Biomaterials, Bioengineering and Sustainability

Volume 9

There is an urgent need to address the current paradigm shifts in bioengineering for Human Health aiming the creation of breakthrough tissue engineered products, manufacturing technologies and effective regenerative treatments for tackling different diseases/disorders in a personalized manner. Yet, the excessive costs and wastes related to the development and production of biomaterials and advanced therapy medicinal products, and the increasing use of plastics in cell culture methods and animal derived reagents has recognized the importance of decreasing the direct carbon footprint and thus, implement sustainable principles and solutions in the innovation ecosystem. The main goal of the volumes in Biomaterials, Bioengineering and Sustainability series is to catapult and consolidate new concepts and solutions towards the development of the next-generation of sustainable and eco-friendly biomaterials and tissue engineering and regenerative medicine approaches. Each volume will focus on the latest developments dealing with the identification of new sources of sustainable or recycled biomaterials, providing ideas for green technologies and methods that can be applied for biomaterials advanced processing and scaffolding strategies, and applications in biofabrication, tissue engineering, regenerative medicine, and drug delivery systems. It also aims to include the exploitation of renewable and sustainable source of human cells applied for cell therapies or in combination with sustainable biomaterials. The develop complex in vitro 3D/4D models and dynamic cell culture systems will be other subjects to be further explored from a sustainable perspective. This series aims to attract the contributions of leading experts in bioengineering, cell biology, materials engineering, and environmental sciences.

Ibrahim Fatih Cengiz • Joaquim M. Oliveira •
Rui L. Reis
Editors

Bioimaging Modalities in Bioengineering

Springer

Editors
Ibrahim Fatih Cengiz
3B's Research Group
University of Minho
Guimarães, Portugal

Joaquim M. Oliveira
3B's Research Group
University of Minho
Guimarães, Portugal

Rui L. Reis
3B's Research Group
University of Minho
Guimarães, Portugal

ISSN 2731-751X ISSN 2731-7528 (electronic)
Biomaterials, Bioengineering and Sustainability
ISBN 978-3-031-99409-8 ISBN 978-3-031-99410-4 (eBook)
https://doi.org/10.1007/978-3-031-99410-4

This work was supported by Università della Svizzera italiana and Radboud University.

This Springer imprint is published by the registered company Springer Nature Switzerland AG
The registered company address is: Gewerbestrasse 11, 6330 Cham, Switzerland

Preface

The curation and development of this book were inspired by the critical roles that imaging plays across the broad and interdisciplinary field of bioengineering. To highlight these roles, this book integrates key concepts, applications, and emerging directions. It also provides a comprehensive overview of bioimaging modalities used in preclinical bioengineering research. It is structured in three main sections, covering topics from the basics of bioimaging to advanced multimodal and computational approaches. Part I is dedicated to the fundamentals and core applications of major imaging modalities, including light microscopy, optical coherence tomography, magnetic resonance imaging (MRI), ultrasound, and micro-computed tomography (micro-CT), with focused chapters on their roles in small animal imaging and preclinical contexts. Part II builds further on this foundation by presenting functional, quantitative, and multimodal approaches, featuring techniques such as photoacoustic imaging, positron emission tomography/single-photon emission computed tomography, and integrated MRI and spectroscopy, highlighting their applications in advanced preclinical imaging. Part III explores emerging technologies and computational advances, including thermoacoustic and Raman-based imaging, second harmonic generation, photon-counting micro-CT, mass spectrometry imaging, laser speckle contrast imaging, and applications of artificial intelligence, computer vision, and motion analysis. Finally, it also provides a concluding perspective on bioimaging in bioengineering. Together, this book serves as a comprehensive, in-depth, and state-of-the-art resource for researchers in the life sciences and engineering fields, including bioimaging, bioengineering, tissue engineering, and regenerative medicine.

Thank you very much to all contributors from diverse geographic regions for supporting our idea and making this book possible. It has been a pleasure to collaborate with you and to bring together leading experts in bioimaging for this unique and concise book, which we are pleased to present after extensive and dedicated efforts.

Guimarães, Portugal

Ibrahim Fatih Cengiz
J. Miguel Oliveira
Rui L. Reis

Acknowledgments IFC thanks Portuguese Foundation for Science and Technology (FCT) and acknowledges the FCT distinction attributed under the Estímulo ao Emprego Científico program (2021.01969.CEECIND). IFC also thanks the TERM RES-Hub, Tissue Engineering and Regenerative Medicine Infrastructure project, funded by FCT. The authors acknowledge the financial support provided by the COMPETE2030 Program and the European Regional Development Fund (FEDER) (Project reference: COMPETE2030-FEDER-00682400). The authors acknowledge the support of CAN-TARGET—Decoding the Complexity of Cancer: Integrative Strategies to Model In vitro and Target the Multi-Cellular Tumor Micro-Environment, reference: NORTE2030-FEDER-02705300, funded by the CCDR-N—Northern Regional Development Coordination Committee, under the Norte 2030 Support System for the Creation of Scientific and Technological Knowledge—Integrated SR&TD Projects. This work has received funding from the European Union's Horizon Europe (2021–2027). The authors thank the financial support provided under UID/50026:3B's-Biomaterials, Biodegradables and Biomimetics Research Group, University of Minho (3B's Res. Group/UM), and the project "HEALTH-UNORTE: Setting-up biobanks and regenerative medicine strategies to boost research in cardiovascular, musculoskeletal, neurological, oncological, immunological and infectious diseases," reference: NORTE-01-0145-FEDER-000039, funded by the Norte Portugal Regional Coordination and Development Commission (CCDR-N), under the NORTE2020 Program.

Contents

Part I

Fundamentals and Applications of Bioimaging Tools in Preclinical Research

Basics of Light Microscopy and Fluorescence Imaging

Rushda Aarif, Sunita Sahoo, and Dillip Kumar Bishi

Abstract

One of the foremost advancing techniques in the domain of scientific research is optical microscopy that enables imaging of cellular components and mechanisms. Modern light microscopy encompasses a variety of techniques, including brightfield, darkfield, and fluorescence microscopy, that have been especially designed to meet specific requirements of research. With the discovery of fluorescence imaging, subcellular and molecular components of live and fixed cells can be visualized with exceptional specificity and resolution. This chapter highlights the fundamentals of light microscopy and advancements in imaging methodologies such as fluorescence lifetime imaging microscopy (FLIM), widefield microscopy, confocal microscopy, and fluorescence resonance energy transfer (FRET), emphasizing its pivotal significance within modern science.

Keywords

Fluorescence microscopy · Widefield microscopy · Confocal microscopy · Electron microscopy · Multi-photon microscopy

Rushda Aarif and Sunita Sahoo contributed equally with all other contributors.

R. Aarif · S. Sahoo · D. K. Bishi (✉)
Department of Biotechnology, Rama Devi Women's University, Bhubaneswar, India
e-mail: dillipkumarbishi@rdwu.ac.in

I. F. Cengiz et al. (eds.), *Bioimaging Modalities in Bioengineering*, Biomaterials, Bioengineering and Sustainability 9,
https://doi.org/10.1007/978-3-031-99410-4_1

1 Introduction

1.1 Overview

Microscopy has been a foundational tool providing insights into the microscopic world, invisible through our naked eyes. It is a simple, basic, and most widely used tool in fundamental and applied scientific research, shaping most groundbreaking discoveries. The light microscope holds a pivotal role due to its simplicity, versatility, accessibility, and historical significance. It has continuously evolved, fostering countless discoveries in fields ranging from cell biology, medicine to material science. It helps the researcher to unfold the hidden truth of life and matter and has a crucial role in developing and enhancing our knowledge related to the intricate details of the microscopic world. From basic brightfield microscopy to more advanced phase-contrast microscopy and fluorescence microscopy have adapted to meet the growing demands of modern science with more precise details.

A light microscope operates using visible light and optical lenses, producing highly magnified, clear, and sharp images. Robert Hooke and Antonie van Leeuwenhoek, the pioneers in the seventeenth century used light microscopes to provide insights into cells and microorganisms leading to the new beginning of cellular biology and microbiology (Somssich 2022). Microscopes have continuously evolved from rudimentary single-lens microscopes to sophisticated compound microscopes that enhance resolution, contrast, and imaging capabilities. The core components of a light microscope are the illumination system and ocular system that produces high-quality images. Over the centuries light microscopy has evolved dramatically, incorporating technological advancements that have pushed the boundaries of imaging. It is used in the study of cells to study synthetic materials and nanostructures.

This technique is based on principles of optical properties such as reflection, refraction, and diffraction which govern the behavior of light as it passes through a condenser lens and interacts with specimens (Lawlor 2019). Enhanced and refined images can be obtained by time-to-time modification of the light source and lens system of an optical microscope. It is a vital tool in various scientific disciplines. In the biological field, light microscopy is a biophysical tool that is used for observing live cells by phase-contrast and interference-contrast microscopes. Fluorescence microscopy opens a whole new path for research in the biomedical field.

1.2 Importance of Light Microscopy in Biology

Microscopy acts as a gateway to discovery, enabling scientists to explore previously unseen microscopic worlds. An optical microscope acts as a fundamental instrument in basic biology research providing an understanding of cellular and tissue level structures. It is cost-effective and user-friendly with a simple setup of lenses, light source, and movable stage, helping in the study of cell morphology, tissue structure, and microbiology. Optical microscopy is the most suitable device for various applications in cell biology especially in distinguishing cell and tissue morphology. A light microscope has been used for the thorough study of the morphology and structure of tendons at different scales. It provides proper imaging of collagen fibers and cells (Xu et al. 2020).

Microscopy can also be modified and developed to enhance image contrast and resolution. It provides detailed visualization of the morphology of cells, tissue architecture, histology, and pathology, enabling the diagnosis of diseases like cancer or infection. Brightfield microscopy is the simplest type of microscopy and remains fundamental in microbiology for observing bacterial morphology, growth patterns, and studying protozoans. Brightfield microscopy enabled the detection of intracellular bacterial communities in urothelial and squamous cells, aiding rapid clinical diagnostics of infections (Martínez-Figueroa et al. 2020).

Integration of technological advancements with light microscopy progress to advanced modern research needs. A fluorescence microscope aids in the visualization of subcellular structures with high specificity by using fluorescent dyes. The visualization and understanding of dynamics and functions of intracellular organelle can be achieved with the rapid advancement of super-resolution microscopy and innovative fluorescent labelling (Zhao and Guo 2023). In neuroscience, light microscopy helps to observe neurons and synaptic connections. Single-cell activity deep in the brain is visualized by multiphoton microscopy. The imaging speed and resolution in real-time can be enhanced using light sheet microscopy (Hillman et al. 2019). In developmental biology, it allows the tracking of cellular differentiation and tissue formation in model organisms in real time. Light microscopes continue to evolve providing more valuable insights into the microscopic world. Techniques like confocal and super-resolution microscopy push the limit of resolution by using lasers and optical sectioning to produce three-dimensional (3D), high-resolution images of samples.

1.3 Advancements in Microscopy Imaging Techniques

Microscope imaging techniques are used to create a visual representation of any biological samples to study the structure, composition, and functions of the sample. The imaging technique provides detailed images with high resolution, magnification, contrast, and dimensionality. The image quality is based on the lens curvature's smoothness and consistency and the lens surface's refractive ability. Aberrations and lack of contrast in the image are due to exposure of the imaging system to too much reflected light. The image is presented onto a camera or photon multiplier tube to display on a computer screen as a real image. The image produced is due to variations in brightness or color. The quality and capability of imaging systems are influenced by the material and precise design of lenses. Images formed are digitalized for display, analysis, interpretation, and storage. The core part of microscopy imaging is digital imaging, that refines the collection of information about biological samples. It is done by digital processing (Merchant and Castleman 2023).

Brightfield microscopy, being the simplest imaging technique, produces images of stained specimens with a dark image on a bright background. It is ineffective for live-cell imaging and produces low contrast for unstained specimens. Darkfield microscopy imaging technique generates an image against a dark background. Phase-contrast imaging technique enables visualization of live, unstained, transparent specimens by enhancing the contrast of the image (Wayne 2019). The advancements in microscopy imaging techniques have transcended the limitations of traditional microscopy, opening new frontiers in science and technology, and allowing for visualization of biological processes at cellular, subcellular, and molecular levels. From nanoscale resolution to real-time, label-free imaging, these innovations have empowered researchers to address complex scientific questions. Super-resolution microscopy has addressed the diffraction limit of traditional optical microscopy, ensuring visualization at nanometer-scale range. Advanced microscopy techniques such as stimulated emission depletion (STED), photoactivated localization microscopy.

(PALM), and stochastic optical reconstruction microscopy (STORM), help in the field of molecular biology and neuroscience as they provide resolution beyond 200 nm, enabling visualization of subcellular organelles. Multiphoton microscopy uses light of longer wavelengths to visualize deep tissue imaging of live cells. Light sheet microscopy, Selective Plane Illumination Microscopy (SPIM), helps in a real-time study of various biological events as it illuminates the sample in a single plane with an increase in imaging speed. The incorporation of AI algorithms with the light phenomenon in the microscope has brought a new dimension to microscopy, reducing background noises, imaging time, reconstructing images, and producing 3D images of live cells. It can be seen in digital holography microscopy (DHM), CARS that produces images of live cells even at a molecular level. These innovations make the microscope more accessible and versatile. The future of microscopy imaging relies on the integration of omics technology and machine learning. Light microscopy can become more efficient and accessible by combining artificial intelligence (AI) allowing researchers to automate the identification of cells, pathogens, or abnormalities, saving time and reducing human error. Moreover, recent trends focus on making light microscopy more sustainable by developing eco-friendly imaging systems with low energy consumption and reusable components (Merchant and Castleman 2023).

2 Principles of Light Microscopy

2.1 Basic Principles of Light, Refraction, and Reflection

2.1.1 Basic Principles of Light

Microscope resolves various unseen queries of biology using two components, light and lens. Properties of light control the visualization of our eyes through the lens. It provides magnified and well-clarity visualization of cell structure and dynamic processes. A lens shapes the trajectory of light as it effectively manipulates light by bending the light rays to focus or spread them, thus lens shares an essential relationship with light and helps in image formation. In optical microscopes, different combinations of convex lenses are used that collect, focus, and magnify light to produce a highly magnified and clear image. It can be stated lens and light work hand-in-hand where light provides the information, and the lens gathers most of the information. Thus, the lens is the path-director of light. In this section, we are going to discuss the simple basic properties of light and their interactions with different matters relevant to a microscope, and how the lens helps in the visualization of such tiny objects. It is amazing to understand how microscope parts are related to our eyes. The human eye is similar to a microscope in an interesting way. The lens of the eye and microscope have the same work of manipulating the light and focusing on the object. The iris of the eye and diaphragm on the microscope controls the intensity of light to pass. Interestingly retina is similar to the CCD of the system and the brain acts as a computer that analyzes the data and produces images.

Light or electromagnetic radiation (EMR) is the form of energy that travels through space in the form of waves or particles. EMR is a broad spectrum of energy with different wavelengths, frequencies, and energy. Wavelength, which is a vital factor for providing contrast, magnified, and highly resolved images, refers to the distance between two successive peaks or troughs of a light wave and is measured in nanometers (nm). In the context of light microscopy, the visible range of EMR is of great importance. The visible spectrum is small, with a broad wavelength range between 400–700 nm. As the wavelength changes the color of the visible spectrum changes. The visible spectrum is a combination of seven colors (VIBGYOR) and when combined it appears white. Any wavelength or color of light can be extracted by using appropriate filters. Monochromatic light with a single wavelength is essential for high-quality imaging. For a proper bright image, intensity of light is very essential. The intensity of light is the amount of energy passing through a specimen. The intensity of light is adjusted by light sources or by using filters high intensity light causes overexposure and low intensity produces a dark image. Incoherent illumination is observed in brightfield microscopy and fluorescence microscopy that uses halogen lamps, mercury vapor bulbs, xenon lamps, and LEDs as light sources where light waves consist of multiple wavelengths, not in a single phase, enhancing contrast. Coherent illumination is observed in confocal microscopy, super-resolution microscopy that uses monochromatic laser light which is highly directional and focused producing high spatial and temporal resolution (J. B. Sanderson 2019)

2.1.2 Interaction of Light with Biological Sample

Vacuum has no air particles so light rays travel straight without being diverted. The bending of light is a phenomenon that occurs due to variations in the speed of light while moving through different mediums. In the context of a microscope, the light rays travel at different speeds from one medium(air) to another (glass or lens) so there is a bending of light. The lens system converges the light rays onto a specimen, forming a magnified image. As the light interacts with the particles in the medium it changes in behavior showing phenomena such as reflection, refraction, absorption, emission, or scattering. Each phenomenon contributes to how light is manipulated and helps to achieve a high-quality image. Besides all these basic phenomena, under specific conditions, interactions like phase-shift, fluorescence, and interference contribute to specialized microscopy techniques (J. B. Sanderson 2019).

Reflection The word "reflection" originates from the Latin prefix "reflex" meaning to bend back. Reflection is the behavior of light where it bounces back from one medium rather than traveling to another medium. It is not absorbed by the medium rather it redirects from the surface of a medium. The conventional light microscopes use mirrors to direct light onto the specimen for illumination. Modern microscopes use direct light sources and various optical components that manipulate and focus light onto a specimen without depending on reflection. Still, some instruments such as darkfield microscopy, fluorescence, and confocal microscope rely on reflection, which is discussed in a later section.

Refraction It originates from the Latin word "refringere" which means "to break up". Refraction is the most important property of light as it helps to focus and produce a magnified image. The light bends as it travels into the lens system and is gathered to a central point. The amount of light bending depends on the curvature and refractive index of the lens. The refractive index quantifies the amount of light that bends down and changes direction while traveling in between mediums. Snell's law defines the refractive index, indicating the ratio of the sine of the angle of incidence to the sine of the angle of refraction remains constant while transitioning between two media:

$$n_1 \sin \theta_1 = n_2 \sin \theta_2$$

where n_1 and n_2 are the refractive indexes of medium 1 (air) and medium 2 (glass), θ_1 is the angle of incidence and θ_2 is the angle of refraction (Gallegos and Stokkermans 2024). The bending of light depends on the refractive index of the lens material and the medium between the lens and specimen. The more the refractive index better the bending of light and a better sharp and focused image will be formed. To avoid reflection of light, the refractive index of the immersion liquid and glass of the lens should be close enough.

Absorption and Emission The absorption phenomenon of light occurs when some of the energy of light is absorbed by molecules of a specimen and converted to other forms of energy. Absorption helps in generating contrast in images and it is useful for brightfield microscopy. Emission is the phenomenon of light where the absorbed energy is re-emitted in a longer wavelength range. Absorption and emission phenomena are important for fluorescence microscopy.

Scattering Scattering occurs when the light is re-directed into different directions as it interacts with the particles within or on the surface of a specimen. Scattered light reduces contrast and sharpness of image making it difficult to distinguish fine details. Scattering of light needs to avoid as much as possible but Scattered light is collected to produce a sharp image in a darkfield microscope.

Phase-shift Phase-shift is a phenomenon of light observed due to a change in the phase of light waves as they interact with a biological sample. It is used in phase-contrast microscopy to generate images of transparent, unstained, live samples having high contrast.

Fluorescence It is the phenomenon observed by fluorophores where excitation light of a specific wavelength is absorbed, and emission of longer wavelengths of light occurs. It is widely used to label specific structures and molecules in a biological sample, producing high-contrast, colored images. Fluorescence is used in various types of fluorescence microscopes which are further discussed in detail in this chapter.

2.2 Anatomy of Light Microscope

A microscope consists of two parts, ocular and mechanical. The mechanical parts provide a structural framework for the whole microscope. There is a well-known saying: "No lens, no microscope". The ocular part of the microscope, which is the instrument's core, is a cylindrical housing with a combination of mirror and lens arranged vertically at precise distances and angles. From the base, there is a condenser lens, objective lens, and eyepiece lens to the detector, which can be an eye or CCD camera. The parallel light rays from a distant object transverse the lens and converge at a point to form the image. The lens system gathers light and provides resolution, magnification, and focus to produce the image. A microscope has thin, double convex lenses that converge light rays and focus them on the object. The lens has an anti-reflective coating that reduces the amount of light reflected by the lens, improves the efficiency of light, reduces glare, and produces better contrast in the image. A condenser lens located between an object and a light source with an adjustable aperture, focuses a uniform illumination as a precise cone onto an object, providing contrast and resolution. A specialized condenser is present in a darkfield and polarized microscope, which will be discussed in a further section. The objective lens, being the primary and most important lens, refracts the light rays from a sample and magnifies the image (J. B. Sanderson 2019). The eyepiece lens has magnification engraved on it (Lawlor 2019). The image travels in the fashion of parallel rays and converges to form a primary image. The primary image plane is formed in the focal plane of the eyepiece lens. The image rays travel parallel through the eye-piece lens and converge on the retina to produce an image. Microscopes with high-quality CCD cameras do not require an eyepiece.

Magnification is the simple enlargement of the primary image allowing the observer to visualize details of the cellular and subcellular structures with precise details that were not visible through our naked eye. The product of the objective and ocular lens magnification is the microscope's overall magnification. The maximum magnification of a light microscope is 1500X. Focal length is an intrinsic, fixed property of the lens responsible for magnification. Focal length is the distance between the lens and the point where light rays converge to form the image. Beyond a certain magnification, a blurry image is produced.

Resolution is the property of a microscope to differentiate two points close together as separate objects producing a clear and sharp image. The resolution of the objective lens helps in the analysis of the details of the object, ocular lens preserves the clarity of a primary image. Better resolution of a microscope relies on a smaller wavelength of incident light and a greater numerical aperture of a lens. The focal plane is the point where the image of the object is most clear and sharp. A sharp image can be achieved by the movement of specimens as it helps to align the specimens with the focal plane of a lens system. Resolution of structures smaller than 200 nanometers is difficult to achieve. Hence, for clear and sharp visualization of intricate structures of an object, there should be a balance between magnification and resolution.

For further distinct visualization of the structural differences of the specimen, a difference in intensity or color can be generated by producing a contrast between the image and background. Contrast is created by differences in light intensity (amplitude-contrast) or differences in the phase of the light wave passing through the specimen (phase-contrast). Phase-contrast mainly helps in generating fine details of live, transparent, unstained specimens. Contrast images can be obtained in darkfield microscopy and phase-contrast microscopy, which are discussed in the following sections. For effective visualization, a balance of contrast with other optical parameters is required.

2.3 Simple and Compound Microscopy

Light microscopes are broadly classified into simple and compound microscopes. A simple light microscope, similar to a magnifying glass, is made up of a single lens focusing light onto a specimen to produce a larger image. It has limitations in magnification and resolution thus being less used in advanced scientific research. A complex microscope is made up of a multiple-lens system (objective and eyepiece) that magnifies images to a greater extent producing more magnified, clear, and sharp images. Thus, compound microscopes are a more versatile and powerful tool in advanced scientific research. Table 1 summarizes the difference between simple and compound microscopes.

3 Types of Light Microscopy

Over time, various types of light microscopy techniques such as brightfield microscopes, darkfield microscopes, fluorescence microscopes, phase-contrast microscopes, confocal microscopes, and most advanced electron microscopes have been

Table 1 Differences between simple and compound microscopes

Characters	Simple microscope	Compound microscope
Number of lenses	Single convex lens	Two or more lenses
Magnification	10–20×	1500× or more
Complexity	Simple design	Complex design with multiple components
Light source	Natural or external light	In-built light sources such as LEDs

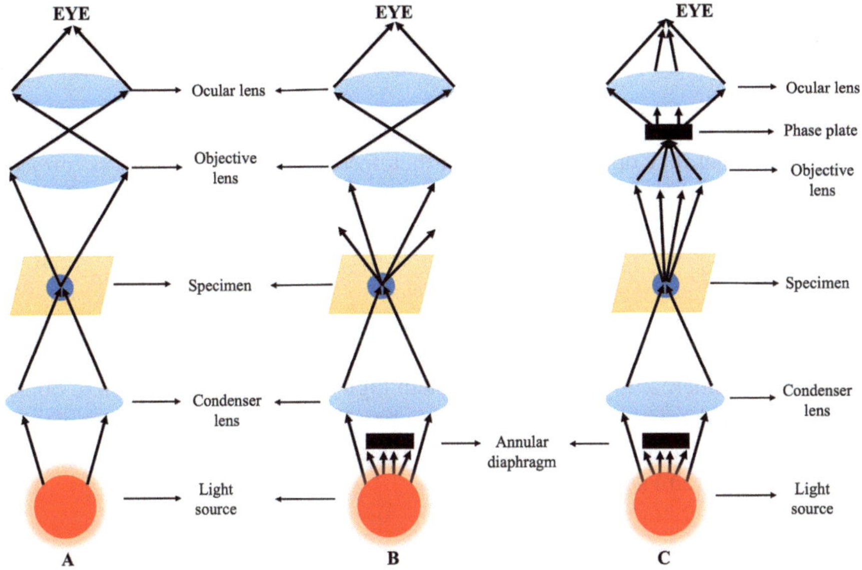

Fig. 1 Schematic representation of image formation in basic optical microscopy techniques. (**a**) Brightfield microscopy: the image is formed when light is absorbed by the specimen and collected by the objective lens, producing an inverted and real image. (**b**) Darkfield microscopy: the condenser collects only scattered light, creating a bright specimen on a dark background. (**c**) Phase-contrast microscopy: the phase plate converts phase shifts caused by refractive index differences in transparent samples into visible contrast for clearer images

developed, each with unique principles and applications to produce better-detailed images. This section explains various light microscopes being used in the biological field and Fig. 1 represents the comparison of the image formation by different microscopes.

3.1 Brightfield Microscopy

Brightfield microscopy helps in the examination and visualization of stained 2D, cellular, and tissue structures. The brightfield microscope uses a visible range of light and produces contrast images based on the transmission and absorption of light by the object. The parallel light rays are focused by the condenser lens onto the object. The object scatters light in all directions and is collected by an objective lens, producing a real, inverted, and magnified image (Mehta et al. 2022). It is simple, rapid, versatile, and inexpensive, but has limitations in contrast and resolution.

Brightfield microscopy is a better option to study cellular morphology and dynamics. Kato, demonstrated the utilization of a brightfield microscope in cell migration through intensity fluctuations in the image, enlightening embryonic development and wound healing (Kato 2024). Brightfield microscopy is used for histological analysis which is helpful in diagnostic validation tests. Serrao et al. utilized an advanced automated brightfield microscope for studying pulmonary tuberculosis smears that promised to enhance diagnostic accuracy (Serrão et al. 2024).

Unstained, transparent, phase objects can also be visualized through a brightfield microscope, but they do not absorb light therefore a low contrast image is formed. However, advancements in this imaging system have also been seen from Zernike phase-contrast and differential phase-contrast to more advanced quantitative phase-contrast that not only makes the object visible but also provides precise information about its physical properties. More advanced brightfield microscopes along with extensive computer modelling can impose multiple images captured from various angles to generate a 3D image. This can be done by introducing optical sectioning in brightfield microscopy (OSBM), to obtain high contrast image of a phase object in a brightfield microscope (Gutiérrez-Medina 2022). Further advancements in brightfield microscopy imaging involve the integration of various machine learning algorithms (Ferreira and Silveira 2024).

3.2 Darkfield Microscopy

Darkfield microscopy is used for observing thin, unstained, transparent living microorganisms, cells, and other specimens with low natural contrast. It enhances the contrast of the image and is based on scattering interaction of light and the specimen. It produces a high-contrast image with a bright specimen with a dark background. A specific component of a Darkfield microscope that allows the study of unstained live specimens is a darkfield condenser that allows direct light at a steep angle onto the specimen, blocking other rays from entering into the objective lens and ensuring only scattered light is observed. The aperture diaphragm controls the light entering through the condenser lens. A light source illuminates the specimen only through the rays travelling through the annular aperture of the condenser lens. The light interacts with the surface of the specimen and scatters in all directions. The unscattered light is blocked, generating a dark background, and the scattered light is gathered by an objective lens to generate an image with a bright specimen. Ongoing advancements continue to expand its capabilities and applications across various scientific disciplines.

3.3 Phase-Contrast Microscopy

Phase-contrast microscopes improve the contrast of transparent, unstained biological samples by converting the difference in the optical phase of light into a difference in light intensity. The amplitude and phase of the object image were reconstructed. An annular light beam is focused on a condenser lens onto the sample that scatters light. Scattered and unscattered light is gathered by an objective lens and passed through the phase-plate. Phase-plate has a phase-shifting region that selectively retards or advances light rays travelling through the specimen. Phase plate introduces phase difference between the two and interference takes place in the detector plane (Mehta et al. 2022).

4 Fluorescence Imaging

The cutting-edge technique of fluorescence microscopy has revolutionized biological and medical imaging. This technology allows for the visualization of individual components in complex biological specimens with high sensitivity and resolution by utilizing the unique properties of fluorescence. This section covers the basic aspects of fluorescence microscopy.

4.1 Principles of Fluorescence

Fluorescence is the property of some substances, known as fluorophores, that, when exposed to light of a given wavelength, re-emit the light at a longer wavelength. Due to the inevitable loss of energy involved in the process, it is red shifted. Absorption of an electron by the fluorophore at an atomic level causes it to move away from the atom nucleus, or into an excited state. However, at an excited state, an electron is unstable, so it returns to its ground state during which a photon of longer wavelength is emitted to dissipate the extra energy, which is known as fluorescence. The shift between the excitation and emission wavelengths allows fluorescence microscopy to distinguish emitted light from the background signal. Fluorescence emission often takes place within nanoseconds of excitation (J. B. Sanderson 2019).

The fundamental concept of a fluorescence microscope is based on filtering, trapping, and detecting the longer wavelength produced by the sample. The excitation wavelength is isolated by the exciter filter and directed towards the dichroic mirror that reflects light only at the excitation wavelength. The objective then focuses the reflected light onto the sample from which the emitted rays are again collected, aggregated, and amplified by the objective lens before being directed to a barrier filter which chooses the emission wavelength and removes any interference from other sources. Finally, selected wavelengths are transmitted to the eyepiece to view or digitize the specimen (Yadav et al. 2022).

4.2 Excitation and Emission Spectra

Fluorescence microscopy typically depends on the excitation and emission spectra of fluorophores to enable imaging. Figure 2 illustrates the Jabłoński diagram (Jablonski 1933), named after Polish physicist Aleksander Jabłoński, that describes the electron states and transitions in a fluorophore. Absorption of a photon by the molecule causes an electron to jump from the ground state (S_o) to a higher energy state (S_1 and S_2) in a fraction of seconds (10^{-15} s). The electron exists at distinct energy levels inside each singlet state due to varying internuclear distances caused by molecular vibrations. Consequently, the electrons can shift into a higher energy level due to the absorption of photons across a spectrum of wavelengths. The *excitation spectrum* denotes the spectrum of wavelengths that a fluorophore absorbs, usually peaking at a particular wavelength. Once excited, the electron can release the surplus energy and return to the ground state through internal conversion and vibrational relaxation, which are non-radiative energy dissipations occurring within picoseconds (10^{-12} s). Internal conversion here refers to the change from one singlet state to another (e.g., S_2 to S_1), while during vibrational relaxation, the electron transfers vibrational energy to the nearby molecules and reverts to the lowest energy level of its present singlet state. Considering the electron can return to any of the ground state's vibrational levels, a range of wavelengths of photons

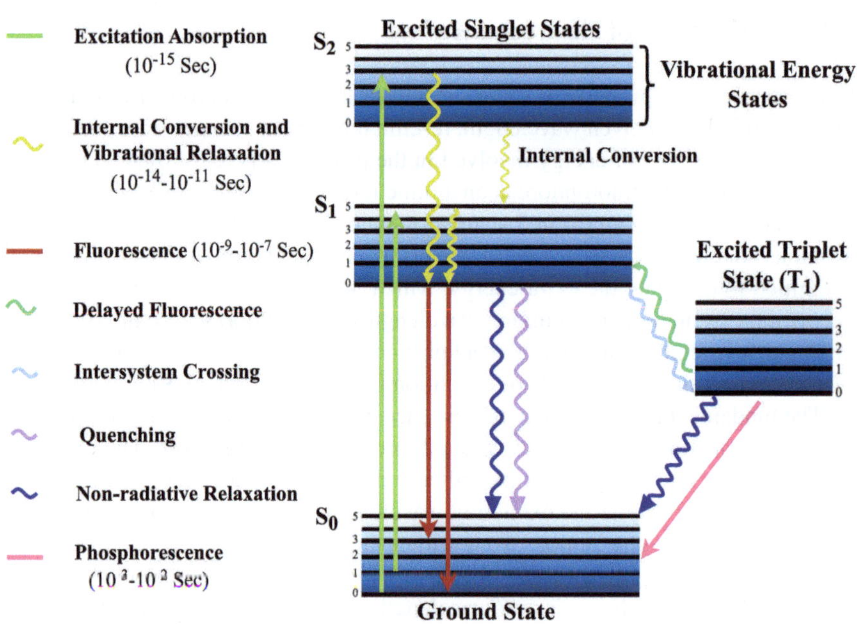

Fig. 2 Jabłoński diagram illustrating the electronic state transitions of a molecule during the fluorescence process. The diagram illustrates how a fluorescent molecule behaves when it interacts with light and the different states (ground state, excited singlet state and excited triplet state) along with both radiative (fluorescence, phosphorescence) and non-radiative pathways

is released which is termed the emission spectra (Chen et al. 2022; Mondal and Diaspro 2014). Ultimately, due to the dissipation of some excitation energy of the electron by non-radiative processes, the released photon will possess lower energy than the absorbed photon. This phenomenon is referred to as the Stokes shift (M. J. Sanderson et al. 2014).

4.3 Fluorophores and Fluorescent Dyes

Compounds that can absorb light at a particular wavelength and emit at a higher wavelength are known as Fluorophores or fluorescent dyes. These are categorized into three types: Green fluorescent proteins, organic dyes, and quantum dots. Table 2 summarizes some of the common fluorophores and their applications.

5 Techniques in Fluorescence Microscopy

5.1 Conventional Fluorescence Microscopy

Fluorescence microscopy encompasses a variety of techniques, exploiting the fluorescence properties of certain molecules to visualize the biological structure and processes, therefore offering insights into dynamic events with high specificity and sensitivity. Figures 3 and 4 represent some of the approaches used in fluorescence microscopy with the details described in the following sections.

Widefield fluorescence microscopy (also known as epifluorescence microscopy) is the simplest and most commonly used technique. Mercury or LED lamps are typically used as a light source which illuminates the entire sample. The optical filters excite the specimen and capture the emitted light through the same objective lens making it ideal to observe labelled cellular structures. While it is the least expensive technique and requires very little amount of excitation light, its widefield illumination produces background fluorescence from out-of-focus planes, restricting resolution in thick specimens (Combs 2010; Danial et al. 2016).

In light sheet fluorescence microscopy (LSFM) a thin sheet of laser illuminates a single optical plane of the specimen, and a perpendicular detection lens captures the fluorescence. This orthogonal arrangement reduces photodamage and photobleaching by restricting light exposure to the target plane. LSFM frequently incorporates Multiview imaging, wherein data is acquired from several perspectives and computationally rebuilt into high-resolution 3D pictures (Buglak et al. 2021; Stelzer et al. 2021).

5.2 Fluorescence-Based Functional Imaging

Fluorescence-based functional imaging includes techniques that reveal molecular interactions, dynamics, and physiological states within cells. Fluorescence reso-

Table 2 Common fluorophores and their applications

Fluorophore/Dye	Structure	Excitation/ Emission Wavelengths (nm)	Applications	References
Cy3	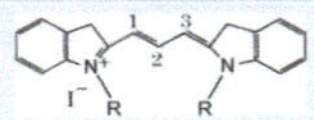	550–570 nm	Nucleic acid labeling, protein interactions	Moreira et al. (2015), Mujumdar et al. (1993)
Fluorescein	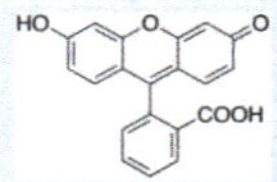	460–521 nm	Western blot analysis, PCR, pH sensing	Johnson (2010), Kobayashi and Choyke (2011)
Rhodamine B	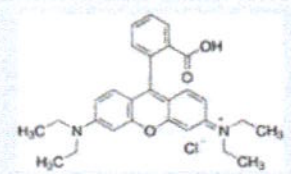	540–590 nm	Tracer dye to determine water flow, membrane dynamics	Lavis and Raines (2008), Preston et al. (2019)
Alexa Fluor 488	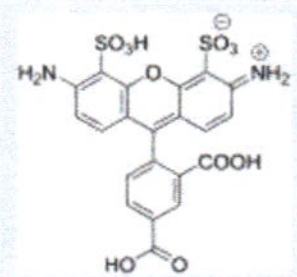	488–496 nm	Immunofluorescence, cell tracing, protein tracking	Lindhoud et al. (2012), Panchuk-Voloshina et al. (1999)
mCherry	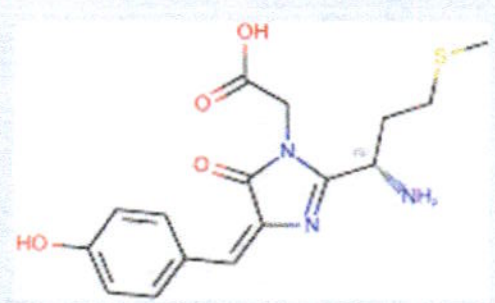	540–650 nm	Fusion protein studies, live-cell imaging	Shaner et al. (2008), Xiao et al. (2018)
GFP (Green fluorescent protein)	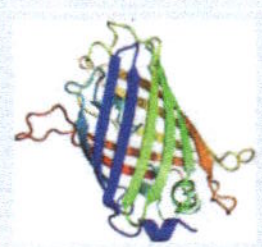	395–509 nm	Protein localization, gene expression studies	Chalfie (1995), Chalfie et al. (1994)

4′,6-diamidino-2-phenylindole (DAPI)	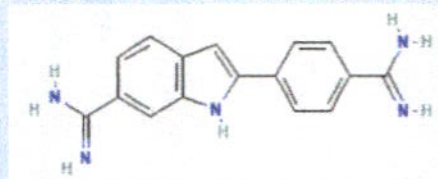	358–461 nm	Cell counting, nuclei and chromosome staining	Kapuscinski (1995), Sanna et al. (1992)
Terbium	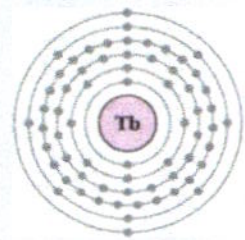	Excitation: ~270–280 nm; Emission: 490–650 nm	Thenostics, track drug delivery	Mohanto et al. (2024), Van Laere et al. (2024)
Europium	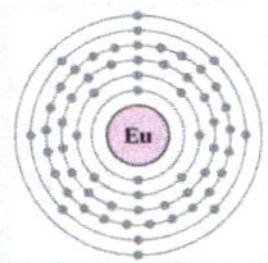	Excitation: ~395 nm; emission: 590–700 nm	Multimodal image, drug delivery, tracking trace materials in cells	Syamchand and Sony (2015)
Cadmium selenide (CdSe)		Size dependent, excitation: 350–400 nm; Emission: 450–650 nm	Drug delivery, tracking single protein	Singh et al. (2023), Song et al. (2005)

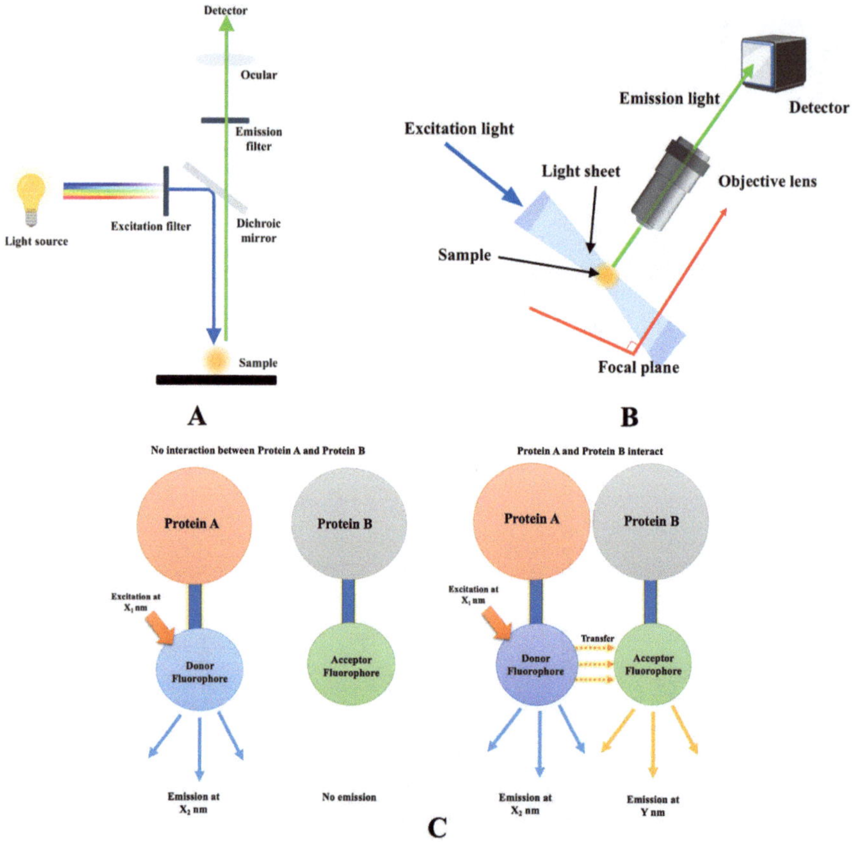

Fig. 3 Schematic representation of advanced fluorescence techniques. (**a**) Widefield fluorescence microscopy illuminates the entire specimen using a broad light source, and the emitted light is captured through the same objective lens. (**b**) Light sheet fluorescence microscopy uses a thin sheet of laser light to illuminate a single plane of the specimen, reducing phototoxicity. (**c**) Fluorescence resonance energy transfer (FRET) measures energy transfer between two fluorophores in close proximity (1–10 nm). (Created with BioRender.com)

nance energy transfer (FRET), a novel fluorescence imaging technique, provides insights into a labeled molecule's immediate environment (typically 1–10 nm) and helps in detecting macromolecular interactions and the proximity of two fluorophores. The procedure entails the radiation-free transfer of energy between the two fluorophores and is only possible when the donor fluorophore's emission spectrum and the acceptor fluorophore's excitation spectrum overlap. Energy transfer efficiency gives quantitative insight into molecular distances and may be assessed using fluorescence intensity or lifetime changes. The applications of FRET include studying enzyme activities, signal transduction, and biomolecular assembly (Liu et al. 2020; Merchant and Castleman 2023).

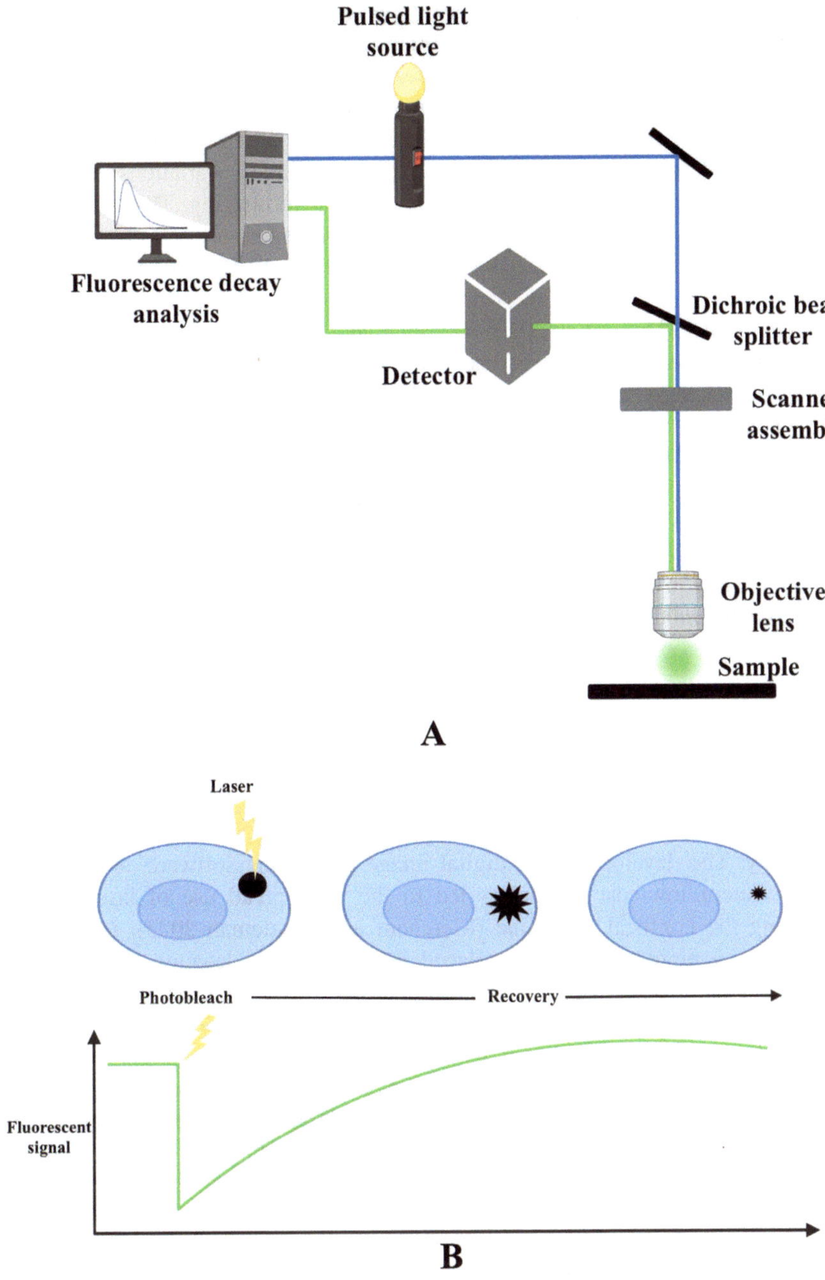

Fig. 4 Schematic representation of further functional fluorescence imaging techniques. (**a**) Fluorescence lifetime imaging microscopy (FLIM) generates images based on fluorescence lifetimes, providing insight into the local environment and molecular interactions. (**b**) Fluorescence recovery after photobleaching (FRAP) involves photobleaching a specific region, followed by monitoring fluorescence recovery over time. (Created with BioRender.com)

Fluorescence lifetime imaging microscopy (FLIM) measures the fluorescence decay time of excited fluorophores to provide spatial and temporal insights into their environment. The fluorescence lifetime refers to the average duration a fluorophore remains in its excited state prior to photon emission and is typically in the range of nanoseconds. In contrast to intensity-based measures, fluorescence lifetime is not influenced by fluorophore concentration, photobleaching, or variations in the light path, rendering it exceptionally reliable for quantitative imaging. This technique is widely used to study metabolic states, molecular interactions, and environmental factors like ion concentrations, oxygen levels, and pH. FLIM can also be used in FRET studies as it enhances the resolution of molecular interaction dynamics at nanometer scales (Datta et al. 2020; Karrobi et al. 2023).

The process in Fluorescence recovery after photobleaching (FRAP) involves photobleaching a specific region of the labeled specimen with a high-intensity laser which results in structural changes of the fluorescent molecules and forfeits their capacity to emit photons. The bleached area is subsequently recovered as the unbleached fluorescent molecules diffuse into the region. The kinetics of fluorescence recovery are analyzed to infer parameters such as diffusion coefficients, binding kinetics, and mobile fraction of molecules. It is also broadly used in cell biology to study protein interactions, intracellular trafficking, protein diffusion, and membrane dynamics (Mantovanelli et al. 2021).

6 Advanced Microscopy in Bioimaging Modalities

6.1 Confocal Microscopy

Confocal microscopy is one of the most notable advancements in 3D imaging technology. The development of digital image processing software, aided by the introduction of low-cost computers led to the widespread use of laser sources supporting the confocal technique (Merchant and Castleman 2023). In confocal microscopy, a pinhole opening is placed in front of the detector, aligning with the image's focal plane. The specimen is illuminated by the laser and is detected through this pinhole resulting in the screening out of background fluorescence. This reduces the haze seen in light microscopy and hence enables optical sectioning. This technique is extensively used in biomedical as well as material sciences. Other fluorescence-based techniques such as FRAP, and FRET are combined with confocal microscopy to study living as well as fixed cells and tissues containing fluorophores (Elliott 2020).

6.2 Electron Microscopy (EM)

In EM, instead of light, a beam of electrons interacts with the sample to generate magnified and extremely high-resolution images. Because electrons have a shorter wavelength than visible light, EM approach allows for precision at an atomic or nanometer scale (Evans and Riley 2021; Houck 2023). EM is of two types: *Scanning Electron Microscope* (SEM) and *Transmission Electron Microscope* (TEM). In SEM, the specimen is coated with a conductive material and the electrons interact with the surface generating a 3D topographic image while in TEM, an ultra-thin section of the specimen is taken through which the electrons pass producing a detailed internal structural image (Carter and Williams 2016). EM is widely used to study micro and macromolecular complexes, viruses, and cellular structures. It also has applications in the field of nanotechnology and is used to study surface properties and nanostructures. Recent advances in EM include cryo-EM which enables the imaging of biomolecules in their native state without any need for crystallization (Joppe et al. 2020).

6.3 Multi-photon Microscopy (MPM)

MPM is an advanced microscopy technique that is based on the possibility that multiple low-energy photons (near-infrared range) interact with the fluorophore. The fluorophore absorbs energy from each photon and goes to an excited state from ground state, the transition akin to that induced by a single high-energy photon. When the fluorophore reverts back to the ground state, it emits the absorbed energy as fluorescence, however, the emission wavelength is shorter compared to that of excitation wavelength. It is a non-linear process that produces optical sectioning, the same as a confocal microscope along with improvising the ability to view live cells. MPM is also ideal for imaging thick, scattering tissues and live specimens as it reduces photodamage and photobleaching effects. The applications of MPM include the study of live animal cells and their cellular processes, tumor microenvironments, tracking nanoparticles, and drug delivery (Eissa et al. 2020; Merchant and Castleman 2023; Yamada et al. 2014).

6.4 Super-Resolution Microscopy (SRM)

The resolution of traditional light microscopy is limited to around 200 nm due to the diffraction limit, which causes the fluorophore to appear blurry as light passes through it. The magnitude of the blur is referred to as point-spread function (PSF). If the diffraction limit is less than the distance between two structures, they look like a single blur instead of two distinct objects. To overcome this limitation, SRM technique is used to attain resolutions down to tens of nanometers, in which the excitation light is temporarily modulated to break the diffraction limit. The benefits of optical microscopy such as target specificity or sample preservation are maintained in the SRM techniques and simultaneously collect data on a number of macromolecules or proteins inside the subcellular components to produce a 3D image of the cellular structures. SRM is further divided into the following three types:

Stimulated Emission Depletion (STED) Microscopy Spatially patterned excitation technique is used in STED. Two lasers are combined on the focal plane which diminishes the effective PSF resulting in a higher resolution. The excited state fluorophore situated near the excitation focal point is suppressed by the STED laser by a process known as stimulated emission, resulting in the diffraction limit higher than the PSF and production of high-resolution images.

Structured Illumination Microscopy (SIM) In SIM, striped illumination patterns are created by using laser beams laser beams with a minimum stripe distance near to resolution limit resulting in Moiré effects (large-scale interferences) that travel through the orifice of the objective lens. A third pattern is produced by this interaction and numerous images are used to obtain an image with twice the resolution of light microscopy.

Single-Molecule Localization Microscopy (SMLM) In SMLM, random activation or deactivation of small subsets of emitters takes place which are segregated spatiotemporally and aggregated over several camera frames. Computational analysis is done based on the quantity of photons detected per emitter to pinpoint their centers with nanometer accuracy and is superimposed to form a binary image on a single plane (Schermelleh et al. 2019; Vangindertael et al. 2018).

7 Conclusions and Future Perspectives

Microscopy has undergone significant transformations in recent years, leading to ground-breaking advancements in the biological and medical field. The continuous evolution of microscopic techniques and imaging processing has transformed the ability to observe and examine biological processes at cellular, subcellular, and molecular levels, producing more accurate results and driving new scientific discoveries across multiple disciplines. The integration of state-of-the-art microscopy techniques enabled researchers to discover the complex dynamics of biological systems by pushing the limits of resolution, contrast, and image depth via the use of techniques like fluorescence microscopy, multi-photon microscopy, super-resolution microscopy, and confocal microscopy. Conventional microscopes had a limit of diffraction due to which a resolution of only 200 nm was possible, but advanced microscopy techniques like Super-resolution microscopy overcome the limit of diffraction and allow visualization of cellular components in nanometer range. Particularly, fluorescence imaging needs development of novel fluorescent probes with better photostability, brightness and specificity to enhance imaging quality. Advances in near-infrared (NIR) and shortwave infrared (SWIR) floresence imaging will allow deeper tissue penetration, aiding for in vivo applications. Although fluorescence imaging techniques have some limitations, these can be overcome using modern microscopy approaches such as super-resolution and multi-photon microscopy. Further advancement and integration of quantum dots, organic dyes and genetically encoded fluorescent proteins may lead to various new insightful discoveries with fluorescent imaging. Label-free imaging without any fluorescent labels has the potential to overcome phototoxicity and increase tissue penetration depth. The continuous advancements of super-resolution techniques will likely push the limitations of microscopic imaging, allowing visualization of finer details of cellular and molecular processes. As the field of nanotechnology progresses, incorporation of nanomaterials into imaging probes may enhance the sensitivity and specificity, enabling highly precise and targeted molecular imaging. Collaboration of interdisciplinary fields has led to significant advancements in microscopy imaging enabling higher resolution, faster analysis, and deeper insights into biological structures. The integration of microscopic imaging with artificial intelligence and machine learning algorithms has enhanced image reconstruction, noise reduction, and automated feature detection. The advancements of multimodal imaging and automated data processing have speed up data acquisition, analysis, and interpretation to produce more precise results. Real-time 3-D imaging and live cell imaging will allow researchers to visualize and monitor dynamic cellular processes. Digital imaging, artificial intelligence, and novel fluorescent labeling combined with light microscopy have enhanced observational precision and specificity as well as reduced human error. The convergence of microscopy with other emerging fields such as quantum imaging and nano-photonics, may produce novel ways to manipulate and visualize images. Such advancements in microscopy techniques and image processing will not only enhance the basic cellular and molecular discoveries but also pave the way for significant discoveries in the field of biomedical research, drug discovery, targeted therapies, and precision medicines. Collectively, these developments underscore the pivotal role of microscopy in modern science, promising continued breakthroughs in biological and material sciences.

Acknowledgments DKB acknowledges the funding support from Science & Technology Department, Government of Odisha, India for the project "Phytonanotechnogical Interventions for Hepatocellular Carcinoma" (ST-BT-MISC-0005-2023).

Conflict of Interest The authors declare that they have no conflict of interest.

References

Buglak NE, Lucitti J, Ariel P, Maiocchi S, Miller FJ, Bahnson ESM (2021) Light sheet fluorescence microscopy as a new method for unbiased three-dimensional analysis of vascular injury. Cardiovasc Res 117(2):520–532. https://doi.org/10.1093/cvr/cvaa037

Carter CB, Williams DB (eds) (2016) Transmission electron microscopy. Springer International Publishing. https://doi.org/10.1007/978-3-319-26651-0

Chalfie M (1995) Green fluorescent protein. Photochem Photobiol 62(4):651–656. https://doi.org/10.1111/j.1751-1097.1995.tb08712.x

Chalfie M, Tu Y, Euskirchen G, Ward WW, Prasher DC (1994) Green fluorescent protein as a marker for gene expression. Science (New York) 263(5148):802–805. https://doi.org/10.1126/science.8303295

Chen K, Yan R, Xiang L, Xu K (2022) Excitation spectral microscopy for highly multiplexed fluorescence imaging and quantitative biosensing. In: Goda K, Tsia KK (eds) High-speed biomedical imaging and spectroscopy VII. SPIE, p 23. https://doi.org/10.1117/12.2612265

Combs CA (2010) Fluorescence microscopy: a concise guide to current imaging methods. Curr Protoc Neurosci 2:Unit2.1. https://doi.org/10.1002/0471142301.ns0201s50

Danial JSH, Aguib Y, Yacoub MH (2016) Advanced fluorescence microscopy techniques for the life sciences. Glob Cardiol Sci Pract 2016(2):e201616. https://doi.org/10.21542/gcsp.2016.16

Datta R, Heaster TM, Sharick JT, Gillette AA, Skala MC (2020) Fluorescence lifetime imaging microscopy: fundamentals and advances in instrumentation, analysis, and applications. J Biomed Opt 25(7):1–43. https://doi.org/10.1117/1.JBO.25.7.071203

Eissa A, Zoeir A, Sighinolfi MC, Puliatti S, Bevilacqua L, Del Prete C, Bertoni L, Azzoni P, Reggiani Bonetti L, Micali S, Bianchi G, Rocco B (2020) "Real-time" assessment of surgical margins during radical prostatectomy: state-of-the-art. Clin Genitourin Cancer 18(2):95–104. https://doi.org/10.1016/j.clgc.2019.07.012

Elliott AD (2020) Confocal microscopy: principles and modern practices. Curr Protoc Cytom 92(1):e68. https://doi.org/10.1002/cpcy.68

Evans TJ, Riley PA (2021) Principles of microscopy, culture and serology-based diagnostics. Medicine 49(10):648–653. https://doi.org/10.1016/j.mpmed.2021.07.009

Ferreira EKGD, Silveira GF (2024) Classification and counting of cells in brightfield microscopy images: an application of convolutional neural networks. Sci Rep 14(1):9031. https://doi.org/10.1038/s41598-024-59625-z

Gallegos J, Stokkermans TJ (2024) Refractive index. In: StatPearls. StatPearls Publishing. http://www.ncbi.nlm.nih.gov/books/NBK592413/

Gutiérrez-Medina B (2022) Optical sectioning of unlabeled samples using bright-field microscopy. Proc Natl Acad Sci USA 119(14):e2122937119. https://doi.org/10.1073/pnas.2122937119

Hillman EMC, Voleti V, Li W, Yu H (2019) Light-sheet microscopy in neuroscience. Annu Rev Neurosci 42(1):295–313. https://doi.org/10.1146/annurev-neuro-070918-050357

Houck MM (2023) Microscopy: Electron. In: Encyclopedia of forensic sciences, 3rd edn. Elsevier, pp 584–589. https://doi.org/10.1016/B978-0-12-823677-2.00032-5

Jablonski A (1933) Efficiency of anti-stokes fluorescence in dyes. Nature 131(3319):839–840. https://doi.org/10.1038/131839b0

Johnson I (ed) (2010) The molecular probes handbook: a guide to fluorescent probes and labeling technologies, 11th edn. Life Technologies Corp

Joppe M, D'Imprima E, Salustros N, Paithankar KS, Vonck J, Grininger M, Kühlbrandt W (2020) The resolution revolution in cryoEM requires high-quality sample preparation: a rapid pipeline to a high-resolution map of yeast fatty acid synthase. IUCrJ 7(2):220–227. https://doi.org/10.1107/S2052252519017366

Kapuscinski J (1995) DAPI: a DNA-specific fluorescent probe. Biotech Histochem 70(5):220–233. https://doi.org/10.3109/10520299509108199

Karrobi K, Tank A, Fuzail MA, Kalidoss M, Tilbury K, Zaman M, Ferruzzi J, Roblyer D (2023) Fluorescence Lifetime Imaging Microscopy (FLIM) reveals spatial-metabolic changes in 3D breast cancer spheroids. Sci Rep 13(1):3624. https://doi.org/10.1038/s41598-023-30403-7

Kato Y (2024) A simple method for visualization and quantitative evaluation of single and collective cell migration in bright field microscopy images, vol 3. Digital Medicine and Healthcare Technology. https://doi.org/10.5772/dmht.20240001

Kobayashi H, Choyke PL (2011) Target-cancer-cell-specific activatable fluorescence imaging probes: rational design and in vivo applications. Acc Chem Res 44(2):83–90. https://doi.org/10.1021/ar1000633

Lavis LD, Raines RT (2008) Bright ideas for chemical biology. ACS Chem Biol 3(3):142–155. https://doi.org/10.1021/cb700248m

Lawlor D (2019) Introduction to light microscopy: tips and tricks for beginners. Springer

Lindhoud S, Westphal AH, Visser AJWG, Borst JW, van Mierlo CPM (2012) Fluorescence of Alexa fluor dye tracks protein folding. PLoS One 7(10):e46838. https://doi.org/10.1371/journal.pone.0046838

Liu P, Wang R, Su W, Qian C, Li X, Gao L, Jiao T (2020) Research advances in preparation and application of chitosan nanofluorescent probes. Int J Biol Macromol 163:1884–1896. https://doi.org/10.1016/j.ijbiomac.2020.09.190

Mantovanelli L, Gaastra BF, Poolman B (2021) Fluorescence-based sensing of the bioenergetic and physicochemical status of the cell. In: Current topics in membranes, vol 88. Elsevier, pp 1–54. https://doi.org/10.1016/bs.ctm.2021.10.002
Martínez-Figueroa C, Cortés-Sarabia K, del Carmen Alarcón-Romero L, Catalán-Nájera HG, Martínez-Alarcón M, Vences-Velázquez A (2020) Observation of intracellular bacterial communities in urinary sediment using brightfield microscopy; a case report. BMC Urol 20(1):89. https://doi.org/10.1186/s12894-020-00661-y
Mehta DS, Butola A, Singh V (2022) Introduction. In: Quantitative phase microscopy and tomography: techniques using partially spatially coherent monochromatic light. IOP Publishing. https://doi.org/10.1088/978-0-7503-3987-2ch1
Merchant FA, Castleman KR (eds) (2023) Microscope image processing, 2nd edn. Academic Press
Mohanto S, Biswas A, Gholap AD, Wahab S, Bhunia A, Nag S, Ahmed MG (2024) Potential biomedical applications of terbium-based nanoparticles (TbNPs): a review on recent advancement. ACS Biomater Sci Eng 10(5):2703–2724. https://doi.org/10.1021/acsbiomaterials.3c01969
Mondal PP, Diaspro A (2014) Basics of fluorescence and photophysics. In: Mondal IPP, Diaspro A (eds) Fundamentals of fluorescence microscopy. Springer, Cham, pp 111–134. https://doi.org/10.1007/978-94-007-7545-9_6
Moreira BG, You Y, Owczarzy R (2015) Cy3 and Cy5 dyes attached to oligonucleotide terminus stabilize DNA duplexes: predictive thermodynamic model. Biophys Chem 198:36–44. https://doi.org/10.1016/j.bpc.2015.01.001
Mujumdar RB, Ernst LA, Mujumdar SR, Lewis CJ, Waggoner AS (1993) Cyanine dye labeling reagents: Sulfoindocyanine succinimidyl esters. Bioconjug Chem 4(2):105–111. https://doi.org/10.1021/bc00020a001
Panchuk-Voloshina N, Haugland RP, Bishop-Stewart J, Bhalgat MK, Millard PJ, Mao F, Leung WY, Haugland RP (1999) Alexa dyes, a series of new fluorescent dyes that yield exceptionally bright, photostable conjugates. J Histochem Cytochem 47(9):1179–1188. https://doi.org/10.1177/002215549904700910
Preston AN, Cervasio DA, Laughlin ST (2019) Visualizing the brain's astrocytes. In: Methods in enzymology, vol 622. Elsevier, pp 129–151. https://doi.org/10.1016/bs.mie.2019.02.006
Sanderson JB (2019) Understanding light microscopy. John Wiley & Sons
Sanderson MJ, Smith I, Parker I, Bootman MD (2014) Fluorescence microscopy. Cold Spring Harb Protoc 2014(10):pdb.top071795. https://doi.org/10.1101/pdb.top071795
Sanna PP, Jirikowski GF, Lewandowski GA, Bloom FE (1992) Applications of DAPI cytochemistry to neurobiology. Biotech Histochem 67(6):346–350. https://doi.org/10.3109/10520299209110047
Schermelleh L, Ferrand A, Huser T, Eggeling C, Sauer M, Biehlmaier O, Drummen GPC (2019) Super-resolution microscopy demystified. Nat Cell Biol 21(1):72–84. https://doi.org/10.1038/s41556-018-0251-8
Serrão MKM, Costa MGF, Fujimoto LBM, Ogusku MM, Costa Filho CFF (2024) Automatic bright-field smear microscopy for diagnosis of pulmonary tuberculosis. Comput Biol Med 172:108167. https://doi.org/10.1016/j.compbiomed.2024.108167
Shaner NC, Lin MZ, McKeown MR, Steinbach PA, Hazelwood KL, Davidson MW, Tsien RY (2008) Improving the photostability of bright monomeric orange and red fluorescent proteins. Nat Methods 5(6):545–551. https://doi.org/10.1038/nmeth.1209
Singh D, Thapa S, Singh KR, Verma R, Singh RP, Singh J (2023) Cadmium selenide quantum dots and its biomedical applications. Mater Lett 18:100200. https://doi.org/10.1016/j.mlblux.2023.100200
Somssich M (2022) A short history of plant light microscopy. Curr Protocols 2(10):e577. https://doi.org/10.1002/cpz1.577
Song J-H, Atay T, Shi S, Urabe H, Nurmikko AV (2005) Large enhancement of fluorescence efficiency from CdSe/ZnS quantum dots induced by resonant coupling to spatially controlled surface Plasmons. Nano Lett 5(8):1557–1561. https://doi.org/10.1021/nl050813r

Stelzer EHK, Strobl F, Chang B-J, Preusser F, Preibisch S, McDole K, Fiolka R (2021) Light sheet fluorescence microscopy. Nat Rev Methods Primers 1(1):73. https://doi.org/10.1038/s43586-021-00069-4

Syamchand SS, Sony G (2015) Europium enabled luminescent nanoparticles for biomedical applications. J Lumin 165:190–215. https://doi.org/10.1016/j.jlumin.2015.04.042

Van Laere C, Koole M, Deroose CM, de Voorde MV, Baete K, Cocolios TE, Duchemin C, Ooms M, Cleeren F (2024) Terbium radionuclides for theranostic applications in nuclear medicine: from atom to bedside. Theranostics 14(4):1720–1743. https://doi.org/10.7150/thno.92775

Vangindertael J, Camacho R, Sempels W, Mizuno H, Dedecker P, Janssen KPF (2018) An introduction to optical super-resolution microscopy for the adventurous biologist. Methods Appl Fluoresc 6(2):022003. https://doi.org/10.1088/2050-6120/aaae0c

Wayne RO (2019) Light and video microscopy. Academic

Xiao C, Li L, Lao L, Liu Y, Wei Q, Ji Q, Sun G, Lin F, Wang J, Bao G (2018) Application of the red fluorescent protein mCherry in mycelial labeling and organelle tracing in the dermatophyte Trichophyton mentagrophytes. FEMS Microbiol Lett 365(6). https://doi.org/10.1093/femsle/fny006

Xu M, Liu J, Sun J, Xu X, Hu Y, Liu B (2020) Optical microscopy and electron microscopy for the morphological evaluation of tendons: a mini review. Orthop Surg 12(2):366–371. https://doi.org/10.1111/os.12637

Yadav A, Kapoor N, Badiye A (2022) Introduction to fluorescence microscope. In: Shukla RK, Kapoor N, Badiye A (eds) Forensic microscopy, 1st edn. CRC Press, pp 123–134. https://doi.org/10.4324/9781003120995-9

Yamada M, Lin LL, Prow TW (2014) Multiphoton microscopy applications in biology. In: Fluorescence microscopy. Elsevier, pp 185–197. https://doi.org/10.1016/B978-0-12-409513-7.00013

Zhao X, Guo S (2023) Methods for visualizing intracellular organelles. J Vis Exp 193. https://doi.org/10.3791/64966

Bioengineering Applications of Light Microscopy and Fluorescence Imaging

Alisha Nayak, Priyadarshini Padhi, and Dillip Kumar Bishi

Abstract

Light microscopy and fluorescence imaging are two powerful and widely used techniques in bioengineering that can be used to visualize biomolecular interactions and cellular dynamics, providing images of high spatial and temporal resolution. Fluorescence imaging, enhanced by fluorescent dyes, proteins, and nanoparticles, offers high specificity and sensitivity for detecting biomolecules and dynamic biological processes. This chapter gives a comprehensive overview of advances fluorescence imaging modalities such as Fluorescence Lifetime Imaging Microscopy (FLIM), Fluorescence Recovery After Photobleaching (FRAP) and Förster resonance energy transfer (FRET) for their contributions towards overcoming classical imaging barriers in imaging complex biological samples and events. This chapter also features the major applications of advanced light microscopy methods with references in bioengineering ranging from cellular and molecular imaging, tissue engineering, and regenerative medicine to drug discovery, disease diagnosis and prognosis, biomaterials research.

Alisha Nayak and Priyadarshini Padhi contributed equally with all other contributors.

A. Nayak
Department of Life Science, Rama Devi Women's University, Bhubaneswar, India

P. Padhi · D. K. Bishi (✉)
Department of Biotechnology, Rama Devi Women's University, Bhubaneswar, India
e-mail: dillipkumarbishi@rdwu.ac.in

I. F. Cengiz et al. (eds.), *Bioimaging Modalities in Bioengineering*,
Biomaterials, Bioengineering and Sustainability 9,
https://doi.org/10.1007/978-3-031-99410-4_2

Keywords

RAP · FLIM · FRET · SIM · STED

1 Introduction

Microscopy plays a vital role in biomedical research. Advances in optics and computer science have greatly improved microscopy technologies, thereby providing biomedical researchers with new insights (Liu et al. 2021). Optical microscopy relies on imaging techniques utilizing the visible light spectrum. This comprises phase contrast microscopy and digital holographic microscopy. The most dominant methodologies used in cell biology are specifically designed to identify one or more fluorophores using fluorescence microscopy (Witte et al. 2018). Fluorescence imaging serves as an essential imaging technique in the realms of biomedical detection and image-guided surgical procedures. It presents several advantages, including noninvasiveness, superior spatiotemporal resolution, the ability for real-time detection, and a cost-effective nature when juxtaposed with conventional imaging modalities such as computed tomography (CT), magnetic resonance imaging (MRI), ultrasound, positron emission tomography, and single photon emission CT (SPECT). Recent decades have witnessed the emergence of novel opportunities for both research and clinical practices (Lawrance et al. 2024). Bioengineering relies on light microscopy, particularly fluorescence imaging, to see biological samples with great contrast. In contrast to multimodal systems, which combine many imaging modalities for more thorough analysis, fluorescence microscopy permits targeted cell labeling and observation (Zhang et al. 2022).

Fluorescence imaging is extensively used in bioengineering, particularly in preclinical applications and medical diagnostics, as a powerful noninvasive optical imaging technique. This technique uses visible, ultraviolet (UV) or near-infrared (NIR) light to detect molecular and physiological information via the use of fluorescence probes. This approach enables real-time image acquisition, has high sensitivity, can achieve subcellular resolution, and has multichannel detection capabilities. In spite of these challenges, including limited tissue penetration, scattering, autofluorescence, and the low photostability from some probes, recent improvements in NIR fluorophores and in nanoparticles (NPs) have substantially enhanced performance or utility. Applications range from image-guided tumor surgery, to probing cellular processes, to whole-animal imaging on the small-animal scale, with three-dimensional fluorescence molecular tomography providing quantitative information on probe distribution (Geraldes 2020). Fluorescence imaging has become a very useful preclinical tool in the past few years. Advancement in the development of fluorescence imaging probes has resulted in contrast agents with greater sophistication and can be detected from extremely low doses (pico- to femto-molar range). These include the conjugation of fluorophores, including fluorescent NPs, to ligands that are unique to various targets, thus greatly enhancing the sensitivity and specificity of fluorescence imaging (Refaat et al. 2022).

2 Applications of Light Microscopy and Fluorescence Imaging

Fluorescence microscopy is the preferred technique for examining essential biological processes, allowing for the imaging of the structure and dynamics of biomolecules in live specimens. This ultimately relies on the advancement of fluorescence probes designed to monitor biological systems with exceptional spatiotemporal precision (Olesińska-Mönch and Deo 2023).

2.1 Cellular and Molecular Imaging

Fluorescent protein tags, live cell dyes, and other approaches may be used to study almost any biological function under a microscope. Minimizing photo damage while keeping a useable signal-to-noise ratio and creating an environment that mimics natural cell dynamics are the two main experimental challenges in live cell imaging (Ettinger and Wittmann 2014). Cellular and subcellular imaging through fluorescence microscopy has enhanced the ability to study dynamic processes. New fluorophores and imaging methods have improved real-time visualization of intracellular structures and metabolic activities. Fluorescence microscopy, particularly with NIR-II fluorophores, allows high-resolution tissue investigation in vivo. This is necessary for studying molecular signaling, cellular dynamics, and other biological processes (T. Wang et al. 2023).

Bioengineering demands noninvasive cellular imaging through fluorescence microscopy, especially with far-red fluorescent proteins (FPs). These proteins facilitate visualization of dynamic biological processes and protein-protein interactions. The development of far-red fluorescent proteins has proven high application value, with its emission spectra peaks at 600–650 nm allows advantages of deep tissue penetration, decreased light scatter and minimal phototoxicity. Multi-color imaging permits detailed investigation of complex cellular structures and functions. This imaging is crucial in the investigation of diseases and the formulations of pharmaceuticals. Fluorescence imaging facilitates a comprehensive analysis of cellular dynamics and interactions in bioengineering research (Shang et al. 2024). Two-photon fluorescence microscopy, in particular high-speed fluorescence lifetime imaging microscopy (2P-FLIM), leads to sensitive detection of metabolic dynamics in the apoptosis. High-resolution imaging of rapid metabolic dynamics is essential for explaining cellular response to stimuli causing apoptosis (Bower et al. 2019). However, intensity-based imaging is inferior to FLIM microscopy since FLIM microscopy enables live real-time quantification of cell proliferation and S phase progression as well as assessment of the effects of anti-proliferative drugs (Okkelman et al. 2016). Smartphone-based fluorescence microscope enables fast and accurate counting of bioparticles such as human leukocytes and fluorescent microparticles through automated image analysis that yields results comparable to those of traditional counting methods but at a substantially lower computational cost (Tahir et al. 2024).

Fluorescence microscopy employs dyes for the visualization and observation of cellular components. 4′,6-diamidino-2-phenylindole (DAPI) is one of the commonly used stains to highlight the nuclear and chromosomal structure of the cells and differentiate viable and non-viable cells. Along with dyes like Annexin V and propidium iodide, this is useful in assessing the cellular apoptosis and necrosis. The dyes present an overall view of cell health and functionality, and their usage is helpful for assessing cell viability, pharmacological testing, and therapies (Wallberg et al. 2016).

Molecular imaging is a field that emerged to meet the demand for improved image contrast in medical diagnostic imaging. This simultaneous application enables a study not only limited to morphological data, but also adds functional modifications within a native tissue at both the cell and molecular level (Geraldes 2020). Forster Resonance Energy Transfer (FRET) and related techniques such as Fluorescence Recovery After Photobleaching (FRAP), and FLIM are widely adopted methods used for the study of biomolecular interactions (Ranjit et al. 2021). FRET is a technique for detecting nanoscale interactions by measuring the energy transfer between fluorophores. FRAP i.e., Fluorescence recovery after photobleaching helps to study the movement and diffusion of the molecules in live cells. On the other hand, FLIM is advantageous to spatial and temporal characterization of molecular habitats through the measurement of fluorescence lifetime changes permitting improved insight in metabolite states and protein interactions (Ishikawa-Ankerhold et al. 2012).

Advanced fluorescence techniques have allowed visualization of mitochondrial dynamics in the context of cellular energy regulation in molecular imaging. We performed a real-time analysis of mitochondrial membrane potential (MMP) and structural dynamics with FRET-based probes. They may also check the relation of MMP loss with cristae fragmentation and implicate processes, such as mitochondrial vacuolization and rupture (Peng et al. 2024). Fluorescence microscopy has become a staple tool in bioengineering for its noninvasive and sensitive imaging capabilities. It allows biologists to obtain intensity-based fluorescence data across multiple biological sample dimensions. FLIM is a powerful approach because it helps to derive this information from a variety of different fluorescent probes and probe the nature of their interactions in the molecular milieu. In the latter the importance of FLIM is especially inherent in the investigation of cellular microenvironments, protein-protein interaction and disease markers (Liu et al. 2019).

FRET is a fundamental method in fluorescence imaging, the most common application of which is used to monitor protein-protein interactions in live cells in the field of bioengineering. This is effective because that method only requires donor and acceptor fluorophores to be within 10 nm of each other, so it can be done in real-time while the molecular is being subjected to forces. Alternatively, FRET can also be measured by FLIM as you can get the changing fluorescence lifetime of your donor/acceptor pair. This capability of FRET allows its usage

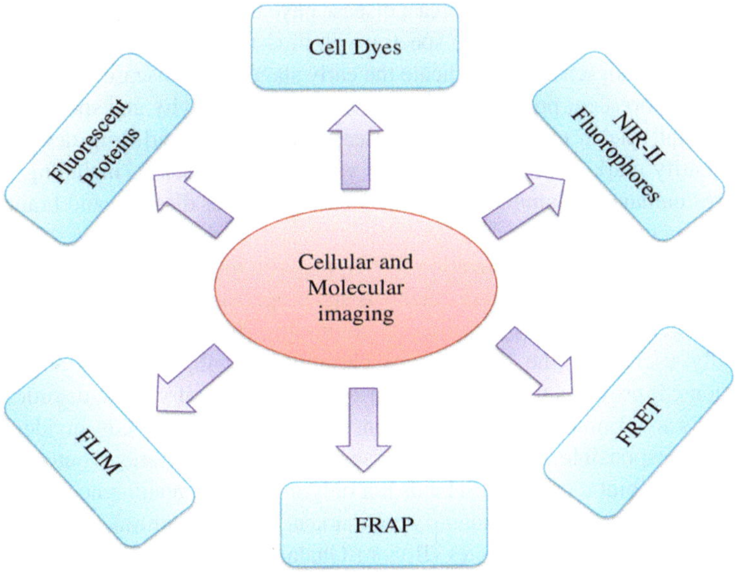

Fig. 1 Overview of various techniques and tools used in cellular and molecular imaging. Cell dyes and fluorescent proteins enable targeted visualization of cellular components, while NIR-II fluorophores enhance deep tissue imaging with improved penetration. Advanced fluorescence-based methods, like FLIM, FRET, FRAP used for monitoring cellular microenvironment, for studying molecular interactions, and for tracking protein mobility, respectively

in visualizing and quantifying protein-protein interactions to help understand cell mechanisms and develop bioengineering remedies. FRAP (Fluorescence Recovery After Photobleaching): Uses photobleaching and lateral diffusion of fluorescently marked molecules (the bleached portion on the right). By applying this method, scientists can quantify the activity and interactions of individual proteins in living cells, providing insights into their dynamic properties and functional significance. FRAP enables quantitation of diffusion coefficients, critical for our understanding of protein mobility and dynamics in the biological systems. It becomes a relevant interpretation for the biological chemistry studies and bioengineering applications, particularly when it is helpful to study the effect of any stimuli that induce the mobilization of the proteins (Ranjit et al. 2021). Figure 1 is an illustration of the role of fluorescence imaging at cellular and molecular level.

2.2 Disease Diagnosis and Prognosis

Fluorescence microscopy becomes a crucial technique for the diagnosis and prognosis of benign breast disease and breast cancer. It allows visualization of biochemical markers for disease activity and progression. Different pathological processes in breast tissues show different fluorescence spectra, and certain benign diseases such as fibroadenoma and fibrocystic breast disease have their own unique characteristics that change a lot of fluorescence spectra with breast cancer samples during tumor growth, which can sensitively indicate the early stage of the disease (Maryakhina et al. 2017). Fluorescent probes, because of their high sensitivity and specificity, play an important role in early diagnosis, disease monitoring, and drug development. They identify markers that are associated with multiple diseases including cellular structures damage, inflammation, cancers, cardiovascular diseases and brain disorders (Wang et al. 2024).

Fluorescence in situ hybridization (FISH) plays an important role in bioengineering, which can be used to disease diagnosis and prognosis. This assists in identifying particular genetic markers for disease, improving diagnostic accuracy. FISH is especially useful for the detection of chromosomal abnormalities and genetic aberrations; it provides the possibility for early diagnosis of genetic disorders such as Down's syndrome and microdeletion syndromes. It also aids in identifying pathogens responsible for infectious diseases, improving patient outcomes and decreasing antibiotic usage. FISH aids not only in the accurate diagnosis of various cancer types but also helps identify biomarkers that are central to personalized medicine and treatment strategies (Bozorg-Ghalati et al. 2019). The dual-modality probes composed of MRI and fluorescence imaging will benefit disease detection and diagnosis through the acquisition of structural information relative to tissues along with insights into the molecular target. In this way, they are capable of imaging diseases on different stages which improves the overall imaging technique (Zhang et al. 2020)

Fluorescence imaging using fluorescent NPs proves to be significantly substantial for early cancer diagnosis. It enables noninvasive imaging of diseased tissues, detecting malignant cancer at an initial stage. Intelligent NPs such as quantum dots and upconversion NPs have improved specificity and sensitivity for cancer imaging, targeting individual cancer cells to improve image resolution and disease staging (Jin et al. 2020). By means of specific response to cancer-associated enzymes, activatable fluorescence probes can utilize high-contrast imaging of cancer lesion during surgery. This leads to the generation of fluorescence signals only in the presence of specific biomarkers, significantly improving both clarity and sensitivity, even in the case of small quantities of enzymes (Fujita and Urano 2024). Fluorescence imaging has vastly improved the sensitivity and versatility of ELISA (enzyme-linked immunosorbent assay) in the detection of antigen-antibody interactions. However, ELISA uses fluorescence labels instead of enzyme substrate reactions, which is more sensitive and specific (Wu et al. 2017). Figure 2 illustrates the diverse applications of fluorescence imaging in disease diagnosis and prognosis.

Fluorescence-based NPs enable early cancer detection and diagnosis. Fluorescent molecular probes aid in detecting specific biomarkers for disease identification and cancer lesion imaging. FISH is critical in detecting genetic markers, identifying chromosomal abnormalities and gene expression patterns. FLISA, Fluorescence-linked immunosorbent assay is extensively used in immunodiagnostics to detect disease specific antigens and antibodies with high sensitivity.

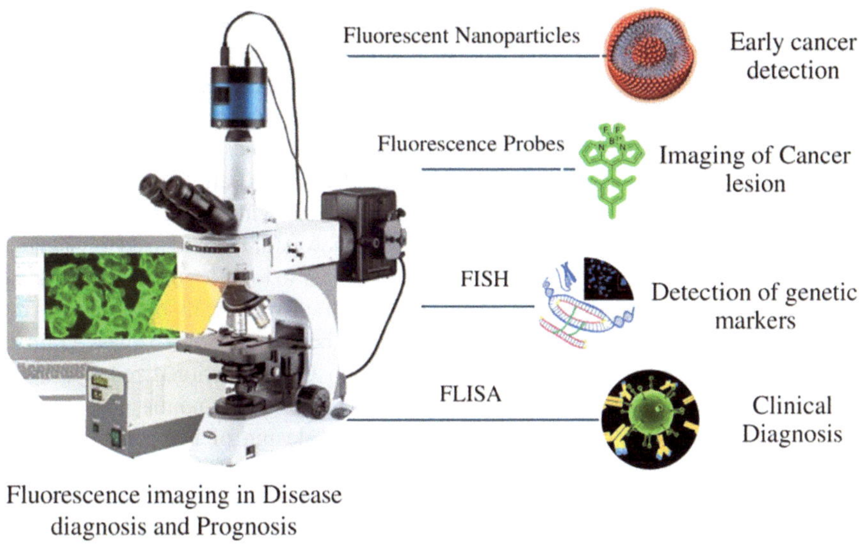

Fig. 2 Fluorescence imaging techniques for disease diagnosis and prognosis

2.3 Biomaterials and Tissue Engineering

Tissue engineering and regenerative medicine processes involving biomaterials pose important challenges to the monitoring and determination of tissue regeneration. The interaction of biomaterials, cells, and tissues in unique environments presents specific challenges for evaluation. Fluorescence microscopy is an indispensable tool in the study of polymers and tissue engineering; it allows the observation of cell interactions, as well as the complex processes of tissue formation at the cellular level. Investigators study engineered tissues in vitro using fluorescent dyes to measure cell viability and to probe the function of tissues enclosed in dense polymeric matrices as thick as 1 mm. Fluorescence microscopy can also use the intrinsic autofluorescence of materials, to image extracellular matrix structures or monitor scaffold degradation (Appel et al. 2013). There is increased relevance of fluorescence microscopy in bioengineering since it offers important information about the behaviors of NPs in contact with cells especially when RapidFLIM enters the picture. RapidFLIM in monitoring the uptake dynamics of NPs offers indispensable information on the localization and clustering of NPs with the close vicinity of the cellular membranes. The improved imaging speed and accuracy facilitate comprehensive examination of subcellular compartments, thereby aiding in the development of targeted delivery systems and strategies in nanomedicine (Ahmed-Cox et al. 2022).

Fluorescence imaging is a noninvasive way to monitor biomaterials such as scaffolds in vivo, making it possible to distantly track stem cells and any deterioration. The developed method has a high sensitivity and spatial resolution and is able to measure the degradation rates of the scaffold and stem cell alive, which facilitates the relationship between scaffold characteristics and tissue (Kim et al. 2020). The fundamentals of fluorescence imaging as discussed in this paper are especially relevant to the design and characterization of nanofibers scaffolds and patches in the fields of biomedicine such as tissue engineering and wound healing. For example, advanced materials to assist in depth (such as NIR probes) allow high demanding imaging of nanofiber structures and their interactions with the respective tissues (Lawrance et al. 2024). Fluorescence imaging is a powerful method used to visualize the nanofibrous patches used in the treatment of cancer to monitor the morphological changes in cancer cell surfaces. This evaluation also aids in determining the feasibility of using drug-eluting patches and its beneficial effectiveness (Jain et al. 2016). Real-time observation of nanofibers in their functionalized hydrogels through fluorescence imaging helps rationally analyze the dynamic properties of such materials, which ultimately contribute to the design of new materials with tailored functionalities (Zhong et al. 2022). Figure 3 depicts the use of fluorescence imaging in tissue engineering for safe and noninvasive monitoring in research in cellular level, in vitro studies, in vivo animal studies and all types of clinical research in human system.

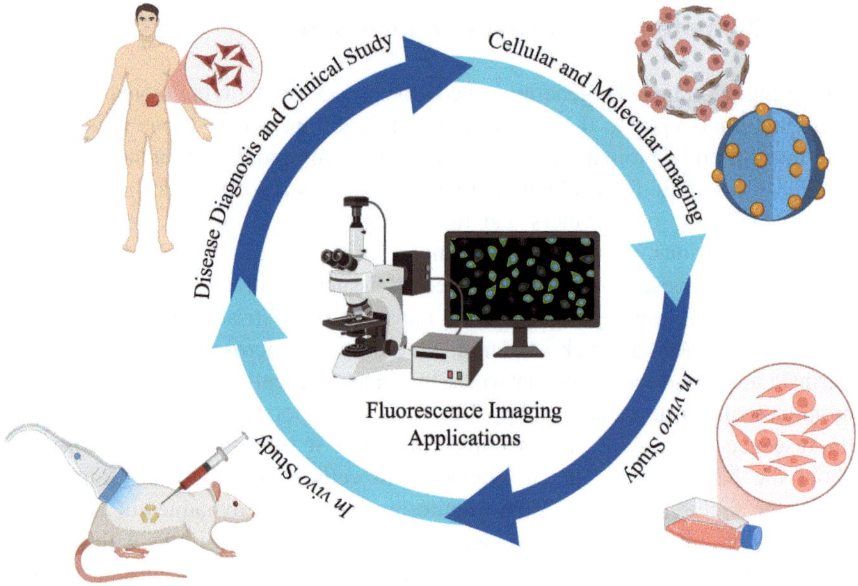

Fig. 3 Applications of Fluorescence imaging in biomedical studies from subcellular level to animal and human studies using safe, quantitative and noninvasive monitoring

2.4 Genetic Engineering

Fluorescence imaging and light microscopy are crucial tools in genetic engineering, enabling precise visualization of cellular events, deoxyribonucleic acid (DNA) modifications, and protein interactions happening during genome editing in real time. Genome editing technologies like clustered regularly interspaced short palindromic repeats-associated protein 9 (CRISPR-Cas9) are often monitored by fluorescence microscopy which gives precise information on the efficiency, specificity, and the required results of the designed genetic modification. Here we overview the use of fluorescence imaging and light microscopy in genome editing with some examples from recent research articles.

2.4.1 Fluorescence Imaging for Analysis of Genome Editing

In fluorescence imaging, fluorophores and fluorescent markers are employed to visualize and track specific genes and their expressions, proteins, or cellular structures. The two widely used fluorescent reporters to mark specific genes or genomic segments are Green Fluorescent Protein (GFP) and Red Fluorescent Protein (RFP). Among different genome editing tools such as CRISPR-Cas9, transcription activator-like effector nucleases, and zinc-finger nucleases, two reporters are mostly used to track and confirm successful gene modifications. In a study, Wimmer et al. have evaluated the mismatch repair mechanism of wild-type CRISPR-Cas9 using a bioluminescence resonance energy transfer (BRET) based reporter system in a single guide ribonucleic acid (sgRNA). GFP2 tagged BRET reported was monitored by fluorescence microscopy (Wimmer et al. 2025). In another study, Bikard et al. have used a GFP-tagged reporter to track the effects of CRISPR-mediated gene knockouts. In their study, the loss of GFP fluorescence indicated successful gene knockouts in both bacterial and mammalian cells (Bikard et al. 2014). CRISPR/Cas9-mediated gene editing is system can easily design large segments of DNA edits in a genome both in vitro and in vivo using homology directed repair and the process can be monitored by fluorescent reporters and can be visualized by fluorescence imaging (Sharma et al. 2018).

Post genome editing, DNA undergoes repair through pathways such as non-homologous end joining (NHEJ) and homology-directed repair to restore the integrity of the genome. The repair process can be visualized in real time using fluorescence microscopy and imaging. In a study, DAPI and AlexaFluor594 were used to visualize FANCD2 gene, a gene required for DNA repair mechanism in NHEJ-mediated repair of CRISPR-Cas9 gene editing in Fanconi anemia – hematopoietic stem cells (Román-Rodríguez et al. 2019).

2.4.2 Light Microscopy in Genome Editing Analysis

The genome editing events and post editing outcomes at cellular and subcellular levels can be tracked and visualized in high resolution images using light microscopy including confocal and super-resolution microscopy. In a study, the cis-cleavage activity of CRISPR/Cas12 on DNA was measured using a confocal microscope and

FRAP to intuitively observe the fluorescence intensity investigated in an aqueous two-phase system comprising polyethylene glycol and dextran, which simulated the intracellular environment (Zhang et al. 2024). Cellular events post genome editing such as cell proliferation, programmed cell death, cell differentiation can be monitored by fluorescent markers. In a study, the transgene expression of plasmids encoding the human histone 2B (H2B), labeled with enhanced GFP, in embryonic stem cells of mouse blastocysts was visualized using an epifluorescence microscope (Yun et al. 2023).

Super-resolution microscopy, including stimulated emission depletion (STED) microscopy, structured illumination microscopy (SIM), Single Molecule Localization Microscopy (SMLM) and stochastic optical reconstruction microscopy (STORM)/photoactivation localization microscopy (PALM) have higher resolution power than conventional light microscopy which can give finer details of chromatin structure and dynamics during and post genome editing. With STORM, a group of researchers have characterized the higher-order chromatin structures at their epigenomic states including histone acetylation, histone methylation and histone methylation in MCF10A cells (Xu et al. 2018). Figure 4 shows the images of microtubules in spread Drosophila macrophages using three techniques SIM, STED, SMLM. While all three techniques resolved the microtubule bundles, STED images gave poorer contrast than the SIM and SMLM images.

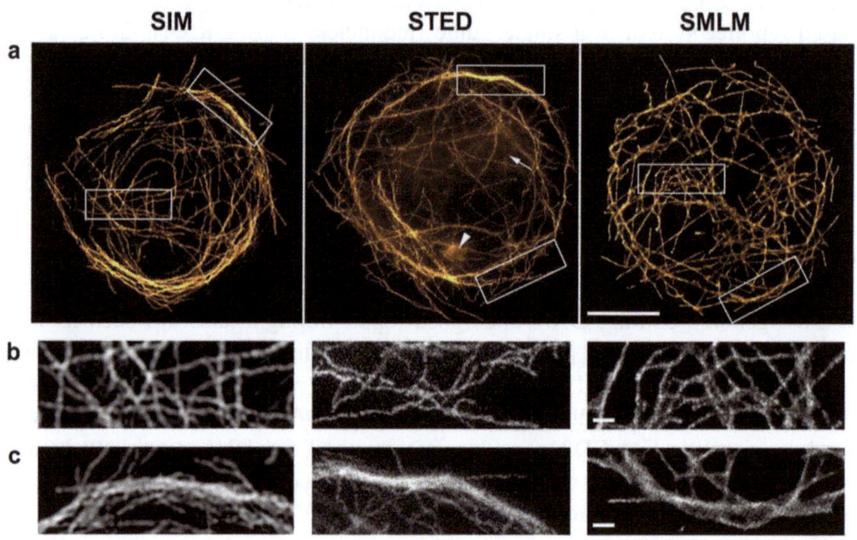

Fig. 4 Microtubules in Drosophila macrophages. Comparison of imaging in SIM, STED, SMLM. (**a**) Single optical section through the microtubule network close to the coverslip. Scale bar, 5 μm. The boxed areas are magnified in (**b**, **c**). In the STED image out of focus blur (arrow) and an artefact caused by high laser intensity (arrowhead) are visible. (**b**) Region showing well separated microtubules. Scale bar, 0.5 μm. (**c**) Region with microtubule bundles. Scale bar, 0.5 μm. (Adapted from Wegel et al. (2016) under a Creative Commons Attribution License (CC BY 4.0))

3 Recent Advances and Challenges in Bioimaging Techniques in Bioengineering Research

The advances in fluorescence imaging and light microscopy in recent years have paved way for the field of bioengineering forward by finer monitoring and analyzing the structural and dynamic studies of biological systems at cellular and molecular level. This has enabled detailed studies on gene editing, drug delivery, tissue engineering, and disease mechanisms. The fluorescence imaging techniques described here are FRAP, the related fluorescence loss in photobleaching (FLIP), fluorescence localization after photobleaching (FLAP), Förster or FRET, widefield or epifluorescence. Nevertheless, the challenges and pitfalls of these techniques required to be addressed for capitalizing their full potential.

Recent advances in FRAP have improved the visualization of physiological transport of biomolecules at cellular and subcellular levels. Where the original version of epifluorescence spot FRAP is not efficient in quantifying interstitial diffusive and convective transport of biomolecules, by using a moderate light source (e.g., Mercury Lamp) instead of a high-power laser beam, researchers have successfully applied FRAP to study diffusion and convection without a prior knowledge of the flow direction (Cai et al. 2022). Similarly, the analysis of interfacial biophysics, i.e., the dynamics of the interaction of biomacromolecules with various surfaces, like ligand-receptor binding on a cell membrane has been improved with recent development of total internal reflection/fluorescence recovery after pattern photobleaching (FRAPP) technique. Recent developments in FRAP technique along with mathematical modelling has improved the analysis of cellular dynamics like cellular and intracellular transport of biomacromolecules (Lippincott-Schwartz et al. 2018). Figure 5 shows the three important steps including the selection of a stable and non-toxic fluorophore for miniaturization of the biosensor, and quantum dots, fluorescent proteins, organic dyes are the commonly used fluorophores; the design of the sensor construct by selecting suitable binding domains for the target and adjusting the distance and orientation between the fluorophores, so, linkers, peptides, or molecular scaffolds may be incorporated to fine-tune the sensor's performance; and the optimization of the sensor performance to design the FRET biosensors.

In recent years, the integration of artificial intelligence (AI) and microfluidic platforms into FRET-based biosensors has improved applications such as disease diagnosis, industrial process control, and environmental monitoring. In a study, researchers have developed machine learning (ML) trained FRET biosensor to detect the changes in dopamine level in brain of a living mice with high specificity and selectivity (Verma et al. 2023). FLIP-based biosensors are efficient in demonstrating connectivity and fluxes of molecules between two cellular or subcellular compartments. FLIP, often used as a control data for FRAP experiments, gives information regarding active or passive transport of biomolecules (Ishikawa-Ankerhold et al. 2012). FLIM is used to map the spatial distribution of the lifetimes within microscopic images of fixed as well as living cells. FLIM is also used in combination with FRET Combination of FLIM and FRET to measure the

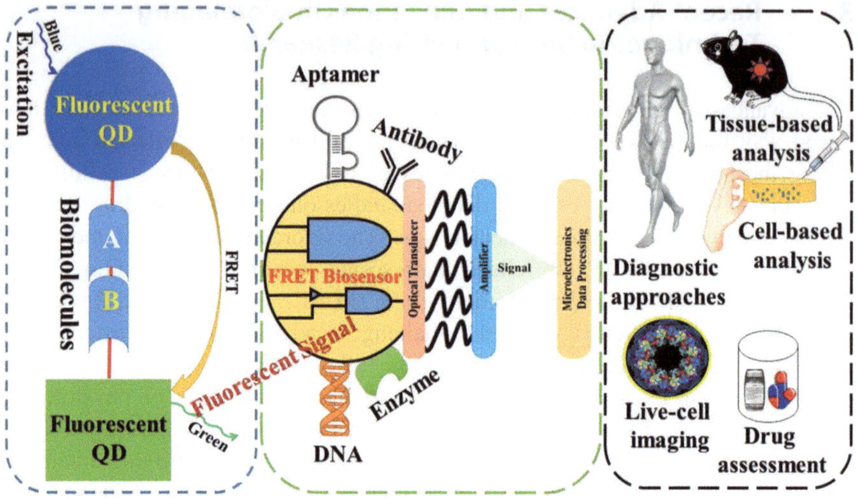

Fig. 5 Schematic of designing and working of FRET biosensor. The key steps include fluorophore selection, sensor construct design, and a through validation of the design to efficiently detect the target molecule. (Reprinted from Verma et al. (2023) under a Creative Commons Attribution License (CC BY 4.0))

lifetime dynamics of biomolecules pixel-by-pixel and thus mapping of its spatial distribution which ultimately quantifies biomolecule concentrations, interactions between biomolecules, and conformational changes with a much higher accuracy than conventional FRET methods (Ishikawa-Ankerhold et al. 2012).

Light microscopy has been evolving significantly since decades to meet the requirements of biologists particularly in the research of cellular dynamics, tissue engineering and organoid development. Light-sheet fluorescence microscopy is well known for providing detailed optically sectioned images by overcoming challenges of photobleaching and phototoxicity. Various methods, such as electrically tunable lenses and tunable acoustic gradient index of refraction lenses, have been adopted to address basic defocusing issues. Hong et al. (2022) has proposed to integrate a deformable mirror into the detection path numerical apertures ranging from 0.72 to 0.75. The study achieved high-speed three-dimensional imaging at a remarkable rate of 26.3 volumes per second, encompassing 35 frames in each volume with a substantial defocus range from -50 to $50\,\mu$m. Again, the combination of STED microscopy and SMLM, proposed by Prakash et al., gives images to cross-validate the observed structures and get more confident results (Prakash et al. 2023).

Advanced methodologies in high-resolution confocal microscopy and in vivo confocal microscopy, multiphoton microscopy and optical coherence tomography have been attempted to obtain high throughput images especially for deeper tissue penetration in thick samples (Wang and Larina 2017). Though bioimaging modalities have come across many advancements, researchers face many challenges due to the pitfalls and artefacts in microscopy and imaging strategies. When it

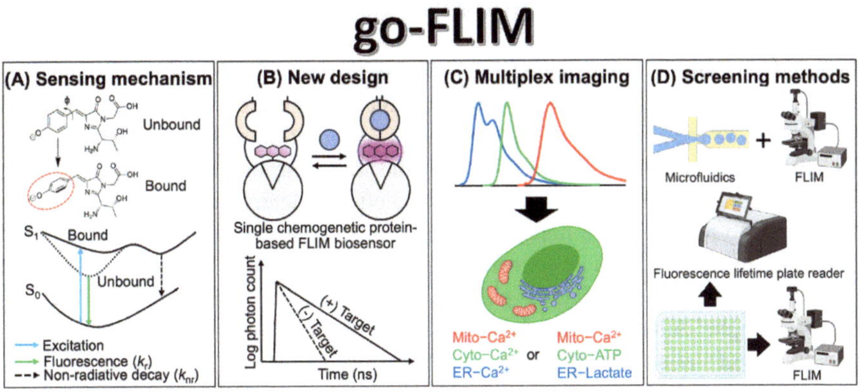

Fig. 6 The promising future of go-FLIM development. (**a**) A schematic representation illustrating the proposed sensing mechanism of single-FP-based FLIM biosensors. A red circle indicates the rotation of chromophores. (**b**) Conceptual design of a single chemogenetic protein-based FLIM biosensor and its FLIM response. (**c**) Representation of multiplex imaging employing go-FLIMs that target various analytes across different organelles such as the mitochondria (mito), cytoplasm (cyto), and endoplasmic reticulum (ER). (**d**) Introduction of methodologies for the screening of go-FLIM. Reprinted from Vu and Arai (2023) under a Creative Commons Attribution License (CC BY 4.0); original illustrations in the figure were created with BioRender.com)

comes to high resolution, the conventional fluorescent probes and light source may face difficulty in deep penetration into thick tissues, dense biological systems like spheroids and organoids, highly scattering tissues, and highly heterogeneous tissue organizations. Long exposure to high-intensity light leads to phototoxicity and photobleaching leading to poor resolution of images. In addition, advanced fluorescence imaging modalities like go-FLIM requires precise calibration, expertise in data analysis as well as high cost with slower data acquisition than intensity-based imaging (Vu and Arai 2023). Figure 6a shows the sensing mechanism behind the fluorescence lifetime changes in an advanced single-FP based FLIM biosensors, Fig. 6b depicts the potential characteristics needed to be considered for the development of a novel class of chemogenetic protein-based FLIM biosensors. Figure 6c shows how multicolored FLIM biosensors enable simultaneous quantification of the same analyte across various organelles or multiple analytes within the same cellular environment. Figure 6d shows the important screening methods for go-FLIMs. With large volume of data generated by advanced high resolution imaging modalities like FRAPP, FRET and multiphoton microscopy, advanced AI-ML techniques can help in robust biomedical imaging.

4 Conclusions and Future Perspectives

Imaging modalities have played a vital role in biomedical research in the last 50 years as we can see almost everything in biology research needs to be validated by visualizing the result under microscope. The evolution of fluorescence imaging and light microscopy has been parred to the mark through each passing decade from basic fluorescent protein to FP-color palette till date. The imaging techniques are poised to revolutionize many scientific fields, from tissue engineering, disease diagnosis to personalized liver medicines. However, in spite of the emergence of advanced fluorescence imaging techniques like FRET, FRAP, FLIM, FLIP, we can see major limitations like photobleaching, poor signal intensity and shallow penetration of the probes still exist. Fluorescent proteins with improved properties, such as enhanced maturation, increased photo-stability and fluorescence intensity will certainly advance future applications of various FRET-based methods.

Super-resolution microscopy techniques such as STORM and PALM will enable researchers to observe cellular structures and molecular interactions at the nanometer scale, providing insights into previously inaccessible biological phenomena. FLIM is a promising technique that can potentially provide information about the local environment of the fluorophore, such as interactions with other molecules or pH or concentration changes in ions. FLIM can remove some of the problems of signal intensity and specificity by measuring the time a fluorophore spends in an excited state. As FLIM technology continues to evolve, it can provide a more stable means of measuring molecular dynamics in real-time with less dependence on fluorescence intensity, which will be better for quantification and signal-to-noise ratio (Park and Gao 2024). New developments in NIR fluorophores and quantum dots are also expected to increase tissue penetration and signal-to-noise ratio, further enhancing the usefulness of fluorescence imaging to deep tissue and live-animal imaging (Gil et al. 2021).

Future perspectives lie in the advancement of high-resolution imaging techniques like SIM, STED microscopy and SMLM to visualize subcellular components at nanoscale range (Wegel et al. 2016). Improvements in Optical coherent tomography and multi-photon microscopy can excel in overcoming problems related to imaging of thick tissues with longer exposure to light source with least damage to the cells. Fluorescence imaging and light microscopy will likely be joined by other imaging modes in the future, such as electron microscopy, MRI, and CT (Refaat et al. 2022). The multimodal imaging will provide complementary information with the high spatial resolution of fluorescence microscopy and wider anatomical context provided by other imaging modes. This fusion will open new paths of in vivo imaging and enable researchers to trace biological occurrences at various scales, from molecular interaction to the dynamics of the whole organ. Additionally advanced AI-ML algorithms associated with high-throughput microscopy can improve large spectral data management, precise unmixed signals, and accurate analysis of pathological samples (Durkee et al. 2021). These computer vision-based techniques will likely be the solution to issues such as noise removal, pattern recognition, and quantification, which will make the accuracy and throughput of fluorescence imaging more efficient in the field of bioengineering. Furthermore, the combination of advanced fluorescence techniques with light microscopy and nanoscopy instruments will help researchers to image structural and functional dynamics of cells and biomolecules in real-time with excellent resolution.

Acknowledgments DKB acknowledges the funding support from Science & Technology Department, Government of Odisha, India for the project "Phytonanotechnogical Interventions for Hepatocellular Carcinoma" (ST-BT-MISC-0005-2023).

Conflict of Interest The authors declare that they have no conflict of interest.

References

Ahmed-Cox A, Macmillan AM, Pandzic E, Whan RM, Kavallaris M (2022) Application of rapid fluorescence lifetime imaging microscopy (RapidFLIM) to examine dynamics of nanoparticle uptake in live cells. Cells 11(4). https://doi.org/10.3390/cells11040642

Appel AA, Anastasio MA, Larson JC, Brey EM (2013) Imaging challenges in biomaterials and tissue engineering. Biomaterials 34(28):6615–6630. https://doi.org/10.1016/j.biomaterials.2013.05.033

Bikard D, Euler CW, Jiang W, Nussenzweig PM, Goldberg GW, Duportet X, Fischetti VA, Marraffini LA (2014) Exploiting CRISPR-Cas nucleases to produce sequence-specific antimicrobials. Nat Biotechnol 32(11):1146–1150. https://doi.org/10.1038/nbt.3043

Bower AJ, Sorrells JE, Li J, Marjanovic M, Barkalifa R, Boppart SA (2019) Tracking metabolic dynamics of apoptosis with high-speed two-photon fluorescence lifetime imaging microscopy. Biomed Opt Express 10(12):6408–6421. https://doi.org/10.1364/BOE.10.006408

Bozorg-Ghalati F, Mohammadpour I, Ranjbaran R (2019) Applications of fluorescence in situ hybridization in detection of disease biomarkers and personalized medicine. Comp Clin Pathol 28(1):3–10. https://doi.org/10.1007/s00580-017-2625-x

Cai N, Lai AC-K, Liao K, Corridon PR, Graves DJ, Chan V (2022) Recent advances in fluorescence recovery after Photobleaching for decoupling transport and kinetics of biomacromolecules in cellular physiology. Polymers 14(9). https://doi.org/10.3390/polym14091913

Durkee MS, Abraham R, Clark MR, Giger ML (2021) Artificial intelligence and cellular segmentation in tissue microscopy images. Am J Pathol 191(10):1693–1701. https://doi.org/10.1016/j.ajpath.2021.05.022

Ettinger A, Wittmann T (2014) Fluorescence live cell imaging. Methods Cell Biol 123:77–94. https://doi.org/10.1016/B978-0-12-420138-5.00005-7

Fujita K, Urano Y (2024) Activity-based fluorescence diagnostics for cancer. Chem Rev 124(7):4021–4078. https://doi.org/10.1021/acs.chemrev.3c00612

Geraldes CFGC (2020) Introduction to infrared and Raman-based biomedical molecular imaging and comparison with other modalities. Molecules 25(23):Article 23. https://doi.org/10.3390/molecules25235547

Gil HM, Price TW, Chelani K, Bouillard JSG, Calaminus SD, Stasiuk GJ (2021) NIR-quantum dots in biomedical imaging and their future. Iscience 24(3):102189. https://doi.org/10.1016/j.isci.2021.102189

Hong W, Wright T, Sparks H, Dvinskikh L, MacLeod K, Paterson C, Dunsby C (2022) Adaptive light-sheet fluorescence microscopy with a deformable mirror for video-rate volumetric imaging. Appl Phys Lett 121(19):193703. https://doi.org/10.1063/5.0125946

Ishikawa-Ankerhold HC, Ankerhold R, Drummen GPC (2012) Advanced fluorescence microscopy techniques—FRAP, FLIP, FLAP, FRET and FLIM. Molecules 17(4):Article 4. https://doi.org/10.3390/molecules17044047

Jain S, Meka SRK, Chatterjee K (2016) Engineering a Piperine eluting nanofibrous patch for cancer treatment. ACS Biomater Sci Eng 2(8):1376–1385. https://doi.org/10.1021/acsbiomaterials.6b00297

Jin K-T, Yao J-Y, Ying X-J, Lin Y, Chen Y-F (2020). Nanomedicine and early cancer diagnosis: molecular imaging using fluorescence nanoparticles. Retrieved December 17, 2024, from https://www.eurekaselect.com/article/110140

Kim SH, Park JH, Kwon JS, Cho JG, Park KG, Park CH, Yoo JJ, Atala A, Choi HS, Kim MS, Lee SJ (2020) NIR fluorescence for monitoring in vivo scaffold degradation along with stem cell tracking in bone tissue engineering. Biomaterials 258:120267. https://doi.org/10.1016/j.biomaterials.2020.120267

Lawrance R, Chowdhury P, Lin H-C, Chan Y-H (2024) The luminous frontier: transformative NIR-IIa fluorescent polymer dots for deep-tissue imaging. RSC Appl Polym 2(5):749–774. https://doi.org/10.1039/D4LP00076E

Lippincott-Schwartz J, Snapp EL, Phair RD (2018) The development and enhancement of FRAP as a key tool for investigating protein dynamics. Biophys J 115(7):1146–1155. https://doi.org/10.1016/j.bpj.2018.08.007

Liu X, Lin D, Becker W, Niu J, Yu B, Liu L, Qu J (2019) Fast fluorescence lifetime imaging techniques: a review on challenge and development. J Innov Opt Health Sci 12(05):1930003. https://doi.org/10.1142/S1793545819300039

Liu Z, Jin L, Chen J, Fang Q, Ablameyko S, Yin Z, Xu Y (2021) A survey on applications of deep learning in microscopy image analysis. Comput Biol Med 134:104523. https://doi.org/10.1016/j.compbiomed.2021.104523

Maryakhina VS, Korneva YS, Chekurov IV, Shisterova OA (2017) Fluorescent diagnostics of benign breast diseases and breast cancer. J Biomed Photon Eng 3(4):Article 4

Okkelman IA, Dmitriev RI, Foley T, Papkovsky DB (2016) Use of fluorescence lifetime imaging microscopy (FLIM) as a timer of cell cycle S phase. PLoS One 11(12):e0167385. https://doi.org/10.1371/journal.pone.0167385

Olesińska-Mönch M, Deo C (2023) Small-molecule photoswitches for fluorescence bioimaging: engineering and applications. Chem Commun 59(6):660–669. https://doi.org/10.1039/D2CC05870G

Park J, Gao L (2024) Advancements in fluorescence lifetime imaging microscopy instrumentation: towards high speed and 3D. Curr Opinion Solid State Mater Sci 30:101147. https://doi.org/10.1016/j.cossms.2024.101147

Peng F, Ai X, Sun J, Ge X, Li M, Xi P, Gao B (2024) Fluorescence lifetime super-resolution imaging unveil the dynamic relationship between mitochondrial membrane potential and cristae structure using the Förster resonance energy transfer strategy. Anal Chem 96(27):11052–11060. https://doi.org/10.1021/acs.analchem.4c01905

Prakash K, Heintzmann R, Manor U, Smith C (2023) Opinions on imaging and cutting-edge developments in light microscopy for biomedical applications. Appl Phys Lett 123(26)

Ranjit S, Lanzanò L, Libby AE, Gratton E, Levi M (2021) Advances in fluorescence microscopy techniques to study kidney function. Nat Rev Nephrol 17(2):128–144. https://doi.org/10.1038/s41581-020-00337-8

Refaat A, Yap ML, Pietersz G, Walsh APG, Zeller J, del Rosal B, Wang X, Peter K (2022) In vivo fluorescence imaging: success in preclinical imaging paves the way for clinical applications. J Nanobiotechnol 20(1):450. https://doi.org/10.1186/s12951-022-01648-7

Román-Rodríguez FJ, Ugalde L, Álvarez L, Díez B, Ramírez MJ, Risueño C, Cortón M, Bogliolo M, Bernal S, March F, Ayuso C, Hanenberg H, Sevilla J, Rodríguez-Perales S, Torres-Ruiz R, Surrallés J, Bueren JA, Río P (2019) NHEJ-mediated repair of CRISPR-Cas9-induced DNA breaks efficiently corrects mutations in HSPCs from patients with Fanconi anemia. Cell Stem Cell 25(5):607–621.e7. https://doi.org/10.1016/j.stem.2019.08.016

Shang A, Shao S, Zhao L, Liu B (2024) Far-red fluorescent proteins: tools for advancing in vivo imaging. Biosensors 14(8):359. https://doi.org/10.3390/bios14080359

Sharma A, Toepfer CN, Ward T, Wasson L, Agarwal R, Conner DA, Hu JH, Seidman CE (2018) CRISPR/Cas9-mediated fluorescent tagging of endogenous proteins in human pluripotent stem cells. Curr Protoc Hum Genet 96:21.11.1–21.11.20. https://doi.org/10.1002/cphg.52

Tahir MN, Xie Y, Sami MA, Punde R, Hassan U (2024) FluoCount: an efficient and accurate cells and bioparticles enumeration mobile application for portable fluorescence microscopes. IEEE Sens Lett 8(4):1–4. https://doi.org/10.1109/LSENS.2024.3371209

Verma AK, Noumani A, Yadav AK, Solanki PR (2023) FRET based biosensor: principle applications recent advances and challenges. Diagnostics 13(8). https://doi.org/10.3390/diagnostics13081375

Vu CQ, Arai S (2023) Quantitative imaging of genetically encoded fluorescence lifetime biosensors. Biosensors 13(10). https://doi.org/10.3390/bios13100939

Wallberg F, Tenev T, Meier P (2016) Analysis of apoptosis and necroptosis by fluorescence-activated cell sorting. Cold Spring Harb Protoc 2016(4):pdb.prot087387. https://doi.org/10.1101/pdb.prot087387

Wang S, Larina IV (2017) 8 - High-resolution imaging techniques in tissue engineering. In: Narayan RJ (ed) Monitoring and evaluation of biomaterials and their performance in vivo. Woodhead Publishing, pp 151–180. https://doi.org/10.1016/B978-0-08-100603-0.00008-0

Wang T, Chen Y, Wang B, Wu M (2023) Recent progress of second near-infrared (NIR-II) fluorescence microscopy in bioimaging. Front Physiol 14. https://doi.org/10.3389/fphys.2023.1126805

Wang X, Ding Q, Groleau RR, Wu L, Mao Y, Che F, Kotova O, Scanlan EM, Lewis SE, Li P, Tang B, James TD, Gunnlaugsson T (2024) Fluorescent probes for disease diagnosis. Chem Rev 124(11):7106–7164. https://doi.org/10.1021/acs.chemrev.3c00776

Wegel E, Göhler A, Lagerholm BC, Wainman A, Uphoff S, Kaufmann R, Dobbie IM (2016) Imaging cellular structures in super-resolution with SIM, STED and Localisation Microscopy: a practical comparison. Sci Rep 6(1):27290. https://doi.org/10.1038/srep27290

Wimmer T, Lorenz A, Hossfeld LT, Goud Ponnam SP, Lytvynchuk L, Stieger K (2025) Evaluation of CRISPR-Cas9 mismatch activity using a BRET-based reporter system. Microchem J 208:112256. https://doi.org/10.1016/j.microc.2024.112256

Witte R, Andriasyan V, Georgi F, Yakimovich A, Greber UF (2018) Concepts in light microscopy of viruses. Viruses 10(4):Article 4. https://doi.org/10.3390/v10040202

Wu Y, Guo W, Peng W, Zhao Q, Piao J, Zhang B, Wu X, Wang H, Gong X, Chang J (2017) Enhanced fluorescence ELISA based on HAT triggering fluorescence "turn-on" with enzyme-antibody dual labeled AuNP probes for ultrasensitive detection of AFP and HBsAg. ACS Appl Mater Interfaces 9(11):9369–9377. https://doi.org/10.1021/acsami.6b16236

Xu J, Ma H, Jin J, Uttam S, Fu R, Huang Y, Liu Y (2018) Super-resolution imaging of higher-order chromatin structures at different Epigenomic states in single mammalian cells. Cell Rep 24(4):873–882. https://doi.org/10.1016/j.celrep.2018.06.085

Yun WS, Cho H, Jeon SI, Lim D-K, Kim K (2023) Fluorescence-based mono- and multimodal imaging for in vivo tracking of mesenchymal stem cells. Biomolecules 13(12). https://doi.org/10.3390/biom13121787

Zhang L, Tong S, Zhang Q, Bao G (2020) Lipid-encapsulated Fe3O4 nanoparticles for multimodal magnetic resonance/fluorescence imaging. ACS Appl Nano Mater 3(7):6785–6797. https://doi.org/10.1021/acsanm.0c01193

Zhang Z, Fan Y, Zuo C (2022) Integrated multimodal and fluorescence imaging microscope based on LED illumination. Adv Opt Imag Technol V 12316:48–54. https://doi.org/10.1117/12.2642266

Zhang Y, Chen J, Wu Z, Zhao C, Wang R, Li Z, Wang J, Wang D (2024). CRISPR / Cas enzyme catalysis in liquid—Liquid phase-separated systems 2407194, pp 1–10. https://doi.org/10.1002/advs.202407194

Zhong J, Zhao T, Liu M (2022) Fluorescence microscopic visualization of functionalized hydrogels. NPG Asia Mater 14(1):1–23. https://doi.org/10.1038/s41427-022-00376-6

Fundamentals of Optical Coherence Tomography

Maitreyee Roy, Rehana Khan, A. Q. M. Sala Uddin Pathan, and Rajiv Raman

Abstract

Optical Coherence Tomography (OCT) is a high-resolution imaging technique that generates cross-sectional images of biological tissue using low-coherence interferometry. OCT revolutionized ophthalmic diagnosis after its invention as it offered real-time structural examination of the retina and anterior segment at micrometer resolution. This chapter introduces the fundamental principles of OCT technology, including light-tissue interactions, axial and lateral resolution trade-offs, and the evolution from Time-Domain to Spectral- and Swept-Source OCT systems. Limitations to OCT, in spite of clinical success, are low penetration in scattering tissue, motion and shadowing artifacts, device cost, inter-platform variation, and the need for skilled operators. Additionally, although AI-powered OCT interpretation and handheld devices show promise, their clinical use is ongoing. The future prospects are geared towards addressing these limitations by creating multimodal, functional OCT systems, incorporating machine learning, and expanding their use to systemic disease monitoring. OCT is also advancing as a useful tool not just in ophthalmology but in multidisciplinary fields of medicine and biomedical research.

Keywords

Low-coherence interferometry · Swept-source OCT · Spectral-domain OCT · Deep learning · Functional imaging

M. Roy · R. Khan · A. Q. M. Sala Uddin Pathan
School of Optometry and Vision Science, University of New South Wales, Sydney, NSW, Australia

R. Raman (✉)
Shri Bhagwan Mahavir Vitreoretinal Services, Sankara Nethralaya, Chennai, India

I. F. Cengiz et al. (eds.), *Bioimaging Modalities in Bioengineering*, Biomaterials, Bioengineering and Sustainability 9,
https://doi.org/10.1007/978-3-031-99410-4_3

1 Introduction

Optical Coherence Tomography (OCT) has transformed ophthalmic imaging by providing unprecedented microscopic detail of the eye with near-histological resolution (Fujimoto et al. 2000). Its high-speed, high-resolution, and non-invasive nature has established it as a valuable asset in the diagnosis and management of ocular conditions. The history of this groundbreaking technology goes back to the initial work of Dr. James G. Fujimoto and his research group at the Massachusetts Institute of Technology in the early 1990s (Fujimoto et al. 2000). In 1991, Fujimoto and others, Carmen Puliafito and David Huang among them, reported a landmark paper in Science that introduced OCT as a new imaging technology based on interferometric principles (Fujimoto et al.; Zeppieri et al. 2023). The technology enabled high-resolution cross-sectional images of biological tissue, with the retina as one of the initial clinical applications. Their publication set the stage for the first application of Time-Domain OCT (TD-OCT) to ophthalmology, which made it possible to achieve precise visualization of retinal layers and structural pathology (Zeppieri et al. 2023; Huang et al. 1991). By 1996, the first commercial versions of OCT machines were released, providing a paradigm shift in retinal diagnosis (Fercher 1996).

Since its advent, OCT technology has come a long way. The shift from TD-OCT to Spectral-Domain OCT (SD-OCT), Swept-Source OCT (SS-OCT), and functional imaging modalities like OCT Angiography (OCTA) has taken it far beyond the reach of what it was in its early stages (Herrera et al. 2025; Xia and Hua 2023; Spaide et al. 2018). Now, OCT is central to the diagnosis and monitoring of conditions like age-related macular degeneration, diabetic retinopathy, and glaucoma, allowing early intervention and tailoring of treatments (Zeppieri et al. 2023). The incorporation of machine learning and artificial intelligence into OCT analysis holds the potential for even more accuracy and efficiency in the detection and prognosis of disease (Kapoor et al. 2019). Even with its common application, cost, accessibility, and standardized interpretation remain areas of continued improvement. The role of OCT in contemporary eye care cannot be overemphasized. With ongoing developments enhancing its functionality, imaging technology continues to lead the way in innovation, driving the future of diagnostics and patient management.

2 Basic Principles of OCT

OCT is a non-invasive, label-free, and interferometric imaging technique that generates high-resolution, cross-sectional images of internal microstructures in biological tissues and materials (Fujimoto and Drexler 2015). OCT generates cross-sectional images by measuring the intensity of backscattered or back-reflected light from different depths within a tissue. This depth-resolved imaging is made possible through the use of low-coherence interferometry, based on the Michelson interferometer (Fujimoto and Drexler 2015).

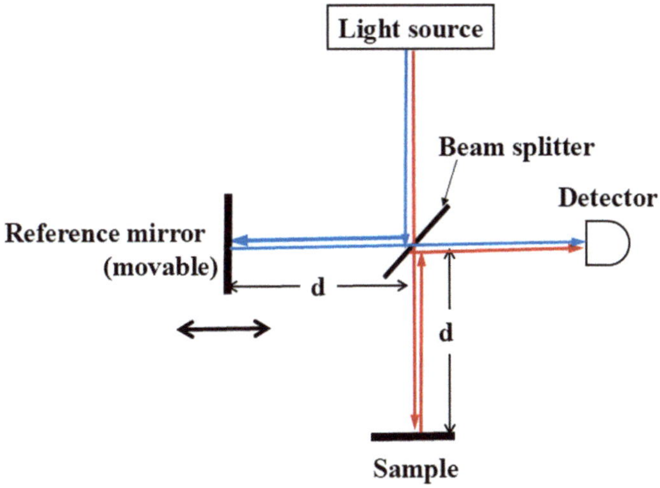

Fig. 1 Schematic of a basic Michelson interferometer setup for low-coherence interferometry. A broadband light source is split into reference and sample arms by a beam splitter. Reflected beams from both arms recombine and interfere at the detector when their optical path lengths match within the coherence length

2.1 Michelson Interferometer

The Michelson interferometer is a fundamental optical setup used to measure small differences in optical path length with high precision, and it forms the core of low-coherence interferometry in OCT (Rollins and Izatt 1999). The Michelson interferometer setup is shown in Fig. 1.

Light from a broadband source is directed toward a beam splitter, which divides the beam into two paths: one toward the reference mirror and the other toward the sample. The light beams reflect back and recombine at the beam splitter. These two beams then interfere, producing an interference pattern that is detected by a photodetector (Maciel et al. 2016). Interference occurs only when the optical path lengths of the two arms are equal or differ by less than the coherence length of the light source (Abdulhalim 2012). The behavior of this interference is strongly influenced by the type of light source. Monochromatic light (e.g., laser) has a long coherence length. It produces sustained and wide-ranging interference fringes, even when there is a substantial difference between the optical path lengths of the sample and reference arms (Aumann et al. 2019). Broadband light (e.g., super luminescent diode or white light) has a short coherence length. This results in interference that is highly localized and occurs only when the optical path lengths of the two arms match very closely, typically within a few micrometers (Aumann et al. 2019). Interference patterns with monochromatic and broadband light sources are illustrated in Fig. 2.

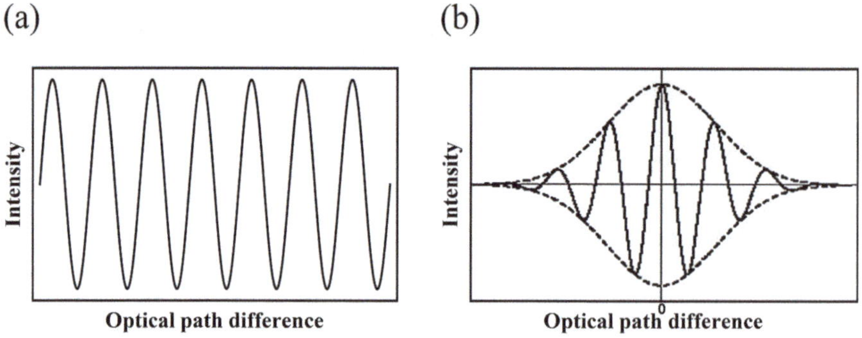

Fig. 2 Interference patterns for different light sources: (**a**) Monochromatic light produces extended, periodic fringes across a wide range of optical path differences, due to its long coherence length. (**b**) Broadband light produces a narrow, localized interference peak centered at zero path difference, due to its short coherence length. This localization enables precise depth resolution in low-coherence interferometry

This distinction is critical to OCT. The use of a low-coherence source allows the system to selectively detect light from specific depths in the sample, enabling depth-resolved imaging (Aumann et al. 2019). Each layer within the tissue generates interference only when it falls within the coherence envelope of the reference arm. This spatial confinement is what grants OCT its high axial resolution and ability to distinguish between microstructural layers in biological tissues.

2.2 Low-Coherence Interferometry

Low-Coherence Interferometry (LCI) uses a broadband light source with a short coherence length, which is essential for enabling depth-resolved imaging in OCT (Sheppard and Roy 2003). When light from the broadband light source is split into two paths, one directed to a movable reference mirror and the other to a layered sample, interference occurs only when the optical path length of the sample arm closely matches that of the reference arm, within the coherence length of the source. This property ensures that only light reflected from a specific depth in the tissue contributes to the interference signal, enabling high axial resolution (Sheppard and Roy 2003).

Figure 3 illustrates how LCI operates when imaging a sample with multiple reflective layers, as is typical in biological tissues. The sample contains three interfaces at different depths; each partially reflects incident light back toward the beam splitter. These reflections recombine with the reference beam, but due to the short coherence length, only those returning from depths matching the reference path generate interference. The result, shown on the right side of Fig. 3, is an interference signal with three distinct peaks, each corresponding to one of the reflective layers. These peaks mark the depth positions of the tissue interfaces,

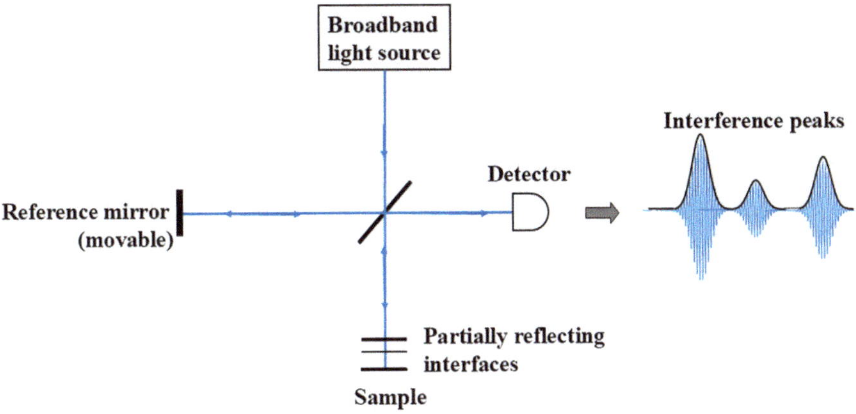

Fig. 3 Schematic representation of low-coherence interferometry: A broadband light source is split into reference and sample arms by a beam splitter. Reflected light from the movable reference mirror and from partially reflecting interfaces within the sample recombines at the detector. Interference occurs when the optical path lengths match within the coherence length of the source, resulting in distinct interference peaks. These peaks correspond to reflections from different sample depths

forming what is known as an A-scan: a one-dimensional profile of reflectivity versus depth (Sheppard and Roy 2003). By scanning the beam laterally across the tissue, a series of A-scans can be acquired and assembled into two-dimensional cross-sectional images (B-scans) or three-dimensional volumetric datasets. This layered, depth-resolved signal generation is the fundamental principle behind OCT imaging.

3 Scanning Protocols and Procedures

OCT scanning protocols are based on fundamental imaging principles that enable high-resolution visualization of ocular structures. These protocols optimize key parameters such as resolution, scan area, and scan density to ensure accurate structural assessment. By tailoring scan methodologies to specific anatomical regions, OCT facilitates precise imaging of the retina, optic nerve, and anterior segment. Advances in OCT technology have led to the development of various scanning approaches, each designed to enhance image acquisition and interpretation while maintaining consistency in fundamental optical principles. An outline of the OCT scanning protocols is detailed in Table 1.

Table 1 Overview of OCT scanning protocols

Scan Type	Fundamental Principles	System Components	Imaging Methodology	Key Considerations
Macular Cube Scan (Massin et al. 2001; Polito et al. 2005; Zhang et al. 2007; Bernardes et al. 2008; Danis et al. 2008)	Utilizes high-resolution volumetric scanning to assess macular structure and detect conditions such as macular degeneration and diabetic macular edema.	Includes a broadband light source, scanning optics, and a high-speed detector.	Acquires volumetric data across a 6×6 mm or 8×8 mm area centered on the fovea, with B-scans composed of multiple A-scans.	High-density modes improve resolution but may introduce motion artifacts and longer scan times.
Optic Nerve Head or Glaucoma Scan (Medeiros et al. 2004; Strouthidis and Garway-Heath 2008)	Measures optic nerve head and retinal nerve fiber layer (RNFL) thickness to monitor glaucomatous changes.	Uses an interferometric system with phase-sensitive detection.	Employs a circular scan pattern (3.4 mm diameter) around the optic disc with high-resolution phase analysis.	Requires precise segmentation algorithms to ensure reproducibility and accuracy in detecting RNFL thinning.
Radial Line or Retinal Thickness Scan (Stanga and Bird 2001; Bhende et al. 2018)	Provides cross-sectional imaging of the retina to assess localized changes in thickness, commonly used for diabetic macular edema.	Consists of a laser light source, a beam splitter, and an interference detector.	Acquires multiple radial scans (12–24 lines) centered on the fovea, optimizing contrast and resolution.	Offers faster acquisition compared to cube scans but lacks full volumetric data.
Anterior Segment Scan (Bhende et al. 2018; Asam et al. 2019)	Captures high-resolution images of the cornea, anterior chamber, and angle structures for evaluating conditions such as glaucoma and corneal abnormalities.	Requires an OCT system with anterior segment imaging capabilities, often with additional hardware.	Uses a high-resolution scan over a 10–16 mm area to capture detailed images of the anterior eye structures.	Some systems require specific attachments or settings for anterior segment imaging.
OCT Angiography (OCTA) Scan (De Carlo et al. 2015; Spaide et al. 2015; Matsunaga et al. 2014)	Visualizes retinal and choroidal microvasculature without requiring contrast agents by detecting motion contrast in repeated OCT scans.	Involves high-speed OCT acquisition and computational motion analysis.	Uses repeated B-scans at the same location to generate vascular maps in scan sizes ranging from 3×3 mm to 8×8 mm.	Highly sensitive to motion artifacts; limited field of view requires advanced image processing techniques.

Widefield or Ultra-Widefield OCT Scan (Kalra et al. 2021; Ripa et al. 2022; Choudhry et al. 2016)	Expands imaging coverage to capture peripheral retina, useful for detecting diabetic retinopathy and peripheral lesions.	Incorporates a specialized optics system with wide-angle optics, swept-source technology, and beam expansion to extend imaging coverage.	Covers areas up to 12 × 12 mm or larger with high-resolution imaging across central and peripheral regions.	Image montaging may be required; peripheral resolution may be lower compared to standard macular scans.
Enhanced Depth Imaging Scan (Wong et al. 2011; Merchant et al. 2013)	Enhances visualization of deeper structures, particularly the choroid, for conditions such as central serous chorioretinopathy.	Uses modified interferometric techniques for improved depth penetration.	Employs a 6 × 6 mm macular scan with an altered reference arm to enhance choroidal contrast.	Requires optimized settings to balance scan time and image clarity, especially for deep-layer analysis.

4 OCT Advancements

4.1 Time-Domain OCT (TD-OCT)

TD-OCT is a cross-sectional imaging technology first introduced by David Huang et al. in 1991 (Huang et al. 1991). It is a non-invasive imaging method that uses the Michelson interferometer technique to obtain high-resolution images of biological tissues. In a typical OCT system based on this technique, light emitted from a low-coherence source is split and directed into two paths: a sample arm equipped with a beam delivery instrument and a reference arm containing a mechanically adjustable mirror for scanning different depths (Huang et al. 1991). The light reflected from both arms is then recombined, and the resulting interference signal is detected by a photodiode. This signal is subsequently demodulated, processed by a computer, and displayed as an image. While TD-OCT is a simple and well-established system with high dynamic range and good axial resolution, it is relatively slow and exhibits lower sensitivity compared to frequency domain OCT (FD-OCT) (Zhang et al. 2023).

4.2 Frequency-Domain OCT (FD-OCT)

FD-OCT is an advanced OCT technique that offers higher speed and sensitivity compared to TD-OCT. (Bouma et al. 2009) FD-OCT captures depth-resolved images by analyzing interference signals using Fourier transformation rather than mechanically scanning a reference mirror (as in TD-OCT). It operates in two main configurations: SD-OCT and SS-OCT. (Bouma et al. 2009) SD-OCT is an advanced cross-sectional imaging technique first demonstrated by Wojtkowski et al. (2002). It is a non-invasive imaging method that leverages the Fourier transformation of the spectral interference signal to obtain depth-resolved images of biological tissue (Wojtkowski et al. 2002). Light from a broadband low-coherence source is split between a stationary reference arm and a sample arm using a fiber-optic coupler. Reflected light from both arms creates interference that is wavelength-dependent and detected by a spectrometer with a line-scan camera. The spectral interferogram is then processed via a Fourier transform to reconstruct a depth profile of the tissue. This eliminates the need for mechanical scanning in the reference arm (Wojtkowski et al. 2002). SD-OCT offers significantly higher imaging speed and sensitivity compared to TD-OCT, but suffers from sensitivity roll-off with depth and is limited by spectrometer resolution (Bouma et al. 2009; Wojtkowski et al. 2002).

4.3 Functional OCT Techniques

4.3.1 Polarization-Sensitive OCT (PS-OCT)

Polarization-Sensitive OCT (PS-OCT) pushes the limit of standard OCT through the assessment of tissue birefringence, enabling anisotropic contrast based on tissue structure. This method involves using a detection platform that is usually combined with a Mach-Zehnder or Michelson interferometer (De Boer et al. 2017). It makes measurements on phase retardation across orthogonal polarizations, thereby enabling visualization of the microstructural parameters of fibrous tissues. The technique is based on a polarization-controlled broadband light source, polarization-diverse detection to measure changes in birefringence, and signal processing algorithms to calculate phase retardation (De Boer et al. 2017; Pircher et al. 2006). PS-OCT is especially useful for imaging tissues with highly ordered collagen structures, enabling contrast mechanisms in addition to standard OCT intensity-based imaging (Yang and Vitkin 2007).

4.3.2 Doppler OCT (D-OCT)

Doppler OCT (D-OCT) is an imaging extension of OCT that employs the Doppler effect to quantify moving particles inside biological tissues (Leitgeb et al. 2014). Through detection of backscattered light frequency changes, the method allows high-resolution visualization of flow. A conventional D-OCT system includes a broadband source, a Michelson interferometer, and an ultra-high-speed detector (Leitgeb et al. 2014; Srinivasan et al. 2012). Velocity measurements are obtained from phase differences between consecutive A-scans, with implementation differing between TD-OCT and FD-OCT. TD-OCT uses a lock-in amplifier to obtain Doppler shifts, whereas FD-OCT obtains phase differences between consecutive scans, providing greater sensitivity and speed (Haavisto et al. 2015). While mainly applied for flow analysis, D-OCT needs to have precise phase stability and motion artifact correction in order to obtain accurate measurements (Morgner et al. 2000).

4.3.3 Spectroscopic OCT (S-OCT)

Spectroscopic OCT (S-OCT) broadens OCT imaging by taking advantage of wavelength-dependent absorption and scattering to retrieve biochemical information from tissues (Nam and Yoo 2018). S-OCT employs a short-time Fourier transform to produce depth-resolved spectral profiles, allowing for improved tissue characterization (Nam and Yoo 2018; Yang 2005). S-OCT examines spectral changes due to endogenous chromophores, and thus it is especially suitable for molecular contrast imaging (Leitgeb et al. 2005). Nevertheless, one of the major challenges is the optimization of the trade-off between spectral and spatial resolution to achieve functional imaging without sacrificing depth penetration (P Mattison et al. 2014). The use of broadband light sources further increases spectral contrast, which offers important information regarding tissue composition at the molecular level (P Mattison et al. 2014).

4.3.4 Molecular Imaging OCT (MI-OCT)

Molecular Imaging OCT improves on the standard OCT with the addition of exogenous contrast agents to promote molecular specificity (Robles et al. 2011). In contrast to S-OCT based on intrinsic optical properties, however, this approach employs labeled fluorophores or nanoparticles to enhance target selectivity (Robles et al. 2011; Oldenburg et al. 2008). The mechanisms of MI-OCT include the interaction of contrast agents with biomolecular targets, the incorporation of pump-probe spectroscopy to increase detection sensitivity, and the application of sophisticated spectral unmixing algorithms to distinguish molecular signatures (Yang 2005). Although it holds promise for accurate molecular imaging, additional innovations are needed to maximize contrast agent delivery and achieve maximum signal specificity within the limitations of OCT imaging (Kirby et al. 2017).

4.3.5 Optical Coherence Elastography (OCE)

Optical Coherence Elastography (OCE) combines mechanical stimulation with OCT imaging to measure tissue stiffness (Kennedy et al. 2013). In contrast to other functional OCT modalities that use endogenous optical contrast, OCE uses an external force to cause mechanical deformation, which is then measured using OCT (Kennedy et al. 2013). The basic principles of OCE include the application of controlled mechanical stress, e.g., acoustic waves, air-puff, or piezoelectric actuators, followed by OCT-based tracking of displacement to quantify the resulting strain response. Computational modeling is then employed to reconstruct tissue elasticity (Sun et al. 2011). OCE is used mainly to measure tissue biomechanical properties, with current research aimed at optimizing force application standardization and enhancing depth-resolved imaging methods (Dubois et al. 2004).

4.3.6 Full-Field OCT (FF-OCT)

Full-Field OCT (FF-OCT) is distinct from traditional point-scanning OCT in that it uses an en-face imaging method to achieve fast, high-resolution imaging in large fields. Rather than sequentially recording depth profiles, FF-OCT records entire cross-sectional planes at once with a broadband light source and high-speed camera (Zheng et al. 2023; Leitgeb 2019). The essential concepts in FF-OCT are interferometric en-face imaging for wide-field visualization, the application of short-coherence light sources for preservation of high axial resolution, and computational phase correction to suppress motion artifacts. The method is best suited for ex vivo tissue imaging with high-throughput optical sectioning properties (Beaurepaire et al. 1998; Roy et al. 2002; Grieve et al. 2004).

5 Challenges and Limitations

While OCT has emerged as the gold standard in biomedical imaging, especially in ophthalmology, its clinical usefulness is also fraught with certain inherent and extrinsic limitations that need to be carefully handled. Perhaps the most critical one is the limited penetration depth. OCT uses near-infrared light, which is ideally adapted for imaging ocular tissue but is hardly able to penetrate highly scattering or pigmented tissue. Therefore, imaging of retinal layers at greater depths or non-ocular tissue becomes especially challenging with pathologies like hemorrhage, fibrosis, or dense cataracts (Aumann et al. 2019).

Beyond depth limitations, OCT image quality is frequently degraded by a broad array of artifacts. These consist of distortions due to eye movement, blinking, poor fixation, and opacities in the media. Motion artifacts can degrade structural integrity significantly in uncooperative patients or handheld scanning situations. Signal dropouts, shadowing secondary to highly reflective structures like blood vessels, and speckle noise can further obscure fine details with accurate interpretation progressively becoming challenging (Lin et al. 2024). Accessibility is another major limitation. High-resolution OCT devices, including SD-OCT and SS-OCT, are costly to procure and maintain. Their high costs are inaccessible in low-resource environments and rural clinics and consequently deny disadvantaged groups access to this beneficial diagnostic tool (Leung et al. 2020; Song et al. 2021). Moreover, while image capture is semi-automatic, high-quality outcomes are still very operator-dependent. OCT image interpretation is specialized, particularly in the differentiation of subtle pathologic changes from normal anatomic variation and requires ongoing training and experience (Mwanza et al. 2011). Another limitation is variability owing to a lack of standardization between OCT platforms. Proprietary algorithms used by manufacturers for image acquisition and processing result in variability of image appearance, resolution, and layer segmentation. This variability limits longitudinal follow-up of patients scanned with different machines as well as meaningful comparison between clinical studies (Barkana et al. 2009; Nouri et al. 2023).

Functionally, conventional OCT is high-resolution structural imaging without dynamic physiological information. While newer developments like D-OCT and OCTA show potential in this aspect, these modalities are yet to be validated and are not included in standard clinical practice workflows. Even with OCTA, important functional parameters like vascular leakage or real-time flow dynamics are yet to be provided compared to traditional angiographic modalities (Srinivasan et al. 2009; Bernardes 2011). Finally, while artificial intelligence has shown much potential in the automation of OCT image analysis, its clinical uptake is plagued by a few challenges. Regulatory clearances are slow, and concerns about algorithmic bias, inability to generalize across heterogeneous populations, and lack of transparency in deep learning models are still daunting hurdles. Furthermore, concerns about data privacy and the need for rigorous real-world validation impede the widespread uptake of AI-augmented OCT interpretation in routine practice (Taribagil et al. 2023; Karalis 2024; Dahrouj and Miller 2021; Djulbegovic et al. 2025).

6 Conclusions and Future Perspectives

OCT has emerged as a revolutionary biomedical optics imaging technology, giving high-resolution, non-invasive images of tissue microstructure. This chapter has surveyed various OCT architectures—ranging from the simple Time-Domain OCT (TD-OCT) to the more advanced Fourier-Domain OCT (FD-OCT), with Spectral-Domain (SD-OCT) and Swept-Source (SS-OCT) variants. All the systems have resolutions, imaging depths, speeds, and complexities with exclusive domains of application ranging from ophthalmology and cardiology to dermatology and oncology.

Among the major highlights are the dramatic improvements in performance achieved with developments in contemporary hardware such as ultrabroadband light sources for sub-micron axial resolution, MHz-scale A-scan rate swept-source lasers, and longer wavelength systems for improved penetration. Enhanced detection sensitivity and speckle noise reduction through balanced detection, polarization diversity, and deep learning techniques have also improved image integrity. Parallelized acquisition (line-field and full-field OCT) and adaptive optics have enhanced acquisition rate and lateral resolution, with real-time 3D imaging now possible.

Apart from structural imaging, functional extensions such as OCT angiography (OCTA), Doppler OCT, polarization-sensitive OCT (PS-OCT), and optical coherence elastography (OCE) are imparting richness to the information content extracted from tissues. These modalities provide rich information regarding vascular networks, flow physics, tissue mechanics, and birefringence characteristics without the need for exogenous contrast agents. Dynamic contrast OCT (DyC-OCT) itself holds high promise as a label-free imaging modality of cells based on detection of temporal signal changes and is therefore of important clinical relevance to cancer diagnosis, organoid tracking, and neuronal activity mapping.

Future OCT systems will be designed and influenced by the intersection of miniaturization, computational imaging, and artificial intelligence. Currently, development is headed towards miniaturized and portable OCT technology powered by photonic integrated circuits, MEMS tunable lasers, and on-chip spectrometers. These systems, combined with real-time GPU/FPGA processing and AI-based reconstruction and interpretation, will enable point-of-care diagnosis, telemedicine, and home monitoring. Along with that, multimodal systems that integrate OCT with fluorescence microscopy, Raman imaging, and photoacoustic modalities are being developed to furnish detailed structural, functional, and biochemical information from one imaging scan.

There has been impressive progress, but many areas still need exploration. Increasing penetration depth in highly scattering tissues, reducing cost for widespread availability, and enhancing dynamic contrast mechanisms are challenges that need to be overcome. In addition, the confirmation of AI models for use in clinical settings, especially in heterogeneous populations and imaging situations, is worth investigating further. Subsequent studies should also investigate the integration of OCT with therapeutic treatments, image-directed surgery, and medication delivery monitoring to maximize its advantage in personalized medicine. OCT has evolved to be a multifaceted and priceless imaging system. With continuous advancements in light sources, detection methods, system integration, and innovative interpretation, the technology is set to take center stage in the future of non-invasive diagnostics as well as translational biomedical investigations.

Acknowledgments N/A

Conflict of Interest The authors declare that they have no conflict of interest.

References

Abdulhalim I (2012) Spatial and temporal coherence effects in interference microscopy and full-field optical coherence tomography. Ann Phys 524(12):787–804

Asam JS, Polzer M, Tafreshi A, Hirnschall N, Findl O (2019) Anterior segment OCT. In: High resolution imaging in microscopy and ophthalmology: new frontiers in biomedical optics. Springer International Publishing, pp 285–299

Aumann S, Donner S, Fischer J, Müller F (2019) Optical coherence tomography (OCT): principle and technical realization. In: High resolution imaging in microscopy and ophthalmology: new frontiers in biomedical optics. Springer International Publishing, pp 59–85

Barkana Y, Burgansky-Eliash Z, Gerber Y, Melamed S, Neudorfer M, Avni I, Bartov E, Morad Y (2009) Inter-device variability of the stratus optical coherence tomography. Am J Ophthalmol 147(2):260–266

Beaurepaire E, Boccara AC, Lebec M, Blanchot L, Saint-Jalmes H (1998) Full-field optical coherence microscopy. Opt Lett 23(4):244–246

Bernardes R (2011) Optical coherence tomography: health information embedded on OCT signal statistics. In: 2011 Annual International Conference of the IEEE Engineering in Medicine and Biology Society. IEEE, pp 6131–6133

Bernardes R, Santos T, Cunha-Vaz J (2008) Increased-resolution OCT thickness mapping of the human macula: a statistically based registration. Invest Ophthalmol Vis Sci 49(5):2046–2052

Bhende M, Shetty S, Parthasarathy MK, Ramya S (2018) Optical coherence tomography: a guide to interpretation of common macular diseases. Indian J Ophthalmol 66(1):20–35

Bouma BE, Yun SH, Vakoc BJ, Suter MJ, Tearney GJ (2009) Fourier-domain optical coherence tomography: recent advances toward clinical utility. Curr Opin Biotechnol 20(1):111–118

Choudhry N, Golding J, Manry MW, Rao RC (2016) Ultra-widefield steering-based spectral-domain optical coherence tomography imaging of the retinal periphery. Ophthalmology 123(6):1368–1374

Dahrouj M, Miller JB. Artificial intelligence (AI) and retinal optical coherence tomography (OCT). In: Seminars in Ophthalmology 2021 36, 4, pp. 341-345). Taylor & Francis

Danis RP, Fisher MR, Lambert E, Goulding A, Wu D, Lee LY (2008) Results and repeatability of retinal thickness measurements from certification submissions. Arch Ophthalmol 126(1):45–50

De Boer JF, Hitzenberger CK, Yasuno Y (2017) Polarization sensitive optical coherence tomography–a review. Biomed Opt Express 8(3):1838–1873

De Carlo TE, Romano A, Waheed NK, Duker JS (2015) A review of optical coherence tomography angiography (OCTA). Int J Retina Vitreous 1:1–5

Djulbegovic MB, Bair H, Gonzalez DJ, Ishikawa H, Wollstein G, Schuman JS (2025) Artificial intelligence for optical coherence tomography in glaucoma. Transl Vision Sci Technol 14(1):27

Dubois A, Grieve K, Moneron G, Lecaque R, Vabre L, Boccara C (2004) Ultrahigh-resolution full-field optical coherence tomography. Appl Opt 43(14):2874–2883

Fercher AF (1996) Optical coherence tomography. J Biomed Opt 1(2):157–173

Fujimoto JG, Drexler W (2015) Introduction to OCT. In: Optical coherence tomography: technology and applications, 2nd edn. Springer International Publishing, pp 3–64

Fujimoto JG, Pitris C, Boppart SA, Brezinski ME (2000) Optical coherence tomography: an emerging technology for biomedical imaging and optical biopsy. Neoplasia 2(1–2):9–25

Fujimoto JG, Brezinski ME, Tearney GJ, Boppart SA, Bouma B, Hee MR, Southern JF, Swanson EA. Optical biopsy and imaging using optical coherence tomography

Grieve K, Paques M, Dubois A, Sahel J, Boccara C, Le Gargasson JF (2004) Ocular tissue imaging using ultrahigh-resolution, full-field optical coherence tomography. Invest Ophthalmol Vis Sci 45(11):4126–4131

Haavisto S, Salmela J, Koponen A (2015) Accurate velocity measurements of boundary-layer flows using Doppler optical coherence tomography. Exp Fluids 56:1–6

Herrera G, Shen M, Trivizki O, Liu J, Shi Y, Hiya FE, Li J, Cheng Y, Lu J, Zhang Q, O'Brien RC (2025) Comparison between spectral-domain and swept-source OCT angiography for the measurement of persistent hypertransmission defects in age-related macular degeneration. Ophthalmol Sci 5(1):100593

Huang D, Swanson EA, Lin CP, Schuman JS, Stinson WG, Chang W, Hee MR, Flotte T, Gregory K, Puliafito CA, Fujimoto JG (1991) Optical coherence tomography. Science 254(5035):1178–1181

Kalra G, Pichi F, Kumar Menia N, Shroff D, Phasukkijwatana N, Aggarwal K, Agarwal A (2021) Recent advances in wide field and ultrawide field optical coherence tomography angiography in retinochoroidal pathologies. Expert Rev Med Devices 18(4):375–386

Kapoor R, Whigham BT, Al-Aswad LA (2019) Artificial intelligence and optical coherence tomography imaging. Asia-Pac J Ophthalmol 8(2):187–194

Karalis VD (2024) The integration of artificial intelligence into clinical practice. Appl Biosci 3(1):14–44

Kennedy BF, Kennedy KM, Sampson DD (2013) A review of optical coherence elastography: fundamentals, techniques and prospects. IEEE J Select Top Quant Electron 20(2):272–288

Kirby MA, Pelivanov I, Song S, Ambrozinski Ł, Yoon SJ, Gao L, Li D, Shen TT, Wang RK, O'Donnell M (2017) Optical coherence elastography in ophthalmology. J Biomed Opt 22(12):121720

Leitgeb RA (2019) En face optical coherence tomography: a technology review. Biomed Opt Express 10(5):2177–2201

Leitgeb RA, Drexler W, Povazay B, Hermann B, Sattmann H, Fercher AF (2005) Spectroscopic Fourier domain optical coherence tomography: principle, limitations, and applications. In: Coherence domain optical methods and optical coherence tomography in biomedicine Ix, vol 5690. SPIE, pp 151–158

Leitgeb RA, Werkmeister RM, Blatter C, Schmetterer L (2014) Doppler optical coherence tomography. Prog Retin Eye Res 41:26–43

Leung EH, Gibbons A, Koch DD (2020) Cost-effectiveness of preoperative OCT in cataract evaluation for multifocal intraocular lens. Ophthalmology 127(7):859–865

Lin WC, Coyner AS, Amankwa CE, Lucero A, Wollstein G, Schuman JS, Ishikawa H (2024) High prevalence of artifacts in optical coherence tomography with adequate signal strength. Transl Vision Sci Technol 13(8):43

Maciel MJ, Costa CG, Silva MF, Peixoto AC, Wolffenbuttel RF, Correia JH (2016) A wafer-level miniaturized Michelson interferometer on glass substrate for optical coherence tomography applications. Sensors Actuators A Phys 242:210–216

Massin P, Vicaut E, Haouchine B, Erginay A, Paques M, Gaudric A (2001) Reproducibility of retinal mapping using optical coherence tomography. Arch Ophthalmol 119(8):1135–1142

Matsunaga D, Yi J, Puliafito CA, Kashani AH (2014) OCT angiography in healthy human subjects. Ophthal Surg Lasers Imag Retina 45(6):510–515

Medeiros FA, Zangwill LM, Bowd C, Weinreb RN (2004) Comparison of the gdx vcc scanning laser polarimeter, hrt ii confocalscanning laser ophthalmoscope, and stratus oct optical coherence tomographfor the detection of glaucoma. Arch Ophthalmol 122(6):827–837

Merchant KY, Su D, Park SC, Qayum S, Banik R, Liebmann JM, Ritch R (2013) Enhanced depth imaging optical coherence tomography of optic nerve head drusen. Ophthalmology 120(7):1409–1414

Morgner U, Drexler W, Kärtner FX, Li XD, Pitris C, Ippen EP, Fujimoto JG (2000) Spectroscopic optical coherence tomography. Opt Lett 25(2):111–113

Mwanza JC, Gendy MG, Feuer WJ, Shi W, Budenz DL (2011) Effects of changing operators and instruments on time-domain and spectral-domain OCT measurements of retinal nerve fiber layer thickness. Ophthal Surg Lasers Imag Retina 42(4):328–337

Nam HS, Yoo H (2018) Spectroscopic optical coherence tomography: a review of concepts and biomedical applications. Appl Spectrosc Rev 53(2–4):91–111

Nouri H, Nasri R, Abtahi SH (2023) Addressing inter-device variations in optical coherence tomography angiography: will image-to-image translation systems help? Int J Retina Vitreous 9(1):51

Oldenburg AL, Applegate BE, Izatt JA, Boppart SA (2008) Molecular OCT Contrast enhancement and imaging. In: Optical coherence tomography: technology and applications. Springer Berlin Heidelberg, Berlin, Heidelberg, pp 713–756

P Mattison S, Kim W, Park J, E Applegate B (2014) Molecular imaging in optical coherence tomography. Curr Mol Imag (Discontinued) 3(2):88–105

Pircher M, Götzinger E, Findl O, Michels S, Geitzenauer W, Leydolt C, Schmidt-Erfurth U, Hitzenberger CK (2006) Human macula investigated in vivo with polarization-sensitive optical coherence tomography. Invest Ophthalmol Vis Sci 47(12):5487–5494

Polito A, Del Borrello M, Isola M, Zemella N, Bandello F (2005) Repeatability and reproducibility of fast macular thickness mapping with stratus optical coherence tomography. Arch Ophthalmol 123(10):1330–1337

Ripa M, Motta L, Florit T, Sahyoun JY, Matello V, Parolini B (2022) The role of widefield and ultra widefield optical coherence tomography in the diagnosis and management of vitreoretinal diseases. Diagnostics 12(9):2247

Robles FE, Wilson C, Grant G, Wax A (2011) Molecular imaging true-colour spectroscopic optical coherence tomography. Nat Photonics 5(12):744–747

Rollins AM, Izatt JA (1999) Optimal interferometer designs for optical coherence tomography. Opt Lett 24(21):1484–1486

Roy M, Svahn P, Cherel L, Sheppard CJR (2002) Geometric phase-shifting for low-coherence interference microscopy. Opt Lasers Eng 37(6):631–641

Sheppard CJR, Roy M (2003) 11 Low-coherence interference microscopy. In: Optical imaging and microscopy. Springer, Berlin, pp 267–273

Song G, Jelly ET, Chu KK, Kendall WY, Wax A (2021) A review of low-cost and portable optical coherence tomography. Progr Biomed Eng 3(3):032002

Spaide RF, Klancnik JM, Cooney MJ (2015) Retinal vascular layers imaged by fluorescein angiography and optical coherence tomography angiography. JAMA Ophthalmol 133(1):45–50

Spaide RF, Fujimoto JG, Waheed NK, Sadda SR, Staurenghi G (2018) Optical coherence tomography angiography. Prog Retin Eye Res 64:1–55

Srinivasan VJ, Chen Y, Duker JS, Fujimoto JG (2009) In vivo functional imaging of intrinsic scattering changes in the human retina with high-speed ultrahigh resolution OCT. Opt Express 17(5):3861–3877

Srinivasan VJ, Radhakrishnan H, Lo EH, Mandeville ET, Jiang JY, Barry S, Cable AE (2012) OCT methods for capillary velocimetry. Biomed Opt Express 3(3):612–629

Stanga PE, Bird AC (2001) Optical coherence tomography (OCT): principles of operation, technology, indications in vitreoretinal imaging and interpretation of results. Laser Scanning: Update 1: First Official Publication of the International Society of Laser Scanning: INSOLAS, pp 9–15

Strouthidis NG, Garway-Heath DF (2008) New developments in Heidelberg retina tomograph for glaucoma. Curr Opin Ophthalmol 19(2):141–148

Sun C, Standish B, Yang VX (2011) Optical coherence elastography: current status and future applications. J Biomed Opt 16(4):043001

Taribagil P, Hogg HJ, Balaskas K, Keane PA (2023) Integrating artificial intelligence into an ophthalmologist's workflow: obstacles and opportunities. Exp Rev Ophthalmol 18(1):45–56

Wojtkowski M, Leitgeb R, Kowalczyk A, Bajraszewski T, Fercher AF (2002) In vivo human retinal imaging by Fourier domain optical coherence tomography. J Biomed Opt 7(3):457–463

Wong IY, Koizumi H, Lai WW (2011) Enhanced depth imaging optical coherence tomography. Ophthal Surg Lasers Imag Retina 42(4):S75–S84

Xia F, Hua R (2023) The latest updates in swept-source optical coherence tomography angiography. Diagnostics 14(1):47

Yang C (2005) Molecular contrast optical coherence tomography: a review¶. Photochem Photobiol 81(2):215–237

Yang VX, Vitkin IA (2007) Principles of Doppler OCT. In: Optical coherence tomography in cardiovascular research. CRC Press, pp 321–334

Zeppieri M, Marsili S, Enaholo ES, Shuaibu AO, Uwagboe N, Salati C, Spadea L, Musa M (2023) Optical coherence tomography (OCT): a brief look at the uses and technological evolution of ophthalmology. Medicina 59(12):2114

Zhang N, Hoffmeyer GC, Young ES, Burns RE, Winter KP, Stinnett SS, Toth CA, Jaffe GJ (2007) Optical coherence tomography reader agreement in neovascular age-related macular degeneration. Am J Ophthalmol 144(1):37–44

Zhang J, Mazlin V, Fei K, Boccara AC, Yuan J, Xiao P (2023) Time-domain full-field optical coherence tomography (TD-FF-OCT) in ophthalmic imaging. Ther Adv Chronic Dis 14:20406223231170146

Zheng W, Kou SS, Sheppard CJ, Roy M (2023) Advancing full-field metrology: rapid 3D imaging with geometric phase ferroelectric liquid crystal technology in full-field optical coherence microscopy. Biomed Opt Express 14(7):3433–3445

Applications of Optical Coherence Tomography (OCT) in Preclinical Practice

Abhi Eathara, Amanda P. Siegel, Maria M. Tsoukas, and Kamran Avanaki

Abstract

Animal models have the potential to probe disease mechanisms and test novel treatments prior to human use, but any model is only as good as the tools available to analyze its output. Non-invasive imaging of animal models enables researchers to visualize and quantify disease processes, evaluate treatments, and refine experimental designs. Optical coherence tomography (OCT), with depth penetration of up to 2 mm and axial and lateral resolution in the range of 10 μm for research applications, gives researchers new tools for their preclinical platforms. Here we present how researchers are using OCT for skin, brain, and cardiovascular research, showing how OCT can enhance each of these fields. Murine models of skin conditions can be visualized non-invasively with greater penetration than dermoscopy. Murine brain imaging by OCT, typically achieved by inserting a cranial optical window prior to experimentation, gives greater penetration and more structural information than confocal microscopy. Non-invasive cardiovascular studies performed by OCT endoscopy enable improved tissue characterization due to OCT's depth penetration. In a complementary fashion, small animals can drive research into enhanced software and hardware

A. Eathara · A. P. Siegel
Richard and Loan Hill Department of Biomedical Engineering, University of Illinois Chicago, Chicago, IL, USA

M. M. Tsoukas
Department of Dermatology, University of Illinois at Chicago, Chicago, IL, USA

K. Avanaki (✉)
Richard and Loan Hill Department of Biomedical Engineering, University of Illinois Chicago, Chicago, IL, USA

Department of Dermatology, University of Illinois at Chicago, Chicago, IL, USA
e-mail: avanaki@uic.edu

I. F. Cengiz et al. (eds.), *Bioimaging Modalities in Bioengineering*,
Biomaterials, Bioengineering and Sustainability 9,
https://doi.org/10.1007/978-3-031-99410-4_4

improvements for OCT imaging. In this chapter, we hope to provide preclinical researchers with a comprehensive summary of how OCT, through its powerful visualization strategies, can be leveraged to enhance our understanding of disease.

Keywords

Skin imaging · Vascular imaging · Brain imaging · Endoscopic OCT · OCT angiography (OCTA)

1 Introduction

Traditionally, preclinical animal research visualization has been dominated by optical imaging microscopy, which enables high resolution, cost-effective, quantitative imaging, but is very limited with regard to tissue penetration. Key characteristics of a good preclinical imaging tool would maintain the benefits of optical microscopy while including greater depth penetration in a system with good temporal resolution. Such a system should also be relatively low-cost, easy to use, and with a straightforward interpretation of the developed images. Optical coherence tomography (OCT) has shown itself to be instrumental in investigating complex biological phenomena in preclinical research. OCT uses low-coherence interferometry to create high resolution, cross-sectional imaging of biological tissues. OCT's ability to assess morphological and architectural changes has been a breakthrough in dermatology for analysis of epidermal and dermal skin layers (Avanaki et al. 2009, 2013; Hessler et al. 2020; O'leary et al. 2018; Hojjatoleslami and Avanaki 2012; Zafar et al. 2024; Akella et al. 2024; Adabi et al. 2017a). In small animal models of the brain, it can capture microvascular changes during pathological events such as stroke and spreading depolarizations, enabling researchers to observe hemodynamic alterations and capillary flow dynamics in real-time. In cardiovascular research, OCT provides detailed visualization of stent healing, vascular calcification, and ischemic injuries. The high-resolution imaging of capillary beds, tubular dynamics, and vascular obstructions has significantly improved the understanding of disease mechanisms and therapeutic responses.

Recent technological advancements enhance OCT's capabilities: line-field confocal OCT (LC-OCT) provides higher axial and lateral resolution than traditional OCT but with lower depth penetration; OCT angiography (OCTA) offers detailed visualization of blood flow; and polarization-sensitive OCT (PS-OCT) provides critical information about tissue birefringence and collagen alignment. Additionally, integrating OCT with other imaging techniques, such as photoacoustic microscopy and fluorescence imaging, enables comprehensive, multimodal data collection. Integration of photoacoustic tomography (Benavides-Lara et al. 2024) and OCT is under study for skin imaging. Portable and robotic OCT systems have expanded OCT's accessibility, allowing for precise imaging in diverse research settings.

This chapter explores cutting-edge applications of OCT in preclinical research and its transformative potential across three main disciplines, imaging of the skin, the brain, and the cardiovascular system. For each subsection, the strengths and weaknesses of OCT for analyzing animal systems of disease are discussed, and key studies are described. By highlighting how OCT bridges the gap between foundational science and clinical applications, this chapter underscores its vital role in advancing biomedical research and translational medicine.

2 Applications of OCT in Skin Imaging

2.1 Overview

OCT enables visualization of skin layers, vascular networks, and microstructural alterations in skin (Schuetzenberger et al. 2019; Lautman et al. 2022; Jacobsen et al. 2023; Pfister et al. 2021). More information has been extracted from OCT images when sophisticated image processing algorithms are used (Hessler et al. 2020; Adabi et al. 2016, 2017b, 2018; Turani et al. 2018; Avanaki and Hojjatoleslami 2013). For this reason, it is very useful for assessing induced pathologies and determining the effect of specific treatments not only topically, but through the epidermis and into the dermis. Significant research has already been conducted in human studies, underscoring the clinical relevance and broad potential of OCT applications in dermatology. Nevertheless, animal models continue to play a valuable role for understanding disease progression, treatment safety and efficacy, monitoring treatment delivery, and developing new implementations of OCT imaging for novel clinical (and preclinical) applications. Figure 1 shows an OCT image of murine skin.

2.2 Assessing Skin Damage and Wound Healing

Researchers have developed many techniques for mimicking skin pathologies, such as optical breakdown caused by fractional picosecond-domain lasers. Because these

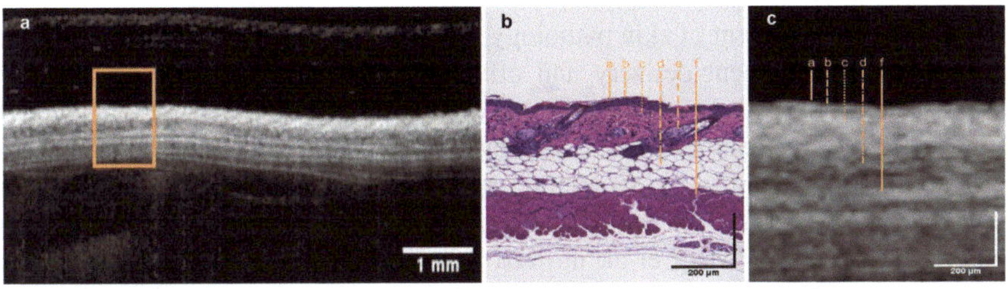

Fig. 1 Cross-sectional image of murine skin. (**a**) OCT full-size image. (**b**) histology (hematoxylin and eosin (H&E) staining). (**c**) Enlarged cross-section from (**a**). Dashed orange lines indicate skin layers and features corresponding to: **a**—epidermis, **b**—papillary dermis, **c**—reticular dermis, **d**—subcutis, **e**—sebaceous gland and hair follicle and **f**—muscle. (Adapted from Schuetzenberger et al. (2019) under a Creative Commons Attribution License (CC BY 4.0))

techniques can cause a variable amount of tissue damage, full characterization of the extent of injury using a non-invasive method is important for preclinical studies (Szczepanik et al. 2022). Jacobsen et al. (2023) analyzed the ability of OCT and LC-OCT to detect melanin-dependent laser-induced optical breakdown including formation of vacuoles and disruption of stratum corneum, and found results were in agreement with histological exam (Jacobsen et al. 2023). Adams et al. utilized polarization-sensitive OCT (PS-OCT) for assessment of fibrosis in a bleomycin-induced skin fibrosis mouse model (Adams et al. 2020). This approach characterizes fibrotic changes based on the measured distribution of optic axis (OA) values measured with PS-OCT, where degree of polarization and OA entropy are significantly decreased in fibrotic tissue, enabling detection of subdural scar tissue as deep as 400 μm below the skin surface. OCT angiography (OCTA) has been used to visualize blood flow and vascular networks in both murine and zebrafish models of diabetes (Pfister et al. 2021; Kim et al. 2023). Diabetic wounds in these animals showed, by OCTA, the postponement of angiogenesis and impairment in tissue remodeling. Ghosh et al. monitored healing of full-thickness excision wounds for 300 days in mice skin utilizing a swept source OCT system (Thorlabs OCT1300SS) with a center wavelength of 1300 nm (Ghosh et al. 2021). By implementing an attenuation-correcting algorithm and monitoring OCT reflectivity, they were able to delineate different stages of wound healing. Uncorrected OCT intensity alone was not able to perform this consistently. OCT, in conjunction with laser Doppler flowmetry (LDF) and diffuse reflectance spectroscopy (DRS), was used to measure autographs and allografts of skin on a burn wound using a murine model (Wistar rats) to detect early non-invasive diagnosis of split skin allograft rejection (Turchin et al. 2023). Allografts are skin taken from a donor and autographs are skin taken from the patient. OCT and OCTA were used for structural (OCT) and functional/angiographic (OCTA) analysis of $3 \times 3 \times 1.5$ mm^3 volumes. While LDF and DRS demonstrated autographs had increased oxygenation, blood content and perfusion, the OCT and OCTA results did now show significant structural or angiographic differences between the two groups.

2.3 Assessing Treatment Delivery and Treatment Efficacy

Important as assessment of skin pathology is for biomedical research, non-invasive assessment of treatment delivery and efficacy can improve longitudinal studies and accelerate hypothesis-testing. For example, OCT has been used in animal studies for testing treatments for hair growth (Liu et al. 2024). OCT was able to visualize improvement in the structural organization of hair follicles and quantify the increase in dermal thickness in mice treated for androgenic alopecia with pilose antler extract. Wang et al. utilized full field OCT (FF-OCT) to assess the efficacy of fractional laser therapy treatment to reverse photoaging in nude mice (photoaging was mimicked by the use of UVB-induced transepidermal water loss and UVA-induced epidermis thickening) (Wang et al. 2022). FF-OCT utilizes a high-brightness broadband-extended source with a reliable speckle suppression method and delivers cellular resolution. Schuetzenberger et al. (2020) performed observations of intradermal hydrogel degradation by OCT, thereby laying the groundwork for assessing therapeutic agents in the management of skin cancer (Schuetzenberger et al. 2020). OCT has also been used to monitor the absorption of gold nanoparticles, applied topically, for the purposes of optimizing the delivery method (Xu et al. 2020).

2.4 Preclinical Research Advancing OCT Technology

OCT is typically considered a structural imaging modality, but it can be used as a functional imaging modality if advanced signal or image processing methods are applied. For example, Gao et al. (2022) developed a deep neural network specifically designed for the segmentation of laser-induced skin injuries (on mice), attaining notable accuracy at injury evaluation and detection of the dermal epidermal junction (Gao et al. 2022). Ji et al. (2022) also utilized semi-supervised learning to modify rodent OCT data to determine epidermal thickness for use in human skin imaging (Ji et al. 2022). Demidov et al. (2021) utilized OCT (1320 central frequency, speckle variance mode) to develop volumetric tumor delineation based on quantitative OCT image speckle analysis using Gamma distribution fits. Accuracy was tested in a murine model of cutaneous melanoma (and also a pancreatic cancer model, through a dorsal skin window) (Demidov et al. 2021). Textural analysis of OCT images has great promise for tumor detection because the optical properties of tissue, which generate the OCT images, are different in tumors compared with healthy tissue. Here, the use of animal models helps train learning-based kernels for eventual implementation in clinical practice.

Preclinical research for improved OCT hardware is also ongoing. Development of an imaging needle for OCT was proposed a quarter century ago (Li et al. 2000), but was abandoned because it was not able to discriminate between different kinds of tissue. Recently, the OCT needle design has been enhanced by incorporating the polarization signal from the OCT fringes of a needle and Doppler-based tracking (Harper et al. 2023). This enhanced design demonstrated the ability of the needle to differentiate skin, subcutaneous tissue, ligament, and epidural spaces during needle insertion, which has many applications for assisting surgical procedures including nerve block administration and for differentiation between muscle and fatty tissue.

In summary, OCT technology can aid skin researchers by: quantifying models of skin pathology; monitoring of skin healing, including wound healing; monitoring of treatment delivery and efficacy, including efficacy of treatments for skin cancer; developing imaging processing methods for functional OCT imaging; and developing new instrumentation for probing cutaneous and subcutaneous tissue. Table 1 summarizes strengths and weaknesses of OCT technologies for different skin applications.

Table 1 Applications of OCT technologies for skin researchers

Applications	Strengths	Shortcomings
Assessment of skin damage and wound healing	Detects subsurface morphological and architectural changes and angiogenesis	Lack of functional/molecular information, cannot measure inflammatory signals directly
Monitoring treatments for skin cancer	Can detect microstructural changes in tumor volume for nonmelanoma skin cancer	Unable to detect or monitor progression of melanoma unless coupled with sophisticated data analysis
Monitoring topical treatment delivery	Useful for materials with structural or reflective properties (e.g., hydrogels, gold nanoparticles)	Cannot monitor delivery of less reflective (and lower concentrations of) nanomaterials
OCT-enhanced needle for drug delivery or fine needle aspiration	Promising ability to differentiate between tissue types including epidermal, dermal, fat and muscle	Limited field of view

3 Applications of OCT in Brain Imaging

3.1 Overview

OCT represents a notable advancement in small animal brain imaging for the examination of the microstructural and microvascular architecture of the brain, enabling visualization of white matter, cortical organization, and dynamic blood flow patterns (Lautman et al. 2022; Park et al. 2018; Yang et al. 2025; Ge et al. 2023). OCT can provide label-free imaging with micron-level precision, enabling researchers to achieve a more profound comprehension of neurological disorders, such as stroke, neurodegeneration, brain injuries, brain tumors, and developmental anomalies (Stamenkovic et al. 2023). OCT is implemented for brain imaging using light sources with a center wavelength of 900 nm, 1300 nm, and even 1700 nm (Park et al. 2018; Zhu et al. 2021), which is also known as optical coherence microscopy (OCM).

3.2 Assessment of Neural Networks, Cerebral Elasticity, and Brain Development

A commercial OCT system (Thorlabs GAN220) was utilized to image a cortical column of 3-month-old C57BL/6 J mice in vivo, using pre-installed cranial windows. The system, with a center wavelength of 900 nm, provides a theoretical axial resolution of 2 μm and lateral resolution of 0.76 μm. A custom-built head-clamp was used to immobilize the mouse head during sedation. Imaging sessions were long (40 min) to image consecutive focal plains spaced 10 μm apart, but after stitching and processing, led to isotropic lateral resolution of 1.1 μm throughout the cortical column (Lautman et al. 2022). The mechanics and functions of cerebrospinal fluid (CSF) circulation in the embryonic tadpole brain were also studied using a Thorlabs Ganymede OCT, which was able to image ventricular and neural development during the onset of CSF circulation (Dur et al. 2020). Thorlabs Ganymede (center wavelength 900 nm, bandwidth 200 nm) was also utilized with a feature segmentation algorithm based on a one-layer convolutional neural network to generate label-free micron-scale 3D imaging of 700 μm depth across the cortex of live mice (Lautman et al. 2022). This tool, which achieved isotropic lateral resolution of 1.1 μm throughout the depth with a 0.7 × 0.7 mm field of view, was able to quantify features such as brain cell bodies' density, volume, and average distance to the nearest cell, which could be used to study microstructural tissue alteration in aging, injury or disease.

To minimize ballistic attenuation and improve visualization of deep cortical and sub-cortical regions, a high numerical aperture OCM approach has been implemented with a center wavelength of 1700 nm (Zhu et al. 2021). There, researchers found significant information regarding disease progression in deep cortical layers, using a murine Alzheimer's disease model, that was not present in superficial cortical layers. Another set of researchers probed longitudinal changes of brain microvascular morphology and flow in mice through a cranial window using a commercial setup (Thorlabs, 1310 nm center wavelength, 170 nm bandwidth). OCT imaging was implemented with different objectives and scan speeds to image the full brain, microvessels, and blood flow via Doppler OCT (Walek et al. 2023). The methodology described enables measurement of vascular properties in a wide range of vessels, from pial vessels to capillaries. In addition to describing imaging methods, the team also developed image processing algorithms to enable longitudinal tracking of vascular features of the brain as it ages.

OCT can also assess mechanical changes in the mouse brain. Ge et al. (2023) utilized OCT reverberant shear wave elastography to assess changes in brain stiffness related to the aging process (Ge et al. 2023). A custom-built OCT system with a center wavelength of 1310 nm and bandwidth of ~140 nm was coupled to a mechanical piezo-electric system with eight points of contact to generate reverberant shear wave fields. The frequency of the piezoelectric device was set to 2000 Hz. Using this setup, the researchers analyzed mouse brain elasticity directly, identifying a relationship between increased stiffness and increased age, which also correlated with reduced glymphatic fluid content (Ge et al. 2023).

OCT, OCTA, and a map of attenuation coefficient by depth can be extracted to develop intricate images of a mouse brain (through a cranial window). Figure 2 shows mouse brain images from a custom-built system with a 1300 nm center wavelength (Choi et al. 2019).

3.3 OCT Imaging of Ischemic Stroke and Traumatic Brain Injury

A range of methods have been developed for mimicking ischemic stroke. One method is photothrombosis (PT), which can be induced by injecting a photo-

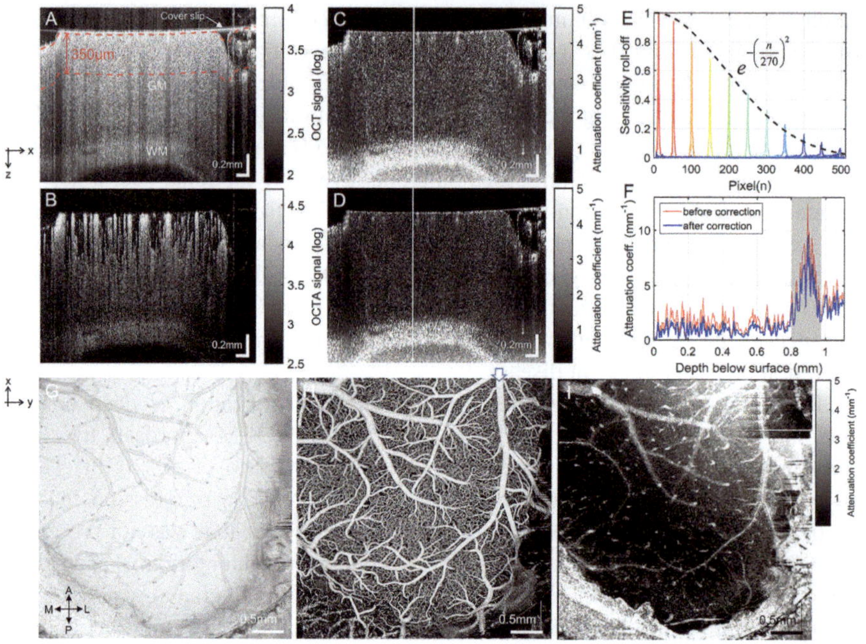

Fig. 2 In vivo OCT imaging of a healthy mouse brain with cranial window preparation. (**a**) Cross-sectional image (log scale), of the cerebral cortex; gray matter (GM) and white matter (WM). (**b**) Corresponding cross-sectional (x–z) OCTA image (log scale), showing blood perfusion in the cortex. Corresponding cross-sectional attenuation coefficient image before compensation for depth-dependent signal decay (**c**) and after compensation (**d**). (**e**) Fitting function (dotted line) of peak amplitudes of normalized sensitivity roll-off used to compensate for signal decay. (**f**) A-lines (while lines in (**c**) and (**d**) comparing the attenuation depth profiles. WM is region marked with gray. Maximum amplitude projections (MAPs) of 3-D OCT, OCTA and attenuation coefficients taken over the depth of 350 μm below the cortical surface (between the red dotted lines in (**a**)) display the structural map (**g**), vascular map (**h**), and attenuation coefficient map (**i**), respectively. (Reprinted with permission from Choi et al. (2019). © 2019 IEEE)

sensitive dye into the brain followed by laser illumination which causes free radical formation leading to endothelial damage, platelet aggregation, and clot formation. In one study, optical resolution photoacoustic microscopy (OR-PAM) was employed to simultaneously monitor vascular and ischemic stroke, while OCT was utilized to image structural and perfusion changes in the brain before and after stroke (Liang et al. 2025). Another team developed a system where a custom-built setup enabled both the green PT-inducing light and the near infrared OCT imaging beam to pass through the same objective lens, and used the set up for long term (weeks) monitoring of a PT stroke (Doleżyczek et al. 2022). This team identified hypo-reflective and hyper-reflective sections of the ischemic area, which, they asserted, corresponded to microglia/macrophage activation and astrocyte activation, respectively. Again, a closed cranial window was constructed for stroke induction and imaging. The temporal hemodynamic response to spreading depolarization (SD) within the cortex is of significant interest for understanding stroke outcome. OCT was used to simultaneously monitor changes in the vascular tree from the capillary bed to the pial arteries through a closed cranial window in mice, where SD was induced by delivering 1 M KCl through an occipital burr hole (Anzabi et al. 2021). The studies showed capillary flow dropped several seconds before pial arteriolar constriction, indicating hemodynamic response to cortical SD originates in the capillary bed.

Traditionally, the stroke core has been evaluated longitudinally with diffusion weighted imaging (DWI), a type of MRI that measures the apparent diffusion coefficient (ADC) of water with brain tissue. While this method is widely accepted, it is difficult to implement in animal models due to high cost and space constraints related to MRI. However, the changes in cellular and intracellular organelle swelling that change ADC also change the optical scattering coefficient within brain tissue. In (Sunil et al. 2021), Sunil et al. demonstrated these changes can be measured by OCT and by spatial frequency domain imaging (SFDI), a wide-field diffuse optical reflectance-based technique that can quantify absorption and scattering coefficients.

Longitudinal OCTA was used to track ischemic tissue repair and microvasculature regeneration at different cortical depths in awake animals following PT-induced ischemia in the mouse cortex (Lu et al. 2020). One team (Nyúl-Tóth et al. 2021a) utilized OCT with two-photon microscopy to *probe* changes in blood brain barrier (BBB) permeability and brain capillarization. Aged mice were given a chronic cranial window. Cortical capillary density, assessed by OCT, decreased in aged mice compared with young mice. Two-photon imaging of fluorescent tracers complemented the research into BBB permeability (Nyúl-Tóth et al. 2021b). Volumetric OCT imaging data enabled capillary orientation analysis to quantity poststroke microvasculature reorganization. OCTA was also used to spatially locate activated regions in the mouse cortex during whisker stimulation using optical microangiography (OMAG), a functional extension of OCTA (Rakymzhan et al. 2020). System specifications included a center wavelength of 1310 nm that provided ~5 μm axial resolution in mouse brain tissue and 2.22 mm imaging depth. Researchers showed that OMAG revealed blood flow changes faster than intrinsic optical signal imaging, and was able to show depth-specific response (Rakymzhan et al. 2020).

A shortcoming of the OCTA acquisition process is imaging speed. OCTA typically requires high sampling density and multiple acquisitions to form a single image frame, but recently deep learning-based methods have been able to accelerate the OCTA acquisition process through AI-enhanced reconstruction techniques that can reconstruct fully-sampled, high-quality angiograms from undersampled, low-quality counterparts by enhancing pixel resolution and image quality (Kim et al. 2022). OCT has also been used to quantify the spatial extent of an infarct caused by an ischemic stroke and measure recovery mechanisms with mice in vivo (Sunil et al. 2021). Imaging results were compared with histological staining. Another application of OCT is the monitoring of cortical changes in the acute and chronic phases of a penetrating traumatic brain injury (TBI) (Osiac et al. 2021). Adult (7–12 months) and aged (20–22 months) Sprague Dawley male rats had partial craniotomies performed to enable precise delivery of a TBI. A Thorlabs

OCT1300SS system, center wavelength 1300 nm was used with a special controller stage for immobilizing the rats. OCT detected age-related differences in response in both the acute and chronic phases of TBI recovery (Osiac et al. 2021).

3.4 OCT Imaging of Dementia and Brain Cancers

Researchers utilized OCTA and two-photon imaging to test their hypothesis that increased capillary stalling is associated with endothelial glycocalyx (cEG) loss in subcortical vascular dementia (Yoon et al. 2022). Here, researchers used OCTA and dynamic OCT to quantify morphological effects of inducing cEG loss on the mice and identify stalled segments near the cortical surface for two-photon imaging. With respect to research into brain cancer, OCT has been coupled with autofluorescence spectroscopy for detection and registration of glioma cells in mice (Zhu et al. 2019). The fluorescent images provided biochemical information and OCT provided structural information. The fused map had higher sensitivity than either single diagnostic modality, with tumor staging accomplished through autofluorescence spectroscopy (Zhu et al. 2019).

3.5 Emerging Neurovascular OCT Technologies

Intravascular endoscopic OCT probes have been developed for clinical use in cardiology (Bouma et al. 2017), gastroenterology (Tsai et al. 2017), real-time colorectal cancer diagnosis (Zeng et al. 2020) use in the clinic, but their utility as a neurosurgical tool has not yet been demonstrated. Depth-resolved polarization sensitive OCT (PS-OCT) was implemented endoscopically using intact porcine brain tissue ex vivo, enabling imaging of white matter structures (DePaoli et al. 2022). Such a tool, if developed clinically, could assist in identifying the placement of critical white matter structures during neurosurgery. Another team is developing an OCT catheter for imaging the cerebral sinus luminal environment, and testing their system in live swine (Pasarikovski et al. 2020). An ultrasound-guided femoral puncture was used to guide an OCT catheter up the right ascending pharyngeal artery to the rete mirabile. Images are acquired at 100 frames per second during a motorized automated pullback. Size of sinus maximum diameter, dural artery diameter, and vein diameter showed fairly good agreement with histology, with just 10% of points outside the 95% confidence intervals (Pasarikovski et al. 2020). Endothelial OCT intrathecal imaging of the spinal cord has also been demonstrated in swine and rabbits (Pasarikovski et al. 2021). In summary, OCT technology can aid neurological research by microstructural assessment of neural networks, cerebral elasticity, and brain development; temporal imaging of ischemic stroke and traumatic brain injury; imaging of indicia of dementia; and for structural registration of brain tumor location, when used with a second modality. Table 2 summarizes strengths and shortcomings of OCT technologies for use with different applications for brain research.

Table 2 Applications of OCT technologies for brain research

Applications	Strengths	Shortcomings
Assessment of neural networks and brain development	Microstructural brain and arterial morphology and architecture detected through implanted window	Requires alternative light source (1700 nm center wavelength) for imaging deep subcortical structures
Monitoring ischemic stroke progression and treatments	Enables quantification of extent of induced stroke and structural and perfusion changes	Sophisticated processing required for temporal measurements as commercial OCT/OCTA volumetric acquisitions can be lengthy
Monitoring progression of traumatic brain injury	Can detect lesions in the brain and changes in lesion volume by analyzing changes in image features such as skewness and kurtosis	Cannot directly measure functional changes in tissue
Assessment of brain tumors	Enables registration (tumor localization) in animal models	A second modality may be useful for tumor staging

4 Applications of OCT in Cardiovascular Research

4.1 Overview

OCT endoscopy, which is used to visualize coronary vascular structures, stent performance, and cardiac dynamics, has become an essential *clinical* imaging technique for detection of coronary atherosclerosis, planning of percutaneous coronary intervention, and detection of plaque (Araki et al. 2022). The clinical use of endoscopic OCT imaging of internal luminal organs is so well-established that it is commonly used as the gold standard for other emerging modalities (Gora et al. 2017). The resolution and depth of field of OCT images markedly enhances the evaluation of stent deployment (Gasior et al. 2020), enables guided interventions, and provides substantial insights into vascular calcification, microvascular pathology, and cardiac development for cardiovascular researchers.

4.2 Basic Research on Cardiovascular Development

OCT can be used in basic cardiovascular research for imaging heart tissue structure and embryonic development. For example, mouse embryos were the focus of studies analyzing embryonic cardiac pumping (Wang and Larina 2020). The same team also demonstrated the ability of optogenetic stimulation to control embryonic heartbeat (Lopez 3rd et al. 2020), measured by OCT and Doppler OCT using software they subsequently published (Faubert et al. 2023). Kim et al. (2023) developed a 9.4 MHz (ultrafast) OCT system to enhance high-speed microvascular flow and cardiac structure imaging, achieving 4D microscopic imaging of a beating *Xenopus* (frog) embryo heart at up to 30 volumes per second (Kim et al. 2020). This system enables the real-time visualization of cardiac dynamics and tissue perfusion.

4.3 Assessment of Novel Treatments

OCT is widely used to assess new cardiovascular treatments in order to demonstrate safety and efficacy. Endoscopic OCT was used for long-term in vivo evaluation of biomechanical performance and vascular healing of a bioresorbable scaffold with a novel thin strut (100 μm) in a porcine model (Gasior et al. 2020). OCT has also been used on ex vivo reanimated porcine hearts to compare two ballooning techniques for percutaneous coronary interventions on bifurcation lesions. OCT imaging of the reanimated swine hearts showed one technique was significantly better than the other (OCT imaging showed equal accuracy as micro-computed tomography, but with lower cost and simpler utilization) (Bianchini et al. 2023). There, researchers analyzed minimum stent expansion, stent eccentricity index at the near end of the main vessel (MV), side branch scaffolding length, and malapposition. Another study tested the hypothesis that a specific oral supplement could suppress coronary stent-induced arterial inflammation and neointimal hyperplasia in a pig model using a combined near infrared fluorescence (NIRF)-OCT endoscopic system (Iwata et al. 2021). Axial and lateral resolution for the OCT images were 10–15 μm and 30–60 μm, respectively; the NIRF data had lateral resolution of 100–200 μm using ProSense VM110, an NIR imaging agent.

4.4 Ex Vivo Assessment of Cardiac Tissue

OCT is also particularly useful for detection of arterial calcification and other pathologies ex vivo. Abrao et al. utilized OCT in swine to assess whether exposure to calcium chloride with a TAPAS catheter could serve as a model for arterial calcification (Abrão et al. 2020). OCT imaging was performed after carotids were harvested. On a smaller scale, OCT and multiphoton microscopy (MPM) were used ex vivo to analyze aortic valve stenosis in healthy and age-matched hyperlipidemic C57BL/6 mice (Jannasch et al. 2021). Using an OCT system with center wavelength of 850 nm and bandwidth of 50 nm, the non-fixed aortic valve tissue was visualized by OCT, which identified early fibrotic thickening in the hyperlipidemic mice with precise spatial resolution. Other techniques were used to assess changes in lipid and collagen content ex vivo. Optical clearing (washing the tissue with glycerol, sucrose, or other clearing agents) is often employed for ex vivo studies (Zhernovaya et al. 2013).

4.5 Emerging Cardiovascular OCT Technologies

A shortcoming of commercial OCT systems is cardiac motion. Cecchetti et al. developed a custom OCT able to synchronize acquisition with the cardiac cycle to enable motion artifact-free OCT imaging of implanted stents, using ECG triggering and a rapid pullback (Cecchetti et al. 2020). The study effectively measured length in reanimated porcine hearts exhibiting minimal motion artifacts. Another team has published an open-source post-acquisition synchronization for imaging the mouse embryonic heart with OCT (Faubert et al. 2023). The assignment process, which operates on a 1D signal, was demonstrated to work on embryos with heart rates varying from 1.24 Hz to 2.13 Hz. The team has also used this synchronization method to report optogenetic control of the embryonic heartbeat, which was discussed above (Lopez 3rd et al. 2020). Wang et al. developed "Heartbeat OCT", a system capable of 4000 frames/s and 100 mm/s pullback speed which can overcome motion-artifacts in existing OCT endoscopic systems and provide near-histology level images of the coronary artery (Wang et al. 2015). Figure 3 shows in vivo images of a swine coronary artery.

Catheter-based ablation is a common intervention for patients with symptomatic atrial fibrillation that does not respond to medication or lifestyle changes. The goal of the intervention is to isolate or destroy areas of heart tissue that fire uncontrollably by converting them to lesions, or scar tissue, that does not conduct electrical impulses. Real-time monitoring of lesion formation during ablation has been demonstrated by integrating an OCT probe into a radio frequency ablation catheter (Tiporlini et al. 2021). The system has been tested ex vivo on tissue, demonstrating real-time lesion imaging, tissue characterization, and catheter tip-tissue contact and orientation. Here, OCT has an advantage over ultrasound or optoacoustic hybrid technologies as the all-optical technology is less susceptible to electric noise from the RF energy delivery. A major limitation, however, is that OCT penetration depth of cardiac tissue (~1 mm) is less than the maximum penetration of irrigated RF ablation (the thickness of a human left atrial wall ranges from 1–5 mm).

In summary, OCT technology can aid cardiovascular research through many types of research, including: advancing basic research on cardiovascular development (particularly when utilized with high speed OCT imaging or a compensatory scheme to synchronize heart beat with image acquisition/analysis); preclinical assessment of new therapies or treatments; and 3D imaging of ex vivo cardiac tissue for pathology. Nevertheless, cardiac OCT imaging faces specific limitations not seen in other forms of OCT imaging. The heart is highly scattering: for systems with a 1300 nm center wavelength, image penetration is limited to around 1 mm (Ambrosi et al. 2009). Moreover, because blood absorption at these wavelengths is high, it is important to either displace the blood with clear buffer (Bianchini et al. 2023) or have the probe come in direct contact with the tissue surface when imaging cardiac tissue in vivo in large animal models (Zhao et al. 2021). Table 3 summarizes strengths and shortcomings of OCT technologies for use with different applications for cardiovascular research.

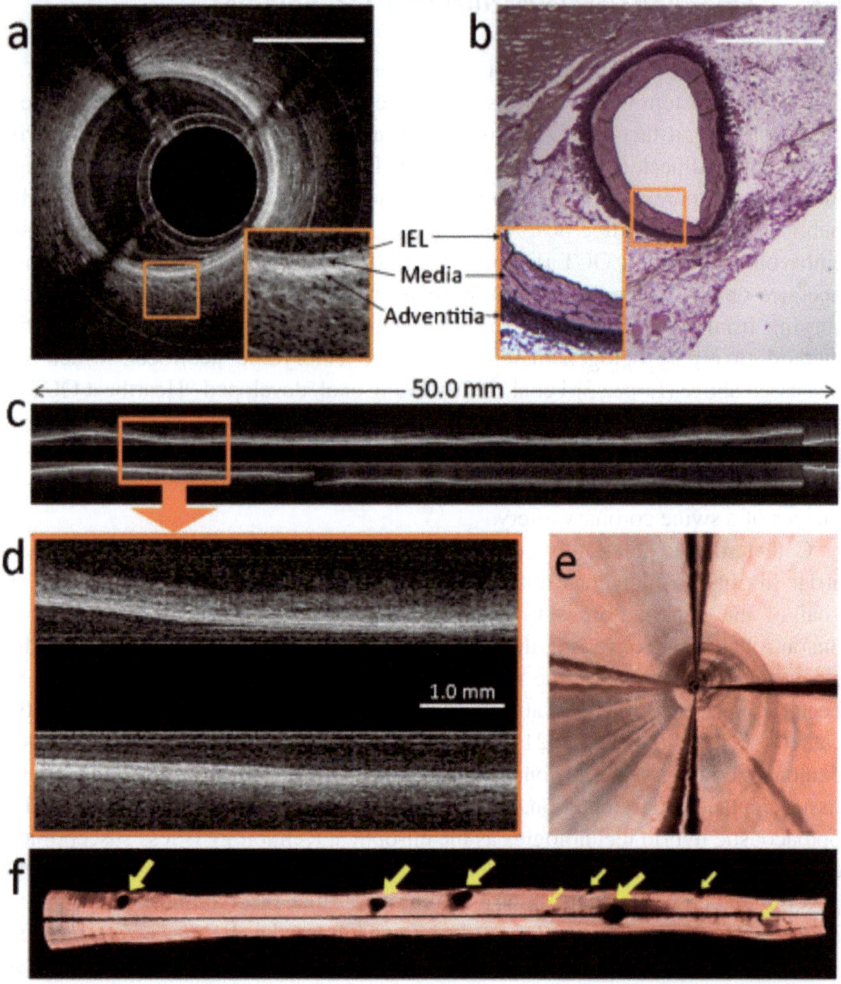

Fig. 3 Images of left anterior descending (LAD) coronary artery from Yorkshire land swine acquired within one cardiac cycle. (**a**) Cross-sectional image of the healthy artery wall (scale bar = 1.0 mm). (**b**) Histological section showing the same layers as in (**a**). (**c**) Longitudinal rendering of the entire artery. (d) Magnification of section of (**c**), showing healthy artery wall and layered structures. (**e**) 3D fly-through inside the artery lumen. (**f**) 3D Image of the artery along the vertical plane. Yellow arrows indicate septal branches. Shadow is caused by the motor lead. (Reprinted from Wang et al. (2015). © 2015 Optical Society of America)

5 Conclusions and Future Perspectives

While OCT has FDA-approved clinical applications for skin, ophthalmologic, and endoscopic imaging, these represent only a fraction of the technology's potential. In small animal studies, OCT has proven invaluable for enabling biomedical

Table 3 Applications of OCT technologies for cardiovascular research

Applications	Strengths	Shortcomings
Assessment of embryonic heart development	Enables assessment of cardiac pumping	Requires image processing or a very fast imaging system to minutely analyze cardiac dynamics and tissue perfusion
Preclinical assessment of novel treatments and technologies	Endoscopic imaging of new treatments in large animal models is similar to endoscopic imaging used clinically, thereby providing relevant preclinical information	Depth penetration is insufficient to fully penetrate cardiac tissue
Ex vivo assessment of cardiac tissue	Excellent at detecting plaque and stenosis	Clearing tissue of blood is needed for improved imaging (greater depth penetration).

discoveries, including assessment of skin damage and wound healing, whole or nearly whole-brain imaging of mice and rats, and evaluation of embryonic cardiac development. The technology has also demonstrated significant value in preclinical studies of large animals, particularly for evaluating surgical treatments, pharmaceutical interventions, and implant performance. The recent integration of OCT with advanced image processing techniques, including deep learning, has expanded its capabilities for functional imaging. Novel OCT devices are being developed to enhance both clinical and preclinical applications. Commercial OCT devices can be purchased from a range of vendors. Thorlabs offers a variety of imaging solutions at different wavelengths, including endoscopic systems. Wasatch Photonics sells OCT systems with patented transmission gratings; these gratings enable a larger bandwidth and better axial resolution compared with competitors. Lumedica uses patented technology to deliver low-cost OCT systems with good imaging depth but slightly reduced axial resolution.

Current technology does face certain limitations with regards to jitter, penetration depth/resolution, functional imaging, and simplified operation. Jitter is a challenge for some applications in small animal research, but can be overcome through hardware stabilization and the use of motion correction algorithms, such as those discussed in Sect. 4.5. OCT's penetration depth is restricted to approximately 2 mm for skin and 1 mm for cardiac tissue. A cranial optical window must be inserted for small animal brain imaging. Additionally, OCT primarily provides structural rather than functional imaging. Exogenous contrast agents can enhance specific regions of interest, but will cause attenuation, effectively blocking the imaging of deeper tissue. One area that needs attention is user experience: standardized protocols and specialized software including analytical tools, particularly for specialized uses of OCT such as OCTA could cause OCT imaging to become more widely used in biomedical research. Several technological advances could also significantly broaden OCT applications. These include improving penetration depth, enhancing signal-to-noise ratio at deeper depths through improved image processing, expanding the field of view, increasing volumetric imaging speed, and reducing system costs.

Acknowledgments This work is financially supported by The Melanoma Research Alliance (Grant# 624320).

Conflict of Interest The authors declare they have no conflict of interest.

References

Abrão SR, Campos CM, Cavalcante R, Eggermont J, Lemos P, Lederman A et al (2020) Percutaneous endovascular delivery of calcium chloride to the intact porcine carotid artery: a novel animal model of arterial calcification. Catheter Cardiovasc Interv 96(4):E484–Ee92. https://doi.org/10.1002/ccd.29070

Adabi S, Conforto S, Clayton A, Podoleanu AG, Hojjat A, Avanaki MR (2016) An intelligent speckle reduction algorithm for optical coherence tomography images. In: 2016 4th International Conference on Photonics, Optics and Laser Technology (PHOTOPTICS). IEEE, pp 1–6

Adabi S, Hosseinzadeh M, Noei S, Conforto S, Daveluy S, Clayton A et al (2017a) Universal in vivo textural model for human skin based on optical coherence tomograms. Sci Rep 7(1):1–11

Adabi S, Turani Z, Fatemizadeh E, Clayton A, Nasiriavanaki M (2017b) Optical coherence tomography technology and quality improvement methods for optical coherence tomography images of skin: a short review. Biomed Eng Comput Biol 8:1179597217713475

Adabi S, Fotouhi A, Xu Q, Daveluy S, Mehregan D, Podoleanu A et al (2018) An overview of methods to mitigate artifacts in optical coherence tomography imaging of the skin. Skin Res Technol 24(2):265–273

Adams DC, Szabari MV, Lagares D, McCrossan AF, Hariri LP, Tager AM et al (2020) Assessing the progression of systemic sclerosis by monitoring the tissue optic axis using PS-OCT. Sci Rep 10(1):2561. https://doi.org/10.1038/s41598-020-59330-7

Akella SS, Lee J, May JR, Puyana C, Kravets S, Dimitropolous V et al (2024) Using optical coherence tomography to optimize Mohs micrographic surgery. Sci Rep 14(1):8900

Ambrosi CM, Moazami N, Rollins AM, Efimov IR (2009) Virtual histology of the human heart using optical coherence tomography. J Biomed Opt 14(5):054002–054007

Anzabi M, Li B, Wang H, Kura S, Sakadžić S, Boas D et al (2021) Optical coherence tomography of arteriolar diameter and capillary perfusion during spreading depolarizations. J Cereb Blood Flow Metab 41(9):2256–2263. https://doi.org/10.1177/0271678x21994013

Araki M, Park S-J, Dauerman HL, Uemura S, Kim J-S, Di Mario C et al (2022) Optical coherence tomography in coronary atherosclerosis assessment and intervention. Nat Rev Cardiol 19(10):684–703. https://doi.org/10.1038/s41569-022-00687-9

Avanaki MR, Hojjatoleslami A (2013) Skin layer detection of optical coherence tomography images. Optik 124(22):5665–5668

Avanaki MR, Hojjat A, Podoleanu AG (2009) Investigation of computer-based skin cancer detection using optical coherence tomography. J Mod Opt 56(13):1536–1544

Avanaki MR, Hojjatoleslami A, Sira M, Schofield JB, Jones C, Podoleanu AG (2013) Investigation of basal cell carcinoma using dynamic focus optical coherence tomography. Appl Opt 52(10):2116–2124

Benavides-Lara J, Siegel AP, Tsoukas MM, Avanaki K (2024) High-frequency photoacoustic and ultrasound imaging for skin evaluation: pilot study for the assessment of a chemical burn. J Biophotonics 17:e202300460

Bianchini F, Cangemi S, DeVos A, Bielecki M, Aurigemma C, Romagnoli E et al (2023) Multimodal comparisons of results achieved by different side branch ballooning techniques for bifurcation provisional stenting. Circ Cardiovasc Interv 16(3):e012908. https://doi.org/10.1161/circinterventions.123.012908

Bouma BE, Villiger M, Otsuka K, Oh W-Y (2017) Intravascular optical coherence tomography. Biomed Opt Express 8(5):2660–2686

Cecchetti L, Wang T, Hoogendoorn A, Witberg KT, Ligthart JMR, Daemen J et al (2020) In-vitro and in-vivo imaging of coronary artery stents with heartbeat OCT. Int J Cardiovasc Imaging 36(6):1021–1029. https://doi.org/10.1007/s10554-020-01796-7

Choi WJ, Li Y, Wang RK (2019) Monitoring acute stroke progression: multi-parametric OCT imaging of cortical perfusion, flow, and tissue scattering in a mouse model of permanent focal ischemia. IEEE Trans Med Imaging 38(6):1427–1437

Demidov V, Demidova N, Pires L, Demidova O, Flueraru C, Wilson BC et al (2021) Volumetric tumor delineation and assessment of its early response to radiotherapy with optical coherence tomography. Biomed Opt Express 12(5):2952–2967. https://doi.org/10.1364/boe.424045

DePaoli D, Côté DC, Bouma BE, Villiger M (2022) Endoscopic imaging of white matter fiber tracts using polarization-sensitive optical coherence tomography. NeuroImage 264:119755. https://doi.org/10.1016/j.neuroimage.2022.119755

Doleżyczek H, Kasprzycki P, Włodarczyk J, Wojtkowski M, Malinowska M (2022) Optical coherence tomography reveals heterogeneity of the brain tissue and vasculature in the ischemic region after photothrombotic stroke in mice. Acta Neurobiol Exp (Wars) 82(1):106–119. https://doi.org/10.55782/ane-2022-010

Dur AH, Tang T, Viviano S, Sekuri A, Willsey HR, Tagare HD et al (2020) In Xenopus ependymal cilia drive embryonic CSF circulation and brain development independently of cardiac pulsatile forces. Fluids Barriers CNS 17(1):72. https://doi.org/10.1186/s12987-020-00234-z

Faubert AC, Larina IV, Wang S (2023) Open-source, highly efficient, post-acquisition synchronization for 4D dual-contrast imaging of the mouse embryonic heart over development with optical coherence tomography. Biomed Opt Express 14(1):163–181. https://doi.org/10.1364/boe.475027

Gao T, Liu S, Gao E, Wang A, Tang X, Fan Y (2022) Automatic segmentation of laser-induced injury OCT images based on a deep neural network model. Int J Mol Sci 23(19). https://doi.org/10.3390/ijms231911079

Gasior P, Cheng Y, Xia J, Conditt GB, McGregor JC, Virmani R et al (2020) Two-year longitudinal evaluation of a second-generation thin-strut sirolimus-eluting bioresorbable coronary scaffold with hybrid cell design in porcine coronary arteries. Cardiol J 27(2):115–125. https://doi.org/10.5603/CJ.a2018.0095

Ge GR, Rolland JP, Song W, Nedergaard M, Parker KJ (2023) Fluid compartments influence elastography of the aging mouse brain. Phys Med Biol 68(9):095004. https://doi.org/10.1088/1361-6560/acc922

Ghosh B, Mandal M, Mitra P, Chatterjee J (2021) Attenuation corrected-optical coherence tomography for quantitative assessment of skin wound healing and scar morphology. J Biophotonics 14(4):e202000357. https://doi.org/10.1002/jbio.202000357

Gora MJ, Suter MJ, Tearney GJ, Li X (2017) Endoscopic optical coherence tomography: technologies and clinical applications [Invited]. Biomed Opt Express 8(5):2405–2444. https://doi.org/10.1364/BOE.8.002405

Harper DJ, Kim Y, Gómez-Ramírez A, Vakoc BJ (2023) Needle guidance with Doppler-tracked polarization-sensitive optical coherence tomography. J Biomed Opt 28(10):102910. https://doi.org/10.1117/1.jbo.28.10.102910

Hessler M, Jalilian E, Xu Q, Reddy S, Horton L, Elkin K et al (2020) Melanoma biomarkers and their potential application for in vivo diagnostic imaging modalities. Int J Mol Sci 21(24):9583

Hojjatoleslami A, Avanaki MR (2012) OCT skin image enhancement through attenuation compensation. Appl Opt 51(21):4927–4935

Iwata H, Osborn EA, Ughi GJ, Murakami K, Goettsch C, Hutcheson JD et al (2021) Highly selective PPARα (peroxisome proliferator-activated receptor α) agonist Pemafibrate inhibits stent inflammation and restenosis assessed by multimodality molecular-microstructural imaging. J Am Heart Assoc 10(20):e020834. https://doi.org/10.1161/jaha.121.020834

Jacobsen K, Ortner VK, Fredman GL, Christensen RL, Dierickx C, Tanghetti E et al (2023) Melanin-dependent tissue interactions induced by a 755-nm picosecond-domain laser: complementary visualization by optical imaging and histology. Lasers Med Sci 38(1):160. https://doi.org/10.1007/s10103-023-03811-4

Jannasch A, Schnabel C, Galli R, Faak S, Büttner P, Dittfeld C et al (2021) Optical coherence tomography and multiphoton microscopy offer new options for the quantification of fibrotic aortic valve disease in ApoE(−/−) mice. Sci Rep 11(1):5834. https://doi.org/10.1038/s41598-021-85142-4

Ji Y, Yang S, Zhou K, Lu J, Wang R, Rocliffe HR et al (2022) Semisupervised representative learning for measuring epidermal thickness in human subjects in optical coherence tomography by leveraging datasets from rodent models. J Biomed Opt 27:27(8). https://doi.org/10.1117/1.jbo.27.8.085002

Kim TS, Joo J, Shin I, Shin P, Kang WJ, Vakoc BJ et al (2020) 9.4 MHz A-line rate optical coherence tomography at 1300 nm using a wavelength-swept laser based on stretched-pulse active mode-locking. Sci Rep 10(1):9328. https://doi.org/10.1038/s41598-020-66322-0

Kim G, Kim J, Choi WJ, Kim C, Lee S (2022) Integrated deep learning framework for accelerated optical coherence tomography angiography. Sci Rep 12(1):1289. https://doi.org/10.1038/s41598-022-05281-0

Kim J, Kim S, Choi WJ (2023) Non-invasive monitoring of cutaneous wound healing in non-diabetic and diabetic model of adult zebrafish using OCT angiography. Bioengineering (Basel) 10(5). https://doi.org/10.3390/bioengineering10050538

Lautman Z, Winetraub Y, Blacher E, Yu C, Terem I, Chibukhchyan A et al (2022) Intravital 3D visualization and segmentation of murine neural networks at micron resolution. Sci Rep 12(1):13130. https://doi.org/10.1038/s41598-022-14450-0

Li X, Chudoba C, Ko T, Pitris C, Fujimoto JG (2000) Imaging needle for optical coherence tomography. Opt Lett 25(20):1520–1522

Liang X, Quan X, Geng X, Huang Y, Zhao Y, Xi L et al (2025) A promising approach for quantifying focal stroke modeling and assessing stroke progression: optical resolution photoacoustic microscopy photothrombosis. Neural Regen Res 20(7):2029–2037. https://doi.org/10.4103/nrr.nrr-d-23-01617

Liu M, He G, Wang F, Sun Y, Ma S, Hao Y et al (2024) Pilose antler extract promotes hair growth in androgenic alopecia mice by promoting the initial anagen phase. Biomed Pharmacother 174:116503. https://doi.org/10.1016/j.biopha.2024.116503

Lopez AL 3rd, Wang S, Larina IV (2020) Optogenetic cardiac pacing in cultured mouse embryos under imaging guidance. J Biophotonics 13(11):e202000223. https://doi.org/10.1002/jbio.202000223

Lu Y, Lu X, Zhang C, Marchand PJ, Lesage F (2020) Longitudinal optical coherence tomography imaging of tissue repair and microvasculature regeneration and function after targeted cerebral ischemia. J Biomed Opt 25(4):1–15. https://doi.org/10.1117/1.jbo.25.4.046002

Nyúl-Tóth Á, Tarantini S, DelFavero J, Yan F, Balasubramanian P, Yabluchanskiy A et al (2021a) Demonstration of age-related blood-brain barrier disruption and cerebromicrovascular rarefaction in mice by longitudinal intravital two-photon microscopy and optical coherence tomography. Am J Physiol Heart Circ Physiol 320:H1370–H1392

Nyúl-Tóth Á, Tarantini S, DelFavero J, Yan F, Balasubramanian P, Yabluchanskiy A et al (2021b) Demonstration of age-related blood-brain barrier disruption and cerebromicrovascular rarefaction in mice by longitudinal intravital two-photon microscopy and optical coherence tomography. Am J Physiol Heart Circ Physiol 320(4):H1370–H1h92. https://doi.org/10.1152/ajpheart.00709.2020

O'leary S, Fotouhi A, Turk D, Sriranga P, Rajabi-Estarabadi A, Nouri K et al (2018) OCT image atlas of healthy skin on sun-exposed areas. Skin Res Technol 24(4):570–586

Osiac E, Mitran SI, Manea CN, Cojocaru A, Rosu G-C, Osiac M et al (2021) Optical coherence tomography microscopy in experimental traumatic brain injury. Microsc Res Tech 84(3):422–431. https://doi.org/10.1002/jemt.23599

Park KS, Shin JG, Qureshi MM, Chung E, Eom TJ (2018) Deep brain optical coherence tomography angiography in mice: in vivo, noninvasive imaging of hippocampal formation. Sci Rep 8(1):11614

Pasarikovski CR, Ku JC, Keith J, Ramjist J, Dobashi Y, Priola SM et al (2020) Endovascular cerebral venous sinus imaging with optical coherence tomography. AJNR Am J Neuroradiol 41(12):2292–2297. https://doi.org/10.3174/ajnr.A6909

Pasarikovski CR, Ku JC, Ramjist J, Dobashi Y, Priola SM, da Costa L et al (2021) Minimally invasive intrathecal spinal cord imaging with optical coherence tomography. J Biomed Opt 26(5). https://doi.org/10.1117/1.jbo.26.5.056002

Pfister M, Stegmann H, Schützenberger K, Schäfer BJ, Hohenadl C, Schmetterer L et al (2021) Deep learning differentiates between healthy and diabetic mouse ears from optical coherence tomography angiography images. Ann N Y Acad Sci 1497(1):15–26. https://doi.org/10.1111/nyas.14582

Rakymzhan A, Li Y, Tang P, Wang RK (2020) Optical microangiography reveals temporal and depth-resolved hemodynamic change in mouse barrel cortex during whisker stimulation. J Biomed Opt 25(9). https://doi.org/10.1117/1.jbo.25.9.096005

Schuetzenberger K, Pfister M, Messner A, Froehlich V, Garhoefer G, Hohenadl C et al (2019) Comparison of optical coherence tomography and high frequency ultrasound imaging in mice for the assessment of skin morphology and intradermal volumes. Sci Rep 9(1):13643. https://doi.org/10.1038/s41598-019-50104-4

Schuetzenberger K, Pfister M, Messner A, Garhöfer G, Hohenadl C, Pfeiffenberger U et al (2020) Cutaneous optical coherence tomography for longitudinal volumetric assessment of intradermal volumes in a mouse model. Sci Rep 10(1):4245. https://doi.org/10.1038/s41598-020-61276-9

Stamenkovic S, Li Y, Waters J, Shih A (2023) Deep imaging to dissect microvascular contributions to white matter degeneration in rodent models of dementia. Stroke 54(5):1403–1415. https://doi.org/10.1161/STROKEAHA.122.037156

Sunil S, Evren Erdener S, Cheng X, Kura S, Tang J, Jiang J et al (2021) Stroke core revealed by tissue scattering using spatial frequency domain imaging. Neuroimage Clin 29:102539. https://doi.org/10.1016/j.nicl.2020.102539

Szczepanik M, Balicki I, Śmiech A, Szadkowski M, Gołyński M, Osęka M et al (2022) The use of optical coherence tomography for skin evaluation in healthy rats. Vet Dermatol 33(4):296–e69. https://doi.org/10.1111/vde.13071

Tiporlini V, Ahderom S, Pratten P, Alameh K (2021) Advanced fully integrated radiofrequency/optical-coherence-tomography irrigated catheter for atrial fibrillation ablation. J Biophotonics 14(2):e202000243. https://doi.org/10.1002/jbio.202000243

Tsai T-H, Leggett CL, Trindade AJ, Sethi A, Swager A-F, Joshi V et al (2017) Optical coherence tomography in gastroenterology: a review and future outlook. J Biomed Opt 22(12):121716

Turani Z, Fatemizadeh E, Xu Q, Daveluy S, Mehregan D, Nasiriavanaki M (2018) Refractive index correction in optical coherence tomography images of multilayer tissues. J Biomed Opt 23(7):070501

Turchin I, Beschastnov V, Peretyagin P, Perekatova V, Kostyuk A, Orlova A et al (2023) Multimodal optical monitoring of auto- and allografts of skin on a burn wound. Biomedicine 11(2). https://doi.org/10.3390/biomedicines11020351

Walek KW, Stefan S, Lee JH, Puttigampala P, Kim AH, Park SW et al (2023) Near-lifespan longitudinal tracking of brain microvascular morphology, topology, and flow in male mice. Nat Commun 14(1):2982. https://doi.org/10.1038/s41467-023-38609-z

Wang S, Larina I (2020) Live mechanistic assessment of localized cardiac pumping in mammalian tubular embryonic heart. J Biomed Opt 25(8):1–19. https://doi.org/10.1117/1.jbo.25.8.086001

Wang T, Pfeiffer T, Regar E, Wieser W, van Beusekom H, Lancee CT et al (2015) Heartbeat OCT: in vivo intravascular megahertz-optical coherence tomography. Biomed Opt Express 6(12):5021–5032. https://doi.org/10.1364/BOE.6.005021

Wang YJ, Chang CC, Lu ME, Wu YH, Shen JW, Chiang HM et al (2022) Photoaging and sequential function reversal with cellular-resolution optical coherence tomography in a nude mice model. Int J Mol Sci 23(13). https://doi.org/10.3390/ijms23137009

Xu Q, Jalilian E, Fakhoury JW, Manwar R, Michniak-Kohn B, Elkin KB et al (2020) Monitoring the topical delivery of ultrasmall gold nanoparticles using optical coherence tomography. Skin Res Technol 26(2):263–268. https://doi.org/10.1111/srt.12789

Yang L, Chen P, Wen X, Zhao Q (2025) Optical coherence tomography (OCT) and OCT angiography: technological development and applications in brain science. Theranostics 15(1):122–140. https://doi.org/10.7150/thno.97192

Yoon JH, Shin P, Joo J, Kim GS, Oh WY, Jeong Y (2022) Increased capillary stalling is associated with endothelial glycocalyx loss in subcortical vascular dementia. J Cereb Blood Flow Metab 42(8):1383–1397. https://doi.org/10.1177/0271678x221076568

Zafar M, Siegel AP, Avanaki K, Manwar R (2024) Skin imaging using optical coherence tomography and photoacoustic imaging: a mini-review. Optics 5(2):248–266

Zeng Y, Xu S, Chapman WC, Li S, Alipour Z, Abdelal H et al (2020) Real-time colorectal cancer diagnosis using PR-OCT with deep learning. Optical Coherence Tomography: Optica Publishing Group, p OW2E. 5

Zhao X, Ziv O, Mohammadpour R, Crosby B, Hoyt WJ, Jenkins MW et al (2021) Polarization-sensitive optical coherence tomography monitoring of percutaneous radiofrequency ablation in left atrium of living swine. Sci Rep 11(1):24330. https://doi.org/10.1038/s41598-021-03724-8

Zhernovaya O, Tuchin VV, Leahy MJ (2013) Blood optical clearing studied by optical coherence tomography. J Biomed Opt 18(2):026014

Zhu M, Chang W, Jing L, Fan Y, Liang P, Zhang X et al (2019) Dual-modality optical diagnosis for precise in vivo identification of tumors in neurosurgery. Theranostics 9(10):2827–2842

Zhu J, Freitas HR, Maezawa I, Jin L-w, Srinivasan VJ (2021) 1700 nm optical coherence microscopy enables minimally invasive, label-free, in vivo optical biopsy deep in the mouse brain. Light: Science & Applications 10(1):145

Basic Principles and Preclinical Applications of Magnetic Resonance Imaging (MRI)

Roland Pilgram and Gerd Melkus

Abstract

Magnetic resonance imaging (MRI) is a non-invasive imaging modality widely used in both clinical practice and research for high-resolution visualization of internal body structures. Its primary advantages include excellent soft tissue contrast, the absence of ionizing radiation, and the ability to acquire both anatomical and functional data. MRI is based on the principles of nuclear magnetic resonance, leveraging the magnetic properties of atomic nuclei, primarily hydrogen. An MRI system comprises several key components: the main magnet, which generates a strong static magnetic field, gradient coils for spatial encoding and radiofrequency (RF) coils for signal transmission and reception. MR signal generation relies on the angular momentum (spin) of nuclei, which align under the magnetic field. The Larmor frequency governs the precession rate of these nuclei and depends on magnetic field strength and the gyromagnetic ratio. Following RF excitation, tissue magnetization returns to equilibrium through relaxation processes, as T1 (longitudinal) and T2 (transverse), which generates contrast in standard imaging sequences. Both T1-weighted and T2-weighted images show detailed anatomical structures. T2-weighted images are also sensitive to fluid and pathology. Image reconstruction is achieved through Fourier transformation of spatially encoded signals. Common pulse sequences

R. Pilgram (✉)
Department of Radiology, Radiation Oncology and Medical Physics, University of Ottawa, Ottawa, ON, Canada
e-mail: rpilgram@uottawa.ca

G. Melkus
Department of Radiology, Radiation Oncology and Medical Physics, University of Ottawa, Ottawa, ON, Canada

Department of Physics, Carleton University, Ottawa, ON, Canada

I. F. Cengiz et al. (eds.), *Bioimaging Modalities in Bioengineering*,
Biomaterials, Bioengineering and Sustainability 9,
https://doi.org/10.1007/978-3-031-99410-4_5

include spin echo, known for high contrast, and gradient echo, valued for faster acquisition and sensitivity to magnetic susceptibility. MRI plays a vital role in neuroimaging for assessing brain structure and function (functional MRI), and in cardiac imaging for evaluating myocardial anatomy and function. These applications underscore MRI's versatility and indispensable role across diverse medical and scientific domains.

Keywords

MRI components · MRI basic principles · MRI relaxation · MRI sequences · MRI applications

1 Introduction

Magnetic resonance imaging (MRI) is a powerful and versatile imaging modality widely used in both clinical and research settings. It provides detailed, non-invasive visualization of internal anatomical structures and physiological processes, making it indispensable in modern medicine and biomedical science. Unlike imaging techniques that rely on ionizing radiation, such as computed tomography (CT) or positron emission tomography (PET), MRI uses non-ionizing radiofrequency (RF) waves, offering a safer alternative particularly suited for longitudinal and functional studies. Proton MRI exploits the quantum properties of hydrogen nuclei, particularly their nuclear spin, and is based on the nuclear magnetic resonance (NMR) phenomenon (Purcell et al. 1946; Bloch 1946). When placed in a strong magnetic field, hydrogen nuclei can absorb and re-emit RF energy as electromagnetic waves at a defined frequency, known as the "Larmor frequency". MRI utilizes the vast abundance of hydrogen atoms in the human body, accounting for up to 70% of body weight, primarily found in water, lipids, carbohydrates, and proteins.

MRI plays a vital role in preclinical research due to its non-destructive nature, superior soft tissue contrast, and ability to provide high-resolution three-dimensional (3D) or four-dimensional (4D) imaging (Marzola et al. 2003; Saito and Ueda 2024). MRI is widely used in neuroscience to study brain structure, function (functional MRI (fMRI)) and disease models (Ma et al. 2024; Mandino et al. 2020). In oncology it is used for tumor detection, characterization, and tumor progression tracking (Hoffmann et al. 2024; Mörchel et al. 2010). MRI allows dynamic visualization of anatomical or physiological processes, such as cardiac motion, to assess e.g., vascularity and myocardial infarction (Wu 2020). It is also popular for studying musculoskeletal systems and internal organs like the liver and kidneys (Kosaka et al. 2000; Samagh et al. 2013; Lindquist et al. 2021). It is an ideal tool for developing novel treatments for translational research. This chapter introduces the components of MRI and explains conceptional the fundamental principle of the MRI signal generation. It provides an overview of MR imaging and image reconstruction techniques, and presents two examples of preclinical MRI applications, namely preclinical brain imaging and cardiac imaging. Conclusions and Future Perspectives are discussed in the end.

2 MRI Scanner Components

The MRI system comprises several major hardware components, namely three major layers within the scanner, and a computer and control system including a data acquisition system (Dance et al. 2014; Schröder and Faber 2011). The three major layers are: the outermost layer, which contains a large and heavy magnet that produces a static, homogeneous high field (B_0, ~Tesla), which is always active (Hanley 1984; Laukien and Tschopp 1994). The second layer includes gradient coils, which can slightly alter the magnetic field (on the order of milli-Tesla) across three dimensions to localize the nuclear spins in space (Golay 1958; Mansfield and Chapman 1986). The innermost layer consists of the RF coils (on the order of micro-Tesla) for sending and receiving RF signals to and from the nuclear spins (Froncisz et al. 1986; Bosch and Ackerman 1992; Gareis et al. 2007).

2.1 Static Magnetic Field

Modern MRI machines use electromagnets, replacing the permanent magnets of earlier models, to create a highly uniform, strong, and stable magnetic field. Higher magnetic fields require increased electric currents and superconducting materials, which exhibit zero electrical resistance near absolute zero (0 K or −273 °C). As a result, an electric current can circulate indefinitely in a superconducting loop due to zero electrical resistance, eliminating the need for a power source. Consequently, the main superconducting coil in an MRI does not consume power directly; instead, most energy is used to keep the coil cold enough to sustain superconductivity, allowing the persistent current to maintain the MRI's magnetic field. The most common superconducting material in MRIs is niobium-titanium (Nb-Ti), maintained at ultra-low temperatures using liquid helium (4 K or −269 °C) (De Graaf 2019). Vacuum-sealed chambers minimize helium evaporation, reducing refill needs and operational costs. Cryo-compressors recondense helium vapor into liquid, sustaining a low-cryogen system during normal operation. This closed-loop refrigeration ensures the helium remains liquid and the magnets stay superconducting. Today, these "zero boil-off" (ZBO) systems are the standard in MRI technology. During a quench, a sudden loss of superconductivity causes the liquid helium to rapidly vaporize. The resulting expansion of helium gas can lead to thermal burns and asphyxiation risks and needs to be released through a quench (venting) pipe outside the building. Newer generations of superconducting magnets do not require liquid helium or liquid nitrogen for the cooling; they are called dry magnets (MRSolutions 2025). Therefore, no quench pipes or valves are necessary, making the installation inexpensive and easy. Such systems can be ramped down and ramped up to different field strengths within a few minutes.

2.2 Gradient and Shimming Coils

Gradient and shimming coils are integral components of MRI systems, each playing distinct yet complementary roles in spatial encoding and magnetic field homogeneity. Gradient coils determine the spatial location of hydrogen nuclei, enabling the conversion of RF frequencies into images. These coils generate linear magnetic field gradients along the x (horizontal, orthogonal to B_0), y (vertical, orthogonal to B_0), or z (aligned with B_0) axes by superimposing additional fields onto B_0 using separate electromagnets (Golay 1958). Nuclei in weaker fields precess at a lower Larmor Frequency, while those in stronger fields precess at a higher, allowing localization of signals and image creation based on their resonant frequencies. Gradients are defined by amplitude, slew rate, and linearity. Rapid switching of these gradients, required for advanced imaging sequences like echo-planar imaging (Mansfield 1977; Ordidge 1999), demands high slew rates, which need to be compensated to reduce gradient settling times (Jensen et al. 1987). Their switching generates eddy currents and the characteristic acoustic noise of MRI systems.

Shimming coils, by contrast, are designed to correct inhomogeneities in the main magnetic field (B_0), which can arise from imperfections in the magnet or the variations in the presence of the animal's susceptibility, particularly in regions with complex tissue-air interfaces (De Graaf 2019). A homogeneous B_0 field is critical for resolution, accurate spatial encoding, and minimizing geometric distortion in functional and diffusion imaging. Shimming is an important step prior to scanning to tailor for individual subjects or specific imaging targets to compensate for both global and localized field inhomogeneities. The integration of high-order shimming coils can further improve the correction for advanced applications such as susceptibility-weighted imaging (Haacke et al. 2004), and quantitative mapping.

2.3 The Radiofrequency (RF) System

The RF system is a fundamental component of the MRI scanner, responsible for generating, transmitting, and receiving the electromagnetic signals necessary for image acquisition. The system typically comprises an RF Synthesizer, RF amplifier, transmit (Tx) and receive coils (Rx), and a receiver chain (De Graaf 2019). The RF synthesizer generates carrier signals at the Larmor frequency, which are modulated into complex RF pulse waveforms (e.g., sinc, Gaussian, adiabatic) for slice selection, excitation, refocusing, or inversion. RF coils transmit precisely timed electromagnetic waves at the Larmor frequency, exciting the nuclei and creating an oscillating magnetic field (B_1) perpendicular to the main static field (B_0) (Link 1992). After the excitation pulse, the system switches rapidly to receive mode to detect the emitted RF signals, which are pre-amplified and converted into digital signals for further processing and image generation. To compensate for the weak RF signals, coils are positioned as close to the body as possible. Separate coils for transmitting and receiving signals are often used to enhance contrast and resolution, with the larger transmit coil generating a homogenous transmit field over the body and the local (and smaller) receiver coils maximize signal strength by being placed closer to the body part of investigation (Hoult and Richards 1976). Coils may be surface, volume (e.g., birdcage coils), or phased-array designs, each optimized for particular anatomical regions and applications, offering improved signal-to-noise ratio (SNR) and spatial coverage (Tropp 1969; Vaughan et al. 2002; Jin et al. 1994). Volume coils are commonly preferred for transmission due to their ability to produce a homogeneous B_1 field, while surface or phased-array coils are often used for reception due to their higher sensitivity, spatial localization and their ability to be used with parallel imaging techniques for scan acceleration (Gareis et al. 2007; Blaimer et al. 2004). The entire RF system operates within an RF-shielded environment (Faraday cage) to prevent external electromagnetic interference, maintaining signal integrity and optimizing SNR for high-resolution imaging.

2.4 Computer and Control System, MRI Safety

The computer and control system of an MRI scanner manages and coordinates all aspects of image acquisition and processing, it ensures synchronization, safety interlocks and precise execution of imaging protocols. The pulse programmer controls the gradient- and RF coils to ensure precise RF signal delivery, encoding, and reception based on a wide range of sequence concepts. The gradient and RF Amplifier Control modulates the magnetic field gradients and transmits RF pulses with high accuracy. The Data Acquisition System receives raw MR signals (echoes) from the receiver coils, digitizes them, and sends the data to the main computer for processing. The image reconstruction unit performs Fourier transformation (Brigham 1988) and other algorithms to convert raw k-space (Mezrich 1995) data into anatomical images. The control system ensures synchronization, safety interlocks, and precise execution of imaging protocols critical for high-quality diagnostic imaging. Users interact with the system through a host computer to change parameters of MRI sequences to generate different image contrasts. Raw data are stored on a computer and processed into images for analysis.

Alongside the technical aspects, MRI safety is a critical consideration. The static magnetic field, always active, poses risks to ferromagnetic materials and sensitive electronics. Key safety concerns include (1) the projectile risk within the MRI scanner room (due to translational and torque forces) and (2) interference of the gradient and RF fields with magnetic field-sensitive implants which can result in local heating and burns (ACR 2024). Metallic objects, such as ear tags or implants in rodents, must be verified or removed for MRI compatibility before experiments. All equipment used near or within the magnet should be tested, and ferromagnetic materials must be strictly excluded. Access to the MRI room is limited to trained, authorized personnel.

3 Basic Principles of MR Signal Generation

The physics underlying MRI has evolved over many years, earning several Nobel Prizes along the way. While a full understanding of magnetic resonance involves complex quantum mechanics and subatomic particle behavior, many key concepts can be grasped through classical physics principles such as precession, resonance, and vector dynamics. The following provides a brief conceptual overview based on Huettel et al. (2014), which also offers a more detailed and quantitative elaboration for interested readers.

MRI begins at the level of the atomic nucleus, which possesses spin, angular momentum, and an associated magnetic moment (Brown et al. 2014). The hydrogen nucleus, consisting of a single proton ^{1}H, has an intrinsic quantum property called "spin," similar to mass and charge (Shaw and Elsken 1950; Odeblad and Lindström 1955). Only atoms with an odd number of protons or neutrons exhibit spin. These spinning nuclei behave like tiny bar magnets, generating magnetic fields. This property gives nuclei a magnetic moment and an intrinsic spin angular momentum. Under normal conditions, the orientation of these magnetic fields is distributed randomly in space. Without an external magnetic field, the tiny bar magnets are unaligned, resulting in a net magnetization vector—the sum of all individual magnetic fields—that is effectively zero (Fig. 1a).

3.1 Magnetization

When hydrogen protons enter a strong external magnetic field, their random orientations align with the field creating a net magnetization used for MRI. Any tissue has a specific net magnetic vector M_z in the direction of the main magnetic field B_0, meaning the magnetic vector aligns with B_0 (by convention, along the z-axis). About half of the hydrogen spins align with the field (N+, parallel in the low-energy state) and others align against the field (N-, antiparallel in the

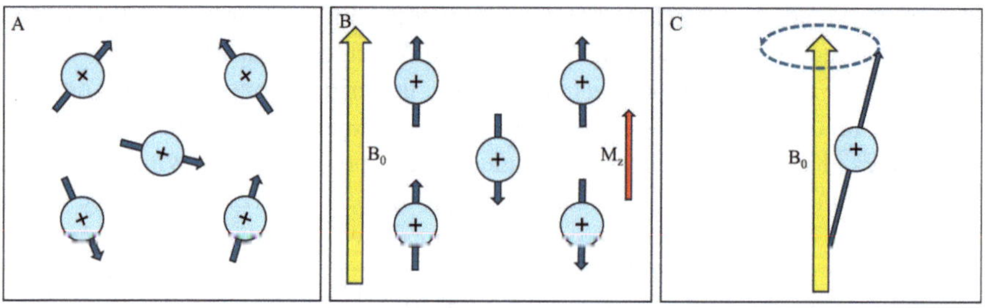

Fig. 1 (**a**) under normal conditions, the orientation of hydrogen protons ^{1}H is distributed randomly in space, resulting in a zero net magnetization vector; (**b**) when hydrogen protons (^{1}H) enter an external magnetic field B_0, their random orientations align with the field creating a net magnetization M_z; (**c**) Hydron protons (^{1}H) precess around B_0 as the protons have an angular momentum

high-energy state) (Fig. 1b). The difference between N+ and N- generates the net longitudinal magnetization component M_Z along the z-axis and is given by the Boltzmann distribution (Dance et al. 2014; Brown et al. 2014):

$$\frac{N_+}{N_-} = e^{\Delta E/kT} = e^{\hbar\omega/kT}$$

where ΔE is the energy difference between two energy states, k is the Boltzmann constant, T is the absolute temperature, ħ is the Planck constant and ω (omega) is the Larmor or resonance frequency. At body temperature in a field of 1.0 T, this equation predicts that the N+ and N- are nearly equal; only a small excess (~3 × 10^{-6}) of spins can be expected to be found in the lower energy (spin-up) state and when measured for 7 T/ 9.4 T it would be ~4 × 10^{-4}/ ~6.4 × 10^{-4}. Although, in effect, only a small fraction of the spins can be detected, the enormous number of spins available in any given voxel—and the whole body—permits the detection of magnetization to generate MR images. Increasing the strength of the external magnetic field (B_0) enhances the net magnetization available for MRI. This boost in field strength increases the received signal, improving the image's SNR. However, rather than aligning exactly, as the protons have an angular momentum, they precess around B_0. Precession is like a wobbling motion around the main magnetic field B_0, described as a cone, in analogs of a spinning top, where the magnetic moment traces a conical path around the main magnetic field (Fig. 1c). While spin is an intrinsic property, precession is the circular movement of that magnetic momentum when placed in an external magnetic field. The specific frequency at which protons rotate is called the Larmor frequency ω and is described by the Larmor equation (Brown et al. 2014):

$$\omega = \gamma B_0$$

which is determined by the magnetic field B_0 (the stronger B_0 the faster the precession frequency is) and on the nuclei itself, which is reflected in the term γ (gamma) or gyromagnetic ratio. A list of NMR nuclei and their properties is shown in Table 1.

Table 1 Common nuclei for MRI

Element	Gyromagnetic ratio γ	Natural abundance	ω at 1.5 T	ω at 7 T	ω at 9.4 T
	MHz T^{-1}	%	MHz	MHz	MHz
^{1}H	42.5774	99.98	63.8661 (Dale et al. 2015)	298.0418	400.2276
^{13}C	10.7084	1.10	16.0626 (Dale et al. 2015)	74.9588	100.659
^{19}F	40.0776	100.00	60.1164 (Dale et al. 2015)	280.5432	376.7294
^{23}Na	11.2686	100.00	16.9029 (Dale et al. 2015)	78.8802	105.9248
^{31}P	17.2514	26.40	25.8771 (Dale et al. 2015)	120.7598	162.1632

Values at 1.5 T is based on data from Dale et al. (2015)

3.2 RF Excitation and Detection

Nuclei align with the main magnetic field (B_0), but their orientation must be manipulated to measure net magnetization. RF transmitter coils generate an oscillating RF field (B_1) perpendicular to B_0. When delivered at the Larmor frequency, the RF pulse disrupts spin equilibrium, enabling energy exchange and tipping the net magnetization vector (M_z) from the longitudinal z-axis to the transverse x–y plane (Fig. 2a-left). This occurs due to resonance between the protons and the RF pulse. The process, called excitation, involves energy absorption and spin phase alteration. At the quantum level, protons transition to a higher energy state (from parallel to anti-parallel alignment with B_0), while macroscopically, the net magnetization spirals into the transverse plane. The RF pulse tips the net magnetization vector by a specific flip angle, which depends on the pulse's intensity, waveform, and duration. A 90° RF pulse tips the net magnetization vector completely into the transverse (x–y) plane. This aligns protons to spin in phase, perpendicular to the

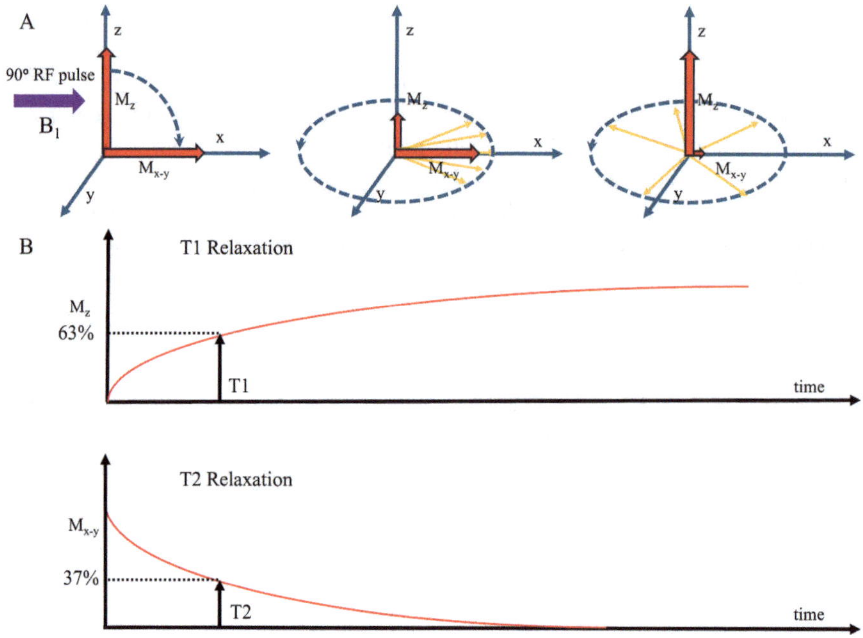

Fig. 2 (**a**) An RF pulse delivered (in this case a 90° pulse) at the Larmor frequency disrupts spin equilibrium, enabling energy exchange and tipping the net magnetization vector (M_z) from the longitudinal z-axis to the transverse x–y plane (left). When the RF pulse ends, net magnetization gradually slowly returns to equilibrium in a process called spin-lattice relaxation, or T1 recovery (middle—starting recovery to right—full recovery); (**b**) Random spin-spin interactions (T2 relaxation) disrupt precession coherence leading to a gradual loss of phase alignment in the transverse plane (see **a**, middle to right) after RF pulse excitation and the decay of transverse magnetization (M_{x-y}) follows an exponential recovery curve (T1 relaxation)

main magnetic field (B_0), with their magnetic moments synchronized. As the system returns to equilibrium, electromagnetic energy is emitted. This RF signal, detected via a receive coil using Faraday's law of induction, is amplified, and processed to produce the MRI image. The signal's strength depends on tissue-specific spin density, flip angle, gyromagnetic ratio, and magnetic field strength.

3.3 Relaxation Types: T1, T2, T2*

When the RF pulse ends, net magnetization gradually returns to equilibrium in a process called relaxation (Fig. 2a-middle and right). The excited spin system, having absorbed RF energy, does not relax instantly. Relaxation time varies based on the microscopic spin environment and differs across tissue types. This process involves two simultaneous mechanisms: T1 relaxation and T2 relaxation (Jung and Weigel 2013). T1 relaxation time, also called longitudinal or spin-lattice relaxation, describes the recovery of longitudinal magnetization after a perturbation. During this process, spins transfer energy to their surrounding chemical and physical environment, known as the 'lattice.' The recovery follows an exponential curve toward equilibrium, characterized by the tissue-specific constant T1. After one T1 period, 63% of the longitudinal magnetization has returned to its equilibrium value (Fig. 2b). Notably, T1 values increase with higher magnetic field strengths.

T2 relaxation, also known as transverse or spin-spin relaxation, is an entropy-driven process where spins lose phase coherence by exchanging energy among themselves without any net energy transfer. Each spinning proton generates a magnetic field that slightly alters the surrounding field. These random spin-spin interactions (T2 relaxation) disrupt precession coherence (indicated in Fig. 2 A-middle and right with orange arrows), leading to a gradual loss of phase alignment in the transverse plane after RF pulse excitation. The decay of transverse magnetization ($M_{x\text{-}y}$) follows an exponential curve and is characterized by the time constant T2, which represents the time required for transverse magnetization to decrease by 63% (Fig. 2b). However, the observed apparent transverse relaxation (T2*) is shorter than true T2 because it includes additional dephasing caused by magnetic field inhomogeneities (Chavhan et al. 2009). These inhomogeneities arise from imperfections in the MRI magnet and susceptibility differences within tissues, such as air-tissue boundaries or the presence of materials like metal and calcium. Variations in the local magnetic field cause protons to precess at different rates, further accelerating dephasing. T2 relaxation is tissue-specific and always shorter than T1 relaxation, reflecting the distinct dynamics of transverse magnetization decay.

With a 90° RF flip angle excitation, turning off the RF transmitter eliminates longitudinal magnetization (M_z) due to the equal distribution of spins between parallel and anti-parallel orientations. Only transverse magnetization remains, characterized by complete phase coherence. This signal, known as free induction decay (FID), oscillates at the Larmor frequency, is detected by the receiver coil, and decays over time due to T2* relaxation. Different tissues have different relaxation times which can be used to generate contrast in the MR image that can reflect not only the net magnetization (the spin density), but also the relaxation properties (Bottomley et al. 1987). T1- and T2-weighting are the two main contrast mechanisms in clinical and pre-clinical anatomical MRI (De Graaf 2019).

To emphasize the T1 signal, excitation pulses are sent rapidly, and the signal is detected immediately, minimizing the effects of T2 dephasing. Conversely, to emphasize the T2 signal, pulses are sent more slowly (to allow T1 relaxation), and the signal is detected after a longer wait time, allowing the dephasing effects (T2 relaxation) to take place. T1 and T2 are intrinsic tissue properties (unless a contrast agent is used). MRI technicians adjust two key time parameters to optimize tissue contrast: repetition time (TR), which is the interval between successive pulse repetitions applied to the same slice, and echo time (TE), which is the time until acquiring the signal. By adjusting these parameters different types of image contrast can be generated.

3.4 MRI Sequences

Anatomical MR imaging is about contrast imaging, maximizing, and measuring the difference of signal intensity between different tissues, or between normal tissue and pathological entities. Tissues have differences in their number of protons (their spin density), T1 and T2 relaxation times. The most common MRI sequences are T1-weighted and T2-weighted scans, each characterized by distinct imaging parameters and contrast mechanisms. T1-weighted images are obtained using short TE and TR times, with contrast primarily influenced by the T1 properties of tissues. For example, water exhibits a long T1 relaxation time, while fat has a short T1. In contrast, T2-weighted images are produced with longer TE and TR times, where the contrast depends on the T2 properties of tissues. In this case, water has a long T2 relaxation time, whereas white matter, with its low water content, has a short T2.

Proton density imaging is achieved with a long TR and a very short TE. Here, neither T1 nor T2 relaxation significantly contributes to contrast, resulting in an image primarily weighted by the tissue's spin density (Boltzmann magnetization). However, in practical applications, tissue signals are typically a composite of T1, T2, and spin density-weighted contributions due to the inherent limitations of real-world conditions. For a quick orientation, T1- and T2-weighted images can typically be distinguished by observing the appearance of cerebrospinal fluid (CSF). On T1-weighted images, CSF appears dark, whereas on T2-weighted images, it appears bright. Additionally, fat appears bright on T1-weighted images, while both fat and water (CSF) appear white on T2-weighted images. A list of relaxation times for different tissues at 1.5T, 7T and 9.4T is shown in Table 2.

T1-weighted imaging can be enhanced with Gadolinium-based contrast agents (GBCAs), which contain paramagnetic Gadolinium in a chelated form (linear or macrocyclic, depending on the molecule). When injected during the scan, GBCAs shorten the T1 relaxation time of nearby hydrogen nuclei, altering MRI signal intensities. The paramagnetic properties of Gadolinium create local magnetic field

Table 2 Tissues and their T1 and T2 values at 1.5 T, 7 T, and 9.4 T

Tissue	T1: 1.5 T/ 7 T/ 9.4 T (ms)	T2: 1.5 T/7 T/9.4 T (ms)
White matter	790 (Dale et al. 2015)/~1500/~1700 (Kuo and Herlihy 2008)	90 (Dale et al. 2015)/~30/~32 (Kuo and Herlihy 2008)
Gray matter	920 (Dale et al. 2015)/~1900/~1900 (Kuo and Herlihy 2008)	100 (Dale et al. 2015)/~40/~40 (Kuo and Herlihy 2008)
Cerebrospinal fluid (CSF)	4000 (Dale et al. 2015)/> 4000/> 4500	2000 (Dale et al. 2015)/~200/~130
Fat	260 (Dale et al. 2015)/~350/~850 (Kuo and Herlihy 2008)	80 (Dale et al. 2015)/~40/~30 (Kuo and Herlihy 2008)
Liver	490 (Dale et al. 2015)/~1000/~1400	40 (Dale et al. 2015)/~40/~30

Values at 1.5 T and certain values at 9.4 T are based on data from Dale et al. (2015), Kuo and Herlihy (2008), respectively

variations, allowing adjacent spins to rapidly release energy to the lattice and shorten T1. As a result, areas with GBCAs appear bright on T1-weighted images. GBCA-enhanced imaging is particularly useful for visualizing vascular structures and assessing blood-brain barrier disruptions.

4 Basic Principles of MR Imaging and Image Reconstruction

Magnetic resonance was first applied in spectroscopy, a major research area in biochemistry (Purcell et al. 1946; Bloch 1946). The development of spatial encoding and Fourier analysis paved the way for MRI, a cornerstone of diagnostic radiology. Key principles behind these advancements are outlined below.

4.1 Spatial Encoding: Slice Selection, Frequency, and Phase Encoding

To generate an MR image, the scanner differentiates RF waves from various spatial locations within the body by applying magnetic field gradients. These gradients adjust the main magnetic field based on location, altering the Larmor frequency. Three primary methods of spatial encodings are used, which can be combined to encode all three dimensions. Slice selection selectively excites spins in a defined plane (typically along the z-axis), while frequency and phase encoding capture the image along the second and third dimensions (x and y directions). Although these encoding steps are commonly used, other techniques and concepts also exist.

Slice Selection A spatial gradient causes a linear variation in resonance frequency along its direction. When an RF pulse at a specific frequency bandwidth is applied, only spins that satisfy the Larmor condition are excited. Protons (^{1}H nuclei) outside the slice plane remain unaffected and do not contribute to the signal. Thus, both slice position and thickness are determined by the gradient strength and RF pulse frequency bandwidth.

Frequency Encoding A method used to localize signals in a spatial direction by applying a magnetic field gradient, known as the frequency encoding or readout gradient (Lauterbur 1973). This creates a spatially dependent Larmor frequency, with the precession frequency varying along the acquisition direction. Frequency encoding typically takes just a few milliseconds to acquire the entire signal.

Phase Encoding Frequency encoding operates in one spatial dimension at a time, and the third dimension is encoded using a phase encoding gradient. This gradient is applied after slice selection and before frequency encoding for a brief period. It modifies the spin resonance frequencies, causing dephasing that persists after the gradient is turned off. Once the phase encoding gradient ends, the spins return to precessing at the same Larmor frequency but with different phases based on their spatial position. Phase encoding typically requires multiple repetitions, each with varying gradient strengths, to capture enough data to encode for the third spatial dimension.

4.2 Image Reconstruction: Raw Data and K-Space

The MRI signal is acquired as a combination of frequencies from all the excited spins in the slice, which must then be transformed into a 2D image. The complex waves received during the encoding process—varying in amplitude, frequency, and phase—are digitized and stored in a data matrix called k-space (Mezrich 1995). K-space represents the spatial frequencies of the MR image, similar to a Fourier plane. A 2D inverse Fourier transform converts the k-space data into image intensities, which are displayed as gray scale values in a pixel matrix. Each point in k-space holds frequency and phase information for every pixel in the final image, with each pixel corresponding to all point in k-space. Data near the center of k-space correspond to low spatial frequencies (representing coarse shapes and contours), while data at the periphery correspond to high spatial frequencies, capturing fine details and edges.

4.3 Basic Pulse Sequences: Gradient Echo (GE) Sequence, Spin Echo (SE) Sequence

Two key parameters are TE, the time from excitation to signal acquisition, and TR, the time between two excitations. The combination of these parameters and the pulse sequence type determines whether the resulting image is T1, T2, (T2*), or spin density weighted as outlined in the chapter above.

Gradient Echo Sequence (GE) The gradient echo sequence is a basic imaging pulse sequence that starts with a single (slice-selective) RF pulse (either 90° or a lower flip angle) (Haase et al. 1986; Markl and Leupold 2012). The signal in the transverse plane is decaying with the apparent transverse relaxation T2* and spatially encoded using the phase and frequency encoding. Each TR this cycle is repeated using different phase encoding gradient strengths (typically 64 to 256 times). Due to the T2* decay, the GE sequence is T2*-weighted but can also be strongly T1-weighted if short TRs and large flip angles are used.

Gradient echo sequences are commonly used in fast data acquisition, such as dynamic imaging, where multiple images are captured rapidly to detect temporal changes.

Spin Echo Sequence (SE) The SE sequence involves two RF pulses before the signal is acquired: a 90° (excitation) pulse followed by a 180° (refocusing) pulse (Jung and Weigel 2013; Hahn 1950). After the 90° pulse, magnetization dephases due to intrinsic spin-spin interactions and magnetic field inhomogeneities (the T2* decay). A 180° refocusing pulse applied TE/2 after excitation refocuses the signal decay due to magnetic field inhomogeneities resulting in a signal only affected by true T2 at time TE. This generated signal, which is formed by a second (180°) pulse is called "spin echo". Spatial encoding is similar to the GE sequence. Each TR this sequence of (slice-selective) 90° pulse, (slice-selective) 180° pulse, phase and frequency encoding are repeated using a different phase encoding strength.

A limitation of the gradient echo sequence is its sensitivity to magnetic field inhomogeneities (sensitivity to T2* decay), which can reduce voxel signal and cause complete signal loss, especially at longer TEs. This issue is mitigated in the SE sequence by the 180° refocusing pulse, which restores the signal loss due to magnetic field inhomogeneities and is only sensitive to T2 decay.

5 MRI for Brain and Cardiac Imaging in Preclinical Research

MRI is the most powerful non-invasive imaging technique in preclinical research. It enables longitudinal studies, reduces the need for invasive procedures, and aids in translating findings from animal models to human clinical applications. The two most popular research areas where MRI is used are brain and cardiac imaging.

5.1 Brain Imaging

Brain imaging focuses on capturing detailed visual representations of the brain's structure and function. Standard MRI scans are usually T1- or T2-weighted images. Figure 3 shows typical examples of these two contrasts for a mouse brain. MR

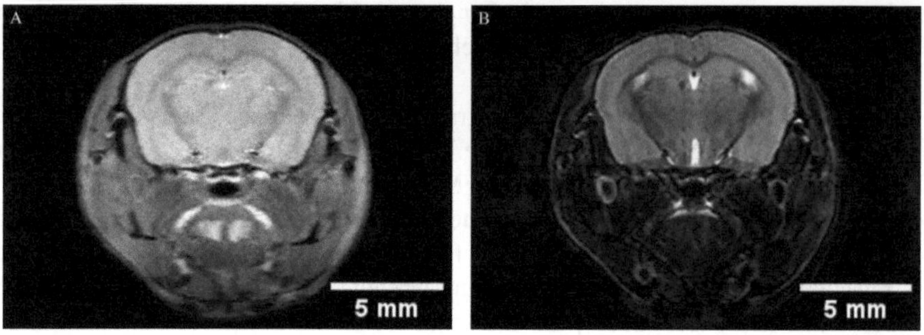

Fig. 3 T1-weighted (**a**) and T2-weighted (**b**) MRI coronal sections of the mouse brain at approximately −1.5 mm relative to Bregma. Grey matter structures (e.g., somatosensory cortex) are brighter than white matter structures (e.g., corpus callosum) in both sequences, but CSF appears very bright on T2-weighted images (e.g., left and right ventricles)

imaging was performed on a 7 T small animal MRS DRYMAG system (MR Solutions, Guildford, UK) using a dedicated mouse volume coil (diameter = 38 mm) (MR Solutions). The mouse was anaesthetized with ~2.0% isoflurane using oxygen, placed in the scanner, with breathing and temperature constantly monitored during the acquisition. Anatomical multi-slice T2-weighted (T2w) turbo-SE scans were performed in the coronal direction using the following sequence parameters (Hennig et al. 1986): TR/TE = 6610/45 ms, flip angle = 90°, echo-train length (ETL) = 8, field of view (FOV) = 20 × 20 mm^2, matrix size = 256 × 256, slice thickness = 0.5 mm, number of slices = 26, number of averages = 2, with a total acquisition time of 7 minutes. For the T1-weighted MRI sequence, a TE of 11 ms and a TR of 800 ms were used, an echo-train length (ETL) = 4, with a total acquisition time of 3.5 minutes.

5.2 Cardiac Imaging

Preclinical cardiac imaging is a highly advanced technique, primarily challenged by heart motion caused by respiratory and cardiac activity (Wu 2020; Phoon and Turnbull 2016). Achieving motion-free cardiac MR images requires synchronizing MRI acquisition with the cardiac cycle while avoiding image sampling during respiration (Price et al. 2010; Nieman et al. 2009).

Traditionally, respiratory and electrocardiogram (ECG) sensors generate the triggering and gating signals. In the prospective approach, the MRI scanner is directly triggered by cardiac pulses (Kinchesh et al. 2018). Alternatively, in the retrospective method, ECG and respiratory data are recorded alongside MRI data and later used to sort and reconstruct cardiac images (Nieman et al. 2009; Bishop et al. 2006; Heijman et al. 2007).

Examples of cardiovascular MRI are shown in Fig. 4. The images were acquired with the system described above. Single-slice, retrospectively gated CINE images

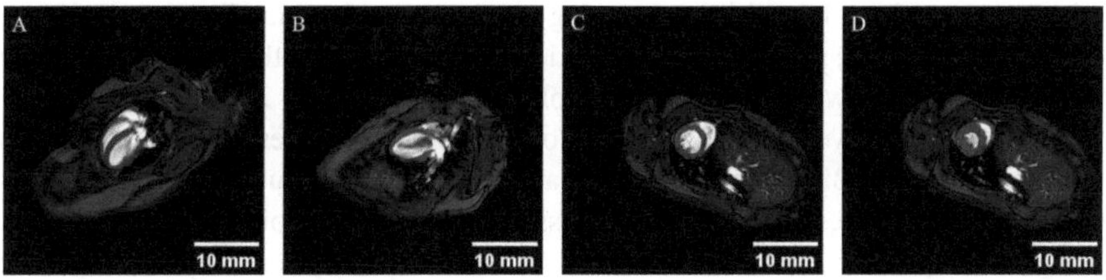

Fig. 4 CINE Mode MRI long-axis scans showing the 4-chamber view with a prominent left and right ventricle and the atria on top (**a**) and the left ventricle including the aortic arch (**b**) of a mouse heart. The cardiac cycle is represented by two CINE mode short-axis scans representing the end diastole phase—filled ventricles with blood (**c**), and the end systole phase—the fully contracted ventricles (**d**) of a mouse heart where the half-moon shape attached to the circular shape—the left ventricle, represents the right ventricle

were acquired in short-axis and long-axis views (two- and four-chamber) using a gradient echo sequence with a slice-excitation-based navigator to detect respiratory and cardiac motion. Acquisition parameters included: TR = 8 ms, TE = 2 ms, flip angle = 15°, FOV = 40 × 40 mm^2, acquisition matrix = 206 × 192, slice thickness = 0.8 mm, k-space repetitions = 100, and scan time = 3.5 minutes. Retrospective triggering enabled higher temporal resolution than the specified TR value. The data were reconstructed into 6 cardiac time frames using the custom open-source software Retrospective, developed by Dr. Gustav J. Strijkers & Dr. Bram F. Coolen (University of Amsterdam, AMC) (Daal et al. 2021), incorporating MATLAB algorithms and compressed sensing tools from the Berkeley Advanced Reconstruction Toolbox (BART).

6 Conclusions and Future Perspectives

As a non-invasive imaging modality that does not use ionizing radiation, MRI offers significant advantages over other radiological techniques. It provides high-resolution anatomical images of soft tissues such as organs and muscles, while also enabling functional, biochemical, and quantitative assessments. MRI allows for repeated measurements in the same subject over time, reducing inter-subject variability and enhancing statistical power. High-field MRI further expands these capabilities by supporting advanced techniques, including fMRI (Ogawa et al. 1990), diffusion tensor imaging (Basser et al. 1994), and MR spectroscopy (Muñoz-Hernández and García-Martín 2018), across diverse biomedical fields such as oncology, neurology, and cardiology.

The interplay between animal physiology and MRI technique plays a crucial role in successful image acquisition. Because most MRI methods are highly sensitive to motion, animals are typically anesthetized during scanning. However, anesthesia significantly influences both animal welfare and the reliability of experimental outcomes (Hildebrandt et al. 2008; Tremoleda et al. 2012). To mitigate these effects, researchers have begun imaging awake animals (Mandino et al. 2024), thereby avoiding the physiological alterations induced by anesthesia and combining technical advances with neuroscience applications (Hike et al. 2025).

MRI itself involves a series of trade-offs among spatial resolution, SNR, scan time, and artifacts (Brown et al. 2014). Spatial resolution is influenced by parameters such as matrix size, FOV, slice thickness, and the type of pulse sequence used. Image quality is further affected by B_0 field homogeneity and sequence-specific factors. SNR depends on voxel size, the number of excitations (or averages), and receiver bandwidth. While SNR can be improved by increasing voxel size (at the cost of resolution), reducing bandwidth, or increasing averages (which lengthens scan time), using specialized hardware can be used to enhance the signal. Dedicated coils, such as local receive coils or phased arrays placed near the region of interest, can significantly enhance SNR compared to larger volume coils (Hoult and Richards 1976). Additionally, higher magnetic field strengths improve SNR, though maintaining both B_0 and B_1 field homogeneity remains essential. To reduce the scan time different parallel imaging techniques like sensitivity encoding (SENSE) (Pruessmann et al. 1999) and generalized autocalibrating partial parallel acquisition (GRAPPA) (Griswold et al. 2002) accelerating phase encoding have been implemented. Scan time can be further reduced using concepts of Compressed Sensing (Lustig et al. 2007; Candès et al. 2006) (especially for 3D sequence) and simultaneous multi-slice imaging (Maudsley 1980; Barth et al. 2016) (for 2D multi-slice imaging). Cardiac or respiratory gating can minimize motion artifacts too, including ghosting and blurring (Russo et al. 2020). Susceptibility artifacts occur at tissue-air interfaces or near metals, causing signal loss or distortion (Hargreaves et al. 2011). Chemical shift artifacts, caused by the 3.5 ppm frequency difference between fat and water protons, can be reduced through different fat suppression techniques or bandwidth adjustments (Delfaut et al. 1999). Aliasing artifacts (wraparound) can be mitigated by expanding the FOV or using antialiasing techniques (Bellon et al. 1986). It is worth noting that ultra-high-field MRI provides better resolution due to higher SNR, but introduces challenges such as increased susceptibility effects, B_1 inhomogeneity, and higher RF energy deposition, which are important for human and animal scans.

Looking ahead, several innovations hold promise for significantly advancing the utility and impact of preclinical MRI. Notably, advancements in hardware, such as ultra-high field magnets exceeding 9.4 T, are enhancing the resolution, speed, and analytical power of MRI for brain research (Eisenstein 2024). These improvements allow for more precise detection of tissue boundaries, vascular structures, plaques, iron deposits, and early-stage pathologies. Additionally, integrating MRI with other imaging modalities like PET or optical imaging enables the simultaneous capture of anatomical, functional, and molecular data, opening new avenues for comprehensive biomedical analysis (Mirshahvalad et al. 2024). Fusion of MRI and CT enables precise in vivo tracking of drug-loaded nanoparticles (Bonlawar et al. 2024).

Artificial intelligence (AI) and machine learning are revolutionizing diagnostics by enhancing accuracy and efficiency (Johnson et al. 2020; Küstner et al. 2024). Leveraging large datasets, these technologies accelerate disease characterization and prognosis, enabling earlier interventions and improved outcomes. On the software front, AI-driven image reconstruction, segmentation, and analysis are streamlining data processing and revealing hidden patterns within complex datasets. New systems are also featured in AI-assisted compressed sensing for fast imaging, and a prospective head motion correction technique for motion free acquisition (Wu et al. 2024). Continued innovation and cross-disciplinary collaboration are vital to overcoming current limitations and unlocking the full potential of preclinical MRI. Establishing standardized protocols, open-access databases, and robust reproducibility frameworks will be key to ensuring consistency and promoting data sharing across institutions. As a cornerstone of preclinical imaging, MRI will continue to drive translational discoveries and accelerate the journey from bench to bedside.

Acknowledgements The authors gratefully acknowledge the support of the Department of Radiology, Dr. Carole Dennie, and Dr. Adam Shuhendler from the Department of Chemistry and Biomolecular Sciences at the University of Ottawa for providing access to the 7 Tesla MRS DRYMAG (MR Solutions, Guildford, UK) at the Ottawa Heart Research Institute.

Conflict of Interest The authors declare no conflicts of interest related to the content of this chapter.

References

ACR (2024) ACR Manual on MR Safety 2024

Barth M et al (2016) Simultaneous multislice (SMS) imaging techniques. Magn Reson Med 75(1):63–81

Basser PJ, Mattiello J, LeBihan D (1994) MR diffusion tensor spectroscopy and imaging. Biophys J 66(1):259–267

Bellon E et al (1986) MR artifacts: a review. AJR Am J Roentgenol 147(6):1271–1281

Bishop J et al (2006) Retrospective gating for mouse cardiac MRI. Magn Reson Med 55(3):472–477

Blaimer M et al (2004) SMASH, SENSE, PILS, GRAPPA: how to choose the optimal method. Top Magn Reson Imaging 15(4):223–236

Bloch F (1946) Nuclear induction. Phys Rev 70(7–8):460–474

Bonlawar J et al (2024) Targeted nanotheranostics: integration of preclinical MRI and CT in the molecular imaging and therapy of advanced diseases. Nano 8(3):401–426

Bosch CS, Ackerman JJ (1992) Surface coil spectroscopy. In: In-vivo magnetic resonance spectroscopy II: Localization and spectral editing, pp 3–44

Bottomley P et al (1987) A review of 1H nuclear magnetic resonance relaxation in pathology: are T1 and T2 diagnostic? Med Phys 14(1):1–37

Brigham EO (1988) The fast Fourier transform and its applications. Prentice-Hall, Inc

Brown RW et al (2014) Magnetic resonance imaging: physical principles and sequence design. John Wiley & Sons

Candès EJ, Romberg J, Tao T (2006) Robust uncertainty principles: exact signal reconstruction from highly incomplete frequency information. IEEE Trans Inf Theory 52(2):489–509

Chavhan GB et al (2009) Principles, techniques, and applications of T2*-based MR imaging and its special applications. Radiographics 29(5):1433–1449

Daal MR et al (2021) Quantification of mouse heart left ventricular function, myocardial strain, and hemodynamic forces by cardiovascular magnetic resonance imaging. J Vis Exp 2021(171)

Dale BM, Brown MA, Semelka RC (2015) MRI: basic principles and applications. John Wiley & Sons
Dance D et al (2014) Diagnostic radiology physics, vol 299. International Atomic Energy Agency, pp 12–14
De Graaf RA (2019) In vivo NMR spectroscopy: principles and techniques. John Wiley & Sons
Delfaut EM et al (1999) Fat suppression in MR imaging: techniques and pitfalls. Radiographics 19(2):373–382
Eisenstein M (2024) Pushing the limits of MRI brain imaging. Nat Methods 21(11):1975–1979
Froncisz W, Jesmanowicz A, Hyde JS (1986) Inductive (flux linkage) coupling to local coils in magnetic resonance imaging and spectroscopy. J Magn Reson (1969) 66(1):135–143
Gareis D et al (2007) Mouse MRI using phased-array coils. NMR Biomed 20(3):326–334
Golay MJ (1958) Field homogenizing coils for nuclear spin resonance instrumentation. Rev Sci Instrum 29(4):313–315
Griswold MA et al (2002) Generalized autocalibrating partially parallel acquisitions (GRAPPA). Magn Reson Med 47(6):1202–1210
Haacke EM et al (2004) Susceptibility weighted imaging (SWI). Magn Reson Med 52(3):612–618
Haase A et al (1986) FLASH imaging. Rapid NMR imaging using low flip-angle pulses. J Magn Reson (1969) 67(2):258–266
Hahn EL (1950) Spin echoes. Phys Rev 80(4):580–594
Hanley P (1984) Magnets for medical applications of NMR. Br Med Bull 40(2):125–131
Hargreaves BA et al (2011) Metal-induced artifacts in MRI. Am J Roentgenol 197(3):547–555
Heijman E et al (2007) Comparison between prospective and retrospective triggering for mouse cardiac MRI. NMR Biomed 20(4):439–447
Hennig J, Nauerth A, Friedburg H (1986) RARE imaging: a fast imaging method for clinical MR. Magn Reson Med 3(6):823–833
Hike D et al (2025) High-resolution awake mouse fMRI at 14 tesla. elife 13:RP95528
Hildebrandt IJ, Su H, Weber WA (2008) Anesthesia and other considerations for in vivo imaging of small animals. ILAR J 49(1):17–26
Hoffmann E et al (2024) Multiparametric MRI for characterization of the tumour microenvironment. Nat Rev Clin Oncol 21(6):428–448
Hoult DI, Richards R (1976) The signal-to-noise ratio of the nuclear magnetic resonance experiment. J Magn Reson (1969) 24(1):71–85
Huettel S, Song A, McCarthy G (2014) Functional magnetic resonance imaging, 3rd edn. Sinauer Associates Inc., Sunderland
Jensen DJ et al (1987) Reduction of pulsed gradient settling time in the superconducting magnet of a magnetic resonance instrument. Med Phys 14(5):859–862
Jin J, Shen G, Perkins T (1994) On the field inhomogeneity of a birdcage coil. Magn Reson Med 32(3):418–422
Johnson PM, Recht MP, Knoll F (2020) Improving the speed of MRI with artificial intelligence. In: Seminars in musculoskeletal radiology. Thieme Medical Publishers
Jung BA, Weigel M (2013) Spin echo magnetic resonance imaging. J Magn Reson Imaging 37(4):805–817
Kinchesh P et al (2018) Prospective gating control for highly efficient cardio-respiratory synchronised short and constant TR MRI in the mouse. Magn Reson Imaging 53:20–27
Kosaka M et al (2000) In vivo NMR micro-imaging of kidney and liver of mouse at 9.4 T. Jpn J Physiol 50(4):463–467
Kuo Y-T, Herlihy AH (2008) Optimization of MRI contrast for pre-clinical studies at high magnetic field. In: Modern magnetic resonance. Springer, pp 759–768
Küstner T et al (2024) The intelligent imaging revolution: artificial intelligence in MRI and MRS acquisition and reconstruction. MAGMA 37(3):329–333
Laukien DD, Tschopp WH (1994) Superconducting NMR magnet design. Concepts Magn Reson 6(4):255–273
Lauterbur PC (1973) Image formation by induced local interactions: examples employing nuclear magnetic resonance. Nature 242(5394):190–191

Lindquist DM et al (2021) MRI measures of murine liver fibrosis. J Magn Reson Imaging 54(3):739–749

Link J (1992) The design of resonator probes with homogeneous radiofrequency fields. In: In-vivo magnetic resonance spectroscopy I: Probeheads and radiofrequency pulses spectrum analysis, pp 3–31

Lustig M, Donoho D, Pauly JM (2007) Sparse MRI: the application of compressed sensing for rapid MR imaging. Magn Reson Med 58(6):1182–1195

Ma X et al (2024) Development and advancements in rodent MRI-based brain atlases. Heliyon 10:e27421

Mandino F et al (2020) Animal functional magnetic resonance imaging: trends and path toward standardization. Front Neuroinform 13:78

Mandino F et al (2024) Where do we stand on fMRI in awake mice? Cereb Cortex 34(1):bhad478

Mansfield P (1977) Multi-planar image formation using NMR spin echoes. J Phys C Solid State Phys 10(3):L55–L58

Mansfield P, Chapman B (1986) Active magnetic screening of coils for static and time-dependent magnetic field generation in NMR imaging. J Phys E 19(7):540–545

Markl M, Leupold J (2012) Gradient echo imaging. J Magn Reson Imaging 35(6):1274–1289

Marzola P, Osculati F, Sbarbati A (2003) High field MRI in preclinical research. Eur J Radiol 48(2):165–170

Maudsley A (1980) Multiple-line-scanning spin density imaging. J Magn Reson (1969) 41(1):112–126

Mezrich R (1995) A perspective on K-space. Radiology 195(2):297–315

Mirshahvalad SA et al (2024) Current applications of PET/MR: part I: technical basics and preclinical/clinical applications. Can Assoc Radiol J 75(4):815–825

Mörchel P et al (2010) Correlating quantitative MR measurements of standardized tumor lines with histological parameters and tumor control dose. Radiother Oncol 96(1):123–130

MRSolutions (2025) MRSDRYMAG: 3T–4.7T-7T-9.4T

Muñoz-Hernández MC, García-Martín ML (2018) In vivo 1 H magnetic resonance spectroscopy. In: Preclinical MRI: methods and protocols, pp 151–167

Nieman BJ, Szulc KU, Turnbull DH (2009) Three-dimensional, in vivo MRI with self-gating and image coregistration in the mouse. Magn Reson Med 61(5):1148–1157

Odeblad E, Lindström G (1955) Some preliminary observations on the proton magnetic resonance in biologic samples. Acta Radiol 6:469–476

Ogawa S et al (1990) Oxygenation-sensitive contrast in magnetic resonance image of rodent brain at high magnetic fields. Magn Reson Med 14(1):68–78

Ordidge R (1999) The development of echo-planar imaging (EPI): 1977–1982. MAGMA 9(3):117–121

Phoon CK, Turnbull DH (2016) Cardiovascular imaging in mice. Curr Protoc Mouse Biol 6(1):15–38

Price AN et al (2010) Cardiovascular magnetic resonance imaging in experimental models. Open Cardiovasc Med J 4:278–292

Pruessmann KP et al (1999) SENSE: sensitivity encoding for fast MRI. Magn Reson Med 42(5):952–962

Purcell EM, Torrey HC, Pound RV (1946) Resonance absorption by nuclear magnetic moments in a solid. Phys Rev 69(1–2):37–38

Russo V, Lovato L, Ligabue G (2020) Cardiac MRI: technical basis. Radiol Med 125(11):1040–1055

Saito S, Ueda J (2024) Preclinical magnetic resonance imaging and spectroscopy in the fields of radiological technology, medical physics, and radiology. Radiol Phys Technol 17(1):47–59

Samagh SP et al (2013) MRI quantification of fatty infiltration and muscle atrophy in a mouse model of rotator cuff tears. J Orthop Res 31(3):421–426

Schröder L, Faber C (2011) Basic contrast mechanisms. In: In vivo NMR imaging: Methods and protocols, pp 45–67

Shaw T, Elsken R (1950) Nuclear magnetic resonance absorption in hygroscopic materials. J Chem Phys 18(8):1113–1114

Tremoleda JL, Kerton A, Gsell W (2012) Anaesthesia and physiological monitoring during in vivo imaging of laboratory rodents: considerations on experimental outcomes and animal welfare. EJNMMI Res 2:1–23

Tropp J (1989) The theory of the bird-cage resonator. J Magn Reson (1969) 82(1):51–62

Vaughan J et al (2002) Detunable transverse electromagnetic (TEM) volume coil for high-field NMR. Magn Reson Med 47(5):990–1000

Wu YL (2020) Cardiac MRI assessment of mouse myocardial infarction and regeneration. In: Cardiac regeneration: methods and protocols. Springer, pp 81–106

Wu D et al (2024) Developing an AI-empowered head-only ultra-high-performance gradient MRI system for high spatiotemporal neuroimaging. NeuroImage 290:120553

Small Animal Preparation and Handling for Magnetic Resonance Imaging (MRI) Applications

Roland Pilgram and Gerd Melkus

Abstract

Magnetic Resonance Imaging (MRI) is a powerful, non-invasive tool widely used in preclinical research to investigate anatomical, functional, and molecular processes in small animal models. Ensuring both high-quality imaging and animal welfare requires adherence to best practices grounded in the 3Rs principles: Replacement, Reduction, and Refinement. Therefore, a schematic outline, a step-by-step workflow, from pre-imaging preparation to post-imaging recovery, is necessary to standardize procedures and promote reproducibility. Key considerations include the careful selection and administration of anesthetic agents. Proper anesthesia management is essential, as it directly affects both animal well-being and the integrity of imaging data by minimizing physiological artifacts. Among others, isoflurane is favored in MRI imaging for its rapid pharmacokinetics, minimal physiological impact, and fine-tuned controllability. Detailed protocols cover experimental setup, including positioning in MRI-compatible cradles, and the use of supportive equipment to maintain physiological stability, such as heating systems, respiratory monitoring, ECG, and temperature regulation. Post-imaging care is equally important, with guidance on recovery monitoring, supportive interventions to ensure ethical treatment. Establishing Standard Operating Procedures (SOPs) for each stage of the entire imaging process,

R. Pilgram (✉)
Department of Radiology, Radiation Oncology and Medical Physics, University of Ottawa, Ottawa, Canada
e-mail: rpilgram@uottawa.ca

G. Melkus
Department of Radiology, Radiation Oncology and Medical Physics, University of Ottawa, Ottawa, Canada

Department of Physics, Carleton University, Ottawa, Canada

I. F. Cengiz et al. (eds.), *Bioimaging Modalities in Bioengineering*, Biomaterials, Bioengineering and Sustainability 9,
https://doi.org/10.1007/978-3-031-99410-4_6

ensures safe and effective small animal handling in MRI studies. This chapter is intended for researchers, veterinary staff, and imaging technicians committed to maintaining ethical standards while achieving scientific rigor in experimental MRI settings.

Keywords

Anesthesia regimes · Experimental setup · Animal handling · Physiological monitoring · Animal recovery

1 Introduction

Small animals, especially rats and mice, are central to translational research, helping to elucidate disease mechanisms and assess therapies (Hormuth et al. 2019). Longitudinal studies provide critical insights over time, both qualitative and quantitative. Ensuring animal welfare is essential, not only ethically, but also to improve data reliability. The 3Rs principle—Replacement, Reduction, and Refinement—introduced by (Russell WMS and Burch 1959; Flecknell 2002), guides humane research practices:

1. **Replacement**: Use non-animal alternatives (e.g., computer models, phantom systems) or substitute vertebrates with less sentient organisms like invertebrates.
2. **Reduction**: Minimize animal use through robust experimental design, genetic homogeneity, and incorporation of in vitro data.
3. **Refinement**: Improve techniques to reduce suffering via enriched housing, species-specific care, and voluntary participation in procedures.

Integrating the 3Rs enhances ethical standards while boosting reproducibility, efficiency, and scientific impact.

Magnetic Resonance Imaging (MRI) is a non-invasive, versatile tool widely used in preclinical and translational research to study anatomical, functional, and metabolic processes in small animals (García-Martín and López-Larrubia 2018). MRI is a safe tool for longitudinal functional and morphological investigations with respect to animal welfare (Baier et al. 2020). High-quality imaging depends on careful animal preparation, optimized protocols, and strict adherence to ethical standards. Key elements include proper anesthesia, continuous monitoring (e.g., heart rate, respiration, temperature), maintenance of normothermia, reproducible positioning, and attentive post-procedure care. Effective integration of MRI into research requires alignment between scientific goals, animal model choice, imaging methods, and ethical compliance. Institutional Animal Care and Use Committee (IACUC) approval is essential, alongside adherence to aseptic techniques and welfare practices. Ethical protocols typically include species and model details, experimental design (per CCAC—Canadian Council on Animal Care: Guidelines (Canadian Council on Animal Care (CCAC) 2025), imaging parameters, timelines, animal number justification, procedural specifics, Standard Operating Procedures (SOPs), anesthesia plans, and monitoring strategies. This chapter focuses on three key aspects of the preclinical MRI workflow: animal preparation mentioning anesthesia and drug administration; the experimental setup including animal handling and physiological monitoring, and animal recovery, followed by a step-by-step animal handling procedure. Challenges and limitations are discussed, followed by conclusions and future perspectives.

2 Animal Preparation

Proper animal preparation is essential for obtaining high-quality, reproducible data in preclinical MRI studies. To reduce motion artifacts, animals must be securely immobilized and maintained under appropriate anesthesia, such as isoflurane, sevoflurane, or injectable agents (McConville 2011; Pinkernelle et al. 2008). Since physiological factors like e.g., obesity can alter anesthetic response, a thorough pre-anesthetic assessment is critical. This evaluation should include verification of the animal's identification, sex, age, weight, hydration status, mucous membrane color, cardiac and respiratory function, and gastrointestinal health. Continuous monitoring and synchronization of physiological parameters with MRI acquisition further minimize motion artifacts and enhance image quality and accuracy.

2.1 Anesthesia

Anesthesia significantly influences both animal welfare and the reliability of experimental outcomes (Hildebrandt et al. 2008; Tremoleda et al. 2012). By suppressing the autonomic nervous system, anesthetics often induce cardiovascular and respiratory depression, hypoxia, and hypothermia, effects that are particularly pronounced in rodents due to their small size and high metabolic rate (Gargiulo et al. 2012). Different anesthetics modulate various targets in the brain and have different impact on animal physiology and neuronal network organizations (Mandino et al. 2020; Paasonen et al. 2018). Commonly used anesthetic agents include isoflurane, ketamine-xylazine, medetomidine, and propofol (see practical comparison in Table 1). Isoflurane is preferred for its rapid onset, easy titration, and minimal cerebral effects at low doses, making it suitable for repeated experiments. However, higher concentrations can cause vasodilation and respiratory suppression (Masamoto and Kanno 2012). The ketamine-xylazine combination provides effective analgesia and muscle relaxation but markedly alters cardiovascular function and cerebral hemodynamics (Hildebrandt et al. 2008; Kober et al. 2004). Medetomidine offers stable and reproducible sedation with minimal variability, making it especially suitable for functional MRI studies (Kober et al. 2004) or in combination with Isoflurane (Grandjean et al. 2014). Propofol allows for rapid induction and recovery but requires vigilant monitoring to maintain physiological stability (Tremoleda et al. 2012).

Table 1 Comparison of anesthetic types

	Inhalant (Isoflurane, Sevoflurane)	Injectable (Medetomidine, Ketamine-Xylazine, Propofol)
Recovery	Fast induction & recovery due to rapid uptake/ elimination via lungs	Slower induction & recovery; propofol can be fast if IV is used and has a rapid recovery
Adjustability	Requires vaporizer & scavenging system	No special equipment needed
Impact	Mild physiological effects if well managed	Greater respiratory & cardiac impact

Combination protocols are also often employed to optimize both induction and maintenance phases, e.g., initiating anesthesia with isoflurane followed by maintenance using ketamine-xylazine. The choice of anesthetic should be tailored to the imaging modality and experimental objectives to ensure minimal physiological interference and high reproducibility. Isoflurane is particularly well-suited for MR-based imaging due to its favorable pharmacokinetics, mild physiological effects, and precise controllability. Further, its non-invasive delivery and rapid recovery reduce animal stress and imaging artifacts. Although isoflurane can still modulate neural activity, using the lowest effective concentration in conjunction with physiological monitoring helps mitigate these effects (Wei et al. 2024).

2.2 Drug Administration

Isoflurane is administered via inhalation using a nose cone or induction chamber, eliminating the need for injections or implants that could compromise imaging quality or alter physiological parameters. This method offers a rapid and efficient route of absorption due to the lungs' extensive alveolar surface area and thin diffusion barrier of just one to two cell layers. Inhalational anesthesia also enables fine control over anesthetic depth, which is essential for maintaining physiological stability during imaging procedures. Dual isoflurane systems are typically used: one for induction and another integrated with the imaging console to provide continuous, regulated delivery throughout the scanning. Key components of the setup include a flowmeter for oxygen or compressed air, a vaporizer connected to a breathing circuit, and a scavenging system to safely remove waste gases.

Injectable anesthetics and agents can be delivered through various routes, depending on pharmacokinetics and experimental objectives (Gargiulo et al. 2012; Kiessling and Pichler 2010). Intravenous (IV) injection, typically via the tail or jugular vein, ensures rapid systemic distribution, intraperitoneal (IP) injection offers slower absorption, and subcutaneous (SC) delivery allows for sustained release. Precision syringes or infusion pumps are used to ensure accurate dosing, while long-term protocols may require catheter implantation for repeated administration. Choosing appropriate anesthetic and drug delivery methods is essential for optimizing both image quality and animal welfare. In MRI studies, gadolinium-based contrast agents (Geelen et al. 2012; Iyad et al. 2023; Zhou and Lu 2013) are commonly used to enhance visualization of specific tissues or pathologies and are typically administered intravenously or intraperitoneally, depending on study design.

3 Experimental Setup

Establishing a robust MRI protocol for preclinical animal studies requires careful attention to both animal welfare and physiological monitoring to ensure consistent, high-quality imaging (Gargiulo et al. 2012). Reliable, high-resolution scans under volatile anesthesia depend on an integrated setup that combines precise temperature control, air circulation, sensitive temperature sensors, and MRI-compatible imaging beds (McConville 2011). These beds are designed to maintain uniform surface temperatures and support the efficient delivery of anesthetic gases. They are typically integrated with MRI-compatible radiofrequency (RF) coils. Volume coils are often preferred over surface coils due to their superior RF homogeneity, particularly when imaging larger fields of view.

To ensure immobilization and reproducibility, species-specific animal cradles are employed. Depending on the model, these may include bite bars, ear bars, stereotaxic frames, or body restraints. Cradles are frequently supplemented with cushions or custom molds to enhance comfort and maintain consistent positioning. While minimizing motion is essential for image quality, care must be taken to avoid excessive restraint that could induce stress or discomfort. A well-integrated setup should support: (a) maintenance of deep anesthesia and complete immobilization; (b) environmental isolation of the subject; (c) precise thermoregulation; (d) flexible and reproducible positioning; (e) seamless integration of ECG, respiratory, and temperature monitoring probes; and (f) smooth transfer of the anesthetized animal from the preparation area to the scanner without disrupting physiological monitoring or environmental control.

3.1 Animal Handling

To ensure physiological stability during imaging, animals must be transported in accordance with established biosafety protocols. Acclimatization, typically lasting 3–7 days, or shorter when using integrated housing and imaging systems, should be completed prior to imaging sessions (Tremoleda et al. 2012). All pre-imaging procedures, including fasting, drug administration, and health assessments (e.g., body weight, general condition), must be thoroughly conducted and documented. Anesthesia induction marks the start of the preparatory phase. Place the animal in an induction chamber connected to the anesthesia station, adhering to standard laboratory protocols to minimize stress. Set the carrier gas flow to approximately ~1 L/min, with an isoflurane concentration of 3–4% (García-Martín and López-

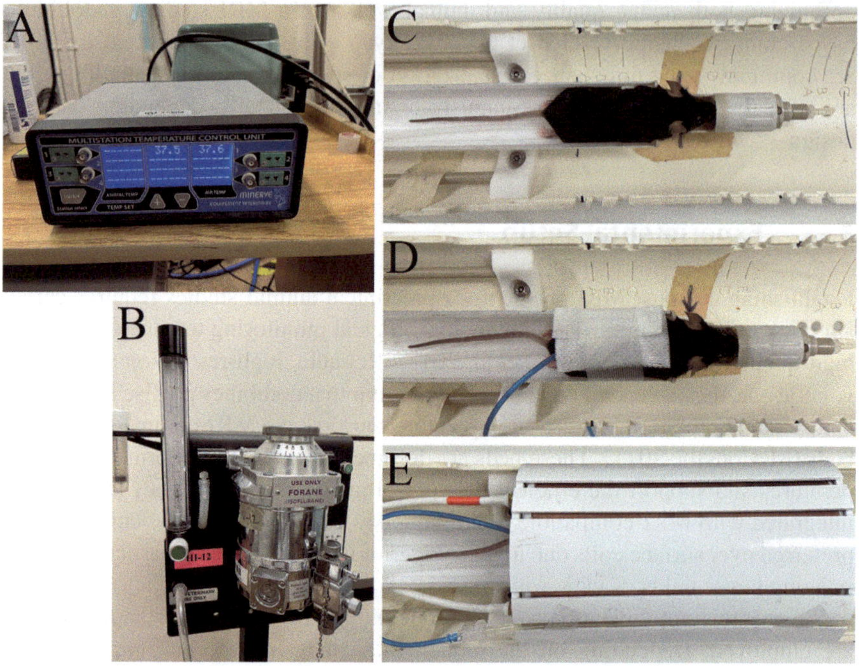

Fig. 1 Typical animal setup, including the temperature control unit with the air pump in the back (**a**), an anesthesia system (**b**), and the mouse positioned in the animal bed (**c**). A small pneumatic respiration pillow sensor is attached to the mouse covered by gauze to protect the skin of the animal (**d**), and the animal within the volume coil (**e**) ready to be placed in the MRI

Larrubia 2018). In rodents, sedation is typically achieved within 2–3 min, indicated by progressive respiratory slowing. Once sedated, reduce the isoflurane concentration to 1–2.5% for maintenance (García-Martín and López-Larrubia 2018), adjusting the gas flow to 0.5 L/min for mice or 0.8 L/min for rats to ensure effective delivery.

Transfer the anesthetized animal to a preheated, thermoregulated bed connected to the temperature control unit (Fig. 1a–c). Preheat the system for 5–10 min beforehand to maintain thermal stability. Preventing hypothermia is critical, as it can introduce physiological variability and pose life-threatening risks. Position the animal so that the region of interest aligns precisely with the imaging field and secure the nose in the anesthesia cone. For brain imaging, stabilize the head using integrated supports: rest the incisors on the tooth bar and gently position the ear bars. Signs of misalignment, seen as closed or bulging eyes, should be corrected immediately. Since isoflurane suppresses the blinking reflex, the cornea is prone to drying. To prevent dehydration, ulceration, and potential vision loss, apply sterile ophthalmic lubricating ointment before imaging. Once the animal is secured and monitoring probes are in place, ensure the target organ is centered within the imaging coil (isocenter) to optimize image quality. During prolonged imaging sessions or extended anesthesia, dehydration can occur. Administer warmed parenteral fluids (e.g., saline) as needed to maintain hydration. For extended or repeated imaging, use minimally metabolized volatile anesthetics such as e.g., isoflurane, which allows for rapid induction, and quick recovery, supporting a swift return to physiological baseline post-procedure.

3.2 Physiological Monitoring

Continuous monitoring of physiological parameters, such as heart rate, respiration, and body temperature, is crucial during imaging sessions (Tremoleda et al. 2012; Gargiulo et al. 2012). This is typically achieved using non-invasive, MRI-compatible equipment to avoid image artifacts and ensure safety. Non-metallic, shielded devices are preferred to reduce magnetic interference. Given that most anesthetic agents influence respiration, cardiovascular function, and thermoregulation, monitoring at least respiratory rate and body temperature is essential for assessing anesthesia depth (McConville 2011). Cardiac monitoring is also strongly recommended, as heart rate and respiratory patterns offer important insights into anesthetic adequacy.

Respiratory Rate Respiratory rate is typically monitored using sensors that detect breathing-associated motion. Tremoleda et al. (2012) provide a comprehensive comparison of respiratory monitoring methods, highlighting pneumatic pressure sensors (Fig. 1d) and optical techniques as optimal for imaging environments due to their sensitivity, ease of use, non-invasive, cost-effective, and minimal interference with imaging modalities. To avoid contamination of the respiratory signal by cardiac motion, placement near the heart should be avoided. Once respiratory monitoring is established, the animal is positioned inside the volume coil for imaging (Fig. 1e). More advanced technologies, such as capnography, utilize infrared spectroscopy to measure carbon dioxide concentrations in inhaled and exhaled air, offering a more detailed evaluation of respiratory function. To minimize motion artifacts during MR imaging, gated acquisition techniques are often used. Since the most pronounced motion from the diaphragm and abdominal wall occurs during inspiration, image acquisition is typically synchronized with expiration, enhancing overall image quality.

Heart Rate Heart rate is most monitored using an electrocardiogram (ECG), a standard tool for assessing cardiovascular stability. ECG provides real-time information on heart rate and rhythm and is essential for detecting arrhythmia, myocardial ischemia, and metabolic disturbances. It is also routinely employed to synchronize cardiac cycles with image acquisition in gated cardiac imaging. For MRI procedures, non-ferromagnetic, MRI-compatible electrodes, such as subdermal needle electrodes, gold disk surface electrodes, or adhesive pads, are used to record ECG signals. Typically, two or three leads are sufficient, with recommended placement on both sides of the chest and near the left hind limb. To minimize electromagnetic interference, lead wires should be twisted, kept as short as possible, and routed away from MRI coil elements, as improper placement can compromise coil tuning and image quality. Pulse oximetry serves as an additional non-invasive modality during anesthesia, offering continuous monitoring of arterial oxygen saturation and pulse rate. Measuring the arterial oxygen saturation can also be valuable for functional MRI of the brain to ensure and monitor the oxygen level while the experiment is performed. Invasive monitoring of heart rate is also possible via arterial catheterization, allowing real-time pulse assessment through blood pressure waveforms.

Body Temperature Anesthetic agents significantly impair thermoregulation (Tremoleda et al. 2012), making rodents, especially nude or hairless mice, highly susceptible to hypothermia due to their high surface area-to-mass ratio and rapid metabolism. Therefore, maintaining normothermia during imaging procedures is essential. Core body temperature is commonly monitored using MRI-compatible rectal thermometers, often integrated with external heating systems. Prior to insertion, the probe tip should be lubricated or fitted with a lubricated sheath to minimize discomfort, and the lead gently taped to the tail for stability. Alternatively, temperature may be monitored by placing the probe between the animal and the imaging bed. Core temperature should be maintained between 36 °C and 38 °C.

4 Animal Recovery

Effective animal recovery and handling in an MRI environment require meticulous planning to ensure safety, minimize stress, and maintain the integrity of imaging procedures (Tremoleda et al. 2012). Continuous monitoring of vital signs is essential to preserve physiological stability under anesthesia and should be sustained until the animal fully regains consciousness and stable functions. Supplemental oxygen should be provided for a few minutes post-scan to facilitate recovery. Animals are generally expected to regain consciousness within 2–3 min; during this time, personnel must remain with the animal until it can support itself in an upright position. Recovery time may vary, depending on factors such as the animal's health status, age, anesthesia type (gaseous or injectable), and duration and dosage of anesthesia.

Post-MRI recovery should occur in a quiet, temperature-controlled environment. Reversal agents may be administered when appropriate. Maintaining normothermia during recovery is especially critical. Hydration and nutrition should be provided once the animal is fully alert, following species-specific post-anesthesia care protocols. In cases of prolonged anesthesia, supplemental oxygen, fluids, and nutritional support may be beneficial. Animals should only be returned to their housing once a full recovery is confirmed. Competent, well-trained personnel are vital to the recovery process, and veterinary support should be readily available for consultation or emergencies.

5 A Step-by-Step Animal Handling Procedure

This is an example of a step-by-step procedure, as part of an SOP of animal preparation and handling for MRI applications:

1. Each animal is anesthetized with inhalant isoflurane throughout the imaging procedure. The physiological monitoring system is used to monitor the animal's vital signs and maintain body temperature via warm air delivery into the MRI bore.
2. Ensure the air within the MRI bore has been warmed using the thermistor control system.
3. Ensure that isoflurane vaporizers, delivery systems and the anesthesia induction chamber are properly setup. Ensure that the isoflurane levels in the vaporizers are filled. When the indicator level drops to less than half, refill the vaporizer using the precision key. Verify that the online gas system is pressurized and ensure the gas delivery is not interrupted.
4. Ensure that the scavenge system has been turned on and is actively providing suction to both the induction station and the MRI bed anesthetic delivery station. Verify scavenge hook-up. Failure to do so may result in personnel having an occupational exposure to anesthetic waste gas.
5. Once animal subjects have been received in the MRI animal preparation area, remove the selected scan subject from its home cage, verify and record the animal identification and place it in the anesthesia induction chamber.
6. Open the oxygen valve for the induction chamber and set the vaporizer to 3–4% isoflurane, at a flow rate of 0.75–1 L/min for mice, 1 L/min for rats.
7. Place the home cage on the heated platform, to warm the cage in preparation for return of the animal.
8. Observe the animal for anesthetic effect/recumbency and turn the vaporizer setting to <OFF> and the oxygen flow to <OFF> for the induction chamber. Remove the animal from the chamber and promptly transfer to the MRI bed. Verify the depth of anesthesia.
9. Set the MRI anesthetic bed station vaporizer to 1–2% (mice), 1.5–2.5% (rats) and flip the Power switch toggle to <ON> and ensure the MRI fresh gas delivery interface is also toggled to <ON>. Adjust the flowmeter to 0.5 L/min for mice and 0.8 L/min for rats.
10. Place the nose of the animal into the isoflurane nose cone on the MRI bed.
11. Ensure that the animal is breathing regularly. Position the animal as per the procedural requirement and apply any required physiological monitoring components.
    a. Arrange/attach the vital sensors to the subject. Standard sensors include: (1) a lubricated rectal temperature probe set up in a feedback loop with the heated ventilation to ensure proper temperature regulation, and (2) a pressure sensor that non-invasively registers the respiration rate, allowing monitoring of the respiration rate and rhythm, SPO_2 and core temperature.

   b. Affix monitoring components as necessary with medical tape and secure the animal into position with gauze and masking tape to ensure that movement of the bed location will not change the orientation of the animal. Note: To prevent injury to the animal, ensure gauze is protecting the fur from the tape.
   c. When monitoring is being conducted, verify that respiration rate and temperatures have stabilized within acceptable ranges (37–38 °C and respiration rates between 40–100 breaths per minute), the scanning procedure may begin. Observe the respiratory rhythm to ensure no aberrant movement that may indicate that the subject is not appropriately anesthetized.
12. Apply eye ointment to prevent corneal drying and damage.
13. Transfer the MRI bed to the MRI bore, verifying that the bore temperature has been appropriately set to maintain optimal body temperature.
14. The length of the procedure will determine the level of monitoring required.
15. Start MRI scanning.
16. When MRI scanning is finished, remove the animal from the scanner and scan bed and return to the recovery home cage. Record the date and procedure and any fluid therapy onto the cage card. NOTE: animals that have been anesthetized for 30 min or longer should receive warmed subcutaneous saline at the time of return to the recovery home cage. Standard fluid administration of 1 mL/mouse, 5 mL/rat.
17. Allow the animal to recover under close observation until fully awake. After the recovery period, make a final assessment to ensure that the animal is healthy and awake before transferring the animal back to its original housing room.

Most consistent anesthetic MRI protocols for rodents rely on inhalant anesthesia, with isoflurane being the most frequently employed inhalant agent. A basic protocol using injectable anesthesia is provided by Oh and Narver (2024).

6 Challenges and Limitations

Animal MRI studies are fraught with methodological challenges that can compromise data quality, reproducibility, and translational value. Common restraint and handling procedures often induce stress responses, triggering physiological changes such as elevated heart and respiratory rates that may introduce imaging artifacts. Although anesthesia is essential for immobilization, it alters cerebral blood flow and neural activity, complicating the interpretation of both functional and structural data. Poor positioning or inadequately designed restraint systems can degrade image quality and negatively affect animal welfare.

Ethical and regulatory considerations add further complexity, demanding a careful balance between scientific rigor and animal welfare. Preclinical MRI requires extensive physiological monitoring, increasing procedural complexity. Advanced respiratory and cardiac gating techniques are often necessary to minimize motion artifacts, complicating imaging setups (Zaitsev et al. 2015). Enhancing spatial resolution typically entails longer scan times, which elevate the risk of stress, anesthetic instability, and physiological drift (Lerch et al. 2012). Moreover, variability in imaging protocols and environmental conditions across laboratories further undermines reproducibility. Ultimately, stress- and anesthesia-induced physiological fluctuations remain major barriers to the translational relevance of animal MRI. Addressing these challenges will require the development of stress-reducing handling methods, non-invasive physiological monitoring, and standardized imaging protocols to improve the reliability, reproducibility, and ethical conduct of preclinical imaging studies.

7 Conclusions and Future Perspectives

Effective animal handling is essential for the success of preclinical MRI studies. Proper techniques help reduce stress in animals and minimize physiological artifacts, leading to more accurate and reproducible imaging results. Best practices include the use of non-invasive restraint systems, acclimation protocols, and anesthetic regimens tailored to maintain physiological stability without compromising image quality. The ability to non-invasively acquire detailed in vivo anatomical and functional data marks a significant step toward refinement and reduction in animal use.

Advancements in imaging technology and handling protocols have greatly improved the quality and translational value of preclinical MRI data. Before initiating any experiment, it is necessary to develop species-specific protocols or SOPs. These should clearly outline the purpose, responsibilities, location, general considerations, and safety warnings (e.g., ensuring animals do not carry metal ear tags or RFID chips near the magnetic field). SOPs should also detail required materials and step-by-step procedures for animal handling and imaging. To further optimize the balance between research goals, animal welfare, imaging modalities, equipment, and strategies, the following considerations are key:

- **Refinement of Handling Protocols**: Continuously improve handling protocols to minimize stress and enhance animal welfare, adhering to the principles of the 3Rs (Replacement, Reduction, Refinement) (Wachsmuth et al. 2021). This includes appropriate anesthetic regimen and monitoring systems to ensure the animal's well-being and to minimize stress-related responses that would compromise the imaging outcomes. It is important to consolidate good protocols for handling, anesthesia and monitoring the animals suitable for all the specific needs of a wide range of imaging experiments.
- **Automation in Handling and Imaging**: Employ automated systems for animal handling, positioning, and imaging to reduce human error, increase efficiency, and improve reproducibility. This ensures that other laboratories can validate findings and advance the field.
- **Advanced Anesthetic Techniques and Standardization**: Develop and implement anesthetics that preserve stable physiological conditions while minimizing interference with MRI outcomes. Standardizing anesthetic and imaging protocols

is critical to enhancing the reproducibility of translational functional brain imaging studies. Minimizing variability in physiological parameters caused by differing anesthetic regimens leads to more consistent and reliable imaging results (Mandino et al. 2020; Celdran-Bonafonte et al. 2025).

- **Ethical and Regulatory Compliance**: Regularly review ethical practices and ensure compliance with evolving regulatory standards to guarantee humane treatment and uphold research integrity.
- **Minimally Invasive Imaging Alternatives**: Explore minimally invasive or non-invasive imaging modalities that complement MRI, further reducing animal stress and enhancing data quality.

By integrating technological advancements with refined animal handling practices, preclinical MRI research can achieve higher-quality, translationally relevant insights while maintaining the highest standards of animal welfare.

Acknowledgements We gratefully acknowledge the support of Dr. Carole Dennie, head of the Department of Radiology, University of Ottawa.

Conflict of Interest The authors declare no conflicts of interest related to the content of this chapter.

References

Baier J et al (2020) Influence of MRI examinations on animal welfare and study results. Investig Radiol 55(8):507–514

Canadian Council on Animal Care (CCAC) (2025) Guidlines and Policies. Available from: https://ccac.ca/en/about/about-the-ccac/

Celdran-Bonafonte D et al (2025) An overview of anesthetic procedures for Ferret (Mustela putorius furo) preclinical brain MRI: a call for standardization. J Am Assoc Lab Anim Sci 64(1):16–28

Flecknell P (2002) Replacement, reduction, refinement. ALTEX 19(2):73–78

García-Martín ML, López-Larrubia P (2018) Preclinical MRI. Springer, New York

Gargiulo S et al (2012) Mice anesthesia, analgesia, and care, Part I: anesthetic considerations in preclinical research. ILAR J 53(1):E55–E69

Geelen T et al (2012) Contrast-enhanced MRI of murine myocardial infarction–Part I. NMR Biomed 25(8):953–968

Grandjean J et al (2014) Optimization of anesthesia protocol for resting-state fMRI in mice based on differential effects of anesthetics on functional connectivity patterns. NeuroImage 102:838–847

Hildebrandt IJ, Su H, Weber WA (2008) Anesthesia and other considerations for in vivo imaging of small animals. ILAR J 49(1):17–26

Hormuth DA et al (2019) Translating preclinical MRI methods to clinical oncology. J Magn Reson Imaging 50(5):1377–1392

Iyad N et al (2023) Gadolinium contrast agents-challenges and opportunities of a multidisciplinary approach: Literature review. Eur J Radiol Open 11:100503

Kiessling F, Pichler BJ (2010) Small animal imaging: basics and practical guide. Springer, Berlin

Kober F et al (2004) Cine-MRI assessment of cardiac function in mice anesthetized with ketamine/xylazine and isoflurane. MAGMA 17:157–161

Lerch JP et al (2012) Wanted dead or alive? The tradeoff between in-vivo versus ex-vivo MR brain imaging in the mouse. Front Neuroinform 6:6

Mandino F et al (2020) Animal functional magnetic resonance imaging: trends and path toward standardization. Front Neuroinform 13:78

Masamoto K, Kanno I (2012) Anesthesia and the quantitative evaluation of neurovascular coupling. J Cereb Blood Flow Metab 32(7):1233–1247

McConville P (2011) Small animal preparation and handling in MRI. Methods Mol Biol 771:89–113

Oh SS, Narver HL (2024) Mouse and rat anesthesia and analgesia. Curr Protoc 4(2):e995

Paasonen J et al (2018) Functional connectivity under six anesthesia protocols and the awake condition in rat brain. NeuroImage 172:9–20

Pinkernelle J et al (2008) Small animal MRI: clinical MRI as an interface to basic biomedical research. Rofo 180(6):505–513

Russell WMS B, Burch R (1959) The principles of humane experimental technique. Special ed, vol 165. Universities Federation for Animal Welfare, Potters Bar, Herts

Tremoleda JL, Kerton A, Gsell W (2012) Anaesthesia and physiological monitoring during in vivo imaging of laboratory rodents: considerations on experimental outcomes and animal welfare. EJNMMI Res 2:1–23

Wachsmuth L et al (2021) Contribution of preclinical MRI to responsible animal research: living up to the 3R principle. MAGMA 34(4):469–474

Wei Z et al (2024) The impact of isoflurane anesthesia on brain metabolism in mice: an MRI and electroencephalography study. NMR Biomed 37(12):e5260

Zaitsev M, Maclaren J, Herbst M (2015) Motion artifacts in MRI: a complex problem with many partial solutions. J Magn Reson Imaging 42(4):887–901

Zhou Z, Lu ZR (2013) Gadolinium-based contrast agents for magnetic resonance cancer imaging. Wiley Interdiscip Rev Nanomed Nanobiotechnol 5(1):1–18

[illegible]

Magnetic Resonance Imaging (MRI) Contrast Agents in Preclinical Oncology and Environmental Impact

Gabriella Baio and Chrit Moonen

Abstract

The extreme progress in high-tech instrumentation has remarkably transformed the way medicine has been practiced making imaging an indispensable foundation for disease diagnosis and treatment. Magnetic Resonance Imaging (MRI) distinguishes itself from other imaging modalities as non-ionizing radiation type of diagnostic imaging, allowing non-invasive visualization and quantification of both anatomical and functional soft tissues features. The introduction of exogenous MR contrast agents has further improved our abilities to target vascular and molecular processes moving from reading images as a result of "an invisible molecular process" to "visualization and quantification of molecular processes" leading to the development of molecular MR imaging. This chapter focuses on the application of MR contrast agents in preclinical studies and how the advancement of their applications has improved our knowledge, especially in the field of oncology. In addition, an overview on the potential environmental impact in the application of MR contrast agents is provided showing how preclinical research could become more environmentally friendly.

Keywords

Molecular imaging · Oncology · Environment · Climate change · Radiology

G. Baio (✉)
Norfolk and Norwich University Hospitals Trust, NNUH, Norwich, UK

University of East Anglia, UEA, Norwich, UK
e-mail: gabriella.baio@nnuh.nhs.uk

C. Moonen
Focused Ultrasound Foundation, Charlottesville, VA, USA

I. F. Cengiz et al. (eds.), *Bioimaging Modalities in Bioengineering*,
Biomaterials, Bioengineering and Sustainability 9,
https://doi.org/10.1007/978-3-031-99410-4_7

1 Introduction

The parallel improvement in our understanding of the molecular and cellular processes related to cancer biology together with the advancement in imaging technology, has increased our ability to identify, visualize and quantify tumor features non-invasively. Magnetic Resonance Imaging (MRI) is a non-invasive, powerful imaging tool with high spatial resolution, and when combined with specific contrast agents, allows visualization of physiological and functional processes in tumors. MR Contrast Agents (MRCAs) are not directly observed in MRI. Instead, they influence the signal strength of water protons because of magnetic interactions. The main advantage is to increase contrast difference from normal and abnormal tissues, improving both sensitivity and specificity of the acquired images (Tang et al. 2013).

The first contrast-enhanced human MRI study was published in 1981 by applying ferric chloride to visualize the gastrointestinal tract (Young et al. 1981). Subsequently, in 1984 it was shown for the first time, that a gadolinium compound could be used as a diagnostic intravascular contrast agent (Carr et al. 1984). The opportunity of combining the anatomical and functional features of MRI with the use of MRCAs combined with more complex compounds able to detect, diagnose and define personalized treatment allowed the development of molecular imaging (MI). Molecular Imaging, especially in the field of oncology, has emerged as an interdisciplinary area under which new targeted contrast agents are developed with the promise for a better understanding and visualization of the molecular biology of cancer. New molecular MR contrast agents are regularly discovered and investigated, especially for preclinical imaging studies. However, it is very difficult for most of these to find a prompt translational clinical application due to the strict safety rules that these must adhere to.

The requirements for applications of MRCAs are mainly safety, stability, low toxicity, and elimination of the compound. Categorization of MRCAs can be done according to their physical and chemical features, such as magnetic properties (e.g., paramagnetic, and superparamagnetic), effect on the MR image (positive or negative), route of administration, or biodistribution (Geraldes and Laurent 2009a) (Fig. 1).

MRCAs 'molecular weight (or the hydrodynamic radius) prevents or allows the passing of the contrast agents from the vascular bed to the extravascular space into the interstitial space. As a consequence, MRCAs able to remain in the blood pool are used for MR angiography studies, to investigate tumor perfusion, or to establish integrity and permeability of the blood–brain barrier. MR tissue-specific contrast agents are instead able to accumulate in a given organ or tissue, such as iron oxide nanoparticles (liver-specific contrast agents) or, when combined with a more complex molecular probe, can specifically target tumor tissues, also named tumor-targeting (Sanchez-Gaytan et al. 2015; Agulla et al. 2014). This chapter provides an overview of the physical and chemical properties of MRCAs, including MR-contrast agent free imaging (Chemical Exchange Saturation Transfer imaging) and their applications in the study of cancer. Environmental impact in the application of

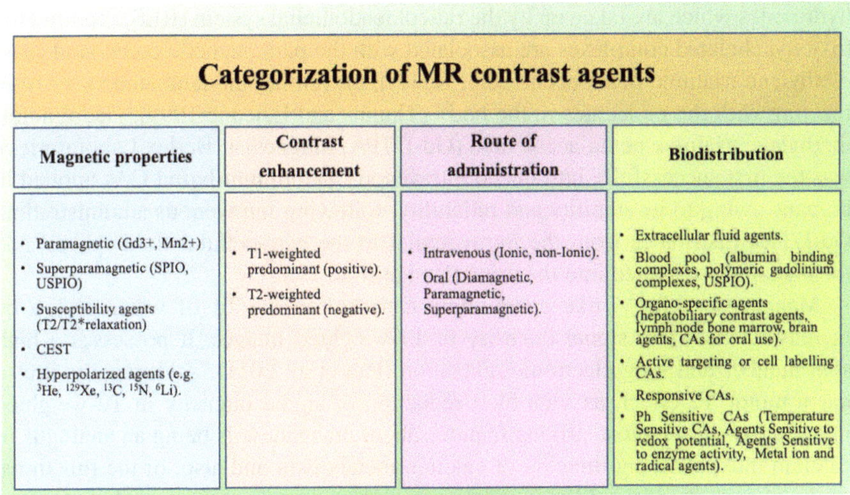

Fig. 1 Magnetic Resonance contrast agent categorization based on magnetic properties, contrast enhancement, route of administration and biodistribution. CAs contrast agents; Gd^{3+} gadolinium; SPIO superparamagnetic iron oxide; USPIO ultrasmall superparamagnetic iron oxides; CEST Chemical Exchange Saturation Transfer; ^{3}He Helium; ^{129}Xe Xenon; ^{13}C Carbon; ^{15}N Nitrogen; ^{6}Li Lithium

MRCAs and MR probes nanoparticles with a view for future directions and on how to become more ecologically responsible, will be explored.

2 T1-Weighted Contrast Agents

2.1 Fundamentals

Two of the most well-known and applied T1-weighed contrast agents are respectively, the lanthanide metal gadolinium (Gd^{3+}) and the transition metal manganese (Mn^{2+}), also classified as paramagnetic contrast agents (CAs) based on their magnetic properties (Fig. 1). Gd^{3+} possesses a high magnetic moment and represents the most stable paramagnetic ion with seven unpaired electrons, while manganese has five. The paramagnetic property of CAs containing gadolinium is to shorten the T1 (or longitudinal) and T2 (or transverse) relaxation time of water protons resulting in an increase of the signal intensity in T1-weighted images (visually a bright contrast) and, reduction of the signal intensity in T2-weighted images (visually a dark contrast) (Mitchell 1996; Wood and Hardy 1993). To obtain a T2 change, a higher concentration of gadolinium is required; however, this increases the risk of toxicity thus rendering T2 shortening of limited clinical use. Therefore, in the clinical environment usually only the T1 effect is considered, where a lower concentration of gadolinium is required (Gandhi et al. 2006). The toxicity of gadolinium complexes is related to the ability of lanthanide salts to hydrolyze into hydroxides which are taken up by the reticuloendothelial system (RES). To limit the toxicity, chelated complexes are associated with the paramagnetic compound (e.g., diethylene triamine penta-acetic acid, DTPA), preventing the lanthanide ions from reacting with the molecules in the body (Thunus and Lejeune 1999). Gadolinium-diethylene triamine penta-acetic acid (Gd-DTPA, Magnevist; Berlex Laboratories), was the first successfully introduced intravenous gadolinium-based CAs applied in humans owing to its stability and reliability: following intravenous administration, Gd-DTPA distributes from the intravascular to the extracellular fluid spaces, and then is rapidly excreted into the urine (Runge et al. 1985).

Manganese (Mn^{2+}) like gadolinium, can shorten the T1 of water proton by increasing the water signal intensity in T1-weighted images. It possesses a high spin number and long electronic relaxation (Pan et al. 2011). Additionally, it also has a minor T2-effect, as seen by a reduction of signal intensity in T2-weighted images. One of the most striking features about manganese is being an analogue of calcium thus, allowing imaging of calcium metabolism and also, of the functional study of mitochondria. Mitochondria are important components of hepatocytes, pancreas, and kidneys and, manganese can target their metabolic functions (high manganese uptake is observed in cells with high density of mitochondria) making it an excellent contrast agent for their investigation (Martin et al. 2010). For this reason, the chelated form of manganese, Manganese dipyridoxyl-diphosphate, has been used for a long time to detect hepatic lesions: once intravenously injected, it dissociates into manganese, which is taken up by the liver cells and excreted into the bile, allowing both imaging of the liver and of the biliary tree, while the chelate component dipyridoxyl-diphosphate is excreted via the kidney (Harisinghani et al. 2001a; Torres et al. 1997) Despite the chelation process is aimed to decrease the toxicity of a contrast agent, in this specific case the dissociation of manganese from the chelated compound dipyridoxyl-diphosphate following injection, has raised concerns around its clinical use, hence preventing clinical applications (Agency 2012).

MRCAs (both T1 and T2-weighted) can also be alternatively delivered by using a carrier. Among these carriers, liposomes have been considered and applied for several years due to their capability of both encapsulating hydrophilic substances in their aqueous inner core or hydrophobic compounds within the bilayer (Torchilin 1996). One of the main advantages of liposomes is their biocompatibility with biological systems making them ideal for both therapeutics and diagnostic in vivo studies. Liposomes can either incorporate into the bilayer gadolinium-conjugated with lipid moieties such as Gd-DTPA-cholesterol or, only Gd-DTPA within the aqueous space (Kamaly and Miller 2010). Encapsulation of gadolinium-chelate avoids gadolinium dispersion into the tissues and interstitial spaces, prolonging the time of circulation (Geraldes and Laurent 2009b). However, a longer circulation time may compromise the stability of the complex due to the risk of leakage of the gadolinium-chelate caused by the circulating serum proteins, thus, affecting stability of the liposome complex. For this reason, sterically stabilized liposomes have been produced by using a covalent bound with PEG (Stealth® liposomes, SEQUUS Pharmaceuticals, Menlo Park, CA, USA). Paramagnetic PEGylated liposomes have a decreased relaxation characteristic ($r_1 = 1.04\ mM^{-1} \cdot s^{-1}$) when compared to Gd-DTPA ($r_1 = 4.0\ mM^{-1} \cdot s^{-1}$), because the liposomal bilayer limits the rate of water exchange between gadolinium-chelate and the bulk of water molecules. If the Gd-DTPA is encapsulated with cholesterol, it will further reduce the water permeability coefficient leading to further decrease in the overserved relaxivity (Unger et al. 1989). This loss of relaxivity is overcome by increasing the concentration of gadolinium in the PEGylated liposomes or by using a smaller size of liposome. Finally, when considering the application of paramagnetic contrast agents, we need to consider that the efficiency of these compounds decreases with the increase of the magnetic field strength: for example, gadolinium-complexes are optimal at fields below 1.5 T; however, at 3 T, T1 relaxivity is reduced to one-third compared with its maximum, while at ultra-high magnetic fields it falls to zero (Caravan 2006).

2.2 Preclinical Applications in the Field of Oncology

One of the main challenges when using T1-weighted contrast agents is given by their toxicity concerns and relatively short circulating time, thus, complex T1-weighted-compounds or nanoparticle-based agents have been developed to be more effective when applied in the MI study of cancer (Raymond and Pierre 2004; Peer et al. 2007). Nanoparticles (NPs) can either target specific molecules expressed by cancer cells or can be internalized into living cells, thus creating a powerful imaging platform not only for diagnosis but also for therapeutic applications.

Macromolecular gadolinium-complexes have been fabricated from polymers, dendrimers, polypeptides, and polysaccharides (Cheng et al. 2010; Nam et al. 2010) by anchoring chelated gadolinium ions into the surface of the NPs (Rieter et al. 2007; Warsi et al. 2010; Manus et al. 2010) Novel MRI and Optical Imaging bimodal probes have also been developed allowing multimodal imaging by using gadolinium combined with optical imaging probes, such as quantum dots (Yang et al. 2020). Encapsulated liposome of Gd-DTPA-bis(steraylamide) has successfully applied allowing the study of cell apoptosis: Gd-DTPA has been incorporated in the lipid bilayer of PEGylated liposomes and covalently coupled with multiple human recombinant annexin A5 molecules.

Manganese compounds similarly to gadolinium can also be combined with other types of nanoparticles such as iron oxides (Zhen and Xie 2012) or carbon nanostructure complexes (e.g., graphene oxide nanoplatelets and graphene oxide nanoribbons) (Paratala et al. 2012), creating a highly effective molecular probe for preclinical molecular imaging studies (Lin and Koretsky 1997; Silva et al. 2004; Guadilla et al. 2023; Wen et al. 2018) Carbonized paramagnetic complexes of manganese, such as Mn-PyC3A, has a promising role as alternative to gadolinium-based contrast agents. Several studies have demonstrated chemical stability with a comparable relaxivity to gadolinium but, more importantly, it may represent a safer type of contrast due to its quick clearance from the body when compared to gadolinium-based agents (Erstad et al. 2019a; Qin et al. 2022). Recent studies have also demonstrated how carbonized polymerized shells could also be used to encapsulate gadolinium ions to minimize leakage and enhance relaxivity with potential impact for both diagnostic and therapy in tumor imaging (Chen et al. 2014, 2016; Chen et al. 2013; Sun et al. 2020; Liu et al. 2015) Manganese enhanced MRI (MEMRI), by using free manganese ions ($MnCl_2$), has been extensively applied for investigating molecular metabolism of cancer based on the study of tumor calcium metabolism (Rallapalli et al. 2020; Bianchi et al. 2017; Wen et al. 2017; Baio et al. 2020).

3 T2-Weighted Contrast Agents

3.1 Fundamentals

Molecular imaging research has found incredible growth with the application of T2-weighted contrast agents, which has been used for over 25 years, especially in studies for cell labelling or cell tracking. While the application of paramagnetic contrast agents, as discussed in the previous paragraphs, implies a certain number of limitations, some of which related to the potential toxicity concerns, when applying T2-weighted contrast agents, these can be overcome due to their low toxicity and stability in physiological conditions (Park et al. 2009; Filippousi et al. 2014). Superparamagnetic iron oxides nanoparticles (SPIONs) are the most extensively studied T2 contrast agents due to their strong magnetic properties, biocompatibility, and versatility. They are made of iron oxide crystals coated with dextran or carboxydextran. The main SPIONs MR effect is to reduce T2 relaxation times in their vicinity, leading to hypointense (dark) regions on both T2- and T2*- weighted images (negative contrast effect, Fig. 1), which corresponds to areas of contrast agent accumulation (Xiao et al. 2016; Orel et al. 2018; Bulte and Kraitchman 2004; Harisinghani et al. 2001b) The negative or dark imaging effect is the result of the strong heterogeneity of the magnetic field caused by these nanoparticles, which increases under high magnetic fields strength, for this reason they have also been named as susceptibility agents (Park et al. 2009; Lam et al. 2013). Despite SPIONs are a negative type of contrast agents, they can also provide T1 shortening effect; however, this becomes less detectable with the increasing of the strength of the magnetic field making these nanoparticles dependent on the strength of the magnetic field applied (Klug et al. 2010). Another important consideration is that the T1- and T2/T2*- effects can change depending on SPIONs conditions in the biological system, such as for example if are cell-bound or free. This feature makes these nanoparticles ideal for in vitro and in vivo studies, allowing to distinguish free from cell-bound SPIONs depending on the T2 or T2*-effect obtained on dedicated spin-echo or gradient-echo sequences (Kuhlpeter et al. 2007).

Another important feature of SPIONs is given by their size. This can span from 2.9–9 nm (Monocrystalline SPIONs—MIOs) to 10–50 nm (Ultra-small SPIONs - USPIONs) and 50–150 nm (Standard SPIONs—SSPIONs and Polydisperse SPIONs), while oral SPIONs can reach 300 nm to 3.5 μm in size (Lam et al.

2013). The diversity of SPIONs 'size is important not only because the saturation magnetization (Ms) of the iron nanoparticles correlates linearly with their size (Bumb et al. 2008; Sun et al. 2008) but also because it impacts on the their behaviors in the biological systems: for example, when the size is less than 100 nm this is preventing uptake of SPIONs from macrophages or from the RES, while when larger than 10 nm in healthy conditions SPIONs are unable to pass through vessels. However, this behavior changes in the presence of pathological conditions (such as inflammation or tumor neoangiogenesis), where instead they can pass through vessels. This phenomenon is called Enhanced Permeability Effect (Mahmoudi et al. 2011). Finally, shape is another key feature of SPIONs. This is important especially when considering the process of synthesis, where the spherical shape represents the most common and simple used because allows control in uniformity of the particle (Zhen et al. 2011); however, other types of shapes have also been explored, such as for example rhombic, ellipsoidal or flower (Mahmoudi et al. 2011). Regarding the impact of shape on their T2-effect in biological systems, this remains a debatable issue which has not completely addressed (Roca et al. 2009).

3.2 Preclinical Applications in the Field of Oncology

T2-weighted MR imaging is valuable for detecting water content and tissue heterogeneity, which are often indicative of malignancy and the use of T2 contrast agents enhances the diagnostic power of T2-weighted imaging, with applications ranging from tumor detection to therapeutic monitoring. T2-based contrast agents have been used preclinically in several oncological applications, as follows:

Tumor detection SPIONs enhance contrast between malignant and healthy tissues, improving tumor visualization and boundary delineation, especially for small or early-stage lesions (Bulte and Kraitchman 2004).

Lymph node imaging SPIONs have been used for mapping sentinel lymph nodes and detecting metastatic spread. Their uptake by macrophages in health lymph nodes creates signal voids, helping identify affected areas, which instead do not uptake iron oxide nanoparticles (Harisinghani et al. 2001b).

Therapeutic monitoring Changes in hypointense regions due to the accumulation of the SPION-labelled compound (tumor targeting) over time can be used to assess tumor responses to treatments like chemotherapy or radiotherapy (Bulte and Kraitchman 2004).

Cell tracking SPION-labelled cells, such as immune cells or stem cells, can be monitored in vivo, aiding the development of cell-based therapies (Harisinghani et al. 2001b; Sun et al. 2008).

Table 1 Comparison of SPIONs and non-SPIONs Agents

Comparison of SPIONs and non-SPIONs agents		
Feature	SPIONs	Non-SPIONs
Primary mechanism	Local magnetic inhomogeneities	Diverse (paramagnetic effects, aggregation)
Specificity	High (targeted nanoparticles)	Variable (depends on design)
Applications	Tumor detection, cell tracking	Tumor microenvironment, vascular imaging
Safety considerations	Low toxicity with modifications	Potential for toxicity with free ions (e.g., manganese)
Multimodality	Well-developed	Emerging

SPIONs superparamagnetic iron oxide nanoparticles

Advancements in SPION design have enlarged the field of oncological applications. Targeted SPIONs have been designed with surface modifications and tumor-specific ligands to improve specificity. Multimodal SPIONs have been described combining magnetic properties with optical or radioactive markers enable multimodal imaging approaches. Biocompatibility enhancements have been shown with modifications to reduce toxicity and improve clearance (Sun et al. 2008; Terreno et al. 2010)

In addition to SPIONs, other classes of contrast agents offer unique mechanisms and applications in T2-weighted imaging. Some of them have been discussed in the previous paragraphs, such as gadolinium-based agents, when used at high concentrations are able to reduce T2 relaxation time and it has been in detecting for example vascular tumors, mapping heterogeneity, and assessing angiogenesis (Xiao et al. 2016; Caravan et al. 1999; McDonald and Watkin 1992; Zhou and Lu 2013) Also, manganese-based nanoparticles or chelates have been effective in visualizing tumor metabolism, microenvironments, and intracellular signaling pathways by using their T2-effect (McDonald and Watkin 1992). Carbon-Based Nanomaterials, such as graphene oxide, carbon nanotubes, and fullerenes doped with paramagnetic elements (e.g., gadolinium or manganese) have intrinsic magnetic properties. They can be useful in imaging hypoxic tumor regions, mapping lymph nodes, and monitoring photothermal therapies (Xu et al. 2016). Aggregation-Induced T2 Agents, are instead agents able to self-assemble at tumor sites, creating localized magnetic field inhomogeneities for enhanced contrast allowing imaging solid tumors, monitoring drug delivery, and evaluating therapeutic responses (Terreno et al. 2010; Neves and Brindle 2006; Ito et al. 2005) Table 1 presents a summary of SPIONs and non-SPIONs features.

A further versatile MR probes with T2 effect are the superparamagnetic liposomes. The first multifunctional hybrid liposome/nanoparticle has been developed in 1988 by De Cuyper M et al. with the development of magnetoliposomes (MLPs) (De Cuyper and Joniau 1988), composed of a core of iron oxide upon which a single phospholipid bilayer was adsorbed (diameter of approximately 20 nm) (De Cuyper 2014). When the MLPs are encapsulated with SPIONs they can be used as T2 contrast agents, if instead USPIONs are encapsulated, they can produce a T1 contrast effect (Park et al. 2008; Gossuin et al. 2009; Soenen et al. 2009) MLPs can be applied to target tumors. Targeting can be achieved either by conjugating antibodies or ligands to the vesicle surface able to identify specific receptors expressed by the tumor cells (biological targeting) (Casals et al. 2003) or by applying an external magnet near the body region affected by the tumor where the MLPs can accumulate (magnetic targeting) (García-Jimeno et al. 2012). Bilabelled contrast agents have been developed by incorporating MNPs with up to 500 gadolinium ions per magnetic vesicle providing both T1 and T2 MR contrast agent in a single system (De Cuyper et al. 2007; Soenen et al. 2007). MLPs can also be applied for theranostics where the liposome contains several cores (or a cluster of cores) of SPIONs, and it can enclose a drug. The encapsulated magnetic nanoparticles allow hyperthermia or magnetic targeting where heat is produced as source of treatment.

In summary, T2-weighted MRI with SPIONs and non-SPIONs contrast agents has transformed oncology imaging by enabling precise tumor detection, characterization, and monitoring. While SPIONs dominate the field due to their strong magnetic effects and proven safety, non-SPIONs agents like gadolinium, manganese-based compounds, and carbon nanomaterials provide unique functionalities that complement SPIONs capabilities. Continued advancements in targeting, biocompatibility, and multifunctionality will enhance the clinical translation of these agents, improving diagnostic and therapeutic outcomes in oncology (Weissleder and Mahmood 2001).

4 Chemical Exchange Saturation Transfer Imaging

4.1 Fundamentals

In the past decade, a novel type of contrast MR imaging, named Chemical Exchange Saturation Transfer (CEST), has been developed for both preclinical and clinical MI. While the rationale of using an injected exogenous CAs is to increase our ability to characterize features of normal or abnormal tissues, CEST is a novel technique which enables imaging of certain molecules which normally are at too low concentration to impact the contrast of a standard MR imaging. This is reflected by exploiting chemical properties of certain metabolites by a process of exchange of water protons (^{1}H) with protons of endogenous molecules, such as cellular proteins (Wu et al. 2016). The use of endogenous processes is also the reason why it is named contrast-free MR imaging. The mechanism of action is based on the principle of the magnetization transfer contrast effects (MTC) which is obtained by using appropriate radiofrequency pulses (at the resonance frequency of labile protons) to transfer magnetization of these labile protons to mobile protons of surrounding water molecules. The amount of MTC effect depends on the chemical and physical properties of the macromolecules used, such as the concentration, chemistry of the surface, and biophysical dynamics. By using different radiofrequency pulses the excitation of different protons can be activated at the same time (multifrequency imaging), resulting in detection of different contrast effects in the same anatomical region (Castelli et al. 2013).

4.2 CEST Applications in the Field of Oncology

Chemical Exchange Saturation Transfer (CEST) imaging has found successful application in several oncological studies; however, it still presents some technical challenges which could potentially limit the application. Amide proton transfer weighted (APT) imaging (a sub-set of CEST), has been applied for the detection and molecular differentiation of glioma tumors (Zhou et al. 2003a, 2008, 2011, 2019; Togao et al. 2014; Sotirios et al. 2020). The mechanism of action of APT is by exploiting saturation exchange from the amide chemical groups, which is sensitive to tissue pH and protein concentration (Zhou et al. 2003b; Ray et al. 2019). Further studies have applied APT allowing molecular prognosis of glioma tumors by predicting mutation of isocitrate dehydrogenase (IDH) or by identifying 1p/19q codeletion status (Yao et al. 2020; Su et al. 2022; Jiang et al. 2017; Paech et al. 2018, 2019) and, also by detecting early tumor response to treatment allowing decision-making, crucial in these types of tumors (Sagiyama et al. 2014). The impact of these studies allowed commercialization of APT imaging making it available for clinical applications (Amide Proton Transfer weighted imaging: Advancement in molecular tumour diagnosis 2018).

CEST imaging has also been tested in the MI of breast tumors. A feasibility study led by Schmitt B et al. (Schmitt et al. 2011) demonstrated the potential of CEST imaging in differentiating normal appearing fibroglandular tissue vs malignant lesions in 50% of the patients. Further application of APTw imaging has been done to assess chemotherapy response in locally advanced breast cancers vs healthy controls (Dula et al. 2013; Zhang et al. 2021; Liu et al. 2022); however, there are some limitations in the interpretation of CEST imaging results due to the high presence of lipids in breast tissues (no fat suppression imaging is done with CEST) and of choline (present in both benign and malignant tissue). To overcome these limitations special fat suppression methods along with dedicated RF coil and sequence optimization have been successfully tested on ultra-high field scanners, such as at 7 T (Klomp et al. 2013) or at 3 T by applying multi-point Dixon acquisition to overcome the limitation due to lipid content (Zhang et al. 2018). An interesting advantage of CEST imaging vs breast contrast-enhanced MRI is that CEST is not affected by menstrual cycle changes, making it more reliable when detecting breast changes related to malignant process rather than to hormone fluctuations (Loi et al. 2022).

Only a limited number of studies have been carried out to investigate the molecular characterization of prostate tumors by using CEST. The rationale of applying CEST in prostate cancer imaging is based on the high cell proliferation and density demonstrated in high-grade prostate cancers compared to more indolent tumors which could be potentially detected by APTw imaging. A study done by Takayama Y et al. (2016) by using a 3 T MR scanner, demonstrated that intermediate-risk prostate tumors had higher APTw signal intensity across all tumors analyzed; however, APT did not show a statistically significant correlation with the non-malignant peripheral zone margins limiting its clinical utility. A further recent study done by Yin et al. (2021) demonstrated higher values of APTw in patients with prostate cancers vs benign prostatic hyperplasia (BPH) with moderate correlation with the Gleason score of tumors. The different results obtained in these studies, could be attributed to the different size of prostate cancers used leading to differences in histological and/or grading features which could have potentially affected CEST imaging quantification.

Further interesting applications of APTw imaging are in the prediction of tumor grade in rectal cancer (Nishie et al. 2018): assessment of response to neoadjuvant chemotherapy in locally advanced rectal cancer and APT values have been assessed (Nishie et al. 2019) demonstrating higher APT values with low response to treatment. Further applications have also demonstrated that advanced and aggressive tumor shows increased APT, like what has been observed in brain, breast, and prostate cancers studies, making CEST imaging a promising MI tool for also clinical imaging applications (Li et al. 2020; Chen et al. 2021; Li et al. 2021).

5 MR Contrast Agents and Environmental Impact

The environmental impact of MR imaging has recently come into focus due to the threat of climate change suggesting that we need to focus on developing a more sustainable type of MR research, especially, when using MRCAs or more complex MR nanoparticles, as described in the previous paragraphs. One of the environmental risks related to MRCAs usage is given by the gadolinium-based contrast agent due to its pharmacokinetics (excreted into the urine) which poses a risk of increasing contamination of global water bodies (i.e., groundwater, lakes and city tap water) (Oluwasola et al. 2022a; Inoue et al. 2020; Boester and Rüde 2020; Oksendal and Hals 1993) Once into wastewater, the gadolinium complex is stable over at least six months under natural conditions and conventional sewage treatment methods cannot adequately remove it (Knappe et al. 2005). As a consequence, this increases the risk of drinking contaminated water and once ingested, could potentially be degraded in free gadolinium (Fig. 2).

More studies are needed to explore the ecological consequences of gadolinium bioaccumulation (Ebrahimi and Barbieri 2019; Souza et al. 2021) and, how this can be monitored and efficiently removed (Brünjes and Hofmann 2020; Lawrence et al. 2010; Cordoba et al. 2021) or how its usage could be limited so, to prevent water contamination (Oluwasola et al. 2022b). Limited usage of MR contrast agents has been proposed by using alternative and more advanced non-contrast MR imaging techniques such as Arterial Spin Labelling (ASL), Time of Fight (TOF), Magnetic Resonance Spectroscopy (MRS) (Delgado et al. 2019) or CEST imaging, as previously described. Further alternatives are offered by the recent introduction of virtual non-contrast (VNC) MRI which thanks to the advancement

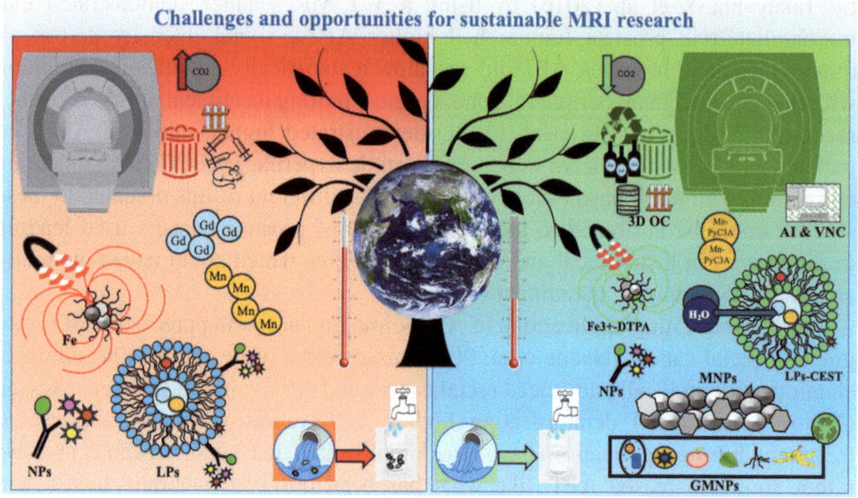

Fig. 2 Challenges and Opportunities for sustainable MRI research. The threat of wastewater contamination related to the usage of gadolinium-based contrast agents, or more complex exogenous and/or non-degradable nanoparticles together with animal and laboratory waste pose a high risk to our environment and health. Opportunities in limiting contrast agents' usage are offered by artificial intelligence (AI) or virtual non-contrast (VNC) MRI to improve non-contrast image quality. When contrast agent is a requirement recycling unused contrast and/or the associated waste material can positively impact our environment. Preclinical imaging research studies by using three-dimensional-organoid cell cultures (3D OC) could also potentially reduce the number of in vivo preclinical imaging experiments resulting in less animal research waste. Generation of contrast agents able to increase relaxivity such as iron-chelated agents (i.e., Fe^{3+}-DTPA) or new generation of manganese-based contrast agents (i.e., MnPyC3A) have the potential to become a new alternative to the commonly used macrocyclic gadolinium-based agents. Development of green magnetic nanoparticles from organic materials or microorganisms (i.e., fruits, roots, leaves, peel, viruses, or yeasts) or MR-contrast free imaging by using CEST could represent a further opportunity to develop a more responsible and sustainable imaging research. Gd gadolinium based-contrast agents; Mn indicates manganese-based contrast agents; Fe indicates iron oxides nanoparticles-based agents (e.g. USPIO, SPIO, MPIO); Fe^{3+}-DTPA indicates an example of chelated iron oxides nanoparticles-based agents; Mn-PyC3A Carbonized paramagnetic complexes of manganese as example of new manganese-based contrast agents; LPs liposomes; LPs-CEST Liposomes plus Chemical Exchange Saturation Transfer imaging; NPs nanoparticles or nanoprobes made of antibodies conjugated with radionuclides or other types of magnetic contrast agents; MNPs magnetic nanoparticles; GMNPs green magnetic nanoparticles; AI&VNC artificial intelligence and Virtual non-contrast imaging; 3D OC three-dimensional organoid cell

of computational methods and sophisticated imaging algorithms, eliminates the need for contrast administration while still providing diagnostically valuable images (Lindner et al. 2021). If the use of MRCAs is a requirement for both research and/or clinical reasons, an alternative could be the use of low gadolinium concentration contrast agents combined with AI tools able to improve imaging quality with smaller dose of contrast (Pasumarthi et al. 2021). Recent and preclinically tested promising alternatives to classic macrocyclic gadolinium compounds are also given by chelated-iron oxide nanoparticles (Boehm-Sturm et al. 2018) or by the new generation of manganese-based contrast agents (Erstad et al. 2019b) (Fig. 2).

The impact of contamination of water bodies as result of a preclinical MRI contrast study (i.e., by using small animal models) has a different impact when compared to a clinical study, due to a vast array of chemicals (not only nanoparticles or probes) typically involved in every step of animal research and testing including chemicals for sanitation, sterilization, animal care and testing procedures (i.e. toxic, infectious, plastic and animal waste disposal), which change the scenario when looking from an environmental perspective (Groff et al. 2014). The recent development from 2D to 3D OC has the potential of reducing disposal of waste related to animal research experiments by using a method which is faithfully mimicking in vivo biologically systems allowing for example, the study of cancer features as we observe in individual patients (Lee et al. 2024; Maharjan et al. 2024; Kim et al. 2024) (Fig. 2).

An alternative type of more sustainable MRCAs is given by the new introduction of biodegradable biologically derived compounds by plants or microorganisms such as green magnetic nanoparticles (GMNPs) (Verma et al. 2024; Ali et al. 2018). Green-magnetic nanoparticles represent a versatile platform which can be used for several type of imaging systems apart from MR, such as computed tomography (CT) or positron emission tomography (PET) and, because they are no-toxic, simple, eco-friendly, and cost effective, increasing demands and widespread application are expected to rise in the future (Obireddy and Wing-Fu Lai 2022; Tran et al. 2022; Dong et al. 2024) (Fig. 2).

Combination of hybrid imaging such as by using PET/MR systems in both clinical and preclinical scenarios could also represent an alternative solution with the potential to mitigate or replace the use of MRCAs, where the functional and advanced imaging provided by contrast-free MR imaging is combined with the sensitivity and specificity given by the introduction of the radionuclides or for example by applying a dual GMNPs (Verma et al. 2024).

Strategies for preventing and minimizing radionuclides waste contamination are required by accurately choosing the type of radionuclides (such as short half-lived radionuclides vs long-lived radionuclides) and by optimizing laboratory procedures in order to reduce the volume of wastes by segregation (Chartier et al. 2014). Most of clinical and research environments are using short half-life radioactive compounds by always taking into consideration radiological, chemical and biological disposal (Shoukat et al. 2010). Hybrid systems might require an increase in energy consumption impacting negatively the carbon footprint; however, they possess a tremendous potential which unfortunately has not been completely appreciated until now. Technological hardware advancement (such as by producing helium free MR system or new sequences/techniques able to consume lower energy with faster automated protocols) combined with the therapeutic effect provided by the new generation of biodegradable green nanoparticles or radionuclides will certainly re-focus our attention on this type of hybrid system finally allowing the generation of one-stop-shop theragnostic imaging.

6 Conclusions and Future Perspectives

When combined with contrast agents, Magnetic Resonance Imaging becomes a powerful imaging tool which allows investigations of the molecular and functional features of tumor, crucial not only for diagnosis but also for planning a better-informed treatment. While gadolinium-based contrast agents represent the first choice for clinical applications, the potential toxicity on global water bodies still requires careful consideration. New and more effective systems to prevent contamination and clearing of wastewater are needed, further implemented and tested. Alternative solution to gadolinium could be offered by carbonized paramagnetic complexes of manganese, such as Mn-PyC3A. T2-weighted contrast agents continue to be the first choice in preclinical MRI studies due to their flexibility and low toxicity, especially, when combined with liposomes. A contrast-free type of MR imaging is provided by CEST which has recently demonstrated a significant impact in both preclinical and clinical oncology imaging studies; however, further implementation of MR systems and hardware are required to overcome some of the previously discussed technical limitations.

Finally, hybrid imaging such as PET/MR possesses tremendous potential thanks to the promising hardware advancement by producing greener scanners combined with the diagnostic and therapeutic effect of new biodegradable green nanoparticles or radionuclides allowing one-stop-shop theragnostic imaging. While thinking at the perseveration of our environment by developing more sustainable systems and a greener future, we should not forget that above all justification remains the principle upon which we base our clinical contrast agents' choices: by relevance, safety, and for lowering costs. Then the approach in the research field should not completely be detached by this reasoning, keeping in mind that new MI contrast agents for preclinical imaging should be developed with the purpose of its translational role to positively impact healthcare by improving diagnosis and treatments for our patients.

Acknowledgments N/A

Conflict of Interest Nothing to declare.

References

Agency (2012) European Medicines Agency. EMA/486286/2012. European Medicines, London

Agulla J, Brea D, Campos F et al (2014) In vivo theranostics at the peri-infarct region in cerebral ischemia. Theranostics 4(1):90–105

Ali I, Peng C, Khan ZM, Naz I, Sultan M (2018) An overview of heavy metal removal from wastewater using magnetotactic bacteria. J Chem Technol Biotechnol 93:2817–2832

Amide Proton Transfer weighted imaging: Advancement in molecular tumour diagnosis (2018) http://images.philips.com/is/content/PhilipsConsumer/Campaigns/HC20140401_DG/HC-Master-White_paper_3D_APT_14062018.PDF

Baio G, Fabbi M, Cilli M et al (2020) Manganese-enhanced MRI (MEMRI) in breast and prostate cancers: preliminary results exploring the potential role of calcium receptors. PLoS One 15(9):e0224414

Bianchi A, Gobbo OL, Dufort S et al (2017) Orotracheal manganese-enhanced MRI (MEMRI): An effective approach for lung tumour detection. NMR Biomed 30(11)
Boehm-Sturm P, Haeckel A, Hauptmann R et al (2018) Low-molecular-weight iron chelates may be an alternative to gadolinium-based contrast agents for T1-weighted contrast-enhanced MR imaging. Radiology 286(2):537–546
Boester U, Rüde TR (2020) Utilize gadolinium as environmental tracer for surface water-groundwater interaction in karst. J Contam Hydrol 235:103710
Brünjes R, Hofmann T (2020) Anthropogenic gadolinium in freshwater and drinking water systems. Water Res 182:115966
Bulte JW, Kraitchman DL (2004) Iron oxide MR contrast agents for molecular and cellular imaging. NMR Biomed 17(7):484–499
Bumb A, Brechbiel MW, Choyke PL, Fugger L, Eggeman A, Prabhakaran D et al (2008) Synthesis and characterization of ultra-small superparamagnetic iron oxide nanoparticles thinly coated with silica. Nanotechnology 19:335601
Caravan P (2006) Strategies for increasing the sensitivity of gadolinium-based MRI contrast agents. Chem Soc Rev 35:512–523
Caravan P, Ellison JJ, McMurry TJ, Lauffer RB (1999) Gadolinium(III) chelates as MRI contrast agents: structure, dynamics, and applications. Chem Rev 99(9):2293–2352
Carr DH, Brown J, Bydder GM et al (1984) Intravenous chelated gadolinium as a contrast agent in NMR imaging of cerebral tumours. Lancet 1:484–486
Casals E, Verdaguer A, Tonda R et al (2003) Atomic force microscopy of liposomes bearing fibrinogen. Bioconjug Chem 14:593–600. https://doi.org/10.1021/bc025641t
Castelli DD, Terreno E, Longo D et al (2013) Nanoparticle-based chemical exchange saturation transfer (CEST) agents. NMR Biomed 26(7):839–849
Chartier Y, Emmanuel J, Pieper U, Prüss A, Rushbrook P, Stringer R, Townend W, Wilburn S, Zghondi R (eds) (2014) Safe management of wastes from health-care activities, 2nd edn. World Health Organization
Chen H, Zhen Z, Tang W et al (2013) Label-free luminescent mesoporous silica nanoparticles for imaging and drug delivery. Theranostics 3(9):650–657
Chen H, Wang GD, Tang W et al (2014) Gd-encapsulated carbonaceous dots with efficient renal clearance for magnetic resonance imaging. Adv Mater 26:6761–6766
Chen H, Wang GD, Sun S et al (2016) Mesoporous silica as nanoreactors to prepare Gd-encapsulated carbon dots of controllable sizes and magnetic properties. Adv Funct Mater 26:3973–3982
Chen W, Li L, Yan Z et al (2021) Three-dimension amide proton transfer MRI of rectal adenocarcinoma: correlation with pathologic prognostic factors and comparison with diffusion kurtosis imaging. Eur Radiol 31(5):3286–3296
Cheng Z, Thorek DLJ, Tsourkas A (2010) Gadolinium-conjugated dendrimer nanoclusters as a tumour-targeted T1 magnetic resonance imaging contrast agent. Angew Chem Int Ed 49:346–350
Cordoba S, Das A, Leon J et al (2021) Double-acting batch reverse osmosis configuration for best-in-class efficiency and low downtime. Desalination 506:114959
De Cuyper M (2014) The original magnetoliposomes: from the physicochemical basics to theranostic nanomedicine. In: Callejas-Fernández J, Estelrich J, Quesada-Pérez M, Forcada J (eds) Soft nanoparticles for biomedical applications. Roya, Cambridge
De Cuyper M, Joniau M (1988) Magnetoliposomes: formation and structural characterization. Eur Biophys J 15:311–319. https://doi.org/10.1007/BF00256482
De Cuyper M, Soenen SJ, Coenegrachts K et al (2007) Surface functionalization of magnetoliposomes in view of improving iron oxide-based magnetic resonance imaging contrast agents: anchoring of gadolinium ions to a lipophilic chelate. Anal Biochem 367:2–273
Delgado AF, Westen DV, Nilsson M et al (2019) Diagnostic value of alternative techniques to gadolinium-based contrast agents in MR neuroimaging—a comprehensive overview. Insights Imaging 10:84

Dong L, Chen G, Liu G et al (2024) A review on recent advances in the applications of composite Fe3O4 magnetic nanoparticles in the food industry. Crit Rev Food Sci Nutr 64(4):1110–1138. https://doi.org/10.1080/10408398.2022.2113363. Epub 2022 Aug 25

Dula AN, Arlinghaus LR, Dortch RD et al (2013) Amide proton transfer imaging of the breast at 3 T: establishing reproducibility and possible feasibility assessing chemotherapy response. Magn Reson Med 70(1):216–224

Ebrahimi P, Barbieri M (2019) Gadolinium as an emerging microcontaminant in water resources: threats and opportunities. Geosciences 9:93

Erstad DJ, Ramsay IA, Jordan VC et al (2019a) Tumour contrast enhancement and whole-body elimination of the manganese-based MRI contrast agent Mn-PyC3A. Invest Radiol 54(11):697–703

Erstad DJ, Ramsay IA, ordan VC, et al. (2019b) Tumour contrast enhancement and whole-body elimination of the manganese-based magnetic resonance imaging contrast agent Mn-PyC3A. Invest Radiol 54(11):697–703

Filippousi M, Angelakeris M, Sikini M et al (2014) Surfactant effects on the structural and magnetic properties of iron oxide nanoparticles. J Phys Chem C 118:16209–16217

Gandhi SN, Brown MA, Wong JG et al (2006) MR contrast agents for liver imaging: what, when, how. Radiographics 26:1621–1636

García-Jimeno S, Escribano E, Queralt J et al (2012) External magnetic field-induced selective biodistribution of magnetoliposomes in mice. Nanoscale Res Lett 7:452. https://doi.org/10.1186/1556-276X-7-452

Geraldes CFGC, Laurent S (2009a) Classification and basic properties of contrast agents for magnetic resonance imaging. Contrast Media Mol Imaging 4:1–23

Geraldes CF, Laurent S (2009b) Classification and basic properties contrast agents for magnetic resonance imaging. Contrast Media Mol Imaging 4:1–23

Gossuin Y, Gillis A, Hocq A et al (2009) Magnetic resonance relaxation properties of superparamagnetic particles. Wiley Interdiscip Rev Nanomed Nanobiotechnol 1:299–310. https://doi.org/10.1002/wnan.36

Groff K, Bachli E, Lansdowne M et al (2014) Review of evidence of environmental impacts of animal research and testing. Environments 1(1):14–30

Guadilla I, González S, Cerdán S et al (2023) Magnetic resonance imaging to assess the brain response to fasting in glioblastoma-bearing rats as a model of cancer anorexia. Cancer Imaging 23(1):36

Harisinghani MG, Jhaveri KS, Weissleder R et al (2001a) MRI contrast agents for evaluating focal hepatic lesions. Clin Radiol 56:714–725

Harisinghani MG, Saini S, Weissleder R (2001b) MR contrast agents for evaluating focal hepatic lesions. Clin Radiol 56(7):547–556

Inoue K, Fukushi M, Furukawa A et al (2020) Impact on gadolinium anomaly in river waters in Tokyo related to the increased number of MRI devices in use. Mar Pollut Bull 154:111148

Ito A, Shinkai M, Honda H, Kobayashi T (2005) Medical application of functionalized magnetic nanoparticles. J Biosci Bioeng 100(1):1–11

Jiang S, Zou T, Eberhart CG et al (2017) Predicting IDH mutation status in grade II gliomas using amide proton transfer-weighted (APTw) MRI. Magn Reson Med 78(3):1100–1109

Kamaly N, Miller AD (2010) Paramagnetic liposome nanoparticles for cellular and tumour imaging. Int J Mol Sci 11:1759–1776

Kim D, Lim H, Youn J et al (2024) Scalable production of uniform and mature organoids in a 3D geometrically engineered permeable membrane. Nat Commun 15(1):9420

Klomp DWJ, Dula AN, Arlinghaus LR et al (2013) Amide proton transfer imaging of the human breast at 7T: development and reproducibility. NMR Biomed 26(10):1271–1277

Klug G, Kampf T, Bloemer S, Bremicker J, Ziener CH, Heymer A et al (2010) Intracellular and extracellularT1andT2relaxivities of magneto-optical nanoparticles at experimental high fields. Magn Reson Med 64:1607–1615

Knappe A, Möller P, Dulski P et al (2005) Positive gadolinium anomaly in surface water and ground water of the urban area Berlin, Germany. Geochemistry 65:167–189

Kuhlpeter R, Dahnke H, Matuszewski L, Persigehl T, von Wallbrunn A, Allkemper T et al (2007) R2 and R2* Mapping for Sensing Cell-bound Superparamagnetic Nanoparticles: In Vitro and Murine in Vivo Testing. Radiology 245:449–457

Lam T, Pouliot P, Avti PK et al (2013) Superparamagnetic iron oxide based nanoprobes for imaging and theranostics. Adv Colloid Interf Sci 199–200:95–113. https://doi.org/10.1016/j.cis.2013.06.007

Lawrence MG, Keller J, Poussade Y et al (2010) Removal of magnetic resonance imaging contrast agents through advanced water treatment plants. Water Sci Technol 61:685–692

Lee SY, Koo IS, Hwang HJ et al (2024) In vitro three-dimensional (3D) cell culture tools for spheroid and organoid models. Adv Drug Deliv Rev 208:115237

Li L, Chen W, Yan Z et al (2020) Comparative analysis of amide proton transfer MRI and diffusion-weighted imaging in assessing p53 and Ki-67 expression of rectal adenocarcinoma. J Magn Reson Imaging 52(5):1487–1496

Li J, Lin L, Gao X, Li S et al (2021) Amide proton transfer weighted and Intravoxel incoherent motion imaging in evaluation of prognostic factors for rectal adenocarcinoma. Front Oncol 11:783544

Lin YJ, Koretsky AP (1997) Manganese ion enhances T1-weighted MRI during brain activation: an approach to direct imaging of brain function. Magn Reson Med 38:378–388

Lindner T, Debus H, Fiehler J (2021) Virtual non-contrast enhanced magnetic resonance imaging (VNC-MRI). Magn Reson Imaging 81:67–74

Liu Y, Chen C, Qian P et al (2015) Gd-metallofullerenol nanomaterial as non-toxic breast cancer stem cell-specific inhibitor. Nat Commun 6:5988

Liu Z, Wen J, Wang M et al (2022) Breast amide proton transfer imaging at 3 T: diagnostic performance and association with pathologic characteristics. J Magn Reson Im 2022:824–833. https://doi.org/10.1002/jmri.28335

Loi L, Goerke S, Zimmermann F et al (2022) Assessing the influence of the menstrual cycle on APT CEST-MRI in the human breast. Magn Reson Imaging:24–31. https://doi.org/10.1016/j.mri.2022.05.006

Maharjan S, Ma C, Singh B et al (2024) Advanced 3D imaging and organoid bioprinting for biomedical research and therapeutic applications. Nat Commun 15(1):9420

Mahmoudi M, Sant S, Wang B, Laurent S, Sen T (2011) Adv Drug Deliv Rev 63:24–46

Manus LM, Mastarone DJ, Waters EA et al (2010) et al. Gd(III)-nanodiamond conjugates for MRI contrast enhancement. Nano Lett 10:484–489

Martin DR, Krishnamoorthy SK, Kalb B et al (2010) Decreased incidence of NSF in patients on dialysis after changing gadolinium contrast-enhanced MRI protocols. J Magn Reson Imaging 31(2):440–446

McDonald MA, Watkin KL (1992) Investigations into the use of manganese as a paramagnetic contrast agent for the gastrointestinal tract. Magn Reson Imaging 10(5):753–759

Mitchell DG (1996) Liver I: currently available gadolinium chelates. Magn Reson Imaging Clin N Am 4:37–51

Nam T, Park S, Lee SY et al (2010) Tumour targeting chitosan nanoparticles for dual-modality optical/MR cancer imaging. Bioconjug Chem 21(2010):578–582

Neves AA, Brindle KM (2006) Assessing responses to cancer therapy using molecular imaging. Biochim Biophys Acta 1766(2):242–261

Nishie A, Takayama Y, Asayama Y et al (2018) Amide proton transfer imaging can predict tumour grade in rectal cancer. Magn Reson Imaging 51:96–103. https://doi.org/10.1016/j.mri.2018.04.017

Nishie A, Asayama Y, Ishigami K et al (2019) Amide proton transfer imaging to predict tumour response to neoadjuvant chemotherapy in locally advanced rectal cancer. J Gastroenterol Hepatol 34(1):140–146. https://doi.org/10.1111/jgh.14315

Obireddy SR, Wing-Fu Lai WF (2022) ROS-generating amine-functionalized magnetic nanoparticles coupled with Carboxymethyl chitosan for pH-responsive release of doxorubicin. Int J Nanomed 17:589–601. https://doi.org/10.2147/IJN.S338897. eCollection 2022

Oksendal AN, Hals P (1993) Biodistribution and toxicity of MR imaging contrast media. J Magn Reson Imaging 3:157–165

Oluwasola IE, Ahmad AL, Shoparwe NF et al (2022a) Ismail S. Gadolinium based contrast agents (GBCAs): uniqueness, aquatic toxicity concerns, and prospective remediation. J Contam Hydrol 250:104057

Oluwasola IE, Ahmad AL, Shoparwe NF et al (2022b) Gadolinium based contrast agents (GBCAs): uniqueness, aquatic toxicity concerns, and prospective remediation. J Contam Hydrol 250:104057

Orel VE, Tselepi M, Mitrelias T et al (2018) Nanomagnetic modulation of tumour redox state. Nanomedicine 14(6):795–807

Paech D, Windschuh J, Oberhollenzer J et al (2018) Assessing the predictability of IDH mutation and MGMT methylation status in glioma patients using relaxation-compensated multipool CEST MRI at 7.0 T. Neuro Oncol 20(12):1661–1671

Paech D, Dreher C, Regnery S et al (2019) Relaxation-compensated amide proton transfer (APT) MRI signal intensity is associated with survival and progression in high-grade glioma patients. Eur Radiol 29(9):4957–4967

Pan D, Schmieder AH, Wickline SA et al (2011) Manganese-based MRI contrast agents: past, present and future. Tetrahedron 67(44):8431–8444

Paratala BS, Jacobson BD, Kanakia S et al (2012) Physicochemical characterization, and relaxometry studies of micro-graphite oxide, graphene nanoplatelets, and nanoribbons. PLoS One 7:e381852012

Park JY, Daksha P, Lee GH et al (2008) Highly dispersible PEG surface modified ultra-small superparamagnetic iron oxide nanoparticles useful for target-specific biomedical applications. Nanotechnology 19:365603. https://doi.org/10.1088/0957-4484/19/36/365603

Park JY, Choi ES, Baek MJ et al (2009) Water-soluble ultra small paramagnetic or superparamagnetic metal oxide nanoparticles for molecular MR Imaging. Eur J Inorg Chem 9:2477–2481

Pasumarthi S, Tamir JI, Christensen S et al (2021) A generic deep learning model for reduced gadolinium dose in contrast-enhanced brain MRI. Magn Reson Med 86:1687–1700

Peer D, Karp JM, Hong S et al (2007) Nanocarriers as an emerging platform for cancer therapy. Nat Nanotechnol 2:751–760

Qin R, Li S, Qiu Y et al (2022) Carbonized paramagnetic complexes of Mn (II) as contrast agents for precise magnetic resonance imaging of sub-millimeter-sized orthotopic tumours. Nat Commun 13(1):1938

Rallapalli H, Tan I-L, Volkova E, Wojcinski A et al (2020) MEMRI-based imaging pipeline for guiding preclinical studies in mouse models of sporadic medulloblastoma. Magn Reson Med 83(1):214–227

Ray KJ, Simard MA, Larkin JR et al (2019) Tumour pH and protein concentration contribute to the signal of amide proton transfer magnetic resonance imaging. Cancer Res 79(7):1343–1352

Raymond KN, Pierre VC (2004) Next generation, high relaxivity gadolinium MRI agents. Bioconjug Chem 16:3–8

Rieter WJ, Kim JS, Taylor KML et al (2007) Hybrid silica nanoparticles for multimodal imaging. Angew Chem Int Ed 46:3680–3682

Roca AG, Costo R, Rebolledo AF, Veintemillas-Verdaguer S, Tartaj P, González-Carreño T et al (2009) Progress in the preparation of magnetic nanoparticles for applications in biomedicine. J Phys D Appl Phys 42:224002

Runge VM, Schoerner W, Niendorf HP et al (1985) Initial clinical evaluation of gadolinium DTPA for contrast-enhanced magnetic resonance imaging. Magn Reson Imaging 3:27–35

Sagiyama K, Mashimo T, Togao O et al (2014) In vivo chemical exchange saturation transfer imaging allows early detection of a therapeutic response in glioblastoma. Proc Natl Acad Sci U S A 111(12):4542–4547

Sanchez-Gaytan BL, Fay F, Lobatto ME et al (2015) HDL-mimetic PLGA nanoparticle to target atherosclerosis plaque macrophages. Bioconjug Chem 26(3):443–451

Schmitt B, Zamecnik P, Zaiss M et al (2011) A new contrast in MR mammography by means of chemical exchange saturation transfer (CEST) imaging at 3 tesla: preliminary results, vol

183. Rofo-Fortschritte Auf Dem Gebiet Der Rontgenstrahlen Und Der Bildgebenden Verf, pp 1030–1036

Shoukat K, Syed AT, Reyaz A et al (2010) Radioactive waste management in a hospital. Int J Health Sci 4:39–46

Silva AC, Lee JH, Aoki L et al (2004) Manganese-enhanced magnetic resonance imaging (MEMRI): methodological and practical considerations. NMR Biomed 17:532–543

Soenen SJ, Desender L, De Cuyper M (2007) Complexation of gadolinium (III) ions on top of nanometer-sized magnetoliposomes. Int J Environ Anal Chem 87:783–796

Soenen SJ, Hodenius M, De Cuyper M (2009) Magnetoliposomes: versatile innovative nanocolloids for use in biotechnology and biomedicine. Nanomedicine (Lond) 4:177–191. https://doi.org/10.2217/17435889.4.2.177

Sotirios B, Demetriou E, Topriceanu CC et al (2020) The role of APT imaging in gliomas grading: a systematic review and meta-analysis. Eur J Radiol 133:109353

Souza LA, Pedreira RMA, Miro M et al (2021) Evidence of high bio-accessibility of gadolinium-contrast agents in natural waters after human oral uptake. Sci Total Environ 793:148506

Su C, Xu S, Lin D et al (2022) Multi-parametric Z-spectral MRI may have a good performance for glioma stratification in clinical patients. Eur Radiol 32(1):101–111

Sun C, Lee J, Zhang M (2008) Magnetic nanoparticles in MR imaging and drug delivery. Adv Drug Deliv Rev 60(11):1252–1265

Sun W, Luo L, Feng Y et al (2020) Gadolinium-rose bengal coordination polymer nanodots for MR–/fluorescence-image-guided radiation and photodynamic therapy. Adv Mater 32(23):e2000377

Takayama Y, Nishie A, Sugimoto M et al (2016) Amide proton transfer (APT) magnetic resonance imaging of prostate cancer: comparison with Gleason scores. MAGMA 29(4):671–679

Tang JB, Sheng YQ, Hu HJ et al (2013) Macromolecular MRI contrast agents: structures, properties and applications. Prog Polym Sci 38:462–502

Terreno E, Castelli DD, Viale A, Aime S (2010) Challenges for molecular magnetic resonance imaging. Chem Rev 110(5):3019–3042

Thunus L, Lejeune R (1999) Overview of transition metal and lanthanide complexes as diagnostic tools. Coordin Chem

Togao O, Yoshiura T, Keupp J et al (2014) Amide proton transfer imaging of adult diffuse gliomas: correlation with histopathological grades. Neuro Oncol 16(3):441–448. https://doi.org/10.1093/neuonc/not158

Torchilin VP (1996) Liposomes as delivery agents for medical imaging. Mol Med 2:242–249

Torres CG, Lundby B, Sterud AT et al (1997) MnDPDP for MR imaging of the liver - results from the European phase III studies. Acta Radiol 38:631–637

Tran HV, Ngo NM, Medhi R et al (2022) Multifunctional iron oxide magnetic nanoparticles for biomedical applications: a review. Materials (Basel) 15(2):503. https://doi.org/10.3390/ma15020503

Unger EC, MacDougall P, Cullis P et al (1989) Liposomal Gd-DTPA: effect of encapsulation on enhancement of hepatoma model by MRI. Magn Reson Imaging 7:417–423

Verma C, Verma DK, Berdimurodov E et al (2024) Green magnetic nanoparticles: a comprehensive review of recent progress in biomedical and environmental applications. J Mater Sci. Review. Published: 21 December 2023, vol 59, pp 325–353

Warsi MF, Adams RW, Duckett SB et al (2010) Gd-functionalised Au nanoparticles as targeted contrast agents in MRI: relaxivity enhancement by polyelectrolyte coating. Chem Commun 46:451–453

Weissleder R, Mahmood U (2001) Molecular imaging. Radiology 219(2):316–333

Wen L, Shi X, He L et al (2017) Manganese-enhanced MRI for the detection of metastatic potential in colorectal cancer. Radiol Exp 1(1):21

Wen L, Shi X, He L et al (2018) Manganese-enhanced magnetic resonance imaging for detection and characterization of colorectal cancers. Tomography 4(2):78–83

Wood ML, Hardy PA (1993) Proton relaxation enhancement. J Magn Reson Imaging 3:149–156

Wu B, Warnock G, Zaiss M et al (2016) An overview of CEST MRI for non-MR physicists. EJNMMI Phys 3(1):19
Xiao YD, Paudel R, Liu J et al (2016) MRI contrast agents: classification and application (Review). Int J Mol Med 38(5):1319–1326
Xu Y, Wang Y, Wang H et al (2016) Recent advancements in graphene-based nanomaterials for T2-weighted magnetic resonance imaging applications. Nanoscale 8(3):1445–1461
Yang CT, Hattiholi A, Tamil Selvan S et al (2020) Gadolinium-based bimodal probes to enhance T1-weighted magnetic resonance/optical imaging. Acta Biomater 110:15–36
Yao J, Hagiwara A, Raymond C et al (2020) Human IDH mutant 1p/19q co-deleted gliomas have low tumour acidity as evidenced by molecular MRI and PET: a retrospective study. Sci Rep 10(1):11922
Yin H, Wang D, Yan R et al (2021) Comparison of diffusion kurtosis imaging and amide proton transfer imaging in the diagnosis and risk assessment of prostate cancer. Front Oncol 11:640906–640906
Young IR, Clarke GJ, Bailes DR et al (1981) Enhancement of relaxation rate with paramagnetic contrast agents in NMR imaging. J Comput Tomogr 5:543–547
Zhang S, Seiler S, Wang X et al (2018) CEST-Dixon for human breast lesion characterization at 3 T: a preliminary study. Magn Reson Med 80(3):895–903
Zhang S, Rauch GM, Adrada BE et al (2021) Assessment of early response to neoadjuvant systemic therapy in triple-negative breast cancer using amide proton transfer-weighted chemical exchange saturation transfer MRI: a pilot study, vol 3. Radiol Imaging Cancer, p e200155
Zhen ZP, Xie J (2012) Development of manganese-based nanoparticles as contrast probes for magnetic resonance imaging. Theranostics 2:45–54
Zhen G, Muir BW, Moffat BA, Harbour P, Murray KS, Moubaraki B et al (2011) Comparative Study of the Magnetic Behavior of Spherical and Cubic Superparamagnetic Iron Oxide Nanoparticles. J Phys Chem C 115:327–334
Zhou Z, Lu ZR (2013) Gadolinium-based contrast agents for magnetic resonance cancer imaging. Wiley Interdiscip Rev Nanomed Nanobiotechnol 5(1):1–18
Zhou J, Lal B, Wilson DA et al (2003a) Amide proton transfer (APT) contrast for imaging of brain tumours. Magn Reson Med 50(6):1120–1126
Zhou J, Payen J-F, Wilson DA et al (2003b) Using the amide proton signals of intracellular propteins and peptides to detect pH effects in MRI. Nat Med 9:1085–1090
Zhou J, Blakeley JO, Hua J et al (2008) Practical data acquisition method for human brain tumour amide proton transfer (APT) imaging. Magn Reson Med 60(4):842–849
Zhou J, Tryggestad E, Wen Z et al (2011) Differentiation between glioma and radiation necrosis using molecular magnetic resonance imaging of endogenous proteins and peptides. Nat Med 17:130–134
Zhou J, Heo HY, Knutsson L et al (2019) MRI: techniques, current neuro applications, and challenging issues. J Magn Reson Imaging 50(2):347–364. https://doi.org/10.1002/jmri.26645

Fundamentals of Ultrasound Imaging

Daria Poul, J. Brian Fowlkes, and Oliver D. Kripfgans

Abstract

In this chapter, we will introduce key topics on fundamentals of ultrasound imaging to provide the basis for understanding how this powerful imaging modality works without going into any sophisticated theories. The primary focus will be on commonly used terminologies and concepts. The chapter begins with a concise overview of the physics characterizing ultrasound waves, followed by a comprehensive introduction of ultrasound transducers. This includes the internal structure of the transducer and their characteristics along with types of transducers frequently employed in research and/or clinics as well as transducer selection criteria. Then, we will move forward by introducing fundamentals of ultrasound image formation, important characteristics of ultrasound B-mode images as well as presenting some advanced ultrasound imaging modes. We will conclude the chapter by discussing important ultrasound imaging artifacts and essential safety considerations.

Keywords

Ultrasound imaging · Ultrasound artifacts · Ultrasound safety indices · Blood flow quantification · Elastography

D. Poul (✉) · J. B. Fowlkes · O. D. Kripfgans
Department of Radiology, Michigan Medicine, University of Michigan, Ann Arbor, MI, USA
e-mail: ssheykho@umich.edu

I. F. Cengiz et al. (eds.), *Bioimaging Modalities in Bioengineering*,
Biomaterials, Bioengineering and Sustainability 9,
https://doi.org/10.1007/978-3-031-99410-4_8

1 Introduction

1.1 Overview

Ultrasound imaging is a non-invasive diagnostic imaging modality to view the internal anatomical structure of the body and soft tissues in real time. As a non-ionizing, portable and relatively inexpensive modality, ultrasound imaging has drawn significant attention for obtaining detailed images of internal structures in various organs from larger ones such as liver and muscles, to smaller sized tissues such as thyroid, tendons, and gum tissues. This non-invasive modality produces a signature gray-scale image, known as B-mode, that provides an image of an organ cross-section, resembling a virtual slice. An example B-mode image of a liver is shown in Fig. 1a. B-mode imaging showcases details of soft tissues and anatomical landmarks in its imaging plane. Such an internal image is acquired using an ultrasound scanner paired with an ultrasound probe (transducer) as illustrated in Fig. 1b, while the probe is held in contact with the skin, targeting the organ of interest from outside of the body. On the gray-scale B-mode, shades (brightness) represent the relative strength of the reflected echoes. Ultrasound B-mode provides a real-time diagnostic tool for distinguishing abnormal changes to anatomical structures of tissues such as alteration of their shape and size or detecting the presence of abnormalities like lesions which are "visible" within the capabilities of ultrasound.

1.2 What Are Ultrasound Waves?

Before defining ultrasound waves, it is helpful to first develop a general understanding of waves. A (mechanical) wave is a form of perturbations or oscillations (like the sinusoid form) that travels through a medium such as soft tissues and carries energy via making its particles oscillate around their equilibrium states without any particle transportation. For instance, when a rope lying on the ground is excited via a left-and-right motion at one end, an oscillating pattern (i.e., wave) is generated at the source. The wave propagates forward along the rope and travels away from the source to the other end. In this example, the rope particles move perpendicular to the direction of the wave along left-and-right directions. Depending on the direction of particle oscillation with respect to the wave propagation, waves are categorized as either transverse or longitudinal. In transverse waves, such as the rope example, particles oscillations are perpendicular to the wave propagation direction whereas in longitudinal waves, these two motions are aligned. In Fig. 1c, an example of longitudinal waves propagating in a medium via its particle oscillations is presented. Here, areas of particle compression and expansion in response to the wave as well as directions of particle oscillation and wave propagation can be observed by comparing two cases of no wave propagation (top) versus longitudinal wave propagation (middle) in a medium. Additionally, a sample sinusoidal plot of amplitude versus time for a longitudinal wave is presented in (c) (bottom) for which

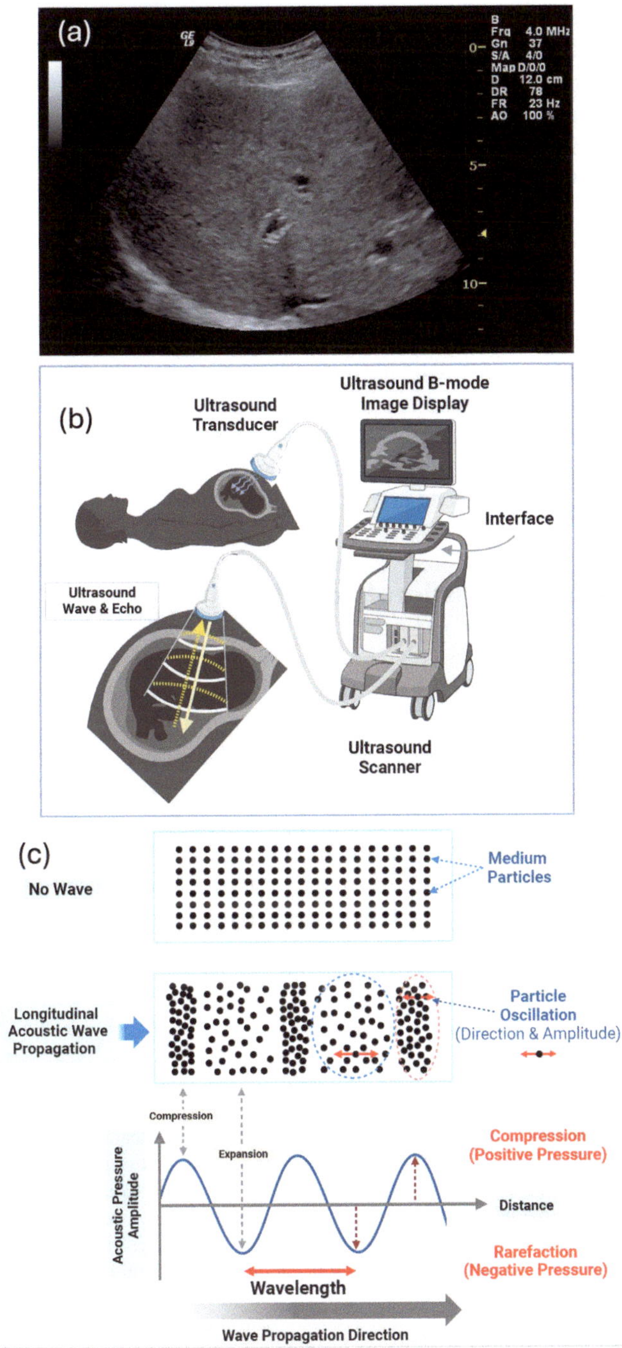

Fig. 1 (**a**) shows a sample of ultrasound gray-scale B-mode image of the human liver acquired at the frequency of 4 megahertz (MHz) (Courtesy of Dr. Jonathan M. Rubin and Dr. Stephen Z. Pinter,

important characteristics of waves such as wavelength, amplitude, compression and rarefaction regions are defined.

An important category of mechanical waves is acoustic waves, which incorporate sound waves and ultrasound waves. Sound waves (such as propagation of air perturbations produced by clapping hands) define the human hearing range while ultrasound waves are acoustic waves beyond the human hearing range which cannot be distinguished by human hearing. More technically, ultrasound waves are longitudinal pressure waves that involve propagations of higher pressure (compression) and lower pressure (rarefaction) fronts within a medium, as illustrated in Fig. 1c. Here, the plot shows how an oscillating pressure wave that undergoes positive and negative pressures impose particle compressions and expansions as it propagates to the right. Ultrasound waves have a wide range of applications, particularly in medical imaging. Ultrasound imaging relies on transmitting ultrasound waves into the body and collecting returning waves (echoes) from internal structures of our body to produce an image of the organ. Acoustic waves are uniquely characterized by key descriptors, introduced in the following section.

1.3 Key Characteristics of Acoustic Waves

Frequency Describes the pitch of the sound and is the number of complete oscillations (cycles) of waves that pass through a point per second. Frequency is shown by f and is measured in the unit of Hertz (Hz). Sound waves are characterized by the frequency range of 20 Hz to 20 kHz (human hearing range) while the sound waves transmitted in medical imaging applications have frequencies greater than 20 kHz (1 kHz $= 10^3$ Hz). For this reason, these high-frequency sound waves are referred to as ultrasound (*ultra-sound*) waves. In medical ultrasound imaging, the range of frequencies employed are commonly in the range of 1–40 megahertz (MHz), 1 MHz $= 10^6$ Hz. Higher frequencies (up to 80 MHz) are also employed less frequently in special applications.

Fig. 1 (continued) University of Michigan, Ann Arbor, MI, USA); (**b**) shows an illustration of ultrasound equipment (a scanner and transducers attached to the scanner) for in vivo ultrasound imaging. Here, like a typical ultrasound scanner, the scanner has a user interface for adjusting imaging acquisition setups, a monitor for displaying ultrasound B-mode image and ports for connecting ultrasound transducers. A schematic of transmitting and receiving ultrasound waves for an imaging case is also depicted to demonstrate basics of the pulse-echo phenomenon; (**c**) shows a detailed description of longitudinal wave propagation (such as an ultrasound wave) in a medium due to particle oscillations. Here, areas of particle compression and expansion in response to the wave, directions of particle oscillation and wave propagation can be observed by comparing two cases of no wave propagation (top) versus longitudinal wave propagation (middle) in a medium. Additionally, an example plot of amplitude variations with respect to time for a longitudinal sinusoidal wave is shown in the bottom of (**c**). Here, important characteristics of waves such as wavelength, amplitude, compression, and rarefaction regions are marked for this wave

Period Describes the time it takes (in seconds, *s*) to complete one wave cycle. Period and frequency of sound waves are inverse of each other ($1\ s = \frac{1}{1\ Hz}$).

Amplitude Describes the loudness of the sound, larger amplitudes basically denote louder sounds (given the frequency of the sound is in the hearing range) or ultrasound. In ultrasound, amplitude is defined as the maximum pressure (P) of waves (or particle displacement) with respect to the undisturbed (equilibrium) state (P_0), or simply as the peak of a sound wave. In acoustics and medical imaging, amplitude is often expressed with the unit of decibel (*dB*). Decibel basically quantifies the relative amplitude of the sound with respect to a reference using a compressed logarithmic scale defined as:

$$\text{Amplitude Level (dB)} = 20\log_{10}\left(\frac{P}{P_{ref}}\right) \tag{1}$$

Intensity It is a measure of the loudness of the sound, proportional to the square of the sound amplitude. Intensity particularly represents the energy carried (in the unit of time) by the sound wave when propagating in space (through a unit cross-sectional area perpendicular to the wave propagation).

Power Describes the total energy carried by an acoustic wave per second.

Sound Speed Describes how fast a sound wave can travel within a medium. It is defined as the distance travelled by a particular point on the wave like its crest (peak) in the unit of time. Sound speed (c) is often given in the unit of meter per second (m/s). Average speed of sound in normal soft tissues and fat are 1540 m/s (approximately 1.5 mm per second) and 1430 m/s, respectively while in bone the sound speed is 3000–4000 m/s (Szabo 2013; Laugier and Haïat 2011).

Wavelength Describes the distance along which the shape of a sound wave repeats (spatial period of sound waves). It is quantified by the distance between two consecutive similar points observed in a wave pattern. For example, the distance between two peaks is shown in the bottom part of Fig. 1c. Wavelength is normally denoted by λ (lambda) and measured in units of meters (m) or millimeter (mm) (1 m = 10^3 mm). Ultrasound waves at megahertz frequencies have wavelengths on the millimeter scale. Wavelength is related to frequency and speed of sound through the following Eq. (2) and inversely changes with frequency; higher frequency ultrasound waves have shorter wavelengths and so have media with slower sound speed.

$$\lambda = \frac{c}{f} \tag{2}$$

In the next section, we will introduce how ultrasound waves with certain characteristics are generated by transducers for medical imaging purposes.

2 Ultrasound Transducers

In ultrasound imaging, a transducer (as shown in Fig. 1b) is basically a device to generate ultrasound waves, transmit them through tissues and collect returning echoes for image formation. More technically, a transducer converts the energy in an electrical pulse into a mechanical pulse (ultrasound wave) and vice versa during the imaging process. Figure 2 showcases a simplified schematic of a transducer with its important internal structures. The energy conversion within a transducer occurs through its arrays of piezoelectric elements. Basically, when an electrical signal (i.e., an oscillating voltage) is applied to a piezoelectric material, it deforms (stretches or compresses). Following the oscillation of applied voltage, length changes of piezoelectric elements are also oscillatory. These spatial motions of elements generate high-frequency ultrasound waves. Conversely, once returning echoes (mechanical pressure waves) from tissues arrive at the transducer, oscillations are induced within piezoelectric elements which are converted back into electrical signals to produce the ultrasound image. Transducers have other essential components including the acoustic matching layer added for an optimized transmission of waves between the transducer and tissues while the backing layer damps any unwanted vibration of piezoelectric elements after the excitation is over. The role of lens in transducers for focusing will be discussed later.

2.1 Imaging Planes of Transducers

A two-dimensional (2D) B-mode image is a cross-sectional representation of how a slice of the underlying tissue appears. A B-mode basically represents the imaging plane of a transducer. However, the ultrasound beam from a transducer has a finite thickness perpendicular to its imaging plane which causes the ultrasound field to be a 3D region in space, as shown in Fig. 2b. ultrasound waves propagate and interact with tissues over this 3D field, which is described by three orientations with respect to the transducer surface: (i) axial and (ii) lateral, and (iii) elevational directions. Axial and lateral are the primary directions defining the transducer's imaging plane. The axial direction is along the propagation direction of the ultrasound beam, that is, the depth of an image, while the lateral direction is defined within the imaging plane perpendicular to the wave propagation direction. The elevational direction describes the orientation perpendicular to the image plane along which the thickness of the transducer imaging plane is measured. These are illustrated in Fig. 2b, with further details to be introduced in the following sections.

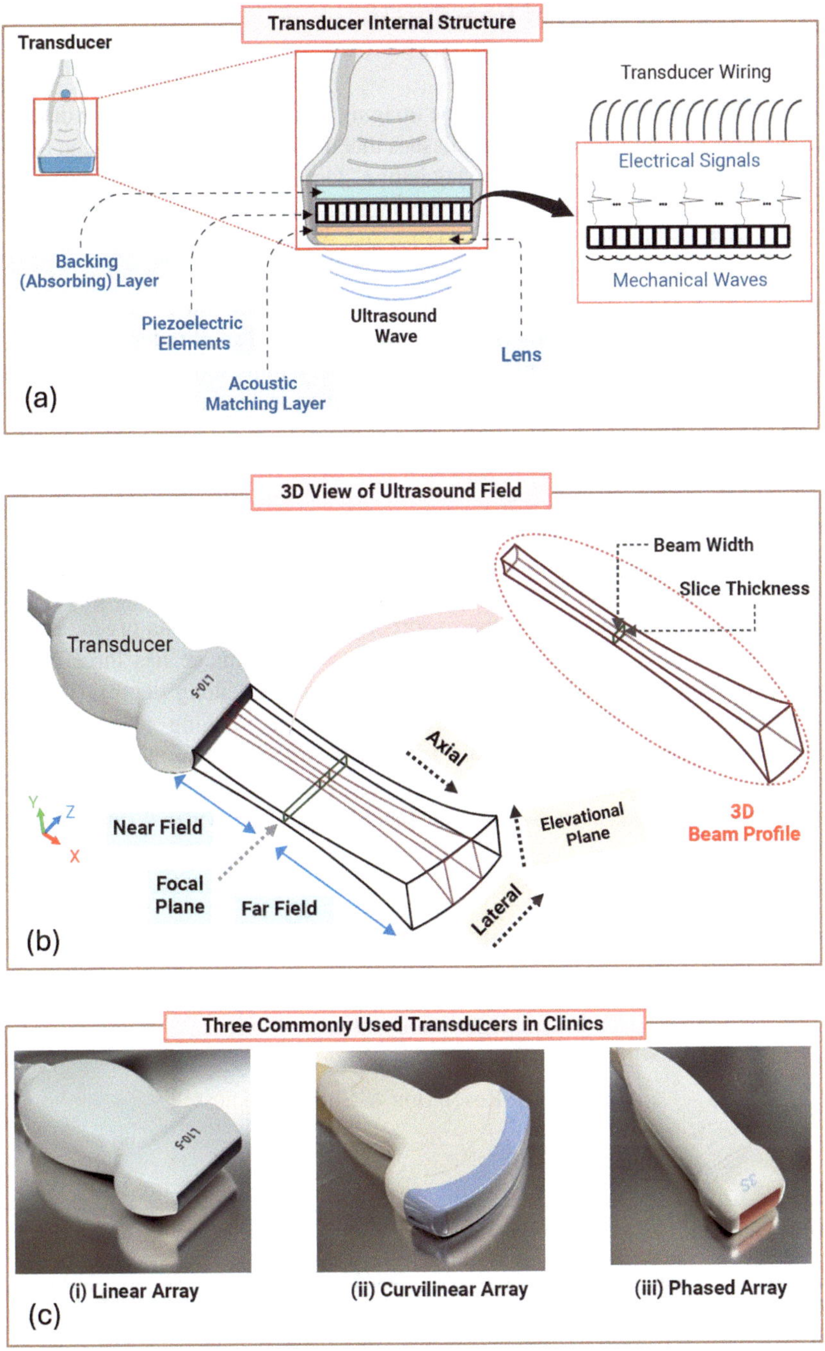

Fig. 2 (**a**) illustrates a simplified schematic showing important internal structures of an ultrasound transducer (backing layer, piezoelectric elements, matching layer, and lens). Piezoelectric elements

2.2 Characteristics of Transducers

Field of View (FOV) It is basically the imaging plane of a transducer and describes the anatomical (spatial) areas within tissues that a transducer can image and display on the scanner in a single image acquisition. Depending on the transducer type, common shapes of the FOV are 2D rectangular, trapezoidal, sector (like a pizza slice!) and 3D volumetric, each suitable for specific applications.

Pulse Length (PL), Beam Width, and Slice Thickness An ultrasound pulse is typically a few cycles of oscillating waves having varying amplitudes with time. The ultrasound pulse is generated by the transducer over a finite short duration, called the PL (or pulse duration). Beam width and slice thickness are two characteristic spatial widths of an ultrasound beam along lateral and elevation direction, respectively, as depicted on the beam profile in Fig. 2b.

Image Resolution It is the minimum distance that two distinct targets can be spaced apart along a specific direction and be distinguishable on the displayed image. Transducers with higher frequencies offer higher resolutions which allow for acquiring ultrasound images that incorporate finer details of an organ, essential for making a more accurate diagnosis. An ultrasound image has three primary spatial resolutions: (i) axial resolution (resolving targets along depth), (ii) lateral resolution (resolving side-by-side targets across the imaging plane), and (iii) elevational resolution which controls how much targets outside the imaging plane are incorrectly included in the displayed image. Lateral and elevational resolutions can vary along depth, narrower around focal regions and wider distal from these regions.

$$\text{Axial resolution} \approx \frac{n\lambda}{2} \ (n : \text{number of cycles in the pulse})$$

$$\text{Lateral Resolution} \approx \text{Beam Width}$$

$$\text{Elevational Resolution} \approx \text{Slice Thickness}$$

Fig. 2 (continued) of transducers convert electrical signals to mechanical waves and vice versa during transmit and receive phenomena; (**b**) shows a 3D view of the ultrasound field for a linear array transducer with its 3D beam profile. The zoomed-in view of the 3D beam profile is also presented on the top corner of (**b**). Here, the imaging orientations of the transducer (axial, lateral and elevational directions), the focal region along the elevational direction as well as the beam width and slice of thickness are defined with respect to this transducer; in (**c**), three types of commonly used ultrasound transducers are shown. (i) linear array, (ii) curvilinear array and, (iii) phased array

Center Frequency and Frequency Bandwidth In ultrasound imaging, a transducer can generate ultrasound pulses within a particular frequency range, termed its bandwidth with the frequency in the middle of this bandwidth called the center frequency (f_C). The center frequency and bandwidth are characteristic features of a transducer affecting its axial resolution. The bandwidth is the range (difference) between the lowest and highest frequency components over which a transducer can work effectively. Units of f_C and bandwidth are MHz. As an example, for a transducer with $f_C = 5$ MHz and a bandwidth of 6 MHz, the operating frequency range is 2 to 8 MHz.

Before introducing different types of transducers, it is essential to note that higher frequency imaging offers a higher axial resolution on B-mode images, because such waves have shorter wavelengths that allow resolving smaller underlying structures of tissues. Also, a transducer producing a shorter PL offers a broader frequency bandwidth and enhanced axial resolution, as *bandwidth* $\approx 1/PL$. However, the imaging depth at higher frequencies and bandwidths is limited due to attenuation. This will be explained further in the Tissue Imaging section.

2.3 Types of Transducers for Imaging

A single piezoelectric element is the building block of various types of transducers for advanced ultrasound imaging of soft tissues. While a single element transducer itself does not provide a 2D ultrasound image of tissues without being physically moved, it has applications in estimating spatial information of targets along the axial direction. These include distance and echo strength of tissue structures along the axial dimension. However, an array (group) of single piezoelectric elements can be assembled into various organized configurations, called array transducers, offering high resolution 2D and 3D imaging with a wide FOV and penetration depths in a single image acquisition without physically moving the transducer. Typical number of elements in a transducer are 64, 128, and 256 (powers of 2) as well as 96 and 192. Three commonly used clinical array transducer configurations are introduced in the following paragraphs and shown in Fig. 2c.

Linear Array Transducer

If multiple piezoelectric elements are aligned in a straight line along a single direction and spaced by the wavelength λ, the configuration is called the linear array transducer. Its clinical applications are primarily cases where image resolution is prioritized over higher penetration depth; that is, high frequency imaging. Important applications include superficial imaging in breast, thyroid, oral tissues, and vascular imaging. These transducers typically offer a rectangular FOV of tissues; however, to achieve a wider FOV, some transducers produce a trapezoidal view from electronically steering ultrasound beams at outer elements. Ultrasound scanners usually allow users to switch between these two views.

Curvilinear Array Transducer
If the alignment of piezoelectric elements is along an arc rather than a straight line and spaced by the wavelength λ, a curvilinear array transducer is formed. Unlike linear arrays, these transducers are particularly employed for deep tissue imaging such as obstetric and abdominal imaging where the organ of interest (such as liver) is located in deeper regions, behind layers of subcutaneous and muscular tissues. Enhanced penetration depth of these transducers comes at the cost of lower image resolution. Such curved alignments of elements enable a broader sector FOV. The center frequency of these transducers is typically lower (1–7 MHz).

Phased Array Transducer
These transducers are designed to image organs such as the heart where the presence of ribs limit the available acoustic window of the transducer to transmit and receive ultrasound signals. A phased array transducer has similar configuration to a linear array transducer; however, its elements are spaced by $\lambda/2$ and it often incorporates fewer elements in its scan plane (i.e., lateral). Phased array transducers use an advanced technique called beam steering, which allows imaging a wide sector-shaped FOV within tissues through a small acoustic window. Beam steering is basically changing the direction of ultrasound beams without moving the transducer by applying time delays to the electrical signals that activate its elements, allowing to image wider lateral regions from a small window without physically moving the transducer. Such time delays can also be used to control the focusing in all of the array configurations described here. It is noted that the linear, curvilinear, and phased array transducers have a fixed elevational focus and are considered 1D arrays, because their elements are organized along a single dimension.

2D Matrix Array Transducer
These are advanced ultrasound transducers, comprised of grids (matrix) of single piezoelectric elements arranged in rows and columns. 2D transducers steer the ultrasound beam in lateral and elevational directions and work based on more advanced image formation techniques which are outside the scope of this chapter. It is noted that there are also some transducers that are not full 2D arrays with varying levels of steering and focusing along the elevational direction but improved spatial resolution.

So far, we have introduced key concepts on physics of ultrasound waves and ultrasound transducers essentials before discussing tissue imaging. In the following sections, we will focus on fundamentals of producing ultrasound images.

3 Tissue Imaging

B-mode image formation basically follows a pulse-echo principle to locate and depict targets, as illustrated for one simple sequence in Fig. 3a. That is, the ultrasound transducer first transmits ultrasound pulses into the body. As each pulse travels through tissues and organ boundaries, it reflects (and scatters) back because

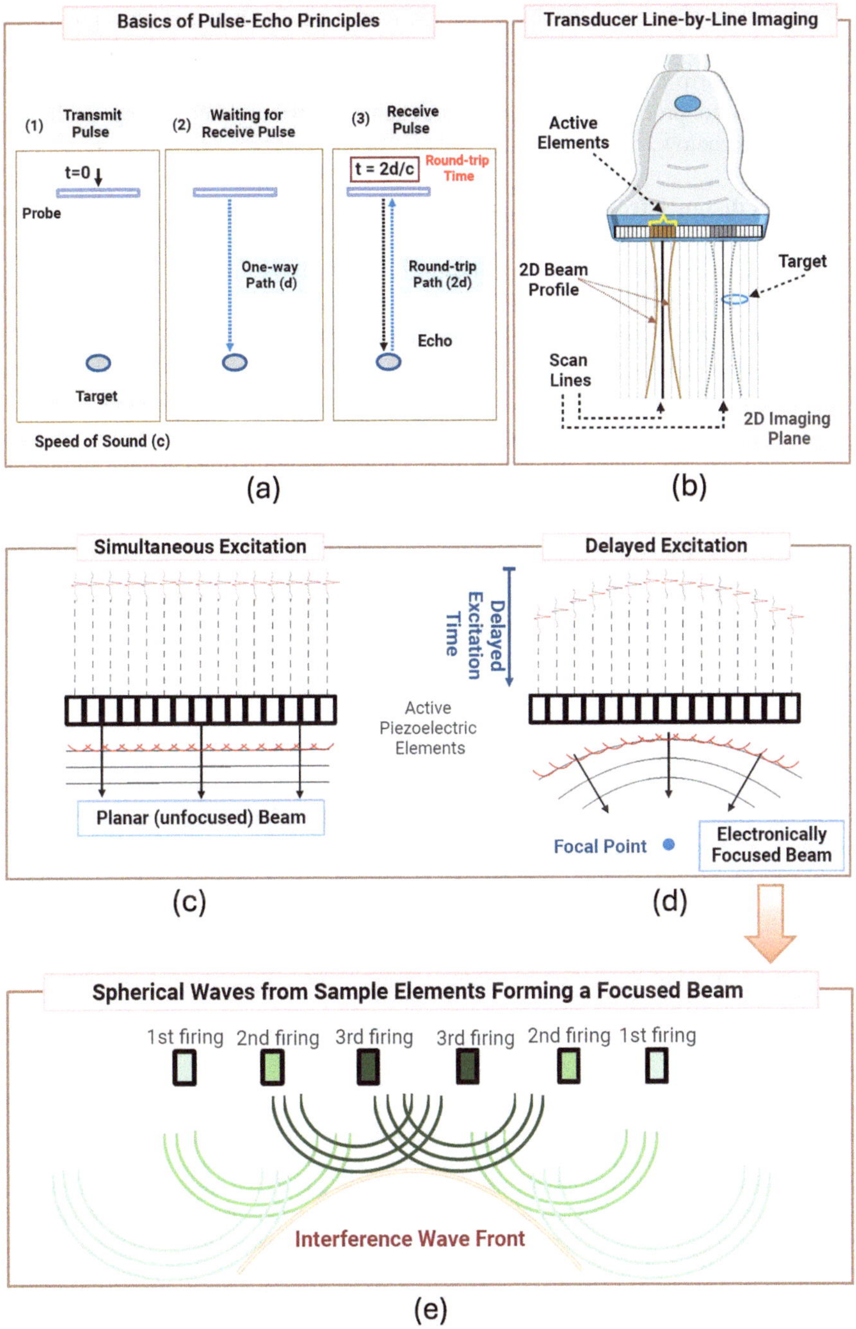

Fig. 3 (**a**) shows basics of the pulse-echo principle in ultrasound image formation. Here, the simplified round-trip of a signal interacting with a target is illustrated in three steps. (1) shows $t = 0$, when the pulse is transmitted from the transducer (probe) to a medium with a speed of sound

of interaction with tissue. Each reflected/scattered pulse, known as an echo, travels back to the ultrasound transducer surface and is received there, completing a "round-trip", (round-trip path $= 2 \times$ target depth). In B-mode image formation through the pulse-echo process, three key components contribute to creating an image, summarized below:

- **Target distance from the transducer:** Recording the time between transmitting a pulse and receiving its echoes ("round-trip" time, t) allows for estimating the distance of the target from the transducer (d), given the speed of sound in tissues (c) is known, according to $d = ct/2$, (reminder: target depth $= 1/2\times$ round-trip path). Practically, it is helpful to remember that it takes approximately 13 microseconds for an ultrasound pulse to complete a 2-centimeter round-trip path.
- **Strength of echoes:** The shade of gray at each spot on a B-mode image is assigned from the relative echo strength, that is, relative magnitude of echo amplitudes received at the probe from their corresponding regions within the tissues. The stronger the echo is, the brighter it appears on the B-mode image.
- **Orientations of transducer scanlines:** Advanced techniques such as electronic focusing and beam steering require knowledge of beam angles to accurately map and scale target locations in the lateral direction on a B-Mode image using individually received echoes from tissues.

It is noteworthy that ultrasound imaging relies on tissues' ability to transmit acoustic waves in the first place, i.e., that tissues allow waves to propagate through them. This capability is assessed by the acoustic impedance of tissues. Impedance, z, is a characteristic feature of a medium such as soft tissue that quantifies its resistance to the propagation of acoustic waves through it. Impedance depends on two bulk properties of tissues: density (ρ) and speed of sound (c), as shown in Eq. (3). For instance, the acoustic impedance of soft tissues is much smaller than bone and relatively larger than that of fat. Soft tissues, fat and bone have acoustic impedances that are orders of magnitudes larger than acoustic impedance of air, as reported in Table 1 (Szabo 2013; Laugier and Haïat 2011; Alvarez-Arenas 2004).

Fig. 3 (continued) denoted with c. (2) depicts the time when the transmitted pulse arrives at a target located within the medium at the distance (depth) of d from the probe surface. (3) shows the time when the reflected echo from the target is received at the transducer surface, completing the round-trip at $t = 2d/c$; (**b**) illustrates the schematic of line-by-line B-mode imaging for a linear array transducer. Here, the 2D imaging plane of the transducer is illustrated which shows individual scanlines forming a B-mode image. Two samples of 2D beam profiles and their associated scanlines are also shown; (**c**) depicts the simultaneous excitation of transducer elements producing a planar unfocused beam compared with the electronic focusing of ultrasound beams shown in (**d**) using delayed excitation (firing) of transducer elements; in (**e**), we have shown a detailed schematic of delayed wave fronts produced by individual elements and their resulting interferences during electronic focusing to produce a focused beam. Here, the case with a limited number of elements is considered for simplicity

Table 1 Typical values for acoustic impedances of some media in the unit of MRayl ($= \frac{kg}{m^2.s} \times 10^{-6}$)

Medium	Soft tissue (Average) [1]	Fat (Szabo 2013)	Bone (Laugier and Haïat 2011)	Air (Alvarez-Arenas 2004)
Acoustic Impedance	1.6	1.327	4 – 8	0.0004

The differences in impedance among tissues is also the property that produces the reflections (echoes) from structures and the proportional strength in these echoes is based on the relative impedance differences. This will be discussed later in Sect. 3.3.

$$z = \rho c \tag{3}$$

3.1 Image Formation

A 2D ultrasound B-mode image is formed from several B-mode scanlines, as shown in Fig. 3b (only limited scanlines are illustrated) for the line-by-line image formation. Each scanline basically represents an image of a small region within tissues located within the 3D beam profile (3D beam profile in Fig. 2b further clarifies this 3D domain). A scanline is typically reconstructed from one sequence of pulse-echo using a group of active piezoelectric elements of the transducer (i.e., a subaperture), where these sub-apertures are then activated sequentially in a certain order to insonify the whole tissue region line by line and produce all the scanlines that form a 2D B-mode image. On the gray-scale B-mode, shades (brightness) represent the relative strength of the reflected echoes from their corresponding region within the tissue. That is, darker (hypo-echoic) regions demonstrate that reflected waves from a specific region are relatively weaker whereas brighter (hyper-echoic) regions on the image represent stronger reflections from internal structures of the body being scanned. This definition led to the introduction of the term B-mode (brightness mode) to refer to the gray-scale ultrasound image.

By producing 2D B-mode scanlines, due to the inclusion of neighboring tissue structures within the 3D ultrasound beam profile along the elevational direction, the reconstructed B-mode image may represent traces of neighboring structures of the desired imaging plane, as well. Moreover, as the spatial resolution of ultrasound imaging systems are finite (i.e., finite lateral and elevational thickness of beam profile as well as a finite wavelength), when a target or an organ boundary is scanned, its appearance on the corresponding ultrasound B-mode is typically a blurred replication of its original shape. These issues are mitigated through advanced beam forming approaches, where the spatial and temporal properties of ultrasound beams are controlled to focus ultrasound energy for an enhanced image quality. To increase image resolution at an area of interest particularly during deep tissue imaging, ultrasound beams are usually focused to reduce beam width and slice thickness at depth as shown in Fig. 2b using focusing mechanisms, giving the beams a concave geometry.

3.2 Ultrasound Beam Focusing

Focusing is basically the narrowing of the ultrasound beam profile and concentrating its energy as it propagates axially. This process ensures a maximum energy density delivery to the focal region while limiting ultrasound interactions with fewer surrounding internal structures at a desired focal depth. Focusing reduces the blurriness of a B-mode image at its focal region due to decreased echo interference from neighboring structures, resulting in enhanced image resolution. The ultrasound beam profile can be focused along the lateral and elevational directions. Two primary approaches are usually employed to achieve ultrasound beam focusing: mechanical and electronic.

Mechanical focusing is implemented by the transducer manufacturer via placing a physical lens in front of the transducer matching layer, as shown in Fig. 2a, or by mounting piezoelectric elements in a curved alignment to physically guide ultrasound beams to converge in a particular direction and at a fixed distance (not adjustable) after leaving the transducer. For example, physical lenses are used on some transducers to provide a fixed focus along the elevational direction.

Electronic focusing is achieved by controlling, namely time delaying, the order at which active piezoelectric elements are electrically excited, creating a converging beam where its focus is adjusted via delay adjustment and aperture size. An ultrasound beam profile is formed from the interference of mechanical waves produced from individual elements grouped together, which is called an aperture. If aperture elements are all activated simultaneously with no time (phase) delays, the resulting interference wave front is planar and unfocused, as shown in Fig. 3c, and the corresponding beam profile does not narrow, at least within a normal imaging depth. However, the excitation order of individual elements can be programmed, as shown in Fig. 3d, where the time each element is fired is calculated from applying time delays, such that an element located away from the center line of the aperture is fired earlier. Two elements at equal distance on two sides are fired simultaneously. The central element firing happens with the longest delay. The consequence is a converging wave as shown in Fig. 3e for delayed excitation. It should be noted that these same ideas of time delays can be used to sum signals from echoes to produce receive beamforming, steering and focusing. Beamforming and electronic focusing are major areas of ultrasound imaging covered by a great body of literature (Szabo 2013; Lu et al. 1994; Lediju et al. 2011; Hyun et al. 2019) and further details on these topics are outside the scope of this chapter.

3.3 Interaction of Ultrasound Waves with Tissues

The interaction of ultrasound waves with tissues occurs through several mechanisms which affect how these acoustic waves propagate within tissues and return as echoes to the transducer, allowing for image formation. Important interactions are introduced in the following section. Being familiar with these interactions is essential in further understanding ultrasound image formation.

Reflection and Transmission

When ultrasound waves travel from medium 1 to medium 2 with different acoustic impedances of Z_1 and Z_2, respectively (namely impedance mismatch), a portion of waves transmits into medium 2 and the remainder reflects at the interface and travels back to medium 1. Depending on the acoustic impedances of these two media, the fraction of wave reflection (R) and transmission (T) as well as the relative amplitude (strength) of reflected acoustic pressure wave are determined, as shown in following equations:

$$\text{Reflection Fraction} : R = \frac{Z_2 - Z_1}{Z_2 + Z_1} \tag{4}$$

$$\text{Transmission Fraction} : T = \frac{2Z_2}{Z_2 + Z_1} \tag{5}$$

$$\frac{p_{reflected}}{p_{incident}} = \frac{Z_2 - Z_1}{Z_2 + Z_1} \tag{6}$$

If ultrasound waves interact with a large interface between two media, having a large impedance mismatch such as soft tissues and bone, waves are strongly reflected into soft tissues and only a weak wave can transmit into bone. Therefore, the corresponding B-mode image of such a structure, i.e., the bone surface, appears hyperechoic (very bright and potentially saturated) with respect to soft tissues located at shallower areas. The echogenicity within the bone quickly reduces due to the weak transmission of acoustic waves into bone and large attenuation within the bone (more about that later). Another important consequence of impedance mismatch is challenges when imaging soft tissues through an air-filled spatial gap between the transducer and tissues. As the impedance of the transducer lens and soft tissues both are significantly larger than air, acoustic waves do not properly transmit from the transducer into air, followed by the same event occurring at the air-tissue interface (given some acoustic waves are transmitted into air at the transducer interface), resulting in no effective acoustic wave transmission to make meaningful images of any tissue structure below the air-filled spatial gap.

Scattering and Refraction

As ultrasound propagates through tissues, it also interacts with smaller interfaces/acoustic targets with acoustic mismatches which have sizes smaller than the wavelength of the incident ultrasound wave. In such scenarios, ultrasound waves scatter seemingly randomly in different directions (rather than reflecting along a certain angle). This phenomenon is called ultrasound backscattering and those small targets with different density and compressibility than the background tissues are called scatterers. Such small scatterers give rise to the granular blurred appearance, termed speckle, visible on ultrasound B-mode images. Speckle will be further discussed in the ultrasound Imaging Artifact section.

Another important event during ultrasound propagation is refraction, which determines the direction of wave transmission when arriving at an interface large compared to λ. If an incident ultrasound wave arrives at an interface at an angle, $\theta_{incident}$ (not perpendicular to the interface), the angle along which portions of wave transmits into the second medium, $\theta_{transmit}$, depends on the relative speed of sound in the two media (c_1 and c_2). This is known as Snell's law and determined by Eq. (7). Also, the incident wave angle is equal to the angle of the reflection with respect to the perpendicular line to the interface. It is noted that if the incident ultrasound wave arrives at an interface at a perpendicular angle, its transmitted portion travels into the second medium without any refraction, regardless of the sound speed ratio. These phenomena are illustrated in Fig. 4.

$$\text{Snell's Law}: \frac{sin\,(\theta_{transmit})}{sin\,(\theta_{incident})} = \frac{c_2}{c_1} \tag{7}$$

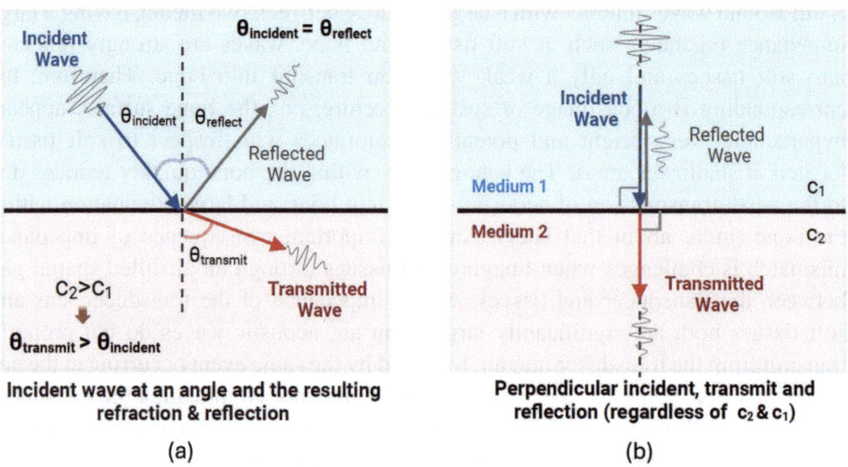

Fig. 4 Interactions of an ultrasound wave at the intersection of two media with different acoustic properties are described for two cases: (**a**) presents the case where the incident wave arrives at the interface at an angle, and (**b**) shows the case for a perpendicular incident wave. The speed of sound is assumed to be higher in the lower medium than the upper medium. Here, the incident wave, as well as transmitted and reflected waves are shown along with their associated orientation with respect to the line perpendicular to the interface of the two media. In (**a**), the angle of incident is equal to the angle of reflection but smaller than the angle of transmitted wave, while in (**b**) for the perpendicular incident wave, the reflected and transmitted waves are also normal to the interface

Absorption and Ultrasound Attenuation

Propagation of an acoustic wave relies on particles in a medium, where particles oscillate back and forth around their equilibrium states while carrying the wave (and its energy) forward along their oscillation direction, as explained earlier in Fig. 1c (longitudinal wave). However, due to the high-frequency (small-wavelength) nature of ultrasound waves, particles fail to transfer along the entire energy content of the wave due to two mechanisms: absorption and scattering. Tissues absorb fractions of the wave energy in the form of their internal viscous heating as well as losing some energy through scattering. This leads to a loss in energy of the ultrasound wave as it propagates.

It is noteworthy to mention that other than these loss mechanisms, an ultrasonic wave will lose energy density through geometrical beam widening, either due to beam refraction or due to divergence after passing the lateral (and/or elevational) focus. This leads to distribution of a particular energy of a wave over a wider spatial volume, resulting in an intensity (energy over area) reduction. In diagnostic ultrasound imaging, the focus is primarily on the first two mechanisms because those are closely related to tissue microstructure. Geometrical loss, on the other hand, is less of a concern due to beam focusing and the relatively small imaging depth.

The intensity (I) of acoustic waves decays exponentially in soft tissues via absorption and scattering, as modelled by Eq. (8). Ultrasound attenuation (α) is the parameter used to measure the rate at which intensity of a wave diminishes with distance (Δx) as it propagates. I_0 is the intensity at the beginning of the distance.

$$\frac{I}{I_0} = e^{-\alpha \Delta x} \tag{8}$$

Ultrasound attenuation is usually expressed in dB/cm as a compressed logarithmic scale over distance. For instance, the rate of attenuation of an ultrasound wave in dB/cm, coming from a target at a depth of d is estimated from the following equations, depending on using intensity (I) or amplitude (A) data ($I \propto A^2$).

$$\alpha_{(dB/cm)} = \frac{-\mathbf{10}\log_{10}\left(\frac{I}{I_0}\right)}{d} \tag{9}$$

$$\alpha_{(dB/cm)} = \frac{-\mathbf{20}\log_{10}\left(\frac{A}{A_0}\right)}{d} \tag{10}$$

A factor of 2 would need to be included in the travel distance if the observed attenuation is in the pulse-echo mode. The definition of attenuation is, however, in distance travelled, i.e., independent of imaging modes. In addition, it is noted that attenuation is also often expressed as $\alpha_{(dB/MHz/cm)}$. The extra division regarding frequency is due to the linear approximation of attenuation increase with frequency. We refer to the definition of conversion into dB provided earlier in this chapter.

3.4 Trade-Off Between Penetration Depth and Image Resolution

In ultrasound imaging, there is a trade-off between frequency (resolution) and imaging depth due to the frequency dependent attenuation mentioned above. At higher frequencies, wavelength becomes shorter, allowing to image (resolve) smaller targets which leads to a higher-resolution imaging (assuming a constant number of excitation cycles, a higher frequency results in a shorter pulse length and a higher resolution). On the other hand, as the frequency increases, absorption, and scattering (and thus, attenuation) will also escalate. This leads to an increased rate of acoustic energy loss with distance, limiting high frequency waves propagating to deeper regions. Therefore, if the ultrasound transmit frequency is increased to enhance image resolution, the effective penetration depth of imaging is decreased accordingly, resulting in loss of deep tissue imaging. The diverse design of transducers is particularly meant to address this limitation.

4 Ultrasound Image Types

Ultrasound imaging has advanced its capabilities in providing diagnostic information beyond what just a gray-scale B-mode image offers for anatomical and morphological imaging. These include but are not limited to techniques for imaging moving structures within body (M-mode), blood flow imaging (color flow, color power, pulse wave (PW) spectra and continuous wave (CW) Doppler), tissue stiffness imaging (elastography) as well as obtaining enhanced B-mode images of difficult-to-scan organs using contrast agents, namely contrast enhanced ultrasound (CEUS) imaging. In the following section elastography, color flow, color power imaging are introduced.

4.1 Elastography

Biomechanical properties of soft tissues, particularly tissue stiffness, may undergo changes in response to certain pathological conditions. Palpation (a manual examination) is a fundamental diagnostic approach practiced in clinics to assess tissue stiffness. It was essentially a qualitative assessment of an associated mechanical biomarker: elasticity.

Tissue elasticity is assessed by measuring Young's modulus (E) or shear modulus (G), where $E = 2G\ (1+\ \nu)$ and ν is Poisson's ratio. These are measures of tissue resistance to normal and shear deformations and allow quantifying tissue deformations in response to two types of forces: normal or shear forces, respectively. These moduli are defined as the ratio of stress to strain (that is, the associated force per unit area divided by corresponding normalized deformation/distortion). While strain is dimensionless, Young's and shear moduli as well as stress are expressed with the unit of pressure, *Pa*.

To assess tissue elasticity quantitatively, in particular for inaccessible tissues in deeper organs, various measures of tissue elastic properties using ultrasound have evolved including sonoelasticity (Parker et al. 1990), strain elastography (Ophir et al. 1991; Doyley et al. 2005), shear wave elasticity imaging (Sarvazyan et al. 1998), and acoustic radiation force impulse imaging (Nightingale et al. 2002). Ultrasound elastography (imaging tissue stiffness) relies on applying controlled mechanical forces to tissues of interest and tracking/measuring resultant tissue responses (displacement or rate of displacement) using an ultrasound system and assess strain or moduli (Gennisson et al. 2013). Such responses provide a measure of tissue stiffness, as stiffer tissues such as a hard lesion usually undergo smaller particle displacements after applying a mechanical force than normal soft tissues normally do. Examples of stiffness measurements for tissue characterization based on ultrasound include those characterizing liver (Barr et al. 2015; Palmeri et al. 2008), prostate (Hoyt et al. 2008; Zhang et al. 2008), brain (Tzschätzsch et al. 2018) and muscles (Eby et al. 2013).

In elastography, there are two primary approaches to excite mechanical forces in tissues:

- Semi-Static methods:

In this approach, tissues are compressed by applying a constant external force, and the corresponding particle displacements as compression is applied, are measured using correlation-based (speckle) tracking algorithms. Particle displacement is expressed as strain which is inversely proportional to Young's modulus. As the quantification of stress in this approach is not practically measured, the strain maps, known as elastograms, are reported as a surrogate for stiffness.

- Dynamic methods:

This approach provides the estimation of shear modulus, an advantage over semi-static methods, by exciting a shear wave (perturbation) within tissues, tracking particle displacement using ultrasound and estimating shear wave propagation speed (c_s). Propagation speed is correlated with shear modulus (G) through density (ρ) interplay according to

$$c_s = \sqrt{G/\rho} \tag{11}$$

The range of *shear* wave speed in soft tissues is typically 1–10 m/s, which is orders of magnitude smaller than propagation speed of the *compressional* wave (~1540 m/s). Here dynamic refers to the time-varying nature of applied force, either transient or steady state, employed in this elastography approach, as opposed to the constant force of semi-static methods. Using an external source such as an actuator

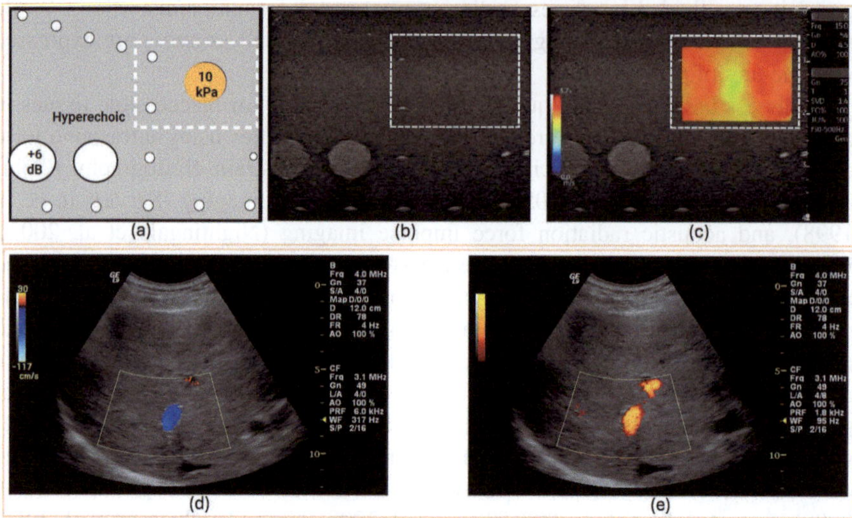

Fig. 5 (**a**)–(**c**) Shear wave speed imaging of a tissue-mimicking phantom: (**a**) shows a phantom incorporating an elasticity target that is softer than the background (phantom model: GSE-040, CIRS Inc., Danville, VA, USA). A rectangular region of interest is also defined on the phantom using white dashed lines; in (**b**), the gray-scale B-mode image of the phantom in (**a**) is shown. The elasticity target is not distinguishable on the B-mode image; (**c**) shows shear wave speed imaging of the phantom. In the region of interest, the target visibly appears softer than its immediate surroundings. Other than the elasticity target, the phantom incorporates various B-mode test objects, which are visible on B-mode imaging; (**d**) and (**e**) are blood flow imaging of a human liver in vivo: (**d**) color flow velocity imaging, and (**e**) color flow power imaging for the liver case shown in Fig. 1a (Panels d and e are courtesy of Dr. Jonathan M. Rubin and Dr. Stephen Z. Pinter, University of Michigan, Ann Arbor, MI, USA)

on the skin or using an acoustic radiation force pulse by the transducer (Nightingale et al. 2002), local perturbations are created in tissues and propagate as shear waves. Figure 5a shows an example of a phantom with an elasticity target that is softer than the background. While the elasticity target does not appear on the B-mode image of the phantom in Fig. 5b, the shear wave speed map in Fig. 5c allows distinguishing it using a region having diminished shear wave speed.

4.2 Blood Flow Imaging

Ultrasound based blood flow imaging provides diagnostic information about blood flow patterns within arteries and veins. Blood contains moving red blood cells that scatter ultrasound and allow for the quantification and visualization of blood flow in vivo through tracking movements of these red blood cells. Ultrasound offers various imaging modes to quantify and/or visualize blood flow (Adler et al. 1990; Rubin et al. 1994; Jensen 1996; Song et al. 2017). Color flow velocity imaging visualizes information about mean velocity and direction of blood flow within lumens with respect to the transducer using a color-coded image (e.g., red-and-blue) superimposed on the B-mode. In color flow velocity images, red and blue indicate blood flowing towards and away from the transducer, respectively while shades of each color represent magnitude of the mean velocity in each pixel. For instance, a region displaying bright red color indicates the presence of blood flow moving towards the transducer with a high mean velocity. An example of color flow velocity image of a liver subject (Pinter et al. 2015) is shown in Fig. 5d.

Color flow power imaging is another Ultrasound-based mode for blood flow quantification which visualizes the power (square of amplitude) of color signals from moving blood cells instead of the velocity using a color-coded image overlaid on B-mode (Rubin et al. 1994). As a single-color echo received at the transducer reflects contributions of many blood cells scattering waves, the strength of backscattered echoes corresponds to the number of red blood cells within the ultrasound beam. A brighter shade on a color power image indicates higher amplitudes and higher red blood cell densities. This mode is particularly helpful in imaging blood flow in small vessels or visualizing cases with small velocities. In Fig. 5e, a sample of color flow power in liver imaging is presented.

Ultrasound systems implement two physical principles in blood flow quantifications to track moving red blood cells: Doppler effect and signal phase shift. When an ultrasound wave interacts with a moving object such as red blood cells, such motions change the frequency of received sound waves at the transducer, which is known as Doppler frequency shift. Additionally, when a moving object interacts with consecutive equally interspaced ultrasound pulses transmitted by a transducer, there will be phase changes among consecutive backscattered echoes because of this relative motion. However, for stationary targets within tissues, such interactions do not cause any frequency or phase shifts. Blood flow measurements with ultrasound works based on quantification of frequency shift or phase shift. Phase shift occurs independent of a Doppler frequency shift and is assessed from signal processing using consecutive transmits/echoes. The Doppler frequency shift Δf of an ultrasound wave is proportional to the speed of red blood cells (v) according to the Eq. (12). By measuring the Doppler frequency shift, blood flow velocity is estimated with respect to time.

$$\Delta f = \frac{2 f_0 v \cos(\theta)}{c} \tag{12}$$

In this equation, c is the speed of sound in the tissue, f_0 is the transmit frequency of ultrasound wave, and θ is the angle between the ultrasound wave propagation and direction of moving object. The term $v\cos(\theta)$ in this equation indicates that for the moving object only the velocity component along the direction of wave direction contributes to the Doppler frequency shift. Moreover, ultrasound systems use two kinds of waves in blood flow quantification: (i) PW for range-gated modes with spatial resolution such as PW spectra (not covered here), color flow velocity and color flow power as well as (ii) CW, without spatial resolution. In range-gated modes short intermittent ultrasound pulses are transmitted by the transducer into the tissue rather than a CW beam. The transducer receives echoes from a specific depth, defined by the range gate, rather than the entire beam path; these echoes are analyzed for their phase. In CW mode, the Doppler frequency shift is assessed using a CW transmit of ultrasound waves at a constant frequency, simultaneously receiving the frequency shifted echoes and estimating the frequency difference between transmitted and received waves from the entire beam depth. CW Doppler imaging does not provide an anatomical image and has limited to no depth information, but is particularly useful for tracking high-velocity flows.

5 Basic Imaging Guide

Figure 6 shows a commercialized ultrasound scanner (ZS3, Mindray Inc., San Jose, CA, USA). A zoomed-in view of the scanner's user interface is shown along with an annotated B-mode image on the scanner display of a tissue-mimicking phantom (CIRS Inc., Danville, VA, USA) obtained by a linear array transducer (L30-8, ZS3, Mindray Inc.). This interface allows adjusting various imaging parameters during imaging such as ultrasound mode, frequency, FOV depth and type, transducer type.

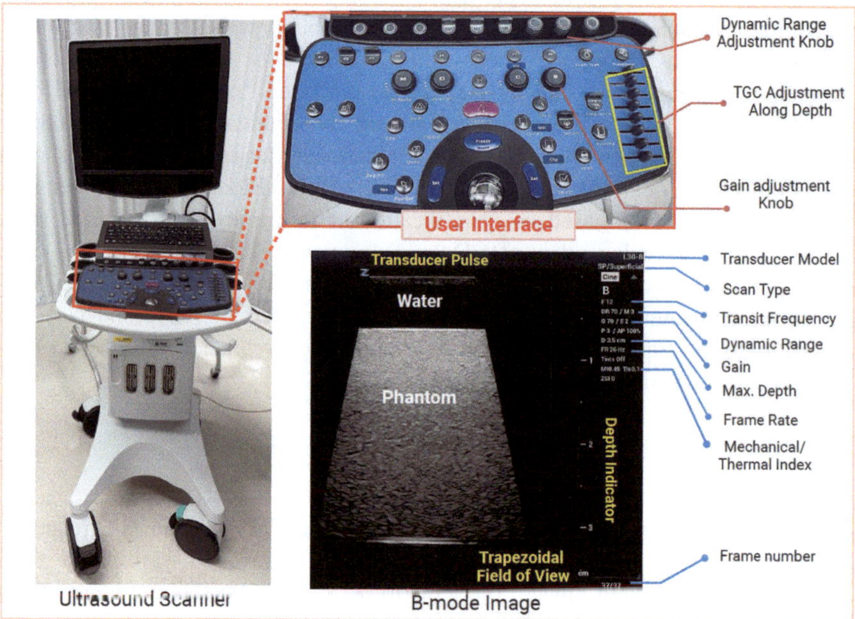

Fig. 6 An example of a commercial ultrasound scanner is shown on the left. A zoomed-in view of the scanner user interface is shown separately on the top corner where two knobs for the gain and dynamic range (DR) adjustments as well as sliders for tuning the TGC are annotated. A sample B-mode image of a phantom submerged in water acquired with this scanner is also presented. On this B-mode, common parameters that are usually present on an ultrasound image display are introduced

In particular, three important setups are also included in ultrasound scanners to allow control of the brightness and contrast of an ultrasound image in real time: time gain compensation (TGC), gain and dynamic range (DR).

- TGC Adjustment: Due to the attenuation of an ultrasound signal as it propagates through tissues, amplitudes of returning echoes from deeper regions become weaker with respect to the depth when received at the transducer. A B-mode image reconstructed from such echoes appears hypoechoic (darker) at deeper regions, deteriorating quality of the image and a loss in ultrasound diagnostic capabilities at deep regions. To compensate for this natural phenomenon, echo amplitudes are amplified proportional to their associated depth via a set of sliders on ultrasound scanners, echoes from deeper regions are amplified more.
- Gain Adjustment: This knob controls the brightness of the entire image by amplifying the received echo. When the overall B-mode image is overly dark or bright, gain adjustment allows for correcting it.
- DR Adjustment: This knob adjusts contrast or number of gray levels used to display an ultrasound image. A higher DR allows displaying subtle details (smaller echoes), leading to a smoother transition between gray levels while a lower DR enhances image contrast by compromising (suppressing) finer details. DR is often expressed in dB units.

6 Ultrasound Imaging Artifacts

For improving an accurate and reliable diagnosis using ultrasound imaging, it is crucial to recognize its common artifacts. Ultrasound artifacts are features that sometimes appear on an ultrasound image that do not correspond to an actual anatomical structure within the tissue being scanned. Artifacts occur due to deviations from underlying assumptions behind implementing an ultrasound imaging system. In the following sections, common artifacts are introduced.

6.1 Mirror Effect

The mirror-image artifact is basically the mirrored (duplicate) appearance of a structure in an ultrasound image at another location other than its original position. This artifact arises from the underlying assumption in the imaging system that sound waves travel directly to and from tissue structures with no accounting of possible multiple reflections back and forth between two structures. To delve more into the occurrence of this artifact from the above assumption, imagine a strong reflecting surface such as diaphragm close to another existing object such a hyperechoic lesion both located within a FOV of an imaging system, as illustrated in Fig. 7a. The imaging system reflects the image of the actual object on B-mode. Additionally, when the ultrasound beam encounters the diaphragm surface at an angle other than 90°, the reflected beam from the diaphragm may travel toward the object instead of

(a)

(b)

(c)

(d)

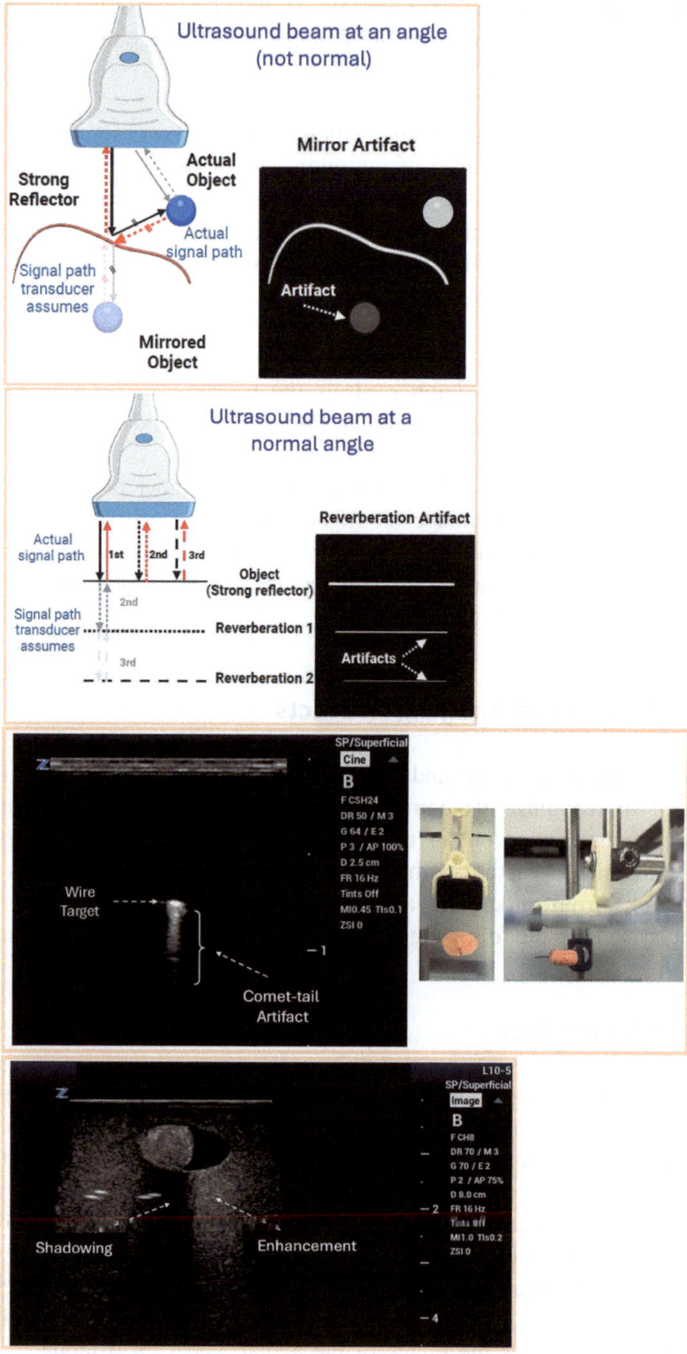

Fig. 7 Five different ultrasound imaging artifacts are presented here. (**a**) shows an illustration of the mirror artifact for a spherical object located close to a strong reflector. The transducer pulse

back to the transducer. This is followed by a second reflection of the beam from the object back to the diaphragm surface and finally, toward the transducer. The imaging system interprets such a delayed echo from the surface as the presence of a lesion structure located at a deeper region along the incident beam at the opposite side of the actual lesion with respect to the reflecting surface. This duplicate mirrored structure does not represent an actual structure beneath the surface and is considered as an artifact. The mirrored object normally appears less echogenic than the actual image of the object.

6.2 Reverberation

The reverberation artifact occurs when sound waves bounce back and forth multiple times between two strong reflecting surfaces parallel to each other and perpendicular to the beam direction as shown in Fig. 7b. This phenomenon results in the appearance of a few false structures in an ultrasound image (equal to the number of back-and-forth reflections) where each reflection artifact is located in equidistance from the next one while they do not represent any corresponding physical structures. Common cases of observing a reverberation artifact in clinical ultrasound imaging are when the transducer surface is aligned parallel to a bone surface or an injected needle or other strong reflecting cases with high impedance mismatch with their surrounding soft tissues such as air.

A special case of reverberation artifact is known as the comet-tail artifact, shown in Fig. 7c. When ultrasound waves are trapped within a small and highly reflective object such as a fluid filled cyst or a highly reflecting needle, repeating back-and-forth reflections occur between internal surfaces. In each of the repeating reflections, a portion of reflecting sound waves escapes back to the transducer while the rest continues traveling back and forth within the object surfaces. The wave received at the transducer weakens after each reflection due to this split of sound energy in leakage and reflection. The imaging system interprets these repeating, delayed and equally spaced reflections as if they are coming from distinct structures at deeper

Fig. 7 (continued) arrives at an angle with respect to the strong reflector. The associated ultrasound image of this structure shows a duplicate (artifact) image of the sphere; in (**b**), an illustration of reverberation artifact is depicted where a strong reflecting surface is located parallel with a transducer surface. Here, the corresponding B-mode illustration incorporates multiple images of one single surface that are being imaged due to multiple reflections of the pulse between the transducer surface and the reflector. The reverberation image decays with depth; (**c**) shows a comet-tail artifact in a B-mode image of a highly reflecting small wire target using a linear array transducer. The two orthogonal views of the imaging setup for this acquisition are also shown in (**c**); in (**d**), the shadowing and enhancement artifacts are presented in a B-mode image of a tissue-mimicking phantom with inclusions. Although the region under these inclusions is homogeneous, these two artifacts resulted in the appearance of two distant (different) regions in the associated B-mode which are considered artifacts tissue regions. This train of reverberations decaying with depth on an ultrasound image produces a comet artifact.

6.3 Shadowing and Enhancement

Acoustic shadowing artifact occurs when a strong reflector or a highly attenuating region within a tissue blocks ultrasound waves from transmitting into deeper region beneath it due to reflecting/absorbing almost the entire ultrasound beam it receives. In other words, the highly reflecting/attenuating object creates a hypo-echoic region below it, imposing a "shadowing" artifact at deeper regions. This phenomenon, shown in Fig. 7d, results in loss of information about distal tissue structures and is considered as an artifact. Bones or calcified regions or air pockets are examples of strong reflectors that could produce this artifact.

The enhancement artifact represents the opposite effect of the shadowing artifact. It is observed when ultrasound waves encounter a low-attenuating region compared to its neighboring regions (left/right neighbors), as shown in Fig. 7d. This results in an increased brightness (hyper-echoic or enhanced appearance) in deeper regions beneath the low-attenuating region in contrast with the neighboring tissue regions which does not necessarily indicate the presence of a hyperechoic region. A fluid-filled cyst or bladder are examples of low-attenuating regions that could create this enhancement artifact.

6.4 Speckle Noise

During ultrasound imaging, when the wave encounters scatterers that are smaller than its characteristic wavelength, λ, ultrasound waves scatter omnidirectionally. Scattered waves from neighboring scatterers interfere with each other, resulting in constructive (stronger) or destructive (weaker) wave interference that ultimately travels back to the transducer. The imaging system cannot distinguish these individual small scatterers from within the backscattered waves, but instead yields a blurred and seemingly random image corresponding to this cloud of scatterers. Such granular blurred patterns (namely speckle patterns) in an ultrasound image are caused by these coherent wave interferences received from tissue microstructures. Speckle degrades image quality by reducing spatial resolution and contrast, considered as an artifact called the speckle noise. This phenomenon is best observed in uniform tissue regions with similar properties with many random scatterers. To reduce speckle noise from ultrasound images for an enhanced image quality and contrast, various techniques have been established over the years and are implemented within ultrasound scanners including post-processing filtering and spatial or frequency compounding (where multiple data are acquired and combined together). Speckle reduction is an advanced ultrasound research area with a substantial body of literature, and we refer to references (Mateo and Fernández-Caballero 2009; Traney et al. 1986; Michailovich and Tannenbaum 2006; Parker 2012) for further reading.

It is noteworthy that although ultrasound speckles are a type of artifact, it usually incorporates statistical information about the microstructure of tissues that could offer diagnostic values for tissue characterization. Speckle modelling is another active area of ultrasound research (Wagner et al. 1983; Mamou and Oelze 2013; Cloutier et al. 2021; Christensen et al. 2024; Parker et al. 2020).

7 Safety Considerations in Ultrasound Imaging

Ultrasound imaging is considered as one of the safest diagnostic imaging modalities due to being a non-invasive method that does not involve any ionizing radiation exposure to patients unlike X-rays or CT scans. In addition to its diagnostic applications, ultrasound offers extensive therapeutic potential where acoustic waves are employed to target energy of waves to a particular region within tissues for treatment purposes. Important therapeutic applications include, but are not limited to, thermal ablation (Madersbacher et al. 1994; Wu et al. 2001); histotripsy, also known as non-invasive surgery for tumor destruction at the cellular level (Xu et al. 2004, 2021); acoustic droplet vaporization (Kripfgans et al. 2000; Zhang et al. 2011; Aliabouzar et al. 2023); and ultrasound -induced blood-brain-barrier opening (Konofagou 2012; Chen and Konofagou 2014), to name a few. Nonetheless, as ultrasonography fundamentally relies on transmitting energy in the form of high-frequency acoustic waves into patients, there are certain safety considerations to prevent any potential negative biological damage to patients at the cellular or structural levels. Biological effects include acoustic cavitation (mechanical effects) and heating within tissues (thermal effects). To monitor the safety of ultrasound, two metrics were introduced, the mechanical index (MI) and the thermal index (TI), along with observing the ultrasound exposure time, also known as dwell time (Fowlkes 2008; Dalecki 2004). MI and TI are discussed in the following section. Their display on ultrasound imaging systems is mandatory depending on the regulatory approval process for the scanner model. During an ultrasound scan, modifying the output power can directly change the values of MI and TI.

7.1 Mechanical Index (MI)

When an ultrasound wave, as an oscillating longitudinal pressure wave with alternating high- and low-pressure fronts shown in Fig. 1c, interacts with gas bubbles in a liquid within the body, it stimulates bubble oscillation with considerable size changes. Gas bubbles, depending on their size and frequency of the ultrasound wave, can undergo rapid size change. This process is driven during the periodic interaction of low-pressure (rarefactional) phases of ultrasound waves and high-pressure (compressional) phases. In this process, bubbles can experience rapid expansions and compressions, which can lead to a state in which the inertia of the fluid around them suddenly drives their collapse against their internal pressure. Such a large imbalance compresses bubbles to an unstable state, and strong collapses.

This phenomenon is termed inertial cavitation. It releases significant localized energy to the surrounding tissue, potentially leading to lasting physical damage to the surrounding tissue. The MI reflects the possibility of physical damages to the tissue because of inertial cavitation forces and is defined using Eq. (13).

$$MI = \frac{P^{-}_{r\,0.3}}{\sqrt{f}} \tag{13}$$

Higher MI indicates higher possibility of cavitation. Note that here, f is the frequency of ultrasound wave in MHz and $P^{-}_{r\,0.3}$, the peak negative pressure in MPa, is the peak amplitude of the ultrasound wave in its low-pressure front derated with an attenuation factor of 0.3 dB/cm/MHz. The choice of $P^{-}_{r\,0.3}$ for MI assessment arises from higher possibility of bubble collapse following bubble expansion in response to lower external pressures. At lower frequencies or lower peak negative pressure, MI and consequently the possibility of cavitation becomes elevated.

The presence of gas bubbles in organs is possible through different mechanisms. They could exist (i) naturally, such as the air trapped in lungs and gas in the intestine, or (ii) deliberately in the form of contrast agents injected for facilitating ultrasound imaging or therapeutic applications. Finally, it could be generated through the cavitation mechanism itself, where gas-filled bubbles form as a result of ultrasound wave interaction with liquids. When the liquid pressure locally drops below its vapor pressure, it leads to vaporization (bubble generation). For safe diagnostic ultrasound imaging, the FDA has approved $MI < 1.9$ as the safe MI range to minimize the chance of any lasting negative mechanical bioeffects on patients from cavitation (Fowlkes 2008).

7.2 Thermal Index (TI)

As an ultrasound wave travels through tissues, some fraction of the acoustic energy it carries is absorbed by tissue and converted into thermal energy. This process leads to a rise in local tissue temperature, particularly with prolonged exposure of tissue to ultrasound during imaging. Such a temperature rise could be negligible or could result in permanent tissue damage. Thermal effects depend on the temperature increase, the duration over which the elevated temperature is maintained, and the sensitivity of tissue being scanned. For instance, bone is considered as a highly absorbing medium and its presence within the FOV results in a higher and faster temperature rise compared to cases without bone. TI, as a dimensionless index obtained from Eq. (14), correlates with the maximum temperature rise somewhere within the scanned area (it is not an actual temperature increase). It reflects the possibility of a 1 °C temperature rise.

$$\text{Thermal Index}: TI = \frac{W}{W_{1\,^{\circ}C}} \tag{14}$$

W is the ultrasound beam power within the tissue while $W_{1\,°C}$ is the required power to induce a maximum temperature rise of 1 °C along the beam under the same imaging setup.

For safety considerations associated with thermal bioeffects, the Bioeffects Committee within the American Institute of Ultrasound in Medicine (AIUM) in an overarching report (Fowlkes 2008) has recommended the joint use of the TI and the dwell time to assess the potential thermal risk. In this report, the maximum safe exposure time is outlined for various levels of temperature increase above the normal physiological temperature (i.e., 37 °C). For example, for a temperature rise of 3.0 °C with the ultrasound exposure duration of 1 h, no significant negative thermal effects are imposed while for an increase of 6.0 °C only allows for an exposure time of 60 s. In conclusion, it is crucial to follow specific MI and TI safety guidelines to ensure patient safety in conducting a clinical ultrasound imaging.

8 Conclusions and Future Perspectives

Ultrasound imaging is a non-invasive, non-ionizing, portable and cost-effective modality that could provide images of internal structures of tissues in real time. Ultrasound also offers advanced imaging mode for diagnosis such as blood flow and tissue stiffness imaging. Ultrasound imaging offers both diagnostic and therapeutic capabilities in clinics which can be performed simultaneously, i.e., tracking the ultrasound-based therapeutic procedure with ultrasound imaging modes.

There are various potential research areas related to ultrasound imaging that offer promise for increasing clinical applications of this non-invasive imaging modality for improving diagnosis. Future research may focus on developing novel ultrasound biomarkers for specific diseases and further enhancing the current biomarkers. Some of more recent research areas include artificial intelligence in ultrasound imaging (Brattain et al. 2018) for image segmentation and disease classification (Amiri et al. 2020; Baek et al. 2021), wearable ultrasound devices (Wang et al. 2022; Lin et al. 2024) for monitoring a condition over an extended time as well as research areas involving quantitative ultrasound for tissue characterization (Cloutier et al. 2021; Poul et al. 2025), advanced beamforming techniques (Lediju et al. 2011; Wiacek et al. 2020) for improving resolution of ultrasound image,, to name a few. Other than diagnostics ultrasound, advanced research on therapeutic ultrasound offers promising clinical potential in areas such as histotripsy for tumor destructions (Xu et al. 2004, 2021), low-intensity ultrasound neuromodulation (Fomenko et al. 2018) to stimulate certain neural activities in the brain (Tufail et al. 2011) and blood-brain barrier opening for the drug delivery (Konofagou 2012; Chen and Konofagou 2014).

With advanced innovations in software and hardware technologies such as system-on-chip, ultrasound imaging systems are becoming smaller and faster which could allow increased access and facilitate more frequent use of this powerful imaging modality. Limitations of this imaging modality are the trade-off between the resolution and depth of imaging as well as being operator dependent.

In this chapter, we have covered fundamental principles of ultrasound imaging to equip readers with essential concepts and definitions to understand how this non-invasive powerful imaging modality works. As advanced innovations in software and hardware designs of the ultrasound imaging system are made, these foundations will still remain at the core of better understanding these advancements.

Acknowledgments Daria Poul is supported by the National Institutes of Health under Award Number F32DE034986. The authors express their gratitude to Drs. Stephen Pinter and Jonathan Rubin for sharing the ultrasound images of the liver from their prior study.

Conflict of Interest The authors declare that they do not have any conflict of interest associated with this work.

References

Adler DD, Carson PL, Rubin JM, Quinn-Reid D (1990) Doppler ultrasound color flow imaging in the study of breast cancer: preliminary findings. Ultrasound Med Biol 16(6):553–559

Aliabouzar M, Kripfgans OD, Fowlkes JB, Fabiilli ML (2023) Bubble nucleation and dynamics in acoustic droplet vaporization: a review of concepts, applications, and new directions. Z Med Phys 33(3):387–406

Alvarez-Arenas TG (2004) Acoustic impedance matching of piezoelectric transducers to the air. IEEE Trans Ultrason Ferroelectr Freq Control 51(5):624–633

Amiri M, Brooks R, Behboodi B, Rivaz H (2020) Two-stage ultrasound image segmentation using U-net and test time augmentation. Int J Comput Assist Radiol Surg 15(6):981–988

Baek J, Poul SS, Basavarajappa L, Reddy S, Tai H, Hoyt K, Parker KJ (2021) Clusters of ultrasound scattering parameters for the classification of steatotic and normal livers. Ultrasound Med Biol 47(10):3014–3027

Barr RG, Ferraioli G, Palmeri ML, Goodman ZD, Garcia-Tsao G, Rubin J, Garra B, Myers RP, Wilson SR, Rubens D (2015) Elastography assessment of liver fibrosis: society of radiologists in ultrasound consensus conference statement. Radiology 276(3):845–861

Brattain LJ, Telfer BA, Dhyani M, Grajo JR, Samir AE (2018) Machine learning for medical ultrasound: status, methods, and future opportunities. Abdominal Radiol 43:786–799

Chen H, Konofagou EE (2014) The size of blood–brain barrier opening induced by focused ultrasound is dictated by the acoustic pressure. J Cereb Blood Flow Metab 34(7):1197–1204

Christensen AM, Rosado-Mendez IM, Hall TJ (2024) A systematized review of quantitative ultrasound based on first-order speckle statistics. IEEE Trans Ultrason Ferroelectr Freq Control 71:872–886

Cloutier G, Destrempes F, Yu F, Tang A (2021) Quantitative ultrasound imaging of soft biological tissues: a primer for radiologists and medical physicists. Insights Imaging 12:1–20

Dalecki D (2004) Mechanical bioeffects of ultrasound. Annu Rev Biomed Eng 6(1):229–248

Doyley MM, Srinivasan S, Pendergrass SA, Wu Z, Ophir J (2005) Comparative evaluation of strain-based and model-based modulus elastography. Ultrasound Med Biol 31(6):787–802

Eby SF, Song P, Chen S, Chen Q, Greenleaf JF, An K-N (2013) Validation of shear wave elastography in skeletal muscle. J Biomech 46(14):2381–2387

Fomenko A, Neudorfer C, Dallapiazza RF, Kalia SK, Lozano AM (2018) Low-intensity ultrasound neuromodulation: An overview of mechanisms and emerging human applications. Brain Stimul 11(6):1209–1217

Fowlkes JB (2008) American Institute of Ultrasound in Medicine consensus report on potential bioeffects of diagnostic ultrasound: executive summary. Journal Ultrasound Med 27(4):503–515

Gennisson J-L, Deffieux T, Fink M, Tanter M (2013) Ultrasound elastography: principles and techniques. Diagn Interv Imaging 94(5):487–495

Hoyt K, Castaneda B, Zhang M, Nigwekar P, di Sant'Agnese PA, Joseph JV, Strang J, Rubens DJ, Parker KJ (2008) Tissue elasticity properties as biomarkers for prostate cancer. Cancer Biomark 4(4–5):213–225

Hyun D, Brickson LL, Looby KT, Dahl JJ (2019) Beamforming and speckle reduction using neural networks. IEEE Trans Ultrason Ferroelectr Freq Control 66(5):898–910

Jensen JA (1996) Estimation of blood velocities using ultrasound: a signal processing approach. Cambridge University Press

Konofagou EE (2012) Optimization of the ultrasound-induced blood-brain barrier opening. Theranostics 2(12):1223–1237

Kripfgans OD, Fowlkes JB, Miller DL, Eldevik OP, Carson PL (2000) Acoustic droplet vaporization for therapeutic and diagnostic applications. Ultrasound Med Biol 26(7):1177–1189

Laugier P, Haïat G (2011) Introduction to the physics of ultrasound. In: Bone quantitative ultrasound. Springer Netherlands, pp 29–45

Lediju MA, Trahey GE, Byram BC, Dahl JJ (2011) Short-lag spatial coherence of backscattered echoes: imaging characteristics. IEEE Trans Ultrason Ferroelectr Freq Control 58(7):1377–1388

Lin M, Zhang Z, Gao X, Bian Y, Wu RS, Park G, Lou Z, Zhang Z, Xu X, Chen X, Kang A, Yang X, Yue W, Yin L, Wang C, Qi B, Zhou S, Hu H, Huang H, Li M, Gu Y, Mu J, Yang A, Yaghi A, Chen Y, Lei Y, Lu C, Wang R, Wang J, Xiang S, Kistler EB, Vasconcelos N, Xu S (2024) A fully integrated wearable ultrasound system to monitor deep tissues in moving subjects. Nat Biotechnol 42(3):448–457

Lu J-Y, Zou H, Greenleaf JF (1994) Biomedical ultrasound beam forming. Ultrasound Med Biol 20(5):403–428

Madersbacher S, Kratzik C, Susani M, Marberger M (1994) Tissue ablation in benign prostatic hyperplasia with high intensity focused ultrasound. J Urol 152(6):1956–1960

Mamou J, Oelze ML (2013) Quantitative ultrasound in soft tissues. Springer

Mateo JL, Fernández-Caballero A (2009) Finding out general tendencies in speckle noise reduction in ultrasound images. Expert Syst Appl 36(4):7786–7797

Michailovich OV, Tannenbaum A (2006) Despeckling of medical ultrasound images. IEEE Trans Ultrason Ferroelectr Freq Control 53(1):64–78

Nightingale K, Soo MS, Nightingale R, Trahey G (2002) Acoustic radiation force impulse imaging: in vivo demonstration of clinical feasibility. Ultrasound Med Biol 28(2):227–235

Ophir J, Cespedes I, Ponnekanti H, Yazdi Y, Li X (1991) Elastography: a quantitative method for imaging the elasticity of biological tissues. Ultrason Imaging 13(2):111–134

Palmeri ML, Wang MH, Dahl JJ, Frinkley KD, Nightingale KR (2008) Quantifying hepatic shear modulus in vivo using acoustic radiation force. Ultrasound Med Biol 34(4):546–558

Parker KJ (2012) Superresolution imaging of scatterers in ultrasound B-scan imaging. J Acoust Soc Am 131(6):4680–4689

Parker K, Huang S, Musulin R, Lerner R (1990) Tissue response to mechanical vibrations for "sonoelasticity imaging". Ultrasound Med Biol 16(3):241–246

Parker KJ, Poul SS, Burr L (2020) Pareto, and logistic distributions from ultrasound speckle. Ultrason Imaging 42(4–5):203–212

Pinter SZ, Rubin JM, Kripfgans OD, Novelli PM, Vargas-Vila M, Hall AL, Fowlkes JB (2015) Volumetric blood flow in transjugular intrahepatic portosystemic shunt revision using 3-dimensional Doppler sonography. J Ultrasound Med 34(2):257–266

Poul D, Samal A, Betancourt AR, Quesada C, Chan H-L, Kripfgans OD (2025) Quantitative ultrasound for periodontal soft tissue characterization. Ultrasound Med Biol 51(2):288–301

Rubin JM, Bude RO, Carson PL, Bree RL, Adler RS (1994) Power Doppler US: a potentially useful alternative to mean frequency-based color Doppler US. Radiology 190(3):853–856

Sarvazyan AP, Rudenko OV, Swanson SD, Fowlkes JB, Emelianov SY (1998) Shear wave elasticity imaging: a new ultrasonic technology of medical diagnostics. Ultrasound Med Biol 24(9):1419–1435

Song P, Trzasko JD, Manduca A, Huang R, Kadirvel R, Kallmes DF, Chen S (2017) Improved super-resolution ultrasound microvessel imaging with spatiotemporal nonlocal means filtering and bipartite graph-based microbubble tracking. IEEE Trans Ultrason Ferroelectr Freq Control 65(2):149–167

Szabo TL (2013) Diagnostic ultrasound imaging: inside out. Academic

Traney G, Allison J, Smith S, Von Ramm O (1986) A quantitative approach to speckle reduction via frequency compounding. Ultrason Imaging 8(3):151–164

Tufail Y, Yoshihiro A, Pati S, Li MM, Tyler WJ (2011) Ultrasonic neuromodulation by brain stimulation with transcranial ultrasound. Nat Protoc 6(9):1453–1470

Tzschätzsch H, Kreft B, Schrank F, Bergs J, Braun J, Sack I (2018) In vivo time-harmonic ultrasound elastography of the human brain detects acute cerebral stiffness changes induced by intracranial pressure variations. Sci Rep 8(1):17888

Wagner RF, Smith SW, Sandrik JM, Lopez H (1983) Statistics of speckle in ultrasound B-scans. IEEE Trans Sonics Ultrason 30(3):156–163

Wang C, Chen X, Wang L, Makihata M, Liu HC, Zhou T, Zhao X (2022) Bioadhesive ultrasound for long-term continuous imaging of diverse organs. Science 377(6605):517–523

Wiacek A, Gonzalez E, Bell MAL (2020) CohereNet: a deep learning architecture for ultrasound spatial correlation estimation and coherence-based beamforming. IEEE Trans Ultrason Ferroelectr Freq Control 67(12):2574–2583

Wu F, Chen W-Z, Bai J, Zou J-Z, Wang Z-L, Zhu H, Wang Z-B (2001) Pathological changes in human malignant carcinoma treated with high-intensity focused ultrasound. Ultrasound Med Biol 27(8):1099–1106

Xu Z, Ludomirsky A, Eun LY, Hall TL, Tran BC, Fowlkes JB, Cain CA (2004) Controlled ultrasound tissue erosion. IEEE Trans Ultrason Ferroelectr Freq Control 51(6):726–736

Xu Z, Hall TL, Vlaisavljevich E, Lee FT Jr (2021) Histotripsy: the first noninvasive, non-ionizing, non-thermal ablation technique based on ultrasound. Int J Hyperth 38(1):561–575

Zhang M, Nigwekar P, Castaneda B, Hoyt K, Joseph JV, di Sant'Agnese A, Messing EM, Strang JG, Rubens DJ, Parker KJ (2008) Quantitative characterization of viscoelastic properties of human prostate correlated with histology. Ultrasound Med Biol 34(7):1033–1042

Zhang M, Fabiilli ML, Haworth KJ, Padilla F, Swanson SD, Kripfgans OD, Carson PL, Fowlkes JB (2011) Acoustic droplet vaporization for enhancement of thermal ablation by high intensity focused ultrasound. Acad Radiol 18(9):1123–1132

Fundamentals and Applications of Micro-computed Tomography

Mert Ocak and Cumali Çatak

Abstract

Micro-computed tomography (micro-CT) has emerged as a transformative imaging technology in dentistry and biomedical research, offering non-invasive, high-resolution, three-dimensional visualization of dental and bone structures. This chapter explores the fundamental principles of micro-CT, its diverse applications in dentistry (including endodontics, restorative dentistry, implantology, orthodontics, and periodontology), and its role in bone tissue analysis, biomaterials research, and cancer metastasis evaluation. The integration of artificial intelligence (AI) and machine learning with micro-CT data enhances automated segmentation, tissue classification, and predictive analysis, while multimodality approaches combining micro-CT with techniques like optical coherence tomography (OCT) and magnetic resonance imaging (MRI) provide comprehensive structural and functional insights. Despite limitations such as beam hardening artifacts and small sample size requirements, advancements in nano-CT and low-dose systems are expanding the clinical potential of micro-CT. Future directions include the development of standardized protocols, real-time 3D imaging during surgical procedures, and the integration of micro-CT into

M. Ocak (✉)
Faculty of Dentistry, Department of Basic Medical Sciences, Anatomy, Ankara University, Ankara, Türkiye
e-mail: mert.ocak@ankara.edu.tr

C. Çatak
Department of Forensic Anthropology, Graduate School of Health Sciences, Ankara University, Ankara, Türkiye

Department of Forensic Anthropology, Institute of Forensic Sciences, Ankara University, Ankara, Türkiye

I. F. Cengiz et al. (eds.), *Bioimaging Modalities in Bioengineering*,
Biomaterials, Bioengineering and Sustainability 9,
https://doi.org/10.1007/978-3-031-99410-4_9

clinical decision-making, paving the way for personalized and precise dental and medical treatments.

Keywords

Micro-computed tomography (micro-CT) · Dentistry applications · Artificial intelligence (AI) · Bone tissue analysis · Multimodality imaging

1 Introduction

The rapid development of imaging technologies has led to radical changes in diagnosis, treatment planning and research processes in medicine and dentistry. Among these developments, micro-computed tomography (micro-CT) has attracted great interest, especially in dentistry, orthodontics, endodontics, periodontology, maxillofacial surgery, implantology and tissue engineering. Similar to conventional medical CT, micro-CT is a non-invasive, high-resolution three-dimensional (3D) image acquisition technique that can provide micrometer-level resolution (Rahardjo et al. 2014; Balta-Uysal et al. 2023). In this way, it is possible to examine the internal structure of tooth and bone tissues, adaptation of restorations, effectiveness of endodontic treatments, implant surface-bone interaction, integration of bone grafts and pathological processes in three dimensions (Demir et al. 2014). One of the most important benefits of micro-CT is the ability to examine specimens without destroying them, thus enabling repeated measurements of the same specimen over time. This is especially useful in experimental studies, monitoring tissue engineering products, monitoring the integration of implants into bone tissue and evaluating the results of various restorative, endodontic and orthodontic procedures (Khalaf et al. 2022; Liu et al. 2015; Lee et al. 2015).

In this chapter, the basic principles of Micro-CT, its common applications in dentistry (endodontics, restorative dentistry, implantology, orthodontics, periodontology), bone tissue analysis, biomaterials research, cancer metastasis evaluation, use of contrast agents, structural analysis methods and future directions will be discussed. In addition, the advantages and limitations of micro-CT, integration of artificial intelligence (AI), multimodality imaging approaches and methodological standards will also be discussed, thus providing the reader with a broad perspective (Carrera et al. 2015; Umanjec-Korac et al. 2016).

2 Fundamentals of Micro-CT

Micro-CT is based on the detection of radiation from an X-ray source passing through the sample, detecting it with a detector, and reconstructing the projections taken from different angles in 3D with computer algorithms (Bakhsh 2021; Zong et al. 2021). Compared to conventional CT, micro-CT offers a much smaller voxel size, allowing for micrometer-level resolution (Gaile et al. 2022). In this way, details such

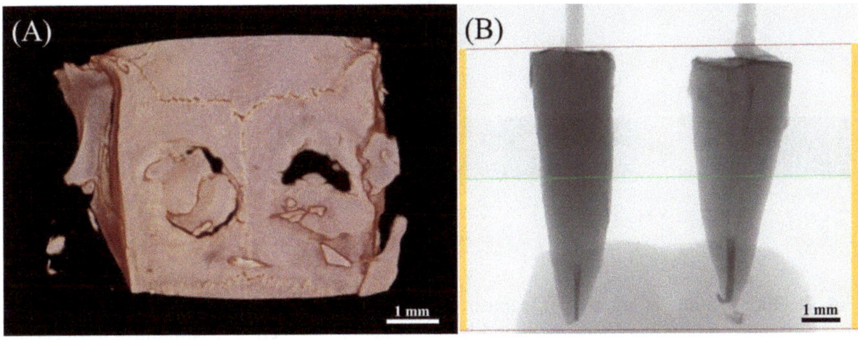

Fig. 1 (**a**) 3D reconstruction of a sample from an animal study evaluating the efficiency of dental grafts, showcasing structural details and integration with bone. (**b**) Initial micro-CT radiographic scan of a human tooth, illustrating alignment, internal geometry, and the enamel-dentin boundary for structural analysis. (Ankara University Faculty of Dentistry Central Research Laboratory, Micro-CT Laboratory, 2024)

as tooth structure, enamel-dentin boundary, pulpalization, canal anatomy, transition areas between restorations and tooth tissue, and bone trabecular architecture can be examined with high accuracy (Mishra et al. 2023; Torres et al. 2021). Figure 1, which represents studies currently being conducted in our laboratory, combines two distinct micro-CT analyses. The first, Fig. 1a, involves the 3D reconstruction of a sample from an animal study evaluating the efficiency of dental grafts, providing detailed visualizations of the sample's alignment, internal geometry, and integration with bone. The second, Fig. 1b, presents the initial micro-CT radiographic scan of a human tooth, showcasing alignment, internal structural details, and the enamel-dentin boundary, serving as a preliminary assessment for further reconstruction. Both images were processed using optimized reconstruction settings such as ring artifact reduction, beam hardening correction, and smoothing, highlighting micro-CT's critical role in the precise analysis of biological structures and material interfaces.

2.1 Technological Components and Factors Affecting Image Quality

The performance of micro-CT devices depends on the focal size of the X-ray tube, resolution of the detector, sample-detector distance, filters used, acquisition voltage, current and exposure time (Rovaris et al. 2019; Aung and Myint 2021). Artifacts such as beam hardening can cause problems such as the absorption of low-energy photons due to the dense structure of the sample and the associated contrast changes (Abdou et al. 2022). Computer-based correction algorithms, calibration phantoms, and appropriate filters are used to reduce such artifacts (Ghavami-Lahiji et al. 2021; Yıldız et al. 2024). In the study presented in Fig. 2 and conducted in our laboratory,

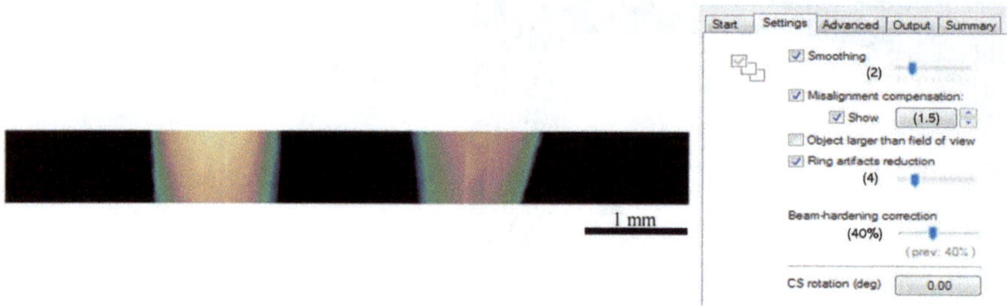

Fig. 2 Optimization of micro-CT reconstruction settings, including smoothing, misalignment compensation, ring artifact reduction, and beam-hardening correction, to minimize artifacts and beam hardening effects for improved image quality. (Ankara University Faculty of Dentistry Central Research Laboratory Micro-CT Laboratory, 2024)

various micro-CT reconstruction settings were optimized to enhance image quality and reduce artifact prevalence. The applied settings included smoothing (level 2) to minimize noise while preserving structural details, misalignment compensation (-1.5) to correct mechanical alignment errors, and ring artifact reduction (level 4) to address circular artifacts in the reconstructed images. Beam-hardening correction was implemented at 40% to improve contrast by mitigating the effects of differential photon absorption. Additionally, CS rotation was set to 0.00° to ensure proper center alignment. These adjustments are critical for producing high-quality and accurate micro-CT images for further analysis.

2.2 Overview of Contrast Enhancement and Soft Tissue Examination

One of the major limitations of micro-CT is the low contrast between soft tissues. Therefore, it can be difficult to clearly distinguish soft tissues such as pulp tissue or connective tissue (Anderson et al. 2014; Salamanca et al. 2022; Özkan et al. 2015). To overcome this problem, contrast agents are used to make vascular structures or soft tissue components more prominent (Bueno et al. 2021; Ahmed et al. 2021). Some studies have shown that combining micro-CT with optical coherence tomography (OCT) or other modalities can provide structural information along with functional information (Jonker 2023).

3 Micro-CT Applications in Dentistry

3.1 Micro-CT in Clinical Diagnostics and Treatment in Dentistry

Micro-CT has revolutionized the detailed examination of dental anatomy. Enamel, dentin, pulp cavity and canal structures can be evaluated in three dimensions and mineral density of different tissues can be determined with precise measurements.

Detection of early caries lesions, monitoring of erosion and remineralization processes can be quantitatively evaluated with micro-CT (Bakhsh 2021; Zong et al. 2021). Beam hardening correction tools can improve the accuracy of diagnosing incipient caries lesions (Rovaris et al. 2019). It has also been shown that the mineral density of enamel and dentin can be measured in 3D in congenital conditions such as ectodermal dysplasia (Yavuz et al. 2022). In this way, changes caused by genetic or environmental influences on dental tissue can be analyzed with high accuracy.

In the endodontic field, micro-CT is widely used to evaluate canal morphology, side canals, isthmuses, curvature, the effectiveness of canal shaping tools and the quality of canal filling (Silva et al. 2020; Alshehri et al. 2016; Carrera et al. 2015). The apical sealing and cavity formation of continuous wave vertical condensation or single cone obturation techniques can be evaluated with micro-CT, improving treatment success (Angerame et al. 2022; Huefner et al. 2022). Micro-CT has also been used to study the effectiveness of minimally invasive endodontic techniques. It is possible to compare the amount of dentin removal of different file systems (such as TruNatomy, XP-endo Shaper) (Rehan 2023). Thus, an objective and quantitative evaluation of the results of endodontic procedures can be made.

In restorative dentistry, micro-CT provides important information on polymerization shrinkage of composite resins, cavity and microleakage formation, and adaptation at the restoration-mine/dentin interface (Balta-Uysal et al. 2023; Kim et al. 2023). The effect of different cavity preparations or access cavity designs on restoration quality can be compared using micro-CT (Silva et al. 2020). Likewise, the effect of different CAD software on the marginal and internal fit of all-ceramic crowns can be evaluated with micro-CT (Akat et al. 2022). Quantitative measurements of leakage, voids, and material adaptation play an important role in developing restorative materials and determining the most clinically relevant restoration techniques (Carrera et al. 2015). In addition, more advanced technologies such as nano-CT through micro-CT are being used to study the internal structure and surface interactions of resin composites, thus opening the door to innovations in materials science (Haugen et al. 2020; Akküç et al. 2023).

3.2 Micro-CT in Implantology, Education, and Forensic Applications

Micro-CT non-invasively evaluates the integration of dental implants into bone tissue and quantitatively measures the bone response to different implant designs, surface modifications and loading protocols. Two- and three-dimensional analysis of platform deformations of tri-channel implants can be performed with micro-CT, and the effect of correct torque application on implant stability can be evaluated (Morais et al. 2023). In addition, the integration of bone grafts, regenerative materials and biomaterials can be monitored with micro-CT and graft resorption, new bone formation and material-bone interface can be examined in detail (Umanjec-Korac et al. 2016; Salamanca et al. 2022; Özkan et al. 2015). Thus, it is possible to design biomaterials that will provide better tissue regeneration.

Micro-CT is an important development in orthodontic treatment, periodontology applications and alveolar bone analysis. This technology allows detailed examination of changes in alveolar bone. In particular, micro-CT is increasingly being used to evaluate bone density changes, microstructural adaptations and tissue-specific healing processes that occur during and after orthodontic tooth movements. The role of micro-CT in orthodontic practice is critical, especially in understanding the effects on alveolar bone during tooth movements. Using micro-CT, the effects of orthodontic forces on alveolar bone can be examined and this technology has been observed to help better understand bone remodeling processes. In this context, the effects of orthodontic forces on alveolar bone are better understood through studies using micro-CT. Studies show the advantages of micro-CT in monitoring bone changes that occur during treatment by evaluating changes in alveolar bone mineral density during orthodontic treatment using micro-CT (Jingtao et al. 2017). These possibilities obtained with micro-CT in the field of periodontology more clearly reveal the negative effects of inflammatory diseases such as periodontitis on alveolar bone loss. The benefit of micro-CT technology is seen in examinations such as examining the accuracy of placement of orthodontic micro-screws and evaluating alveolar bone loss. The data obtained with micro-CT during treatment processes constitute an important source for monitoring tissue healing in regenerative applications such as bone grafts or the use of growth factors (Deglow et al. 2021; Aristizábal-Elejalde 2022).

Micro-CT is also used in the field of education as it enables non-destructive examination of specimens. Students can examine dental anatomy, restorations, or root canals in three dimensions, thus obtaining detailed information without the need for traditional sectioning methods (Liao et al. 2018; Shahmoradi et al. 2016; Campioni et al. 2020). In forensic dentistry, micro-CT enables detailed examination of dental samples, age determination studies, and analysis of the structural properties of dental tissues (Yıldız et al. 2024). This is useful for the examination of archaeological samples or identification studies.

4 Non-dentistry Applications

Micro-CT is not limited to dentistry. The potential of micro-CT as a decision-making tool to support pathologic examinations in patients with non-small cell lung cancer is being explored to distinguish between metastatic and non-metastatic lymph nodes (Cangır et al. 2023). Similarly, in tissue engineering, orthopedics and cardiovascular research, micro-CT is used to evaluate vascular structures, bone healing processes and tissue integrity (Anderson et al. 2014; Salamanca et al. 2022; Özkan et al. 2015). In tissue engineering, the porosity, internal structure integrity and cell-material interaction of scaffold structures are evaluated with micro-CT to develop optimal biomaterials (Bueno et al. 2021; Ahmed et al. 2021). Likewise, in the cardiovascular field, micro-CT can be used to visualize vascular structures in 3D, thereby monitoring angiogenesis, vascular regeneration, and integration of implants into the vascular bed (Huefner et al. 2022; Jiang et al. 2021).

5 Multimodality Approaches and Use of AI

5.1 Multimodality Approaches and Functional Imaging

The integration of micro-CT with other imaging techniques, such as optical coherence tomography (OCT), micro magnetic resonance imaging (MRI), scanning electron microscopy (SEM) and confocal microscopy, allows both structural and functional information to be combined. This multimodal approach allows simultaneous assessment of mineral distribution and organic matrix structure. The imaging capabilities of micro-CT, when combined with other techniques, offer a more comprehensive analysis. The data obtained with micro-CT, when combined with the soft tissue contrast of MRI, helps to more accurately identify tumors and other pathological structures. Dynamic monitoring of healing processes and regenerative therapies is possible using time-lapse (4D) micro-CT. This plays a critical role, especially in tissue engineering and repair processes, as micro-CT offers the ability to observe changes in the treatment process in real time. The three-dimensional images obtained with micro-CT allow the monitoring of morphological changes that occur during tissue regeneration, which helps to optimize treatment strategies (Wang et al. 2023a, b) (Table 1).

5.2 AI and Machine Learning

Data obtained with micro-CT is difficult to analyze due to their complex structure and large size. In this context, AI and machine learning applications offer significant

Table 1 Advantages and application areas of integrating various imaging techniques with micro-CT

Imaging method	Advantages	Application areas
Micro-CT and OCT	Integration of both structural and functional information (Metscher 2009)	Multimodal analyses (Clark and Badea 2014)
Micro-CT and micro MRI	More accurate tumor and pathological structure identification with soft tissue contrast (Senter-Zapata et al. 2016)	Tumor identification and pathology studies (Teplov et al. 2019)
Micro-CT and SEM	Simultaneous analysis of mineral distribution and organic matrix structure (Li et al. 2021)	Tissue engineering and regeneration (Bartoš et al. 2018)
Micro-CT and confocal microscopy	More comprehensive morphological analysis capabilities (Ball et al. 2016)	Detailed cellular imaging (Marr et al. 2020)
Time-lapse (4D) micro-CT	Real-time monitoring of dynamic processes and treatment optimization (Liu et al. 2023)	Dynamic monitoring of treatment processes and optimization (Christen and Müller 2017)

advantages in processes such as automatic segmentation, tissue classification, lesion detection, canal morphology analysis and implant stability prediction (Clark and Badea 2014; Gong et al. 2019). Deep learning-based models can be trained on large data sets, reducing operator dependency, and facilitating more quantitative, reproducible results (Thenmozhi and Reddy 2019). A model observer based on deep learning and partial least square regression has also been developed for low-contrast lesion detection. Such systems provide more reliable results by minimizing human error in medical imaging (Shen).

Similar approaches also yield successful results in the analysis of histopathological images. Alon et al. (2021) conducted a study aiming to synthetically generate three-dimensional cancer cell models from histopathological images. Such AI-based methods accelerate clinical diagnosis and treatment planning processes by automatically classifying the morphological characteristics of cells (Giri and Cheng 2023). Similarly, Liang and He (2021) opened new horizons in dentistry applications by reconstructing the three-dimensional structure of the oral cavity from panoramic X-ray data. On the other hand, Oluwadare et al. (2019) illuminated dynamic aspects of cellular morphology, focusing on the construction of 3D chromosome and genome structures from Hi-C data.

AI and machine learning, as in other fields, also play a transformative role in dentistry and medical studies (Báskay et al. 2024). Studies utilizing 3D images obtained through micro-CT, such as the work depicted in Fig. 3, showcase the significant potential of these technologies in processing and analyzing complex imaging data, driving advancements in dental and medical imaging analysis. This

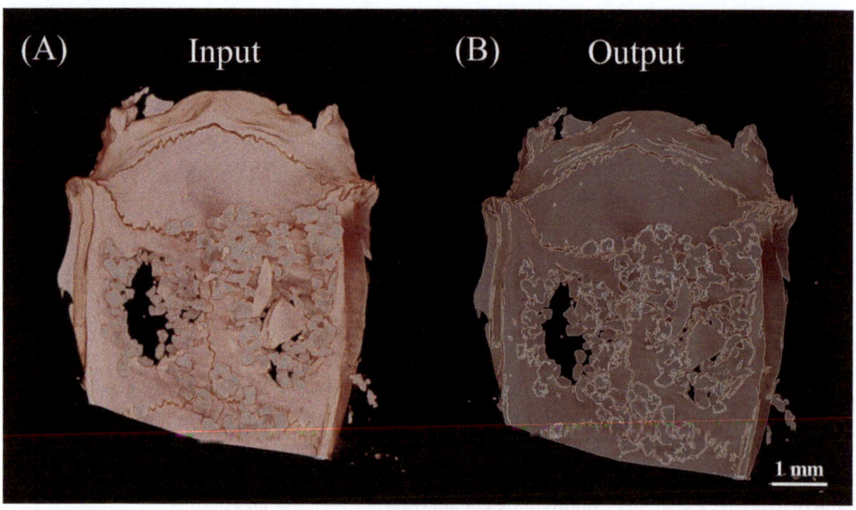

Fig. 3 Segmentation Analysis of Dental Morphology: (**a**) Raw 3D micro-CT reconstruction; (**b**) Segmented image using U-Net, enhancing morphological analysis for precise diagnostics and treatment planning. (Ankara University Faculty of Dentistry Central Research Laboratory Micro-CT Laboratory, 2024)

figure demonstrates the application of advanced segmentation techniques on micro-CT images, emphasizing anatomical details and potential pathological changes in dental structures, as part of our current research efforts.

The application of deep learning-based approaches to topics such as seismic performance and building durability analysis (Kazemi and Jankowski 2023) or pest classification in agriculture (Thenmozhi and Reddy 2019) allows the evaluation of large amounts of data in a short time and with high accuracy (Friedrich et al. 2014). In addition, micro-CT data opens new horizons in dentistry applications such as canal morphology and root canal filling methods (Bhandi et al. 2021). All these developments indicate that fast and accurate analysis of micro-CT data will become more widespread in the future in many fields, especially in medical diagnosis and biomedical research.

6 Conclusions and Future Perspectives

Limitations of micro-CT include long scan times, small sample size requirement to achieve high resolution, beam hardening and motion artifacts (Yavuz et al. 2022; Kim et al. 2021). In addition, radiation dose may adversely affect cell viability in living samples. Therefore, it is important to optimize protocols in animal experiments or biological samples and to follow the "As Low As Reasonably Achievable" (ALARA) principle (Gaile et al. 2022; Dudhe et al. 2024). The fact that standard protocols and reporting criteria have not yet been fully established makes comparability in studies difficult. Therefore, developing methodological standards is necessary to facilitate harmonization and data sharing between different research groups (Cangır et al. 2021; Torres et al. 2021).

Advances in micro-CT technology are moving towards nano-CT level resolution. Nano-CT allows for more detailed examination of dental hard tissues, collagen fibril orientation and the restoration-tissue interface. Micro-CT and nano-CT can more clearly visualize the internal structure of teeth and tissue interactions, enabling more accurate diagnosis and treatment planning in clinical applications (Campioni et al. 2020). In addition, these technologies provide high quality images with less radiation exposure compared to other imaging methods used in dentistry (Xu et al. 2023).

The nano-CT technique contributes to a higher level of understanding of material-biological tissue interactions. This is especially critical for assessing the biocompatibility of dental implants and restorative materials. For example, by studying the interactions of various materials used in dentistry with dental tissue using nano-CT, more information can be obtained about the long-term performance of these materials. Such information helps to develop more effective and reliable treatment methods in clinical applications (Özdemir and Kopac 2022).

Currently, micro-CT is mainly used in research laboratories. However, in the future, with the development of more compact, fast, and low-dose systems, micro-CT is likely to be integrated into clinical applications (Liao et al. 2018; Shahmoradi et al. 2016). Applications such as instant 3D evaluation of resected tissues during surgical procedures or detailed analysis of bone quality before implant surgery are conceivable. With the widespread adoption of CBCT in clinical dentistry, similar adaptation of micro-CT capabilities to the clinic will be an important step in patient-specific diagnosis and treatment planning (Shahmoradi et al. 2016).

Micro-CT has become an indispensable imaging tool in dentistry and related biomedical fields. The possibility of three-dimensional, high-resolution, non-invasive, and non-destructive examination enables detailed analysis of dental structures, restorations, endodontic treatments, implant-bone integration, orthodontic and periodontic changes, biomaterial evaluations and even cancer metastasis research. Although there are some limitations (beam hardening, soft tissue contrast, high data volume), advancing technology, AI -assisted analysis, nano-CT, and multi-modality imaging approaches aim to overcome these problems. Thus, micro-CT is expected to offer a wider range of applications in the future, take part in clinical decision support systems and contribute to the development of more effective, individualized treatments.

Acknowledgements The authors would like to acknowledge the contributions of their respective institutions and funding bodies that supported this research.

Conflict of Interest The authors declare no conflict of interest related to this study.

References

Abdou A, Matoug-Elwerfelli M, Nagendrababu V et al (2022) Tooth auto-transplantation: a bibliometric analysis of the top 100 most-cited articles. Dent Traumatol. https://doi.org/10.1111/edt.8060

Ahmed M, Ikram Y, Qureshi F et al (2021) Assessment of jaw bone density in terms of Hounsfield units using cone beam computed tomography for dental implant treatment planning. PAFMJ. https://doi.org/10.5124/PAFMJ.3000

Akat B, Şentürk A, Ocak M, Kılıçarslan M, Özcan M (2022) Does CAD software affect the marginal and internal fit of milled full ceramic crowns? Braz Oral Res 36. https://doi.org/10.1590/1807-3107bor-2022.vol36

Akküç S, Duruk G, Keleş A (2023) Remineralization effect of three different agents on initial caries and erosive lesions: a micro-computed tomography and scanning electron microscopy analysis. BMC Oral Health 23(1):4020. https://doi.org/10.1186/s12903-023-04020-x

Alon Y, Yu X, Zhou H (2021) Synthetic generation of three-dimensional cancer cell models from histopathological images. arXiv preprint arXiv:210111600

Alshehri M, Alamri HM, Alshwaimi E, Kujan O (2016) Micro-computed tomographic assessment of quality of obturation in the apical third with continuous wave vertical compaction and single match taper sized cone obturation techniques. Scanning 38(4):352–356

Anderson PJ, Yong R, Surman TL et al (2014) Application of three-dimensional computed tomography in craniofacial clinical practice and research. Aust Dent J 59(4):601–410. https://doi.org/10.1111/adj.12218

Angerame D, De Biasi M, Lenhardt M, Porrelli D, Bevilacqua L, Generali L et al (2022) Root-end resection with or without retrograde obturation after orthograde filling with two techniques: a micro-CT study. Aust Endod J 48(3):423–430

Aristizábal-Elejalde D (2022) Concrescence in anterior teeth assessed by cone beam computed tomography. A case report. Revista Científica Odontológica 10(1):e102. https://doi.org/10.21142/2523-2754-1001-2022-102

Aung NM, Myint KK (2021) Diagnostic accuracy of CBCT for detection of second canal of permanent teeth: a systematic review and meta-analysis. Int J Dent 2021:11060. https://doi.org/10.1155/2021/11060

Bakhsh TA (2021) Optical comparison between micro-CT and OCT in imaging of marginal composite adaptation: observational study. J Microsc 284(2):60–110. https://doi.org/10.1111/jmi.13000

Ball AD, Goral T, Kamanli SA (2016) Confocal microscopy applied to paleontological specimens. Paleontol Soc Pap 22:39–55

Balta-Uysal VM, Orhan K, Oguz EI, Guzeldemir-Akcakanat E (2023) Analysing the effect of sodium bicarbonate and glycine air polishing on tooth surfaces with two different imaging methods. J Microsc 289(3):180–186

Bartoš M, Suchý T, Foltán R (2018) Note on the use of different approaches to determine the pore sizes of tissue engineering scaffolds: what do we measure? Biomed Eng Online 17:1–15

Báskay J, Kivovics M, Pénzes D, Kontsek E, Pesti A, Kiss A et al (2024) Reconstructing 3D histological structures using machine learning (artificial intelligence) algorithms. Die Pathol:1–7

Bhandi S, Mashyakhy M, AbuMelha AS, Alkahtany MF, Jamal M, Others (2021) Complete obturation—cold lateral condensation vs. thermoplastic techniques: a systematic review of micro-CT studies. Materials 14(14):4013

Bueno CRE, Sumida DH, Duarte MAH, Ordinola-Zapata R, Azuma MM, Guimarães G et al (2021) Accuracy of radiographic pixel linear analysis in detecting bone loss in periodontal disease: study in diabetic rats. Saudi Dent J 33(8):987–996

Campioni I, Pecci R, Bedini R (2020) Ten years of micro-CT in dentistry and maxillofacial surgery: a literature overview. Appl Sci 10(19):80190. https://doi.org/10.3390/app10198019

Cangır A, Dizbay Sak S, Güneş G et al (2021) Differentiation of benign and malignant regions in paraffin-embedded tissue blocks of pulmonary adenocarcinoma using micro CT scanning of paraffin tissue blocks: a pilot study for method validation. Surg Today 51(5):7050. https://doi.org/10.1007/s00595-021-02203-3

Cangır A, Orhan K, Güneş SG et al (2023) Could microcomputed tomography be a new method to distinguish between metastatic and non-metastatic lymph nodes in patients with non-small cell lung cancer as a decision support tool for pathological examination? [Manuscript in preparation]

Carrera CA, Lan C, Escobar-Sanabria D, Li Y, Rudney J, Aparicio C, Fok A (2015) The use of micro-CT with image segmentation to quantify leakage in dental restorations. Dent Mater 31(4):382–390

Christen P, Müller R (2017) In vivo visualisation and quantification of bone resorption and bone formation from time-lapse imaging. Curr Osteoporos Rep 15:311–317

Clark DP, Badea CT (2014) Micro-CT of rodents: state-of-the-art and future perspectives. Phys Med 30(6):619–634

Deglow R, Toledano Gil S, Zubizarreta-Macho Á, Bufalá Pérez M, Rodríguez Torres P, Tzironi G et al (2021) Influence of the computer-aided static navigation technique and mixed reality technology on the accuracy of the orthodontic micro-screws placement: an in vitro study. J Pers Med 11(10):964

Demir N, Ozturk AN, Malkoc MA (2014) Evaluation of the marginal fit of full ceramic crowns by the microcomputed tomography (micro-CT) technique. Eur J Dent 8(04):437–444

Dudhe SS, Mishra G, Parihar P, Nimodia D, Kumari A (2024) Radiation dose optimization in radiology: a comprehensive review of safeguarding patients and preserving image fidelity. Cureus 16(5):e60846

Friedrich F, Matsumura Y, Pohl H, Bai M, Hörnschemeyer T, Beutel RG (2014) Insect morphology in the age of phylogenomics: innovative techniques and its future role in systematics. Entomol Sci 17(1):1–24

Gaile M, Papia E, Zalite V, Locs J, Soboleva U (2022) Resin cement residue removal techniques: in vitro analysis of marginal defects and discoloration intensity using micro-CT and stereomicroscopy. Dent J 10(4):55

Ghavami-Lahiji M, Tayefeh Davalloo R, Tajziehchi G et al (2021) Micro-computed tomography in preventive and restorative dental research: a review. Imaging Sci Dent 51(3):110–190. https://doi.org/10.5624/isd.20210223

Giri N, Cheng J (2023) Deep learning for reconstructing protein structures from cryo-EM density maps: recent advances and future directions. Curr Opin Struct Biol 79:102536

Gong H, Yu L, Leng S, Dilger SK, Ren L, Zhou W, McCollough CH (2019) A deep learning-and partial least square regression-based model observer for a low-contrast lesion detection task in CT. Med Phys 46(5):2052–2063

Haugen HJ, Bin Qasim SS, Matinlinna JP et al (2020) Nano-CT as tool for characterization of dental resin composites. Sci Rep 10(1):210–140. https://doi.org/10.1038/s41598-020-69192-2

Huefner M, Rauch N, Schwarz-Herzke B, Knorr IJ, Sager M, Drescher D, Becker K (2022) Micro-angiogenic patterns around orthodontic implants migrating in bone: a micro-CT study in the rat tail model. J Clin Periodontol 49(2):188–197

Jiang J, Xiao J, Wang D et al (2021) Application of implantable polylactic-co-glycolic acid microcapsule in repairing alveolar bone defects. Evid Based Complement Alternat Med 2021:1–10. https://doi.org/10.1155/2021/1102346

Jingtao A, Liu Y, Liu Z, Wang R, Zhang B (2017) A micro-ct study of microstructure change of alveolar bone during orthodontic tooth movement under different force magnitudes in rats. Exp Ther Med 13(5):1793–1798. https://doi.org/10.3892/etm.2017.4186

Jonker CH (2023) Root and canal morphology of the mandibular first molar: a micro-computed tomography-focused observation of literature with illustrative cases. Part 1: external root morphology. South African. Dent J 78(2):45–52. https://doi.org/10.17159/2519-0105/2023/v78no2a2

Kazemi F, Jankowski R (2023) Machine learning-based seismic response and performance assessment of reinforced concrete buildings. Arch Civ Mech Eng 23(2):94

Khalaf K, Mustafa S, Maarouf TB (2022) Is orthodontic treatment a risk factor of cervical root resorption? A systematic review. Eur J Dent 16(4):1–10. https://doi.org/10.1055/s-0042-1750593

Kim Y, Jung GI, Jeon D, Wijesinghe RE, Seong D, Lee J et al (2021) Non-invasive optical coherence tomography data-based quantitative algorithm for the assessment of residual adhesive on bracket-removed dental surface. Sensors 21(14):4670

Kim HJ, Lee JS, Gwak DH, Ko YS, Lim CI, Lee SY (2023) In vitro comparison of differences in setting time of premixed calcium silicate-based mineral trioxide aggregate according to moisture content of gypsum. Materials 17(1):35

Lee S, Lee DK, Chung HJ et al (2015) Dental optical coherence tomography: new potential diagnostic system for cracked-tooth syndrome. Surg Radiol Anat 37(5):412–480. https://doi.org/10.1007/s00276-015-1464-6

Li Z, Du T, Ruan C, Niu X (2021) Bioinspired mineralized collagen scaffolds for bone tissue engineering. Bioact Mater 6(5):1491–1511

Liang Y, He L (2021) Oral-3D: reconstructing the 3D structure of oral cavity from panoramic X-ray. Proc AAAI Conf Artif Intell 35(1):566–573

Liao CW, Fuh LJ, Shen YW, Huang HL, Kuo CW, Tsai MT, Hsu JT (2018) Self-assembled micro-computed tomography for dental education. PLoS One 13(12):e0209698

Liu Z, Yan C, Kang C, Zhang B, Li Y (2015) Distributional variations in trabecular architecture of the mandibular bone: an in vivo micro-CT analysis in rats. PLoS One 10(1):e0116194

Liu Q, Sun M, Sun X, Liu B, Ostadhassan M, Huang W et al (2023) Pore network characterization of shale reservoirs through state-of-the-art X-ray computed tomography: a review. Gas Sci Eng 113:204967

Marr N, Hopkinson M, Hibbert AP, Pitsillides AA, Thorpe CT (2020) Bimodal whole-mount imaging of tendon using confocal microscopy and x-ray micro-computed tomography. Biol Proced Online 22:1–14

Metscher BD (2009) MicroCT for developmental biology: a versatile tool for high-contrast 3D imaging at histological resolutions. Dev Dyn 238(3):632–640

Mishra S, Kumar M, Mishra L, Panda S, Panda S, Lewkowicz N, Lapinska B (2023) Estimation of cancellous changes using fractal analysis in patients with periodontitis. Biomedicine 2023(11):2547

Morais RC, Simionato AA, Moris ICM et al (2023) Influence of torque on platform deformity of the tri-channel implant: two- and three-dimensional analysis using micro-computed tomography. Medicina 59(3):402. https://doi.org/10.3390/medicina59030402

Oluwadare O, Highsmith M, Cheng J (2019) An overview of methods for reconstructing 3-D chromosome and genome structures from Hi-C data. Biol Proced Online 21:1–20

Özdemir O, Kopac T (2022) Recent progress on the applications of nanomaterials and nano-characterization techniques in endodontics: a review. Materials 15(15):5109

Özkan G, Kanli A, Başeren NM, Arslan U, Tatar İ (2015) Validation of micro-computed tomography for occlusal caries detection: an in vitro study. Braz Oral Res 29:01–07

Rahardjo A, Sahertian RD, Ramadhani SA, Maharani DA, Latief FD (2014) The effect of milk or its combination with tea and 0.2% NaF on dental enamel demineralization analyzed by micro computed tomography. J Dent Indones 21(2):53–56

Rehan EA (2023) In-vitro comparison of TruNatomy and XP-endo Shaper on amount of dentin removal: a micro-computed tomography study. Manuscript submitted for publication

Rovaris K, Sousa Melo SLS, Queiroz PM et al (2019) Beam hardening correction tool improves the diagnosis of incipient caries lesions in micro-CT images. Dentomaxillofac Radiol 48(1):301–310. https://doi.org/10.1259/dmfr.20180259

Salamanca E, Pan Y-H, Sun Y-S et al (2022) Magnesium-modified β-tricalcium phosphate induces cell osteogenic differentiation in vitro and bone regeneration in vivo. Int J Mol Sci 23(4):8040. https://doi.org/10.3390/ijms23048040

Senter-Zapata M, Patel K, Bautista PA, Griffin M, Michaelson J, Yagi Y (2016) The role of micro-CT in 3D histology imaging. Pathobiology 83(2–3):140–147

Shahmoradi M, Lashgari M, Rabbani H et al (2016) A comparative study of new and current methods for dental micro-CT image denoising. Dentomaxillofac Radiol 45(4):220–160. https://doi.org/10.1259/dmfr.20150271

Shen D Computer-aided diagnosis with deep learning architecture: applications to breast lesions in US images and pulmonary nodules in CT scans

Silva EJNL, Oliveira VB, Silva AA, Belladonna FG, Prado M, Antunes HS, De-Deus G (2020) Effect of access cavity design on gaps and void formation in resin composite restorations following root canal treatment on extracted teeth. Int Endod J 53(11):1540–1548

Teplov A, Tabata K, Fu X, Uraoka N, Roehrl MH, Ntiamoah P et al (2019) Development of standard operating procedure (SOP) of micro-computed tomography (micro-CT) in pathology. Diagn Pathol 5(1):273

Thenmozhi K, Reddy US (2019) Crop pest classification based on deep convolutional neural network and transfer learning. Comput Electron Agric 164:104906

Torres FFE, Guerreiro-Tanomaru JM, Pinto JC et al (2021) Effect of different dimensions of test samples on the volumetric change assessment of endodontic materials. Braz Dent J 32(2):200. https://doi.org/10.1590/0103-6440202104124

Umanjec-Korac S, Parsa A, Nikoozad AD et al (2016) Accuracy of cone beam computed tomography in following simulated autogenous graft resorption in maxillary sinus augmentation procedure: an ex vivo study. Dentomaxillofac Radiol 45(8):706–760. https://doi.org/10.1259/dmfr.20150405

Wang J, Zhang R, Zhang Z, Geng C, Zhang Y (2023a) Micro-computed tomography evaluation of the effects of orthodontic force on immature maxillary first molars and alveolar bone mineral density of Sprague–dawley rats. Korean J Orthod 53(3):205–216. https://doi.org/10.4041/kjod22.209

Wang Q, Sheng Y, Jiang Z, Liu H, Lu H, Xing W (2023b) What imaging modality is more effective in predicting early recurrence of hepatocellular carcinoma after hepatectomy using Radiomics analysis: CT or MRI or both? Diagnostics 13(12):2012

Xu S, Hu Y, Li B, Inscoe CR, Tyndall DA, Lee YZ et al (2023) Volumetric computed tomography with carbon nanotube x-ray source array for improved image quality and accuracy. Communications. Engineering 2(1). https://doi.org/10.1038/s44172-023-00123-x

Yavuz Y, Akleyin E, Doğan MS, Goncharuk-Khomyn M, Akkus Z (2022) Can the density of mineralized dental tissues (dentin and enamel) be measured and compared with 3D cone beam computed tomography in cases of ectodermal dysplasia? Med Sci Monit 28:e937003-1

Yıldız H, Bagis N, Camgoz M, Karacaoglu F, Ocak M, Orhan K (2024) Micro-CT evaluation of the effect of various ScRp instrumentation methods on cement loss, porosity and micro-crack formation. Int J Dent Hyg 22(3):588–595

Zong C, Bronckaers A, Vande Velde G et al (2021) In vivo micro-computerized tomography tracking of human periodontal ligament stem cells labeled with gold nanocomplexes. Adv Healthc Mater 10(21):e210111. https://doi.org/10.1002/adhm.202101111

Part II

Functional, Quantitative, and Advanced Multimodal Approaches in Preclinical Bioimaging

3D Histopathology of Musculoskeletal Tissues Using Contrast-Enhanced Micro-computed Tomography

Shuvashis Das Gupta, Sami Kauppinen, and Mikko A. J. Finnilä

Abstract

Traditional two-dimensional (2D) histology, while foundational in anatomic pathology for understanding the complex composition and microarchitecture of biological tissues, suffers from inherent limitations. This method is time-consuming, prone to artifacts, and fails to capture the intricate 3D tissue architecture, thereby limiting volumetric analysis and accurate quantification of diverse tissues. These limitations restrict our understanding of complex biological systems, particularly in the field of musculoskeletal research. X-ray micro-computed tomography (μCT) emerges as a promising alternative for 3D visualization of heterogeneous tissues, offering complete three-dimensional (3D) information, high spatial resolution, extensive field of view, and reduced sample preparation workload. Initially limited to mineralized tissues due to the weak X-ray absorbing properties of nonmineralized or soft biological tissues, the application of μCT has been revolutionized by the development of radiopaque contrast-enhancing staining agents (CESAs) and X-ray imaging techniques. These advancements enable high-resolution 3D visualization of both mineralized and soft tissues, offering a comprehensive view of tissue microarchitecture and quantitative virtual 3D anatomical pathology. This chapter focuses on contrast-enhanced μCT (CE-μCT) ex vivo imaging for 3D anatomical pathology of musculoskeletal tissues. We review key studies that showcase the efficacy of CESAs and discuss innovative techniques, such as chemical drying and phase contrast μCT, highlighting their applications in musculoskeletal tissues. We believe that integrating CE-μCT imaging techniques provides a comprehensive

S. Das Gupta (✉) · S. Kauppinen · M. A. J. Finnilä
Research unit of Health Sciences and Technology, University of Oulu, Oulu, Finland
e-mail: shuvashis.dasgupta@oulu.fi

I. F. Cengiz et al. (eds.), *Bioimaging Modalities in Bioengineering*,
Biomaterials, Bioengineering and Sustainability 9,
https://doi.org/10.1007/978-3-031-99410-4_10

approach to studying the 3D anatomical pathology of musculoskeletal tissues, offering significant improvements over traditional 2D histopathology methods.

Keywords

Micro-computed tomography · Contrast-enhanced micro-CT · Phase-contrast micro-CT · Soft tissue imaging · 3D histopathology

1 Introduction

The current gold standard in clinical diagnostics, prognostics, and preclinical research of anatomic pathology is the decades-proven slide-based two-dimensional (2D) histology. This microscopic examination provides a crucial understanding of the complex composition and microarchitecture of tissues, ranging from the cellular to the tissue scale, in both healthy and diseased conditions. The histopathological method involves tissue embedding and sectioning, followed by colorimetric, immunohistochemical, or immunofluorescent staining, and microscopic evaluation. A broad selection of dyes is available to stain nucleic acids, proteins, lipids, and carbohydrates, allowing visualization of specific components. However, conventional 2D histopathological processes are often time-consuming and costly. For instance, mineralized tissues typically undergo a lengthy decalcification process before sectioning. The sectioning process is also susceptible to artifacts and distortions. Additionally, there might be a need to process multiple sub-samples or possibly reorient the sample to thoroughly analyze its spatial structure. The microscopic examination of stained sections leads to a limited 2D representation of the anatomic pathology. Any perception beyond the sections depends on the experience and subjectivity of the pathologist. While there are attempts to transfer this existing 2D technique to create 3D reconstructions from the serial-stained tissue sections, the 3D whole slide imaging is still limited to a smaller sample size and resolution in the third dimension (Farahani et al. 2017; Roberts et al. 2012). These drawbacks greatly hinder thorough spatial analysis of highly diverse tissues, such as trabecular bone and vascular networks, and obstruct accurate quantification of spatial relationships among various tissue types within the sample.

X-ray microfocus computed tomography (μCT) is an excellent candidate for 3D visualization of heterogeneous tissues. It provides complete 3D information with adequate spatial resolution (<1 μm), sufficient contrast for the structures of interest, and an extensive field of view in any direction within the volume. Additionally, it has a reduced workload for sample preparation, which should enhance throughput. Being non-destructive by nature, μCT imaging can preserve tissue integrity and be compatible with downstream histology or other biochemical assays. Early μCT studies mostly focused on imaging mineralized tissues, e.g., bone (Layton et al. 1988). Mineralized tissues provide excellent contrast, whereas soft and demineralized hard tissues, particularly when embedded in paraffin, exhibit poor contrast, complicating routine μCT imaging. To overcome this limitation, several μCT techniques often depend on contrast-enhancing stains or specialized X-ray systems for soft tissue imaging. Over the last two decades, the introduction of radiopaque contrast-enhancing staining agents (CESAs), innovative sample preparation methods, and advancement in X-ray imaging techniques have broadened the applicability of μCT imaging to non-mineralized soft tissues for 3D histopathology with sufficient contrast. This chapter focuses on contrast-enhanced-μCT (CE-μCT) ex vivo imaging for quantitative virtual 3D anatomical pathology, specifically for musculoskeletal tissues. We highlight several studies that contributed to the development of CESA-based virtual 3D anatomical pathology and explore different chemical drying techniques and phase contrast μCT (PhC-μCT) techniques that were employed for studying the 3D anatomical pathology of musculoskeletal tissues.

2 Applications of Contrast-Enhancing Staining Agents (CESAs) in Soft Tissue Imaging

At a specific X-ray energy, the absorption coefficient depends on the atomic number of the target, resulting in higher X-ray absorption in heavier elements compared to lighter ones. This principle is utilized in CE-μCT techniques, where CESAs containing elements with high atomic numbers are employed to improve image contrast and detail (Rawson et al. 2020). Osmium tetroxide (OsO_4) is the first reported commercially available CESA applied for 3D anatomical ex vivo visualization of mouse embryos (Johnson et al. 2006) and pig lungs (Litzlbauer et al. 2006), respectively. Although effective, the extreme toxicity of OsO_4 limited its widespread use. Metscher et al. significantly advanced the field by performing quantitative comparisons of CE-μCT images of chick embryos treated with several CESAs [Gallocyanin-chromalum, 1% iodine in ethanol, iodine potassium iodide (I_2KI, Lugol's iodine), and phosphotungstic acid (PTA)] (B. D. Metscher 2009). Their research demonstrated the staining potential of readily available and less toxic CESAs, such as I_2KI and PTA, for soft tissue CE-μCT imaging. Subsequently, a comprehensive screening study with 28 different chemical substances identified suitable CESAs for larger soft tissue samples (more than 1 cm^3) from bacon strips and mouse legs (Pauwels et al. 2013). These exploratory studies mostly focused on the feasibility of soft tissue CE-μCT imaging and did not explore the mechanisms of their tissue interactions or their potentially destructive effects due to the high osmolality of CESAs compared to the soft tissue. (Pauwels et al. 2013). The literature review conducted by De Bournonville et al. in 2019 provides a valuable resource for the comparisons between CESAs that are commonly used for various musculoskeletal tissues (de Bournonville et al. 2019). Recent reviews have summarized applications of CESAs for visualizing the fine microarchitecture of collagen with CE-μCT imaging (Balcaen et al. 2025; Glancy et al. 2023) and reported that I_2KI, PTA, mercury(II) chloride ($HgCl_2$), and Wells–Dawson polyoxometalates (WD-POMs) all can provide sufficient image contrast for the illustration of collagenous tissues.

PTA and phosphomolybdic acid (PMA) are widely used POMs in histochemistry. Particularly, PTA has been extensively used in μCT-based studies due to its ability to enhance collagen contrast (Karhula et al. 2017b; Lesciotto et al. 2020). PTA binds to collagen due to its high negative charge density and affinity for positively charged collagen in acidic environments. However, due to their slow diffusion rate, PTA and PMA are primarily limited to smaller tissue specimens. Moreover, they are both highly acidic, causing tissue damage and shrinkage. Other POMs include Monolacunary WD-POM, Parent-WD-POM, and 1:2 Hafnium-substituted WD-POM (Hf-WD 1:2 POM) (Glancy et al. 2023). WD-POMs can minimize tissue damage since they can be dissolved in near-physiological pH solutions. They offer comparable contrast enhancement to PTA for imaging soft skeletal tissues (Kerckhofs et al. 2018; Pascart et al. 2022). However, WD-POMs are not commercially available and must be synthesized for use.

Recent research has explored using conventional histological stains as novel CESAs for high-resolution imaging of soft tissues. Eosin, commonly used for its histopathological applications, has Bromine within its molecular structure and has been shown to generate high contrast in mouse kidneys (Busse et al. 2018). The eosin-based, cytoplasm-specific staining protocol is nontoxic, fast for whole-organ staining, and highly compatible with conventional histological examinations. Similarly, a hematein-lead(II)-based staining method was developed for CE-μCT imaging of cell nuclei within soft-tissue samples (Müller et al. 2018). Subsequently, a lead(II) acetate-based CESA was developed for nuclear staining, enabling CE-μCT imaging in aqueous solutions and subsequent removal of the CESA using ethylenediaminetetraacetic acid for further histological examinations (Metscher 2021).

Although the applications of CESA mostly focus on imaging of soft tissues in a similar manner to classical histology, it is still a major challenge to visualize specific molecules or genes. An immunohistochemical approach involving peroxidase-assisted reduction of silver ions was successfully employed in 2011 to map the 3D distribution of collagen type II and acetylated alpha-tubulin proteins in chick embryos (Metscher and Müller 2011). A recent study by Väänänen et al. adopted a similar and accessible methodology aimed at detecting and illustrating the distribution of expression patterns of collagen type II alpha one and sonic hedgehog in developing mouse tissues, utilizing peroxidase-assisted reduction of silver in *in situ* hybridization (Väänänen et al. 2023). They further visualized 3D gene expression with the soft tissue morphology using a staining protocol combining PTA with Gold enhancement of silver *in situ* hybridization.

3 Alternative Contrast Enhancement Methods

3.1 Sample Preparation Methods

CE-μCT methods based on CESAs are affected by the electrostatic repulsion or attraction between the tissue constituents and the contrast agent molecule. Moreover, CESAs often have an inhomogeneous distribution in samples due to their diffusion limits into the tissue and leakage from the tissue in the surrounding medium during imaging (Docter et al. 2024; Li et al. 2015). Therefore, CESA-based CE-μCT faces challenges in the high-resolution imaging of 3D cell morphometry and collagen organization. Chemical drying of soft tissues as a CE-μCT method has become relatively common in recent years. This method utilizes the contrast within the tissue without needing any contrast agent in the μCT images. Moreover, the drying process of tissues is usually very reproducible and straightforward. Besides the conventional air-drying method (with ascending alcohol series), critical-point drying (CPD) and hexamethyldisilazane (HMDS)-based air-drying are mostly used for soft tissue preparation.

CPD is commonly used during sample preparation in scanning electron microscopy (SEM) to minimize tissue artifacts that occur during the air-drying process. During the air-drying process, the transition or crossing from liquid to gas phase can induce significant tangential forces due to surface tension, potentially altering the tissue's nano- and microstructures. At the critical point, the distinction between liquid and gas phases vanishes. During the CPD process, water in the tissue is replaced by ethanol or acetone and then with liquid carbon dioxide (CO_2). The critical point (31 °C, 74 bar) of CO_2 is suitable for biological tissues, and at this point, the transition from interfaces between liquid and gaseous phases avoids the tissue-damaging effects. CPD has been successfully applied to enhance μCT imaging of soft-tissue samples and can be considered the state-of-the-art method to preserve sample morphology (Busse et al. 2018; Karjalainen et al. 2024; Runge et al. 2022).

One of the major disadvantages of CPD is that it needs a specialized device to hold liquid CO_2 under a vacuum. HMDS-based air-drying is a low-cost alternative to CPD that also saves considerable sample preparation time (Bhattacharya et al. 2020). HMDS was initially introduced for SEM in 1983 (Nation 1983) and has been adopted for preparing biological specimens in CE-μCT imaging (Chen et al. 2018; Kestilä et al. 2018, 2019; Parapar et al. 2017; Verleden et al. 2014). The drying process includes dehydration of the tissue sample using an ethanol series, immersion in HMDS, and air drying. While the precise mechanism remains unclear, HMDS can induce protein cross-linking and reduce surface tension to strengthen the tissue sample, which eventually reduces the collapse and distortion of tissue during the air-drying process (Nation 1983). Although CPD and HMDS-based drying processes introduce fragility and potential tissue shrinkage, they have proven valuable for visualizing the intricate microstructures of musculoskeletal soft tissue samples, especially for examining collagen fiber orientation and chondron morphology (V. Karjalainen et al. 2024; Kestilä et al. 2018, 2019; Marr et al. 2020).

3.2 Phase-Contrast Micro-computed Tomography (PhC-μCT)

PhC-μCT is another emerging technique, both laboratory-based and synchrotron-based, suitable for 3D structural characterization of soft tissues at the microscale. This methodology utilizes the phase shift encountered by X-rays as they pass through biological tissue and provides structural characteristics that have traditionally been regarded as invisible to absorption-based settings (Disney et al. 2018). Phase contrast is more beneficial for soft tissues with low atomic numbers since the phase shift has a stronger effect than absorption on the materials with low atomic numbers in the relevant energy range (Häggmark et al. 2024).

The most common PhC-μCT setups are (1) analyzer-based imaging, (2) X-ray grating interferometry imaging, and (3) propagation-based tomography (Furlani et al. 2023; Quenot et al. 2022). The analyzer-based imaging setting uses Bragg's geometry, where a monochromator is placed before the sample, and a highly selective analyzer crystal is placed between the sample and the detector. This setting produces high sensitivity and low X-ray dose deposition to samples; still, it suffers from low beam stability and requires longer exposure time due to the low X-ray flux after monochromatizing. This makes the analyzer-based imaging setting challenging to implement with the conventional laboratory-based X-ray sources. X-ray grating interferometry setting uses gratings (usually made of Au or Si) to create periodic interference patterns and photon-counting pixel detectors placed at characteristic distances of high peak intensity (Furlani et al. 2023). Moreover, an attenuation grating that matches the periodicity of the original interference pattern is placed in front of the detector. Thus, this comprehensive configuration can retrieve distortions in the interference patterns due to the attenuation, refraction, and scattering of the X-ray from the sample. X-ray grating interferometry imaging can be used with conventional X-ray sources (Pfeiffer et al. 2006) and has good sensitivity. Nevertheless, it requires complicated and expensive experimental arrangements, particularly for large fields of view. Additionally, the time taken for acquisition remains relatively lengthy, and further research is needed to minimize noise levels, making it not yet suitable for widespread use in clinical or research settings.

Propagation-based PhC-μCT is the most frequently employed experimental set-up at synchrotron facilities, largely due to its simplicity, efficient use of photons, and simple numerical phase retrieval processing. This configuration leverages the spatial coherence of synchrotron X-rays to generate distinctive interference patterns. Unlike other far-field phase-contrast methods that demand highly monochromatic X-ray beams, the propagation-based set-up uses large bandwidth beams generated by a wiggler or a bending magnet. This minimizes photon flux loss, which is a critical advantage in maximizing signal strength. Experimentally, the propagation-based set-up closely resembles the conventional radiography set-up, with the key difference being the adjustable distance between the sample and the detector. This propagation distance is crucial for controlling the desired spatial resolution. Propagation-based PhC-μCT focuses on capturing the effects of local transversal phase derivatives of X-ray wavefront caused by variations in the sample's density that enhance the contrast of otherwise transparent features within the sample (lensing effect). As propagation distance increases, small phase gradients become more visible, provided their induced angular deviation exceeds the angular diameters of both the source and detector pixels from the sample's perspective (Mirone et al. 2023). Propagation-based PhC-μCT amplifies the signal above Poisson noise and systematic errors (detection irregularities or beam inhomogeneities) without requiring an increased radiation dose, which is often detrimental to the samples.

The decoupling of phase and absorption effects simplifies the image reconstruction process, reducing Propagation-based PhC-μCT to standard absorption tomography. The phase information is retrieved from the interference fringes of the sample's edge using phase retrieval techniques. The most common phase retrieval technique is the 'Paganin method' (Paganin et al. 2002), based on the transport-of-intensity equation under the paraxial approximation (Zuo et al. 2020). Moreover, the recently developed 'Eikonal phase retrieval' algorithm, based on the *eikonal* approximation, performs the phase retrieval task over the entire near-field range, producing images of exceptional quality for mixed attenuation objects compared to the Paganin phase-retrieval algorithm (Mirone et al. 2023).

PhC-μCT imaging methods can distinguish the extracellular matrix and even cellular alterations in soft tissues to approximate histopathology, which is otherwise not obtainable with conventional absorption-based settings. Recently, these techniques have been used to examine collagen fibers in various soft musculoskeletal tissues, including tendons (Silva Barreto et al. 2024; Zhang et al. 2022), menisci (Einarsson et al. 2022), intervertebral discs (Disney et al. 2017, 2019), and articular cartilage (Dejea et al. 2024; Horng et al. 2021; Xu et al. 2024).

4 Contrast-Enhanced Micro-computed Tomography (CE-μCT) of Musculoskeletal Tissues

4.1 Articular Cartilage Imaging

Cartilage covers the entire articulating surface of long bones, and it comprises a dense extracellular matrix with sparsely distributed chondrocytes (Mow et al. 1992). The extracellular matrix primarily contains water, collagen fibrils (mainly type II), proteoglycans with negatively charged glycosaminoglycans (GAGs), other non-collagenous proteins, and glycoproteins. Healthy hyaline articular cartilage is structurally inhomogeneous and can be divided into the surface, middle, and deep zones (Lane and Weiss 1975; Mow et al. 1992; Redler and Zimny 1970). The distributions of extracellular matrix materials, the orientations of collagen fibril and chondrocyte, and the chondrocyte density and morphology all depend on the tissue depth. Musculoskeletal disease like osteoarthritis (OA) leads to progressive heterogeneity of the extracellular matrix, including loss of proteoglycans and degeneration of the collagen network (Das Gupta et al. 2020; Karhula et al. 2017b; Oinas et al. 2016).

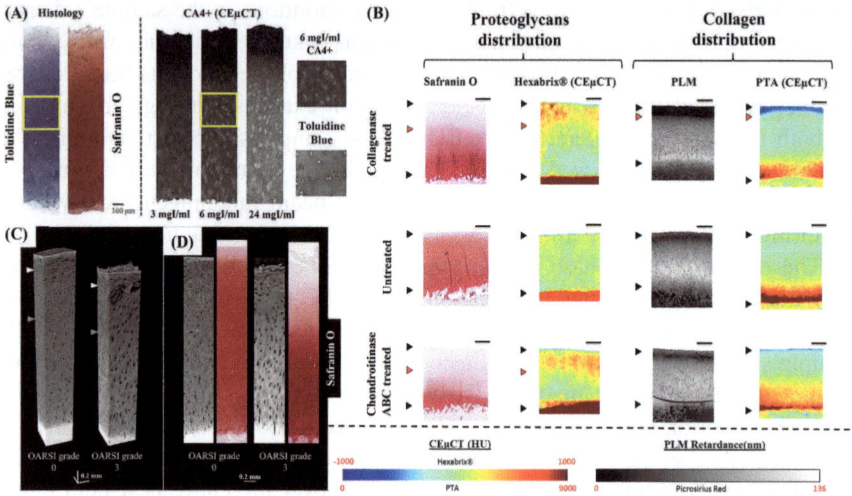

Fig. 1 CE-μCT imaging of articular cartilage. (**a**) Safranin O and Toluidine Blue indicate similar depth-wise GAG distribution as in the different concentrations of the CA4+ stained cartilage sample. Magnified areas (restricted by the yellow square) show that CA4+ accumulation exceeds the cell areas according to the GAG content around the chondrons (yellow dotted line circle). Adapted from (Karhula et al. 2017a) under a Creative Commons Attribution License (CC BY 4.0); (**b**) 2D CE-μCT images of Hexabrix- and PTA-labeled cartilage compared to the corresponding histological collagen (polarized light microscopy, PLM of Picrosirius red stain) and PG (Safranin O) stains. The histological stains showed clear effects of the enzymes, which can be seen as insufficient staining on the surface of the sample as compared to the control samples (red arrows). While the Picrosirius red stain and the PTA showed highly similar collagen distributions, Safranin O and Hexabrix showed highly similar PG distributions. Black arrows indicate the cartilage surface and the uncalcified cartilage-CC interface. Scale bars equal to 500 μm. Adapted from (Karhula et al. 2017b) under a Creative Commons Attribution License (CC BY 4.0); (**c**) Volumetric visualizations of HMDS-dried normal, intact cartilage (histopathological severity, Osteoarthritis Research Society International (OARSI) grade 0) and osteoarthritic cartilage with vertical fissures (OARSI grades 3). The white arrowhead refers to the border between the superficial zone (0–10% from the cartilage surface) and the transitional zone (10–40% from the cartilage surface), and the gray arrowhead refers to the border between the transitional and deep zones (40–100% from the cartilage surface); (**d**) 2D images of intact (OARSI grade 0) and osteoarthritic (OARSI grade 3) cartilage samples imaged with HMDS-based CE-μCT and their adjacent Safranin O-stained histological sections. Images in panels C and D were adapted from (Kestilä et al. 2018) under a Creative Commons Attribution License (CC BY 4.0)

Ionic CESAs (either anionic or cationic) are commonly used to examine the loss of GAGs within cartilage (Fig. 1). Based on the Donnan equilibrium ion distribution law, the negatively charged anionic CESAs are repelled, and the positively charged cationic CESAs are attracted by the fixed charge density (FCD) generated by the highly negatively charged GAGs (Mow et al. 1992). This results in decreased (in the case of anionic CESAs) or increased (in the case of cationic CESAs) X-ray attenuation within the cartilage, making it appear darker or brighter in the reconstructed images, respectively. Cationic CESAs are more strongly associated with GAG content than anionic CESAs (Bansal et al. 2011).

Palmer et al. demonstrated the first application of an anionic CESA to monitor the GAG distribution in an in vitro model of IL-1-induced cartilage degradation using equilibrium partitioning of an ionic contrast agent *via* μCT (EPIC-μCT) with Hexabrix® 320, containing ioxaglate, a negatively charged hexaiodinated dimer (Palmer et al. 2006). However, since the X-ray attenuation of subchondral bone and Hexabrix-equilibrated cartilage can overlap in EPIC-μCT, the authors failed to distinguish them accurately from the 3D images, hindering a detailed evaluation of the cartilage during OA progression. Cationic contrast agents, particularly CA4+, show a strong positive correlation with GAG content compared to anionic agents, effectively revealing GAG distribution in articular cartilage even at low concentrations (Bansal et al. 2011). High-resolution CE-μCT imaging with CA4+ found that CA4+ accumulates in chondrocytes, and it affects the low-resolution ex vivo μCT imaging studies of CA4+ stained human cartilage with different OA severities (Karhula et al. 2017a). Moreover, CE-μCT imaging with inorganic CESAs, such as PTA and PMA, for characterizing collagen distribution within the entire cartilage volume was validated using Fourier-transform infrared imaging, histology, and biochemical analyses (Karhula et al. 2017b; Nieminen et al. 2015).

The feasibility of an analyzer-based PhC-μCT setup for visualizing articular cartilage at high resolution has been demonstrated on ex vivo human patella (Coan et al. 2010). The PhC-μCT imaging showed an excellent depiction of the complete volume and of the 3D architecture of the cartilage. It enabled the differentiation of OA and healthy cartilage samples using the histologic characteristics of cartilage structures. Horng et al. explored the feasibility of high-resolution propagation-based PhC-μCT imaging of hyaline cartilage from human cadaveric knees and compared it with conventional CT, synchrotron radiation absorption-based CT, and magnetic resonance imaging (MRI), and corresponding histology (Horng et al. 2014). The propagation-based setting achieved sufficient contrast and resolution for evaluating cartilage, as well as for analyzing the underlying subchondral bone and its morphological alterations in osteoarthritic conditions. Subsequently, state-of-the-art propagation-based setups and image-processing methods were implemented by the same research group in a laboratory setting for high-resolution imaging of healthy and degenerated cartilage samples (Horng et al. 2021). The PhC-μCT setting achieved the necessary resolution to visualize cellular and extracellular variation in cartilage matrix, approximating histology and TEM analyses, making it a non-invasive virtual histopathology technique.

4.2 Bone Marrow Imaging

The adult bone marrow is a complex, adipocyte-rich environment where adipose tissue is a loose connective tissue that has a distinct pattern throughout the marrow cavity and is dynamically regulated by a variety of physiological and pathological conditions (Li et al. 2018). Bone marrow adipocytes are actively engaged in both local and systemic metabolic processes (e.g., lipid metabolism, energy storage, thermogenesis) that are significant in maintaining skeletal and glucose homeostasis (Li et al. 2018; Suchacki et al. 2020). High-resolution 3D visualization of bone marrow adipose tissues, with their spatial distribution and structural properties, can provide a better understanding of their impact on metabolic and skeletal disorders.

OsO_4 staining specifically interacts with unsaturated double bonds of fatty acids and has been applied as a CESA in μCT visualization of bone marrow adipose tissue in decalcified long bones of small animal models (Scheller et al. 2015; Styner et al. 2014). However, OsO4 is highly toxic, has limited tissue penetration, and limits subsequent 2D histological assessment. Kerckhofs et al. introduced a novel CE-μCT imaging protocol with Hf-WD 1:2 POM staining, simultaneously allowing visualization of bone, bone marrow vasculature, and adipocytes (Kerckhofs et al. 2018). Hf-WD 1:2 POM staining resulted in μCT images with bone remains as the highest attenuating tissue, followed by bone marrow vasculature, and adipocytes having the lowest grey value. They also performed Raman spectroscopy to understand binding mechanisms and affinity of the CESA and reported that Hf-WD 1:2 POM does not chemically alter several collagen type II and fibrin proteins, unlike PTA, highlighting its non-destructive nature (Kerckhofs et al. 2018). The Hf-WD 1:2 POM staining allows bone marrow adipocytes to be visualized at the single-cell level to quantify adipocyte number and diameter, rather than only limited to volumetric tissue level imaging. Additionally, the vascular network in the marrow space could be visualized and analyzed. The Hf-WD 1:2 POM staining provides a valuable solution for 3D cellular-level quantitative and spatial analysis of bone marrow adipose tissues, offering complementary multi-tissue insights compared to standard 2D histomorphometry, thereby enhancing our understanding of the interactions between tissues in bone.

4.3 Meniscus Imaging

The menisci are essential for proper functioning and the long-term health of the knee joint. They are semi-lunar-shaped fibrocartilaginous structures between the bones that significantly contribute to the load distribution of the joint by enhancing the contact area (Fox et al. 2015). Menisci are made up of tightly organized collagen fibers (primarily type I) that form thicker bundles arranged in a circumferential direction. The outer zone of a meniscus has vascularity and innervation, while the inner zone is avascular and more similar to articular cartilage, containing collagen type II (Fox et al. 2015; Karjalainen et al. 2024).

CE-μCT imaging of ex vivo menisci using CA4+ and Ioxaglate was compared to Safranin O staining images, and the enhanced attenuation by CA4+ was found to be strongly correlated with the GAGs (Lakin et al. 2013). Moreover, the volumetric shrinkage and radiodensity of menisci for iodine-based CESAs (water and saline-based I_2KI) staining were investigated using sheep and pig menisci (Orellana et al. 2024). HMDS drying-based CE-μCT was successfully applied to effectively visualize and quantitatively assess the microstructures of human menisci (Fig. 2) (Karjalainen et al. 2021; Kestilä et al. 2019). CE-μCT images revealed various morphological features on the meniscal surface and within the collagen network similar to the traditional 2D histopathological assessments, as well as assessments of local collagen fiber orientation analysis. Furthermore, this method allowed visualization of meniscus cells, calcifications, vascular channels, and collagen layers. Alternatively, the CPD drying process was utilized to quantitatively analyze the 3D vascular network in an ex vivo porcine meniscus model of obesity and early OA (Karjalainen et al. 2024).

Synchrotron radiation-based PhC-μCT imaging was also applied for high-resolution 3D imaging of fresh frozen human menisci (Einarsson et al. 2022). Crimping pattern characterization and orientation analysis of collagen fiber from the PhC-μCT images were performed to assess the degeneration of meniscus tissue. Subsequently, the same research group investigated the simultaneous microstructural and mechanical behavior of meniscus tissues using a time-resolved synchrotron radiation-based PhC-μCT imaging setup (Dejea et al. 2024). Authors reported that during compression, the fibers in the meniscus reorient perpendicularly to the loading direction, which then partially returns during relaxation. These studies suggest that quantitative 3D parameters calculated from the CE-μCT techniques (e.g., analyses of the vascular network, collagen fiber orientation, and crimping pattern) could be used as potential markers of meniscus degeneration.

4.4 Tendons, Ligaments, and Muscle Fiber Imaging

Tendons and ligaments facilitate locomotion in the musculoskeletal system and are vital to transmitting force and stabilization. They are structured, unidirectional fibrous composites of mainly collagen type I molecules. The mechanical properties of collagen arise from its hierarchical organization and assembly. Each level of the hierarchy is combined with a non-collagenous matrix rich in proteoglycan (Handsfield et al. 2016). To study the changes in the structure of tendon fibers before attachment to the bone, an iodine-based contrast agent, Imeron 300, was used as a CESA for high-resolution CE-μCT imaging (Rossetti et al. 2017). Tendon fibers were found to unravel into thinner interface fibers upon entering the tendon–bone insertion region. Balint et al. compared I_2KI, PTA, and PMA as optimal CESA for CE-μCT imaging of porcine knee ligaments and tendons (Balint et al. 2016). They reported that I_2KI had rapid diffusion and caused the least sample shrinkage. Murine Achilles tendon entheses were visualized using CE-μCT imaging by PTA staining or by a combination of cell-maceration, demineralization, and CPD with propagation-based PhC-μCT (Sartori et al. 2018). CE-μCT images were then compared with the classical histological sections to validate the method as a 3D histological tool. While PTA-based CE-μCT imaging introduced tissue shrinkage, it successfully visualized the collagen fiber bundles within the tendon. PhC-μCT imaging enabled the tracking of tendon fiber courses and characterization of fiber shapes, as well as the quantitative assessment of fiber thickness distribution and

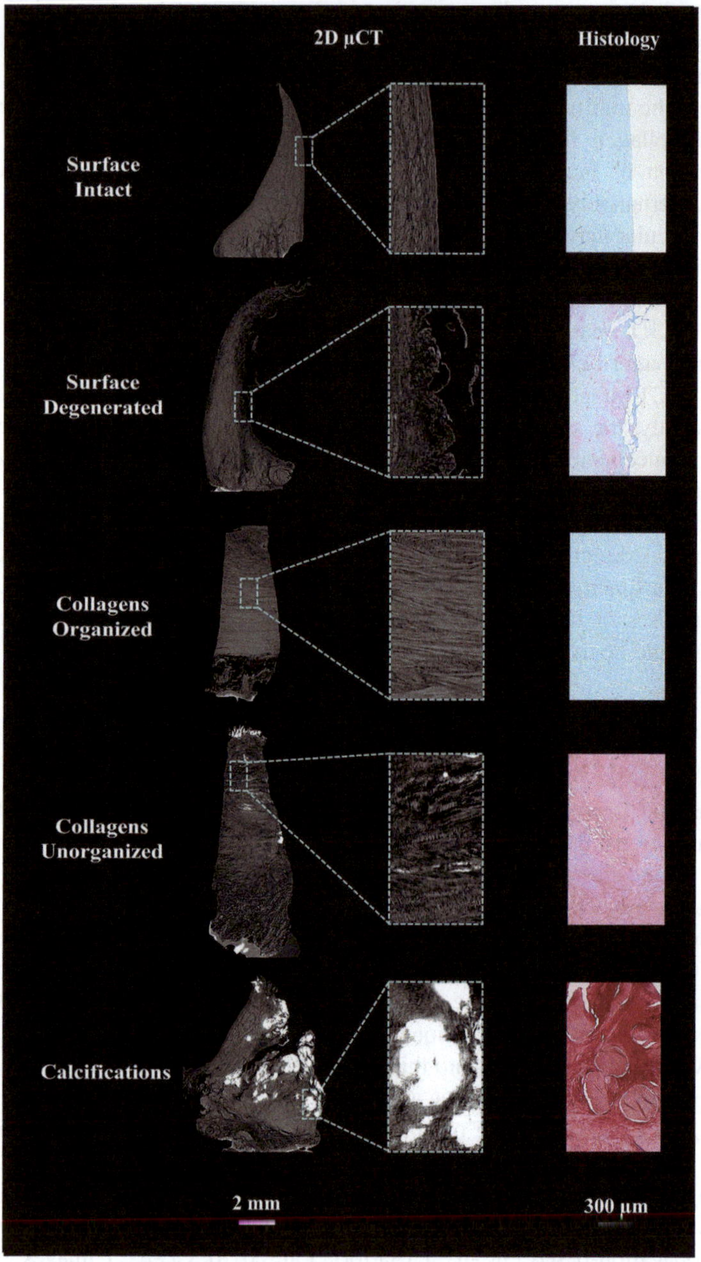

Fig. 2 A comparative visualization of various meniscal structural features using HMDS drying-based CE-μCT and Safranin O Fast Green stained images. CE-μCT images can effectively visualize several meniscal structural features, such as surface morphology and collagen network organization, that are comparable to traditional histopathological assessment. Moreover, meniscus calcifications are visualized in the HMDS drying-based CE-μCT imaging. (Adapted from (Kestilä et al. 2019) under a Creative Commons Attribution License (CC BY 4.0))

volume within the tendon (Sartori et al. 2018). Pierantoni et al. reported an optimized protocol for synchrotron PhC-μCT imaging of the human Achilles tendon by comparing four types of sample preparations (fresh frozen, fixed, embedded, and stained) and two imaging setups (sub-micrometric and micrometric final pixel sizes) (Pierantoni et al. 2021). They found that fresh-frozen tendon samples resulted in high-quality PhC-μCT images capturing the complex 3D organization of fibers in nearly native states.

The HMDS-based drying protocol enabled 3D CE-μCT visualization of human ACL bundles and their bone insertions (Parrilli et al. 2024). This imaging protocol facilitated a virtual macroscopic analysis of the shape of ACL insertions at the femoral and tibial levels, offering valuable insights into how the conformation of the ACL bundles relates to their physiological function. Moreover, this protocol allowed the authors to perform subsequent histology with Safranin O staining on the same samples after rehydrating. By registering the 2D histology sections with μCT images, the authors achieved precise identification of the calcified fibrocartilage area, providing a more in-depth understanding of the calcification of the ACL attachment system at the tibial level (Parrilli et al. 2024). Previously, a bimodal protocol for imaging adult rats' Achilles tendons was developed using whole-mount confocal microscopy and HMDS-based CE-μCT to visualize the cellular and microstructural properties in 3D (Marr et al. 2020). The optical clearing approach rendered the tendon transparent for confocal imaging, followed by reverse clarification and HMDS-based drying that permits CE-μCT. This provides independent fluorescence-based immunolabeled markers at the cellular level and X-ray-based volumetric measurements for dense connective tissues.

3D imaging of skeletal muscle tissues and analysis of muscle fiber structure have been the subjects of numerous CE-μCT-based research studies, which utilize staining with inorganic iodine solutions, e.g., I_2KI (Dickinson et al. 2018; Gignac et al. 2016; Kupczik et al. 2015). However, I_2KI can introduce substantial tissue shrinkage, which may presumably affect the visualization of individual fibers. Recently, Maes et al. demonstrated the nondestructive 3D visualization of individual muscle fibers using cryogenic CE-μCT (Maes et al. 2022). By combining sample staining with contrast-enhancing Hf-WD 1:2 POM, with freezing the sample at the optimal freezing rate, cryogenic CE-μCT enabled 3D visualization and structural analysis of muscle and tendon fibers.

5 Practical Challenges

Several important practical challenges related to CE-μCT imaging in 3D anatomical pathology of musculoskeletal tissues must be considered. It is essential to optimize the CESA staining protocol and ensure consistent sample preparation for μCT imaging (e.g., maintaining the appropriate pH of the staining solution) to minimize significant tissue shrinkage and regulate CESA diffusion within the tissue. Additionally, for ex vivo CE-μCT imaging, the removal of CESA staining is crucial for subsequent tissue experiments. The removal efficiency largely depends on the type and concentration of CESA, the size and permeability properties of tissue, and the staining duration. Routine washing with PBS or saline can eliminate weakly covalently bonded CESA, while chemical treatments may be necessary to remove tightly bonded CESA from the tissue. The electrostatic interactions between biomolecules and CESA may require several washing cycles, depending on the CESA's concentration and the ionic strength of the washing solution. In some situations, complete removal of CESA from the tissues may not be feasible due to prolonged staining times and high concentrations.

A significant limitation of μCT systems is the ratio of sample size to voxel size, which restricts the visualization of small structures, such as individual cells, within larger tissues. Laboratory μCT systems often lack the capability to provide a sufficiently wide field of view for capturing the entire tissue sample at high resolution. This issue can be addressed through synchrotron imaging, which offers a broader field of view. Additionally, synchrotron X-rays are significantly brighter, resulting in an improved signal-to-noise ratio and reduced scan times. For instance, the high coherence and tomography beam-line configurations facilitate long propagation distances, making them suitable for in-line PhC-μCT imaging of soft tissues. Nonetheless, it is essential to strike a balance between the radiation dose and the scanning parameters, including beam filters, the number of projections, and exposure time.

6 Conclusions and Future Perspectives

The importance of conventional 2D histopathology in characterizing pathology-induced alterations in microstructure cannot be understated. Nevertheless, there remains an unmet need for imaging tools to facilitate 3D microstructural analysis for anatomical pathology. CE-μCT imaging could provide unprecedented insight into the 3D microstructural organization of various soft biological tissue types, including musculoskeletal tissues. This chapter explores various contrast enhancement techniques for ex vivo μCT imaging, with a focus on the application of CE-μCT imaging to analyze different musculoskeletal tissues in both healthy and diseased states. Although considerable progress has been made in the applicability of CE-μCT imaging methods, several limitations and practical considerations must be taken into account before quantitative virtual 3D anatomical pathology based on CE-μCT imaging becomes a routinely used complementary technique to 2D histochemistry or immuno-histomorphometry. Developing tissue-specific CESA (potentially via immunodetection), understanding their binding mechanisms, optimizing the staining protocol, and minimizing potential tissue shrinkage remain crucial. Moreover, the use of PhC-μCT in non-synchrotron settings presents certain challenges due to the relatively large focus spot sizes and broad energy spectra of commercially available conventional X-ray tube beams. Addressing these limitations will be crucial for advancing CE-μCT imaging towards becoming a standard tool for quantitative virtual 3D anatomical pathology, facilitating significant progress in basic and preclinical research.

Acknowledgments We would like to acknowledge the funding from the Research Council of Finland (Academy research fellows, Grant: 347445) and Funding from Jane and Erkko Aatos Foundation.

Conflict of Interest The authors report no conflict of interest.

References

Balcaen T, Vangrunderbeeck S, De Borggraeve WM, Kerckhofs G (2025) Contrast-enhancing staining agents for ex vivo contrast-enhanced computed tomography: a review. Tomography of. Mater Struct 7(June 2024):100052. https://doi.org/10.1016/j.tmater.2025.100052

Balint R, Lowe T, Shearer T (2016) Optimal contrast agent staining of ligaments and tendons for X-ray computed tomography. PLoS One 11(4):e0153552. https://doi.org/10.1371/journal.pone.0153552

Bansal PN, Joshi NS, Entezari V, Malone BC, Stewart RC, Snyder BD, Grinstaff MW (2011) Cationic contrast agents improve quantification of glycosaminoglycan (GAG) content by contrast enhanced CT imaging of cartilage. J Orthop Res 29(5):704–709. https://doi.org/10.1002/jor.21312

Bhattacharya R, Saha S, Kostina O, Muravnik L, Mitra A (2020) Replacing critical point drying with a low-cost chemical drying provides comparable surface image quality of glandular trichomes from leaves of Millingtonia hortensis L. f. in scanning electron micrograph. Appl Microsc 50(1):15. https://doi.org/10.1186/s42649-020-00035-6

Busse M, Müller M, Kimm MA, Ferstl S, Allner S, Achterhold K, Herzen J, Pfeiffer F (2018) Three-dimensional virtual histology enabled through cytoplasm-specific X-ray stain for microscopic and nanoscopic computed tomography. Proc Natl Acad Sci 115(10):2293–2298. https://doi.org/10.1073/pnas.1720862115

Chen W-C, Chen H-Y, Liao P-C, Wang S-J, Tsai M-Y, Chen Y-H, Lin W-Y (2018) Toward a new insight of calcium oxalate stones in drosophila by micro-computerized tomography. Urolithiasis 46(2):149–155. https://doi.org/10.1007/s00240-017-0967-0

Coan P, Bamberg F, Diemoz PC, Bravin A, Timpert K, Mützel E, Raya JG, Adam-Neumair S, Reiser MF, Glaser C (2010) Characterization of osteoarthritic and normal human patella cartilage by computed tomography X-ray phase-contrast imaging: a feasibility study. Investig Radiol 45(7):437–444. https://doi.org/10.1097/RLI.0b013e3181e193bd

Das Gupta S, Finnilä MAJ, Karhula SS, Kauppinen S, Joukainen A, Kröger H, Korhonen RK, Thambyah A, Rieppo L, Saarakkala S (2020) Raman microspectroscopic analysis of the tissue-specific composition of the human osteochondral junction in osteoarthritis: a pilot study. Acta Biomater 106:145–155. https://doi.org/10.1016/j.actbio.2020.02.020

de Bournonville S, Vangrunderbeeck S, Kerckhofs G (2019) Contrast-enhanced MicroCT for virtual 3D anatomical pathology of biological tissues: a literature review. Contrast Media Mol Imaging 2019:1–9. https://doi.org/10.1155/2019/8617406

Dejea H, Pierantoni M, Orozco GA, Wrammerfors ETB, Gstöhl SJ, Schlepütz CM, Isaksson H (2024) In situ loading and time-resolved synchrotron-based phase contrast tomography for the mechanical investigation of connective knee tissues: a proof-of-concept study. Adv Sci:1–14. https://doi.org/10.1002/advs.202308811

Dickinson E, Stark H, Kupczik K (2018) Non-destructive determination of muscle architectural variables through the use of DiceCT. Anat Rec 301(2):363–377. https://doi.org/10.1002/ar.23716

Disney CM, Madi K, Bodey AJ, Lee PD, Hoyland JA, Sherratt MJ (2017) Visualising the 3D microstructure of stained and native intervertebral discs using X-ray microtomography. Sci Rep 7(1):1–11. https://doi.org/10.1038/s41598-017-16354-w

Disney CM, Lee PD, Hoyland JA, Sherratt MJ, Bay BK (2018) A review of techniques for visualising soft tissue microstructure deformation and quantifying strain ex vivo. J Microsc 272(3):165–179. https://doi.org/10.1111/jmi.12701

Disney CM, Eckersley A, McConnell JC, Geng H, Bodey AJ, Hoyland JA, Lee PD, Sherratt MJ, Bay BK (2019) Synchrotron tomography of intervertebral disc deformation quantified by digital volume correlation reveals microstructural influence on strain patterns. Acta Biomater 92:290–304. https://doi.org/10.1016/j.actbio.2019.05.021

Docter D, Timmerman M, Dawood Y, Hagoort J, Lobe N, van Heurn E, Gorter R, Jacobs K, Pyka G, Kerckhofs G, van den Hoff MJB, de Bakker B (2024) Scaling up contrast-enhanced micro-CT imaging: optimizing contrast and acquisition for large ex-vivo human samples. Forensic Imaging 37(April):200596. https://doi.org/10.1016/j.fri.2024.200596

Einarsson E, Pierantoni M, Novak V, Svensson J, Isaksson H, Englund M (2022) Phase-contrast enhanced synchrotron micro-tomography of human meniscus tissue. Osteoarthr Cartil 30(9):1222–1233. https://doi.org/10.1016/j.joca.2022.06.003

Farahani N, Braun A, Jutt D, Huffman T, Reder N, Liu Z, Yagi Y, Pantanowitz L (2017) Three-dimensional imaging and scanning: current and future applications for pathology. J Pathol Inf 8(1):36. https://doi.org/10.4103/jpi.jpi_32_17

Fox AJS, Wanivenhaus F, Burge AJ, Warren RF, Rodeo SA (2015) The human meniscus: a review of anatomy, function, injury, and advances in treatment. Clin Anat 28(2):269–287

Furlani M, Riberti N, Gatto ML, Giuliani A (2023) High-resolution phase-contrast tomography on human collagenous tissues: a comprehensive review. Tomography 9(6):2116–2133. https://doi.org/10.3390/tomography9060166

Gignac PM, Kley NJ, Clarke JA, Colbert MW, Morhardt AC, Cerio D, Cost IN, Cox PG, Daza JD, Early CM, Echols MS, Henkelman RM, Herdina AN, Holliday CM, Li Z, Mahlow K, Merchant S, Müller J, Orsbon CP et al (2016) Diffusible iodine-based contrast-enhanced computed tomography (diceCT): an emerging tool for rapid, high-resolution, 3-D imaging of metazoan soft tissues. J Anat 228(6):889–909. https://doi.org/10.1111/joa.12449

Glancy SB, Morris HD, Ho VB, Klarmann GJ (2023) Optimal agents for visualizing collagen tissue microarchitecture using contrast-enhanced MicroCT. Pharmaceuticals 16(12):1719. https://doi.org/10.3390/ph16121719

Häggmark I, Hoshino M, Uesugi K, Sasaki T (2024) X-ray phase contrast reveals soft tissue and shell growth lines in mollusks. Commun Biol 7(1):17. https://doi.org/10.1038/s42003-023-05457-y

Handsfield GG, Slane LC, Screen HRC (2016) Nomenclature of the tendon hierarchy: an overview of inconsistent terminology and a proposed size-based naming scheme with terminology for multi-muscle tendons. J Biomech 49(13):3122–3124. https://doi.org/10.1016/j.jbiomech.2016.06.028

Horng A, Brun E, Mittone A, Gasilov S, Weber L, Geith T, Adam-Neumair S, Auweter SD, Bravin A, Reiser MF, Coan P (2014) Cartilage and soft tissue imaging using X-rays: propagation-based phase-contrast computed tomography of the Human Knee in comparison with clinical imaging techniques and histology. Investig Radiol 49(9):627–634. https://doi.org/10.1097/RLI.0000000000000063

Horng A, Stroebel J, Geith T, Milz S, Pacureanu A, Yang Y, Cloetens P, Lovric G, Mittone A, Bravin A, Coan P (2021) Multiscale X-ray phase contrast imaging of human cartilage for investigating osteoarthritis formation. J Biomed Sci 28(1):1–14. https://doi.org/10.1186/s12929-021-00739-1

Johnson JT, Hansen MS, Wu I, Healy LJ, Johnson CR, Jones GM, Capecchi MR, Keller C (2006) Virtual histology of transgenic mouse embryos for high-throughput phenotyping. PLoS Genet 2(4):471–477. https://doi.org/10.1371/journal.pgen.0020061

Karhula SS, Finnilä MA, Freedman JD, Kauppinen S, Valkealahti M, Lehenkari P, Pritzker KPH, Nieminen HJ, Snyder BD, Grinstaff MW, Saarakkala S (2017a) Micro-scale distribution of CA4+ in ex vivo human articular cartilage detected with contrast-enhanced micro-computed tomography imaging. Frontiers. Physics 5(August):1–8. https://doi.org/10.3389/fphy.2017.00038

Karhula SS, Finnilä MA, Lammi MJ, Ylärinne JH, Kauppinen S, Rieppo L, Pritzker KPH, Nieminen HJ, Saarakkala S (2017b) Effects of articular cartilage constituents on Phosphotungstic acid enhanced micro-computed tomography. PLoS One 12(1):e0171075. https://doi.org/10.1371/journal.pone.0171075

Karjalainen VP, Kestilä I, Finnilä MA, Folkesson E, Turkiewicz A, Önnerfjord P, Hughes V, Tjörnstrand J, Englund M, Saarakkala S (2021) Quantitative three-dimensional collagen orientation analysis of human meniscus posterior horn in health and osteoarthritis using micro-computed tomography. Osteoarthr Cartil 29(5):762–772. https://doi.org/10.1016/j.joca.2021.01.009

Karjalainen V, Herrera Millar VR, Modina S, Peretti GM, Pallaoro M, Elkhouly K, Saarakkala S, Mobasheri A, Di Giancamillo A, Finnilä MAJ (2024) Age and anatomical region-related differences in vascularization of the porcine meniscus using microcomputed tomography imaging. J Orthop Res 42(9):2095–2105. https://doi.org/10.1002/jor.25862

Kerckhofs G, Stegen S, van Gastel N, Sap A, Falgayrac G, Penel G, Durand M, Luyten FP, Geris L, Vandamme K, Parac-Vogt T, Carmeliet G (2018) Simultaneous three-dimensional visualization of mineralized and soft skeletal tissues by a novel microCT contrast agent with polyoxometalate structure. Biomaterials 159:1–12. https://doi.org/10.1016/j.biomaterials.2017.12.016

Kestilä I, Thevenot J, Finnilä MA, Karhula SS, Hadjab I, Kauppinen S, Garon M, Quenneville E, Haapea M, Rieppo L, Pritzker KP, Buschmann MD, Nieminen HJ, Saarakkala S (2018) In vitro method for 3D morphometry of human articular cartilage chondrons based on micro-computed tomography. Osteoarthr Cartil 26:1118–1126. https://doi.org/10.1016/j.joca.2018.05.012

Kestilä I, Folkesson E, Finnilä MA, Turkiewicz A, Önnerfjord P, Hughes V, Tjörnstrand J, Englund M, Saarakkala S (2019) Three-dimensional microstructure of human meniscus posterior horn in health and osteoarthritis. Osteoarthr Cartil 27(12):1790–1799. https://doi.org/10.1016/j.joca.2019.07.003

Kupczik K, Stark H, Mundry R, Neininger FT, Heidlauf T, Röhrle O (2015) Reconstruction of muscle fascicle architecture from iodine-enhanced microCT images: a combined texture mapping and streamline approach. J Theor Biol 382:34–43. https://doi.org/10.1016/j.jtbi.2015.06.034

Lakin BA, Grasso DJ, Stewart RC, Freedman JD, Snyder BD, Grinstaff MW (2013) Contrast enhanced CT attenuation correlates with the GAG content of bovine meniscus. J Orthop Res 31(11):1765–1771. https://doi.org/10.1002/jor.22421

Lane JM, Weiss C (1975) Review of articular cartilage collagen research. Arthritis Rheum 18(6):553–562. https://doi.org/10.1002/art.1780180605

Layton MW, Goldstein SA, Goulet RW, Feldkamp LA, Kubinski DJ, Bole GG (1988) Examination of subchondral bone architecture in experimental osteoarthritis by microscopic computed axial tomography. Arthritis Rheum 31(11):1400–1405. https://doi.org/10.1002/art.1780311109

Lesciotto KM, Motch Perrine SM, Kawasaki M, Stecko T, Ryan TM, Kawasaki K, Richtsmeier JT (2020) Phosphotungstic acid-enhanced microCT: optimized protocols for embryonic and early postnatal mice. Dev Dyn 249(4):573–585. https://doi.org/10.1002/dvdy.136

Li Z, Clarke JA, Ketcham RA, Colbert MW, Yan F (2015) An investigation of the efficacy and mechanism of contrast-enhanced X-ray computed tomography utilizing iodine for large specimens through experimental and simulation approaches. BMC Physiol 15(1):1–16. https://doi.org/10.1186/s12899-015-0019-3

Li Z, Hardij J, Bagchi DP, Scheller EL, MacDougald OA (2018) Development, regulation, metabolism and function of bone marrow adipose tissues. Bone 110:134–140. https://doi.org/10.1016/j.bone.2018.01.008

Litzlbauer HD, Neuhaeuser C, Moell A, Greschus S, Breithecker A, Franke FE, Kummer W, Rau WS (2006) Three-dimensional imaging and morphometric analysis of alveolar tissue from microfocal X-ray-computed tomography. Am J Phys Lung Cell Mol Phys 291(3):L535–L545. https://doi.org/10.1152/ajplung.00088.2005

Maes A, Pestiaux C, Marino A, Balcaen T, Leyssens L, Vangrunderbeeck S, Pyka G, De Borggraeve WM, Bertrand L, Beauloye C, Horman S, Wevers M, Kerckhofs G (2022) Cryogenic contrast-enhanced microCT enables nondestructive 3D quantitative histopathology of soft biological tissues. Nat Commun 13(1):6207. https://doi.org/10.1038/s41467-022-34048-4

Marr N, Hopkinson M, Hibbert AP, Pitsillides AA, Thorpe CT (2020) Bimodal whole-mount imaging of tendon using confocal microscopy and X-ray micro-computed tomography. Biol Proced Online 22(1):13. https://doi.org/10.1186/s12575-020-00126-4

Metscher BD (2009) MicroCT for developmental biology: a versatile tool for high-contrast 3D imaging at histological resolutions. Dev Dyn 238(3):632–640. https://doi.org/10.1002/dvdy.21857

Metscher B (2021) A simple nuclear contrast staining method for microCT-based 3D histology using lead(II) acetate. J Anat 238(4):1036–1041. https://doi.org/10.1111/joa.13351

Metscher BD, Müller GB (2011) MicroCT for molecular imaging: quantitative visualization of complete three-dimensional distributions of gene products in embryonic limbs. Dev Dyn 240(10):2301–2308. https://doi.org/10.1002/dvdy.22733

Mirone A, Brunet J, Admans L, Boistel R, Sowinski M, Berruyer C, Payno H, Boller E, Paleo P, Walsh CL (2023) Eikonal phase retrieval: unleashing the fourth generation sources potential for enhanced propagation based tomography on biological samples. ArXiv Preprint ArXiv:2311.14745

Mow VC, Ratcliffe A, Robin Poole A (1992) Cartilage and diarthrodial joints as paradigms for hierarchical materials and structures. Biomaterials 13(2):67–97. https://doi.org/10.1016/0142-9612(92)90001-5

Müller M, Kimm MA, Ferstl S, Allner S, Achterhold K, Herzen J, Pfeiffer F, Busse M (2018) Nucleus-specific X-ray stain for 3D virtual histology. Sci Rep 8(1):17855. https://doi.org/10.1038/s41598-018-36067-y

Nation JL (1983) A new method using hexamethyldisilazane for preparation of soft insect tissues for scanning electron microscopy. Stain Technol 58(6):347–351

Nieminen HJ, Ylitalo T, Karhula S, Suuronen JP, Kauppinen S, Serimaa R, Hæggström E, Pritzker KPH, Valkealahti M, Lehenkari P, Finnilä M, Saarakkala S (2015) Determining collagen distribution in articular cartilage using contrast-enhanced micro-computed tomography. Osteoarthr Cartil 23(9):1613–1621. https://doi.org/10.1016/j.joca.2015.05.004

Oinas J, Rieppo L, Finnilä MAJ, Valkealahti M, Lehenkari P, Saarakkala S (2016) Imaging of osteoarthritic human articular cartilage using Fourier transform infrared microspectroscopy combined with multivariate and univariate analysis. Sci Rep 6(1):30008. https://doi.org/10.1038/srep30008

Orellana F, Grassi A, Nuss KM, Wahl P, Neels A, Zaffagnini S, Parrilli A (2024) Spatial and temporal evaluation of iodine uptake and radiodensity in meniscus tissue using contrast-enhanced micro-CT. Heliyon 10(24):e41080. https://doi.org/10.1016/j.heliyon.2024.e41080

Paganin D, Mayo SC, Gureyev TE, Miller PR, Wilkins SW (2002) Simultaneous phase and amplitude extraction from a single defocused image of a homogeneous object. J Microsc 206(1):33–40. https://doi.org/10.1046/j.1365-2818.2002.01010.x

Palmer AW, Guldberg RE, Levenston ME (2006) Analysis of cartilage matrix fixed charge density and three-dimensional morphology via contrast-enhanced microcomputed tomography. Proc Natl Acad Sci USA 103(51):19255–19260. https://doi.org/10.1073/pnas.0606406103

Parapar J, Candás M, Cunha-Veira X, Moreira J (2017) Exploring annelid anatomy using micro-computed tomography: a taxonomic approach. Zool Anz 270:19–42. https://doi.org/10.1016/j.jcz.2017.09.001

Parrilli A, Grassi A, Orellana F, Lolli R, Marchiori G, Berni M, Fini M, Lopomo NF, Zaffagnini S (2024) 3D visualization of the human anterior cruciate ligament combining micro-CT and histological analysis. Surg Radiol Anat 46(2):249–258. https://doi.org/10.1007/s00276-023-03295-5

Pascart T, Falgayrac G, Cortet B, Paccou J, Bleuse M, Coursier R, Putman S, Quinchon JF, Bertheaume N, Delattre J, Marchandise P, Cultot A, Norberciak L, Kerckhofs G, Budzik JF (2022) Subchondral involvement in osteonecrosis of the femoral head: insight on local composition, microstructure and vascularization. Osteoarthr Cartil 30(8):1103–1115. https://doi.org/10.1016/j.joca.2022.05.003

Pauwels E, Van Loo D, Cornillie P, Brabant L, Van Hoorebeke L (2013) An exploratory study of contrast agents for soft tissue visualization by means of high resolution X-ray computed tomography imaging. J Microsc 250(1):21–31. https://doi.org/10.1111/jmi.12013

Pfeiffer F, Weitkamp T, Bunk O, David C (2006) Phase retrieval and differential phase-contrast imaging with low-brilliance X-ray sources. Nat Phys 2(4):258–261. https://doi.org/10.1038/nphys265

Pierantoni M, Silva Barreto I, Hammerman M, Verhoeven L, Törnquist E, Novak V, Mokso R, Eliasson P, Isaksson H (2021) A quality optimization approach to image Achilles tendon microstructure by phase-contrast enhanced synchrotron micro-tomography. Sci Rep 11(1):17313. https://doi.org/10.1038/s41598-021-96589-w

Quenot L, Bohic S, Brun E (2022) X-ray phase contrast imaging from synchrotron to conventional sources: a review of the existing techniques for biological applications. Appl Sci (Switzerland) 12(19). https://doi.org/10.3390/app12199539

Rawson SD, Maksimcuka J, Withers PJ, Cartmell SH (2020) X-ray computed tomography in life sciences. BMC Biol 18(1):21. https://doi.org/10.1186/s12915-020-0753-2

Redler I, Zimny ML (1970) Scanning electron microscopy of normal and abnormal articular cartilage and synovium. J Bone Joint Surg Am 52(7):1395–1404

Roberts N, Magee D, Song Y, Brabazon K, Shires M, Crellin D, Orsi NM, Quirke R, Quirke P, Treanor D (2012) Toward routine use of 3D histopathology as a research tool. Am J Pathol 180(5):1835–1842. https://doi.org/10.1016/j.ajpath.2012.01.033

Rossetti L, Kuntz LA, Kunold E, Schock J, Müller KW, Grabmayr H, Stolberg-Stolberg J, Pfeiffer F, Sieber SA, Burgkart R, Bausch AR (2017) The microstructure and micromechanics of the tendon–bone insertion. Nat Mater 16(6):664–670. https://doi.org/10.1038/nmat4863

Runge J, Stahnke T, Guthoff RF, Wree A, Keiler J (2022) Micro-CT in ophthalmology: ex vivo preparation and contrasting methods for detailed 3D-visualization of eye anatomy with special emphasis on critical point drying. Quant Imaging Med Surg 12(9):4361–4376. https://doi.org/10.21037/qims-22-109

Sartori J, Köhring S, Witte H, Fischer MS, Löffler M (2018) Three-dimensional imaging of the fibrous microstructure of Achilles tendon entheses in Mus musculus. J Anat 233(3):370–380. https://doi.org/10.1111/joa.12837

Scheller EL, Doucette CR, Learman BS, Cawthorn WP, Khandaker S, Schell B, Wu B, Ding SY, Bredella MA, Fazeli PK, Khoury B, Jepsen KJ, Pilch PF, Klibanski A, Rosen CJ, MacDougald OA (2015) Region-specific variation in the properties of skeletal adipocytes reveals regulated and constitutive marrow adipose tissues. Nat Commun 6:7808. https://doi.org/10.1038/ncomms8808

Silva Barreto I, Pierantoni M, Nielsen LC, Hammerman M, Diaz A, Novak V, Eliasson P, Liebi M, Isaksson H (2024) Micro- and nanostructure specific X-ray tomography reveals less matrix formation and altered collagen organization following reduced loading during Achilles tendon healing. Acta Biomater 174:245–257. https://doi.org/10.1016/j.actbio.2023.12.015

Styner M, Thompson WR, Galior K, Uzer G, Wu X, Kadari S, Case N, Xie Z, Sen B, Romaine A, Pagnotti GM, Rubin CT, Styner MA, Horowitz MC, Rubin J (2014) Bone marrow fat accumulation accelerated by high fat diet is suppressed by exercise. Bone 64:39–46. https://doi.org/10.1016/j.bone.2014.03.044

Suchacki KJ, Tavares AAS, Mattiucci D, Scheller EL, Papanastasiou G, Gray C, Sinton MC, Ramage LE, McDougald WA, Lovdel A, Sulston RJ, Thomas BJ, Nicholson BM, Drake AJ, Alcaide-Corral CJ, Said D, Poloni A, Cinti S, Macpherson GJ et al (2020) Bone marrow adipose tissue is a unique adipose subtype with distinct roles in glucose homeostasis. Nat Commun 11(1):3097. https://doi.org/10.1038/s41467-020-16878-2

Väänänen V, Christensen MM, Suhonen H, Jernvall J (2023) Gene expression detection in developing mouse tissue using in situ hybridization and μCT imaging. Proc Natl Acad Sci 120(24):2017. https://doi.org/10.1073/pnas.2301876120

Verleden SE, Vasilescu DM, Willems S, Ruttens D, Vos R, Vandermeulen E, Hostens J, McDonough JE, Verbeken EK, Verschakelen J (2014) The site and nature of airway obstruction after lung transplantation. Am J Respir Crit Care Med 189(3):292–300

Xu H, Olivier C, Sajidy H, Pallu S, Portier H, Peyrin F, Chappard C (2024) Cell quantification at the osteochondral interface from synchrotron radiation phase contrast micro-computed tomography images using a deep learning approach. Sci Rep 14(1):29619. https://doi.org/10.1038/s41598-024-81333-x

Zhang T, Li S, Chen Y, Xiao H, Wang L, Hu J, Xu D, Lu H (2022) Characterize the microstructure change after tendon enthesis injury using synchrotron radiation μCT. J Orthop Res 40(11):2678–2687. https://doi.org/10.1002/jor.25289

Zuo C, Li J, Sun J, Fan Y, Zhang J, Lu L, Zhang R, Wang B, Huang L, Chen Q (2020) Transport of intensity equation: a tutorial. Opt Lasers Eng 135(April):106187. https://doi.org/10.1016/j.optlaseng.2020.106187

Comprehensive Analysis of Cartilage Integrity with Cartilage Roughness Score from Micro-computed Tomography

Sami Kauppinen, Lucia Baixauli Marin, and Mikko A. J. Finnilä

Abstract

Osteoarthritis (OA) is a degenerative joint disease affecting a significant portion of the population. No approved therapies for OA exist, and advanced methods to evaluate preclinical outcomes are urgently needed. OA is a whole joint disease, and micro-computed tomography is frequently used to analyze bone structures in preclinical models. Still, there is a limited number of methods for cartilage analysis. Accurate detection of cartilage surface defects is critical when assessing OA progression and therapeutic outcomes. Current methods like histological grading are time-intensive and limited by their two-dimensional nature. This chapter introduces the Cartilage Roughness Score (CRS) from a contrast-enhanced micro-computed tomography (CEμCT) as a detailed, three-dimensional analysis of cartilage surfaces.

This chapter will provide an outline of the sample preparation and actual analysis. Possible sample preparation needs to maintain surface integrity. Most of the processing steps are automated, including cartilage segmentation and iterative polynomial surface fitting to approximate healthy cartilage. Localized angular deviations are calculated within defined windows, providing a spatially detailed map of surface irregularities. Outputs such as CRS(mean), CRS > X°, and sum CRS are discussed as measures for assessing surface degradation and

S. Kauppinen (✉) · L. B. Marin
Research Unit of Health Sciences and Technology, Faculty of Medicine, University of Oulu, Oulu, Finland
e-mail: sami.kauppinen@oulu.fi

M. A. J. Finnilä
Research Unit of Health Sciences and Technology, Faculty of Medicine, University of Oulu, Oulu, Finland

Biocenter Oulu, University of Oulu, Oulu, Finland

I. F. Cengiz et al. (eds.), *Bioimaging Modalities in Bioengineering*, Biomaterials, Bioengineering and Sustainability 9,
https://doi.org/10.1007/978-3-031-99410-4_11

tracking lesion progression. The advantages of CRS include its ability to analyze the entire joint surface, reducing the bias introduced by histological sampling. The method is adaptable to various imaging modalities as long as resolution criteria are met. This chapter details the practical steps for implementing CRS, highlights its potential in OA research, and discusses areas for further refinement and automation.

Keywords

Cartilage roughness score · Micro-computed tomography · Surface analysis · Osteoarthritis · MATLAB

1 Introduction

Osteoarthritis (OA) is a chronic, progressive, degenerative joint disease affecting the whole joint. The rate of OA is increasing with numerous risk factors identified, the most significant being age, previous injury, female gender, and obesity (Palazzo et al. 2016). Recent evidence indicates that rather than being a single well-defined disease, OA is an umbrella term for the state of a joint where increasing symptoms have resulted in patients seeking medical help and, therefore, receiving a diagnosis (Bruyère et al. 2015; Roemer et al. 2023). This further refers to different OA phenotypes that develop through specific pathophysiological mechanics (endotypes) that require specific therapy (Hunter and Deveza 2024).

Preclinical studies using different animal models are vital in understanding the pathophysiological mechanisms of different OA phenotypes. They also play a key role in developing therapeutic biomaterials or pharmaceutical products. Cartilage Roughness Score (CRS) is highly relevant in these studies as it allows for a detailed evaluation of the cartilage surface condition. It provides valuable quantitative data by capturing subtle changes that conventional histological approaches might miss. In drug development, accurate preclinical assessment is crucial because it ensures the identification of effective therapies, thereby reducing the risk of advancing ineffective drugs to clinical trials. The ability to detect early-stage cartilage degradation makes it possible to evaluate therapeutical interventions that try to prevent progression in the early phase, where a so-called treatment window is still open.

At the time of writing this paragraph, no successful disease-modifying drugs are clinically available, and current guidelines are still focused on alleviating pain only (Rodriguez-Merchan 2023). Although changes could be detected before becoming irreversible and within the treatment window, actionable interventions remain limited. The treatment window is crucial because, once cartilage loses the Benninghoff architecture (Benninghoff 1925), biomechanical changes in the joint can trigger a cascading process of degeneration. There also begins a cycle of a self-feeding loop where the particles from breaking cartilage start to feed an inflammatory process because these particles need to be cleaned out. The inflammatory process releases more particles, which leads to maintaining the inflammatory process (Sokolove and Lepus 2013).

Many studies investigating OA pathophysiology have focused on bone or cartilage, but both tissues and all other knee tissues should be considered together for a comprehensive understanding (Findlay and Kuliwaba 2016). Micro-computed X-ray tomography (μCT) is highly effective for analyzing bone changes in 3D due to the natural X-ray contrast in bone. Several tools are available for bone analysis, with BoneJ (Doube et al. 2010) being a commonly used free software.

Cartilage imaging, however, requires contrast-enhancing techniques, such as using negatively charged phosphotungstic acid (PTA), which binds to positively charged collagen (Karhula et al. 2017b; Lesciotto et al. 2019). The tungsten atom in PTA provides the necessary X-ray contrast, enabling clear visualization and analysis of cartilage surfaces. Since the cartilage surface often serves as the initiation site of OA progression, accurately identifying the earliest surface defects is critical.

Traditionally, OA progression is evaluated using subjective histopathological methods like the Osteoarthritis Research Society International (OARSI) histopathology grading system (Gerwin et al. 2010; Pritzker et al. 2006) or Mankin grading system (Mankin et al. 1971), which is resource-intensive, requires both experienced personnel and substantial time, ultimately increasing development costs. Another major limitation of histology is its reliance on slice-wise representation, which captures only a small subsection of the joint surface. For instance, while a typical histological slice is 5 μm thick, the anteroposterior length of a Wistar rat's joint surface ranges from 4000 μm to 6000 μm. Even 10 evenly spaced coronal sections represent merely ~1% of the surface.

The most accurate way to analyze cartilage is to examine the entire surface rather than isolated subsections. High-resolution 3D data addresses this limitation, offering a comprehensive view of the surface and the ability to detect early degradation signs, such as minor fibrillation. This approach eliminates uncertainties inherent in conventional methods and provides critical insights into the earliest stages of OA.

2 Essential Data Requirements

2.1 Sample Preparation

Cartilage and connective tissues do not provide sufficient contrast in absorption-based μCT imaging. Currently, the most utilized methods for contrasting cartilage are PTA (Karhula et al. 2017b; Metscher 2009), hafnium-substituted Wells-Dawson polyoxometalates (de Bournonville et al. 2020) and CA4+ (Bansal et al. 2011; Karhula et al. 2017a). The CRS method has been used to analyze separate surfaces where joints have been dissected open. When scanning an intact knee using a contrast agent, one should apply a slight pull to open joint space and avoid contact between cartilage surfaces. Furthermore, distinct contrast for cartilage will allow straightforward segmentation, and the results will reflect the actual cartilage condition. Some contrast agents like PTA have limited diffusion and require sample preparation. In such cases, the cartilage surface must remain intact and free from any artifacts, as even the smallest incision from a scalpel can impact the results.

Additionally, any soft tissues that may fold on top of the cartilage if not adequately extracted, such as the ACL, PCL, synovium, or other ligaments, must be carefully removed to avoid interfering with the analysis. Please refer to Supplementary Data 2 in the published article for further details (Kauppinen et al. 2025).

2.2 Noise Level

Surface analysis requires sufficient resolution and contrast to identify micro-scale surface fibrillation features. Noise level is also a key factor in the data; therefore, imaging settings should be set to minimize noise. This will always increase the scanning time of the sample. The best noise removal is achieved by adding frame averaging, increasing exposure time, or increasing projections. The noise should be minimized because the analysis aims to see minimal changes in the surface, and it will also see the noise. If the fibrillations in the cartilage surface are the same size as the image noise, the analysis loses sensitivity in detecting the earliest changes. This reduced sensitivity could delay identifying initial cartilage degradation, potentially impacting early intervention efforts and therapeutic evaluations. An efficient way to deal with noise after the scan is The Perona-Malik filter, also known as anisotropic diffusion, which effectively reduces noise while preserving edges by diffusing the image selectively along regions of low gradient. This makes it particularly useful in enhancing cartilage surface details without blurring crucial features (Perona and Malik 1990).

2.3 Sample Orientation and Data Format

Samples need to be oriented for the analysis to work correctly. The way that the samples need to be oriented is shown in (Fig. 1). The surface of interest must be at the bottom of the image stack and have a proper orientation where the surface that needs analysis is facing the bottom without significant tilts. The rotation of the sample does not matter, but we suggest using the same orientation to get result images in a similar orientation.

The analysis requires data in the form of an image stack or sequence of 8-bit image files, such as .png, .tiff, or .bmp. Although scanners typically save files in their proprietary formats, these can be easily converted to image sequences using proprietary software or third-party tools.

3 Cartilage Roughness Score

3.1 Surface Segmentation

The cartilage surface is segmented using the automatic Otsu method. Otsu separates data into two classes by maximizing inter-class variance (Otsu 1979) (background

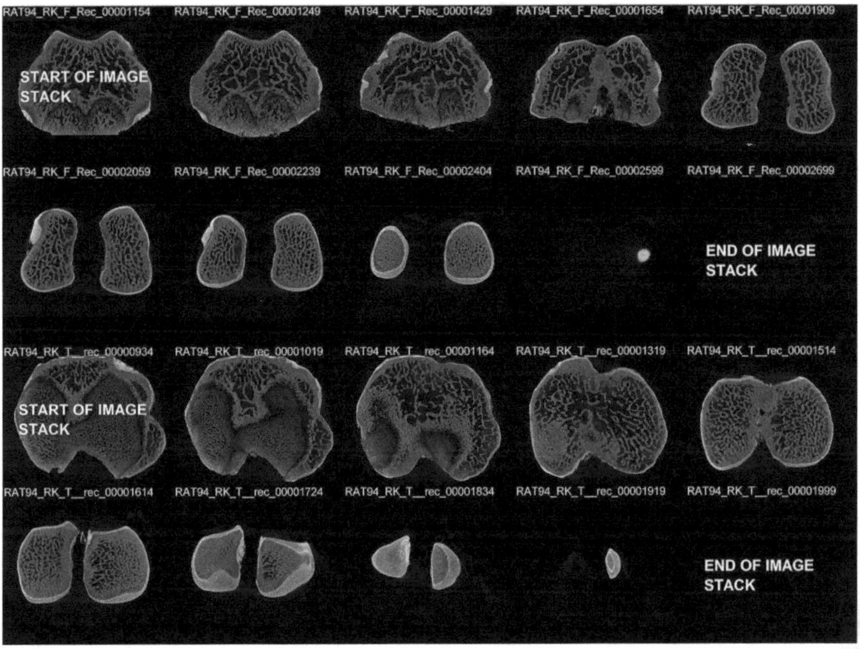

Fig. 1 Example slice sequences from a Femur (top) and Tibia (bottom). Example of what data that is ready for analysis should look like. The example data from femur has slices from 1154 to 2699 and the data from tibia has slices from 934 to 1999 (ten slices shown here from both). Image sequence must have ascending numbering where the surface that needs analysis is found from the end of the image stack

and sample). We have added a threshold multiplier to the settings that may need to be fine-tuned based on the staining intensity of the cartilage in relation to the bone because the bone has a higher volume fraction; otherwise, segmentation would aim to segment bone from the background. The modifiable threshold multiplier is applied to the automatic threshold level. If the staining intensity of the cartilage is less than that of the bone, then the multiplier should be less than one, and if the intensity is higher than that of the bone, then the multiplier should be more than one. The algorithm outputs example images from the initial segmentation where different multipliers are visualized. Viewing these output images will help decide the correct multiplier by examining how well the background and cartilage are separated. The image is then binarized and despeckled. The surface is recognized by running a 2D plane through the volume, identifying the first nonzero voxel representing the surface's outermost boundary, and saving its height to a 2D matrix. This approach ensures that only the topmost layer of the surface is captured, which is critical for accurate surface analysis. The surface is thus represented by a height matrix $Z(x, y)$ where Z is the height at position x, y.

3.2 Choosing a Region of Interest

An essential step in the analysis is choosing the region of interest (ROI), as it directly impacts the accuracy and relevance of the results by ensuring that only the cartilage surface is analyzed. This is currently done using ImageJ and false color image output from MATLAB. This part of the analysis pipeline is the only portion that requires manual sample-specific input from the user. The user must delineate the lateral and medial condyle from the images using the polygon tool and save them to the ImageJ ROI manager:

- Open the ROI manager by pressing ctrl+T.
- Rename both ROIs with the full filename (without file extension), appending "_lat" for the lateral condyle and "_med" for the medial condyle.
- Save the ROIs using the same filename without extension, adding "_RoiSet" at the end.

This structured naming scheme ensures that the ROIs match the correct files, facilitating accurate extraction of lateral and medial volumes. The sub-volumes are placed in new folders, "lateral" and "medial," in the full-volume folder. More detailed instructions on ROI placement, including step-by-step visuals and examples of correctly labeled ROIs to guide users through the process, can be found on GitHub (Kauppinen 2024a, b).

3.3 Estimation of Healthy Surface

A healthy surface is estimated using an iterative 5-degree polynomial surface fit. The primary purpose of this fit is to try to achieve a surface that mimics the crude shape of the joint surface and overlooks sharp changes in the surface. To achieve this, we apply the fit to the detected surface Z(x, y) and then compare the fitted surface to the original surface. Next, possible artifacts are removed (CRS4.removal_threshold) by comparing the fit to the target surface. The default value for this function is ten pixels, and locations where the distance between the fit and the target surface is more than ten are set to not a number (NaN). These areas are excluded from the subsequent iterative surface fits. This iterative process continues until there are less than 10 locations where fit does not meet the given criteria. Ten is used as the default cutoff to make the fit converge faster. Before each fit, the borders of the sample are set back from the original surface in case too many coordinates are removed from the edges, which causes the fitted surface to lose its curve at the edge and results in high angular values at the edges. As a result, we have a fitted surface that has a tendency towards smooth/even areas in the sample. One possible limitation of this approach is when there is a severe erosion of articular cartilage and only a little normal cartilage left. In such a case, the fit might represent an average shape not weighted towards the healthy part but just the parts with the least variation rather than a healthy surface. Nevertheless, even this type of surface will suffice in local comparison when sharp changes in the surface are of interest because the 5th-degree polynomial will have a smooth overall shape and sharp changes in cartilage surface are still detected.

3.4 Local Comparison

The local comparison or CRS is calculated using a moving rectangular window. The pixel size should be small enough to see structural changes visually. In the original dataset (Kauppinen 2024a, b), an isotropic voxel size of 2.8 μm was used for developing the CRS. To reduce noise, a Gaussian smoothing with a kernel size of half the moving window size is applied before CRS analysis. Each coordinate in the surface and the fitted surface is analyzed based on its neighborhood, where the neighborhood size is the moving window. A 3D plane is fitted to both surfaces at each coordinate. Surface normals are extracted from these planes, which give two-unit vectors. The angle between these unit vectors is calculated using the basic dot product of 3D vectors. The angle is transformed to degrees and saved to surface CRS(x, y).

The local window size and minimum angle will affect the results. The window size is set automatically based on the pixel size. We found that the diameter of the rectangular window should be close to 30 μm x 30 μm to catch features such as early fibrillation of the surface. These fibrillations cause changes in surface roughness that may be just a few degrees depending on window size, and it is important to optimize corresponding parameters for final analysis. This effect is visualized in Fig. 2, where CRS analysis is performed on "healthy," "moderate OA," and "advanced OA" samples. The healthy surface was very smooth and had no signs of OA in either histopathological OARSI grading (Gerwin et al. 2010) or CRS analysis using the default window size (28 μm * 28 μm). The analysis was run with five different window sizes where half width of the local moving window was 1, 3, 5, 7 and 9 pixels using 2.8 μm data. These numbers equal to rectangular squares with side lengths of 5.6 μm, 16.8 μm, 28 μm, 39.2 μm and 50.4 μm, respectively. The data was also visualized using different minimum thresholds from 3° to 15° using 3° steps. Figure 2 shows the various outputs from these analyses. When the smallest possible window and threshold (5.6 μm × 5.6 μm, 3°-45°) is used, the shape of the 5-polynomial fit also becomes visible. Furthermore, the CRS appears rather noisy, making actual surface features challenging to identify. Therefore, the lowest possible window size should not be used.

At the center of the three example panels is the default size (28 μm × 28 μm). This window size was deemed most suitable to capture relevant features without excess noise (Kauppinen et al. 2025) and has also been used in studies (Asadikorayem et al. 2025; Weber et al. 2024). One may use bigger or smaller windows when analyzing their data. However, to keep results from studies comparable to each other, we suggest that the default window size is used when cartilage defects are analyzed from rat joints. In the case of mouse models, where the defects may be

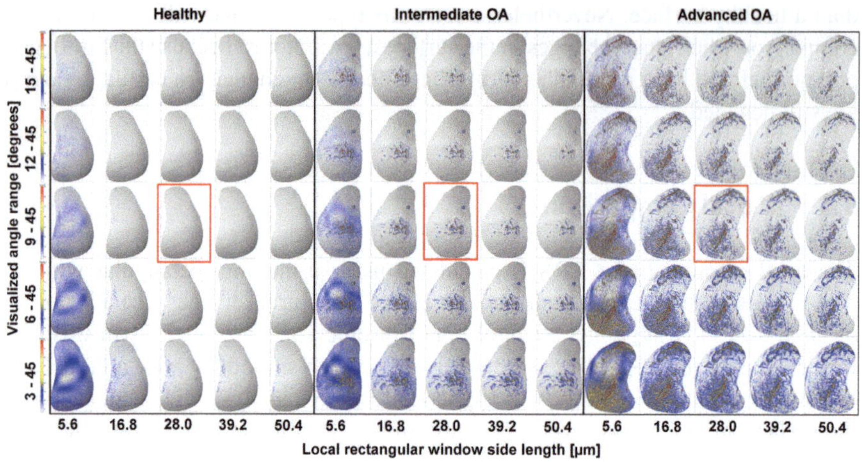

Fig. 2 Image panels to visually demonstrate the effects of different sizes of moving windows and angular thresholds. The impact of settings is demonstrated for three different samples representing healthy, intermediate OA (local defect), Advanced stage OA (spread cartilage damage) from left to right, respectively. Each sample was analyzed with 5 different moving window sizes: 5.6 μm, 16.8 μm, 28.0 μm, 39.2 μm, and 50.4 μm. The low limit threshold for visualization varies from 3° to 15° with 3° steps, while the high limit is kept at 45°. The default visualization range and window size are highlighted with the red rectangle where the samples appear most different from each other

even more minor, the data should be first acquired with a higher resolution in order to see the earliest changes. Then, a smaller window may be needed to see groupwise differences. Selected parameters should allow for a convincing representation of surface integrity, both visually and numerically. The color scheme helps to capture how surface degradation occurs spatially.

3.5 Result Outputs

CRS analysis provides comprehensive statistics on surface integrity. The basic outputs are CRS(mean) and CRS(std), which describe the joint state by looking at average CRS and the variation of CRS values in the joint surface. Another important parameter to describe cartilage degradation over the whole joint affected area. This is the area percentage after the minimum angular threshold (CRS>X° %). Figure 3 contains example images and parameters that can be expected from successful CRS analysis.

These are effective measures for tracking the severity and progression of cartilage lesions because the earliest fibrillation will affect the local angular value. Cracks and lesions will further increase the CRS values, especially on edges. CRS>X° % can be used to express overall cartilage degradation, which is visually also evident. Overall degradation can also be estimated with "sum CRS." This parameter exists

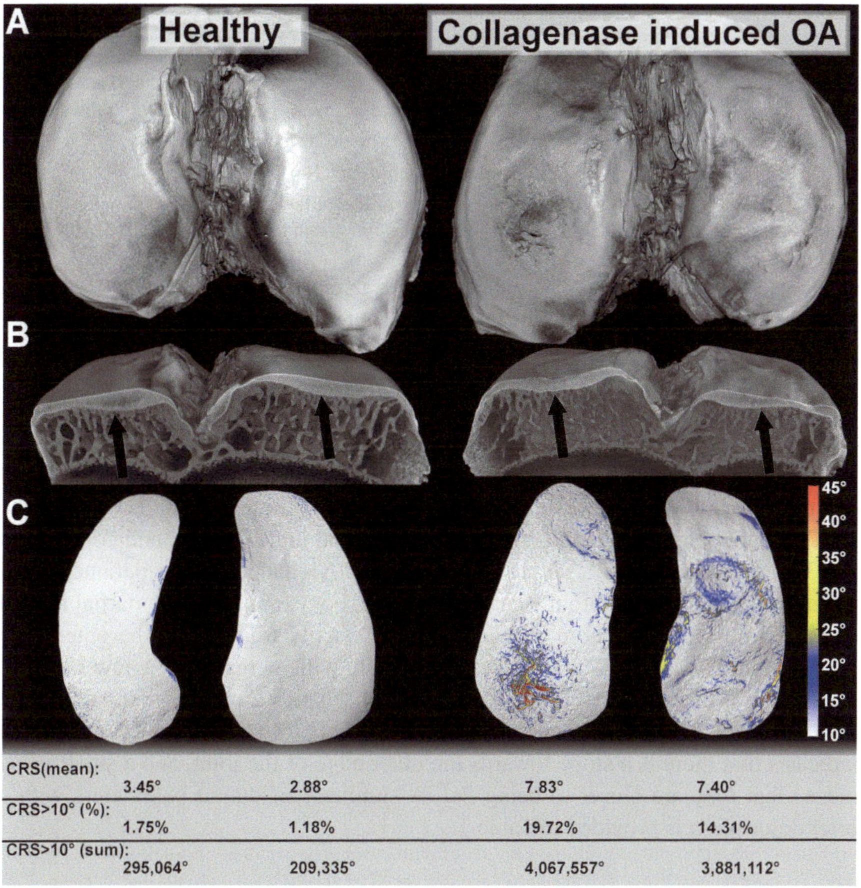

Fig. 3 Example analysis output from a healthy and Collagenase induced OA rat tibia. (**a**) 3D visualizations of phosphotungstic acid-stained tibial surfaces from μCT-data. (**b**) Same volumes in an angle showing well-stained articular cartilage tissue (Black arrows) with subchondral bone tissue underneath. Cartilage loss is not evident, but the surface of the cartilage appears to have minor damage on the collagenase induced OA sample. (**c**) CRS maps from corresponding tibial plateaus with color showing where angles between 10° and 45° are. White color means that the surface is locally smooth (angels are less than 10°) and blue to yellow to red indicates an increase in the local roughness (angles are higher than 10°). Relevant outputs and the actual values are shown below each CRS map. Collagenase-induced OA has much higher values in all outputs compared to the healthy sample

to differentiate between samples that may have similar CRS > X° but a difference in overall roughness. By summing together all the CRS values exceeding a certain threshold, we can further estimate the degeneration that the samples may have. This output was designed to have a parameter that can better differentiate the late stage (where many of the CRS values may be high) from samples where damage is spread but is not very extensive yet.

Table 1 Output parameters with descriptions

Parameter	Description
CRS (mean)	The average "state" of the joint surface. Local angular comparison of surface normals averaged throughout the surface
CRS (median)	Median CRS reflects how skewed the CRS is.
CRS (std)	The standard deviation of CRS
CRS > X° %	Gives a percentage of the "affected" area in the joint based on the threshold. Threshold X can be between 5° to 20° with 1° steps
CRSsum > X°	The sum of CRS after threshold. X (5° to 20° with 1° steps)
Relative tilt (lateral vs. medial)	Crude anatomy of the lateral and medial joint surfaces in one simple measure

The final parameter in the current version is the average angular tilt between the lateral and medial condyles. It is calculated using the same algorithm as the local comparison, but instead of using a local window, the whole condyle is used for the fit. This parameter reflects the crude anatomy of the joint surfaces and compresses the 3D information into a single angular value. A change in this parameter can provide relevant insight into the analyzed model. For example, in the partial medial meniscectomy model, where cartilage wear is usually found from the outer edge of the joint (Doan et al. 2021) the relative tilt will increase by a few degrees. Furthermore, the joint shape may be a factor that plays a part in what shape of a joint is more susceptible to cartilage damage. A bigger relative angle in the tibia indicates that there is a slope towards the outer edge of the joint, and a small value means that they are closer to being flat. A summary of output parameters with a brief description is provided in Table 1.

3.6 How Settings Affect Results

The algorithm is simple to use, and a detailed and updated guide is found on GitHub (Kauppinen 2024a, b). However, one should modify the settings file before analysis. Table 2 lists the settings that can be altered and describes how they affect the final analysis.

4 Conclusions and Future Perspectives

The CRS method is an excellent tool for assessing the state of the cartilage surface in OA animal models. It can be run on any data where the surface segmentation is feasible. Therefore, MRI, multiphoton, and other microscopic imaging modalities could be used with applications beyond cartilage integrity. Practically any imaging modality that can capture the surface features of interest with at least twice the resolution is suitable.

Table 2 Algorithm settings and their impact on analysis

Setting name	Default	Effect on analysis
CRS1.binning	0	Option to bin the data with inputs 0, 2, 3, or 4. A quick test with a binning of 4 can be helpful if the data size is huge
CRS1.dataformat	*.png	It is recommended to use 8-bit .png, .bmp, or .tiff data. 16-bit images are transformed to 8-bit upon loading
CRS1.flipH	0	Horizontal flip of the data. To keep the right and left legs looking the correct way from different scanners. Bruker scanners: 0, Scanco and Zeiss: 1. Only affects the visualization
CRS1.ThreeD	0	Removing excess particles to create surface projection images. 0: a slice-by-slice despeckle to keep memory sufficient. 1: a 3D 26-connected component analysis to extract only the largest object. (memory heavy)
CRS1.threshold_mult	1	Otsu threshold modifier. Keep it as one if the cartilage intensity matches the bone. Increase if cartilage is brighter than bone, decrease if bone is brighter than cartilage
CRS3.flipH	0	Use the same as in CRS1.flipH
CRS3.ThreeD	1	Keep this as 1. The 26-connected component analysis works better as you now analyze sub-volumes
CRS3.filetype	*.png	Matching filetype from CRS1.dataformat
CRS3.file_extension	.png	Filetype to save the images
CRS3.common_nom	RAT	Common name of image files that all samples included in a batch have. This is used to get only the correct files inside the folders
CRS4.voxel_size	2.8	Input for the isotropic voxel size. Make sure that you set this correctly. The analysis is based on the pixel size
CRS4.polX	5	Surface fitting polynomial. 1, 2, 3, 4 or 5. Use 5 for the rat joint surface to allow a more precise fit
CRS4.polY	5	Surface fitting polynomial. 1, 2, 3, 4 or 5. Use 5 for the rat joint surface to allow a more precise fit
CRS4.removal_threshold	10	An outlier modifier for iterative fit. Locations where the fit difference is more than this value are removed from the next fit (set as NaN) until the fitting converges
CRS4.holefill	150000	Default value for filling closed holes inside binarized 3D volume. Holes larger than 150,000/voxel size are filled. (does not affect surface results)
CRS6.Low_limit	10	The lower limit in the result images. Angles less than this are shown as white
CRS6.High_limit	45	Upper CRS limit in the result images. Angles between low limit and high limit are visualized. Keep it as narrow as possible to get the best dynamic range for the colors but the same for the whole study
CRS5.radius	Automatic	This parameter should not be modified. It is appended to the result file based on the pixel size

When writing this chapter, the analysis can be run on MATLAB, and a Python implementation is expected in the future for more accessible use. Additionally, improving user-friendliness and adding new features will be implemented based on user feedback. The next version will also include an automated ROI selection for analysis. The current hand-drawn ROIs are always user dependent and our earlier repeatability analysis (Kauppinen et al. 2025) showed that some variation will be introduced from different users drawing the same ROIs.

The current estimation of the healthy surface is rather a mathematical approximation of a smooth surface and is not a "real" healthy surface. The possible solution would be to use a surface fit with more freedom to be any shape, allowing normal anatomical variation on the healthy surface. Such fits could, for example, be a MATLAB function (Gridfit (Surface Fitting Using Gridfit – File Exchange – MATLAB Central)) that is very efficient, and the settings for the fits are easy to change. Combining the current iterative approach with *gridfit* may improve fit on degraded areas. Furthermore, it would allow for the estimation of the surface based on the shape of the healthy areas of the joint. The joint structure is subjected to various factors of used models, including weight, gender, breed, and age. However, combining multiple studies would allow for building more sophisticated surface approximations, e.g., a statistical shape model. However, the current approach provides valuable insight into cartilage surface integrity and works even on smaller studies, where relevant features can be visualized and quantified.

Acknowledgments We want to acknowledge people who have made the development of the CRS algorithm possible. Especially the Tissue Engineering and Biofabrication group (ETH Zurich), with special mentions to Prof. Marcy Zenobi-Wong and David Fercher and to Prof. Simo Saarakkala and Tuomas Frondelius from the Research Unit of Health Sciences and Technology (University of Oulu). We would like to acknowledge the funding that has made this work possible. Grants or funding sources that have helped during the development of this algorithm are as follows: Research Council of Finland (Academy research fellows, Grant: 347445), Funding from the University of Oulu graduate school (UniOGS), Swiss National Science Foundation (Grant 315230_192656), Finnish Cultural Foundation (Grant: 60231705) and Funding from Jane and Erkko Aatos Foundation.

Conflict of Interest The authors report no conflicts of interest.

References

Asadikorayem M, Weber P, Zhang S, Surman F, Fercher D, Fonti M, Bevc K, Kauppinen S, Frondelius T, Finnilä MAJ, Zenobi-Wong M (2025) In-situ-forming zwitterionic hydrogel does not ameliorate osteoarthritis in vivo, despite protective effects ex vivo. Biomater Adv 169:214151. https://doi.org/10.1016/J.BIOADV.2024.214151

Bansal PN, Stewart RC, Entezari V, Snyder BD, Grinstaff MW (2011) Contrast agent electrostatic attraction rather than repulsion to glycosaminoglycans affords a greater contrast uptake ratio and improved quantitative CT imaging in cartilage. Osteoarthr Cartil 19(8):970–976. https://doi.org/10.1016/J.JOCA.2011.04.004

Benninghoff A (1925) Form und Bau der Gelenkknorpel in ihren Beziehungen zur Funktion. Z Anat Entwicklungsgesch 76(1):43–63. https://doi.org/10.1007/BF02134417

Bruyère O, Cooper C, Arden N, Branco J, Brandi ML, Herrero-Beaumont G, Berenbaum F, Dennison E, Devogelaer JP, Hochberg M, Kanis J, Laslop A, McAlindon T, Reiter S, Richette P, Rizzoli R, Reginster JY (2015) Can we identify patients with high risk of osteoarthritis progression who will respond to treatment? A focus on epidemiology and phenotype of osteoarthritis. Drugs Aging 32(3):179–187. https://doi.org/10.1007/S40266-015-0243-3/TABLES/3

de Bournonville S, Vangrunderbeeck S, Ly HGT, Geeroms C, De Borggraeve WM, Parac-Vogt TN, Kerckhofs G (2020) Exploring polyoxometalates as non-destructive staining agents for contrast-enhanced microfocus computed tomography of biological tissues. Acta Biomater 105:253–262. https://doi.org/10.1016/J.ACTBIO.2020.01.038

Doan TN, Mckinney JM, Pucha KA, Bernard FC, Willett NJ, Mckinney J (2021) Osteoarthritis early-, mid- and late-stage progression in the rat medial meniscus transection model. BioRxiv, 2021.03.11.434909. https://doi.org/10.1101/2021.03.11.434909

Doube M, Klosowski MM, Arganda-Carreras I, Cordelières FP, Dougherty RP, Jackson JS, Schmid B, Hutchinson JR, Shefelbine SJ (2010) BoneJ: free and extensible bone image analysis in ImageJ. Bone 47(6):1076–1079. https://doi.org/10.1016/j.bone.2010.08.023

Findlay DM, Kuliwaba JS (2016) Bone–cartilage crosstalk: a conversation for understanding osteoarthritis. Bone Res 4(1):1–12. https://doi.org/10.1038/boneres.2016.28

Gerwin N, Bendele AM, Glasson S, Carlson CS (2010) The OARSI histopathology initiative—recommendations for histological assessments of osteoarthritis in the rat. Osteoarthr Cartil 18(SUPPL. 3):S24–S34. https://doi.org/10.1016/J.JOCA.2010.05.030

Hunter DJ, Deveza LA (2024) Deconstructing the "types" of osteoarthritis. Osteoarthritis Imaging:100257. https://doi.org/10.1016/J.OSTIMA.2024.100257

Karhula SS, Finnilä MA, Freedman JD, Kauppinen S, Valkealahti M, Lehenkari P, Pritzker KPH, Nieminen HJ, Snyder BD, Grinstaff MW, Saarakkala S (2017a) Micro-scale distribution of CA4+ in ex vivo human articular cartilage detected with contrast-enhanced micro-computed tomography imaging. Front Phys 5(AUG). https://doi.org/10.3389/fphy.2017.00038

Karhula SS, Finnilä MA, Lammi MJ, Ylärinne JH, Kauppinen S, Rieppo L, Pritzker KPH, Nieminen HJ, Saarakkala S (2017b) Effects of articular cartilage constituents on Phosphotungstic acid enhanced micro-computed tomography. PLoS One 12(1):e0171075. https://doi.org/10.1371/JOURNAL.PONE.0171075

Kauppinen S (2024a) CRS Wiki. https://github.com/sambacaramba/CRS/wiki

Kauppinen S (2024b) μCt and virtual histology data of rat hind legs, collagenase induced osteoarthritis. University of Oulu, Health Sciences and Technology. https://doi.org/10.23729/f33d3ba8-01b8-47af-90dd-94b7c7861247

Kauppinen S, Fercher D, Barreto G, Karjalainen VP, Virtanen V, Baixauli-Marin L, Fonti M, Zhang S, Frondelius T, Weber P, Saarakkala S, Zenobi-Wong M, Finnilä MAJ (2025) Assessment of whole cartilage surface damage in an osteoarthritis rat model: the cartilage roughness score (CRS) utilizing microcomputed tomography. Osteoarthr Cartil 33(1):134–145. https://doi.org/10.1016/J.JOCA.2024.09.008

Lesciotto KM, Motch Perrine SM, Kawasaki M, Stecko T, Ryan TM, Kawasaki K, Richtsmeier JT (2019) Phosphotungstic acid-enhanced microCT: Optimized protocols for embryonic and early postnatal mice. Dev Dyn 249:573–585. https://doi.org/10.1002/dvdy.136

Mankin HJ, Dorfman H, Lippiello L, Zarins A (1971) Biochemical and metabolic abnormalities in articular cartilage from osteo-arthritic human hips. II. Correlation of morphology with biochemical and metabolic data. J Bone Joint Surg Am 53(3):523–537. https://doi.org/10.2106/00004623-197153030-00009

Metscher BD (2009) Micro CT for comparative morphology: simple staining methods allow high-contrast 3D imaging of diverse non-mineralized animal tissues. BMC Physiol 9(1):1–14. https://doi.org/10.1186/1472-6793-9-11/FIGURES/15

Otsu N (1979) Threshold selection method from gray-level histograms. IEEE Trans Syst Man Cybern SMC-9(1):62–66. https://doi.org/10.1109/TSMC.1979.4310076

Palazzo C, Nguyen C, Lefevre-Colau MM, Rannou F, Poiraudeau S (2016) Risk factors and burden of osteoarthritis. Ann Phys Rehabil Med 59(3):134–138. https://doi.org/10.1016/J.REHAB.2016.01.006

Perona P, Malik J (1990) Scale-space and edge detection using anisotropic diffusion. IEEE Trans Pattern Anal Mach Intell 12(7):629–639. https://doi.org/10.1109/34.56205

Pritzker KPH, Gay S, Jimenez SA, Ostergaard K, Pelletier JP, Revell K, Salter D, van den Berg WB (2006) Osteoarthritis cartilage histopathology: grading and staging. Osteoarthr Cartil 14(1):13–29. https://doi.org/10.1016/j.joca.2005.07.014

Rodriguez-Merchan EC (2023) The current role of disease-modifying osteoarthritis drugs. Arch Bone Joint Surg 11(1):11. https://doi.org/10.22038/ABJS.2021.56530.2807

Roemer FW, Jarraya M, Collins JE, Kwoh CK, Hayashi D, Hunter DJ, Guermazi A (2023) Structural phenotypes of knee osteoarthritis: potential clinical and research relevance. Skeletal Radiol 52(11):2021–2030. https://doi.org/10.1007/S00256-022-04191-6/FIGURES/2

Sokolove J, Lepus CM (2013) Role of inflammation in the pathogenesis of osteoarthritis: latest findings and interpretations. Ther Adv Musculoskelet Dis 5(2):77–94. https://doi.org/10.1177/1759720X12467868

Surface Fitting using gridfit – File Exchange – MATLAB Central Retrieved December 21, 2024, from https://se.mathworks.com/matlabcentral/fileexchange/8998-surface-fitting-using-gridfit

Weber P, Bevc K, Fercher D, Kauppinen S, Zhang S, Asadikorayem M, Marin LB, Zhang T, Frondelius T, Salzmann G, Bruhin V, Hax J, Barreto G, Finnilä MAJ, Zenobi-Wong M (2024) The collagenase-induced osteoarthritis (CIOA) model: where mechanical damage meets inflammation. Osteoarthritis Cartilage Open 6(4):100539. https://doi.org/10.1016/J.OCARTO.2024.100539

Fundamentals of Photoacoustic Imaging

Mohammad Reza Fouladi, Parastoo Farnia, Reza Naghne, Niuosha Ghadesi, Tina Anjomshoa, and Ebrahim Najafzadeh

Abstract

Photoacoustic Imaging (PAI), also known as optoacoustic imaging, is a promising hybrid technology that integrates the advantages of optical and ultrasound imaging to achieve high-resolution, label-free imaging of deep biological tissues.

M. R. Fouladi
Department of Medical Physics and Biomedical Engineering, Tehran University of Medical Sciences (TUMS), Tehran, Iran

P. Farnia · R. Naghne
Department of Medical Physics and Biomedical Engineering, Tehran University of Medical Sciences (TUMS), Tehran, Iran

Research Center for Intelligent Technologies in Medicine (RCITM), Advanced Medical Technologies and Equipment Research Institute (AMTEI), TUMS, Tehran, Iran

Research Center of Biomedical Technology and Robotics (RCBTR), Advanced Medical Technologies and Equipment Research Institute (AMTEI), TUMS, Tehran, Iran

N. Ghadesi · T. Anjomshoa
Department of Medical Physics and Biomedical Engineering, Tehran University of Medical Sciences (TUMS), Tehran, Iran

Research Center for Intelligent Technologies in Medicine (RCITM), Advanced Medical Technologies and Equipment Research Institute (AMTEI), TUMS, Tehran, Iran

E. Najafzadeh (✉)
Finetech in Medicine Research Center, School of Medicine, Iran University of Medical Sciences, Tehran, Iran

Department of Medical Physics, School of Medicine, Iran University of Medical Sciences, Tehran, Iran

Department of Molecular Imaging, Faculty of Advanced Technologies in Medicine, Iran University of Medical Sciences, Tehran, Iran
e-mail: najafzadeh.e@iums.ac.ir

I. F. Cengiz et al. (eds.), *Bioimaging Modalities in Bioengineering*,
Biomaterials, Bioengineering and Sustainability 9,
https://doi.org/10.1007/978-3-031-99410-4_12

Utilizing the PA effect, PAI converts absorbed optical energy into ultrasound signals, enabling detailed structural and functional assessment of tissues with minimal invasiveness. This chapter explores the fundamental principles of PAI, focusing on its signal generation and detection mechanisms, the roles of endogenous and exogenous contrast agents, and various imaging methodologies, including PA spectroscopy, microscopy, remote sensing, and tomography. Additionally, it examines advanced image reconstruction techniques critical for enhancing the quality of PA images, emphasizing the need for practical algorithms due to the complexities of acoustic signal propagation within heterogeneous biological tissues. As PAI technology advances, integrating artificial intelligence into imaging analyses and employing novel detectors and light sources will undoubtedly enhance its applications and diagnostic capabilities. The discussion culminates with insights into ongoing challenges and future directions, particularly concerning developing portable and cost-effective systems that can augment the clinical utility of PAI in diverse healthcare settings. Overall, this chapter provides a comprehensive overview of the fundamentals, technological innovations, and prospective advancements in the field of PAI, highlighting its transformative potential in biomedical research and clinical practice.

Keywords

Photoacoustic imaging · Photoacoustic tomography · Photoacoustic signal generation · Photoacoustic contrast agents · Photoacoustic remote sensing

1 Introduction

Photoacoustic Imaging (PAI), also known as optoacoustic imaging, is a hybrid technique combining the advantageous characteristics of optical and ultrasound imaging. This method derives high contrast property through optical imaging, while ultrasound imaging mainly enhances depth penetration and contributes to overall image resolution. PAI works based on the PA effect, which transforms the absorbed optical energy into acoustic pressure waves in ultrasound range. Since acoustic waves are attenuated less than optical waves in soft tissues, higher-resolution imaging in tissues which are located further below the skin's surface can be performed. In the recent decades, PAI has drawn attention as it can provide high-quality structural, functional, and molecular images non-invasively. Labelling a cell seems essential in most optical imaging modalities, which is not necessary for PAI. PAI shows minimal to no dependency on labels (label-free), which makes the imaging process easier and more accessible. However, contrast agents can be used to enable imaging of deeper structures while providing targeted contrasts (Park et al. 2024; Zhao et al. 2019; Attia et al. 2019; Lin and Wang 2021).

The subsequent sections of this chapter will provide the intricate background of PAI, beginning with the fundamental principles underlying the PA effect, followed by an exploration of signal generation and detection mechanisms. In addition, the chapter will highlight the role of contrast agents, which are essential for enhancing image quality and specificity. Methodological approaches to PAI will be examined in detail, covering both single-modality and multi-modality techniques, thus providing a comprehensive perspective on the technological diversity within the field. The chapter will also address PA tomography due to its unique capabilities of deep tissue imaging, high spatial resolution, and scalability over other modalities, focusing on irradiation and detection mechanisms, as well as advanced image reconstruction techniques that facilitate volumetric imaging. Finally, the chapter will conclude with insights into ongoing challenges and potential future directions that may shape the landscape of PAI in the coming years.

2 Principles of PAI

2.1 The PA Effect

PA effect refers to the phenomenon in which a target is exposed to electromagnetic wave (in optical domain), and consequently, generate acoustic waves (Barbosa and Mendes 2022). This phenomenon was first discovered by Alexander Graham Bell during his photophone experiments, which transmitted speech to a remote listener (Barbosa and Mendes 2022). The photophone used vibrating mirrors to modulate a sunlight beam with voice signals, then transformed the signals back into sound at the receiving mirror. Following this invention, he stated that illuminating a solid material via sunlight produced acoustic signals at modulated frequency. However, this phenomenon did not come into account till the discovery of laser (Barbosa and Mendes 2022).

The PA effect primarily operates based on the absorption of light and the subsequent pressure changes in the material. This involves several stages, including light absorption, thermal expansion, and acoustic wave generation. The first step involves the absorption of light by a material and this absorbance depends on their respective absorption coefficients. The absorbed light energy is transformed into heat, increasing the local temperature. As the temperature rises, the material experience rapid structural change. This rapid thermal expansion generates localized pressure changes in the medium surrounding the absorbing material. This rapid structural change occurs due to the excitation at molecular level originating from absorbing optical energy. The pressure changes caused by thermal expansion generate acoustic waves. These waves propagate through the material and can be detected by appropriate sensors. The characteristics of these acoustic waves depend on several parameters, including the intensity of the light and the acoustic and mechanical properties of the imaging target and the surrounding medium (Barbosa and Mendes 2022; Illienko et al. 2024; Wang et al. 2022; Gao et al. 2021).

2.2 Signal Generation and Detection

For signal generation in PAI, the pulsed optical source with a suitable pulse duration illuminates the tissue of interest. The laser pulse duration must be shorter than the tissue or medium's thermal and stress relaxation times. This condition ensures that the absorbed energy does not disperse away before it can generate a pressure wave. Thermal relaxation time is the time it takes for the absorbed heat to diffuse through the medium, and stress relaxation time is the time it takes for the mechanical stress caused by the absorption of the laser energy to return to its equilibrium state.

Since biological tissues have higher absorption coefficients in the near-infrared or visible part of the optical spectrum, and light penetrates more in these windows (Lane et al. 2018), PAI primarily works well in this part of the spectrum. Since each tissue has a unique optical absorption in a specified wavelength, the contrast in PA images can be defined. When a pulsed laser, typically with pulse widths ranging from nanoseconds to hundreds of nanoseconds, irradiates the tissue, photons scatter and are absorbed in chromophores (Liu et al. 2024). For PA wave generation, the condition $\tau \mu_a v_s \ll 1$ must be satisfied (Tsubouchi et al. 2020), where τ is the laser pulse duration, μ_a is the optical absorption coefficient, and v_s is the speed of sound in the medium. This condition ensures that the energy is confined locally and can generate a strong pressure wave (Moore and Jokerst 2019). Absorbed energy at the location r of a chromophore, $H(r)$ depends on μ_a and illuminated fluence (F) at the location r (Eq. 1);

$$H(r) = \mu_a(r) F\left(r, \mu_a, \mu_s, g\right) \tag{1}$$

Where $F(r, \mu_a, \mu_s, g)$ is light fluence (Han et al. 2022). Fluence is defined as the total radiant energy per target's area. Under the mentioned circumstances, the absorbed energy at location r leads to the generation of an initial pressure wave P_0 which then propagates through the tissue. These acoustic waves are eventually detected by a sensor, which can be a typical ultrasound transducer (Du et al. 2021).

The magnitude of the generated initial PA pressure can be described using Eq. 2;

$$P_0 = \frac{\beta {v_s}^2}{c_p} H(r) \tag{2}$$

Where β is the thermoelastic coefficient of volume expansion, and c_p is the specific heat capacity. Generate acoustic waves propagate through the medium and are detected by a transducer. Considering a homogenous medium, the pressure detected by the transducer can be calculated via Eq. 3.

$$p\left(r, t\right) = \frac{1}{4\pi v_s} \frac{\partial}{\partial t} \int_{s'(t)}^{11} \frac{p_0\left(r'\right)}{\left|r - r'\right|} ds'(t) \tag{3}$$

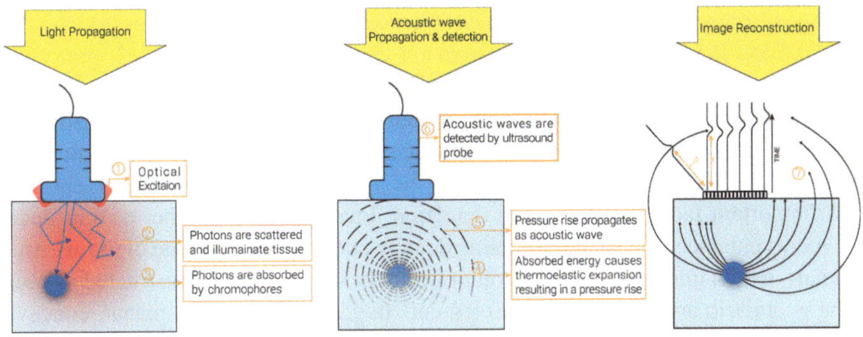

Fig. 1 Schematic diagram of the PA signal generation, detection, and reconstruction mechanism. Different stages of producing PA signals including: (1) Optical excitation, (2) Scattering of photons and tissue illumination, (3) Absorption of photons by chromophores, (4) Thermoelastic expansion due to absorbed energy, (5) Acoustic wave propagation, (6) Detection of acoustic waves by ultrasound probe, and (7) Reconstruction of acquired signals and image generation

Where t represents the time and $|r - r'|$ represents the distance between the initial pressure source and the transducer surface, which can be equal to $v_s\ t$ (Liu et al. 2024). As it mentioned before v_s defines the speed of sound in medium. Briefly, the pulsed laser with an appropriate wavelength corresponding to the peak optical absorption of the target tissue illuminates the tissue. The light absorption leads to subsequent heating, which, due to the thermoelastic effect, causes ultrasound waves to be produced and propagate through the medium to reach a transducer. Then detected signals can be reconstructed, and the PA image be generated; these processes have been shown in Fig. 1.

There are several types of excitation sources such as Light Diodes, LEDs, Xenon Flash Lamp (XFL) and continuous wave lasers. However a common excitation source for PAI is solid-state lasers, such as Q-switched rare-earth-doped crystals (e.g., Nd:YAG) (Fatima et al. 2019). These sources provide high pulse energy, resulting in higher Signal-to-Noise Ratios (SNRs). Despite their perfect ability to provide high-quality images, they tend to be expensive, bulky and require cooling, making them impractical for clinical use. Consequently, semiconductor-based sources are increasingly being adopted. Semiconductor sources, including diode lasers and LEDs, that can be used in surgical units or at the patient's bedside as they are cheap, lightweight, and portable. These lasers almost share the same physical principle. However, LEDs use a broader optical range. Although these sources provide flexibility in the output wave shape, their energy is less than Nd:YAG lasers (Najafzadeh et al. 2020). Fiber lasers can be an alternative to diode lasers and LEDs. They operate based on the amplification method, are more compact, and provide higher output (Riksen et al. 2023).

Generated PA waves are mostly detected by a typical ultrasound probe, which is mostly known as conventional probe and translates the pressure of detected waves into electrical current based on the piezoelectric effect. The conventional probes are readily available and cost effective than other means of detection (Manwar et al. 2020a).; however, new probes based on capacitive micromachined ultrasound transducers exhibit wider bandwidths, which causes the imaging system to be more sensitive and provides better resolution. Moreover, other types of transducers, such as ultra-wideband or focused transducers, have also recently drawn attention (Liu et al. 2024). Also, geometry and the number of transducer elements are two important factors in PA signal detection. Increasing the number of transducer elements enhances the spatial sampling density and angular coverage of the acoustic signals, which improves spatial resolution and the accuracy of reconstructed images. The geometric arrangement of the elements (e.g., linear, circular, planar, concave, arc, ring-shaped, hemispherical arrays, etc.) also affects the sensitivity and angular coverage, which in turn impacts image quality. Higher numbers of transducer arrays with optimized configurations provide broad acoustic detection coverage, enhancing image fidelity and resolution (Moore and Jokerst 2019). Although linear array transducers are the most widely used geometric configuration of transducer elements, due to their clinical prevalence and affordability, ease of use, and practicality, they can cause some artifacts like out-of-plane artifacts and limited view problems. In out-of-plane artifacts, signals generated by absorbers located outside the imaging plane are detected due to the sensitivity of the transducer, causing features outside the actual imaging plane to appear in images. Also, the linear array's finite aperture means it cannot capture PA signals from all directions, especially missing signals from structures oriented perpendicular to the array surface, causing incomplete or distorted imaging of some vascular features. Moreover, detectors with unique shapes exist for specific applications; for example, the developed hockey stick-shaped sensor for dental, Hemispherical for breast and brain, and catheter shaped for colon imaging (Manwar et al. 2020a; Li et al. 2018; Jiang et al. 2024).

3 PAI Methodologies

This section addresses various single-modality PA techniques (Fig. 2), including PA Spectroscopy (Fig. 2a), PA Microscopy (Fig. 2b), PA Tomography (Fig. 2c), and PA Remote Sensing (Fig. 2d). Additionally, it explores the concept of multi-modality PAI, highlighting its significance and advantages.

3.1 Single-Modality PAI Techniques

PAI techniques have emerged as cutting-edge methods in the realm of biomedical imaging. The following section will introduce the fundamental aspects of various PAI techniques including PA Spectroscopy (PAS) and microscopy, along with tomography and remote sensing. As this section explores each technique, it will highlight their principles, applications, advantages, and limitations, thus providing a comprehensive understanding of their roles within the broader context of biomedical imaging.

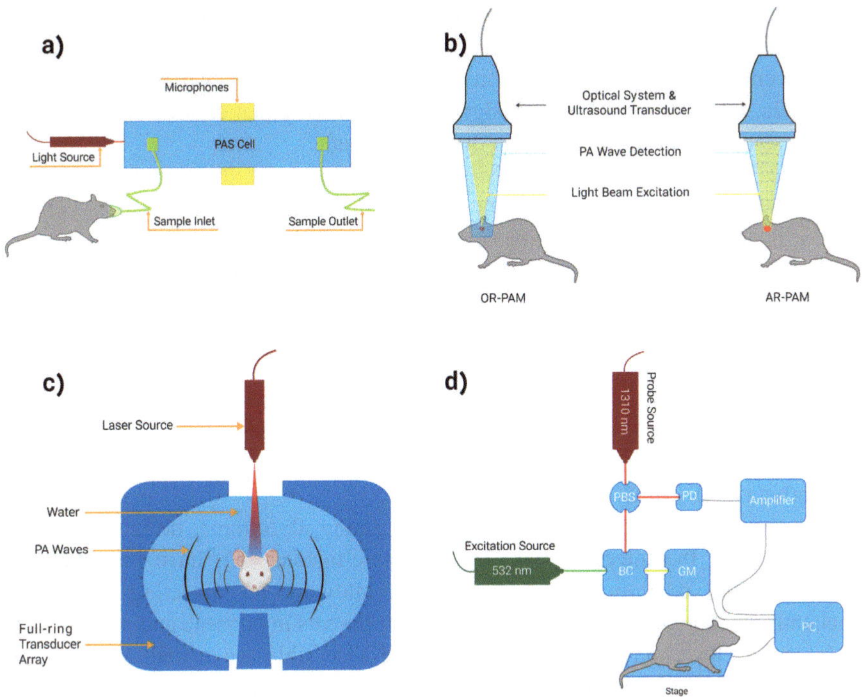

Fig. 2 Schematic of various single-modality PA techniques. (**a**) A simple schematic of the PAS setup with a gas handling system for mice breath analysis. (**b**) Left setup: OR-PAM that captures high spatial resolution by utilizing optical focusing, and the right setup: AR-PAM that captures deeper tissue images by utilizing acoustical focusing in optical diffusive regime. (**c**) A PAT system utilizing a full-ring circular/cylindrical ultrasonic transducer array. (**d**) A PARS system, where both excitation and detection components are optical. (PBS polarized beamsplitter, PD photodiode, BC beam combiner, GM galvo-scanner mirrors, PC personal computer)

PA Spectroscopy (PAS)

PAS is a technique that measures the acoustic waves generated when a sample target absorbs periodically modulated light energy. This method utilizes light with various wavelengths, from a few hertz to kilohertz ranges, facilitating analysis across a broad wavelength spectrum for the same object. The light source in PAS is usually a tunable laser or a laser diode that emits light at various wavelengths (Kaniyala Melanthota et al. 2022). A typical PAS system comprises a light source, a PAS cell where the sample is positioned and is designed to absorb light and convert it into acoustic waves, acoustic detectors, amplifiers for enhancing detection, and a data acquisition unit that collects and processes the detected signals for further analysis (Jin et al. 2022). Figure 2a shows the schematic of a PAS system which consists of light source, PAS cell, and microphones as three main components of a PAS system. PAS cell is the detection volume that contains the gas or sample being analyzed and ensures that the environment is controlled for precise interaction between light and molecules. Also, the cell should be designed to optimize the detection of the PA signals by housing sensitive acoustic detectors (like microphones or ultrasound transducers) that measure the pressure waves induced by the absorbed light energy. In Fig. 2a, PAS cells have been designed in such a way that collects mouse breath using a sample inlet for the analysis of mouse breath.

While PAS offers good sensitivity, its penetration depth is moderate, about millimeters or less, due to significant medium scattering (Jeng et al. 2021). Therefore, it is particularly effective for analyzing surface tissues, making it extensively used for detecting liquid, solid, plasma states, and especially gases, even with low concentrations, in fields such as biology, medicine, chemistry, physics, and environmental monitoring, all with high accuracy and sensitivity (Jin et al. 2022). Additionally, it is widely used in revealing the molecular composition and material properties of the sample (Kaniyala Melanthota et al. 2022). For example, it is employed in cancer detection, blood glucose monitoring (Jin et al. 2022), skin lesion diagnosis (Jeng et al. 2021), and the detection of harmful gases in the air (Kaniyala Melanthota et al. 2022).

Innovations such as enhanced signal processing algorithms, utilizing acoustic resonators and the integration of Artificial Intelligence (AI) have improved the sensitivity and resolution of PAS systems, allowing for detecting even trace concentrations of analytes. Furthermore, developments in portable and miniaturized PAS devices have made the technology more accessible for in-field applications, particularly in environmental monitoring and biomedical diagnostics (Liu et al. 2024).

PA Microscopy (PAM)

PA Microscopy (PAM) is another approach to PAI that provides structural, functional, and molecular insights for both preclinical and clinical research, which employs focused lasers to achieve high spatial resolution imaging at depths typically up to 1–2 mm below the surface. Also, PAM leverages the rich optical contrast of various chromophores, such as hemoglobin and melanin, allowing it to distinguish between different types of tissues and structures based on their specific characteristics; thus, it provides detailed functional and structural information at the cellular and sub-cellular levels. PAM can be categorized as two main techniques based on utilized focusing methods: Acoustic Resolution PA Microscopy (AR-PAM) and Optical Resolution PA Microscopy (OR-PAM), each employing different optical and acoustic focusing methods, which are shown in Fig. 2b (Zafar et al. 2019, 2023, 2024). AR-PAM employs wide-field optical excitation, enabling it to achieve penetration depths 3–5 times greater than OR-PAM. Conversely, OR-PAM provides a markedly narrower optical focus, approximately 10 times smaller than the acoustic focus, resulting in higher-resolution imaging, which is helpful for imaging fine details (Cho et al. 2024; Huang and Wong 2024; Ghadiri et al. 2016; Ghadiri et al. 2018). These capabilities make PAM a valuable tool for biomedical research. However, PAM has some limitations, including high costs and complexity due to its requirement for powerful lasers (Kaniyala Melanthota et al. 2022).

PA Tomography (PAT)

PA Tomography (PAT) or PA Computed Tomography (PACT) is an advanced imaging technique that creates images of tissues or objects by measuring the amount of light absorbed at various depths, thus enabling the extraction of information from different tissue layers such as the epidermis (outer layer of the skin, mainly made up of keratinocytes), dermis (which contains blood vessels, connective tissue, and diverse cell types), and subcutaneous fat (Pandey et al. 2020), as well as deeper structures like muscles and organs. This technique, referred to as depth profiling, evaluates information from various tissue depths.

Similar to X-ray Computed Tomography (CT), PAT acquires images from multiple viewing angles. It employs various geometrical configurations of ultrasound transducer arrays around the target surface to produce either 2D cross-sectional or 3D volumetric images (Choi et al. 2020). A schematic of the PAT system has been shown in Fig. 2c. PAT is the most commonly utilized PAI system for in vivo imaging (Benavides-Lara et al. 2023; Manwar et al. 2022; Nasiriavanaki et al. 2014) and preclinical research (Kratkiewicz et al. 2021; Mahmoodkalayeh et al. 2021; Manwar et al. 2020b, 2023a). Therefore, it has been explained in more detail in Sect. 5.

PA Remote Sensing (PARS)

PA Remote Sensing (PARS) is a biomedical imaging technique that achieves sub-micron resolution through a reflection-mode setup. Unlike other PA methods that use acoustic transducers, this method employs a continuous-wave detection laser to measure the initial acoustic pressures near the absorber. It utilizes a co-focused detection beam that captures thermal and pressure effects as intensity changes in reflected light. It is not restricted by the speed of ultrasound waves, allowing for high-quality optical-resolution PA imaging. Unlike many other optical-resolution PAI systems requiring physical contact, PARS can capture images without any connections to the sample, making it advantageous for non-contact imaging applications.

The all-optical design of PARS (illustrated in Fig. 2d) eliminates the need for any additional components between the imaging lens (objective) and the target (tissue) such as Fabry-Perot (Ma et al. 2022) and optical resonators, resulting in a more straightforward system that makes PARS highly versatile and capable of achieving improved resolution and SNR. While PARS does not need physical contact, the imaging process can still induce localized stress within the tissue due to the absorption of laser energy. This stress limits capturing depth information because it requires short acquisition times, typically hundreds of nanoseconds, which fail to gather depth data. However, this capability allows fast imaging, which enables real-time video feedback and rapid 3D visualizations. Notably, PARS has been used for retinal imaging in vivo (Abbasi et al. 2020, 2021; Hosseinaee et al. 2021). The principles, applications, advantages, limitations, and imaging depths of mentioned single-modality PAI systems have been summarized in Table 1.

Table 1 Summary of essential characteristics of single-modality PAI techniques

Technique	Principle	Applications	Advantages	Limitations	Imaging depth
Photoacoustic tomography (PAT)	Images are generated from multiple angles and depths	Cancer detection, brain imaging, etc.	High spatial resolution, deeper tissue imaging	Requires advanced imaging setups, hardware complexity and cost, reconstruction challenges	Several centimeters (up to ~7 cm)
Photoacoustic spectroscopy (PAS)	Multiple wavelengths are used to identify specific tissues	Molecular composition analysis, biomedical diagnostics, evaluation of blood oxygen saturation, etc.	High sensitivity, broad wavelength analysis capacity	Sensitivity to environmental noise, limited penetration depth	From sub-micrometer up to few hundred micrometers
Photoacoustic Microscopy (PAM)	Optical/acoustic microscopy principles are combined with PAI for high-resolution/high-penetration imaging	Cellular and sub-cellular imaging, imaging of blood vessels, etc.	High spatial resolution, detailed structural and functional information	Limited penetration depth	Up to about 1–1.2 mm
Photoacoustic Remote Sensing (PARS)	An all-optical approach is employed to capture PA signals	In vivo imaging, retinal imaging, etc.	Non-contact, fast imaging, simplicity	Low scanning speed and field of view, limited penetration depth	Approximately 1 mm

3.2 Multi-modality PAI

Multimodal imaging involves integrating various imaging techniques to gather complementary information about biological structures, ranging from single molecules to entire organisms. PAI primarily focuses on light absorption, which limits its sensitivity to other contrast types. Researchers combine PAI with different modalities to overcome these challenges and enhance image quality. Initially, PAI was integrated with ultrasound and fluorescence (FL). More recently, multimodal PAI has incorporated additional technologies, including X-ray CT, Magnetic Resonance Imaging (MRI), Single-Photon Emission Computed Tomography (SPECT), and Positron Emission Tomography (PET). These integrations utilize diverse anatomical, functional, and molecular contrast forms to improve the researcher's knowledge. However, the design of imaging systems becomes more complicated.

The ultrasound offers essential insights about tissue structure and blood flow while PAI provides valuable information on blood oxygen levels and molecular characteristics. Together, these modalities confer numerous benefits, such as the absence of ionizing radiation, the ability to capture images in real time, high spatial and temporal resolution, and ease of portability (Kratkiewicz et al. 2019). Also, PA and ultrasound are the two most utilized modalities together due to their similar detection systems, enhancing diagnostic capabilities in various applications. FL imaging is widely used in biology because of its sensitivity for detecting molecular labels. Still, it has limitations in imaging depths greater than 1 mm due to photon scattering within tissues. In contrast, PAI can achieve greater depth. As a result, combining FL and PAI has been frequently utilized in clinical settings, such as ophthalmology and surgery.

MRI offers excellent soft tissue contrast, serving as a detailed anatomical guide for PAI. In contrast, PAI provides essential hemodynamic and molecular data that enhances the blood-oxygen-level-dependent (BOLD) signal from MRI, which mainly illustrates deoxygenated hemoglobin levels. Therefore, integrating these two techniques can significantly advance multiple fields of biomedical research, improving diagnostic and monitoring capabilities for conditions such as breast cancer, cardiovascular diseases, and brain imaging. Nuclear medicine imaging techniques, such as PET and SPECT, visualize the distribution of radioactive tracers in the body, providing valuable information about organ function and metabolism. However, these methods suffer from limitations in anatomical contrast and spatial resolution. To overcome these issues, they are frequently combined with CT, which provides excellent spatial resolution. However, it involves ionizing radiation and lacks sufficient molecular sensitivity for deeper insights. Combining PAI with these techniques can merge functional data on blood oxygenation from PA with high-quality structural information from CT and functional details from PET or SPECT. This multimodal strategy significantly improves cancer diagnosis and treatment monitoring, particularly by offering crucial insights into tumor hypoxia (Chen et al. 2024).

Integrating PAI with other imaging modalities such as CT, PET, SPECT, ultrasound, and MRI present several challenges in the fusion and registration of images (Zhang et al. 2022, 2024a, b; Zhong et al. 2023).

4 PA Tomography

4.1 Irradiation and Detection Mechanisms

In PAT, the source light is unfocused; in other words, it illuminates a broad region of interest (ROI) instead of concentrating on a small, precise region. This unfocused illumination facilitates the excitation of a larger tissue volume, enhancing the likelihood of detecting PA signals from diverse anatomical structures. The array of detectors can then capture the signals from a wide area at various depths and angles, which is beneficial for detecting the diverse and distributed nature of PA sources within the tissue. In its most straightforward configuration, the PAT system utilized a single-element transducer to generate cross-sectional images. This process involves scanning with the transducer, utilizing either a translational stage (which moves in a linear trajectory) or a rotational stage to acquire images from various angles. As technology has developed, more complex configurations began to incorporate sophisticated transducer arrays to enhance the system's capabilities, such as deeper tissue penetration, higher resolution, and higher imaging speed (Neprokin et al. 2022).

In the first generation, linear array transducers were commonly employed in PAT because they are highly compatible with existing ultrasound systems and cost-effective. However, linear arrays have limited resolution in the elevation direction compared to their axial and lateral directions, which makes them less suitable for three-dimensional imaging. This limitation causes artifacts in the PA images. Therefore, various transducer geometries have been developed to address this issue, including circular/cylindrical, convex, planar, or hemispherical-shaped and curved transducers that provide better access to the generated PA waves from multiple angles (Zheng et al. 2022). Since the detectors are not focused on specific points and capture signals from multiple angles, the relationship between the received acoustic signals and their sources in the tissue becomes more complex. Therefore, it is essential to use efficient reconstruction algorithms in PAT.

4.2 Image Reconstruction Techniques

The primary aim of PAT is to reconstruct the distribution of optical absorption within the biological tissues. It means determining the extent of light absorption at different depths and angles within the tissue. Reconstruction algorithms, known as beam-formers, interpret the observed pressure signals so that the signal's amplitude at time t is related to the integral of pressure distribution (P_0). It is calculated over a spherical surface in the center of a detector, with a radius of $v_s\, t$ (where v_s represents the speed of sound, and t is time), as outlined in Eq. 3. The effectiveness of these algorithms directly impacts the quality of the resulting images and the accuracy of the diagnostic information derived from them.

The principle behind image reconstruction is reversing the propagated acoustic signals to the initial state and predicting the locations of their origins by measuring the arrival times, which depend on the distance from the acoustic source and angles of these signals detected across multiple detectors (Choi et al. 2020; Prakash et al. 2023). The general inverse problem utilized in the reconstruction algorithm is defined by Eq. 4 (Hsu et al. 2021):

$$\mathrm{g} = \mathrm{A}\ \left(\mathrm{f}^{*}\right) + \delta \mathrm{g} \tag{4}$$

It involves two main parameters: 1. Unknown optical absorption coefficient (f^*) that represents the amount of light absorbed by tissue, so it identifies initial pressure distribution, and 2. Measured acoustic pressure time series (g) that reflects the acoustic signals collected from the tissue over time, obtained from the aforementioned inverse problem. The function $A: X \rightarrow Y$ is a forward operator that models how the absorption coefficient f^* produces the measured acoustic signals. Additionally, δg accounts for measurement errors, noise affecting the data, or any inaccuracies.

Several conventional image reconstruction techniques have been employed in PAT, such as Filtered Back Projection (FBP), delay-and-sum (DAS) beamforming, delay multiply and sum (DMAS), time-reversal (TR), and Fourier transform-based algorithms (Dange 2021; Yang et al. 2021). FBP is a common technique for image reconstruction that takes projections from various angles, applies a filter to enhance the edges and reduce blurring, and then back-projects the filtered data onto an image space. However, this method operates under the assumption that the medium is acoustically homogeneous, which can lead to artifacts in heterogeneous tissues. DAS beamforming combines signals from multiple transducer elements by delaying and summing them to construct an image; however, it can suffer from low contrast and is sensitive to noise. DMAS is a modified version of DAS that has been shown to provide improved image quality in linear array PAT. However, DMAS continues to face challenges related to noise. TR methods reverse the detected acoustic signals back to their sources and consist of two main components: 1. The acoustic inverse problem, which estimates the initial acoustic pressure distribution within the object from the measured PA time series, and 2. The optical inverse problem estimates the distributions of optical parameters based on the absorbed optical energy density or the initial pressure distribution obtained from the acoustic inversion. TR methods can focus on specific ROI and demonstrate robustness to certain types of noise; nevertheless, they increase computational complexity (Tarvainen and Cox 2024). Fourier transform-based algorithms utilize the frequency domain characteristics of the signals to construct images. These algorithms can enhance image quality through frequency-domain filtering; however, they may be computationally expensive and often require significant data pre-processing to ensure accurate reconstructions. All these algorithms can provide accurate images when all captured data are ideal; however, in real-world applications, they can be time-consuming and costly because of the dense sampling of objects in full view. Specifically, in vivo applications require a high frame rate, which is restricted by the time necessary to acquire the acoustic boundary data and produce an image.

Several methods have been developed to address the aforementioned challenges and mitigate artifacts, including interpolation-based methods, optimization-based iterative reconstruction methods, compressed sensing reconstruction techniques, and deep learning reconstruction techniques (Wang et al. 2024; WangLihong 2021; Shen and Tian 2021). In interpolation-based methods, unmeasured signals are obtained from measured signals via interpolation. This technique reduces unwanted artifacts and enhances image quality; however, it can be low in accuracy and computationally intensive. Iterative reconstruction methods are essential for generating high-quality images from sparsely sampled acoustic measurements. One such method is the Algebraic Reconstruction Technique (ART), developed for handling incomplete and noisy data; however, if the projections from certain angles are missing, it can result in blurred edges and artifacts. So, Simultaneous Iterative Reconstruction Technique (SIRT) and its modified version (MSIRT) were developed to enhance image reconstruction. They calculate the correct value for each pixel by averaging across all iterations, resulting in high-precision outcomes with fewer iterations than ART and faster convergence. However, they are computationally more expensive compared to the FBP method. The Filtered Mean-Back-Projection Iterative (FMBPI) technique further improves image reconstruction quality by utilizing a weight factor and an estimated relationship between the initial and actual distributions of acoustic pressure, thereby decreasing the number of iterations required.

Compressed sensing (CS) techniques are effective methods for sparse-data PAT imaging. They enable reconstructing whole acoustic signals from partial measurements by employing a sparse-constrained algorithm, allowing for high-quality imaging with less acoustic data. CS-PAT in the frequency domain offers better Contrast-to-Noise Ratio (CNR) and faster reconstruction than time-domain CS. CS-PAT provides superior image quality compared to TR and traditional analytic approaches for acoustic inversion. Furthermore, frequency-domain CS outperforms time-domain CS in enhancing the CNR and accelerating reconstruction speed (Tarvainen and Cox 2024; Zheng and Xiangyang 2020). However, these methods for PA reconstruction face challenges due to the use of a sparse sensing strategy for data acquisition and limited view geometry, have shown in Fig. 3 (Chen et al. 2024; Hsu et al. 2021).

To address these challenges, deep learning-based methods, particularly convolutional neural networks (CNNs) for extracting spatial features and recurrent neural networks (RNNs) for utilizing temporal information, are being developed to produce high-quality image reconstructions. Integrating these techniques with conventional methods can yield results comparable to or even exceed those of existing approaches. For example, model-based learned iterative reconstruction enhances PAT image quality by combining deep learning techniques, which help reduce artifacts. However, this approach faces challenges such as computational costs and memory usage. Additionally, using deep learning methods for in vivo imaging encounters limitations due to a lack of training and testing data (Hsu et al. 2021; Zheng and Xiangyang 2020; Gröhl et al. 2021; Kim et al. 2020; Lan et al. 2020). As a result, selecting the most suitable method depends on specific applications and requires consideration of both time and accuracy.

5 Contrast Agents in PAI

In PAI, contrast agents play a pivotal role in enhancing image quality and specificity that are classified into two main categories: endogenous and exogenous contrast agents. Endogenous agents are naturally existing substances within biological tissues. Notable examples include hemoglobin (which monitors blood oxygen

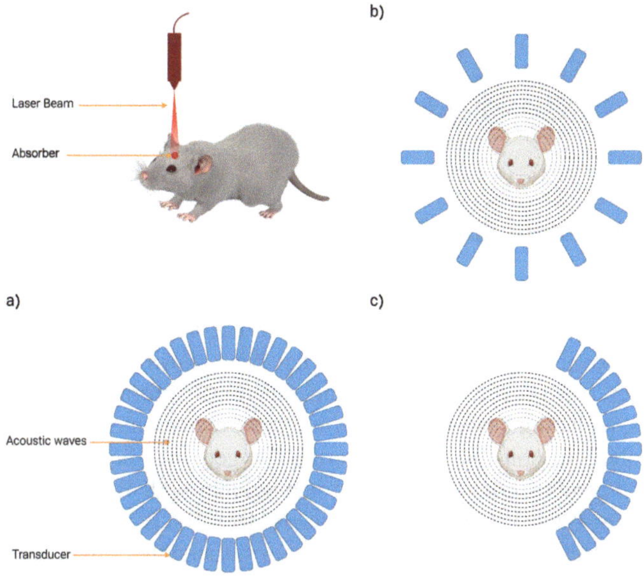

Fig. 3 Illustration of various PA signal acquisition techniques. (**a**) PAI process with complete data; PA signals from 360° around the sample are acquired. (**b**) Sparse sensing; in this technique, number of US transducer elements have been reduced. (**c**) limited view geometry; in this technique, PA waves from limited-angle are collected

levels), and melanin (which can indicate certain skin conditions). These agents provide inherent contrast due to their unique optical absorption properties, allowing for label-free imaging. On the other hand, exogenous agents are manufactured substances designed to improve imaging performance when introduced at the imaging site. They also make molecular imaging possible. In many cases, these agents outperform endogenous agents. Also, exogenous contrast agents can be used for theranostic and tumor imaging. Exogenous contrast agents include nanoparticles, dyes, and various compounds specifically engineered to absorb light at specified wavelengths, thus improving the specificity and contrast of the PAI. The key difference lies in their origin; endogenous agents utilize the body's natural chromophores, while exogenous agents are synthetic and can be tailored for specific applications, allowing for advanced imaging techniques that surpass the limitations imposed by endogenous agents (Tsang et al. 2020).

Generally exogenous agents can be used when endogenous contrast is insufficient for imaging needs, for example in the case of low signal from early-stage tumors, when endogenous agents like hemoglobin and melanin provide natural contrast but often fail to deliver sufficient sensitivity or specificity for detecting early-stage tumors or small lesions, exogenous agents enhance contrast and improve visualization. Also, exogenous agents are being used to achieve enhanced imaging depth and contrast. They are engineered to absorb strongly in the near-infrared windows (NIR-I: 650–900 nm, NIR-II: 1000–1700 nm) where tissue absorption and scattering are minimal. Also, exogenous agents are useful for functional and multi-modality imaging, where exogenous agents can be functionalized to enable simultaneous imaging with complementary modalities (e.g., MRI, fluorescence, ultrasound), providing comprehensive diagnostic information beyond PA imaging alone.

5.1 Endogenous Contrast Agents

Natural chromophores such as melanin, lipids, deoxyhemoglobin (HbR), and oxyhemoglobin (HbO_2) fall into this category. The absorption spectrum of some significant molecules in biological tissue has been shown in Fig. 4. Endogenous chromophores can provide structural and functional information for biomedical applications, including cancer prognosis and brain activity measurement.

Hemoglobin, responsible for delivering oxygen via the bloodstream, is an excellent endogenous contrast. Imaging hemoglobin with PAI can help visualize the angiogenesis, a crucial parameter for early detection of tumor development. In addition to providing anatomical information of vasculature, functional information associated with blood oxygen saturation (SO_2) can also be extracted in vivo (Manwar et al. 2022). SO_2 is defined as the fraction of HbO_2 relative to total hemoglobin concentration in blood. According to Fig. 4, at wavelengths between 650 nm and 900 nm, both HbR and HbO_2 have an absorption coefficient at least one magnitude larger than those of other chromophores. Although melanin has a

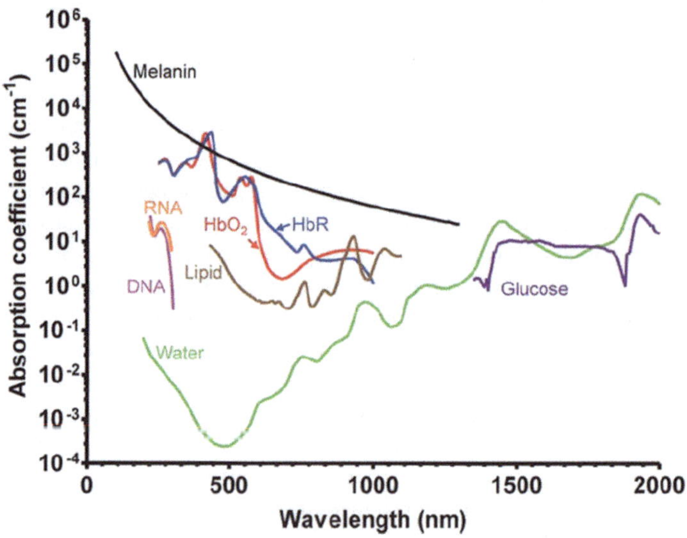

Fig. 4 The optical absorption coefficients of different natural chromophores in various wavelengths of irradiation (RNA, DNA, Lipid, HbO2, HbR, water, Glucose, and Melanin). (Reproduced with permission from Yao and Wang 2013)

higher absorption coefficient within this wavelength range than HbR and HbO2, it is highly localized in the skin or retina and thus does not contribute to oxygenation quantification (Han et al. 2022). Although NIR is a suitable range for SO_2 mapping, an LED array including 750 nm and 850 nm elements can be utilized in a dual-wavelength technique for oxygen saturation imaging (Bulsink et al. 2021).

5.2 Exogenous Contrast Agents

Tissue differentiation can be challenging due to the close absorption coefficient of tissue of interest to the surrounding tissues, so using an extra substance seems necessary. Exogenous contrast agents refer to artificial contrast media made that are introduced to the body as additional substances. In addition to providing better contrast of deeper tissues, it provides the ideal opportunity for molecular imaging (Nasri et al. 2023). Several factors should be considered when selecting exogenous agents. The most crucial factors are biocompatibility and degradability. The second factor is the ability of the contrast agent to reach the target tissue without deposition by kidneys (Farooq et al. 2022).

In addition to being non-toxic and cost-effective, they should have an adequate quantum yield and high absorption peak in the near-infrared range. Exogenous agents can be categorized into three subgroups: organic-based, metal-based, and dye-based. Organic-based agents' main benefit is their low toxicity. Also, some modifications can enhance their efficiency. Metal-based agents work based on plasmon surface resonance. The urge to use these agents lies behind the fact that they are photostable, have enhanced brightness, and have manageable photopeaks. Also, these agents have already been used in PET and CT, so that they can be a good option for dual-modality imaging. Dye-based agents, such as indocyanine green, methylene blue, and Evans blue, can be used as contrast agents in PAI that can produce high signal intensities. These contrast agents are typically combined with other materials (Han et al. 2022).

6 Conclusions and Future Perspectives

PAI represents a significant advancement in biomedical imaging, offering a unique combination of high resolution, deep tissue penetration, and label-free imaging capabilities (Kratkiewicz et al. 2021). Through the PA effect, PAI excels in providing structural and functional information about biological tissues in a non-invasive manner. As this chapter illustrates, signal generation and detection methodologies, various hardware setups, sampling, and image reconstruction techniques form the backbone of PAI, facilitating its application across various fields, from cancer diagnostics to functional brain imaging. However, despite its myriad advantages, PAI still faces notable challenges, particularly in achieving high frame rates necessary for in vivo applications and overcoming limitations imposed by dense sampling requirements. As we look to the future, the integration of AI and the developments in detectors and light sources will likely play pivotal roles in addressing these challenges and expanding the capabilities of PAI.

A promising direction involves the increasing use of AI, particularly deep learning technologies, to improve image reconstruction and analysis in PAI frameworks. Emerging research indicates that these algorithms can markedly enhance the quality of images reconstructed from limited or noisy datasets, facilitating faster and more accurate interpretation of imaging results. As large datasets for biomedical imaging become more accessible, the potential for transfer learning further enhances model performance and reliability in real-world preclinical applications (Gröhl et al. 2021). Furthermore, recent advances in novel detectors and light sources are poised to revolutionize PAI technology. Innovations in laser technology, particularly tunable laser sources and fiber lasers, enable tailored wavelength selection for specific tissue types, leading to improved imaging contrasts and depths. Developing novel ultrasound detectors enhances sensitivity and resolution while allowing seamless integration with PAI systems. Similarly, incorporating novel contrast agents, including biocompatible and targeted nanoparticles, promises to enhance PAI's specificity and diagnostic power.

The future landscape for PAI in preclinical research presents both exciting opportunities and challenges. Interdisciplinary research is crucial to overcoming existing technological limitations and facilitating the integration of PAI with complementary imaging modalities, thereby enhancing its diagnostic repertoire. Developing portable and cost-effective imaging systems remains essential to making PAI broadly accessible for various research initiatives involving small animals. In conclusion, leveraging AI and advancing detection and light source technologies are two critical strategies that can significantly enhance the capabilities of PAI. As these fields continue to evolve, PAI is likely to witness transformations that enhance not only its diagnostic accuracy and efficiency but also its role in personalized medicine and innovative therapeutic approaches.

Acknowledgments N/A

Conflict of Interest The authors have no conflicts of interest to declare. All co-authors have seen and agree with the contents of the manuscript, and there is no financial interest to report. We certify that the submission is an original manuscript and is not under review at any other publication.

References

Abbasi S, Bell K, Haji Reza P (2020) Rapid high-resolution mosaic acquisition for photoacoustic remote sensing. Sensors 20(4):1027

Abbasi S, Bell K, Ecclestone B, Reza PH (2021) Live feedback and 3D photoacoustic remote sensing. Quant Imaging Med Surg 11(3):1033–1045

Attia ABE et al (2019) A review of clinical photoacoustic imaging: current and future trends. Photoacoustics 16:100144

Barbosa RCS, Mendes PM (2022) A comprehensive review on photoacoustic-based devices for biomedical applications. Sensors 22(23):9541

Benavides-Lara J et al (2023) Transfontanelle photoacoustic imaging of intraventricular brain hemorrhages in live sheep. Photoacoustics 33:100549

Bulsink R, Kuniyil Ajith Singh M, Xavierselvan M, Mallidi S, Steenbergen W, Francis KJ (2021) Oxygen saturation imaging using LED-based photoacoustic system. Sensors 21(1):283

Chen Z, Gezginer I, Zhou Q, Tang L, Deán-Ben XL, Razansky D (2024) Multimodal optoacoustic imaging: methods and contrast materials. Chem Soc Rev. https://doi.org/10.1039/d3cs00565h

Cho S-W, Nguyen VT, DiSpirito A, Yang J, Kim C-S, Yao J (2024) Sounding out the dynamics: a concise review of high-speed photoacoustic microscopy. J Biomed Opt 29(S1):S11521–S11521

Choi W, Oh D, Kim C (2020) Practical photoacoustic tomography: realistic limitations and technical solutions. J Appl Phys 127(23)

Dange S (2021) Time reversal reconstruction algorithm for photoacoustic imaging. In: AIJR proceedings, pp 405–413

Du J, Yang S, Qiao Y, Lu H, Dong H (2021) Recent progress in near-infrared photoacoustic imaging. Biosens Bioelectron 191:113478

Farooq A, Sabah S, Dhou S, Alsawaftah N, Husseini G (2022) Exogenous contrast agents in photoacoustic imaging: an in vivo review for tumor imaging. Nanomaterials 12(3):393

Fatima A et al (2019) Review of cost reduction methods in photoacoustic computed tomography. Photoacoustics 15:100137

Gao R, Xu Z, Ren Y, Song L, Liu C (2021) Nonlinear mechanisms in photoacoustics—powerful tools in photoacoustic imaging. Photoacoustics 22:100243

Ghadiri H, Fouladi M, Rahmim A (2016) Analysis of various parameters affecting signals in resting-state mouse brain photoacoustic microscopy: A simulation study. In: 2016 2nd International Conference of Signal Processing and Intelligent Systems (ICSPIS). IEEE, pp 1–5

Ghadiri H, Fouladi MR, Rahmim A (2018) An analysis scheme for investigation of effects of various parameters on signals in acoustic-resolution photoacoustic microscopy of mice brain: a simulation study. arXiv preprint arXiv:1805.06236

Gröhl J, Schellenberg M, Dreher K, Maier-Hein L (2021) Deep learning for biomedical photoacoustic imaging: a review. Photoacoustics 22:100241

Han S, Lee D, Kim S, Kim H-H, Jeong S, Kim J (2022) Contrast agents for photoacoustic imaging: a review focusing on the wavelength range. Biosensors 12(8):594

Hosseinaee Z, Abbasi N, Pellegrino N, Khalili L, Mukhangaliyeva L, Haji Reza P (2021) Functional and structural ophthalmic imaging using noncontact multimodal photoacoustic remote sensing microscopy and optical coherence tomography. Sci Rep 11(1):11466

Hsu K-T, Guan S, Chitnis PV (2021) Comparing deep learning frameworks for photoacoustic tomography image reconstruction. Photoacoustics 23:100271

Huang B, Wong TT (2024) Review of low-cost light sources and miniaturized designs in photoacoustic microscopy. J Biomed Opt 29(S1):S11503–S11503

Illienko M, Velsink MC, Witte S (2024) Understanding photoacoustic signal formation in the presence of transparent thin films. Photoacoustics 38:100617

Jeng G-S et al (2021) Real-time interleaved spectroscopic photoacoustic and ultrasound (PAUS) scanning with simultaneous fluence compensation and motion correction. Nat Commun 12(1):716

Jiang D, Zhu L, Tong S, Shen Y, Gao F, Gao F (2024) Photoacoustic imaging plus X: a review. J Biomed Opt 29(S1):S11513–S11513

Jin Y, Yin Y, Li C, Liu H, Shi J (2022) Non-invasive monitoring of human health by photoacoustic spectroscopy. Sensors 22(3):1155

Kaniyala Melanthota S et al (2022) Types of spectroscopy and microscopy techniques for cancer diagnosis: a review. Lasers Med Sci 37(8):3067–3084

Kim M, Jeng G-S, Pelivanov I, O'Donnell M (2020) Deep-learning image reconstruction for real-time photoacoustic system. IEEE Trans Med Imaging 39(11):3379–3390

Kratkiewicz K et al (2019) Photoacoustic/ultrasound/optical coherence tomography evaluation of melanoma lesion and healthy skin in a swine model. Sensors 19(12):2815

Kratkiewicz K, Manwar R, Zhou Y, Mozaffarzadeh M, Avanaki K (2021) Technical considerations in the Verasonics research ultrasound platform for developing a photoacoustic imaging system. Biomed Opt Express 12(2):1050–1084

Lan H, Jiang D, Yang C, Gao F, Gao F (2020) Y-net: hybrid deep learning image reconstruction for photoacoustic tomography in vivo. Photoacoustics 20:100197

Lane LA, Xue R, Nie S (2018) Emergence of two near-infrared windows for in vivo and intraoperative SERS. Curr Opin Chem Biol 45:95–103

Li Y et al (2018) In vivo photoacoustic/ultrasonic dual-modality endoscopy with a miniaturized full field-of-view catheter. J Biophotonics 11(10):e201800034

Lin L, Wang LV (2021) Photoacoustic imaging. In: Optical imaging in human disease and biological research. Springer, pp 147–175

Liu H, Teng X, Yu S, Yang W, Kong T, Liu T (2024) Recent advances in photoacoustic imaging: current status and future perspectives. Micromachines 15(8):1007

Ma X, Fan M, Cai Y, Xu L, Ma J (2022) A Fabry–Perot fiber-optic array for photoacoustic imaging. IEEE Trans Instrum Meas 71:1–8

Mahmoodkalayeh S et al (2021) Wavelength and pulse energy optimization for detecting hypoxia in photoacoustic imaging of the neonatal brain: a simulation study. Biomed Opt Express 12(12):7458–7477

Manwar R, Kratkiewicz K, Avanaki K (2020a) Overview of ultrasound detection technologies for photoacoustic imaging. Micromachines 11(7):692

Manwar R, Kratkiewicz K, Avanaki K (2020b) Investigation of the effect of the skull in transcranial photoacoustic imaging: a preliminary ex vivo study. Sensors 20(15):4189

Manwar R et al (2022) Transfontanelle photoacoustic imaging for in-vivo cerebral oxygenation measurement. Sci Rep 12(1):1–7

Manwar R, Gelovani JG, Avanaki K (2023a) Bilirubin-biliverdin concentration measurement using photoacoustic spectroscopic analysis for determining hemorrhage age. J Biophoton 16:e202200316

Manwar R, Islam MT, Shoo A, Pillers DA, Avanaki K (2023b) Development of ex-vivo brain hemorrhage phantom for photoacoustic imaging. J Biophotonics 16:e202200313

Moore C, Jokerst JV (2019) Strategies for image-guided therapy, surgery, and drug delivery using photoacoustic imaging. Theranostics 9(6):1550–1571

Najafzadeh E, Ghadiri H, Alimohamadi M, Farnia P, Mehrmohammadi M, Ahmadian A (2020) Evaluation of multi-wavelengths LED-based photoacoustic imaging for maximum safe resection of glioma: a proof of concept study. Int J Comput Assist Radiol Surg 15:1053–1062

Nasiriavanaki M, Xia J, Wan H, Bauer AQ, Culver JP, Wang LV (2014) High-resolution photoacoustic tomography of resting-state functional connectivity in the mouse brain. Proc Natl Acad Sci 111(1):21–26

Nasri D, Manwar R, Kaushik A, Er EE, Avanaki K (2023) Photoacoustic imaging for investigating tumor hypoxia: a strategic assessment. Theranostics 13(10):3346–3367

Neprokin A, Broadway C, Myllylä T, Bykov A, Meglinski I (2022) Photoacoustic imaging in biomedicine and life sciences. Life 12(4):588

Pandey P et al (2020) Emerging trends in nanomedicine for topical delivery in skin disorders: current and translational approaches. Dermatol Ther 33(3):e13292

Park J et al (2024) Clinical translation of photoacoustic imaging. Nat Rev Bioeng:1–20. https://doi.org/10.1038/s44222-024-00240-y

Prakash R, Manwar R, Avanaki K (2023) Evaluation of 10 current image reconstruction algorithms for linear array photoacoustic imaging. J Biophotonics 17:e202300117

Riksen JJ, Nikolaev AV, van Soest G (2023) Photoacoustic imaging on its way toward clinical utility: a tutorial review focusing on practical application in medicine. J Biomed Opt 28(12):121205–121205

Shen K, Tian C (2021) Deep filtered back projection for photoacoustic image reconstruction. In: M4G.3: Asia communications and photonics conference. Optica Publishing Group

Tarvainen T, Cox B (2024) Quantitative photoacoustic tomography: modeling and inverse problems. J Biomed Opt 29(S1):S11509–S11509

Tsang VT, Li X, Wong TT (2020) A review of endogenous and exogenous contrast agents used in photoacoustic tomography with different sensing configurations. Sensors 20(19):5595
Tsubouchi M, Hoshina H, Nagai M, Isoyama G (2020) Plane photoacoustic wave generation in liquid water using irradiation of terahertz pulses. Sci Rep 10(1):18537
Wang H, Chen Y-S, Zhao Y (2022) Understanding the near-field photoacoustic spatiotemporal profile from nanostructures. Photoacoustics 28:100425
Wang Y, Chen Y, Zhao Y, Liu S (2024) Compressed sensing for biomedical photoacoustic imaging: a review. Sensors 24(9):2670
WangLihong V (2021) Recent advances in photoacoustic tomography. BME Front 2021:9823268
Yang G, Amidi E, Zhu Q (2021) Photoacoustic tomography reconstruction using lag-based delay multiply and sum with a coherence factor improves in vivo ovarian cancer diagnosis. Biomed Opt Express 12(4):2250–2263
Yao J, Wang LV (2013) Photoacoustic microscopy. Laser Photonics Rev 7(5):758–778
Zafar M, Kratkiewicz K, Manwar R, Avanaki M (2019) Development of low-cost fast photoacoustic computed tomography: system characterization and phantom study. Appl Sci 9(3):374
Zafar M, Manwar R, McGuire LS, Charbel FT, Avanaki K (2023) Ultra-Widefield and high-speed spiral laser scanning OR-PAM: system development and characterization. J Biophotonics 16:e202200383
Zafar M et al (2024) Spiral laser scanning photoacoustic microscopy for functional brain imaging in rats. Neurophotonics 11(1):015007–015007
Zhang S et al (2022) MRI information-based correction and restoration of photoacoustic tomography. IEEE Trans Med Imaging 41(9):2543–2555
Zhang R et al (2024a) Navigating challenges and solutions in quantitative photoacoustic imaging. Appl Phys Rev 11(3)
Zhang S, Miao J, Li LS (2024b) Challenges and advances in two-dimensional photoacoustic computed tomography: a review. J Biomed Opt 29(7):070901–070901
Zhao T, Desjardins AE, Ourselin S, Vercauteren T, Xia W (2019) Minimally invasive photoacoustic imaging: current status and future perspectives. Photoacoustics 16:100146
Zheng S, Xiangyang Y (2020) Image reconstruction based on compressed sensing for sparse-data endoscopic photoacoustic tomography. Comput Biol Med 116:103587
Zheng W, Huang C, Zhang H, Xia J (2022) Slit-based photoacoustic tomography with co-planar light illumination and acoustic detection for high-resolution vascular imaging in human using a linear transducer array. Biomed Eng Lett 12(2):125–133
Zhong Y et al (2023) Unsupervised fusion of misaligned PAT and MRI images via mutually reinforcing cross-modality image generation and registration. IEEE Trans Med Imaging 43:1702–1714

Applications of Photoacoustic Imaging in Preclinical Research

Ebrahim Najafzadeh, Parastoo Farnia, Reza Naghne, Niuosha Ghadesi, Tina Anjomshoa, and Mohammad Reza Fouladi

Abstract

Photoacoustic Imaging (PAI) is a cutting-edge imaging modality that combines the advantages of optical and ultrasound imaging, offering high-contrast and deeper penetration in order to visualize and extract anatomical, functional, and molecular information of the biological tissues. Due to the noninvasive nature

E. Najafzadeh
Finetech in Medicine Research Center, School of Medicine, Iran University of Medical Sciences, Tehran, Iran

Department of Medical Physics, School of Medicine, Iran University of Medical Sciences, Tehran, Iran

Department of Molecular Imaging, Faculty of Advanced Technologies in Medicine, Iran University of Medical Sciences, Tehran, Iran

P. Farnia · R. Naghne
Department of Medical Physics and Biomedical Engineering, Tehran University of Medical Sciences (TUMS), Tehran, Iran

Research Center for Intelligent Technologies in Medicine (RCITM), Advanced Medical Technologies and Equipment Research Institute (AMTEI), TUMS, Tehran, Iran

Research Center of Biomedical Technology and Robotics (RCBTR), Advanced Medical Technologies and Equipment Research Institute (AMTEI), TUMS, Tehran, Iran

N. Ghadesi · T. Anjomshoa
Department of Medical Physics and Biomedical Engineering, Tehran University of Medical Sciences (TUMS), Tehran, Iran

Research Center for Intelligent Technologies in Medicine (RCITM), Advanced Medical Technologies and Equipment Research Institute (AMTEI), TUMS, Tehran, Iran

M. R. Fouladi (✉)
Department of Medical Physics and Biomedical Engineering, Tehran University of Medical Sciences (TUMS), Tehran, Iran

I. F. Cengiz et al. (eds.), *Bioimaging Modalities in Bioengineering*,
Biomaterials, Bioengineering and Sustainability 9,
https://doi.org/10.1007/978-3-031-99410-4_13

and easy integration with popular ultrasound diagnostic imaging, PAI has been utilized for several applications in preclinical studies, such as vascular imaging, utilizing endogenous contrast agents' characteristics like hemoglobin to measure parameters such as blood oxygenation and flow velocity. Major configurations of PAI i.e., photoacoustic tomography and photoacoustic microscopy excel in imaging small vessels and microvasculature, enabling the study of angiogenesis, tumor malignancy, ischemic conditions, etc. PAI's ability to differentiate oxygenated and deoxygenated hemoglobin provides critical insights into tissue states, particularly for stroke and tumor imaging. Advanced systems such as multi-spectral optoacoustic tomography and spiral volumetric optoacoustic tomography enable 3D imaging with enhanced resolution and speed. PAI also integrates exogenous contrast agents, including nanoparticles and dyes, to enhance imaging specificity for targeted molecular and functional studies. Its applications range from tumor imaging, wound healing monitoring to drug delivery assessment. By merging functional and molecular imaging capabilities, PAI demonstrates its versatility in preclinical research and its potential for clinical application.

Keywords

Photoacoustic imaging (PAI) · Vascular imaging · Hemoglobin oxygen saturation · Angiogenesis · Exogenous contrast agents

1 Introduction

PAI has emerged as a revolutionary modality for biomedical imaging, integrating optical and acoustic capabilities to achieve detailed images of biological tissues. By utilizing endogenous contrast agents such as hemoglobin and melanin, PAI enables label-free noninvasive imaging of vascular structures and tissue oxygenation without requiring exogenous contrast agents (Farajollahi and Baharvand 2024; Liu and Yao 2018; Zafar et al. 2024). Its unique ability to integrate structural, functional, and molecular imaging has positioned PAI as a promising tool for various applications, including characterization of tumors and ischemic stroke research (Ivankovic et al. 2024).

Two types of PAI techniques, photoacoustic tomography (PAT) and photoacoustic microscopy (PAM), have shown exceptional utility in preclinical and clinical research. PAT is a particularly effective method for imaging large structures and assessing hemodynamic parameters such as blood flow velocity. In contrast, PAM offers superior spatial resolution in microvascular imaging (Zhu et al. 2024; Zhang et al. 2018). Furthermore, multi-spectral optoacoustic tomography (MSOT) and 3D volumetric imaging approaches such as spiral volumetric optoacoustic tomography (SVOT) have significantly enhanced the PAI's capabilities by providing high-resolution, multi-spectral, and real-time imaging of complex tissue structures in animal models (Chen et al. 2020; Yan et al. 2023).

Recent advancements have introduced exogenous contrast agents such as nanoparticles, engineered biomolecules, and molecular dyes, enhancing PA imaging specificity and sensitivity. These developments have facilitated targeted molecular imaging, providing deeper insights into pathological changes, the microenvironments of tumors, and drug delivery dynamics (Farajollahi and Baharvand 2024). Gold nanoparticles have been engineered to amplify photoacoustic signals for tumor imaging, and indocyanine green (ICG) has demonstrated utility in monitoring therapeutic responses (Yao and Wang 2018).

A pivotal frontier in PAI research is the integration of artificial intelligence (AI). AI-driven algorithms have demonstrated great promise in optimizing photoacoustic data acquisition, improving image reconstruction, etc. For instance, machine learning techniques are utilized to un-mix complex multi-spectral data, identify subtle vascular abnormalities, and automate the classification of pathological features in PAI images (Liu et al. 2024; Kim et al. 2020). Deep learning approaches have further enabled real-time image reconstruction and noise reduction, enhancing the resolution (Grasso et al. 2023; Rodrigues et al. 2024).

In this chapter, recent applications of PAI in anatomical, functional, and molecular studies have been reviewed. In the anatomical PAI section, capabilities of PAI for label-free vascular imaging have been described. Also, MSOT and SVOT, as remarkable volumetric PAI systems, have been introduced. In the functional PAI section, applications and methodologies of hemoglobin concentration, blood flow, and hemoglobin oxygen saturation (sO_2) assessment have been described, and further developments have been reviewed. In the molecular PAI section, utilization of endogenous and exogenous contrast agents for enabling molecular features using PAT or PAM systems have been investigated.

2 Anatomical PAI

PAI has emerged as a powerful tool for anatomical, structural, and morphological imaging in preclinical research, offering unique advantages in visualizing biological tissues with high spatial resolution and deep penetration rather than pure optical imaging techniques. This section introduces anatomical PAI, focusing on its ability to provide detailed images of tissue architecture and vascular structures. Unlike purely optical or ultrasound imaging, PAI capitalizes on the optical absorption properties of tissues, translating absorbed pulsed light into ultrasonic waves, thus combining optical contrast with ultrasonic resolution. These characteristics make PAI particularly well-suited for imaging complex anatomical features such as microvasculature, organ boundaries, and tissue morphology with exceptional clarity and depth.

A key strength of anatomical PAI lies in its capability to image vascular networks in vivo, which is critical for understanding physiological and pathological processes. Blood vessels are among the most optically absorbing endogenous chromophores in tissue due to the presence of hemoglobin. This enables non-invasive visualization and mapping of vasculature with precise delineation of vessel morphology, branching patterns, and density. Such detailed vascular imaging plays an essential role in preclinical studies of cancer angiogenesis, cardiovascular disease, and tissue regeneration. PAT and PAM are the two primary PAI configurations in vascular imaging of animal models. Also, PAT and PAM systems can independently measure functional parameters, including blood flow velocity, which can also be assessed using Doppler imaging. However, unlike PAI, Doppler imaging is less applicable for small vessels due to its limited resolution (Farajollahi and Baharvand 2024; Li et al. 2021).

PAI does not require exogenous contrast agents for vascular imaging, making this modality a unique approach for imaging and quantifying vascular networks (Brown et al. 2022). There are endogenous contrast agents used for vascular imaging, such as hemoglobin, which has an optical absorption peak in the near-infrared (NIR) and visible spectrum range (John et al. 2023). Figure 1 shows PAM images of the mouse brain before and after photothrombotic stroke induction. The figure includes control and stroke conditions, with vessel-level imaging at 1-minute intervals for 5 minutes. These sequential images illustrate structural changes in the cerebral vasculature and highlight ability of anatomical PAI to monitor vascular responses in real time (Seong et al. 2022).

The concentrations of oxyhemoglobin (HbO_2) and deoxyhemoglobin (HbR) are utilized to calculate the sO_2 that will be discussed in the functional PAI section. Besides enabling functional imaging, these concentrations also reflect important factors such as angiogenesis, the process of forming new blood vessels, specifically showing tumor growth. In tumor imaging, malignancy is often associated with the presence of abnormal new vessels and increased vascular density. PAT is a valuable modality for visualizing vascular shape and density changes both with and without exogenous contrast agents. Additionally, it provides detailed information on oxygenation and blood flow velocity, that besides structural information, functional data can be obtained for further assessments (Gu et al. 2023).

Thus far, we have established that PAI is a promising modality for assessing cardiovascular diseases, like stroke, as it could demonstrate vascular malfunction and hemodynamic parameters. It could also be utilized for heart muscle imaging as it can use myoglobin as an endogenous contrast agent, which can be found in muscle (Karlas et al. 2019). Also, PAI has advantages in imaging microvasculature and peripheral vessel imaging, specifically in skin imaging. It provides the opportunity of imaging vasculature and melanin distribution, concurrently. As a demonstrated application, in order to facilitate early identification and quantitative evaluation of an immune-mediated condition known as systemic sclerosis, which causes internal organ and skin fibrosis and vasculopathy, PAI has been used as a technique to identify alterations in skin micro-vessels (Ahn et al. 2024).

Both endogenous and exogenous contrast agents are utilized to enhance the performance of anatomical PAI. Endogenous agents, primarily hemoglobin and melanin, intrinsically provide rich contrast for vascular and tissue imaging, facilitating label-free visualization of blood vessel structures and anatomical features. However, to achieve enhanced contrast, deeper tissue penetration, or targeted imaging of specific structures, exogenous contrast agents such as nanoparticles

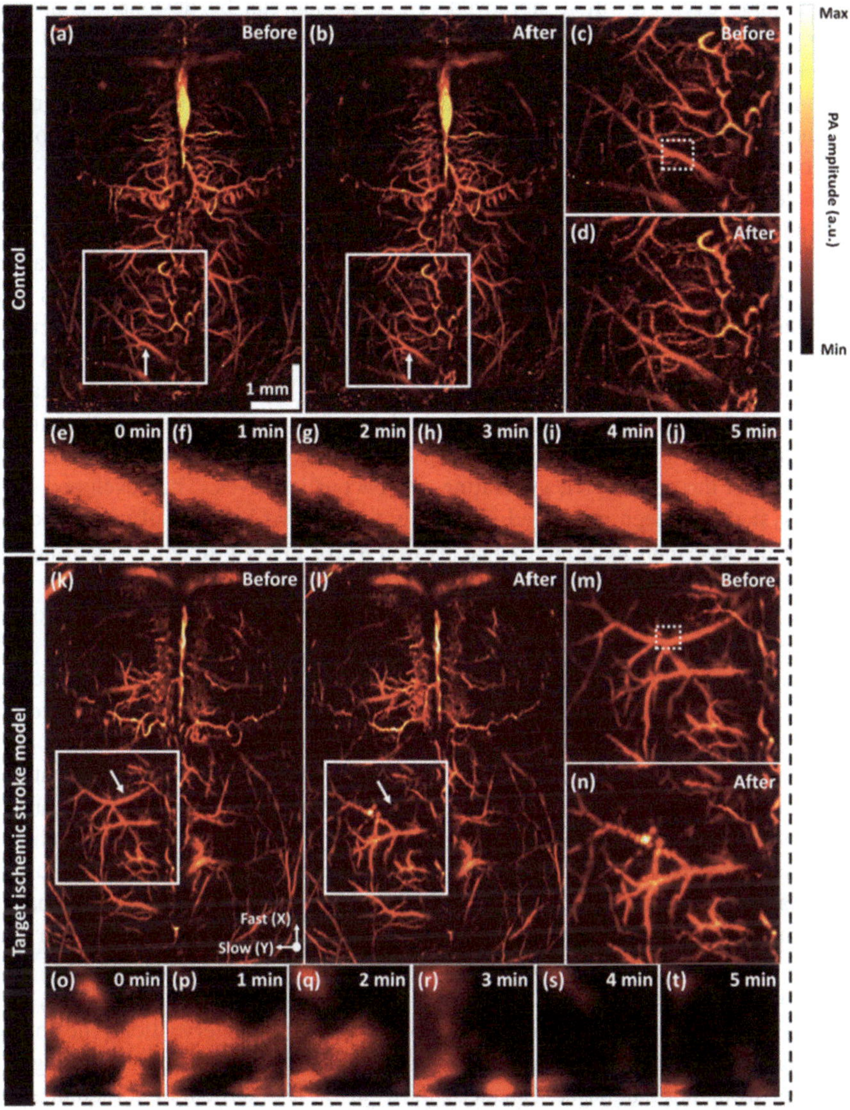

Fig. 1 PAM imaging of cerebral vessels before and after photothrombotic stroke induction in a mouse model. Panels (**a**) and (**b**) show maximum amplitude projection images of the control mouse brain before and after laser illumination. Panels (**c**) and (**d**) show enlarged views of the boxed region in (**a**) and (**b**), respectively. Panels (**e**) to (**j**) show changes in the selected vessel (white dotted box in (**c**)), imaged every minute during a 5-minute period. Panels (**k**) and (**l**) show maximum amplitude projection images of the brain in the stroke model before and after illumination. Panels (**m**) and (**n**) show magnified views of the boxed regions in (**k**) and (**l**), while panels (**o**) to (**t**) illustrate vessel-level changes in the selected area (white dotted box in (**m**)), imaged every minute during 5 minutes. (Reprinted with permission from (Seong et al. 2022), Copyright (2022), with permission from Elsevier)

and organic dyes are often employed. These agents can be designed to absorb strongly at specific wavelengths suitable for PAI, improving image contrast and enabling molecular targeting within anatomical imaging frameworks. The judicious use of these contrast agents in anatomical PAI expands its versatility for preclinical research, allowing more detailed and specific imaging of structural and morphological characteristics (Karlas et al. 2019; Attia et al. 2019; Choi et al. 2018; Deán-Ben et al. 2017).

Endogenous contrast agents are natural chromophores in the animal's body that act as absorbers, providing contrast in the resulting image. Hemoglobin, melanin, water, collagen, DNA, RNA, and lipids are some endogenous contrast agents that have been utilized for PAI. For instance, Imaging of cellular nuclei is essential for the diagnosis and prognosis of cancer. Research has shown, imaging DNA and RNA with the help of an Ultraviolet PAM (UV-PAM) system provides the chance to achieve noninvasive imaging of cell nuclei in vivo without the use of staining techniques. This system utilizes 266 nm UV light to excite unlabeled DNA and RNA within the cell nuclei, generating photoacoustic waves. The developed UV-PAM system has been employed effectively for ex vivo imaging of cell nuclei in murine lip and small intestine tissues, as well as for in vivo imaging of cell nuclei in mouse skin (Hosseinaee et al. 2020).

The qualitative assessment of morphological features using PAI, whether through endogenous or exogenous contrast agents, has advanced the visualization and characterization of tissue architecture in preclinical research. Qualitative parameters commonly extracted include vessel morphology, structure, shape, and the delineation of tissue or organ boundaries—attributes crucial for the characterization of normal and pathological states in living organisms. PAI's hybrid imaging nature, combining the high contrast of optical imaging with the spatial resolution of ultrasound, allows researchers to discern complex patterns in vascular branching, tortuosity, and distribution, features that are especially relevant in the study of tumor vasculature, inflammatory processes, and neurovascular anatomy. Additionally, innovations in image acquisition and analysis, such as advanced segmentation and deep learning algorithms, have further empowered PAI to extract detailed and descriptive visual representations of tissue organization, spatial heterogeneity, and changes associated with disease progression or therapeutic intervention. The ability to monitor the dynamic perfusion of contrast agents and to differentiate pathological features from normal tissue enhances the role of PAI as a tool for qualitative morphological assessment across a diverse array of preclinical applications.

In addition to qualitative descriptions, PAI provides a robust platform for quantitative extraction of morphological parameters that objectively characterize biological tissue structure and function. Quantitative measurements possible with PAI include vessel diameter, vessel density, branching indices, spatial distribution, blood volume fraction, and tissue oxygen saturation, metrics that are indispensable for rigorous evaluation in vascular biology, oncology, and tissue engineering. With endogenous agents, quantitative PAI enables precise measurement of hemoglobin concentration, sO_2, and microvascular density, informing on physiological states such as tissue hypoxia and responses to therapies. The introduction of exogenous contrast agents such as nanoparticles, organic dyes, and functionalized probes extends the quantitative capabilities of PAI by enabling the measurement of agent accumulation, bio-distribution, binding efficiency, and targeting specificity within specific tissue compartments. Contemporary methodologies leverage quantitative PAI for applications including the assessment of molecular targeting strategies, evaluation of therapy responses, and validation of in vivo imaging with direct correlation to histopathological analyses, ensuring that quantitative morphological metrics derived from PAI are both accurate and biologically relevant.

Volumetric PAI has established itself as a transformative technology in preclinical research, allowing for the comprehensive, non-invasive evaluation of small animal models across multiple biological dimensions. Among the cutting-edge PAT systems, MSOT has garnered notable attention for its integration of multi wavelength laser excitation with sophisticated ultrasound detector arrays, enabling high-resolution whole-body and organ-level imaging in small animals. MSOT's core strength lies in its ability to exploit the unique optical absorption spectra of both endogenous tissue chromophores (such as hemoglobin and melanin) and a wide range of exogenous contrast agents, delivering simultaneous anatomical, functional, and molecular PAI. This is achieved through multi-spectral acquisition and advanced spectral un-mixing algorithms, which allow for the distinctive mapping of various tissue components and molecular markers within physiologically relevant depths. The volumetric capability of MSOT, often realized via curved or hemispherical transducer arrays combined with mechanical or electronic scanning, facilitates comprehensive 3D visualization for dynamic studies, providing a powerful platform for investigating disease progression, therapeutic response, and molecular interactions in vivo.

Complementing the advancements offered by MSOT, SVOT represents a new paradigm in volumetric PAI, characterized by its unique spiral (helical) scanning trajectory and the use of spherical matrix ultrasound detection arrays. This innovative system allows for rapid, panoramic head-to-tail imaging of small animals, attaining isotropic spatial resolution frequently in the range of 90–200 micrometers and achieving volumetric frame rates up to 100 volumes per second—capabilities that are unprecedented in other preclinical imaging modalities. SVOT's design not only supports high-quality anatomical imaging but also facilitates the quantification of dynamic physiological and molecular processes, such as tracking bio-distribution, pharmacokinetics, and tissue perfusion with remarkable temporal and spatial fidelity. Through fast multispectral excitation and real-time three-dimensional reconstruction, SVOT enables whole-body anatomical, functional, and molecular PAI.

2.1 Multi-Spectral Optoacoustic Tomography (MSOT)

For volumetric imaging of preclinical models, a multiple-wavelength basis called MSOT is required. By using multiple wavelengths, spectral absorption contrast and information on different chromophores will be provided (Najafzadeh et al. 2019).

MSOT is suitable for imaging large and small vessels and can be used in diagnosis of diseases which cause changes in blood vessels. In addition to vascular imaging, MSOT can be used in Pilosebaceous units imaging and can be helpful for hair care products and gastrointestinal imaging. Another application for MSOT is assessing the metabolic as it can differentiate different types of adipose tissue (Attia et al. 2019). Same as other PA techniques, MSOT consists of an optical source, mainly an adjustable laser to confirm multiple wavelength imaging and a focused acoustic transducer. The laser must provide a fast repetition rate to acquire cross-sectional images (Gu et al. 2023). MSOT uses a high-powered excitation source such as a neodymium-doped yttrium-aluminum-garnet solid-state laser (MacCuaig et al. 2020). A schematic illustration of the excitation and detection mechanism of the MSOT system, besides images provided by an MSOT system have been shown in Fig. 2. Figure 2a shows the schematic of an MSOT system and Fig. 2b shows a reconstructed MSOT image, at 800 nm, with linear un-mixing data of Hb, HbO_2 and lipid from chow- and high fat diet-fed mice. The color bar shows the color coding of MSOT from 0 to maximum (bottom to top). The utilization of MSOT in this context exemplifies its strength as a label-free, functional, and molecular imaging tool, enabling the quantitative assessment of lipid content and tissue oxygenation with spatial resolution suitable for preclinical studies. Such capabilities are particularly critical in studying metabolic diseases like non-alcoholic fatty liver disease, where lipid quantification and oxygenation dynamics serve as important biomarkers for disease progression and therapeutic response (Huang et al. 2021).

The transducer is an important means to be a low-frequency transducer as it can detect low frequency signals containing signals from deeper structures due to less attenuation of lower frequencies. The transducers are mostly one-dimensional or two-dimensional linear arrays. The linear array transducers can have a curved structure, which allows for acquiring data from multiple angles and creating 3D or volumetric images. The curved structure has superiority over the planar transducer in case of higher resolution and sensitivity. Other 3D configurations, such as hemispheric transducers, provide a larger field of view and higher image quality as there are more signals for image reconstruction (MacCuaig et al. 2020). After image reconstruction, each pixel in a PA image demonstrates a mixture of signals from multiple chromophores in varying concentrations that necessitates a subpixel detection approach, where the algorithm must discern the individual contributions of each chromophore from the composite signal. The un-mixing or material decomposition process can be done using linear regression; however, due to some issues in this method that cause less accuracy in the resulting image, other methods, such as adaptive match filters or adaptive cosine estimators, have been proposed.

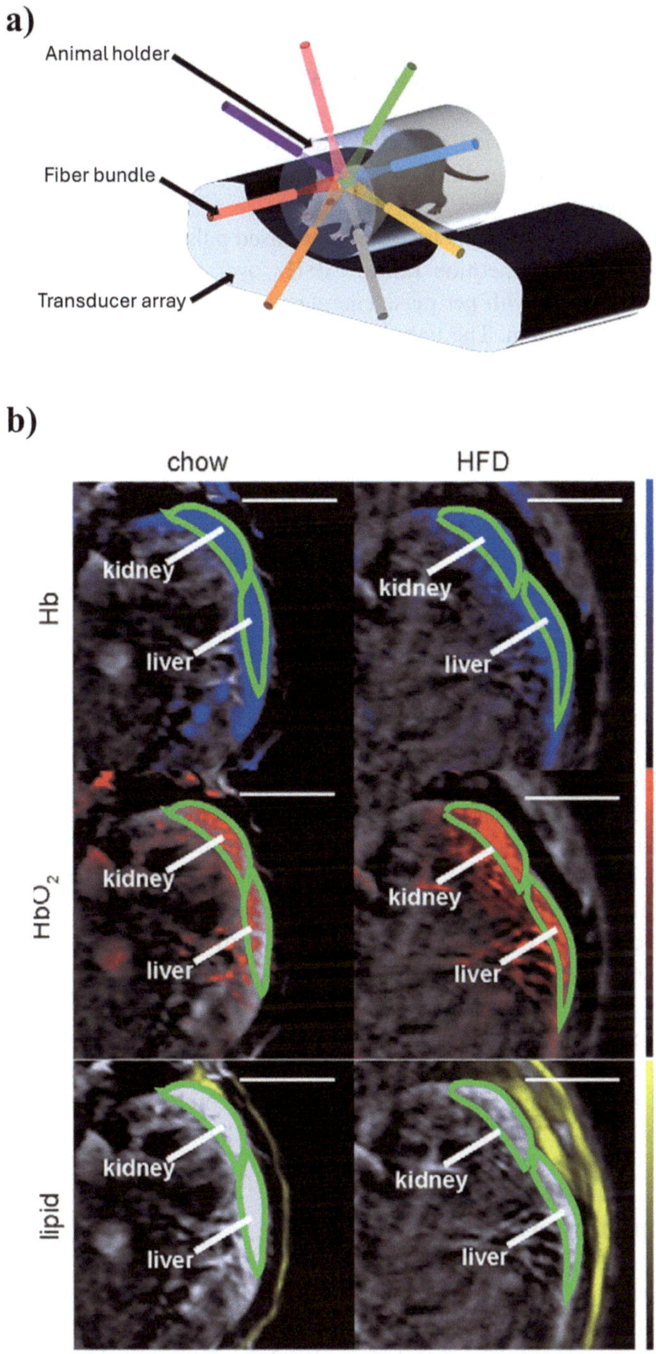

Fig. 2 Multispectral optoacoustic tomography (MSOT) system and reconstructed MSOT image. (**a**) A schematic illustrates the MSOT setup, where an optical fiber uniformly illuminates the animal, and a cylindrically focused curved transducer array detects the generated photoacoustic signals; (**b**) A reconstructed MSOT image at 800 nm shows linear unmixing of hemoglobin (Hb), oxygenated hemoglobin (HbO_2), and lipid in mice fed either a chow or high-fat diet (HFD). Vertical color bars indicate signal intensity from minimum (bottom) to maximum (top). (Image in panel (b) is adapted from (Huang et al. 2021) under a Creative Commons Attribution License (CC BY 4.0))

2.2 Spiral Volumetric Optoacoustic Tomography (SVOT)

SVOT is another volumetric PAT method, as illustrated in Fig. 3a, which consists of a pulsed laser and fiber bundles, a spherical matrix transducer array, animal holder, and a data acquisition system.

A broadband optical parametric oscillator-based pulsed laser that delivers <10 ns duration pulses at a repetition rate of 10 Hz over a broad tunable wavelength range (680–1250 nm) with per-pulse energies up to ∼180 mJ has been used in the developed SVOT system. The light beam has been guided through a custom-made fiber bundle that bifurcates into five individual outputs with an equal number of fibers. All five outputs were placed at the same radial distance of 40 mm from the center of a custom-made spherical array. One of these outputs was inserted into a central cavity of the array, while the other four were arranged on both sides of the array.

A custom-made spherical array consisting of 512 individual piezoelectric sensor elements, having 7 mm^2 area, 7 MHz central detection frequency, and 85% bandwidth was used to collect the PA responses. The PA signals were simultaneously digitized at 40 mega samples per second by a custom-made parallel data acquisition unit (Kalva et al. 2023a). This configuration can accomplish scanning in three modes. 1-step and go, 2- continuous, and 3- single sweep (Kalva et al. 2023b). These scanning modes provide a trade-off between image quality and duration of imaging. SVOT provides high spatial and temporal resolution; however, its main problem relies on animal positioning. Anything between the animal and the transducer, such as animal holder, interfere the measurements, which is one of SVOT systems design challenges (Kalva et al. 2023b). Figure 3b shows in vivo head-to-tail 3D imaging of a mouse using SVOT system. Maximum intensity projections of the rendered volume are shown (left to right) from the back, right, left-front, and left views. The capability of SVOT in resolving fine structure, such as superior sagittal and transverse sinuses, superior temporal veins, jugular veins, inferior vena cava, thoracic and femoral vessels, in whole body mouse imaging is obvious in Fig. 3b (Kalva et al. 2023a). SVOT's multiscale imaging capability can be utilized to measure perfusion, visualize rapid biodynamics, track reactions to stimuli and treatments, and quantify the dynamics of drug and molecular agent accumulation and clearance throughout the body (Kalva et al. 2023b).

3 Functional PAI

PAI provides high-resolution imaging of tissue oxygenation, hemoglobin concentration, and vascular dynamics without relying on ionizing radiation or exogenous contrast agents, which are essential in Positron Emission Tomography (PET) and single-photon emission computed tomography (SPECT), the predominant techniques in functional and molecular imaging which are not often portable (Yu et al. 2020; Park et al. 2024). PAI enables multiscale imaging, from cellular to

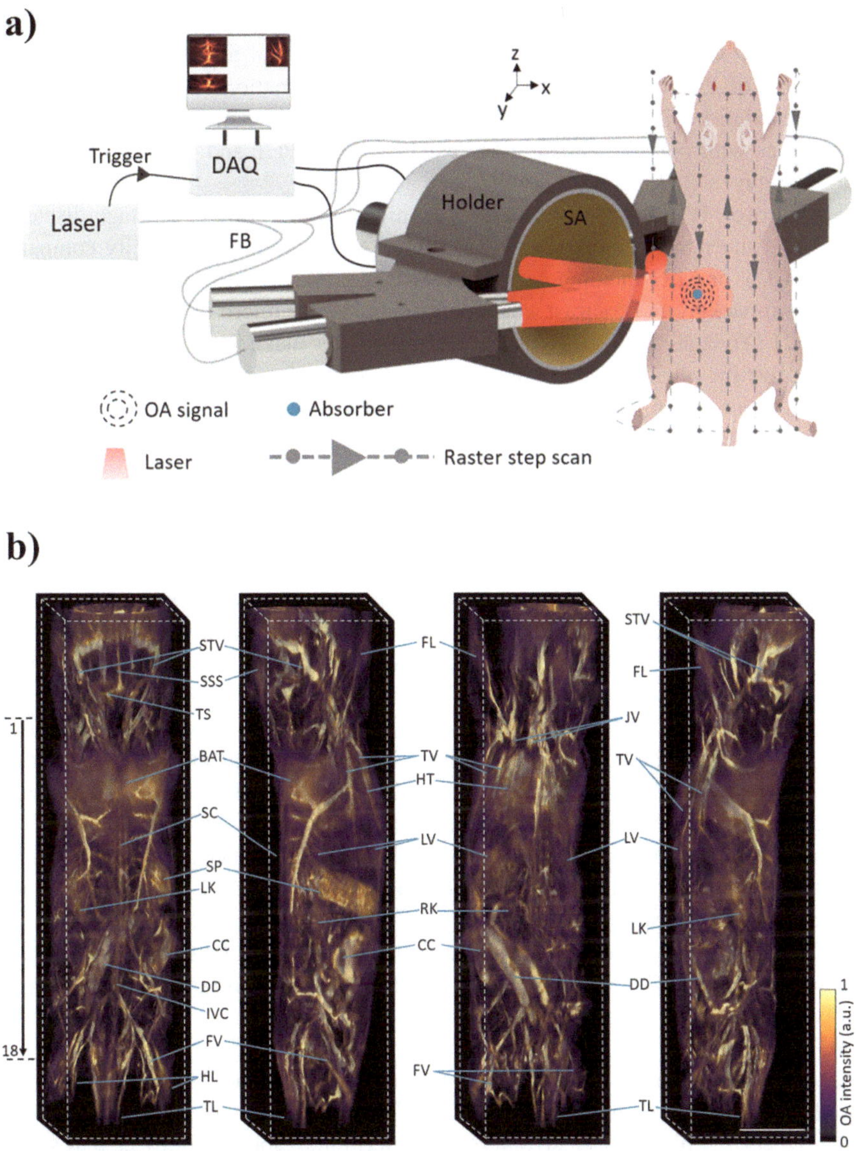

Fig. 3 Spiral volumetric optoacoustic tomography (SVOT) system, and whole-body imaging of a mouse. (**a**) The schematic illustrates the main components of the SVOT system, including a fiber bundle (FB), spherical matrix transducer array (SA), data acquisition system (DAQ), and detection of optoacoustic (OA) signals; (**b**) In vivo 3D imaging of a mouse using SVOT, with maximum intensity projections of the reconstructed volume shown (left to right) from the back, right, left-front, left views. Labeled anatomical structures include the superficial temporal vein (STV), superior sagittal sinus (SSS), transverse sinus (TS), brown adipose tissue (BAT), spinal cord (SC), spleen (SP), liver (LV), left kidney (LK), right kidney (RK), fore limb (FL), hind limb (HL), tail (TL), heart (HT), jugular vein (JV), thoracic vessels (TV), caecum (CC), duodenum (DD), inferior vena cava (IVC), and femoral vessels (FV). (Image in panel (a) is adapted, and image in panel (b) is reproduced, both from (Kalva et al. 2023a) under a Creative Commons Attribution License (CC BY 4.0))

organ resolution, and can be effectively integrated with complementary imaging modalities, such as ultrasound, enhancing the functional and molecular imaging strengths of PAI by adding further anatomical information. The combination of these modalities is particularly practical in the real-time monitoring of biologic processes with portable capabilities, less feasible in PET and SPECT (Park et al. 2024; Nguyen et al. 2023).

Moreover, PAI represents superior sensitivity in detecting early-stage diseases, such as cancer micro-metastases, with higher sensitivity and specificity compared to PET/CT or MRI in some cases. For instance, PAI has demonstrated improved tumor-to-background contrast when identifying hepatic micro-metastases relative to PET/CT (Yu et al. 2020). As a result, these attributes make PAI a versatile and safer alternative for a wide range of clinical and preclinical applications. As mentioned earlier, hemoglobin, as an endogenous contrast agent, can be utilized to illustrate blood arteries. Oxygen can bind to it and as a result, the absorption peak is varied. Thus, hemoglobin can be used as an appropriate chromophore for both anatomical and functional imaging. Hemoglobin, HbO_2, and HbR are attractive chromophores for PA structural and functional studies and are suitable for various applications like early detection of cancer (Tsang et al. 2020).

PAI has gained significant attention in preclinical ischemic stroke imaging due to its capability to show tissue states and its superiority over other modalities. While PAT is valuable for stroke imaging, using PAM is particularly effective for imaging small vessels due to its higher resolution. Also, high-speed PAM can be used to track the incidence of mini-stroke at the capillary level in mice models (Yang et al. 2021). Microvasculature can be examined with high temporal and spatial resolution using a high pulse repetition rate laser combined with a micro-electro-mechanical system (MEMS) scanning mirror (Zafar et al. 2022, 2023, 2024). This innovative approach in using high-speed PAM with MEMS enables measuring the variations in microvascular blood flow induced by the photothrombosis and allows for the observation of the blockage site (Menozzi et al. 2022).

In this section, three specific quantitative parameters that show vascular and neural function have been introduced. Hemoglobin concentration, blood flow, and sO_2 are three main quantitative parameters for assessing tissue's functions using PAI. Each provides a different point of view for studying functional states. For instance, assessment of hemoglobin concentration is suitable for disease research, such as in models of chronic anemia, blood flow is an important factor for assessing angiogenesis, which plays a pivotal role in determining abnormalities such as stroke or diabetes, and sO_2 assessment is primarily suitable for brain studies.

3.1 Hemoglobin Concentration Assessment Using PAI

The measurement of hemoglobin concentration in mouse models is particularly useful in the context of disease research, such as in models of chronic anemia. A robust approach for assessing the hemoglobin concentration is normalizing by PA signals amplitude. In this regard, an isosbestic wavelength should be chosen.

An isosbestic wavelength is a specific wavelength at which different forms of hemoglobin (oxygenated and deoxygenated) exhibit the same optical absorption coefficients. This unique property allows the measurement of total hemoglobin concentration (C_{Hb}) independently of sO_2, because the photoacoustic signal at this wavelength reflects hemoglobin concentration regardless of its oxygenation state. Mathematically, the optical absorption coefficient (μ_a), which is the final value of a PA image, can be defined by Eq. 1;

$$\mu_a = \beta C_{Hb} \left[sO_2.\varepsilon^{HbO_2} + (1 - sO_2)\, \varepsilon^{HbR} \right] \tag{1}$$

Where ε^{HbO_2} and ε^{HbR} are molar distinction coefficients of oxy- and deoxy-hemoglobins, β is a fixed constant, and C_{Hb} is hemoglobin concentration. By choosing an isosbestic wavelength of hemoglobin, ε^{HbO_2} and ε^{HbR} will have equal values, named ε^{Hb}, and Eq. 1 transforms to Eq. 2;

$$\mu_a = \beta C_{Hb} \varepsilon^{Hb} \tag{2}$$

Therefore, the PA signal amplitude (μ_a) is directly dependent on hemoglobin concentration (C_{Hb}). The typical isosbestic wavelengths for hemoglobin near the 500 nm range are 498, 545, 570, and 584 nm. By selecting such isosbestic wavelengths, the total hemoglobin concentration can be acquired by single-wavelength PAI.

3.2 Blood Flow Assessment Using PAI

Blood flow is an important factor for assessing angiogenesis, which plays a key role in determining abnormalities such as stroke or diabetes. Blood flow can be calculated in various ways using PAI. One direct approach involves computing the distribution of PA signals within the flow speed distribution. Considering both the moving target and probe, the blood flow velocity at a specific depth can be determined. This method also allows the calculation of the flow speed of a single red blood cell. Despite its real-time measurement, this method might be dependent on depth.

Another method for calculating blood flow relies on modulating optical properties in spatial or temporal domains. Temporal modulation of the continuous light generates time-invariant PA signals, which encode blood flow information, and blood flow velocities can be calculated. Also, light can be modulated in the spatial domain so that a modulated PA signal will be received. This can provide some advantages, like higher sensitivity and higher signal-to-noise ratio (SNR) (Liu and Wang 2022). It is worth noting that angiogenesis is not exclusive to malignant tumors; it also plays a critical role in wound healing. In fact, forming new vessels supplies sufficient oxygen to the wound and accelerates the healing process. This process involves several steps, each with unique properties that can be effectively studied using PAM. The high resolution of PAM makes it a good candidate for demonstrating the microvasculature. Besides changes in sO_2 and blood flow (for example, blood flow is above the baseline in feeding arteries in primary days of wounding), two major factors can be calculated, which are oxygen extraction fraction (OEF) and metabolic rate of oxygen (MRO_2). These factors can maintain different values during different healing steps. Thus, PAM can help evaluate the wound stage and the healing progress. OEF is typically a measure of tissue's capacity to absorb oxygen from the surrounding circulation to preserve its structural integrity and shape. This factor utilizes the average saturation of feeding arteries and draining veins and can also be used to assess conditions such as stroke and tumor growth. Calculating OEF via optical-resolution-PAM (OR-PAM) shows a decreased value over baseline and provides a quantitative criterion, which can demonstrate the blocked vessel in stroke. OEF also decreases in tumors, indicating that tumors have a reduced capacity to receive oxygen. MRO_2 can accurately reflect the state of oxygen deprivation in the targeted tissue. Researches show that during the early stages of tumor growth, MRO_2 is elevated; however after the tumor's core necrosis, this value drops significantly (Liu and Wang 2022).

3.3 Hemoglobin Oxygen Saturation (sO_2) Assessment Using PAI

PAI is a powerful modality for functional imaging of biological tissues with high contrast and resolution. By using the optical characteristics of HbO_2 and HbR, as endogenous contrast agents, SO_2 in tissues can be calculated, which is an indirect measure of function in biological tissues. In addition to the capability of PAI in demonstrating the structure of the vasculature, resulting PA images can provide valuable insights into sO_2 levels because the presence of oxygen in blood changes the hemoglobin absorption spectra. Therefore, oxygenated and deoxygenated hemoglobin in blood can be differentiated using PAI. However, achieving this requires a broader spectrum of wavelengths. This procedure can help assess various factors, such as deoxygenated tissue, which is significant in tumor imaging as it may indicate malignancy or a complex response to treatment (John et al. 2023). For calculation of the sO_2, which is essential for the PAI system to work in at least two wavelengths, using Eq. 3, concentrations of hemoglobin and HbO_2 can be formulated;

$$\left.\begin{aligned} C_{Hb} &= \frac{\varepsilon_{HbO_2}^{\lambda 1} * \mu_a^{\lambda 2} - \varepsilon_{HbO_2}^{\lambda 2} * \mu_a^{\lambda 1}}{\varepsilon_{Hb}^{\lambda 2} * \varepsilon_{HbO_2}^{\lambda 1} - \varepsilon_{Hb}^{\lambda 1} * \varepsilon_{HbO_2}^{\lambda 2}} \\ C_{HbO2} &= \frac{\varepsilon_{Hb}^{\lambda 2} * \mu_a^{\lambda 1} - \varepsilon_{Hb}^{\lambda 1} * \mu_a^{\lambda 1}}{\varepsilon_{Hb}^{\lambda 2} * \varepsilon_{HbO_2}^{\lambda 1} - \varepsilon_{Hb}^{\lambda 1} * \varepsilon_{HbO_2}^{\lambda 2}} \end{aligned}\right\} \tag{3}$$

In Eq. 3, ε represents the molar extinction coefficient of hemoglobin and HbO_2 at wavelengths λ_1 and λ_2, while μ represents the measure of optical absorption coefficients at wavelengths λ_1 andλ_2 during PAI.

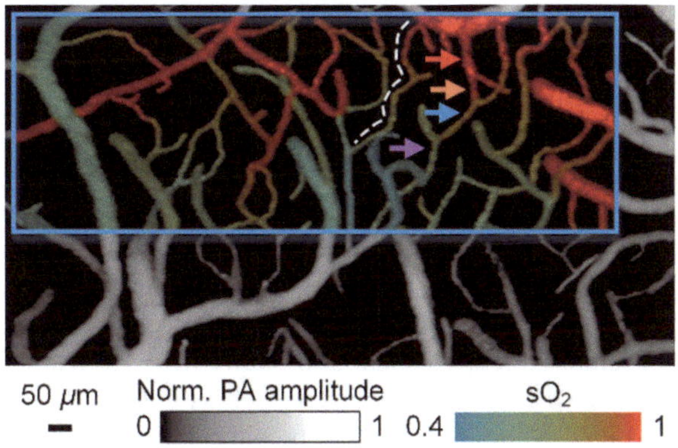

Fig. 4 Hemoglobin oxygen saturation (sO_2) map of a brain region (highlighted by the blue box) obtained after electrical stimulation. The map indicates the localized changes in sO_2 that can be used for functional image analysis. (Adapted from (He et al. 2020) under a Creative Commons Attribution License (CC BY 4.0))

sO_2 can be determined using Eq. 4, by using the measured values of C_{Hb} and C_{HbO_2};

$$sO_2 = \frac{C_{HbO_2}}{C_{HbO_2} - C_{Hb}} \tag{4}$$

Then, based on acquired μ maps at two wavelengths of the PAI system, concentrations of hemoglobin and oxy-hemoglobins can be calculated in each pixel/voxel of the image and SO_2 map can be concluded (John et al. 2023). Figure 4 shows a map of sO_2. In the sO_2 map of Fig. 4, higher saturations after electrical stimulation have been shown with red and lower saturations around 0.4 have been shown with a color near the blue (He et al. 2020).

Another method for sO_2 calculation is based on the concept of saturation. However, this method is different as it uses two laser sources with different pulse widths: a nanosecond pulsed laser and a picosecond laser source. This method works by analyzing the photoacoustic signals generated by both pulse widths. The relative concentrations of HbO_2 and HbR can be inferred based on how each chromophore responds to the different saturation levels. It should be noted that these methods are not error-free and potentially lead to inaccuracies. To address this issue, novel compensation techniques have been developed, which can be utilized based on optical or acoustics approaches (Liu and Wang 2022).

One of the most significant challenges in functional imaging is the time required for the PAI system to acquire PA signals. In this regard, different setups using various transducer geometries have been assessed. Single-element US transducers, with a step angle of 1.5°, have been utilized to capture cross-sectional brain images of rats. The mentioned setup has reached a spatial resolution of approximately 60 μm and an imaging depth of about 1 cm. However, each whole brain image required around 24 minutes to be scanned. Recently, a fast 3D PAI system that utilizes four combined curved array modules, each with 256 elements, has been proposed that decreases the required scan time. The mentioned design completes a full scan in only 5 seconds by connecting all elements directly and results in volumetric images of the mouse brain at a resolution of approximately 390 μm and a depth of 10 mm (Kye et al. 2022).

Slow imaging speeds can affect the accuracy of 3D PAI due to breathing-related distortions. To overcome this issue, an integrated simultaneous US and PAI system has been proposed that enables breath-compensated 3D photoacoustic whole-body imaging of mice. The breathing-related distortions were eliminated by segmenting the skin profile in US images and realigning the signals, which also enabled the creation of detailed and accurate 3D sO_2 maps (Menozzi et al. 2022).

Mice and rats are the two most widely used animals in PAI for studies of ischemic stroke, as their thin skulls help reduce scattering and improve image quality when using techniques such as OR-PAM, acoustic-resolution (AR-PAM), and PAT. Although OR-PAM provides high-resolution images, invasive procedures like scalp removal or cranial windows may be needed. In contrast, AR-PAM has better depth penetration while it has a lower resolution than OR-PAM. In general, PAT is usually the best technique for whole-brain imaging, as it reaches spatial resolutions of several hundred micrometers and offers good depth penetration.

PAI has been effectively used for imaging larger animals both in vivo and ex vivo, which better represents humans' brain size and skull thickness for clinical translations. In this regard the capability of PAI for doing studies on pig models in vivo has been evaluated. Neonatal pigs within 2 hours after perinatal stroke surgery were studied, and significant blood flow and oxygen level variations between the injured and unaffected brain regions were observed. These results highlight the promise of PAI as a technique for identifying perinatal ischemic stroke. The development of PAI studies in large animals demonstrates promise for researching ischemic stroke in humans, but additional investigation is required.

The application of functional imaging using PAI systems extends beyond brain studies in preclinical research; several investigations have done functional PAI of the abdominal region. In this regard, a slip-ring-based system has been proposed to enhance imaging speed by applying two symmetrically arranged single-element focused transducers. The mentioned setup has enabled the acquisition of complete cross-sectional images of mice's abdominal areas with a spatial resolution of approximately 129 μm in 9 minutes (Menozzi et al. 2022).

4 Molecular PAI

The unique strength of molecular PAI lies in its versatility to probe physiologically and pathologically relevant molecular processes either using endogenous contrast agents such as hemoglobin, melanin, and lipids that naturally occur within tissues, or by using a vast array of exogenous contrast agents, including engineered nanoparticles, organic dyes, and probes, designed to target specific pathways or cellular markers. Employing endogenous contrast agents enables label-free functional and molecular imaging by leveraging the intrinsic absorption signatures of key biomolecules, facilitating noninvasive evaluation of tissue composition, oxygenation, metabolic activity, and disease microenvironments. In parallel, employing exogenous contrast agents builds upon advances in contrast agent chemistry and nanotechnology to provide enhanced signal, tailored targeting, multiplexing capabilities, and integrated therapeutic functionalities, greatly expanding the molecular imaging toolbox for cancer research, drug delivery monitoring, and the study of molecular interactions in living systems. The applications of both endogenous and exogenous contrast agents in molecular PAI have been described in this section.

4.1 Molecular PAI Using Endogenous Contrast Agents

Molecular PAI utilizing endogenous contrast agents such as melanin, lipids, and collagen empowers label-free, in vivo characterization of tissue composition and pathophysiological states across various biomedical applications, providing critical molecular insights into tissue function, disease mechanisms, and prognosis without the need for exogenous labels. Melanin is present in various tissues, including the skin, eyes, and hair, and it has the potential to give rise to malignant cells known as melanoma. Quantitative applications of PAI in the investigation of cutaneous melanoma include the assessment of tumor depth and thickness, evaluation of cancer-related angiogenesis, detection of metastases to lymph nodes, monitoring of circulating tumor cells, and application of virtual histology. Numerous studies have been conducted to characterize melanoma quantitatively using preclinical models. Dual-wavelength AR-PAM systems have been employed to detect melanoma in mice inoculated with highly invasive B16 skin melanoma cells. By exploiting the differences in peak optical absorption between melanin and hemoglobin, NIR light at 764 nm and visible light at 584 nm have been utilized to visualize melanoma and the adjacent vascular structures. Additionally, a handheld AR-PAM system has been developed to measure melanoma depth. This system employs an annular-shaped light illumination method, where light with an inner diameter of 8 mm and an outer diameter of 20 mm bypasses the tumor center and is delivered perpendicularly to the skin surface (Fakhoury et al. 2024).

Lipids are defined as structures that have a high hydrocarbon content. Imaging lipids is valuable for evaluating vascular diseases. There is also potential for myelin imaging using variational PAM. Despite its high resolution, this technique faces limitations when it comes to imaging deep structures. It is noteworthy that the excitation source can operate at wavelengths of either 1210 nm or 1720 nm. The observed absorption peak at these higher wavelengths is attributed to the second overtone of hydrocarbon, indicating that the second frequency exceeds the fundamental frequency (Tsang et al. 2020). Collagen constitutes a vital component of atherosclerotic lesions. The discrepancies in optical absorption coefficients among standard plaque constituents, such as lipids, water, blood, and collagen, facilitate the characterization of various atherosclerotic plaques. The lipid-to-collagen ratio can be utilized to classify plaques, thereby informing the development of targeted therapeutic strategies. Advancements in multi-wavelength PAI have enabled the mapping of lipid and collagen content. Specifically, PA images captured at a total of 60 wavelengths within the range of 1650–1850 nm were analyzed using alternating least squares analysis to un-mix the spectral information pertaining to lipids and collagen. Furthermore, multi-wavelength intravascular PA (IVPA) imaging, conducted within the 680–900 nm range, has been employed to assess the vulnerability of atherosclerotic plaques. IVPA imaging of a normal aorta and aortic tissue from a rabbit model of atherosclerosis demonstrated a strong correlation between collagen deposits in the thickened fibrous cap of the plaque and histological images obtained from the corresponding cross-section (John et al. 2023).

These endogenous absorbers enable the simultaneous, label-free visualization and quantification of multiple biomolecules, supporting multi-parametric imaging for monitoring disease progression, angiogenesis, and therapeutic response in oncology, cardiovascular medicine, dermatology, and metabolic studies; they provide functional molecular insight unique to the intrinsic biochemistry of tissues, sparing patients the risk associated with exogenous agents and facilitating broader, safer adoption of molecular PAI in clinical and preclinical practice. The specificity and reliability of endogenous contrast-based PAI have been significantly elevated by recent advances in multispectral and machine-learning-assisted analysis, spectral un-mixing, and quantitative image reconstruction, which allow discrimination of even less prominent biomarkers and overcome technical limitations related to depth and heterogeneity, thereby solidifying the central role of melanin, lipids, collagen, hemoglobin, and other intrinsic absorbers in the expanding landscape of molecular PAI applications.

4.2 Molecular PAI Using Exogenous Contrast Agents

Molecular PAI is an evolving technique that synergizes the principles of PAI technology with molecular biology to provide unprecedented insights into the complex biological processes at the cellular and molecular levels. The integration of exogenous contrast agents plays a pivotal role in enhancing the sensitivity and specificity of molecular PAI, facilitating the visual detection of targeted biomolecules, such as proteins and nucleic acids, within various biological processes. These contrast agents improve the contrast-to-background ratio, thereby enabling high-resolution imaging of pathological changes, tumor microenvironments, and therapeutic responses. By utilizing specific wavelengths for excitation, typically within the NIR spectrum, molecular PAI can penetrate deeper into tissues while minimizing photo-toxicity, making it a powerful tool for preclinical applications (Farooq et al. 2022).

PAI has emerged as a powerful modality for enhancing the visualization of biological tissues, particularly when combined with exogenous contrast agents. These agents are designed to improve the sensitivity and specificity of PAI by providing additional optical absorption contrasts that can be targeted to specific cellular and molecular markers. The development and synthesis of various exogenous contrast agents entail sophisticated chemical and biophysical approaches, enabling fine-tuning their absorption characteristics to match the wavelengths utilized in PAI systems. Diverse types of exogenous contrast agents, including nanoparticles, dyes, and engineered biomolecules, have been developed to facilitate a wide range of applications, from tumor detection to molecular imaging of diseases. By augmenting the inherent contrast of tissues, exogenous agents significantly enhance the diagnostic capabilities of PAI, paving the way for more precise and targeted imaging strategies in both preclinical and clinical settings (James et al. 2021; Sridharan and Lim 2023; Upputuri and Pramanik 2020; Han et al. 2022).

Nanoparticle contrast agents have gained significant attention in PAI due to their unique properties, enabling various applications across preclinical research. These agents typically possess tailored optical absorption characteristics, enhancing contrast and specificity in imaging targeted tissues. One notable application is in detecting tumors; for instance, gold nanoparticles have been utilized in a dual-wavelength PAI system to differentiate malignant tissues by exploiting the distinct absorption profiles of the nanoparticles. Additionally, PAI systems equipped with silica-coated iron oxide nanoparticles have been employed to assess the depth and vascularization of tumors, providing crucial insights into tumor biology and response to therapy. Furthermore, polymeric nanoparticles have demonstrated efficacy in imaging lymphatic vessels and assessing lymphatic metastasis using multi-wavelength PAI systems, which enhance imaging resolution and facilitate the study of metastatic pathways (Jung et al. 2019).

Dyes are specifically designed to absorb light at wavelengths suitable for PAI. One prominent application of dyes is in monitoring tumor dynamics; for instance, ICG has been extensively utilized in PAI systems to visualize tumor vascularization and assess drug delivery dynamics in preclinical models. In studies utilizing a multi-wavelength PAI system, ICG has demonstrated its capacity to provide real-time feedback on tumor responses to therapeutic interventions, thereby facilitating the evaluation of treatment efficacy (Tsang et al. 2020; James et al. 2021; Shinoda et al. 2022). Moreover, ICG's NIR absorption properties facilitate deeper tissue penetration and minimize autofluorescence interference, making it suitable for in vivo imaging applications. In recent research, ICG conjugated with single-wall carbon nanotubes has exhibited remarkable improvements in PA signal intensity, demonstrating a peak absorption enhancement that provides valuable metabolic and physiological information about the tumor microenvironment (Chaudhary et al. 2019).

Engineered biomolecule contrast agents have emerged as pivotal tools in enhancing the capabilities of PAI for preclinical research. These agents, often designed to target specific biological markers, allow for the visualization of various physiological processes at a molecular level. For instance, the use of engineered antibodies or peptides conjugated with chromophores has enabled the detection of tumor-associated antigens, facilitating early cancer diagnosis and monitoring of therapeutic responses. The PAI systems employed in these applications typically operate within the NIR spectrum, particularly around 700 to 900 nm, where tissue absorption is minimized, allowing for deeper tissue penetration and improved imaging resolution (Chaudhary et al. 2019; Moore et al. 2019).

In this regard, a recent study has showcased the use of a subcutaneous lacZ-expressing tumor model in conjunction with AR-PAM to visualize tumor-associated microvasculature by utilizing a molecular probe, X-gal, which absorbs light in the wavelength range between 605 and 655 nm. This study highlights the capacity of engineered biomolecules to provide spatially resolved images of tumor tissues, thus aiding in the early detection of cancers (Kim et al. 2024; Shrestha et al. 2020).

Regarding vascular imaging, engineered biomolecules such as antibodies and peptides conjugated with photoacoustic absorbers have been employed to provide real-time assessment of blood flow and tissue oxygenation. Applications have included the real-time monitoring of venous return during occlusion tests using a photoacoustic patch that utilizes screened laser sources to monitor blood flow variations. The results have indicated significant changes in vein dynamics during the occlusion, illustrating the biomolecule's role in assessing vascular function and pathophysiology (Karlas et al. 2019; Maturi et al. 2019). Additionally, the emergence of engineered nanoparticles, such as gold nanocages and other metallic nanostructures, presents novel approaches for enhancing the sensitivity of PAI systems. These engineered biomolecules exhibit tunable optical properties, allowing for selective targeting of specific cells or tissues at particular wavelengths, where they exhibit high absorption coefficients. The application of such engineered biomolecules in PAI has not only facilitated enhanced tumor imaging but also improved microvasculature imaging, providing insights into the vascular architecture crucial for understanding tumor biology and response to treatment (Kye et al. 2022).

Gold nanoparticles have recently been modified to generate gas-generating laser-activatable nanorods (GLANCE), which create nitrogen gas bubbles upon NIR laser irradiation to improve the performance of PAI. These nanoparticles were injected into the mice, and PAI was used to monitor the distribution and effects of the nanoparticles. The NIR laser activated the nanoparticles, producing gas bubbles that enhanced the imaging contrast (Kim et al. 2024). In another study, researchers developed nanoparticles made from a unique polymer known as poly N-isopropyl acrylamide (PNIPAM), which reduces its size when heated. They incorporated gold nanorods, made of gold that have a rod-like shape, into these particles. As the temperature rises, the PNIPAM contracts, causing the gold nanorods to move closer together, subsequently changing their light absorption properties. These particles were tested on mice tumors to facilitate real-time monitoring during photothermal therapy. This approach allowed for an assessment of treatment effectiveness while minimizing interference from blood by decreasing background noise from hemoglobin, resulting in more explicit images and improving the accuracy of cancer therapy evaluation (Moore et al. 2019). Also, both functional and molecular aspects of PAI have been utilized for monitoring myocardial infarction and associated inflammation in mice models. The researchers induced a heart attack in the mice and employed PAI, which gathers data across multiple wavelengths of light to visualize the resulting inflammation in the heart. After that, they injected a specialized dye, dPGS-NIR, which targets inflammation-related molecules (P- and L-selectins) to enhance the imaging results. The outcomes demonstrated that PAI can effectively achieve images of inflamed myocardial tissue, which provides essential information for better understanding and potentially treating heart attacks (Karlas et al. 2019).

PAI can be used as a molecular imaging technique to assess drug delivery responses by detecting variations in the vascular system. One study aimed to investigate how blood vessels react to different drugs using PAM without using a contrast agent, monitoring changes such as narrowing (vasoconstriction) or widening (vasodilation). The researchers monitored the effect of corticosteroid treatment on mouse skin, utilizing corticosteroid (a class of steroid hormones) medications to reduce inflammation and suppress the immune response. They understood that the topical application of corticosteroids led to vasoconstriction of blood vessels in the skin, in contrast to the effects observed with subcutaneously injected corticosteroids. Furthermore, many studies apply PAI to track the biodistribution and pharmacokinetics of administered contrast agents.

5 Conclusions and Future Perspectives

PAI has demonstrated remarkable versatility across various fields of preclinical research, including cardiovascular imaging, renal and hepatic disease models, drug delivery and pharmacological studies, drug delivery, oncology, and neuroscience. Its ability to provide high-resolution, noninvasive insights into biological tissues makes it an invaluable tool for studying angiogenesis, ischemic conditions, tumor microenvironments, and functional brain activity. By integrating optical and acoustic modalities, PAI offers unique advantages in both structural and functional imaging, making it a cornerstone of modern biomedical research.

PAI, despite all its advantages and applications, has some limitations that need further investigation. These limitations include depth penetration constrained by tissue optical attenuation, limiting its ability to image deep tissue with good contrast and high SNR (Choi et al. 2023). In addition, the acoustic waves generated during the photoacoustic process encounter significant attenuation and phase distortion when passing through dense bone structures. This phenomenon leads to a restriction in the overall quality of the resulting images (Ye et al. 2024). Another limitation of PAI is technical constraints, which include restricted detection angles and view, hardware complexity, or lack of the biometric information required for image reconstruction (Wang et al. 2023).

Incorporating AI has revolutionized PAI by addressing challenges in data acquisition, image reconstruction, etc. The ability of AI-driven algorithms to enhance image quality, improve speed by mitigating the limitation of spatial sampling and limited view (Wang et al. 2023), and diagnostic precision has further expanded the utility of PAI in both clinical and preclinical settings. With advancements in deep learning, machine learning, and predictive modeling, AI has become a pivotal component in PAI, transforming it into a more powerful, efficient, and accessible imaging modality. Some novel algorithms, such as DiffPam, utilize diffusion models to accelerate PAI without compromising accuracy. This is particularly beneficial for large-area scans, making the technology more efficient, faster, and user-friendly (Loc and Unlu 2023). Deep learning approaches have been particularly impactful in tackling key challenges in PAI, such as low SNR, inaccurate quantification, and limitations in numerical reconstruction. For instance, convolutional neural networks are used to enhance image reconstruction by reducing noise and correcting artifacts, even in limited-view and sparse data scenarios. This ensures high-resolution and artifact-free imaging, which is critical for clinical applications (Yang et al. 2023).

AI also facilitates real-time processing, enabling faster imaging systems suitable for dynamic physiological studies such as blood flow analysis and functional brain imaging. Multi-spectral PAI data, which captures complex optical absorption characteristics of tissues, presents significant challenges for manual analysis. AI has addressed this by automating spectral un-mixing processes and extracting key biomarkers like oxygenation levels, hemoglobin concentration, and molecular signatures. This capability is crucial for studying tumor hypoxia, vascular abnormalities, and inflammation, offering deeper insights into disease mechanisms (Yang et al. 2023).

AI-powered predictive models are transforming PAI's role in preclinical research. For instance, machine learning algorithms have successfully assessed wound healing dynamics, and monitored drug delivery in real time. Furthermore, AI enables the development of automated systems for detecting subtle pathological changes, improving diagnostic accuracy in conditions such as cancer, stroke, and cardiovascular diseases. For instance, an AI model named YOLOv8 was applied for classifying brain tumors using PAI which has shown promising results, achieving high accuracy in distinguishing between benign and malignant tumors. This highlights the potential of AI to optimize diagnostic strategies in oncology with PAI (Chen et al. 2024).

Although, the computational complexity of AI algorithms can limit their application in preclinical PAI. To overcome these limitations, future research is focused on developing lightweight AI models suitable for real-time utilization in PAI systems. Another promising direction is the combination of AI with multimodal imaging systems, such as integrating PAI with ultrasound or MRI. This strategy leverages the advantages of each modality while employing AI to fuse and analyze the data, offering a more comprehensive view of tissue structure and function (Rana et al. 2024).

Acknowledgments N/A

Conflict of Interest The authors have no conflicts of interest to declare. All co-authors have seen and agree with the contents of the manuscript, and there is no financial interest to report. We certify that the submission is an original manuscript and is not under review at any other publication.

References

Ahn J, Kim M, Kim C, Choi W (2024) In vivo multi-scale clinical photoacoustic imaging for analysis of skin vasculature and pigmentation: a comparative review. Adv Imaging 1(3):032002

Attia ABE et al (2019) A review of clinical photoacoustic imaging: current and future trends. Photoacoustics 16:100144

Brown EL et al (2022) Quantification of vascular networks in photoacoustic mesoscopy. Photoacoustics 26:100357

Chaudhary Z et al (2019) Efficient photoacoustic imaging using indocyanine green (ICG) loaded functionalized mesoporous silica nanoparticles. Biomater Sci 7(12):5002–5015

Chen Z, Özbek A, Rebling J, Zhou Q, Deán-Ben XL, Razansky D (2020) Multifocal structured illumination optoacoustic microscopy. Light Sci Appl 9(1):152

Chen Y et al (2024) Adaptive detection and classification of brain tumour images based on photoacoustic imaging. Appl Sci 14(12):5270

Choi W, Park E-Y, Jeon S, Kim C (2018) Clinical photoacoustic imaging platforms. Biomed Eng Lett 8:139–155

Choi W, Park B, Choi S, Oh D, Kim J, Kim C (2023) Recent advances in contrast-enhanced photoacoustic imaging: overcoming the physical and practical challenges. Chem Rev 123(11):7379–7419

Deán-Ben XL, Fehm TF, Ford SJ, Gottschalk S, Razansky D (2017) Spiral volumetric optoacoustic tomography visualizes multi-scale dynamics in mice. Light Sci Appl 6(4):e16247

Fakhoury JW et al (2024) Photoacoustic imaging for cutaneous melanoma assessment: a comprehensive review. J Biomed Opt 29(S1):S11518

Farajollahi A, Baharvand M (2024) Advancements in photoacoustic imaging for cancer diagnosis and treatment. Int J Pharm 665:124736

Farooq A, Sabah S, Dhou S, Alsawaftah N, Husseini G (2022) Exogenous contrast agents in photoacoustic imaging: an in vivo review for tumor imaging. Nanomaterials 12(3):393

Grasso V, Willumeit-Roemer R, Jose J (2023) Development of an AI-assisted multi-spectral photoacoustic imaging for volumetric molecular tissue composition: a multi-frequency translational approach. In: Photons plus ultrasound: imaging and sensing 2023, vol 12379. SPIE, p 1237902

Gu Y, Sun Y, Wang X, Li H, Qiu J, Lu W (2023) Application of photoacoustic computed tomography in biomedical imaging: a literature review. Bioeng Transl Med 8(2):e10419

Han S, Lee D, Kim S, Kim H-H, Jeong S, Kim J (2022) Contrast agents for photoacoustic imaging: a review focusing on the wavelength range. Biosensors 12(8):594

He Y, Shi J, Maslov KI, Cao R, Wang LV (2020) Wave of single-impulse-stimulated fast initial dip in single vessels of mouse brains imaged by high-speed functional photoacoustic microscopy. J Biomed Opt 25(6):066501

Hosseinaee Z, Le M, Bell K, Reza PH (2020) Towards non-contact photoacoustic imaging. Photoacoustics 20:100207

Huang S et al (2021) Functional multispectral optoacoustic tomography imaging of hepatic steatosis development in mice. EMBO Mol Med 13(9):e13490

Ivankovic I, Lin HCA, Özbek A, Orive A, Deán-Ben XL, Razansky D (2024) Multispectral optoacoustic tomography enables in vivo anatomical and functional assessment of human tendons. Adv Sci 11(18):2308336

James S, Neuhaus K, Murphy M, Leahy M (2021) Contrast agents for photoacoustic imaging: a review of stem cell tracking. Stem Cell Res Ther 12:1–19

John S et al (2023) Niche preclinical and clinical applications of photoacoustic imaging with endogenous contrast. Photoacoustics 32:100533

Jung D, Park S, Lee C, Kim H (2019) Recent progress on near-infrared photoacoustic imaging: imaging modality and organic semiconducting agents. Polymers 11(10):1693

Kalva SK, Deán-Ben XL, Reiss M, Razansky D (2023a) Head-to-tail imaging of mice with spiral volumetric optoacoustic tomography. Photoacoustics 30:100480

Kalva SK, Deán-Ben XL, Reiss M, Razansky D (2023b) Spiral volumetric optoacoustic tomography for imaging whole-body biodynamics in small animals. Nat Protoc 18(7):2124–2142

Karlas A et al (2019) Cardiovascular optoacoustics: from mice to men–A review. *Photoacoustics* 14:19–30

Kim M, Jeng G-S, Pelivanov I, O'Donnell M (2020) Deep-learning image reconstruction for real-time photoacoustic system. IEEE Trans Med Imaging 39(11):3379–3390

Kim J, Choi S, Kim C, Kim J, Park B (2024) Review on photoacoustic monitoring after drug delivery: from label-free biomarkers to pharmacokinetics agents. Pharmaceutics 16(10):1240

Kye H, Song Y, Ninjbadgar T, Kim C, Kim J (2022) Whole-body photoacoustic imaging techniques for preclinical small animal studies. Sensors 22(14):5130

Li D, Humayun L, Vienneau E, Vu T, Yao J (2021) Seeing through the skin: photoacoustic tomography of skin vasculature and beyond. JID Innov 1(3):100039

Liu C, Wang L (2022) Functional photoacoustic microscopy of hemodynamics: a review. Biomed Eng Lett 12(2):97–124

Liu W, Yao J (2018) Photoacoustic microscopy: principles and biomedical applications. Biomed Eng Lett 8:203–213

Liu H, Teng X, Yu S, Yang W, Kong T, Liu T (2024) Recent advances in photoacoustic imaging: current status and future perspectives. Micromachines 15(8):1007

Loc I, Unlu MB. Speeding up photoacoustic imaging using diffusion models, arXiv preprint arXiv:2312.08834, 2023.

MacCuaig WM, Jones MA, Abeyakoon O, McNally LR (2020) Development of multispectral optoacoustic tomography as a clinically translatable modality for cancer imaging. Radiol Imaging Cancer 2(6):e200066

Maturi M, Locatelli E, Monaco I, Franchini MC (2019) Current concepts in nanostructured contrast media development for in vivo photoacoustic imaging. Biomater Sci 7(5):1746–1775

Menozzi L, Yang W, Feng W, Yao J (2022) Sound out the impaired perfusion: Photoacoustic imaging in preclinical ischemic stroke. Front Neurosci 16:1055552

Moore C, Chen F, Wang J, Jokerst JV (2019) Listening for the therapeutic window: advances in drug delivery utilizing photoacoustic imaging. Adv Drug Deliv Rev 144:78–89

Najafzadeh E, Ghadiri H, Alimohamadi M, Farnia P, Mehrmohammadi M, Ahmadian A (2019) Application of multi-wavelength technique for photoacoustic imaging to delineate tumor margins during maximum-safe resection of glioma: A preliminary simulation study. J Clin Neurosci 70:242–246

Nguyen VP, Zhe J, Hu J, Ahmed U, Paulus YM (2023) Molecular and cellular imaging of the eye. Biomed Opt Express 15(1):360–386

Park J et al (2024) Clinical translation of photoacoustic imaging. Nat Rev Bioeng:1–20

Rana M, Sall S, Bijoor VS, Gaiwad V, Gaikwad UV, Patil P, Meher K (2024) Obstacles to the full realization and adoption of Artificial Intelligence (AI). South East Eur J Public Health:1003–1016. https://doi.org/10.70135/seejph.vi.2251

Rodrigues J et al (2024) Machine learning enabled photoacoustic spectroscopy for noninvasive assessment of breast tumor progression in vivo: a preclinical study. ACS Sens 9(2):589–601

Seong D et al (2022) Target ischemic stroke model creation method using photoacoustic microscopy with simultaneous vessel monitoring and dynamic photothrombosis induction. Photoacoustics 27:100376

Shinoda K et al (2022) Indocyanine green conjugated phototheranostic nanoparticle for photodiagnosis and photodynamic therapy. Photodiagn Photodyn Ther 39:103041

Shrestha B, DeLuna F, Anastasio MA, Yong Ye J, Brey EM (2020) Photoacoustic imaging in tissue engineering and regenerative medicine. Tissue Eng B Rev 26(1):79–102

Sridharan B, Lim HG (2023) Advances in photoacoustic imaging aided by nano contrast agents: special focus on role of lymphatic system imaging for cancer theranostics. J Nanobiotechnol 21(1):437

Tsang VT, Li X, Wong TT (2020) A review of endogenous and exogenous contrast agents used in photoacoustic tomography with different sensing configurations. Sensors 20(19):5595

Upputuri PK, Pramanik M (2020) Recent advances in photoacoustic contrast agents for in vivo imaging. Wiley Interdiscip Rev Nanomed Nanobiotechnol 12(4):e1618

Wang R, Zhu J, Xia J, Yao J, Shi J, Li C (2023) Photoacoustic imaging with limited sampling: a review of machine learning approaches. Biomed Opt Express 14(4):1777–1799

Yan B, Song B, Mu G, Fan Y, Zhao Y (2023) Compressed single-shot 3D photoacoustic imaging with a single-element transducer. Photoacoustics 34:100570

Yang X, Chen Y-H, Xia F, Sawan M (2021) Photoacoustic imaging for monitoring of stroke diseases: a review. Photoacoustics 23:100287

Yang J, Choi S, Kim J, Park B, Kim C (2023) Recent advances in deep-learning-enhanced photoacoustic imaging. Adv Photonics Nexus 2(5):054001–054001

Yao J, Wang LV (2018) Recent progress in photoacoustic molecular imaging. Curr Opin Chem Biol 45:104–112

Ye L, Chang C-C, Li Q, Tintut Y, Hsu JJ (2024) Advanced imaging techniques for atherosclerosis and cardiovascular calcification in animal models. J Cardiovasc Dev Dis 11(12):410

Yu Q, Huang S, Wu Z, Zheng J, Chen X, Nie L (2020) Label-free visualization of early cancer hepatic micrometastasis and intraoperative image-guided surgery by photoacoustic imaging. J Nucl Med 61(7):1079–1085

Zafar M, Manwar R, Avanaki K (2022) High-fidelity compression for high-throughput photoacoustic microscopy systems. J Biophotonics 15:e202100350

Zafar M, Manwar R, McGuire LS, Charbel FT, Avanaki K (2023) Ultra-widefield and high-speed spiral laser scanning OR-PAM: system development and characterization. J Biophotonics 16:e202200383

Zafar M et al (2024) Spiral laser scanning photoacoustic microscopy for functional brain imaging in rats. Neurophotonics 11(1):015007

Zhang P et al (2018) High-resolution deep functional imaging of the whole mouse brain by photoacoustic computed tomography in vivo. J Biophotonics 11(1):e201700024

Zhu X, Menozzi L, Cho S-W, Yao J (2024) High speed innovations in photoacoustic microscopy. npj Imaging 2(1):46

Fundamentals of Positron Emission Tomography (PET) and Its Integration with Magnetic Resonance Imaging (MRI)

Usevalad Ustsinau and Cécile Philippe

Abstract

Positron Emission Tomography (PET) combined with Magnetic Resonance Imaging (MRI) is a state-of-the-art hybrid imaging modality that merges PET's functional imaging with MRI's exceptional anatomical resolution. This integration enables the simultaneous acquisition of metabolic, structural, and functional data, offering unique diagnostic and research potential. PET/MRI systems are particularly advantageous in oncology, neurology, and cardiology, as they provide high soft-tissue contrast, enhanced lesion detection, and precise localization—with MRI contributing no additional ionizing radiation, making the combined modality comparatively safe. The following chapter explains the fundamentals of PET, including its principles, data corrections, and radiotracer production, such as the radiolabeling with fluorine-18. Coupled with MRI's imaging basics, the chapter offers background knowledge about in vivo imaging, quantification methods such as standardized uptake values (SUVs), and an introduction to more advanced graphical analysis and compartmental models. The ability to analyze metabolic and functional changes alongside detailed anatomical imaging opens new research opportunities, improves early diagnosis, optimizes treatment planning, and supports therapy monitoring. Despite challenges like high costs and operational complexities, PET/MRI systems continue to evolve, with advancements in hybrid technology and quantification methods pushing the boundaries of personalized and precise medicine. These innovations highlight PET/MRI's transformative potential, not only in research settings

U. Ustsinau · C. Philippe (✉)
Division of Nuclear Medicine, Department of Biomedical Imaging and Image-Guided Therapy, Medical University of Vienna, Vienna, Austria
e-mail: cecile.philippe@meduniwien.ac.at

I. F. Cengiz et al. (eds.), *Bioimaging Modalities in Bioengineering*, Biomaterials, Bioengineering and Sustainability 9, https://doi.org/10.1007/978-3-031-99410-4_14

but also in routine clinical practice, paving the way for more effective and comprehensive diagnostic solutions.

Keywords

Hybrid imaging · Radiotracer · Fluorine-18 · FDG · Image quantification

1 Introduction to Radiological Imaging Modalities

The current level of technological advancement enables the non-invasive study of physiological and anatomical alterations using a variety of radiological imaging modalities. These methods allow researchers to monitor changes in organ functions and related physiological parameters, providing tests for analysis of disease progression, responses to treatments, and potential therapeutic targets. Radiological imaging includes several key modalities, each offering unique data. X-ray radiography provides images primarily for bone and chest evaluations, while computed tomography (CT) offers detailed cross-sectional views useful for complex diseases and injuries. Magnetic resonance imaging (MRI) excels at imaging soft tissues, such as the brain and heart, without ionizing radiation. Ultrasound, using sound waves, is ideal for real-time imaging in obstetrics and abdominal exams. Fluoroscopy enables real-time imaging of moving structures, often enhanced with contrast agents. Positron emission tomography (PET) and nuclear medicine imaging reveal metabolic and biochemical activity, valuable in e.g., oncology and cardiology. Together, these modalities support comprehensive patient care across various medical fields.

This chapter covers the most advanced and molecular-sensitive hybrid imaging technique: PET combined with MRI. The standing out power of PET/MRI is in a combination of functional and quantitative data from PET scanners and detailed anatomical images captured by MRI. As a result, it can visualize and quantify the metabolic activity, volume, and distribution of organs in vivo with a high level of certainty. A PET scan is a cornerstone of nuclear medicine, already implemented in many clinical practices, and complements blood tests and other biomarkers associated with various health conditions.

2 Fundamentals of PET

2.1 Principles of PET

PET imaging relies on the principle of coincidence counting, which was described by Wrenn et al. (1951). Wrenn and colleagues were among the first to demonstrate that the simultaneous detection of annihilation photons could replace the need for lead collimators, providing spatial information about the radiation's origin through "electronic collimation". During a PET scan, a radiopharmaceutical, often referred to as a tracer, is usually injected into the patient's bloodstream (although in certain cases oral administration or inhalation of gaseous radiotracers is possible), where it is distributed and accumulated at the target tissue or organ. PET-radiopharmaceuticals are labeled with positron-emitting radionuclides such as fluorine-18 (^{18}F), carbon-11 (^{11}C), gallium-68 (^{68}Ga), and others (see Sect. 2.4). As the radionuclide undergoes radioactive β^+-decay, it emits a positron, which is the antiparticle of an electron. When a positron encounters an electron, they annihilate each other. During this process, the mass energy of both particles is converted into two photons, each with an energy of 511 keV (reflecting mass conservation), which are emitted in nearly opposite directions (~180°), ensuring momentum conservation. Figure 1 illustrates this principle. When two opposite detectors spot the photons, the origin of the radiation can be located along the line connecting the detectors, known as the line of response (LOR).

Scintillation detectors are used for the detection of the positron-electron annihilation events. These detectors are scintillation crystals, such as lutetium oxyorthosilicate (LSO, Lu_2SiO_5:Ce) (Stickel et al. 2007), lutetium yttrium oxyorthosilicate (LYSO, $(Lu\text{-}Y)_2SiO_5$:Ce) (Korzhik et al. 2007) or bismuth germanate (BGO, $Bi_4Ge_3O_{12}$) (Kwon et al. 2016), which convert the energy of incoming gamma rays into flashes of light. Further, to detect and amplify the flashlights, pho-

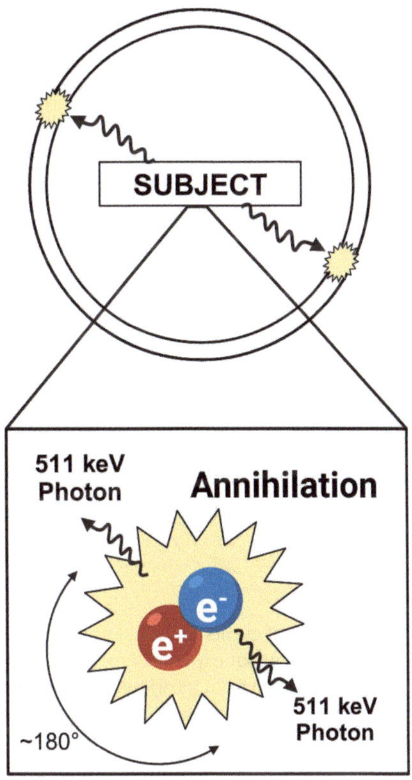

Fig. 1 Positron annihilation process. In a nucleus with a neutron deficiency, a proton is converted to a neutron, creating a positron. It is emitted from the nucleus into the surrounding tissue, losing its kinetic energy before pairing with an electron. The electron and positron pair transforms its mass into two photons of 511 keV each

tomultiplier tubes (PMT) or avalanche photodiodes (APD) are used to convert the light signals into electrical signals. Modern PET systems increasingly utilize silicon photomultipliers (SiPMs) (Roncali and Cherry 2011), which offer higher sensitivity and improved timing resolution. Operating in Geiger mode, they are providing exceptional photon detection efficiency and spatial resolution. Additionally, being silicon-based, SiPMs exhibit reduced sensitivity to magnetic fields, enabling compatibility and integration with MRI (see more in Sect. 3.2). Due to the ring of detectors, data collected in PET imaging can be reconstructed into a 3D representation using advanced computational algorithms, creating a three-dimensional image of metabolic activity within the subject. By analyzing the distribution and intensity of detected events, PET imaging provides information about metabolic rates, cellular functions, and disease processes.

2.2 PET Data Corrections

To accurately evaluate PET measurements and assess physiological processes, several quantitative corrections are essential. They typically include decay correction, scatter correction, normalization, and randoms subtraction. In a PET system, when two photons are detected almost simultaneously within a predefined time window (typically around 4 ns (Rausch et al. 2015)), they are assumed to originate from the same annihilation event. The known positions of these detectors allow the signal to be mapped to a specific LOR. Photon pairs from an annihilation event, that reach the detectors without interference, are termed “true” events (Fig. 2). However, in practice, positron annihilation occurs within an object, and photons may also interact with surrounding material, causing scattering or absorption. Scattering deflects the photon from its original path and results in a loss of energy (Fig. 2). Although in theory, scattered events can be distinguished from unscattered photons

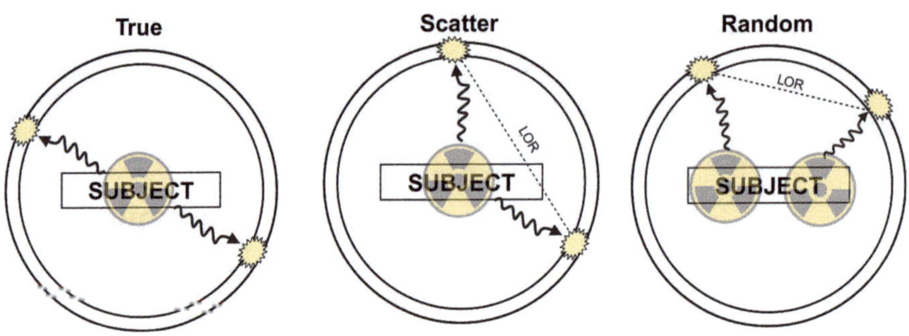

Fig. 2 Types of events in PET. A true event occurs when two photons from the same annihilation are detected without experiencing scatter or attenuation. In a scattered event, one of the photons has been deflected from its original path due to scattering. A random event involves the detection of two photons from separate positron annihilations that are registered within the coincidence timing window

by their energy, PET detectors have limited energy resolution, making it difficult to resolve small energy differences from the scatter. Consequently, modern PET systems typically use energy windows of 430 keV to 650 keV to capture true events, which include approximately 30% scatter events in 3D-PET (Rausch et al. 2015). Another situation involves the simultaneous detection of two photons from different annihilation events, known as "randoms" (Fig. 2). Their frequency can be reduced, though not completely eliminated, by using a narrow coincidence time window. This window must be at least as long as the time required for photons to travel across the detector ring's diameter.

Since scatter and random events do not represent true events, they both must be accounted for and corrected. Random events are estimated by measuring delayed timing windows and subtracting these counts from the total. Scattered events are generally approximated using simulated scatter distributions. In these simulations, the origin and count of annihilation photons are estimated from the uncorrected emission measurements, while the scattering characteristics of the surrounding medium (e.g., the patient) are derived from attenuation correction maps. This information is then used to calculate the scatter distribution, often using a single scatter simulation technique, such as the method proposed by Watson et al. (1996).

2.3 Principles of PET Radiotracers

Radiopharmaceuticals are drugs that act as tracers and include or are combined with radioactive chemical elements, known as radionuclides. Once administered to a subject, these substances are rapidly absorbed by tissues, where they participate in biochemical interactions.

The radiopharmaceuticals must consist of two components:

- A radioactive nuclide responsible for emitting a detectable signal outside the organism, allowing visualization through nuclear medical methods (Fig. 3. Radioactive compound).
- A molecular structure (vehicle, ligand) that determines the radiopharmaceutical's behavior within the organism, responsible for aspects of pharmacokinetics and pharmacodynamics (Fig. 3. Targeting Molecule).

A linker (Fig. 3. Linker) may be needed for the chemical fitting together of two parts, where the vehicle plays a vital role in defining the biological characteristics of the radiopharmaceutical. The vehicle is responsible for facilitating chemical and biochemical interactions within the living organism, ensuring proper uptake, distribution, and retention of the radiopharmaceutical in target tissues or organs.

PET is classified as the molecular imaging modality. This term reflects the fact that radiopharmaceuticals are chemically identical to their non-radioactive forms, meaning the body processes them in the same way and incorporates the radiolabeled molecules into most biochemical functions. As a result, PET enables direct visualization of these functional processes in vivo. This is particularly

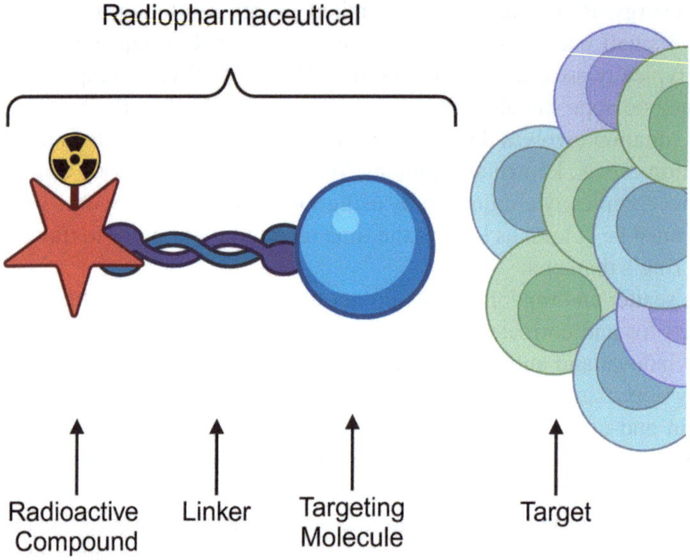

Fig. 3 Representation of the interaction mechanism of a PET radiopharmaceutical within the target. The most sophisticated radiopharmaceutical consists of a radioactive compound (nuclide), a targeting molecule (or vessel), responsible for affinity with the target, and, if chemically required, a linker between the radioactive compound and the targeting molecule

valuable for detecting pathologies at the molecular level, where abnormal function can be observed using PET radiopharmaceuticals before any structural changes become apparent.

2.4 PET Radionuclides

In PET, ligand molecules are labeled with positron-emitting nuclides. Table 1 summarizes the most commonly used nuclides for PET imaging.

It is challenging for chemists to develop rapid synthetic techniques to incorporate short-lived positron-emitting nuclides into target molecules. The synthesis, purification, analysis, and formulation of the labeled probe must typically be completed within minutes. To ensure an adequate supply of radiolabeled material for a PET scan, the total synthesis time is generally kept within approximately two half-lives of the radionuclide. Due to the short half-lives of the nuclides shown in Table 1, labeled probes must be synthesized close to the production site and used immediately after the synthesis. To support this, many modern PET facilities have in-house cyclotrons (for radionuclide production), radiochemical labs, and PET scanners in one location, enabling efficient production and quick transfer of short-lived PET probes from the lab to the scanner.

Table 1 Common PET radionuclides

Nuclide	Half-Life, min	Preparation methods	Application
Fluorine-18 (^{18}F)	109.8	Cyclotron production using $^{18}O(p,n)^{18}F$ reaction	Incorporation into organic molecules via nucleophilic substitution (Haveman et al. 2023) or electrophilic addition (Kuik et al. 2015)
Carbon-11 (^{11}C)	20.4	Cyclotron production using $^{14}N(p,\alpha)^{11}C$ reaction	Synthesis of [^{11}C]methyl iodide followed by nucleophilic substitution or [^{11}C]CO and [^{11}C]CO_2 for carboxylation reactions (Eriksson et al. 2021)
Oxygen-15 (^{15}O)	2.0	Cyclotron production using $^{14}N(d,n)^{15}O$ or $^{16}O(p,\alpha)^{15}N$ reaction	Incorporation into water or labeled gases for physiological studies (Pulagam et al. 2020)
Nitrogen-13 (^{13}N)	9.97	Cyclotron production using $^{16}O(d,\alpha)^{13}N$ or $^{14}N(d,p)^{15}N$ reaction	Synthesis of [^{13}N]ammonia for myocardial perfusion imaging (Pulagam et al. 2020)
Gallium-68 (^{68}Ga)	67.7	Generator-produced from the decay of germanium-68	Complexation with chelators for radiolabeling peptides and other molecules (Synowiecki et al. 2018)

PET centers with their in-house cyclotron have the biggest advantage as it requires a high level of investment and scientific expertise to handle them. Cyclotrons are essential for producing radiopharmaceuticals used in medical imaging and research, providing a consistent supply of radionuclides for various diagnostic and therapeutic applications. A cyclotron functions by accelerating charged particles, usually protons, to high energies using powerful magnetic fields. Its core structure includes two hollow, D-shaped electrodes called "dees" positioned within a vacuum chamber. An alternating high-frequency voltage applied to the dees creates an electric field that propels charged particles along a spiral path. A strong magnetic field perpendicular to their motion guides these particles outward in a spiral. When the particles reach the outer edge of the dees, they pass through a gap in the magnetic field and are directed toward a target material. Upon collision with the target, nuclear reactions occur, producing desired radioactive nuclides (Friesel and Antaya 2009). For instance, fluorine-18 is produced by accelerating protons onto an oxygen-18 target, resulting in the reaction $^{18}O(p,n)^{18}F$. Other PET nuclides, such as carbon-11, nitrogen-13, and oxygen-15, can also be produced in cyclotrons via specific nuclear reactions. These nuclides are then incorporated into target molecules to create radiotracers for PET imaging (refer to Table 1 Application). Selecting the appropriate radionuclide necessitates striking a balance between availability, economic feasibility, radiolabeling feasibility, and radiopharmacological considerations.

2.5 Radiolabeling with Fluorine-18 and Quality Control

Fluorine-18 is a leading radionuclide for PET imaging. In medicinal chemistry and radiopharmacy, it is often termed the "radionuclide of choice". It possesses favorable physical and nuclear characteristics, including an optimal half-life that allows sufficient time for complex synthesis processes. With a high proportion of β^+ decay (97%, with 3% electron capture), low positron energy (maximum 0.635 MeV), and a short positron range in tissue (~2.3 mm), fluorine-18 enables high-resolution PET images. Compared to other small positron emitters like carbon-11, oxygen-15, or nitrogen-13, fluorine-18 has a relatively long half-life of 109.8 min (see Table 1), which supports multistep synthesis and prolonged imaging sessions. Its extended half-life also allows for the transportation of doses over hours, enabling distribution to remote locations. Additionally, ^{18}F-labeled compounds are efficient, allowing multiple patient scans from a single dose, making them highly practical for clinical PET imaging facilities.

The widespread use of fluorine-18 is largely due to the development and broad application of the PET tracer 2-[^{18}F]fluoro-2-deoxy-D-glucose ([^{18}F]FDG), a glucose analog that has greatly enhanced cancer diagnosis, treatment planning, and monitoring. Due to the Warburg effect (Potter et al. 2016), which involves changes in cell metabolism through increased glucose uptake, PET systems are now widely used in oncology to detect and stage tumors.

2.5.1 Radiosynthesis of [^{18}F]FDG

The production of [^{18}F]FDG can be achieved via two chemical reactions: electrophilic or nucleophilic fluorination. Electrophilic fluorination, first conducted by Wolf et al. in 1976 (Fowler and Ido 2002), is based on [^{18}F]F_2 gas and usually is limited with the radiochemical yield (RCY) of a maximum of 50%. The presently favored synthesis method utilizes nucleophilic fluorination with the addition of Kryptofix 222 as a catalyst, a technique introduced by Hamacher et al. in 1986 (Hamacher et al. 1986). Acetylated mannose triflate serves as the precursor in this approach and undergoes nucleophilic substitution by $^{18}F^-$ to form FTAG (2-[^{18}F]-fluoro-1,3,4,6-tetra-O-acetyl-D-glucose) (presented in Fig. 4). Such advancement has yielded efficiencies exceeding 50% with reaction times under 50 min. Further enhancements in reaction conditions and the implementation of automated synthesis modules have resulted in rapid reactions (<25 min) with first-rate yields (>70%,

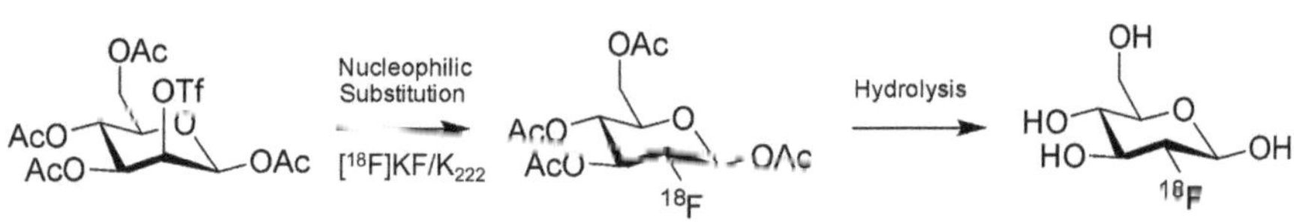

Fig. 4 Schematic description of the [^{18}F]FDG synthesis by nucleophilic fluorination: an initial nucleophilic substitution step, facilitated by Kryptofix 222 as a phase-transfer catalyst, followed by a hydrolysis step to yield the final product

uncorrected for decay), high purity (>95%), and perfect reproducibility, establishing [^{18}F]FDG as a widely applicable tracer for routine use.

2.5.2 Radioligand Dose and Quality Control

Radiation exposure levels are directly related to the amount of tracer injected. Adhering to the ALARA principle ("As Low as Reasonably Achievable"), outlined in Article 67 of the Euratom Directive, radiopharmaceuticals must be applied with minimal exposure while ensuring sufficient image quality for clinical assessment. This principle emphasizes using the lowest possible tracer dose that still provides clear, clinically useful images. In most countries, PET radiopharmaceuticals are classified as drugs, subjecting them to strict regulatory approval processes for human use. Furthermore, nearly all clinical applications involve intravenous administration of these radiopharmaceuticals, necessitating rigorous safety and quality controls. Careful consideration is given to factors such as appropriate glass, rubber stopper materials, and sterility, given PET tracers' short half-life and unique properties. Prior to administration, each radiopharmaceutical must be tested for safety, radiochemical purity, and regulatory compliance (see Table 2). A radiochemist evaluates the total radioactivity and molar activity, with decay corrections applied to ensure these values are accurate at the time of administration. The proper injection volume is calculated to keep both the tracer dose (in MBq) and the radiopharmaceutical mass (in μg or nmol) within safe, predefined limits, preventing "mass effects" that could disrupt the tracer's distribution or cause unintended pharmacological effects. An insufficient dose may yield unreliable tissue and blood radioactivity measurements,

Table 2 Necessary quality control tests of fluorine-18 labeled PET tracers before application

Parameters	Method	Acceptance criterion
Radiochemical Purity, %	Analytical HPLC[a]	>95
Radiochemical Identity	Analytical HPLC[a], Radioactivity detector	Matches retention time of the standard
Radionuclidic Purity	Gamma spectrometer	Presence of peak at 511 keV
Chemical Purity	Analytical HPLC[a], UV-detector	Matches retention time of the standard
Kryptofix, μg/mL	Kryptofix test	≤50
Residual solvent, ppm	Gas-chromatographer	Acetonitrile < 410; Methanol <3000 and others
pH	pH indicator strip	4.0–8.5
Osmolality, mosm/kg	Osmometer	200–400
Sterility	Inoculation method	No microbiological growth
Endotoxins, EU/mL	Kinetic-turbidimetric or kinetic-chromogenic assay	<1.0

[a] *HPLC* High-performance liquid chromatography

while an excessively high dose increases radiation exposure for both staff and patients, potentially limiting the number of allowable PET studies.

3 Hybrid PET/MRI Systems

Providing functional and quantitative data from PET scanners and detailed anatomical images captured by CT or MRI, hybrid PET systems are standing out as the most powerful and molecular-sensitive imaging techniques nowadays. Such integration allows the precise localization of metabolic abnormalities, enhances diagnostic accuracy, and facilitates treatment planning. PET/MR systems are relatively new to the market, with clinical systems introduced in 2010 and preclinical prototypes emerging in 2006 (Herzog 2012). They enhance hybrid imaging by providing superior soft tissue contrast, multi-parametric capabilities for comprehensive tissue characterization, and the advantage of no ionizing radiation exposure compared to CT (Ehman et al. 2017). This subchapter aims to briefly introduce the concept of MRI and highlight the technical challenges of combining PET and MRI.

3.1 Basics of MRI

MRI operates by detecting nuclei with a magnetic moment. When placed in an external magnetic field B_0, these nuclei align their magnetic moments with the field. A radiofrequency pulse at a specific frequency is then used to excite the nuclei. The frequency required for this excitation is determined by the gyromagnetic ratio γ of the nuclei being studied and the strength of the external magnetic field B_0:

$$\omega = \gamma * B_0 \tag{1}$$

The magnetic moment of the stimulated nuclei precesses around the external magnetic field at a frequency ω, known as the Larmor frequency. These nuclei are considered to be in an excited energy state and will gradually return to their ground state over time. During this process, the precessing magnetic moment realigns with the external magnetic field, emitting electromagnetic radiation at a frequency equal to the precession frequency. This emitted radiofrequency is detected using specialized coils, producing a signal proportional to the number of stimulated nuclei (Rausch et al. 2017).

In routine MRI imaging (excluding MR spectroscopy), the nuclei of interest are typically hydrogen atoms, which have a gyromagnetic ratio of $\gamma_H = 42.6$ MHz/T. In a clinical PET/MR system with a static magnetic field strength of $B_0 = 3$ T, this results in a Larmor frequency of $\omega = 127.8$ MHz (Rausch et al. 2017). The static magnetic field is generated by a superconducting coil and is highly homogeneous within the MRI's field of view (FOV). However, if only this homogeneous field were used, the radiofrequency pulse would excite nuclei throughout the entire volume of the homogeneous field. Consequently, the signal obtained from such a setup would lack spatial information about the stimulated nuclei. To achieve spatial localization, additional external magnetic fields, known as gradient fields, are superimposed onto the static field. These gradient fields are generated by specialized gradient coils in the MRI system and can be adjusted to vary the magnetic field strength at different points within the FOV. This controlled variation enables spatially selective excitation of nuclei, allowing the received signal to be mapped to specific locations.

3.2 Combining PET and MRI

Integrating PET and MRI into a single system presented significant technical challenges due to the close proximity of the two modalities. Traditional PET detectors, such as photomultiplier tubes (PMTs), cannot function within strong magnetic fields (Delso et al. 2015). Additionally, the switching gradient fields and radiofrequency signals used in MRI can induce Foucault's currents in the PET electronics, potentially causing inaccurate readouts and overheating of the PET detectors, which can compromise system calibration (Spanoudaki et al. 2008). Furthermore, the homogeneity of the MRI's magnetic field is crucial for accurate imaging, as it may be disrupted by the presence of PET components and Foucault's currents they generate.

Early solutions to these issues focused on the physical separation of the PET and MRI components. For example, the Philips Gemini TF PET/MRI system featured a tandem arrangement where the PET and MRI components were separated by a rotating patient handling system (Ojha et al. 2010). The PET gantry was shielded from the magnetic field, and its electronics were housed outside the examination room to minimize electromagnetic interference with the MRI (Ojha et al. 2010). A similar approach was seen in GE's triple-modality PET/CT-MRI system, where the PET/CT and MRI were located in separate rooms. This setup utilized a specialized mobile patient handling system that docks with both devices, enabling patient positioning on both systems with minimal repositioning between scans (Veit-Haibach et al. 2013).

For applications requiring simultaneous PET and MRI acquisition, significant modifications to the individual systems are necessary. A key advancement is the replacement of traditional photomultiplier tubes (PMTs) with solid-state photodetectors that are insensitive to magnetic fields. A notable example of such detectors is specialized avalanche photodiodes (APDs). When photons enter the APD, they cause ionization, generating electron-hole pairs. The applied potential accelerates these electrons, triggering secondary ionizations and resulting in a charge "avalanche," which amplifies the signal. Although APDs have certain disadvantages compared to PMTs, such as lower gain, slower rise times, and sensitivity to temperature, these limitations can be mitigated through advanced electronics and cooling systems. Crucially, APDs are unaffected by magnetic fields, making them well-suited for use in integrated PET/MRI systems.

A more advanced design solution for solid-state PET detectors is the silicon photomultiplier (SiPM), which is essentially an array of miniature avalanche photodiodes (APDs) operated in Geiger mode (Roncali and Cherry 2011). When a photon strikes an APD, it triggers a Geiger-type discharge, with each discharge representing the detection of a single photon. By summing the signals from all APDs in the array, the resulting output is proportional to the number of detected photons. This allows SiPMs to measure the energy of gamma photons, as the number of photons produced in a scintillating material is directly related to the energy deposited.

Compared to APDs, SiPMs offer several advantages, including a gain comparable to PMTs, on the order of 10^6, and excellent timing resolution. This timing capability supports the use of time-of-flight techniques, which measure the time difference between the detection of two annihilation photons to estimate their origin along the line of response. Incorporating this TOF information into image reconstruction improves precision and reduces noise. Currently, both APDs and SiPMs are employed in clinical PET/MRI systems (Siemens Biograph mMR, GE Signa, and others), often paired with effective electromagnetic shielding and active cooling systems. Figure 5 provides a schematic representation of a fully integrated PET/MRI design.

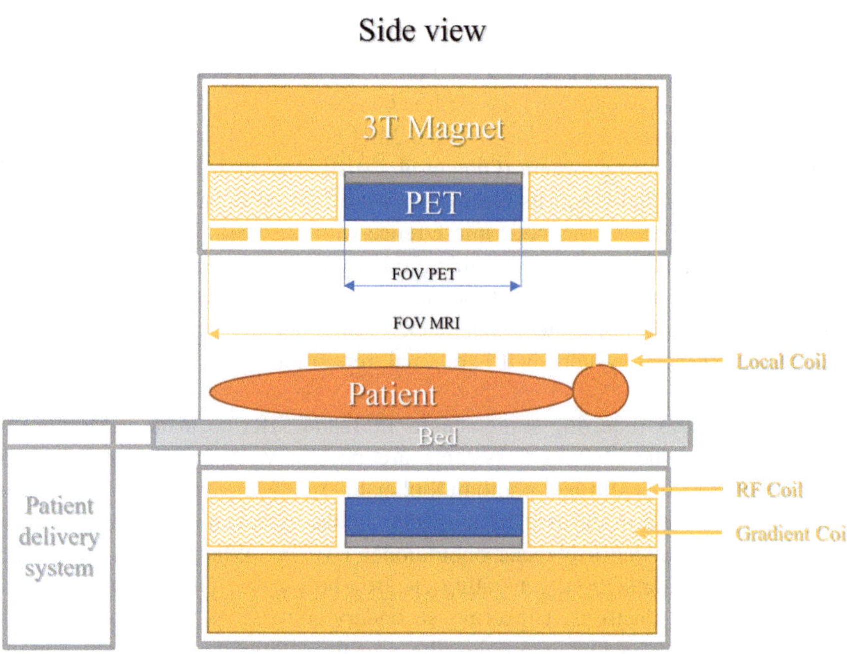

Fig. 5 Schematic concept of a fully integrated PET/MR system. The PET detectors are arranged within the MRI gantry to allow simultaneous acquisition of PET and MR data. Gradient MR coils are placed next to PET detectors, radiofrequency (RF) coils are in the inner circle and the local coil is above the patient

3.3 In Vivo Imaging

In clinical practice, imaging modalities like MRI, CT, PET, ultrasound, or any physiological measurement conducted directly on a patient are considered in vivo methods. In vivo refers to studies or experiments conducted within a living organism, such as an animal or human. These measurements assess bodily processes, structures, or functions in their natural state within the patient. In a clinical context, terms like "non-invasive" or "real-time imaging" are often used alongside or instead of in vivo to describe routine patient measurements more specifically.

Non-invasive PET/MR measurements are crucial in medical imaging because they combine the functional insights of PET with the high-resolution anatomical detail of MRI, enabling early and accurate diagnosis for conditions such as cancer, neurological, and cardiovascular diseases. This method enhances patient comfort and safety, avoiding the risks of surgical procedures or biopsies, while also reducing radiation exposure by replacing CT with MRI for structural imaging. Simultaneous PET/MR provides both functional and structural information in a single scan procedure. Additionally, PET/MR facilitates personalized medicine by allowing researchers to study disease mechanisms and responses over time in the same individual, promoting more tailored treatment strategies.

4 Quantification of PET Scans

Radioactivity measurements are quantified in units like Becquerel (Bq), which represents one decay per second, or 1 Curie, which is equivalent to 37 Gigabecquerels (GBq). Since radioactive decay occurs randomly, statistical methods are important for precise measurement. Reliable count statistics are achieved by collecting data over time and adjusting the sample size or duration. In in vivo imaging, however, measurement time is limited by the dynamic changes in radioactivity concentration, requiring data collection within brief intervals to capture accurate count statistics.

Modern PET systems are able to generate 4D data consisting of pixels in dynamic images, where counts are collected over specific time frames. Dynamic PET studies require capturing multiple images over time (about an hour) to create time-activity curves (TACs) for specific organs or structures, along with simultaneous blood sampling to measure tracer concentration. Concentration rates in equilibrium with blood are then calculated using specific metabolic models—multiple-time graphical analysis or compartmental modeling. However, these studies are time and labor-intensive and require extensive post-processing, limiting the axial examination area to a single FOV. Due to these demands, kinetic modeling is generally confined to research settings and is rarely used in clinical practice.

4.1 Standardized Uptake Values

In clinical practice, PET measurements are typically conducted as static scans, capturing a snapshot of activity distribution at a specific time point. This allows for relative comparisons of functional activity in selected areas, such as organs or lesions. However, the activity concentration in tissues depends on the tracer available within the patient. To enable comparisons across multiple scans of the same patient or between different patients, these values must be normalized.

The standardized uptake value (SUV) is a commonly employed PET metric. It is calculated as the ratio of tissue radioactivity concentration in units of kBq or MBq per mL at time T, denoted as $C_{PET}(T)$, to the administered dose of kBq or MBq at the injection time, divided by body weight (BW), usually in kg or g (see Eq. 2). In both clinical and preclinical research, SUV is commonly expressed in g/ml.

$$SUV_{BW} = \frac{C_{PET}(T)}{Dose/Weight} \tag{2}$$

Both tissue radioactivity and dose require decay correction to align with the same time point. The denominator of the SUV equation represents the average radioactivity concentration per unit weight in the entire body, rendering SUV as the regional to whole body ratio of radioactivity concentration. The average SUV of the whole body is the same as the density of the body and ideal conditions should be $\lesssim 1$.

Instead of using body weight alone, the administered dose can be adjusted based on lean body mass (LBM):

$$SUV_{LBM} = \frac{C_{PET}(T)}{Dose/LBM} \tag{3}$$

where it is divided for LBM for men:

$$LBM_{men} = \frac{9.27 \times 10^3 \times BW}{6.68 \times 10^3 + 216 \times BMI} \tag{4}$$

and LBM for women:

$$LBM_{women} = \frac{9.27 \times 10^3 \times BW}{8.78 \times 10^3 + 244 \times BMI} \tag{5}$$

here BW is denoted for body weight and BMI for body mass index.

Specifically for [^{18}F]FDG-PET studies, it is often recommended to use SUV_{LBM}, since [^{18}F]FDG uptake in white adipose tissue is minimal. Additionally, for some research purposes, normalization to the body surface area (BSA) may be used. The equations BSA may vary from selected models and parameters and more precisely are gathered and compared with BW and LBM in the work of Ustsinau et al. (2024b).

SUVs can also be defined by the statistics of the regions of interest. SUV_{mean} or SUV_{aver}, is the most common parameter for measuring regional uptake, representing the average value of voxels within a defined volume or region of interest. Additional metrics like SUV_{max} and SUV_{min}, representing the maximum and minimum uptake values, are also used. SUV_{max} is commonly applied in assessing cancer treatment response with [^{18}F]FDG-PET due to its low inter-observer variability, though SUV_{mean} has shown better reproducibility and repeatability. SUV_{max} tends to be higher due to factors like point-spread-function and time-of-flight corrections. Another metric, SUV_{peak}, uses a spherical VOI to capture the maximal average, reducing noise bias from SUV_{max} but being more affected by partial volume effects. A combination of SUV_{max} and SUV_{peak} has been suggested for multicenter PET studies to improve the consistency of datasets.

The European Association of Nuclear Medicine (EANM) provides specific guidelines on SUV ranges and acceptable variations in PET imaging: tumors (Boellaard et al. 2015), brain (Guedj et al. 2022), parathyroid (Mottet et al. 2021), prostate (Petranović Ovčariček et al. 2021), and bone (Van den Wyngaert et al. 2016). However, the criteria for visually interpreting these SUV values must be tailored to each study protocol. This customization is necessary due to variations in organ or tumor locations, disease stages, and others, which can influence radiotracer uptake. Therefore, while general SUV guidelines are available, individual protocols require specific interpretative criteria to ensure accurate assessment and diagnostic consistency across different applications.

4.2 Multiple-Time Graphical Analysis and Compartmental Models

For research needs more sophisticated quantification methods have been developed, which is why SUVs are sometimes referred to as a semi-quantitative approach. These models consider the activity in the blood pool or a reference region and typically require dynamic PET imaging over a specified period. They are commonly represented as multi-time graphical analyses (MTGA) and compartmental models.

In multiple-time graphical analysis, concentration curves of the radiopharmaceutical in tissue regions of interest and arterial plasma are transformed and combined into a single curve, with the goal of achieving linearity under specific conditions. This involves plotting the data and fitting a line to the linear phase, where the slope represents the net uptake rate (also known as the net influx rate—K_i) of the radiopharmaceutical or its volume of distribution. In some cases, for example, areas of the brain with non-specific uptake, a reference region curve can be used instead of the arterial plasma input function. These graphical methods are model-independent, although the slope can be interpreted in terms of model parameters for specific structures. They have been developed for both irreversibly binding (Patlak plot (Patlak and Blasberg 1985)) and reversibly binding (Logan plot (Logan 2000)) radiopharmaceuticals. However, these methods are vulnerable to errors in plasma metabolite analysis and non-specific binding to plasma proteins. For cases where the blood curve is measured or known, fractional uptake rate (FUR) offers a simple alternative to SUV or MTGA. FUR, which approximates the Patlak slope (K_i), does not require dynamic scans, is proportional to SUV, and is related to plasma clearance rate.

Compartmental modeling is the most precise and complicated quantification method. Here the physiological system of dynamic processes within the target tissue is divided into multiple interacting compartments. Each compartment contains the positron-emitting nuclide label at specific concentrations. It is important to calculate the concentration in the target tissue in equilibrium. For that tracking not only the original radioligand but also any labeled molecules formed within the tissue (e.g., phosphorylated [^{18}F]FDG) or transported there from other organs via the bloodstream is essential. The tissue and blood interactions typically require the solution of complicated differential or integral calculus. The simplest form of this model is the two-compartment model, or one-tissue compartment model (Fig. 6), where the first compartment represents the input function (plasma or blood curve)

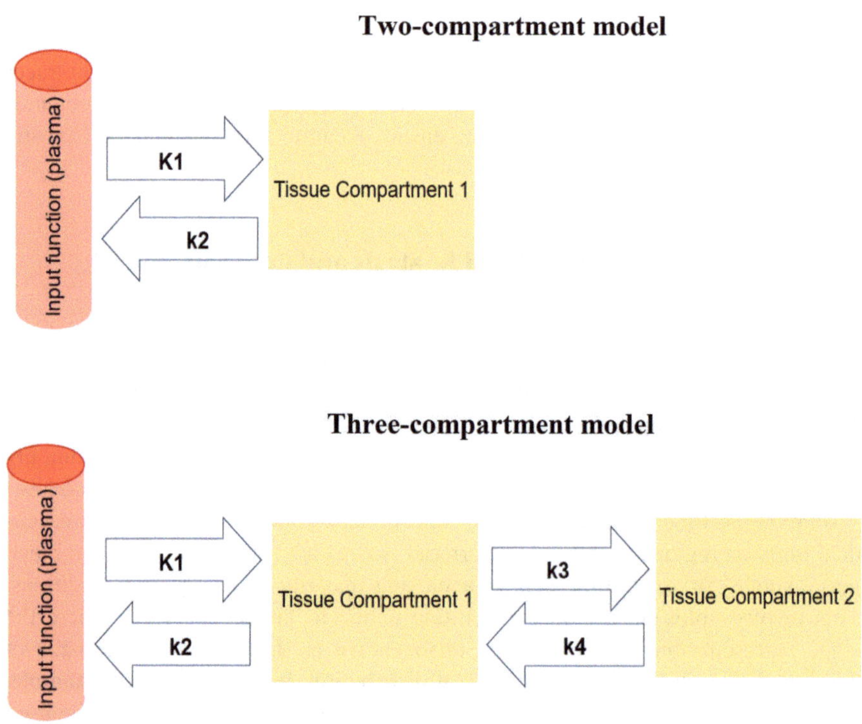

Fig. 6 Schematic representation of compartmental modeling: two and three-compartment models, where K_1 is constant from blood to tissue, k_2 constant from tissue to blood, and k3 and k4 are representatives of the further tracer distribution

and the second represents the labeled nuclide in the tissue, with these compartments connected by two rate constants, K_1 (constant from blood to tissue) and k_2 (constant from tissue to blood). More complex versions, like the three-compartment (two-tissue) (Fig. 6) and four-compartment (three-tissue) models, are also frequently used in research. When the model is sufficiently detailed, it is considered robust, with minimal vulnerabilities, though systematic errors in plasma metabolite analysis may still affect plasma input-based methods.

5 Conclusions and Future Perspectives

A PET/MR system is an integrated radiological medical imaging technology that combines positron emission tomography with MRI in a single scanner. PET provides functional information by detecting the decay of positron-emitting nuclides from a radioactive tracer injected into the body, revealing metabolic and biochemical processes. MRI, on the other hand, offers detailed anatomical images using magnetic fields. Together, PET/MR systems provide simultaneous functional and structural information, minimize patient movement, and enable more precise diagnostics, particularly useful in oncology, neurology, and cardiology.

Various radiopharmaceuticals provide a big potential for detecting the small and early changes in metabolism within the body and extending the research questions and diagnostic accuracy. By now PET is among the best techniques for assessing treatment efficacy and visualizing biological processes before the presence of any morphological changes. Measurements with the glucose analog [^{18}F]FDG are a great example that is especially beneficial for complex cases in clinical practice. A wide range of available tracers is extending the future application list. As a non-invasive technique, PET becomes suitable for frequent monitoring of in vivo conditions.

MRI is instrumental for visualizing various soft tissues, especially in the brain, allowing the precise quantification of tissue uptake in PET. Current novel techniques of MRI are opening the horizons for research and diagnosis. Anatomical DIXON images are suited for attenuation correction of PET signals and for quantifying the fat content of various tissues. This data can further be analyzed and implemented into SUV assessment (Ustsinau et al. 2024b). Among the most well-established MR research tools is magnetic resonance spectroscopy (MRS), which measures the chemical composition of tissues by analyzing the frequency spectrum of specific metabolites, allowing for the detection and quantification of biochemical compounds in tissues. Together with PET, MRS can effectively assess the metabolic cycle of the body and help to differentiate tracer metabolites or to phenotype the diseases (Ustsinau et al. 2023). In the future deuterium metabolic imaging, also a part of MRS, can extend the nutrient assessment and together offer the full picture of molecular interaction during metabolic disorders (Ehret et al. 2024). Another technique, CEST MRI can potentially be combined with PET and reveal chemical exchange during metabolic processes (Vinogradov et al. 2023). BOLD MRI, a specialized imaging technique used to measure changes in blood oxygen levels, is another candidate for complementing PET research in the future (Zhou et al. 2019).

The hybrid system enables the simultaneous acquisition of metabolic, structural, and functional data, which is especially beneficial for complex cases, such as neurology and cardiology. The integration of PET's quantitative metabolic data with MRI's superior tissue characterization is expected to enhance early diagnosis, improve treatment planning, and aid in monitoring therapy response more effectively than either modality alone. Although PET/MRI systems are currently more common for research purposes due to high costs and operational complexities, their potential to provide better results with reduced radiation exposure suggests a promising future in clinical practice for personalized, precise medical imaging.

Acknowledgments This work was supported by the Medical Imaging Cluster (MIC) of the Medical University of Vienna and has been funded by the Vienna Science and Technology Fund (WWTF) [https://doi.org/10.47379/LS19046].

Conflict of Interest The authors declare that they have no known competing financial interests or personal relationships that could have appeared to influence the work reported in this work.

References

Boellaard R, Delgado-Bolton R, Oyen WJG, Giammarile F, Tatsch K, Eschner W, Verzijlbergen FJ, Barrington SF, Pike LC, Weber WA et al (2015) FDG PET/CT: EANM procedure guidelines for tumour imaging: version 2.0. Eur J Nucl Med Mol Imaging 42:328–354

Delso G, Ter Voert E, Veit-Haibach P (2015) How does PET/MR work? Basic physics for physicians. Abdom Imaging 40:1352–1357

Ehman EC, Johnson GB, Villanueva-Meyer JE, Cha S, Leynes AP, Larson PEZ, Hope TA (2017) PET/MRI: where might it replace PET/CT? J Magn Reson Imaging 46:1247–1262

Ehret V, Ustsinau U, Friske J, Scherer T, Fürnsinn C, Helbich TH, Philippe C, Krššák M (2024) Evaluation of hepatic glucose and palmitic acid metabolism in rodents on high-fat diet using deuterium metabolic imaging. J Magn Reson Imaging

Eriksson J, Antoni G, Långström B, Itsenko O (2021) The development of 11C-carbonylation chemistry: a systematic view. Nucl Med Biol 92:115–137

Fowler JS, Ido T (2002) Initial and subsequent approach for the synthesis of 18FDG. Semin Nucl Med 32:6–12

Friesel DL, Antaya TA (2009) Medical Cyclotrons. Rev Accl Sci Tech 02:133–156

Guedj E, Varrone A, Boellaard R, Albert NL, Barthel H, van Berckel B, Brendel M, Cecchin D, Ekmekcioglu O, Garibotto V et al (2022) EANM procedure guidelines for brain PET imaging using [18F]FDG, version 3. Eur J Nucl Med Mol Imaging 49:632–651

Hamacher K, Coenen HH, Stöcklin G (1986) Efficient stereospecific synthesis of no-carrier-added 2-[18F]-fluoro-2-deoxy-D-glucose using aminopolyether supported nucleophilic substitution. J Nucl Med 27:235–238

Haveman LYF, Vugts DJ, Windhorst AD (2023) State of the art procedures towards reactive [18F]fluoride in PET tracer synthesis. EJNMMI Radiopharm Chem 8:28

Herzog H (2012) PET/MRI: challenges, solutions and perspectives. Z Med Phys 22:281–298

Korzhik M, Fedorov A, Annenkov A, Borissevitch A, Dossovitski A, Missevitch O, Lecoq P (2007) Development of scintillation materials for PET scanners. Nucl Instrum Methods Phys Res A 571:122–125

Kuik W-J, Kema IP, Brouwers AH, Zijlma R, Neumann KD, Dierckx RAJO, DiMagno SG, Elsinga PH (2015) In vivo biodistribution of no-carrier-added 6-18F-fluoro-3,4-dihydroxy-L-phenylalanine (18F-DOPA), produced by a new nucleophilic substitution approach, compared with carrier-added 18F-DOPA, prepared by conventional electrophilic substitution. J Nucl Med 56:106–112

Kwon SI, Gola A, Ferri A, Piemonte C, Cherry SR (2016) Bismuth germanate coupled to near ultraviolet silicon photomultipliers for time-of-flight PET. Phys Med Biol 61:L38–L47

Logan J (2000) Graphical analysis of PET data applied to reversible and irreversible tracers. Nucl Med Biol 27:661–670

Mottet N, van den Bergh RCN, Briers E, Van den Broeck T, Cumberbatch MG, De Santis M, Fanti S, Fossati N, Gandaglia G, Gillessen S et al (2021) EAU-EANM-ESTRO-ESUR-SIOG guidelines on prostate cancer-2020 update. Part 1: screening, diagnosis, and local treatment with curative intent. Eur Urol 79:243–262

Ojha N, Griesmer J, Hu Z, Shao L, Izquierdo D, Machac J, Ratib O, Zaidi H, Fuster V, Fayad ZA (2010) PET performance of the GEMINI TF PET—MR: the world's first whole body PET—MRI scanner. In: IEEE nuclear science symposuim & medical imaging conference. IEEE, pp 2013–2015

Patlak CS, Blasberg RG (1985) Graphical evaluation of blood-to-brain transfer constants from multiple-time uptake data. Generalizations. J Cereb Blood Flow Metab 5:584–590

Petranović Ovčariček P, Giovanella L, Carrió Gasset I, Hindié E, Huellner MW, Luster M, Piccardo A, Weber T, Talbot J-N, Verburg FA (2021) The EANM practice guidelines for parathyroid imaging. Eur J Nucl Med Mol Imaging 48:2801–2822

Potter M, Newport E, Morten KJ (2016) The Warburg effect: 80 years on. Biochem Soc Trans 44:1499–1505

Pulagam KR, Gómez-Vallejo V, López-Gallego F, Rejc L, Llop J (2020) Synthesis of 13 N- and 15 O-Labeled Radiopharmaceuticals. In: Handbook of radiopharmaceuticals, pp 109–142

Rausch I, Cal-González J, Dapra D, Gallowitsch HJ, Lind P, Beyer T, Minear G (2015) Performance evaluation of the Biograph mCT Flow PET/CT system according to the NEMA NU2-2012 standard. EJNMMI Phys 2:26

Rausch I, Quick HH, Cal-Gonzalez J, Sattler B, Boellaard R, Beyer T (2017) Technical and instrumentational foundations of PET/MRI. Eur J Radiol 94:A3–A13

Roncali E, Cherry SR (2011) Application of silicon photomultipliers to positron emission tomography. Ann Biomed Eng 39:1358–1377

Spanoudaki VC, McElroy DP, Torres-Espallardo I, Ziegler SI (2008) Effect of temperature on the performance of proportional APD-based modules for gamma ray detection in positron emission tomography. IEEE Trans Nucl Sci 55:469–480

Stickel JR, Qi J, Cherry SR (2007) Fabrication and characterization of a 0.5-mm lutetium oxyorthosilicate detector array for high-resolution PET applications. J Nucl Med 48:115–121

Synowiecki MA, Perk LR, Nijsen JFW (2018) Production of novel diagnostic radionuclides in small medical cyclotrons. EJNMMI Radiopharm Chem 3:3

Ustsinau U, Ehret V, Fürnsinn C, Scherer T, Helbich TH, Hacker M, Krššák M, Philippe C (2023) Novel approach using [18F]FTHA-PET and de novo synthesized VLDL for assessment of FFA metabolism in a rat model of diet induced NAFLD. Clin Nutr 42:1839–1848

Ustsinau U, Nics L, Hacker M, Philippe C (2024a) A fully automated synthesis of 14-(R,S)-[18F]fluoro-6-thia-heptadecanoic Acid ([18F]FTHA) on the Elixys radiosynthesizer. Pharmaceuticals (Basel) 17:318

Ustsinau U, Kulterer OC, Rausch I, Krššák M, Kiefer FW, Hacker M, Philippe C (2024b) A PET/MRI study on the effect of obesity and NAFLD on hepatic [18F]FDG uptake. Eur J Radiol 177:111552

Van den Wyngaert T, Strobel K, Kampen WU, Kuwert T, van der Bruggen W, Mohan HK, Gnanasegaran G, Delgado-Bolton R, Weber WA, Beheshti M et al (2016) The EANM practice guidelines for bone scintigraphy. Eur J Nucl Med Mol Imaging 43:1723–1738

Veit-Haibach P, Kuhn FP, Wiesinger F, Delso G, von Schulthess G (2013) PET-MR imaging using a tri-modality PET/CT-MR system with a dedicated shuttle in clinical routine. MAGMA 26:25–35

Vinogradov E, Keupp J, Dimitrov IE, Seiler S, Pedrosa I (2023) CEST-MRI for body oncologic imaging: are we there yet? NMR Biomed 36:e4906

Watson CC, Newport D, Casey ME (1996) A single scatter simulation technique for scatter correction in 3D PET. In: Computational imaging and vision. Springer Netherlands, Dordrecht, pp 255–268

Wrenn FR Jr, Good ML, Handler P (1951) The use of positron-emitting radioisotopes for the localization of brain tumors. Science 113:525–527

Zhou H, Chiguru S, Hallac RR, Yang D, Hao G, Peschke P, Mason RP (2019) Examining correlations of oxygen sensitive MRI (BOLD/TOLD) with [18F]FMISO PET in rat prostate tumors. Am J Nucl Med Mol Imaging 9:156–167

Applications of Positron Emission Tomography (PET) in Preclinical Research

Junlong Wang and Ruiqing Ni

Abstract

Positron emission tomography (PET) has become an important tool in preclinical research, offering insights into molecular and physiological processes with 3-dimensional in vivo imaging at around 1 mm spatial resolution. This chapter highlights the applications of PET across diverse fields of preclinical research, emphasizing its unique role in drug development, cancer research, neurological studies, and cardiovascular investigations. In drug development, PET facilitates the evaluation of target engagement and pharmacokinetics, enabling the precise characterization of drug behavior in living systems. Within the domain of cancer research, PET has advanced our understanding of tumor metabolism and supported the monitoring of immunotherapy efficacy, paving the way for theranostic approaches. In neurological research, PET is instrumental in studying brain metabolism, neurotransmitter systems, neuroinflammation, and pathological protein aggregation, which are central to brain diseases such as neurodegenerative diseases. Similarly, in cardiovascular studies, PET provides valuable insights into cardiac metabolism and atherosclerotic plaque characterization, aiding in the early detection and management of cardiovascular diseases. By combining robust quantitative capabilities with high sensitivity, PET continues to enhance our understanding of disease mechanisms, accelerating the

---

J. Wang
Department of Nuclear Medicine, Inselspital, University of Bern, Bern, Switzerland

R. Ni (✉)
Department of Nuclear Medicine, Inselspital, University of Bern, Bern, Switzerland

Institute for Regenerative Medicine, University of Zurich, Zurich, Switzerland

Institute for Biomedical Engineering, University of Zurich & ETH Zurich, Zurich, Switzerland
e-mail: ruiqing.ni@uzh.ch

I. F. Cengiz et al. (eds.), *Bioimaging Modalities in Bioengineering*,
Biomaterials, Bioengineering and Sustainability 9,
https://doi.org/10.1007/978-3-031-99410-4_15

translation of preclinical discoveries into clinical advancements. This chapter reviews the current state of preclinical PET applications and highlights emerging trends and future perspectives.

Keywords

Positron emission tomography · Radiotracers · Preclinical research · Cancer imaging · Neuroimaging

1 Introduction

Positron emission tomography (PET) has fundamentally advanced the field of molecular imaging, offering researchers the ability to monitor biochemical and physiological processes in vivo with exceptional precision. By employing radiolabeled tracers, PET provides insights into the cellular functions and disease mechanisms that are otherwise inaccessible through conventional imaging techniques (Nwabufo and Aigbogun 2022). Its high sensitivity and ability to quantify molecular interactions noninvasively make PET indispensable in both clinical and preclinical research (Amirrashedi et al. 2020a). The processes of imaging with PET include the generation of positron-emitting radionuclides, radiochemical synthesis, preparation of suitable radiopharmaceutical formulations, PET imaging data acquisition, PET data pre- and postprocessing, radiometabolites analysis, compartmental analysis, kinetic modeling, and PET images generation (Fig. 1) (Nerella et al. 2022).

In preclinical settings, PET plays a critical role in translational research, bridging the gap between laboratory discoveries and clinical applications. This technology has revolutionized drug development by enabling the real-time evaluation of pharmacokinetics, biodistribution, and drug–target interactions (Matthews et al. 2012). For example, PET imaging is instrumental in assessing whether a candidate drug reaches its intended target and produces the desired biological effect (Ametamey et al. 2008). Additionally, PET aids in oncology research by revealing alterations in tumor metabolism and evaluating therapeutic responses, including immunotherapy and combination treatments (Jensėn and Kjaer 2015). Beyond oncology, PET has demonstrated extensive applications in neuroscience, where it helps to unravel complex neurobiological processes. Researchers have utilized PET to study neurotransmitter dynamics, synaptic density, neuroinflammation, and pathological protein aggregates in models of neurodegenerative diseases such as Alzheimer's disease (AD) and Parkinson's disease (Chen et al. 2022). Similarly, in cardiovascular research, PET facilitates the investigation of myocardial metabolism and early detection of atherosclerosis, contributing to the development of targeted interventions (Yao et al. 2012). Despite these advances, several challenges remain. The development of novel radiotracers tailored to specific research needs, coupled with the high costs of PET scanners and operational logistics, limits broader

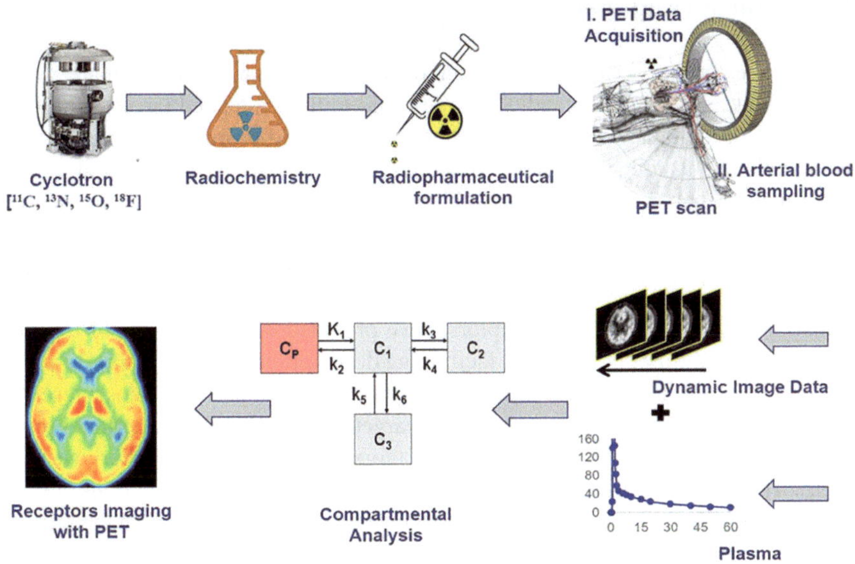

Fig. 1 Schematic representation of receptor imaging with PET (Nerella et al. 2022). First step: generation of positron-emitting radionuclides; Second step: radiochemical synthesis, Third step: preparation of suitable radiopharmaceutical formulations; Fourth step: PET imaging data acquisition; Fifth step: PET data pre- and postprocessing via software technologies, and Sixth step: output of PET images based on radiotracer uptake. (Reprinted from (Nerella et al. 2022) under a Creative Commons Attribution License (CC BY 4.0))

adoption in preclinical studies (Fig. 1). This chapter provides a comprehensive overview of the application of PET in preclinical research, with a focus on its roles in drug development, cancer studies, neurological disorders, and cardiovascular diseases. Through these examples, we highlight the potential of PET in accelerating the transition from preclinical to clinical application.

2 Preclinical PET in Drug Development

The integration of PET in preclinical drug development has significantly advanced the understanding of drug–target interactions and pharmacokinetics. PET enables real-time, quantitative 3-dimensional imaging of biological processes, providing valuable data on how candidate drugs interact with their targets and behave within biological systems. This has made PET an indispensable tool for optimizing drug development pipelines and improving translational outcomes.

2.1 Target Engagement

One of the primary roles of PET in drug development is to validate target engagement, ensuring that candidate drugs bind to their intended targets with high specificity and efficacy. Radiolabeled tracers engineered to mimic or interact with therapeutic agents are administered to animal models (Fig. 1) (Nerella et al. 2022). PET imaging then tracks the tracer's distribution, enabling researchers to visualize and quantify receptor or enzyme occupancy in real time. For example, PET imaging has been widely used to evaluate receptor–ligand interactions in drug candidates that target the central nervous system. Radiolabeled tracers such as [^{11}C]raclopride, which binds to dopamine D2 receptors, have provided direct evidence of receptor occupancy by antipsychotic drugs, facilitating dose optimization during early-phase studies (Jacobs et al. 2003). Similarly, PET has been applied in oncology research by using radiotracers specific to molecular targets such as human epidermal growth factor receptor 2, aiding the development of targeted cancer therapies (Serkova et al. 2020). This ability to directly observe and quantify drug–target interactions ensures that candidate drugs are advancing through the development pipeline on the basis of robust evidence. In addition, PET facilitates the identification of off-target binding of a drug / molecule, i.e., the binding to a target other than the intended primary target. or suboptimal properties, enabling early termination of unsuitable candidates and saving time and resources.

2.2 Pharmacokinetics

PET also provides unparalleled insights into the pharmacokinetics (PK) of drug candidates by visualizing and quantifying the absorption, distribution, metabolism, and excretion (ADME) processes. The use of radiolabeled compounds allows researchers to track the biodistribution of drugs throughout the body in real time, offering precise data on how drugs reach and persist in their target tissues. For example, PET imaging has been utilized to study the hepatic and renal clearance of small molecules, by assessing the glomerular filtration rates and effective renal plasma flow that can guide whether structural modifications of molecules are necessary (Toyama et al. 2021). Radiolabeled analogs such as [^{18}F]fluorothymidine (FLT) have also been employed to study drug uptake and metabolism in specific tissues, such as the liver or brain, providing critical insights into organ-specific dynamics (Cunha et al. 2014). One strength of PET is its ability to generate time-resolved imaging data, enabling the creation of kinetic models that describe drug behavior quantitatively. These models inform dose selection, predict therapeutic windows, and identify potential risks associated with prolonged exposure or insufficient target delivery. For example, in longitudinal studies, PET has been used to assess the sustained release profiles of drug delivery systems, ensuring that the therapeutic concentrations are maintained over time. Furthermore, PET aids in understanding the interaction of drugs with biological barriers such as the blood–brain barrier.

Radiolabeled tracers are used to assess a drug's ability to penetrate these barriers and accumulate in target tissues, a critical step in the development of therapies for central nervous system disorders (Amirrashedi et al. 2020b). In summary, the ability of PET to provide real-time, quantitative data on target engagement and pharmacokinetics makes it an indispensable tool in preclinical drug development. Its applications have facilitated the drug discovery process and increased the likelihood of successful clinical translation.

3 PET in Preclinical Cancer Research

PET imaging has become a cornerstone in preclinical cancer research, providing invaluable insights into tumor biology, metabolic alterations, and therapeutic responses. By leveraging its ability to detect molecular and functional changes in vivo, PET facilitates the development of early detection tools, the understanding of progression of normal cells to cancerous states, and the development of novel treatment strategies.

3.1 Metabolism Imaging

Metabolic imaging is one of the foundational applications of PET in cancer research, leveraging its ability to noninvasively measure altered biochemical processes in tumors. The radiotracers most commonly target at glucose metabolism, amino acid metabolism, nucleoside synthesis, hypoxia (tumor microenvironment) (Table 1). [^{18}F]fluorodeoxyglucose (FDG) remains the most widely used tracer, exploiting the increased glycolysis observed in cancer cells, a hallmark of malignancy. This increased glucose uptake and utilization make [^{18}F]FDG PET an essential tool for studying tumor growth, progression, and response to therapy (Pantel et al. 2018). Recent advancements have expanded metabolic imaging beyond glycolysis. Novel tracers such as [^{11}C]methionine and [^{18}F]FLT allow researchers to explore amino acid metabolism and nucleoside synthesis, providing a broader perspective on tumor proliferation and biosynthetic activity (Abrantes et al. 2020). Imaging of tumor microenvironment features such as hypoxia via tracers such as [^{18}F]fluoroazomycin arabinoside ([^{18}F]FAZA) has provided critical insights into resistance mechanisms, guiding therapeutic strategies targeting hypoxic cells (Chang et al. 2015). In addition

Table 1 Tracers for metabolic imaging in cancer research

Tracer	Target	Reference
[^{18}F]fluorodeoxyglucose	Glucose metabolism	Pantel et al. (2018)
[^{11}C]methionine	Amino acid metabolism	Abrantes et al. (2020)
[^{18}F]fluorothymidine	Nucleoside synthesis	Abrantes et al. (2020)
[^{18}F]fluoroazomycin arabinoside	Hypoxia (tumor microenvironment)	Chang et al. (2015)

to preclinical research, PET-based metabolic imaging is instrumental in translational studies, assisting in the evaluation of novel therapies and identifying biomarkers for clinical trials. For example, combining metabolic imaging with therapeutic interventions has elucidated the metabolic dependencies of specific cancer subtypes, driving the development of precision oncology strategies.

3.2 Immunotherapy Monitoring & Theragnostics

The rapid advancement of immunotherapy has created a demand for imaging techniques that can monitor immune responses and predict therapeutic outcomes. Immuno-PET employs radiolabeled antibodies or immune-specific tracers, offering a noninvasive solution to visualize and quantify immune cell dynamics within the tumor microenvironment (Wei et al. 2020; Kong et al. 2023). Radiotracers targeting immune checkpoint molecules, such as programmed cell death protein 1 (PD-1) and programmed cell death-ligand 1 (PD-L1), have been successfully used to evaluate the presence and activity of immune cells. For example, [^{89}Zr]labeled anti-PD-L1 antibodies enable longitudinal monitoring of checkpoint inhibitor responses, offering critical insights into resistance mechanisms and guiding therapeutic decisions (Bouleau et al. 2022). Theragnostic PET imaging, which integrates diagnostic imaging with targeted therapy, is another transformative approach. Radiopharmaceuticals such as [^{68}Ga]DOTA-D-Phe1-Tyr3-octreotate (DOTA-TATE) and [^{177}Lu]DOTATATE) allow simultaneous tumor visualization and treatment delivery, exemplifying the dual capabilities of theranostic agents (Pantel et al. 2018). In preclinical cancer models, PET imaging has also been used to track the trafficking and efficacy of engineered immune cells, such as chimeric antigen receptor-T cells (CAR-T cells). Radiolabeled tracers enable real-time visualization of the localization and activity of CAR-T cells, providing critical data for optimizing dosing strategies and therapeutic protocols (Abrantes et al. 2020).

4 PET in Preclinical Neurological Research

PET is a powerful tool in neuroscience, enabling precise imaging of brain metabolism, neurotransmitter systems, neuroinflammation, and pathological proteins in neurodegenerative diseases. This section explores the role of PET in preclinical neurological research, focusing on its major applications.

4.1 Brain Metabolism Imaging

Brain metabolism imaging with PET has been fundamental for studying energy utilization and neuronal activity under various physiological and pathological conditions. [^{18}F]FDG PET, a gold standard for glucose metabolism imaging, reveals patterns of altered metabolism in conditions such as AD, ischemic stroke, epilepsy, and traumatic brain injury. Preclinical studies in rodent models of amyloidosis have shown early glucose hypometabolism in the hippocampus and cortex (Tanner et al. 2022). Recent advancements include the use of alternative tracers targeting metabolic molecules and pathways, such as acetate and lactate, to investigate brain energy metabolism in novel contexts. These tracers are expanding the utility of PET in understanding non-glucose-dependent energy pathways and their relevance to diseases such as Parkinson's disease and Huntington's disease.

4.2 Neurotransmitter and Synaptic Density Imaging

PET imaging of neurotransmitter systems has revolutionized the study of brain signaling pathways, particularly in preclinical models of neuropsychiatric and neurodegenerative diseases. Radiolabeled tracers targeting dopamine, serotonin, and glutamate receptors allow for the quantification of neurotransmitter dynamics in the brain under physiological and diseased conditions (Kong et al. 2024a; Wang et al. 2024). For example, PET using [^{11}C]raclopride, or [^{18}F]fallypride has been extensively used to study the in dopamine D2 receptors in animal models (Vučković et al. 2010). Additionally, novel tracers such as [^{11}C]UCB-J, [^{18}F]SDM-8, which target synaptic vesicle protein 2A (SV2A), enable the quantification of synaptic density (Nabulsi et al. 2016). SV2A PET providing valuable biomarkers for identifying changes in brain diseases such as AD (Wang et al. 2024) and schizophrenia (Finnema et al. 2016). These innovations offer a deeper understanding of synaptic integrity and loss, which are critical for studying disease mechanisms (Friston 1998).

4.3 Neuroinflammation Imaging

Neuroinflammation plays a central role in many neurological disorders. Translocator protein (TSPO) PET tracers such as first-generation [^{11}C]PK11195 is mostly widely used but has limitations including high nonspecific binding, low brain penetration, signal-to-background ratio and sensitivity (Papadopoulos et al. 2006; Kreisl et al. 2020). Second-generation TSPO tracers such as [^{18}F]DPA714 and [^{11}C]PBR28 improved upon the limitations of [^{11}C]PK11195 but were affected by TSPO rs6971 polymorphism which affected the binding (Papadopoulos et al. 2006; Kreisl et al. 2020). Third-generation tracers such as [^{18}F]GE180, [^{11}C]ER176, and [^{18}F]CB251 that circumvent the limitation in TSPO rs6971 polymorphism, have been developed and applied in preclinical and clinical research (Papadopoulos et al. 2006; Kreisl et al. 2020). Preclinical models of multiple sclerosis and traumatic brain injury have demonstrated the ability of TSPO-PET to correlate neuroinflammation with disease progression (Jain et al. 2020; Zhou et al. 2021). However challenges remain for targeting TSPO as TSPO is expressed not only as microglia as well as on astrocytes and endothelial cells and vascular smooth muscle cells, and is difficult to associate with a subtypes of microglia (Jain et al. 2020; Zhou et al. 2021; Gui et al. 2020).

Given the limitations of TSPO tracers, other targets, such as the purinergic P2X7 receptor (P2X7R) and triggering receptor expressed on myeloid cells 2 (TREM2), are emerging. P2X7R is a receptor highly involved in the regulation of microglial activation and neuroinflammatory responses. PET imaging using P2X7R-specific tracers has shown promise in mapping neuroinflammation with greater specificity to activated microglia and their signaling pathways (Illes et al. 2019; Kong et al. 2024b). Similarly, TREM2, which is expressed on microglia, plays a critical role in regulating microglial responses to injury and maintaining homeostasis. Mutations in TREM2 are associated with increased susceptibility to late-onset AD. PET imaging of TREM2 provides insights into its role in modulating microglial activity during neurodegeneration and inflammation (Beaino et al. 2021). These alternative targets highlight the potential for complementing or replacing TSPO-PET imaging to better understand the dynamic and complex nature of neuroinflammation. Emerging tracers, such as the monoamine oxidase B tracer (MAO-B) tracers [^{11}C]deuterium-l-deprenyl ([^{11}C-DED) (Rodriguez-Vieitez et al. 2015), [^{18}F]F-DED (Ballweg et al. 2023), [^{18}F]SMBT-1 (Kong et al. 2024c, d), allows researchers to comprehensively evaluate the dynamic astrocytic activation in vivo.

4.4 Amyloid and Tau Imaging in Neurodegenerative Disease Models

PET imaging of amyloid-beta (Aβ) and tau proteins has transformed preclinical research in AD and other neurodegenerative diseases. Amyloid tracers such as [^{11}C]pittsburgh compound B (PIB) and [^{18}F]florbetapir, [^{18}F]florbetaben, and [^{18}F]flutemetamol enable the visualization of Aβ plaques, providing powerful tools to investigate Aβ deposition and its role in disease progression (Chisholm and Hunter 2024). These tracers provide insights into the temporal and spatial dynamics of protein aggregation in disease models, such as the widely-used 3 × Tg, 5 × FAD, Tg2576 and APP/PS1 mice (Fig. 2) as well as rat models (Ballweg et al. 2023; Gnörich et al. 2024; Meier et al. 2022). Using [^{18}F]florbetaben, sufficient difference in the brain uptake can be detected in APP/PS1 and WT mice aged 13–16 months across different scanners and imaging sites (Fig. 2) (Gnörich et al. 2024), indicating the potential for multicenter studies in the future.

Moreover, tau tracers such as [^{18}F]flortaucipir, [^{18}F]THK-5117, [^{18}F]PM-PBB3, and [^{18}F]PI-2620 allow imaging of tau pathology in AD patients as well as at early stages (Saint-Aubert et al. 2017). Moreover, [^{18}F]PM-PBB3 exhibits high sensitivity for identifying tau deposits in AD as well as in primary tauopathies; [^{18}F]PI-2620 provides strong binding characteristics for 4-repeat tau pathologies; [^{18}F]MK6240 offers superior selectivity for tau fibrils in AD, making it an excellent tracer for tau detection and progression monitoring, however not yet been reported in disease animal models (Maschio and Ni 2022). Tau PET using [^{18}F]PM-PBB3, [^{11}C]PBB3, [^{18}F]PI-2620 has revealed the spread of tau pathology in relation to neuronal loss

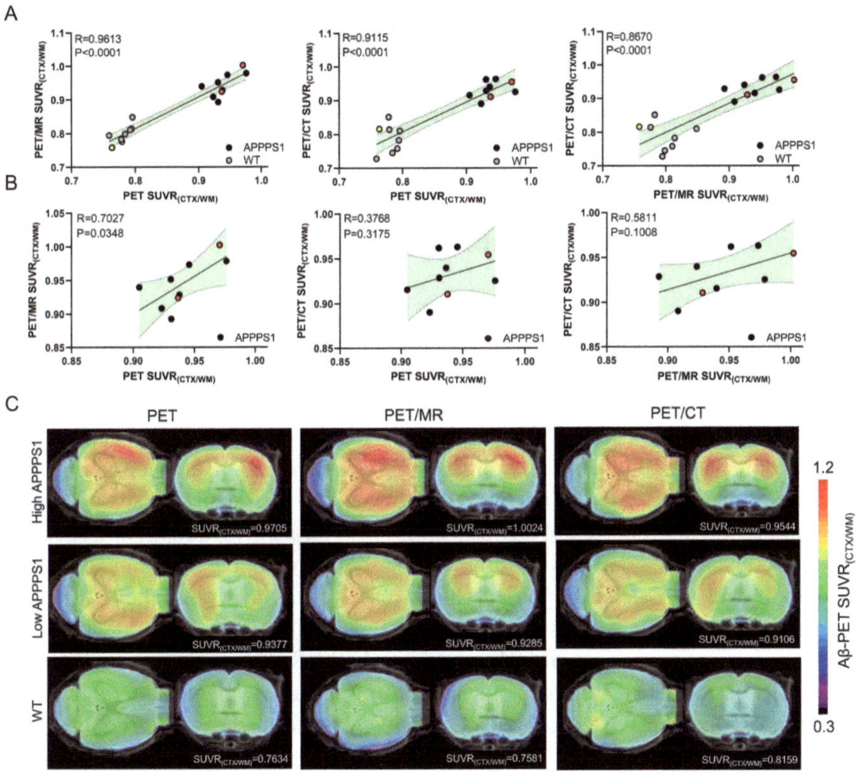

Fig. 2 Amyloid imaging in mouse model of amyloidosis (**a**) Correlation analysis between the [18F]florbetaben $SUVR_{(CTX/WM)}$ values determined by the three different scanners in APP/PS1 ($n = 9$) and WT mice ($n = 8$) aged 13–16 months. Scatter plots show strong correlations between all the modalities. (**b**) Dedicated APP/PS1 analysis shows significant correlation between PET vs. PET/MR ($p < 0.05$), whereas PET vs. PET/CT and PET/MR vs. PET/CT did not meet the 5% significance level. R indicates Pearson's coefficient of correlation. Red dots highlight two distinct APPPS1 mice, and yellow dots highlight one respective WT mouse visualized in panel c. (**c**) Representative [18F]florbetaben PET SUVR images upon an MR template of two transgenic APP/PS1 mice compared to one WT mouse show similar $SUVR_{(CTX/WM)}$ values and visual image impressions across the three PET systems. (Reprinted from (Gnörich et al. 2024) under a Creative Commons Attribution License (CC BY 4.0))

in transgenic mouse models of 4-repeat tau, such as P301L and P301S mice (Saint-Aubert et al. 2017; Ni et al. 2018; Ishikawa et al. 2018), providing insights into the disease mechanisms and potential therapeutic targets. Combining these tracers with metabolic and neuroinflammation imaging provides a comprehensive view of the disease process and supports therapeutic development.

5 PET in Preclinical Cardiovascular Research

PET has been established as an essential tool in cardiovascular research, providing detailed insights into cardiac metabolism and the pathological processes underlying atherosclerosis.

5.1 Cardiac Metabolism Imaging

PET imaging of cardiac metabolism provides critical information on myocardial energy utilization, which is pivotal in understanding cardiac pathophysiology. The use of [^{18}F]FDG PET allows researchers to assess glucose metabolism in the myocardium, a process often altered in diseases such as heart failure and myocardial infarction (Ribeiro et al. 2022). High-resolution cardiac PET/CT scanners have enabled precise imaging of left ventricular glucose metabolism, facilitating evaluations of the ejection fraction and stroke volume in rodent models. Dynamic imaging using [^{11}C]acetate PET has also been utilized to quantify myocardial oxygen consumption and cardiac efficiency in both preclinical and clinical settings (Hansson et al. 2018). This approach helps researchers understand the relationship between metabolic alterations and contractile function, particularly under conditions such as ischemia or hypoxia (Hansson et al. 2018). Furthermore, multimodal PET/MRI has provided simultaneous insights into metabolism and cardiac function, enabling comprehensive assessments of the response of the heart to stress or injury. This integration is particularly valuable for studying dynamic metabolic changes alongside structural alterations in the myocardium (Barton et al. 2018).

5.2 Atherosclerosis Imaging

PET imaging has advanced the understanding of atherosclerosis by enabling the visualization of key biological processes, such as inflammation, microcalcification, and hypoxia, within vulnerable plaques, that are atherosclerotic plaques in arteries prone to rupture, leading to serious cardiovascular events. [^{18}F]FDG PET, widely used as a surrogate marker for plaque inflammation, has been instrumental in characterizing high-risk plaques prone to rupture. Preclinical studies have shown that [^{18}F]FDG uptake is strongly correlated with macrophage activity and plaque vulnerability (Sriranjan et al. 2019). Emerging tracers, such as [^{18}F]sodium fluoride (NaF), [^{18}F]FLT for proliferation, [^{68}Ga]pentixafor (Hyafil et al. 2017), [^{18}F]AlF-NOTA-SC (Spahn et al. 2024) and [^{68}Ga]mCXCL12 (Zacherl et al. 2021) (Fig. 3) for chemokine receptor CXCR4, offers a noninvasive approach to study alterations in the infarct region or early atherosclerotic changes. In addition, recent developments in macrophage-targeted PET tracers have provided deeper insights into the inflammatory processes driving atherosclerosis (Li et al. 2020). These tracers

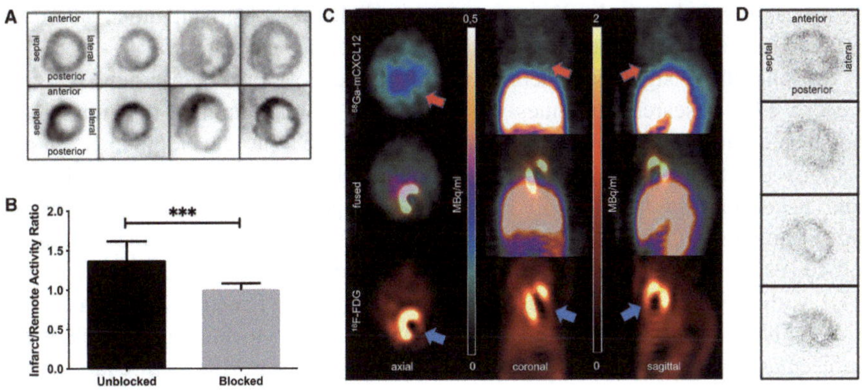

Fig. 3 Specificity of [^{68}Ga]mCXCL12. (**a**) Unblocked slices with [^{68}Ga]mCXCL12 accumulation in the infarcted heart in the anterior and antero-septal wall (bottom row). After blocking with CXCR4 antibody sections show no visible [^{68}Ga]mCXCL12 uptake anymore (top row). (**b**) Significantly higher [^{68}Ga]mCXCL12 myocardial infarct-to-remote activity ratios as compared to the unblocked slices. All data represent mean ± SD. (**c**) PET image on day 2 shows no visible [^{68}Ga]mCXCL12 uptake in the infarcted myocardium after Co-Injection with excess of native mCXCL12. [^{18}F]FDG PET showing the viable myocardium (red arrow: no [^{68}Ga]mCXCL12 uptake; blue arrow: infarcted myocardium). (**d**) Autoradiography on day 2 after Co-Injection with excess of native mCXCL12 shows no significant [^{68}Ga]mCXCL12 uptake (top to bottom shows representative slides from the base to the apex of the heart). (Reprinted from (Zacherl et al. 2021) under a Creative Commons Attribution License (CC BY 4.0))

have been critical in preclinical studies exploring the efficacy of anti-inflammatory therapies and their potential to reduce plaque vulnerability (Li et al. 2020).

6 Conclusions and Future Perspectives

PET has proven to be an indispensable tool in preclinical biomedical research, offering unparalleled insights into a wide array of physiological, metabolic, and pathological processes. By providing precise and quantitative imaging at the molecular level, PET has revolutionized research in drug development, cancer biology, neuroscience, and cardiovascular disease, bridging the gap between basic science and clinical applications.

In drug development, PET has significantly enhanced our ability to validate drug-target engagement, optimize dosing strategies, and monitor pharmacokinetics. Its application has accelerated the development pipeline, offering a robust platform for evaluating therapeutic efficacy with precision. In cancer research, PET has transformed our understanding of tumor metabolism, immune responses, and theranostic applications, which integrate diagnosis and treatment to improve outcomes. PET has also been pivotal in neuroscience, enabling breakthroughs in brain metabolism, neurotransmitter dynamics, neuroinflammation, and the imaging of amyloid and tau proteins in neurodegenerative disease models. Similarly, in cardiovascular research,

PET has provided critical insights into myocardial metabolism and atherosclerosis, guiding innovative therapeutic strategies and preventive measures.

Despite these significant achievements in the development of new imaging tracers, PET research continues to evolve, and holding immense potential for advancing biomedical research and clinical practice. The development of novel radiotracers that target specific immune cell subsets, rare metabolic pathways, and novel protein targets will expand the scope of PET applications. These advancements will enable the exploration of previously inaccessible biological processes, further enhancing our understanding of complex diseases. Integrating PET with other imaging modalities, such as MRI and CT, will also play a crucial role in the future. The complementary strengths of these technologies, such as the molecular specificity of PET with anatomical detail from MRI, will enable more comprehensive assessments of disease processes and therapeutic responses. These multimodal systems are expected to become increasingly accessible, transforming how researchers and clinicians approach diagnostics and treatment planning.

AI and machine learning are poised to revolutionize PET imaging by improving image reconstruction, quantification accuracy, and the detection of subtle patterns in complex datasets. AI-driven tools can accelerate data analysis and uncover previously unrecognized insights, making PET imaging faster, more reliable, and more widely applicable. Additionally, the miniaturization of PET systems and the development of cost-effective compact scanners for small animal imaging will democratize access to this powerful technology, enabling more laboratories to contribute to cutting-edge research. Another exciting area of growth is in theranostics, which combines diagnostic imaging with targeted radiopharmaceutical therapy. PET-guided theranostic agents are promising in oncology, and their applications are likely to expand into other areas, such as neurology and cardiology. These innovations hold the potential to deliver highly personalized treatment strategies, minimizing side effects while maximizing therapeutic benefits. There are a few remaining challenges in the widespread adoption of PET in preclinical research, including high costs, the complexity of radiotracer production, and the need for specialized infrastructure and expertise. Addressing these barriers will require advances in radiochemistry, automation, and scanner technology, along with collaborative efforts among academia, industry, and policymakers to promote accessibility and innovation.

In conclusion, PET has become a cornerstone of modern biomedical research and clinical translation. Its ability to provide real-time, noninvasive, and quantitative imaging of molecular processes makes it an unparalleled tool for advancing our understanding of human biology and disease. As technology continues to evolve, the role of PET in driving precision medicine, developing novel therapies, and improving patient outcomes is set to expand further, ensuring its place as a transformative force in the life sciences for years to come.

Acknowledgement RN acknowledges Swiss Center for applied human toxicology (SCAHT-AP22-01).

Conflict of Interest The authors have no conflict of interest.

References

Abrantes AM, Pires AS, Monteiro L, Teixo R, Neves AR, Tavares NT et al (2020) Tumour functional imaging by PET. Biochim Biophys Acta (BBA)-Mol Basis Dis 1866(6):165717

Ametamey SM, Honer M, Schubiger PA (2008) Molecular imaging with PET. Chem Rev 108(5):1501–1516

Amirrashedi M, Zaidi H, Ay MR (2020a) Advances in preclinical PET instrumentation. PET Clin 15(4):403–426

Amirrashedi M, Zaidi H, Ay MR (2020b) Towards quantitative small-animal imaging on hybrid PET/CT and PET/MRI systems. Clin Translat Imaging 8:243–263

Ballweg A, Klaus C, Vogler L, Katzdobler S, Wind K, Zatcepin A et al (2023) [(18)F]F-DED PET imaging of reactive astrogliosis in neurodegenerative diseases: preclinical proof of concept and first-in-human data. J Neuroinflammation 20(1):68

Barton G, Vildberg L, Goss K, Aggarwal N, Eldridge M, McMillan A (2018) Simultaneous determination of dynamic cardiac metabolism and function using PET/MRI. J Nucl Cardiol:1–12

Beaino W, Janssen B, Vugts DJ, de Vries HE, Windhorst AD (2021) Towards PET imaging of the dynamic phenotypes of microglia. Clin Exp Immunol 206(3):282–300

Bouleau A, Nozach H, Dubois S, Kereselidze D, Chevaleyre C, Wang CI et al (2022) Optimizing immuno-PET imaging of tumor PD-L1 expression: pharmacokinetic, biodistribution, and dosimetric comparisons of (89)Zr-labeled anti-PD-L1 antibody formats. J Nucl Med 63(8):1259–1265

Chang E, Liu H, Unterschemmann K, Ellinghaus P, Liu S, Gekeler V et al (2015) 18F-FAZA PET imaging response tracks the reoxygenation of tumors in mice upon treatment with the mitochondrial complex I inhibitor BAY 87-2243. Clin Cancer Res 21(2):335–346

Chen B, Marquez-Nostra B, Belitzky E, Toyonaga T, Tong J, Huang Y et al (2022) PET imaging in animal models of Alzheimer's disease. Front Neurosci 16:872509

Chisholm TS, Hunter CA (2024) A closer look at amyloid ligands, and what they tell us about protein aggregates. Chem Soc Rev 53(3):1354–1374

Cunha L, Horvath I, Ferreira S, Lemos J, Costa P, Vieira D et al (2014) Preclinical imaging: an essential ally in modern biosciences. Mol Diagn Ther 18:153–173

Finnema SJ, Nabulsi NB, Eid T, Detyniecki K, Lin S-f, Chen M-K et al (2016) Imaging synaptic density in the living human brain. Sci Transl Med 8(348):348ra396

Friston KJ (1998) The disconnection hypothesis. Schizophr Res 30(2):115–125

Gnörich J, Koehler M, Wind-Mark K, Klaus C, Zatcepin A, Palumbo G et al (2024) Towards multicenter β-amyloid PET imaging in mouse models: a triple scanner head-to-head comparison. NeuroImage 297:120748

Gui Y, Marks JD, Das S, Hyman BT, Serrano-Pozo A (2020) Characterization of the 18 kDa translocator protein (TSPO) expression in post-mortem normal and Alzheimer's disease brains. Brain Pathol 30(1):151–164

Hansson NH, Harms HJ, Kim WY, Nielsen R, Tolbod LP, Frøkiær J et al (2018) Test–retest repeatability of myocardial oxidative metabolism and efficiency using standalone dynamic 11C-acetate PET and multimodality approaches in healthy controls. J Nucl Cardiol 25(6):1929–1936

Hyafil F, Pelisek J, Laitinen I, Schottelius M, Mohring M, Döring Y et al (2017) Imaging the cytokine receptor CXCR4 in atherosclerotic plaques with the radiotracer (68)Ga-Pentixafor for PET. J Nucl Med 58(3):499–506

Illes P, Rubini P, Huang L, Tang Y (2019) The P2X7 receptor: a new therapeutic target in Alzheimer's disease. Expert Opin Ther Targets 23(3):165–176

Ishikawa A, Tokunaga M, Maeda J, Minamihisamatsu T, Shimojo M, Takuwa H et al (2018) In vivo visualization of tau accumulation, microglial activation, and brain atrophy in a mouse model of tauopathy rTg4510. J Alzheimers Dis 61(3):1037–1052

Jacobs AH, Li H, Winkeler A, Hilker R, Knoess C, Rueger A et al (2003) PET-based molecular imaging in neuroscience. Eur J Nucl Med Mol Imaging 30:1051–1065

Jain P, Chaney AM, Carlson ML, Jackson IM, Rao A, James ML (2020) Neuroinflammation PET imaging: current opinion and future directions. J Nucl Med 61(8):1107–1112

Jensen MM, Kjaer A (2015) Monitoring of anti-cancer treatment with 18F-FDG and 18F-FLT PET: a comprehensive review of pre-clinical studies. Am J Nucl Med Mol Imaging 5(5):431–456

Kong Y, Xie F, Zhang Z, Wang S, Zhang Y, Di Y et al (2023) Evaluation of novel anti-CEACAM6 antibody-based conjugates for radioimmunotheranostics of pancreatic ductal adenocarcinoma. Eur Radiol 33(10):7077–7088

Kong Y, Cao L, Xie F, Wang X, Zuo C, Shi K et al (2024a) Reduced SV2A and GABAA receptor levels in the brains of type 2 diabetic rats revealed by [18F]SDM-8 and [18F]flumazenil PET. Biomed Pharmacother 172:116252

Kong Y, Cao L, Wang J, Zhuang J, Liu Y, Bi L et al (2024b) Increased cerebral level of P2X7R in a Tauopathy mouse model by PET using [(18)F]GSK1482160, ACS Chem Neurosci 15(11):2112–2120

Kong Y, Maschio CA, Shi X, Xie F, Zuo C, Konietzko U et al (2024c) Relationship between reactive astrocytes, by [(18)F]SMBT-1 imaging, with amyloid-beta, tau, glucose metabolism, and TSPO in mouse models of Alzheimer's disease. Mol Neurobiol 61(10):8387–8401

Kong Y, Cao L, Wang J, Zhuang J, Xie F, Zuo C et al (2024d) In vivo reactive astrocyte imaging using [18F]SMBT-1 in tauopathy and familial Alzheimer's disease mouse models: a multi-tracer study. J Neurol Sci 123079:123079

Kreisl WC, Kim M-J, Coughlin JM, Henter ID, Owen DR, Innis RB (2020) PET imaging of neuroinflammation in neurological disorders. Lancet Neurol 19(11):940–950

Li X, Rosenkrans Z, Wang J, Cai W (2020) PET imaging of macrophages in cardiovascular diseases. Am J Transl Res 12(5):1491–1514

Maschio C, Ni R (2022) Amyloid and tau positron emission tomography imaging in Alzheimer's disease and other tauopathies. Front Aging Neurosci 14:838034

Matthews PM, Rabiner EA, Passchier J, Gunn RN (2012) Positron emission tomography molecular imaging for drug development. Br J Clin Pharmacol 73(2):175–186

Meier SR, Sehlin D, Roshanbin S, Falk VL, Saito T, Saido TC et al (2022) (11)C-PiB and (124)I-antibody PET provide differing estimates of brain amyloid-β after therapeutic intervention. J Nucl Med 63(2):302–309

Nabulsi NB, Mercier J, Holden D, Carré S, Najafzadeh S, Vandergeten M-C et al (2016) Synthesis and preclinical evaluation of 11C-UCB-J as a PET tracer for imaging the synaptic vesicle glycoprotein 2A in the brain. J Nucl Med 57(5):777–784

Nerella SG, Singh P, Sanam T, Digwal CS (2022) PET molecular imaging in drug development: the imaging and chemistry perspective. Front Med 9:812270

Ni R, Ji B, Ono M, Sahara N, Zhang MR, Aoki I et al (2018) Comparative in vitro and in vivo quantifications of pathologic tau deposits and their association with neurodegeneration in tauopathy mouse models. J Nucl Med 59(6):960–966

Nwabufo CK, Aigbogun OP (2022) Diagnostic and therapeutic agents that target alpha-synuclein in Parkinson's disease. J Neurol 269(11):5762–5786

Pantel AR, Ackerman D, Lee S-C, Mankoff DA, Gade TP (2018) Imaging cancer metabolism: underlying biology and emerging strategies. J Nucl Med 59(9):1340–1349

Papadopoulos V, Baraldi M, Guilarte TR, Knudsen TB, Lacapère J-J, Lindemann P et al (2006) Translocator protein (18 kDa): new nomenclature for the peripheral-type benzodiazepine receptor based on its structure and molecular function. Trends Pharmacol Sci 27(8):402–409

Ribeiro FM, Correia PMM, Santos AC, Veloso JFCA (2022) A guideline proposal for mice preparation and care in 18F-FDG PET imaging. EJNMMI Res 12(1):49

Rodriguez-Vieitez E, Ni R, Gulyás B, Tóth M, Häggkvist J, Halldin C et al (2015) Astrocytosis precedes amyloid plaque deposition in Alzheimer APPswe transgenic mouse brain: a correlative positron emission tomography and in vitro imaging study. Eur J Nucl Med Mol Imaging 42(7):1119–1132

Saint-Aubert L, Lemoine L, Chiotis K, Leuzy A, Rodriguez-Vieitez E, Nordberg A (2017) Tau PET imaging: present and future directions. Mol Neurodegener 12:1–21

Serkova N, Glunde K, Haney C, Farhoud M, De Lille A, Redente E et al (2020) Preclinical applications of multi-platform imaging in animal models of cancer. Cancer Res 81:1189–1200

Spahn MA, Loy TV, Celen S, Koole M, Deroose CM, Cawthorne C et al (2024) Selective PET imaging of CXCR4 using the Al(18)F-labeled antagonist LY2510924. Eur J Nucl Med Mol Imaging 52:1723–1738

Sriranjan R, Tarkin J, Evans N, Le E, Chowdhury M, Rudd J (2019) Atherosclerosis imaging using PET: insights and applications. Br J Pharmacol 178:2186–2203

Tanner J, Iaccarino L, Edwards L, Asken B, Gorno-Tempini M, Kramer J et al (2022) Amyloid, tau and metabolic PET correlates of cognition in early and late-onset Alzheimer's disease. Brain 145:4489–4505

Toyama Y, Werner RA, Ruiz-Bedoya CA, Ordonez AA, Takase K, Lapa C et al (2021) Current and future perspectives on functional molecular imaging in nephro-urology: theranostics on the horizon. Theranostics 11(12):6105–6119

Vučković MG, Li Q, Fisher B, Nacca A, Leahy RM, Walsh JP et al (2010) Exercise elevates dopamine D2 receptor in a mouse model of Parkinson's disease: in vivo imaging with [^{18}F]fallypride. Mov Disord 25(16):2777–2784

Wang J, Huang Q, He K, Li J, Guo T, Yang Y et al (2024) Presynaptic density determined by SV2A PET is closely associated with postsynaptic metabotropic glutamate receptor 5 availability and independent of amyloid pathology in early cognitive impairment. Alzheimers Dement 20:3876–3888

Wei W, Rosenkrans ZT, Liu J, Huang G, Luo QY, Cai W (2020) ImmunoPET: concept, design, and applications. Chem Rev 120(8):3787–3851

Yao R, Lecomte R, Crawford ES (2012) Small-animal PET: what is it, and why do we need it? J Nucl Med Technol 40(3):157–165

Zacherl MJ, Todica A, Wängler C, Schirrmacher R, Hajebrahimi MA, Pircher J et al (2021) Molecular imaging of cardiac CXCR4 expression in a mouse model of acute myocardial infarction using a novel (68)Ga-mCXCL12 PET tracer. J Nucl Cardiol 28(6):2965–2975

Zhou R, Ji B, Kong Y, Qin L, Ren W, Guan Y et al (2021) PET imaging of neuroinflammation in Alzheimer's disease. Front Immunol 12:739130

Applications of Single-Photon Emission Computed Tomography (SPECT) in Preclinical Research

Alan Miranda, Stijn De Schepper, and Filipe Elvas

Abstract

Preclinical Single-Photon Emission Computed Tomography (SPECT) is an imaging tool that has gained importance over the years thanks to technological improvements and to the development of novel radiotracers to image a wide variety of biologically relevant targets. Preclinical SPECT scanners offer improved spatial resolution and sensitivity compared with their clinical counterparts and allow in vivo experimentation in animal models of disease. Research using several radiotracers targeting neurological and oncological applications has provided important insight into several diseases and advanced drug development. With the advent of radiotheranostics, the role of preclinical SPECT has become even more significant for providing radiation dosimetry estimates.

Keywords

SPECT · Preclinical · Neurology · Oncology · Theranostics

A. Miranda · F. Elvas (✉)
Molecular Imaging Center Antwerp, University of Antwerp, Antwerp, Belgium
e-mail: filipe.elvas@uantwerpen.be

S. De Schepper
Molecular Imaging Center Antwerp, University of Antwerp, Antwerp, Belgium

Department of Nuclear Medicine, Antwerp University Hospital, Antwerp, Belgium

I. F. Cengiz et al. (eds.), *Bioimaging Modalities in Bioengineering*,
Biomaterials, Bioengineering and Sustainability 9,
https://doi.org/10.1007/978-3-031-99410-4_16

1 Introduction

1.1 General Principle

Molecular imaging provides a direct probe into physiological processes at work by attaching a radioactive isotope to a targeting vector. A popular analogy is that the isotope is the package, and the vector molecule is the GPS guiding the package to its destination, the binding receptors. If the vector molecule is injected in a trace amount, there is no noticeable change in binding properties due to the injected molecule which is in competition with the normal physiological process. Therefore, the binding of the molecules and thus the isotope distribution can be considered proportional to the true physiological conditions.

1.2 Decay Characteristics

X-ray emission is governed by the quantum mechanics of the atomic electrons. Similarly, gamma emission is governed by quantum mechanics of the atomic nuclei. The photon energy is determined by the difference in energy between the original state and the one to which it decays. The energy for a given decay is thus quantized and the resulting gamma photons always have the same energy. For ^{99m}Tc—the most used isotope in Single-Photon Emission Computed Tomography (SPECT) imaging—this is always 140.5 keV. Therefore, we can acquire an image of the ^{99m}Tc distribution if we discriminate the photons from the nuclear decay from others by only considering photons of that particular energy, reducing the noise from other photons that might reach the detector.

Each nucleus of an isotope has at any moment an equal and constant probability of undergoing decay. This results in an exponential decay in the radioactivity of an ensemble of nuclei:

$$A(t) = A(0) \times e^{-\lambda t}, \tag{1}$$

where *A(t)*, the activity at time *t*, is proportional to the activity at time 0 and decays exponentially with λ the decay constant. The time after which half of the activity is decayed is called the half-life, $T_{1/2} = -\ln 2 \times \lambda$. The reconstruction process requires sufficient statistics to obtain a qualitative image. If we need to acquire an image after a certain time post injection, it is essential that the decay is taken into account so that enough activity is available to acquire sufficient photons. The emission of photons has no preferred direction. The resulting flux of photons from the decay of a large number of nuclei follows the inverse-square law where the number of photons that can be measured is proportional to $\frac{1}{R^2}$. It is therefore important to minimize the distance between source and detector to obtain optimal statistics.

1.3 Interaction with Matter

Photons travel through matter with a constant probability of interacting with their environment. The equation for the number of particles after a distance x traveled in a homogeneous medium is similar to Eq. (1):

$$I(x) = I(0) \times e^{-\mu x}, \tag{2}$$

where $I(x)$ is the beam intensity at thickness x and μ is the attenuation coefficient. The attenuation coefficient depends on both the photon and the matter through which the photons move. The attenuation coefficient is the sum of the three interactions: photo-electric, (in)coherent scattering and pair production.

At low energy, the photo-electric effect is dominant. At these energies, the photons have the required energy to directly interact with individual orbital electrons and send them to a higher energy orbital or freeing them completely. At intermediate energy, the photons mostly scatter off of electrons either elastic (Rayleigh scatter) or inelastic (Compton scatter). When photon energy exceeds 1.022 MeV, an electron-positron pair can be created when the photon interacts with either the electrons or the nuclei. All the previous interactions depend on the electrons of the absorber through which the photons travel. It should therefore not come as a surprise that the atomic number is a large influence on the attenuation. Materials are generally electrically neutral, therefore higher atomic number absorbers have a larger density of associated electrons.

1.4 Detection

The goal of SPECT imaging is to collect the gamma photons and convert these photons to an electronic signal which can be subsequently converted into an image. The first step towards conversion is stopping the incoming photons. A detector must therefore have a high stopping power, which correlates with the attenuation coefficient. The conversion can be done either indirectly or directly. Some materials have the property that they can absorb high energy photons, while emitting lower energy photons. This process is called scintillation. The most used scintillator in SPECT imaging is NaI(Tl), where a NaI crystal is doped with Tl-atoms. When electrons are excited by the incoming photons, they become trapped in the electron traps created by the Tl. When they decay from these traps, low energy photons are emitted. These photons can be collected by photomultiplier tubes (PMT's) and converted to electrical signals. It is obvious why this detection mode is considered indirect.

Direct conversion is possible using semiconductors. In these semiconductors the photons directly create charge carriers (electron-hole pairs) which lead to an electrical signal. CdZnTe (CZT) has seen a large increase in popularity in clinical systems (Van den Wyngaert et al. 2020). Still questions remain regarding the cost-effectiveness of such systems. In the preclinical ecosystem there has yet to be a breakthrough of commercial CZT systems. For deeper knowledge on radiation detection, we recommend the reference textbook on the matter by Knoll (2010).

1.5 Shielding

There is a continuous background of ionizing radiation (and non-ionizing photons) around us, even when no tracer is present. This radiation does not contribute to the image and must be blocked from reaching the detector. For this application, high atomic number and high-density materials are best suited. Most common are lead and tungsten. The ability of a material to be a shield for photons of a certain energy is captured in the half-value thickness (HVT) or tenth-value thickness (TVT), the thickness after which the beam intensity is half or one tenth of its original intensity. For example, when comparing the HVT of water to the TVT of lead, 4.6 cm and 0.9 mm respectively, you can appreciate the compactness of Pb shielding and understand its omnipresence in radiation protection.

1.6 Collimation

Spatial information about the origin of the photons reaching the detector is essential to obtain an image. In this energy range it is impossible to have a lens, therefore the alternative choice is made to block all photons but the ones from certain directions. A classic example is the parallel-hole collimator where the idea is that only photons originating perpendicular to the detector surface reach the detector (Fig. 1a). Collimation is a suboptimal solution to the problem. First, the high-density material causes a shower of scattered photons. Second, the finite size of the holes exposes more than only the line perpendicular to the detector surface. This can be improved by increasing the height of the septa in between holes (Fig. 1b). Less

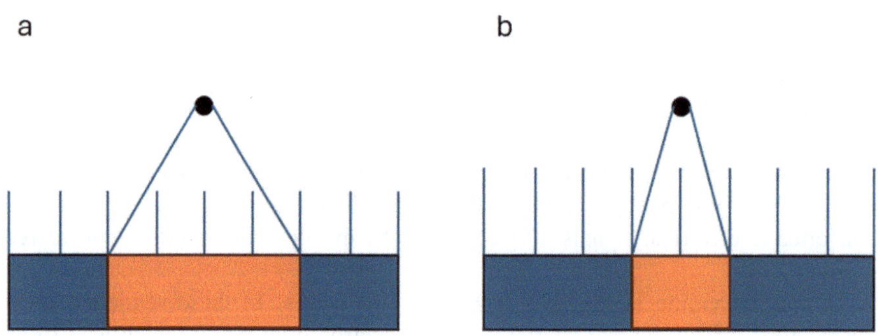

Fig. 1 (**a**) The cross-section of a parallel-hole collimator with a point source. (**b**) The same source-detector configuration, but the height of the septa is increased which reduces the part of the detector that is exposed

detector is exposed, and spatial information is more accurate as a result. However, this will lower the photons reaching the detector and thus the system sensitivity. Collimator design will always be a compromise of sensitivity and spatial resolution: decreasing sensitivity will increase noise and/or acquisition time, while decreasing spatial resolution will increase photons being misplaced outside the true source region. This spill-out effect is called the partial volume effect. The collimator design is often characterized by its response to exposure to a point source: the point spread function (PSF). The PSF of the collimator in Fig. 1b is narrower than that of the collimator in Fig. 1a.

1.7 Radiochemistry

With regard to the preparation of SPECT radiopharmaceuticals, radiochemistry approaches include both the use of bifunctional chelating agents to incorporate metallic radionuclides (e.g., ^{99m}Tc, ^{111}In, and ^{67}Ga), and covalent labeling methods to incorporate nonmetallic radionuclides (e.g., ^{123}I). Single-photon emitters with half-lives longer than those of commonly used PET radionuclides, enable the use of more complex radiolabeling procedures, as well as prolonged imaging protocols. These aspects are of particular relevance for the radiolabeling of antibodies, antibody fragments and peptides, radioligand vectors that benefit from longer reaction times under milder conditions, and imaging at longer time points to accommodate their long circulation half-life.

2 Preclinical SPECT Systems

In order to perform preclinical research in small animals with SPECT radiotracers, scanners must be adapted to the needs of small animal imaging. The requirements of preclinical SPECT scanners are more demanding than in clinical SPECT scanners. To observe smaller structures, e.g., in rodents, the spatial resolution should be in the order of ~1 mm or lower. Partial volume effect is directly related to spatial resolution and therefore its improvement also produces images with less partial volume effect. This in turn improves quantification in the small structures of small animals. This also offers the opportunity to precisely measure the expression and function of specific molecules in small tumors with greater detail. Sensitivity should also be higher in order to obtain good quality images with lower injected doses, often required to avoid excessive radiation to the animal, or to avoid a pharmacological effect of the radiotracer. These requirements have been handled in several ways by different scanner vendors and research labs, using, for example, different collimators, different detector materials and different detector geometries.

The typical SPECT scanner configuration used in clinical devices consists of parallel hole collimators coupled to scintillation crystal detectors. The signal from the scintillator crystal is then converted to an electric signal using photomultipliers tubes. The field of view (FOV) of the detectors usually covers the subject partially, and they must be rotated around the patient to cover the complete set of projections needed to reconstruct the tomographic image. Different types of collimators, scintillator crystals, and photomultipliers technologies exist that can be used to construct SPECT systems with different spatial resolutions and sensitivities. The geometry of the setup can also be designed in several ways to optimize sensitivity and spatial resolution given the chosen hardware. Below different technologies used in preclinical SPECT scanners are detailed.

Most preclinical studies combine the use of SPECT with CT or MRI to provide an anatomical reference and for attenuation and scatter correction. However, CT and MRI can offer many advanced features beyond anatomical information. The development of advanced CT and MRI techniques has enabled to obtain functional and physiological tumor information such as oxygenation levels, cellularity and vascularity that can be related to the expression and function of specific biomarkers as measured by SPECT. Hybrid imaging, including the incorporation of CT or MRI in small animal SPECT, can additionally contribute to correlate focal radiopharmaceutical uptake on the SPECT image within a specific organ or tissue location. Importantly, the incorporation of CT or MRI in SPECT image acquisition brings several advantages such as a more accurate interpretation of the SPECT data. When compared with CT, MRI provides superior soft tissue contrast and imposes no ionizing radiation. Improved soft-tissue contrast would enable improved delineation and follow up of tumor growth in orthotopic or genetically engineered spontaneous tumor models, in which the tumor is either implanted in the original anatomical region of the body or spontaneously grows in an unknown region of the body.

2.1 Hardware

The parallel hole collimator from Fig. 1 is usually suboptimal to obtain the desired spatial resolution and sensitivity in small animal imaging. Pinhole collimators on the other hand, thanks to their amplification capabilities, are a better alternative for small-animal SPECT imaging. Sensitivity of pinhole collimators is lower than parallel hole collimators, but this can be improved by using multiple (clusters) pinholes. (Multi)Pinhole collimators are used in all modern preclinical SPECT scanners.

Typically, preclinical system vendors offer multiple collimators designs which can be changed for different applications, such as for high resolution, low sensitivity scans, or large field of view applications (Matsusaka et al. 2022), for example. Depending on the diameter of the pinhole(s), the sensitivity and spatial resolution can be modified, with higher sensitivity and lower spatial resolution with larger pinhole diameter, and vice-versa. For a complete review on the properties of different SPECT collimator configurations please see the work published by Van Audenhaege et al. (2015).

Scintillation crystal detectors are currently the most common type of detectors used in preclinical SPECT scanners. Among these, thallium-doped sodium iodide (NaI(Tl)) is the most common material. Most of the systems use monolithic crystals, but pixelated crystals can also be used. A few systems make use of semiconductor detectors such as CZT or cadmium telluride (CdTe). The X-SPECT (Mok et al. 2012) (Gamma Medica-Ideas, Inc., Northridge, CA, USA) and the eXplore speCZT (Matsunari et al. 2014) (GE Healthcare, Milwaukee, WI, USA) are examples of systems which use CZT detectors, while the FullSPECT 3D (Modry et al. 2022) use CdTe detectors. Semiconductor detectors are less used than scintillation crystal due to their higher cost and lower detection efficiency for higher energy gamma rays, but with recent fabrication improvements the use of this technology is increasing.

Conversion of scintillation crystals light to electric signal is typically done coupling multiple PMT's to the crystal. The signals from the PMT's are then combined to calculate the location of the event using Anger logic (Peterson and Furenlid 2011). To improve the calculation of the event location, position sensitive PMT's can be used, which offer discrete position of the event without recurring to Anger logic. Solid state light transducers, such as avalanche photodiodes or silicon photomultipliers (SiPM's), can also be used to convert the scintillation light into an electric pulse. PMT's are still being regularly used, but SiPM's are increasingly being used in medical imaging applications.

The geometry of the system can also have a great impact on the image quality. Depending on the type of collimator and detectors used, some geometries might result in better image quality than others. Similar to clinical SPECT systems, many preclinical systems acquire a full set of projections by rotation of the detectors. A disadvantage of this type of acquisition is that artifacts can result from the change of activity distribution between projections acquired at different time points. This is particularly problematic for tracers with fast kinetics. Additionally, dynamic scans are difficult to perform. To solve this limitation, some preclinical scanners have a set of fixed collimators and detectors to perform the scan without rotation of the detectors. Preclinical SPECT scanners from MILabs (Utrecht, The Netherlands) are an example of systems with this acquisition type. Some other scanners offer both possibilities of detection, either by rotation or with fixed detectors, such as the nanoSPECT scanner (Mediso Medical Imaging Systems, Budapest, Hungary). For this scanner, it was found that static detectors acquisition produces lower quality images than acquisition with rotation of the detectors (Lange et al. 2014).

In order to obtain anatomical information of the subject, preclinical SPECT systems are often coupled to small animal CT scanners. The CT image also serves to obtain an accurate attenuation map used for attenuation and scatter correction of the SPECT image. Some systems also offer the possibility to perform SPECT and positron emission tomography (PET) scans. For example, the Trimodal Albira Si (Bruker BioSpin, E., Germany) offers a PET subsystem next to the SPECT and CT hardware. Another approach is to use the same SPECT detectors and collimators for PET tracers gamma detection, such as in the U-SPECT Vector system from MILabs. This allows simultaneous acquisition of PET and SPECT tracers (with non-overlapping gamma energy), but with the drawback of lower sensitivity and smaller field of view compared with dedicated PET scanners.

Simultaneous SPECT and MRI has also been investigated. Especially for neurology applications, the excellent soft tissue contrast offered by MRI can be used for better anatomical localization of SPECT tracer distribution. However, designing SPECT hardware technology compatible with MRI scanners is a challenging task, in which several research centers are continuously involved in. For example, simulations of a special multipin-hole configuration, to allow a compact design for insertion in typical MRI scanner bores, has been performed (Lai and Meng 2018). Additionally, an EU consortium termed INSERT has dedicated efforts to construct an MRI compatible SPECT insert for mouse brain imaging, which has been evaluated in a Philips MRI scanner (Carminati et al. 2019). One of the few commercial preclinical SPECT/MRI scanners is offered by Mediso.

2.2 Commercial SPECT Scanners

Some preclinical SPECT scanners have been discontinued over the years, such as the (Triumph) X-SPECT and the eXplore speCZT. Currently available preclinical SPECT systems include the MILabs U-SPECT, the Bruker Albira Si, the Molecubes γ-cube, and the Mediso nanoSPECT (Fig. 2). The Molecubes' SPECT γ-cube consists of a compact scanner with NaI(Tl) detectors coupled to SiPM's (Decuyper et al. 2022). This SPECT system can be used in complementarity with the PET β-cube and/or the CT X-cube, from the same vendor, to obtain multimodality images. The MILabs U-SPECT/CT systems comprise fixed NaI(Tl) detectors with interchangeable collimators. These systems have been updated over the years.

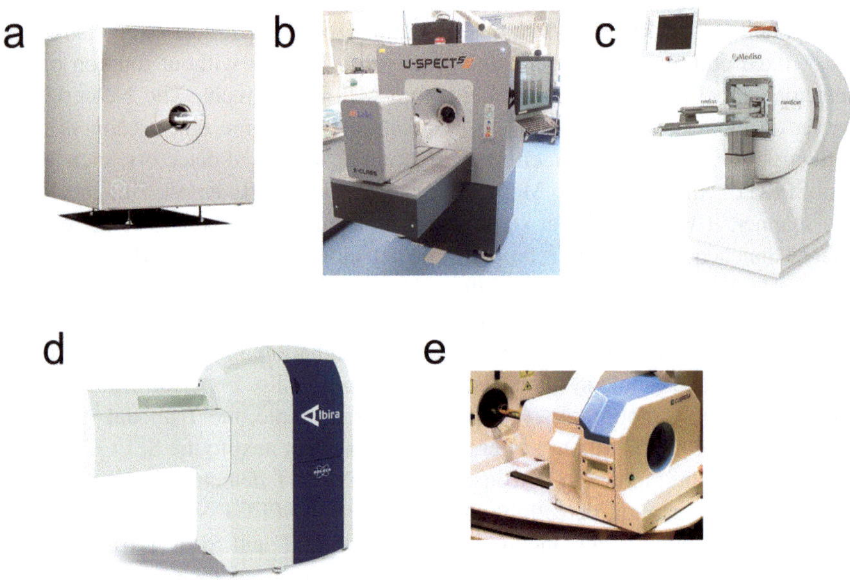

Fig. 2 Preclinical scanners (**a**) Molecubes SPECT γ-cube, (**b**) MILabs U-SPECT/CT E-class, (**c**) Mediso nanoSPECT/CT, (**d**) Bruker Albira Si, and (**e**) Cubresa Spark

The U-SPECT-II was the second iteration of the U-SPECT system, with larger detectors surface and higher sensitivity (van der Have et al. 2009). The U-SPECT$^+$ is the successor of the U-SPECT-II system, with improved collimators, image reconstruction and quantification (Ivashchenko et al. 2015a, b). The U-SPECT E-class is a reduced cost system with only 2 detector heads, instead of 3 as in the other systems (Matsusaka et al. 2022; Hoffmann et al. 2020; Prieto et al. 2022).

The Mediso nanoSPECT/CT plus comprises NaI(Tl) detectors coupled to position sensitive PMT's (Lukas et al. 2020). The system offers the possibility to perform acquisitions with rotating or stationary detectors (Lange et al. 2014). It also offers parallel hole or pinhole collimators for different applications.

The Bruker Albira Si is a trimodal SPECT/PET/CT system, where the SPECT sub system has CsI(Na) detectors coupled to position sensitive PMT's ((Bruker BioSpin, E., Germany; Attarwala et al. 2021). The detectors rotate around the subject to obtain a complete set of projections. The Cubresa Spark (Strugari et al. 2023; Cubresa) (Cubresa Inc., Winnipeg, Canada) is another benchtop commercial scanner, which is intended to be used together with existing installed scanners to perform multi-modal imaging. The system has CsI(Na) monolithic scintillation crystals coupled to SiPM's.

2.3 Standardization

The European Society for Molecular Imaging (ESMI) and the European Association of Nuclear Medicine (EANM) have published guidelines for quality control of SPECT preclinical laboratories (Vanhove et al. 2024). Similar to clinical SPECT, where quantification and standardization can be challenging due to the nature of the acquisition and/or difference in technology between systems (Dickson et al. 2019), standardization and quality control of preclinical SPECT is currently suboptimal (Vanhove et al. 2024). For example, preclinical SPECT systems can use a wide variety of collimators for different detection goals, and acquisition can either be performed with rotation or static detectors. The guidelines published by the ESMI and the EANM aim to tackle this standardization gap by proposing methods to, for example, quantify image quality and verify that detectors are in optimal state (Vanhove et al. 2024). Table 1 presents a summary of these guidelines.

3 SPECT Preclinical Applications on Biomedical Research

Preclinical SPECT can be used to test novel radiotracers before being used in humans. Radiotracer biodistribution, as well as dosimetry calculations can be extrapolated from rodents to humans (Stabin et al. 2006), keeping in mind possible species differences. Recently, with the advent of radiotheranostics, preclinical studies started exploring dosimetry calculations in small animals (Gupta et al. 2020).

Once SPECT tracers have been validated for preclinical studies, they can be used to investigate the status of their respective target/receptor under different conditions.

Table 1 Overview of recommended QC tests for preclinical SPECT. The purpose of the QC tests, the recommended frequency, the source required, and the suggested radioactivity are listed

Routine test	Physical inspection	Detector check and photopeak drift	Image uniformity	Quantitative accuracy	Recovery coefficients	SPECT/CT alignment
Purpose	Check for collimator damage and collimator mountings to detector head	To check that gamma-detectors are working	To estimate voxel-value uniformity by imaging a uniformly filled object	Determine that calibration constant is constant over time	Determine that spatial resolution is constant over time using the recovery coefficients concept	Determine possible misalignments between PET and CT component
Comments	Inspect for mechanical and other defects that may compromise safety of animals. If collimator damage is detected or suspected, a uniformity test should be performed	When using a single-photon emitter that is not frequently used, detector check and photopeak drift should be performed prior to scanning	When using a collimator-isotope combinations that is not frequently used, image uniformity should be checked prior to scanning	When using a collimator-isotope combinations that is not frequently used, quantitative accuracy should be checked prior to scanning	Can be used to harmonize SPECT acquisitions and reconstruction	When using a collimator-isotope combinations that is not frequently used, SPECT/CT alignment should be checked prior to scanning
Activity	Not applicable	± 1 MBq	10–20 MBq	10–20 MBq	37 MBq	10–20 MBq
Source	Not applicable	Point source	Cylindrical phantom	Cylindrical phantom with known activity concentration	NEMA NU 4–2008 image quality phantom (or alternative)	Cylindrical phantom, capillary tubes or zeolite beads
Frequency	Daily (recommended) Certainly on a scan day (minimal)	Daily (recommended) Certainly on a scan day (minimal)	Monthly, using the collimator that is most frequently used in daily practice, and a single-photon emitter that is easily available	Monthly, using the collimator that is most frequently used in daily practice, and a single-photon emitter that is easily available	Yearly, using the collimator that is most frequently used in daily practice, and a single-photon emitter that is easily available	Yearly, using the collimator that is most frequently used in daily practice, and a single-photon emitter that is easily available

Reproduced from Vanhove et al. (2024), with permission from Springer Nature

For example, receptor density in health and disease, drug-receptor occupancy, change in physiological parameters due to therapy, and tumor progression after therapy, can be investigated with a wide variety of SPECT tracers. Below specific applications in neurology and oncology research are detailed.

3.1 Preclinical SPECT on Neurological Research

3.1.1 Dopamine Transporter

Several SPECT radiotracers have been developed to image the dopamine transporter (DAT). Diseases showing abnormal DAT expression, such as Parkinson's disease (Lin et al. 2018) and dementia with Lewy bodies (McKeith et al. 2007), have been investigated with SPECT DAT tracers. [^{123}I]FP-CIT is one of the most frequently used DAT tracers. For example, this tracer has been used to determine the density of DAT in young and aged rats (Ninerola-Baizan et al. 2015), showing an age-related decline in DAT integrity. It also has been used to measure the binding of L-DOPA to DAT at different doses in rats (Nikolaus et al. 2013), showing reduced [^{123}I]FP-CIT binding possibly due to competition with elevated endogenous dopamine. In a Monte Carlo simulation study the effect of different reconstruction and image correction methods was evaluated in rat studies using [^{123}I]FP-CIT (Roe-Vellve et al. 2014), where it was found that correction for point spread function and partial volume effects are important to improve quantification accuracy. A pharmacological study with the antipsychotic drug haloperidol was also performed using [^{123}I]FP-CIT (Park et al. 2019), where changes in endogenous dopamine concentrations due to administration of haloperidol were detected with SPECT imaging. Another SPECT DAT tracer is [^{123}I]-β-CIT. This tracer was thoroughly validated to perform DAT imaging studies in mice (Pitkonen et al. 2012), showing good test-retest performance and stable specific to non-specific binding ratios after 2 h post-injection. [^{99m}Tc]TRODAT-1 can also be used to image DAT. This tracer has been used in a preclinical Parkinson's disease rat model, in which rats with striatal lesions showed significantly lower tracer binding (Lin et al. 2018).

3.1.2 Dopamine $D_{2/3}$

[^{123}I]Epidepride is a dopamine $D_{2/3}$ receptor SPECT tracer. It has been used in rat schizophrenia models to detect alterations in dopamine $D_{2/3}$ receptors (Huang et al. 2012). Improved kinetic modelling has also been recently developed for [^{123}I]Epidepride (Tsartsalis et al. 2020). [^{123}I]IBZM has also been used to quantify dopamine $D_{2/3}$ receptor binding in preclinical studies (Tsartsalis et al. 2017). [^{123}I]IBF is a dopamine D_2 receptor tracer which has been used in drug occupancy studies with SPECT (Acton et al. 2002).

3.1.3 Translocator Protein

[^{125}I]CLINDE and [^{123}I]CLINME are 2 SPECT tracers used for translocator protein (TSPO) imaging. [^{125}I]CLINDE was used in a rat model of microglial activation where radiotracer biding was correlated with neuroinflammation (Arlicot et al.

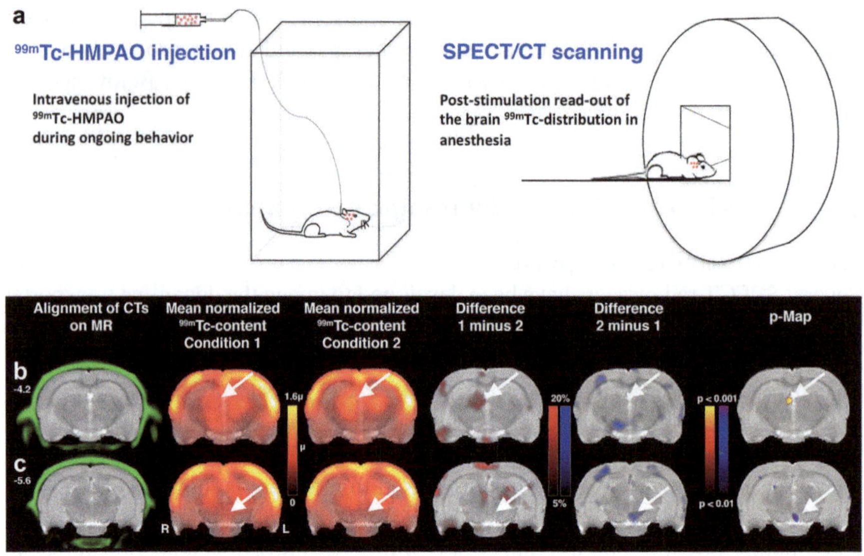

Fig. 3 Top row: example of behavioral rat scans performed with [^{99m}Tc]HMPAO (Oelschlegel and Goldschmidt 2020). The animal is injected with the radiotracer outside the scanner while performing a task. The tracer is trapped in the brain reflecting cerebral blood flow (CBF) during the task, which can then be imaged with SPECT. Bottom row: example rat brain images of CBF obtained with [^{99m}Tc]HMPAO in different conditions to calculate differences between groups. (Adapted from Oelschlegel and Goldschmidt (2020) under a Creative Commons Attribution License (CC BY 4.0))

2008). [^{123}I]CLINME was also validated in rat studies where it was found to have high selectivity for TSPO (Mattner et al. 2015).

3.1.4 Cerebral Blood Flow

[^{123}I]IMP, [^{99m}Tc]HMPAO, and [^{99m}Tc]ECD are SPECT tracers used to measure cerebral blood flow (CBF) (Holschneider and Maarek 2004; Oelschlegel and Goldschmidt 2020; Apostolova et al. 2012). Among these, [^{99m}Tc]HMPAO (Fig. 3) is the most commonly used. Due to the high retention of the tracer in brain, it can be used in behavioral rat studies to then relate brain activation with CBF. For example, the animal injected with [^{99m}Tc]HMPAO outside the scanner performs a defined task, and is then anesthetized and scanned in a SPECT system to measure the distribution of the tracer which will reflect CBF/activation (Oelschlegel and Goldschmidt 2020) during the task. Making use of the ability of SPECT scanners to measure 2 radioisotopes with different gamma energies, rats were injected with [^{99m}Tc]HMPAO and [^{123}I]FP-CIT to simultaneously measure CBF and dopamine transporter binding in the brain (Kobayashi et al. 2016).

3.1.5 Beta-Amyloid and Tau

In order to image the proteins expressed in Alzheimer's disease, i.e., beta-amyloid and hyperphosphorylated tau protein aggregates, several SPECT radiotracers have been developed. SPECT tracers that bind to beta-amyloid include [$^{123/125}$I]DRM106, [^{125}I]IMPY, and [^{123}I]ABC577 (Valotassiou et al. 2018), while tau tracers include [^{125}I]BIP (Ono et al. 2016), and [^{125}I]PDB-3 (Watanabe et al. 2015).

3.2 Preclinical SPECT on Oncology Research

Preclinical SPECT is widely used in oncology research to study the biological and molecular mechanisms of cancer in animal models. In this section we focus on applications of preclinical SPECT, used in combination with CT or MRI, for studying tumor development and progression, for monitoring treatment efficacy, for developing and validating novel radiopharmaceuticals and drugs, and in theranostic approaches. These applications make preclinical SPECT a powerful tool in translational cancer research, bridging the gap between basic research and clinical practice.

3.2.1 Angiogenesis

Tumors are known to display an aberrant vascular network and microcirculation, which are responsible for the rapid growth, invasion and spread of malignant cells (Carmeliet 2003). The new blood vessels formed by the process of tumor angiogenesis ensure the transport of oxygen and nutrients that support tumor growth, enabling fast tumor growth once a well-developed vascularization network has been established within the tumor. The angiogenic process depends on the migration and invasion of vascular endothelial cells and is regulated by vascular endothelial growth factor (VEGF) receptors and cell adhesion receptors, called integrins (Mezu-Ndubuisi and Maheshwari 2021). Several angiogenesis-targeting SPECT imaging agents have been developed targeting integrin $\alpha_v\beta_3$, which is upregulated on activated endothelial cells of tumor neovasculature during angiogenesis but not in quiescent endothelial cells. Importantly, the expression level of integrin $\alpha_v\beta_3$ is associated with the potential for tumor metastasis and aggressiveness. Therefore, imaging $\alpha_v\beta_3$ integrin expression can potentially be used as an approach for early detection of tumors with active neovascularization and as a marker of effectiveness of antiangiogenic therapies. Imaging agents have been developed based on radiolabeled cyclic Arg-Gly-Asp (RGD) peptides due to their high binding affinity to integrin $\alpha_v\beta_3$. For example, radiolabeled cyclic RGD peptide dimers, such as E[c(RGDfK)]$_2$ and 3P-RGD$_2$, have been used as diagnostic radiotracers (^{99m}Tc and ^{111}In) for tumor SPECT imaging in human breast, glioma, lung, colorectal and prostate cancer xenografts, demonstrating rapid biodistribution and excellent tumor contrast as early as 1 h post injection (Liu et al. 2007; Jia et al. 2006, 2008; Zhou et al. 2011). Importantly, a correlation was demonstrated between the radiotracer uptake and $\alpha_v\beta_3$ expression levels on the tumor cells and tumor neovasculature, demonstrating the utility of these imaging agents to visualize integrin $\alpha_v\beta_3$ expression in tumors.

One important feature of SPECT is the ability to image very small primary tumors and their metastases. [^{99m}Tc]3P-RGD$_2$ SPECT/CT has been used to monitor the tumor growth and progression of metastasis in an orthotopic breast cancer model, enabling in vivo monitoring of the burden of lung metastasis (Zhou et al. 2012).

The approach of targeting tumor angiogenesis with anti-angiogenic agents represents a promising strategy for controlling and eradicating tumor growth. SPECT/CT imaging with [^{99m}Tc]3P-RGD$_2$ has been used to monitor (Ji et al. 2013a). Tumors responding to linifanib treatment (a multi-targeted tyrosine kinase inhibitor against VEGF and platelet-derived growth factor receptors) showed a decrease in radiotracer uptake, while no changes were observed (Ji et al. 2013b). Importantly, tumors with a higher radiotracer uptake at baseline, corresponding to a higher vascularization and expression level of integrin $\alpha_v\beta_3$, showed better tumor response to linifanib therapy then tumors showing low uptake of [^{99m}Tc]3P-RGD$_2$ (poorly vascularized tumors with low expression level of integrin $\alpha_v\beta_3$). In summary, SPECT/CT imaging of angiogenesis can be used to characterize tumor angiogenesis, identify patients who may benefit from anti-angiogenic treatment and to monitor the tumor response to anti-angiogenic therapies.

3.2.2 Cell Death

Tumor development involves specific mutations that lead to deregulated cellular proliferation and suppression of the mechanisms that control cell death (Herr and Debatin 2001). Many anti-cancer therapeutic strategies are directed at promoting tumor cell death. Therefore, molecular imaging of cell death can be used as a useful approach for treatment response evaluation in patients and in the development of novel anti-cancer drugs. Apoptosis and necrosis are two of the main mechanisms of cell death induced by anti-cancer treatments. The loss of the asymmetric distribution of phospholipids in the plasma membrane is an early event of apoptosis. In particular, the exposure of phosphatidylserine (PS) and phosphatidylethanolamine (PE) in the outer leaflet of the cell membrane are considered universal biomarkers of apoptosis. Because the rupture of the plasma membrane during necrosis also results in accessible PS and PE, both phospholipids represent high abundance targets for imaging both apoptosis and necrosis.

The most explored cell death SPECT imaging agents include ^{99m}Tc- and ^{111}In-labeled annexin V, a 36-kDa protein that binds to PS, and [^{99m}Tc]duramycin, a small peptide (2 kDa) that binds to PE. Radiolabeled annexin V has been used to image therapy-induced cell death in tumors, where it showed more uptake in treated tumors when compared with non-treated tumors early after the start of chemotherapy and radiotherapy (Ke et al. 2004; Kuge et al. 2004; Guo et al. 2009). Other PS-binding probes include the C2A domain of synaptotagmin I (16 kDa). ^{99m}Tc- and ^{111}In-labeled C2Am have been used to detect response to chemotherapy in mouse models of lymphoma and colorectal cancer, respectively, where accumulation of these tracers was found to correlate well with histological evidence of cell death in the tumors (Neves et al. 2017). [^{99m}Tc]duramycin has been shown to have avid uptake in apoptotic cells, and rapid renal clearance with low blood pool and hepatic background, which represents an advantage when compared with radiolabeled PS-binding proteins. Molecular imaging of cell death using [^{99m}Tc]duramycin was capable of providing an early visualization and quantification of the response to chemotherapy, radiotherapy and targeted therapy regimens in human colorectal adenocarcinoma xenograft models (Elvas et al. 2015, 2016, 2017). The high apoptotic response induced 24 h after a single treatment with conatumumab (a fully human monoclonal agonist antibody directed against the death receptor 5) was specifically detected in therapy-sensitive Colo205 tumors, while therapy-resistant HT29 tumors showed low levels of [^{99m}Tc]duramycin uptake after treatment which corresponded to a lack of cell death induction. In comparison, [^{18}F]FDG was not able to detect tumor response in the first 24 h after the onset of therapy (Fig. 4). Another promising application of this approach is the monitoring of chemotherapy-induced drug toxicity. Toxicity to healthy tissues is a known limitation of chemotherapy regimens. In a preclinical proof-of-concept study, [^{99m}Tc]duramycin enabled whole-body longitudinal imaging of toxicity following a single dose of various chemotherapeutic agents, successfully identifying heterogeneous toxic effects across different tissues and individual subjects (Johnson et al. 2018).

3.2.3 Other Cancer Biomarkers

Many other SPECT imaging agents specific for tumor targets that contribute to cancer development have been evaluated for tumor detection, localization and

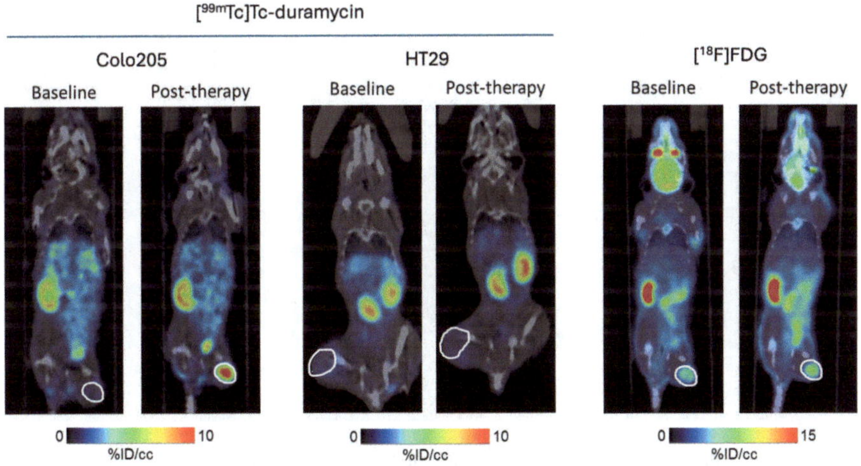

Fig. 4 [^{99m}Tc]duramycin SPECT/CT and [^{18}F]FDG PET/CT images of COLO205 and HT29 xenografts at baseline and 24 h after conatumumab treatment in mice. Tumors are encircled. (Adapted from Elvas et al. (2017). This research was originally published in *JNM*. Elvas et al. (2017). © SNMMI.)

treatment response evaluation. Such radiotracers include antibodies, peptides and small molecules targeting the epidermal growth factor receptor 2 (HER2), the folate receptor (FR), the prostate specific membrane antigen (PSMA), the somatostatin receptors (SSTRs), and the fibroblast activation protein (FAP) (Wu et al. 2020; Zhao et al. 2021; Müller et al. 2008, 2014; Luan et al. 2024).

3.2.4 Immune System

The immune system plays an important role in the development and treatment response of metastatic cancer. Immunotherapies are now established as an essential weapon for treating several types of cancer, with a rapidly growing list of approved drugs and disease indications. However, despite impressive anti-tumor effects and extended patient survival, only a minority of all patients respond, and life-threatening toxicity can occur, highlighting the need for predictive imaging biomarkers. To facilitate the development and assessment of tumor response to novel immunotherapies, a tumor imaging method based on the expression of murine sodium iodide symporter (mNIS) was developed (Merrill et al. 2023). This approach allowed the detection of early metastatic hepatic lesions and *in vivo* longitudinal assessment of response to immunotherapy in an immune-competent pancreatic ductal adenocarcinoma mouse model using SPECT/CT imaging following injection of [^{99m}Tc]pertechnetate (Fig. 5).

Ideal monitoring of adaptive immune responses involving T lymphocytes requires the ability to track T-cell migration to the tumor site, engagement with the antigen-bearing tumor cells, as well as T-cell- expansion and persistence at the tumor site—all essential steps for therapeutic effect. In vivo tracking of CAR T cells

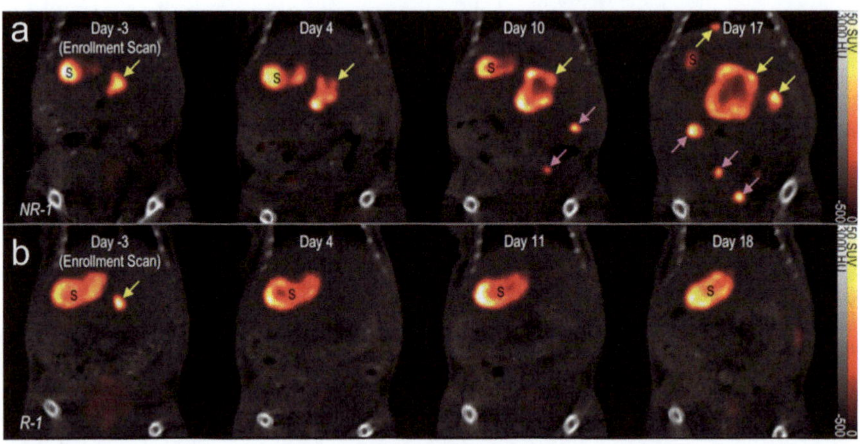

Fig. 5 Treatment response monitoring of metastatic PDAC tumors using mNIS-based SPECT/CT. Tumor response to anti-PD-1 immunotherapy treatment was followed by weekly mNIS-SPECT imaging and allowed differentiation of nonresponsive (upper panel) and responding animals (lower panel). (Adapted from Merrill et al. (2023) under a Creative Commons Attribution License (CC BY 4.0))

and autologous T cells can be realized by direct cell labeling with [^{111}In]tropolonate, [^{111}In]oxine or [^{99m}Tc]HMPAO followed by SPECT/CT imaging (Parente-Pereira et al. 2011; Cussó et al. 2014; Sharif-Paghaleh et al. 2014). However, this strategy faces some inherent shortcomings for longitudinal cell tracking due to imaging signal dilution and decay of the radionuclide, providing limited information at later stages of the treatment cycle. Therefore, transduction of CAR T cells with reporter genes has been employed for pairing with SPECT imaging agents. To date, many such reporters in combination with [^{123}I]MIBG or ^{99m}Tc-pertechnetate have been developed to achieve serial monitoring of CAR T-cells in preclinical mouse models (Volpe et al. 2020, 2022; Moroz et al. 2007; Doubrovin et al. 2007).

In addition, cell surface markers can provide useful information regarding migration, tumor retention and the functional status of T cells. For example, programmed death 1 (PD-1) is a cell surface receptor present on T cells and has a role in regulating immune response. When bound to its ligand, PD-L1, PD-1 inhibits T-cell activation and promotes effector T-cell exhaustion. Therefore, ^{111}In-labeled antibodies targeting PD-L1 have been developed to evaluate the expression levels of PD-L1 and reflect the immune status for the tumor (Chatterjee et al. 2016; Heskamp et al. 2015, 2019; Josefsson et al. 2016). This is a promising approach to identify patients who would be eligible to undergo PD-L1 inhibition as a therapeutic strategy.

3.2.5 Theranostics

Radiotheranostics involves the combination of targeted radiopharmaceuticals with diagnostic or therapeutic radionuclides. Therapeutic radiopharmaceuticals often have matched diagnostic pairs for pre-therapy SPECT imaging, enabling diagnostic information on disease spread and assisting in patient selection and dosimetry calculations. Some radionuclides, given their individual decay characteristics, can have a dual role, emitting particle radiation useful for radionuclide therapy as well as gamma radiation for SPECT imaging allowing an accurate assessment of the resulting radiation dose (Nelson et al. 2024). These radionuclides can be divided into three groups by the therapeutic emission type, such as electrons (^{177}Lu, ^{67}Cu, ^{90}Y, ^{188}Re, ^{131}I, ^{161}Tb), Auger electrons (^{67}Ga, ^{111}In, ^{123}I), and alpha particles (^{211}At, ^{213}Bi, ^{225}Ac, ^{212}Pb).

The feasibility of quantitative SPECT imaging of the theranostic radionuclide pair, ^{155}Tb/^{161}Tb, has been shown with high spatial resolution using a preclinical SPECT/CT scanner (Koniar et al. 2024). This imaging method can provide critical quantitative information for performing image-based dosimetry calculations using terbium-labeled radiopharmaceuticals. A study using [^{161}Tb]PSMA-617 in an LNCaP tumor bearing mouse showed accurate quantification of tumor activity (Wright et al. 2024). In an element-matched radiotheranostic study using [$^{155/161}$Tb]Crown-TATE, a ligand targeting SSTR2-positive neuroendocrine tumors, longitudinal SPECT/CT imaging was used to show equivalent biodistribution and pharmacokinetic profile and AR42J tumor imaging (Wharton et al. 2023). Imaging tumor uptake and biodistribution of [^{212}Pb]NG001, a PSMA-binding ligand, was demonstrated in mice using a clinical SPECT/CT scanner (Kvassheim et al. 2023).

Although it should be noted that a preclinical SPECT scanner should be preferred over a clinical SPECT scanner to decrease quantification uncertainties in small tumor volumes.

Small animal data are required to obtain successful clinical translation of new radiotheranostics. Small animal-to-human radiation dosimetry extrapolations based on SPECT images have been attempted with therapeutic radioligands (Gnesin et al. 2023; Sachindra et al. 2021). The most commonly used methods are extrapolation based on weight or time, taking into account either the difference in relative weight of the individual organs compared to the bodyweight or the different metabolic rates (Stabin et al. 2022). However, both approaches still present some challenges which are not easy to overcome, and no method is preferred over the other.

3.2.6 Dual Isotope Imaging

SPECT offers the unique advantage of allowing simultaneous imaging of two different radionuclides. Dual-isotope SPECT enables the simultaneous detection of radiotracer distribution with spatial and temporal resolution, thus enabling the detection of multiple radiotracers in a single imaging session. In preclinical research, this imaging approach has showed its value for optimization of the development of novel anti-cancer therapies, for example by combining the use of a NIS reporter/^{125}I pair and [^{99m}Tc]duramycin SPECT imaging for simultaneous monitoring of oncolytic virus therapy and the induction of treatment-associated cell death (Zhang et al. 2019). Additionally, this technique can assist in the development and validation of new molecular imaging agents, since it enables the true quantification of target expression levels, by comparison of the biodistribution of a specific peptide or antibody and its respective non-specific control ligand labeled with two different radionuclides simultaneously in the same animal (Hijnen et al. 2012; Knight et al. 2019). This paired design improves statistical analysis of results, while decreasing the number of animals needed to obtain statistically relevant results. Similarly dual-isotope ^{111}In/^{177}Lu SPECT imaging has been used as a tool to preclinically evaluate theranostic nanobodies (Beltrán Hernández et al. 2019).

4 Conclusions and Future Perspectives

Recent advances in preclinical SPECT are shaped by ongoing improvement in hardware, radiolabeling and SPECT radiotracer development, which together with greater accessibility to these tools have facilitated a wider application and growth of this technique in small animal imaging. Technological advancements have contributed to handling the demanding requirements of small animal imaging. Innovation in detector materials (e.g., CZT) and optimized multipinhole collimator design have contributed to improved spatial resolution and sensitivity, enabling the visualization of smaller structures in animal models. These exciting technological developments have fueled the development of novel SPECT radiotracers, creating new radiolabeling strategies, such as new bifunctional chelators and prosthetic groups, and new carrier vectors that can be labeled with SPECT radionuclides.

The development of new radiotracers targeting specific molecular pathways (e.g., brain receptors and cancer biomarkers) will expand the applications of preclinical SPECT, accelerating the introduction of new radiopharmaceuticals into clinical practice for studying neurodegenerative diseases and cancer. The development of innovative SPECT radionuclide pairs has also opened new possibilities in the field of theranostics, integrating diagnostic imaging with targeted radioligand therapy. In this strategy, gamma-emitting radionuclides allow visualization of disease sites and quantification of radiotracer distribution while, beta- or alpha-emitting radionuclides deliver cytotoxic radiation to the same biological target. In recent years, with the surge of radiotheranostics, it became of utmost importance assessing the biodistribution, pharmacokinetics, and dosimetry of therapeutic radioligands. The challenges in imaging the emissions of therapeutic radionuclides have been increasingly addressed using preclinical SPECT scanners. These images can be used to perform lesion dosimetry analysis, such as estimating the absorbed radiation dose in the tumors or lesions. Improved spatial resolution and accuracy of SPECT could facilitate more efficient development of novel radioligand therapies in the preclinical phase by providing accurate quantitative in vivo pharmacokinetic information essential for estimating dosimetry, assess therapeutic efficacy and estimate potential radiation-induced toxicities in radionuclide therapy.

Improvements in quantification and standardization are currently being addressed. The surge in the field of theranostics and associated requirements for dosimetry have provided a boost to research of quantitative accuracy of SPECT imaging. Several nuclear medicine and molecular imaging societies have made efforts on SPECT quality control standardization.

Acknowledgments Filipe Elvas is supported by a tenure-track fellowship provided by the University of Antwerp (BOF: TTZAPBOF).

Conflict of Interest The authors declare that they have no conflict of interest.

References

Acton PD et al (2002) Occupancy of dopamine D2 receptors in the mouse brain measured using ultra-high-resolution single-photon emission tomography and [123]IBF. Eur J Nucl Med Mol Imaging 29:1507–1515

Apostolova I et al (2012) Brain perfusion SPECT in the mouse: normal pattern according to gender and age. NeuroImage 63:1807–1817

Arlicot N et al (2008) Evaluation of CLINDE as potent translocator protein (18 kDa) SPECT radiotracer reflecting the degree of neuroinflammation in a rat model of microglial activation. Eur J Nucl Med Mol I(35):2203–2211

Attarwala AA et al (2021) Performance assessment of the ALBIRA II pre-clinical SPECT S102 system for (99m)Tc imaging. Ann Nucl Med 35:111–120

Beltrán Hernández I et al (2019) Imaging of tumor spheroids, dual-isotope SPECT, and autoradiographic analysis to assess the tumor uptake and distribution of different nanobodies. Mol Imaging Biol 21:1079–1088

Bruker BioSpin, E., Germany. https://www.bruker.com/en/products-and-solutions/preclinical-imaging/nmi/albirasi.html
Carmeliet P (2003) Angiogenesis in health and disease. Nat Med 9:653–660
Carminati M et al (2019) Validation and performance assessment of a preclinical SiPM-based SPECT/MRI insert. Ieee T Radiat Plasma 3:483–490
Chatterjee S et al (2016) A humanized antibody for imaging immune checkpoint ligand PD-L1 expression in tumors. Oncotarget 7:10215–10227
Cubresa. https://www.cubresa.com/
Cussó L et al (2014) Combination of single-photon emission computed tomography and magnetic resonance imaging to track 111In-Oxine–labeled human mesenchymal stem cells in neuroblastoma-bearing mice. Mol Imaging 13:7290.2014.00033
Decuyper M, Deprez K, Mollet P (2022) Neural networks for gamma interaction positioning in continuous SPECT detectors. In: 2022 IEEE Nuclear Science Symposium and Medical Imaging Conference (NSS/MIC), pp 1–3
Dickson J, Ross J, Vöö S (2019) Quantitative SPECT: the time is now. EJNMMI Phys 6:4
Doubrovin MM et al (2007) In vivo imaging and quantitation of adoptively transferred human antigen-specific T cells transduced to express a human norepinephrine transporter gene. Cancer Res 67:11959–11969
Elvas F et al (2015) Characterization of [(99m)Tc]duramycin as a SPECT imaging agent for early assessment of tumor apoptosis. Mol Imaging Biol 17:838–847
Elvas F et al (2016) Early prediction of tumor response to treatment: preclinical validation of 99mTc-duramycin. J Nucl Med 57:805–811
Elvas F et al (2017) (99m)Tc-duramycin SPECT imaging of early tumor response to targeted therapy: a comparison with (18)F-FDG PET. J Nucl Med 58:665–670
Gnesin S et al (2023) From bench to bedside: 64Cu/177Lu 1C1m-Fc anti TEM-1: mice-to-human dosimetry extrapolations for future theranostic applications. EJNMMI Res 13:59
Guo M-F et al (2009) In vivo99mTc-HYNIC-annexin V imaging of early tumor apoptosis in mice after single dose irradiation. J Exp Clin Cancer Res 28:136
Gupta A, Lee MS, Kim JH, Lee DS, Lee JS (2020) Preclinical voxel-based dosimetry in theranostics: a review. Nucl Med Mol Imaging 54:86–97
Herr I, Debatin K-M (2001) Cellular stress response and apoptosis in cancer therapy. Blood 98:2603–2614
Heskamp S et al (2015) Noninvasive imaging of tumor PD-L1 expression using radiolabeled anti–PD-L1 antibodies. Cancer Res 75:2928–2936
Heskamp S et al (2019) PD-L1 microSPECT/CT imaging for longitudinal monitoring of PD-L1 expression in syngeneic and humanized mouse models for cancer. Cancer Immunol Res 7:150–161
Hijnen NM, de Vries A, Nicolay K, Grüll H (2012) Dual-isotope 111In/177Lu SPECT imaging as a tool in molecular imaging tracer design. Contrast Media Mol Imaging 7:214–222
Hoffmann JV et al (2020) Performance evaluation of fifth-generation ultra-high-resolution SPECT system with two stationary detectors and multi-pinhole imaging. EJNMMI Phys 7:64
Holschneider DP, Maarek JM (2004) Mapping brain function in freely moving subjects. Neurosci Biobehav Rev 28:449–461
Huang YR et al (2012) [123I]Epidepride neuroimaging of dopamine D receptor in chronic MK-801-induced rat schizophrenia model. Nucl Med Biol 39:826–832
Ivashchenko O et al (2015a) Quarter-millimeter-resolution molecular mouse imaging with U-SPECT. Mol Imaging 14
Ivashchenko O, van der Have F, Goorden MC, Ramakers RM, Beekman FJ (2015b) Ultra-high-sensitivity submillimeter mouse SPECT. J Nucl Med 56:470–475
Ji S, Zheng Y, Shao G, Zhou Y, Liu S (2013a) Integrin alpha(v)beta(3)-targeted radiotracer (99m)Tc-3P-RGD(2) useful for noninvasive monitoring of breast tumor response to antiangiogenic linifanib therapy but not anti-integrin alpha(v)beta(3) RGD(2) therapy. Theranostics 3:816–830

Ji S et al (2013b) Monitoring tumor response to linifanib therapy with SPECT/CT using the integrin alphavbeta3-targeted radiotracer 99mTc-3P-RGD2. J Pharmacol Exp Ther 346:251–258
Jia B et al (2006) 99mTc-labeled cyclic RGDfK dimer: initial evaluation for SPECT imaging of glioma integrin αvβ3 expression. Bioconjug Chem 17:1069–1076
Jia B et al (2008) Linker effects on biological properties of 111In-labeled DTPA conjugates of a cyclic RGDfK dimer. Bioconjug Chem 19:201–210
Johnson SE et al (2018) Whole-body imaging of cell death provides a systemic, minimally invasive, dynamic, and near-real time indicator for chemotherapeutic drug toxicity, vol 25. Clin Cancer Res, pp 1331–1342
Josefsson A et al (2016) Imaging, biodistribution, and dosimetry of radionuclide-labeled PD-L1 antibody in an immunocompetent mouse model of breast cancer. Cancer Res 76:472–479
Ke S et al (2004) Imaging Taxane-induced tumor apoptosis using PEGylated, 111In-labeled annexin V. J Nucl Med 45:108–115
Knight JC et al (2019) Dual-isotope imaging allows in vivo immunohistochemistry using radiolabelled antibodies in tumours. Nucl Med Biol 70:14–22
Knoll GF (2010) Radiation detection and measurement. Wiley
Kobayashi M et al (2016) Simultaneous acquisition of (99m)Tc- and (123)I-labeled radiotracers using a preclinical SPECT scanner with CZT detectors. Ann Nucl Med 30:263–271
Koniar H et al (2024) Quantitative SPECT imaging of 155Tb and 161Tb for preclinical theranostic radiopharmaceutical development. EJNMMI Phys 11:77
Kuge Y et al (2004) Feasibility of 99mTc-annexin V for repetitive detection of apoptotic tumor response to chemotherapy: an experimental study using a rat tumor model. J Nucl Med 45:309–312
Kvassheim M, Tornes AJK, Juzeniene A, Stokke C, Revheim M-ER (2023) Imaging of 212Pb in mice with a clinical SPECT/CT. EJNMMI Phys 10:47
Lai X, Meng LJ (2018) Simulation study of the second-generation MR-compatible SPECT system based on the inverted compound-eye gamma camera design. Phys Med Biol 63:045008
Lange C et al (2014) Performance evaluation of stationary and semi-stationary acquisition with a non-stationary small animal multi-pinhole SPECT system. Mol Imaging Biol 16:311–316
Lin L, Ye J, Zhang H, Han ZF, Zheng ZH (2018) Degree of dopaminergic degeneration measured by (99m)Tc-TRODAT-1 SPECT/CT imaging. Neural Regen Res 13:1281–1287
Liu S et al (2007) Evaluation of a 99mTc-labeled cyclic RGD tetramer for noninvasive imaging integrin αvβ3-positive breast cancer. Bioconjug Chem 18:438–446
Luan X et al (2024) The first-in-human preclinical evaluation of the new probe [123I]I-PSMA-7 for real-time intraoperative targeted biopsy and SPECT/CT imaging in prostate cancer. Eur J Nucl Med Mol I(51):4141–4150
Lukas M, Kluge A, Beindorff N, Brenner W (2020) Multi-isotope capabilities of a small-animal multi-pinhole SPECT system. J Nucl Med 61:152–161
Matsunari I et al (2014) Performance evaluation of the eXplore speCZT preclinical imaging system. Ann Nucl Med 28:484–497
Matsusaka Y et al (2022) Performance evaluation of a preclinical SPECT scanner with a collimator designed for medium-sized animals. Mol Imaging 2022
Mattner F et al (2015) Radiosynthesis, in vivo biological evaluation, and imaging of brain lesions with [123I]-CLINME, a new SPECT tracer for the translocator protein. Dis Markers 2015:729698
McKeith I et al (2007) Sensitivity and specificity of dopamine transporter imaging with 123I-FP-CIT SPECT in dementia with Lewy bodies: a phase III, multicentre study. Lancet Neurol 6:305–313
Merrill JR et al (2023) Sensitive, non-immunogenic in vivo imaging of cancer metastases and immunotherapy response. Cell Stress 7:59–68
Mezu-Ndubuisi OJ, Maheshwari A (2021) The role of integrins in inflammation and angiogenesis. Pediatr Res 89:1619–1626
MILabs. https://www.milabs.com/u-pet/

Modry A et al (2022) SPECT imaging properties of the CdTe-Timepix3 detector used in a new prototype FullSPECT 3D multimodal preclinical imaging system: comparison with scintillation gamma camera. Nucl Instrum Meth A 1032:166531
Mok GS, Yu J, Du Y, Wang Y, Tsui BM (2012) Evaluation of a multi-pinhole collimator for imaging small animals with different sizes. Mol Imaging Biol 14:60–69
Moroz MA et al (2007) Imaging hNET reporter gene expression with 124I-MIBG. J Nucl Med 48:827–836
Müller C, Forrer F, Schibli R, Krenning EP, de Jong M (2008) SPECT study of folate receptor-positive malignant and normal tissues in mice using a novel 99mTc-Radiofolate. J Nucl Med 49:310–317
Müller C et al (2014) Future prospects for SPECT imaging using the radiolanthanide terbium-155—production and preclinical evaluation in tumor-bearing mice. Nucl Med Biol 41:e58–e65
Nelson BJB et al (2024) Aspects and prospects of preclinical theranostic radiopharmaceutical development. Theranostics 14:6446–6470
Neves AA et al (2017) Rapid imaging of tumor cell death in vivo using the C2A domain of Synaptotagmin-I. J Nucl Med 58:881–887
Nikolaus S et al (2013) Effects of L-DOPA on striatal iodine-123-FP-CIT binding and behavioral parameters in the rat. Nucl Med Commun 34:1223–1232
Ninerola-Baizan A et al (2015) Dopamine transporter imaging in the aged rat: a [(1)(2)(3)I]FP-CIT SPECT study. Nucl Med Biol 42:395–398
Oelschlegel AM, Goldschmidt J (2020) Functional neuroimaging in rodents using cerebral blood flow SPECT. Front Phys-Lausanne 8:152
Ono M et al (2016) Highly selective tau-SPECT imaging probes for detection of neurofibrillary tangles in Alzheimer's disease. Sci Rep 6:34197
Parente-Pereira AC et al (2011) Trafficking of CAR-engineered human T cells following regional or systemic adoptive transfer in SCID beige mice. J Clin Immunol 31:710–718
Park SH et al (2019) Combination of in vivo [(123)I]FP-CIT SPECT and microdialysis reveals an antipsychotic drug haloperidol-induced synaptic dopamine availability in the rat midbrain and striatum. Exp Neurobiol 28:602–611
Peterson TE, Furenlid LR (2011) SPECT detectors: the anger camera and beyond. Phys Med Biol 56:R145–R182
Pitkonen M et al (2012) Advanced brain dopamine transporter imaging in mice using small-animal SPECT/CT. EJNMMI Res 2:55
Prieto E et al (2022) Performance evaluation of a preclinical SPECT/CT system for multi-animal and multi-isotope quantitative experiments. Sci Rep-UK 12
Roe-Vellve N et al (2014) Quantification of rat brain SPECT with (123)I-ioflupane: evaluation of different reconstruction methods and image degradation compensations using Monte Carlo simulation. Phys Med Biol 59:4567–4582
Sachindra S et al (2021) SPECT/CT imaging, biodistribution and radiation dosimetry of a 177Lu-DOTA-integrin αvβ6 cystine knot peptide in a pancreatic cancer xenograft model. Front Oncol 11:684713
Sharif-Paghaleh E et al (2014) Monitoring the efficacy of dendritic cell vaccination by early detection of (99m) Tc-HMPAO-labelled CD4(+) T cells. Eur J Immunol 44:2188–2191
Stabin MG, Peterson TE, Holburn GE, Emmons MA (2006) Voxel-based mouse and rat models for internal dose calculations. J Nucl Med 47:655–659
Stabin MG, Wendt RE, Flux GD (2022) RADAR guide: standard methods for calculating radiation doses for radiopharmaceuticals, part 1—collection of data for radiopharmaceutical dosimetry. J Nucl Med 63:316–322
Strugari ME, DeBay DR, Beyea SD, Brewer KD (2023) NEMA NU 1-2018 performance characterization and Monte Carlo model validation of the Cubresa Spark SiPM-based preclinical SPECT scanner. EJNMMI Phys 10:35
Tsartsalis S et al (2017) A single-scan protocol for absolute D(2/3) receptor quantification with [(123)I]IBZM SPECT. NeuroImage 147:461–472

Tsartsalis S, Tournier BB, Millet P (2020) In vivo absolute quantification of striatal and extrastriatal D receptors with [I]epidepride SPECT. EJNMMI Res 10:66
Valotassiou V et al (2018) SPECT and PET imaging in Alzheimer's disease. Ann Nucl Med 32:583–593
Van Audenhaege K et al (2015) Review of SPECT collimator selection, optimization, and fabrication for clinical and preclinical imaging. Med Phys 42:4796–4813
Van den Wyngaert T, Elvas F, De Schepper S, Kennedy JA, Israel O (2020) SPECT/CT: standing on the shoulders of giants, it is time to reach for the sky! J Nucl Med 61:1284–1291
van der Have F et al (2009) U-SPECT-II: an ultra-high-resolution device for molecular small-animal imaging. J Nucl Med 50:599–605
Vanhove C et al (2024) Preclinical SPECT and PET: joint EANM and ESMI procedure guideline for implementing an efficient quality control programme. Eur J Nucl Med Mol Imaging 51:3822–3839
Volpe A et al (2020) Spatiotemporal PET imaging reveals differences in CAR-T tumor retention in triple-negative breast cancer models. Mol Ther 28:2271–2285
Volpe A, Adusumilli PS, Schoder H, Ponomarev V (2022) Imaging cellular immunotherapies and immune cell biomarkers: from preclinical studies to patients. J Immunother Cancer 10:e004902
Watanabe H, Ono M, Saji H (2015) Novel PET/SPECT probes for imaging of tau in Alzheimer's disease. ScientificWorldJournal 2015:124192
Wharton L et al (2023) Preclinical evaluation of [155/161Tb]Tb-crown-TATE—A novel SPECT imaging theranostic agent targeting neuroendocrine tumours. Molecules 28:3155
Wright J, Renard I, Middleton I, Lusby P, Archibald S (2024) Quantitative dynamic whole body planar scintigraphy in mice using 99mTc and 161Tb. J Nucl Med 65:242331
Wu Y et al (2020) Imaging and monitoring HER2 expression in breast cancer during trastuzumab therapy with a peptide probe 99mTc-HYNIC-H10F. Eur J Nucl Med Mol I(47):2613–2623
Zhang L et al (2019) Dual-isotope SPECT imaging with NIS reporter gene and duramycin to visualize tumor susceptibility to oncolytic virus infection. Mol Ther Oncolytics 15:178–185
Zhao L et al (2021) Development of a 99mTc-labeled single-domain antibody for SPECT/CT assessment of HER2 expression in breast cancer. Mol Pharm 18:3616–3622
Zhou Y et al (2011) 99mTc-labeled cyclic RGD peptides for noninvasive monitoring of tumor integrin αvβ3 expression. Mol Imaging 10:7290.2011.00006–7290.2011.00397
Zhou Y, Shao G, Liu S (2012) Monitoring breast tumor lung metastasis by U-SPECT-II/CT with an integrin αvβ3-targeted radiotracer 99mTc-3P-RGD2. Theranostics 2:577–588

Applications of Ultrasound (US) Imaging in Preclinical Research

Nimit L. Patel, Nicola Boccella, and Joseph D. Kalen

Abstract

Ultrasound (US) imaging has become an essential modality in preclinical research due to its numerous advantages in the study of biological systems and disease models. This non-invasive, real-time imaging technique enables high-resolution, dynamic visualization of soft tissues, vasculature, and organ structures, positioning it as a critical radiological tool in the fields of oncology and cardiology. This book chapter reviews the application of various US imaging techniques in preclinical research. Specifically, it examines anatomical imaging modes, including 2D and 3D B-mode US, for assessing disease progression, evaluating therapeutic interventions, monitoring the effects of novel drugs or treatments, and guiding image-assisted injections. The chapter further explores contrast-enhanced imaging techniques for evaluating tissue perfusion. Additionally, M-mode, Strain and Strain Rate imaging, 4D echocardiography, and Doppler modalities are discussed in the context of assessing heart chamber dimensions, contractility, blood flow velocities, and valve function, contributing to a comprehensive evaluation of cardiac health. Collectively, US imaging serves as a versatile, cost-effective, and reproducible platform, enhancing our understanding of disease mechanisms, supporting drug development, and advancing translational research.

N. L. Patel (✉) · J. D. Kalen
Small Animal Imaging Program, Frederick National Laboratory for Cancer Research, Leidos Biomedical Research Inc., Frederick, MD, USA
e-mail: nimit.patel@astrazeneca.com

N. Boccella
Mouse Cancer Genetics Program, Center for Cancer Research, National Cancer Institute, Frederick, MD, USA

I. F. Cengiz et al. (eds.), *Bioimaging Modalities in Bioengineering*, Biomaterials, Bioengineering and Sustainability 9,
https://doi.org/10.1007/978-3-031-99410-4_17

Keywords

B-mode · Doppler ultrasound · Ultrasound guided injections · Echocardiography · Contrast-enhanced imaging

1 Introduction

Ultrasound (US) imaging has been extensively used in the study of preclinical animal models since its inception in the early 2000s. It provides superior spatial resolution [i.e., US (30 μm); MRI (125 μm) and optical and nuclear (>1000 μm)] without the use of ionizing radiation which allows for longitudinal imaging feasible. Overtime, US imaging techniques have steadily evolved as a rapid, reliable and comparatively inexpensive imaging research tool (Foster et al. 2011).Initially, much of the research utilized clinical transducers which were operating at relatively lower frequency range (~10 to 20 MHz) which limited spatial resolution to several hundred micrometers and their effectiveness in detecting abnormalities in rodent preclinical models (Greco et al. 2012). Additionally, these earlier (mechanical) transducers lacked temporal resolution (frame rate) to capture the rapid heart movements (400–600 bpm) of preclinical models (Coatney 2001). The innovations in new solid-state transducer technology and scanner architecture helped overcome these challenges (Foster et al. 2011; Coatney 2001). Current preclinical US systems can achieve spatial resolution of ~30 micrometers with frame rates high enough to detect rapid rodent heart movements to image cardiac structure and physiologic function. In contrast to clinical US imaging, where the sonographer moves the transducer in real-time over the patient's skin for correct orientation of the transducer to organ of interest, preclinical US scanning typically employs a transducer mounted on a robotic arm which translates the transducer in small increments for acquiring 3D images and positioning the transducer at precise angles in relation to an organ. To achieve high spatial image resolution requires the reduction of rodent motion (anesthesia) and gating (pulmonary or cardiac). The rodent loses body heat to the imaging stage and therefore the stage must be heated (37 °C). The US detects changes in the echogenicity of the media and to reduce image artifacts due to the skin-hair-transducer interfaces, the rodent hair/fur is removed with a depilatory cream and a warmed (37 °C) coupling gel applied for improving the efficiency of the US wave transmission. Most imaging modes provide real-time readout, allowing operators to freeze the image acquisition and review a set number of recent captures, either on the scanner or exported for offline analysis.

US imaging provides accurate images of various target organs, including tumor masses and plays a key role in preclinical cancer research. Use of doppler imaging enables real-time measurement of blood flow in cardiac chambers and in the vessels of the kidneys, liver, and uterus, while elastography techniques can assess tissue stiffness. Emerging innovative applications include the use of ultrasonic contrast microbubbles, engineered to target specific biological markers, enabling the visualization and qualitative analysis of blood perfusion and therapeutic distribution in animal models. As a research tool, noninvasive US imaging represents a significant advancement in evaluating and quantifying the structure and function of target organs in rodent species. This book chapter primarily focuses on specific research applications to demonstrate how US imaging is used in studies involving rodents, particularly in cancer and cardiac research.

2 B-mode Imaging

The most common imaging format for general US use is B-mode or brightness-mode, producing real-time 2D cross-sectional images of target structure. It operates in transmit-receive mode, where the transducer emits a US pulse, then captures reflected signals to create gray-scale images. The appearance of organs as brighter (hyperechoic) or darker (hypoechoic) on B-mode US is determined by the structural characteristics and tissue composition, which influence the reflection and transmission of sound waves. Typically, calcified structures (bones, gall and kidney stones) and dense tissues appear brighter as most of the transmitted US is backscattered (reflected) to the transducer. Fluid-filled chambers (blood vessels, cysts, bladder, gallbladder) poorly reflect sound waves resulting in a darker appearance. By reducing the size of field of view (FOV) and number of focal zones, 2D B-mode can achieve high frame rates which is useful to obtain imaging kinetics. High frame rate 2D B-mode assists in the guidance for injections and implementation of various imaging techniques, such as M-mode, color doppler and contrast enhanced physiological imaging.

Although 2D B-mode imaging is primarily used to evaluate cardiac structure and function due to fast frame rates, its 3D extension is widely applied in preclinical cancer research to assess volumetric changes in specific organs. To capture the volume of a target organ, the US transducer is moved in precise increments across the X-Y plane using a stepper motor which the visualization software then renders the stack of 2D slices into a 3D volume. Traditionally, caliper measurements have been used to assess tumor burden in cancer research; however, these are typically limited to subcutaneous lesions, the 3D formula assumes a spheroid shape and exhibit significant variability among technicians (Jensen et al. 2008). These challenges are effectively addressed by 3D B-mode volumetric measurements, which offer enhanced accuracy and consistency compared to caliper methods (Roth et al. 2024) and has been extensively utilized in the study of disease biology, with particular emphasis on cancer research, where it facilitates the monitoring of tumor progression and the evaluation of therapeutic efficacy. The next subsections provide specific examples in B-mode imaging.

2.1 Ovary

The ovary is a female reproductive organ responsible for oocyte production and secretion of hormones that regulate the menstrual cycle and fertility. Polycystic ovary syndrome (PCOS), the most common hormonal disorder in women, is characterized by the presence of ovarian cysts and dysregulation of reproductive hormones. In preclinical models, B-mode US is utilized to assess ovarian morphology, enabling the identification of large antral follicles and fluid-filled cysts (Feng et al. 2018). Follicular and luteal dynamics can also be studied using this method (Jaiswal et al. 2009). Additionally, B-mode imaging is used to characterize and longitudinally monitor development of cystic follicles in chemically induced PCOS models (Ren et al. 2024) and to study spontaneous preclinical ovarian cancer models (Fig. 1a, b), which closely mimic the pathophysiology of the disease (Lyshchik et al. 2007).

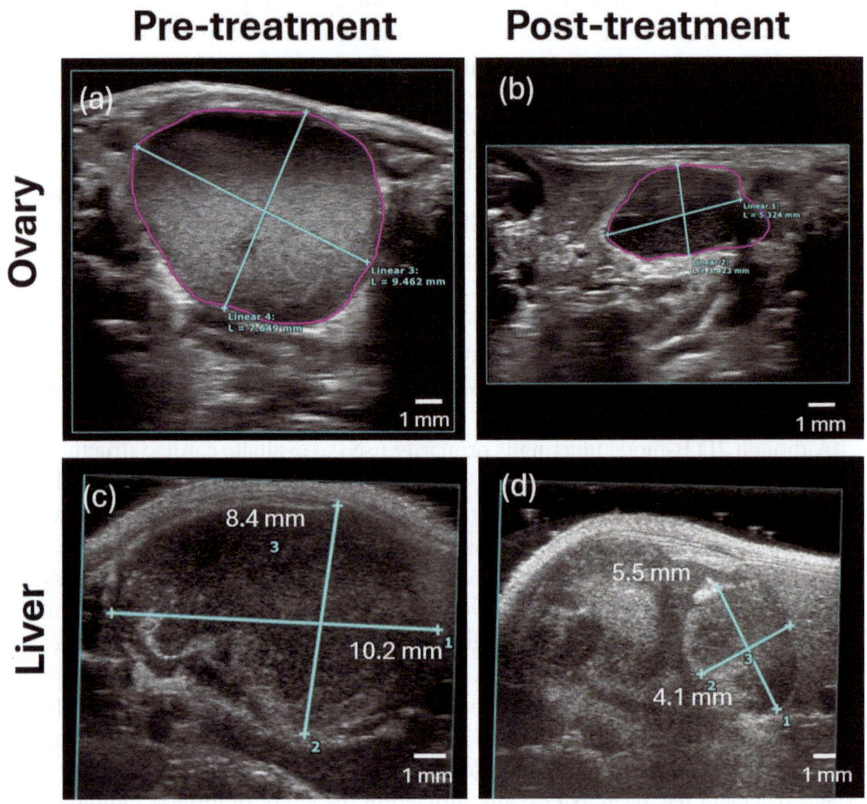

Fig. 1 3D B-mode tumor imaging. Representative slice (longest tumor diameter) of a murine syngeneic allograft ovarian tumor: (**a**) before and (**b**) after 6 weeks of treatment. Representative slice (longest tumor diameter) of a Neuroblastoma PDX liver metastatic tumor: (**c**) before and (**d**) after 7 weeks of treatment

2.2 Liver

B-mode US is commonly utilized to assess liver structure, detect conditions such as hepatic steatosis, fibrosis, and tumor growth (Fig. 1c, d), and evaluate the response to treatments in models of nonalcoholic fatty liver disease (NAFLD), nonalcoholic steatohepatitis (NASH), and cirrhosis (Hayakawa et al. 2023; D'Souza et al. 2019; Kuroda et al. 2012). Comparing the brightness of the liver to the renal cortex, B-mode US imaging can assess liver fibrosis and cirrhosis (D'Souza et al. 2019). Elastography is a specialized US technique used to quantify liver stiffness (Czernuszewicz et al. 2022), measuring the change in velocity of the shearwave generated by an external mechanical pulse, allowing for longitudinal monitoring of disease progression and evaluation of the impact of potential therapeutic interventions (D'Souza et al. 2019; Lessa et al. 2010).

2.3 Pancreas

B-mode US has become an invaluable tool which allows real-time monitoring of the pancreas, offering a cost-effective alternative to more expensive and time-consuming techniques like MRI. Despite challenges posed by the pancreas's undefined shape, small size, and deep location in murine models, 3D B-mode US enables the study of pancreatic pathophysiology and treatment efficacy. B-mode US can provide detailed images of the pancreatic structure in head, body and tail locations (Fig. 2a-c) (Sastra and Olive 2013), allowing researchers to observe changes in size, echogenicity, and morphology indicative of disease progression or therapeutic response (Goetze et al. 2018). For instance, in genetically engineered mouse models (GEMM) and orthotopic pancreatic cancer models, 3D B-mode US imaging has been a primary imaging modality to non-invasively monitor tumor onset and growth (Sastra and Olive 2013; Goetze et al. 2018; Rojas et al. 2022; Cook et al. 2008). In GEMMs, abnormalities such as large cysts, enlarged gallbladder,

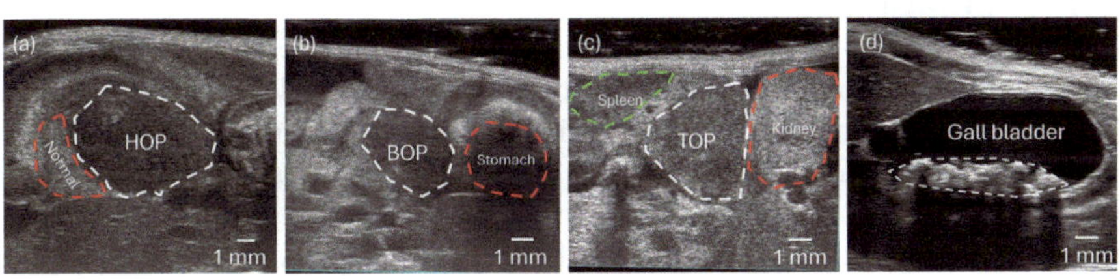

Fig. 2 3D B-mode imaging of pancreatic GEMM tumor model. (**a**) Representative slice showing tumor in Head of Pancreas (shown with white dotted line), normal pancreas is shown with red dotted line. (**b**) Representative image showing tumor in Body of Pancreas (shown with white dotted line), stomach (anatomical reference) is shown with red dotted line. (**c**) Representative image showing tumor in Tail of Pancreas (shown with white dotted line), spleen and kidney (anatomical reference) are shown with green and red dotted lines, respectively. (**d**) Enlarged gallbladder with stones (shown with white dotted line)

spleen, or abnormal liver can be detected by B-mode US during prescreening (Fig. 2d), thereby reducing potential biases in drug efficacy data.

3 US Guided Injections

Murine models of human diseases are essential in the translational research process. Orthotopic tumor mouse models are becoming increasingly popular in cancer research due to their greater clinical relevance compared to subcutaneous xenograft models (Stribbling et al. 2024). The use of image guidance significantly enhances the accuracy of orthotopic injections compared to traditional blind methods (Price et al. 2021), enabling more precise targeting while reducing the need for extensive technical experience. Specifically, US-guided injections are commonly employed in the development of cancer and cardiovascular murine models, as this technique improves injection precision and accuracy, resulting in higher success rates and a marked reduction in mortality compared to conventional manual methods. Additionally, this approach reduces pain, stress, and recovery time in animals, thereby improving the ethical standards of research (Camara Serrano 2022). After the injection, B-mode US imaging serves as a reliable tool for monitoring disease progression (McVeigh et al. 2019). In cases where the tumor is not detectable via US due to its location (i.e. lung), alternative imaging methods such as bioluminescence or magnetic resonance imaging (MRI) can be used to verify the accuracy of the injections and assess disease progression (Balathasan et al. 2013; Labora et al. 2023; Choi et al. 2023; Zhou and Zhao 2014). This technique is broadly applicable to any organ visible via B-mode US, a comprehensive review of the organs suitable for injection is available elsewhere (Camara Serrano 2022). The following subsections describe the two commonly implemented image guided injection routes. In conclusion, the integration of US-guided injections represents a significant advancement in the creation of reliable and consistent animal models of cancer.

3.1 Intracardiac

Cell injections via intracardiac route has been utilized for the development of rodent metastatic cancer models, where injections via the left ventricle (LV) result in metastasis to the brain (Balathasan et al. 2013) and bone (Chang et al. 2022) while right ventricle (RV) injections develop metastasis to the lung (Labora et al. 2023). When these injections are administered using the blind (non-imaging) method, their success is largely dependent on the skills and experience of the technician. US-guided cardiac injections (Fig. 3a, b) mitigate this dependency, ensuring real-time accuracy and consistency in injection placement. This approach decreases the number of mis-injected mice included in the study, enhances the distribution of tumor cells, and reduces the pulmonary metastatic burden typically observed with tail vein injections (Price et al. 2021).

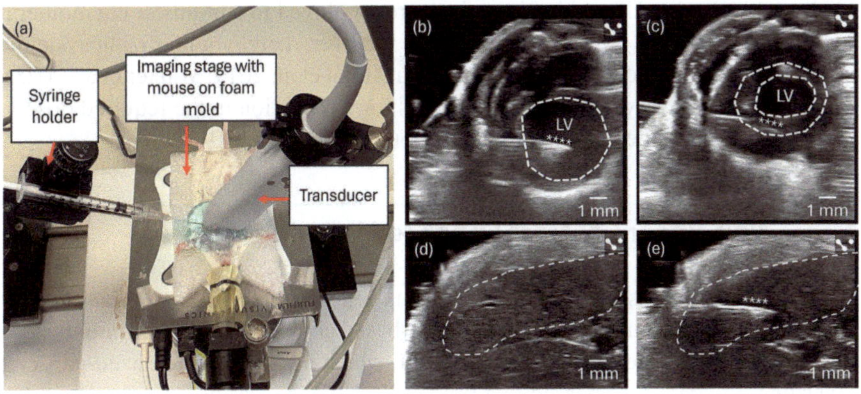

Fig. 3 Image guided injections. (**a**) Typical imaging setup for intracardiac injection. Foam mold is used to restrict animal movement and to provide reproducible placement of mouse during injections. (**b**) B-mode US image of left ventricle (area inside the dotted line) with needle (tip marked by "*"). (**c**) B-mode US image showing needle (tip marked by "*") placement in myocardium (area between the two dotted line). (**d**) B-mode US image showing liver lobe (area inside the dotted line). (**e**) Needle placement (tip marked by "*") in liver lobe

In cardiovascular research, murine models are essential for elucidating pathophysiological mechanisms and evaluating therapeutic interventions. Traditional intramyocardial injection methods typically require invasive procedures, such as sternotomy, which can limit experimental design and lead to complications, including fibrosis. The US-guided intramyocardial injection technique offers a minimally invasive approach that enables precise delivery of agents to the myocardium without the need for thoracotomy (Fig. 3c), minimizes associated complications and enhances flexibility in experimental protocols. This provides a technically straightforward and efficient method for myocardium manipulation in murine studies.

3.2 Intrahepatic

The tumor microenvironment (TME) is considered a critical factor influencing tumor growth, therapeutic response, and immune interactions in vivo (McVeigh et al. 2019). Tumors grown subcutaneously fail to replicate the complex, naturally occurring TME. Hence, subcutaneous models, in general, exhibit limited predictive value for drug efficacy in humans, highlighting the need for more clinically relevant orthotopic models of human cancers. To address this need, US-guided cancer cell injections are used to establish orthotopic liver tumor/metastatic models (Fig. 3d, e). Implementing B-mode US, liver lobes can be identified, and cells injected directly into the hepatic parenchyma or portal vein for developing rodent models for Hepatocellular carcinoma (HCC) (Bou About et al. 2019), and liver metastatic model of melanoma and colorectal cancers (McVeigh et al. 2019; Labora et al.

2023). Additionally, US-guided injections are employed to administer chemotherapeutic agents and cell therapies directly into the liver, thereby minimizing systemic exposure and associated adverse effects (Bou About et al. 2019; Walker et al. 2023). This localized delivery approach also allows for a reduction in the required dose to achieve the desired therapeutic outcome (Bou About et al. 2019).

4 Non-linear Contrast Enhanced Imaging

Molecular imaging using US contrast agents relies on principles similar to other imaging modalities, such as X-ray and MRI-based contrast agents. Commonly used US contrast agents are microbubbles (MBs), which are micrometer-sized particles consisting of a gas core encased in a lipid shell that enhances backscatter of the US pulse, improving image contrast. MBs can be non-targeted, allowing visualization of blood flow and vascularity changes, or functionalized with polysaccharides, antibodies, or peptides for targeted molecular imaging (Alzaraa et al. 2012; Rychak et al. 2006; Takalkar et al. 2004). Functionalized MBs selectively accumulate at target sites, enhancing imaging of specific biomarkers or receptors associated with disease. In general, Contrast-enhanced US (CEUS) is effective for assessing tissue perfusion, molecular biomarker quantification, necrotic regions, and angiogenesis. The following subsections provide a detailed description of the imaging techniques employed for non-targeted and targeted microbubbles.

4.1 Non-targeted Microbubbles

When using non-targeted contrast agents, the focus is on analyzing the perfusion kinetics of a tissue, organ, or tumor. The contrast agent can be administered via bolus injection (Fig. 4a), generating a time versus intensity curve (Fig. 4b), or continuously infused with a destructive pulse (Fig. 4c) to produce a replenishment curve (Foster et al. 2011). Both methods are incorporated in the instrument-specific software, which uses proprietary filtering to remove tissue signal responses, isolating the microbubble response for quantification (Liang and Blomley 2003). Perfusion modeling combines amplitude-related parameters (echo power) and time-related parameters (in seconds) to derive blood flow metrics. Fitting a general equation for describing the uptake and washout of the microbubbles and displaying the fitted parameters (Parametric map) in a false color images provide for easier visualization (Fig. 4d) of the organ/tumor perfusion. Studies have shown that non-targeted CEUS imaging is more sensitive than B-mode imaging in diagnosing early-stage fatty infiltration and vascular changes in liver parenchyma and provides a more accurate diagnosis of steatohepatitis in mouse models (Pandit et al. 2019). Additionally, sub-micrometer-sized non-targeted nanobubble contrast agents predict disease progression and therapeutic reversal early in asymptomatic type 1 diabetes by accumulating in pancreatic islets, correlating with insulitis. This accumulation

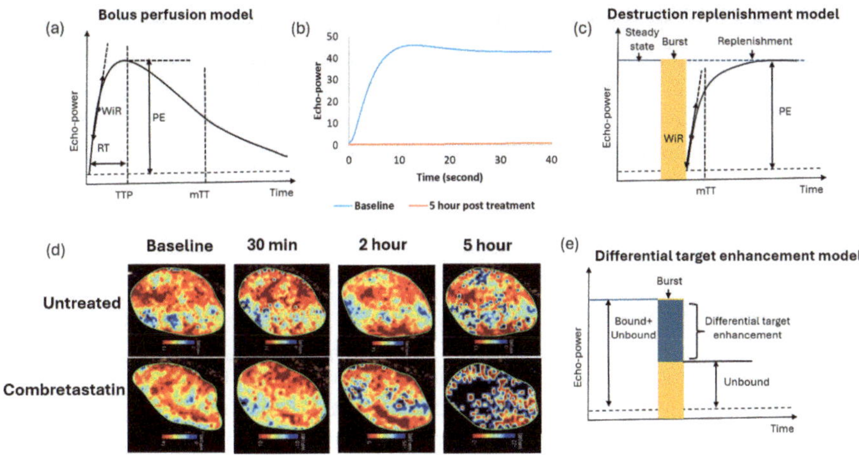

Fig. 4 Contrast Enhanced US. (**a**) Schematic of bolus perfusion model with derived parameters. (**b**) Representative time-intensity plot before and after therapy. (**c**) Schematic of destruction replenishment model with derived parameters. (**d**) Representative parametric map of WiR for untreated and treated alveolar soft-part sarcoma tumors at various timepoints. (**e**) Schematic of Differential target enhancement model. *TTP* time to peak, *RT* rise time, *PE* peak enhancement, *WiR* wash-in-rate, *mTT* mean transit time. (The images in panels **a**, **c** and **e** are modified with permission from Fujifilm VisualSonics, Inc. (https://www.visualsonics.com/sites/default/files/PB_2100_Cb_VevoCQ.pdf))

can be detected early in disease progression and decreases with therapeutic intervention (Ramirez et al. 2020).

4.2 Targeted Microbubbles

To assess targeted MBs, a late-phase contrast enhancement technique is utilized. In this approach, targeted MBs are injected intravenously, and imaging is paused for 3–4 min to allow the MBs to bind to the target. Following this period, a couple hundred frames are acquired, capturing signals from both bound and freely circulating MBs. A high-intensity US pulse is then applied to destroy the MBs within the imaging plane, after which additional frames are captured that reflect only the signals from the freely circulating MBs. The differential signal, derived from the data obtained before and after the destruction pulse, provides an estimate of the quantity of bound contrast agent (Fig. 4e) (El Kaffas et al. 2017). When generated for both a specific target and a negative control, this semi-quantitative value allows assessment of the relative expression for the target and color-coded parametric images are created for qualitative assessment of the spatial distribution of the bound contrast agent. This method is widely used to detect and quantify tumor angiogenesis in murine models (Willmann et al. 2010; Liu et al. 2020). VEGFR1-targeted MBs are commonly used (Liu et al. 2020), while efforts continue to develop novel MBs targeting various in vivo cell surface markers (Willmann et al. 2010). These site-specific markers enable target validation, blocking studies, and drug efficacy evaluation (Willmann et al. 2010; Liu et al. 2020). MBs bearing antibodies targeting lymphocyte CD4, endothelial P-selectin, and E-selectin expand the use of this technique to study inflammatory diseases, such as chronic inflammatory bowel disease, myocardial inflammation, colitis, and treatment efficacy (El Kaffas et al. 2017; Wang et al. 2018; Steinl et al. 2016).

5 Echocardiography

US imaging is widely utilized in preclinical cardiac research (echocardiography) as a non-invasive technique for evaluating cardiac function and morphology in animal models. Echocardiography offers real-time, quantitative measurements of heart chamber dimensions, contractility, blood flow velocities, and valve function, thereby providing critical insights into cardiovascular pathophysiology. Various echo modes offer distinct information regarding various aspects of cardiac health, and the integration of multiple modes facilitates a comprehensive assessment of cardiac function. The following subsections provide detailed descriptions of commonly employed cardiac imaging modes.

5.1 M-mode

M-mode (motion) imaging is a specialized US technique for assessing rapidly moving cardiac structures. This technique captures data along a single line in the B-mode image, providing high temporal resolution for observing rapid cardiac movements such as the heart wall, offering valuable insights into dynamic cardiac function (Ram et al. 2011).

Rodent echocardiography requires a solid-state transducer for imaging elevated heart rates (>500 BPM) in conscious mice which requires prior acclimatization (handling of the mouse over several sessions) to reduce stress and to minimize parasympathetic nervous system activation before imaging (Mitchell et al. 1963; Lindsey et al. 2018; Yang et al. 1999). Standard preclinical imaging implements anesthesia to minimize motion artifacts but for echocardiography, rodents can be imaged in the conscious state to avoid the effects of anesthesia on the cardiovascular system (Pachon et al. 2015). Cardiac images are acquired in a specific orientation from parasternal long-axis (PLAX) or short-axis (SAX), which is perpendicular to the long-axis, to evaluate Left Ventricle (LV) function (Fig. 5a, b). In M-mode, depth is represented along the vertical axis, while time is displayed along the horizontal axis, yielding a high-resolution, temporal representation of cardiac motion, for assessing LV structural and functional properties during the cardiac cycle (end-systole and end-diastole). Quantitative measurements of the interventricular septum (IVS), LV internal diameter (LVID), and LV posterior wall (LVPW) are facilitated using echocardiography software (Fig. 5a) and are critical for calculating ejection

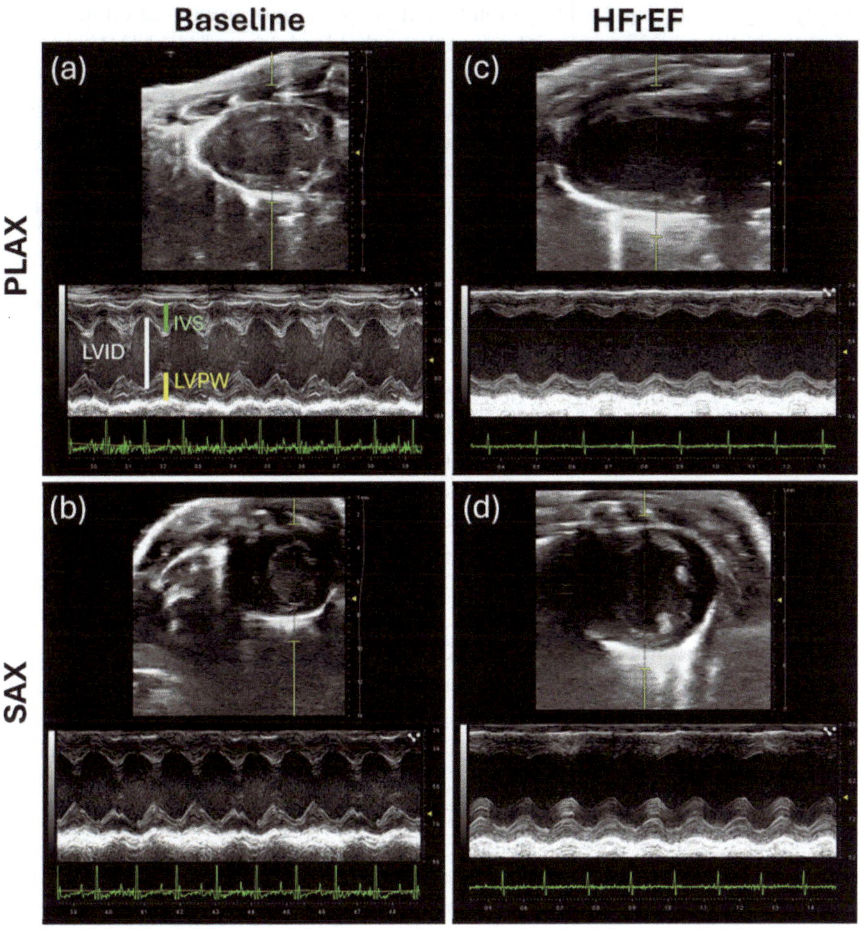

Fig. 5 M-mode imaging. (**a**) Parasternal long axis (PLAX) M-mode image in healthy mouse with geometric measurements. (**b**) Parasternal short axis (SAX) M-mode image in healthy mouse. (**c**) PLAX M-mode image in HFrEF model. (**d**) SAX M-mode image in HFrEF model. *IVS* interventricular septal, *LVID* Left ventricle internal dimensions, *LVPW* left ventricle posterior wall

fraction (EF) and fractional shortening (FS), which serve as key indicators of myocardial performance and contractility (Gao et al. 2011).

These measurements provide an accurate technique to investigate rodent heart disease models, including those that examine adverse cardiac remodeling following interventions such as left anterior descending artery (LAD) ligation (Yang et al. 2002), transverse aortic constriction (TAC), or pharmacological stress (Patten and Hall-Porter 2009). These models are crucial for elucidating the pathophysiology of heart failure, a major cause of global morbidity and mortality (Murphy et al. 2020). A hallmark of these models is Heart Failure with reduced Ejection Fraction (HFrEF), indicative of impaired LV contractile function and inadequate blood supply (Fig. 5c, d). The LAD ligation model, induces myocardial infarction (MI) in mice, providing significant insights into the pathophysiology of HFrEF (Yang et al. 2002). However, M-mode imaging may yield inaccuracies when quantifying infarct size as the assessment is only limited to a single slice on the imaging plane (Yuan et al. 2011; Dann et al. 2022; Dawson et al. 2004). To overcome this issue, recent advanced imaging techniques, such as 4D US, can provide a more comprehensive and accurate assessment, enabling detailed evaluation of overall cardiac health in these models (discussed in Sect. 5.3).

5.2 Global Longitudinal Strain and Strain Rate Imaging

LV ejection fraction (LVEF), derived from M-mode imaging, has long been considered a critical parameter for assessing LV systolic function in echocardiography. However, LVEF has several limitations; it serves as an indirect measure of myocardial contractility, rather than directly assessing myocardial function, and it can be influenced by external factors, such as heart rate, potentially leading to inaccurate interpretations. More importantly, LVEF lacks the sensitivity to detect subtle changes in myocardial contractility, making it less effective in identifying early or subclinical myocardial damage, which can have significant therapeutic and prognostic implications (Bansal and Kasliwal 2013). In contrast, strain imaging directly quantifies myocardial deformation, overcoming many of the limitations associated with M-mode derived LVEF.

Echocardiographic strain and strain rate imaging, also referred to as deformation imaging, are based on speckle tracking technology, which tracks the movement of naturally occurring speckle patterns within the myocardium on a slice-by-slice basis (Bansal and Kasliwal 2013). Strain is a dimensionless parameter that quantifies myocardial deformation relative to the heart's original shape, while strain rate represents the speed at which this deformation occurs over time (Dandel et al. 2009). This allows for precise quantification of myocardial deformation (strain) and the rate of deformation (strain rate), providing valuable insights into both regional and global cardiac function.

Global Longitudinal Strain (GLS) is used for evaluating LV systolic function, measuring longitudinal myocardial shortening throughout the cardiac cycle (Yang et al. 2002). GLS is typically calculated by averaging the peak strain values obtained from the apical views, spanning from the apex to the aortic root, thereby providing a comprehensive measure of overall LV contractility (Fig. 6a, b) (Bansal and Kasliwal 2013). A reduction in the GLS value is indicative of impaired myocardial function and is commonly observed in conditions such as heart failure, ischemia, and cardiomyopathy (Fig. 6c, d) (Sławiński et al. 2023; Bauer et al. 2011). Unlike conventional echocardiographic measurements, which are influenced by operator variability, GLS is primarily derived through automated image processing software, thereby minimizing inter-operator variability (Karlsen et al. 2019). Furthermore, GLS can detect early subclinical myocardial dysfunction, like HFpEF (Daniels et al. 2023) which demonstrates GLS as a valuable tool for identifying myocardial

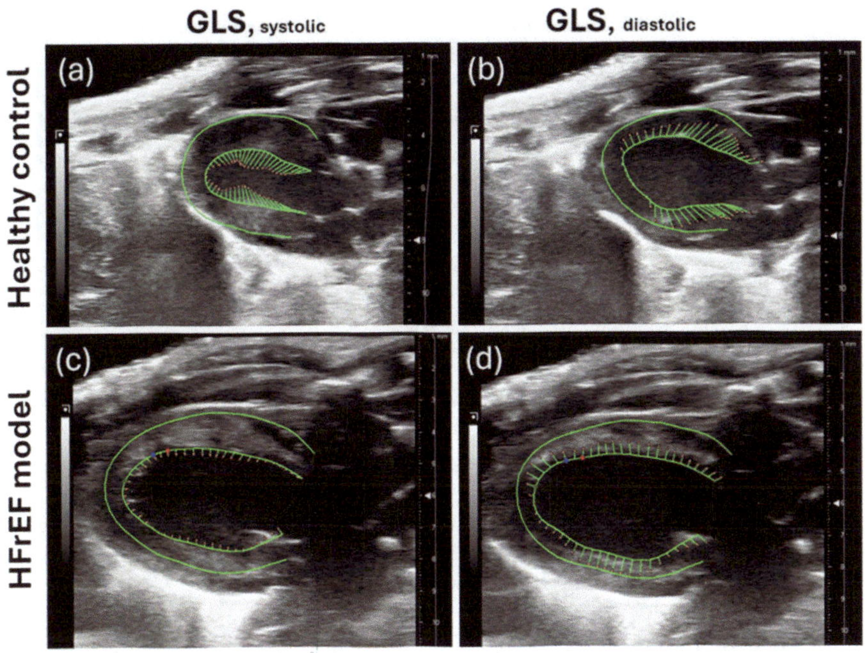

Fig. 6 Strain rate imaging. Image showing Global longitudinal strain (GLS) in healthy mouse with radial traces during (**a**) systolic phase and (**b**) diastolic phase. Representative image of GLS in HFrEF model with abnormal radial traces during (**c**) systolic phase and (**d**) diastolic phase

dysfunction in rodent models, often revealing abnormalities not detectable by conventional M-mode imaging techniques.

5.3 4D Echocardiography

Four-dimensional (4D) echocardiography is an advanced imaging technique that enables the visualization of myocardial dynamics throughout the cardiac cycle. Unlike traditional two-dimensional (2D) echocardiography (Fig. 7a, b), which captures a single cardiac gated plane, 4D echocardiography incorporates the heart's three-dimensional geometry (stitching multiple 2D gated scans), providing an improved, accurate and comprehensive, real-time 3D-gated depiction of cardiac function (Dann et al. 2022; Rutledge et al. 2020).

A key advantage of 4D echocardiography is its ability to visualize myocardial motion in a 3D-gated format, facilitating an enhanced and more accurate assessment of myocardial deformation, alterations in chamber volume, and the interactions between cardiac structures throughout the cardiac cycle. This capability is especially valuable in evaluating LAD-induced MI, as it provides critical information on the extent of myocardial damage, functional impairment, and potential recovery or

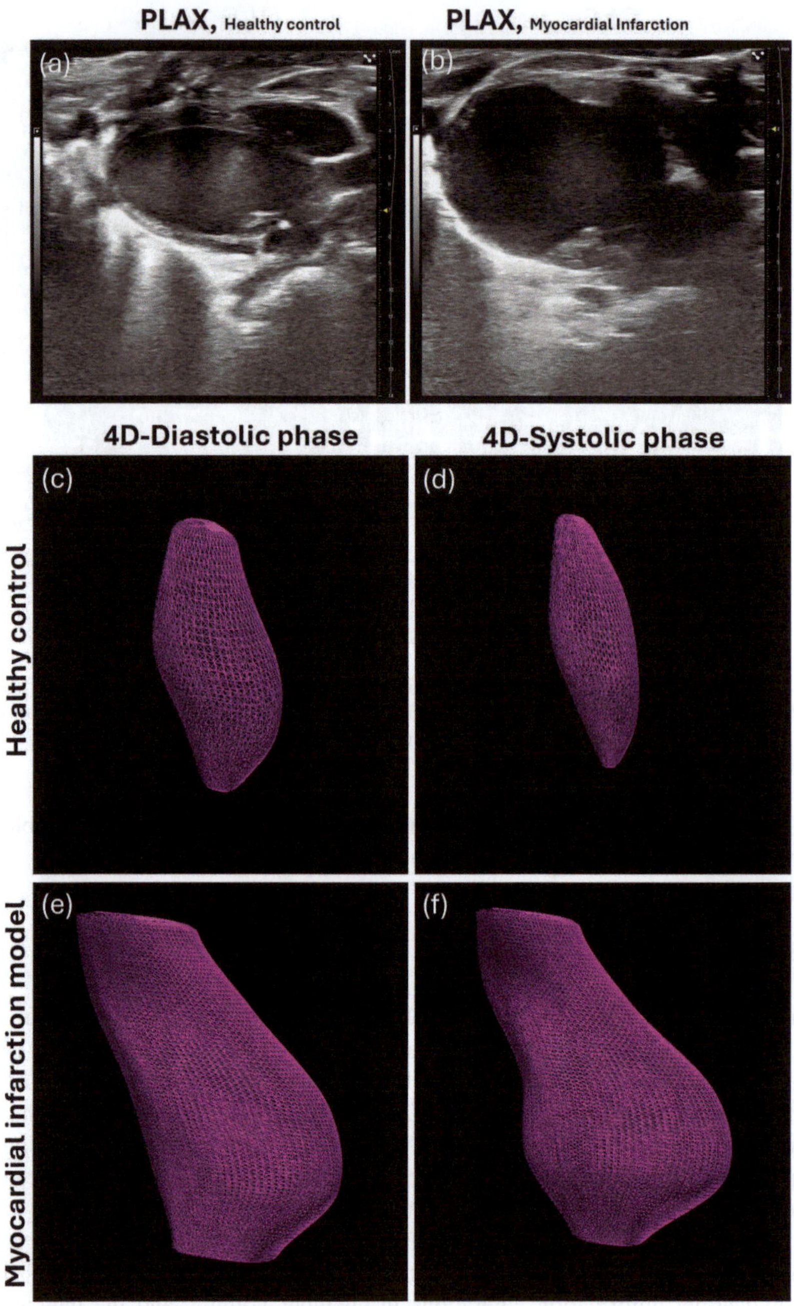

Fig. 7 B-mode and 4D imaging. (**a**) Parasternal long axis (PLAX) B-mode image in healthy mouse. (**b**) PLAX B-mode image after inducing myocardial infarction (MI). (**c**) 4-D image of heart in healthy mouse during diastolic phase. (**d**) 4-D image of heart in healthy mouse during systolic phase. (**e**) 4-D image of heart in MI model during diastolic phase. (**f**) 4-D image of heart in MI model during systolic phase

remodeling following infarction (Fig. 7c–f) (Rutledge et al. 2020; Omoto et al. 2024). Traditional echocardiographic techniques such as M-mode imaging may be less reliable when assessing infarction or post-infarction remodeling, as they rely on 2D-gated images which do not account for 3D-gated regional abnormalities. Furthermore, synchronizing the electrocardiogram (ECG) of the 4D-gated echocardiogram with respiratory gating, which minimizes motion artifacts, the 4D-gated technique ensures high-quality data (Tsujita et al. 2005).

This enhanced technique provides significant advantages in evaluating myocardial damage and ventricular remodeling such as evaluating therapeutic strategies, especially when monitoring post-MI recovery and assessing responses to interventions such as revascularization, pharmacological treatments and cardiac toxicities (Omoto et al. 2024). 4D-gated technique offers real-time data on chamber volumes, wall motion, and LV remodeling, ultimately leading to improved diagnostic and prognostic outcomes in animal models following LAD ligation driven injury.

5.4 Doppler Mode

Color Doppler (CD), Pulsed Wave Velocity (PWV), and Tissue Doppler Imaging (TDI) are echocardiographic techniques that utilizes Doppler to assess distinct physiological parameters (Tsujita et al. 2005). Utilizing the Doppler principle, CD provides real-time, quantitative data on blood flow, its velocity (direction and speed) and turbulence (Viola et al. 2024). Utilizing the CD mode, PWV mode measures blood flow velocity at specific points in the heart or vasculature by detecting Doppler shifts from moving red blood cells. In contrast, TDI evaluates the velocity of myocardial tissue, providing insights into the dynamics of heart muscle rather than blood flow (Stypmann 2007).

Doppler imaging modes are extensively employed to evaluate diastolic dysfunction, a pathological condition characterized by impaired myocardial relaxation and inadequate ventricular filling during diastole (Viola et al. 2024; Stypmann 2007). This dysfunction is particularly prevalent in individuals with HFpEF, a condition where the ejection fraction remains within normal limits, but elevated filling pressures are observed due to compromised diastolic relaxation. To investigate HFpEF in vivo, murine models incorporating risk factors such as a high-fat diet (HFD) and N(γ)-nitro-L-arginine methyl ester (L-NAME) have been established (Zhang et al. 2024). Doppler imaging modes are pivotal in detecting early signs of diastolic dysfunction in this model by quantifying the velocity and timing of blood flow during various phases of the cardiac cycle (Zhang et al. 2024).

To evaluate the HFpEF murine model, the apical four-chamber (B-mode) view is commonly utilized (Stypmann 2007), which provides a comprehensive visualization of both the atria and ventricles, as well as the mitral valve (Fig. 8a). When coupled with CD imaging, it enables the evaluation of blood flow through the mitral valve (Fig. 8b) (Brunner et al. 2020). Diastolic function is characterized by four distinct phases, each of which generates unique signature peaks in PWV and TDI: (1) isovolumic relaxation time (IVRT) (Fig. 8c), (2) rapid filling (E wave in PWV

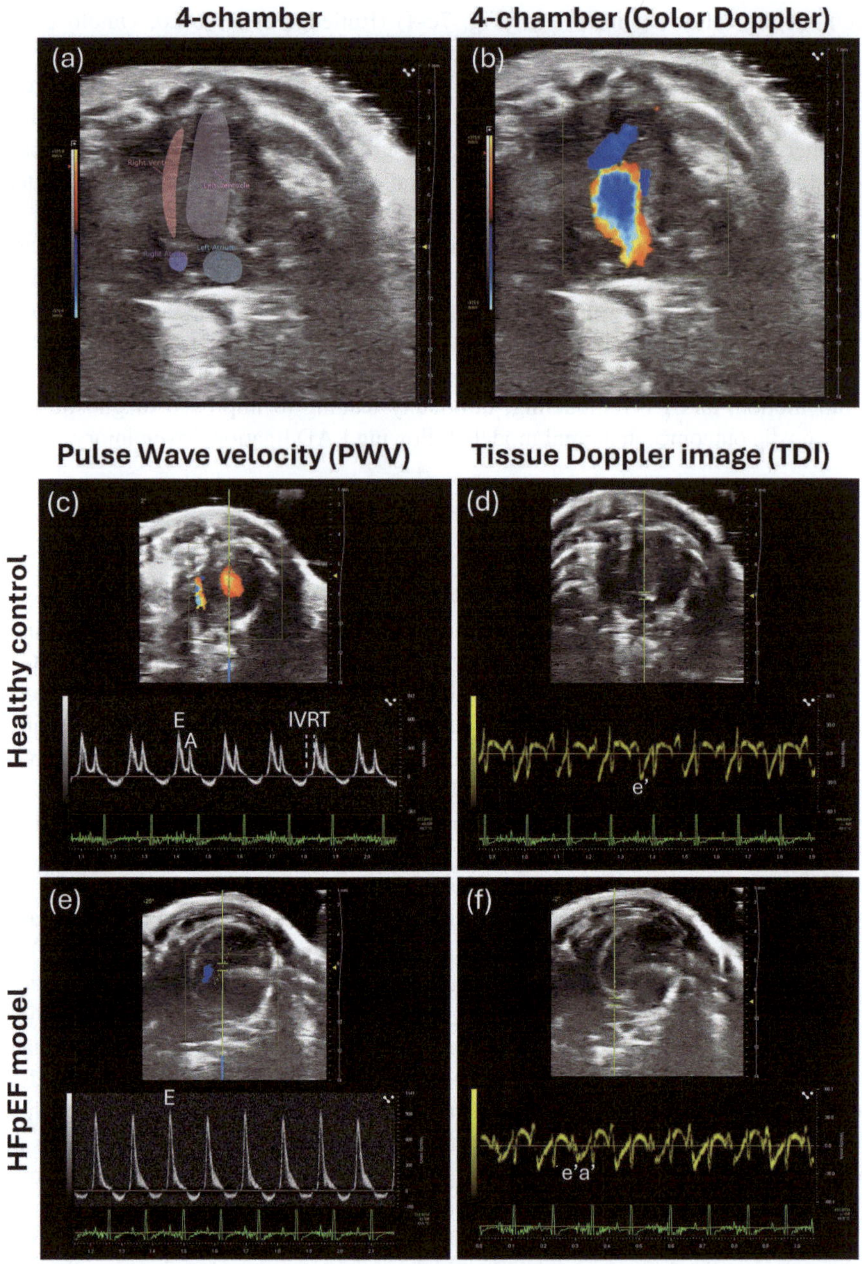

Fig. 8 Four-chamber view and doppler imaging. (**a**) B-mode four-chamber view. (**b**) Color Doppler mode image showing blood through mitral valve in healthy mouse. (**c**) Pulse Wave Velocity mode in healthy mouse. (**d**) Tissue Doppler mode in healthy mouse. (**e**) Pulse Wave Velocity mode in HFpEF model. (**f**) Tissue Doppler mode in HFpEF model. *E* Early diastolic flow peak velocity of the mitral valve, *A* Late flow peak velocity of the mitral valve, *IVRT* isovolumic relaxation time, e' early diastolic motion, a' late diastolic motion

or e' wave in TDI) (Fig. 8c, d), (3) slow filling, and (4) active filling (A wave in PWV and a' wave in TDI) (Fig. 8c, d) (Macé et al. 2018). In animal models with diastolic dysfunction, these signature patterns are modified, indicating the presence of pathological changes (Schnelle et al. 2018). For instance, in mice with HFpEF, an increased ratio of early (E-wave) to late (A-wave) blood flow velocities across the mitral valve is a common indicator of diastolic dysfunction (Fig. 8e). Similarly, an elevated E/e' ratio suggests increased LV filling pressures and serves as a critical marker of diastolic dysfunction (Fig. 8e, f). Furthermore, an extended IVRT implies elevated filling pressures. These alterations are strongly correlated with impaired relaxation and increased LV filling pressures, which are hallmark features of diastolic dysfunction in models of HFpEF (Schnelle et al. 2018).

6 Conclusions and Future Perspectives

US imaging is a widely utilized, non-invasive in vivo technique used in preclinical, translational, and clinical research; particularly in cancer, and cardiovascular modeling. It provides real-time, high-resolution readout of anatomical structures and physiological processes and allows longitudinal monitoring with minimal disruption to the subject. Compared to other radiological modalities, preclinical US systems are cheaper to purchase and maintain, and daily quality control tests are not necessary to assure the scanner is operating within manufacturer specifications. B-mode imaging is used to visualize anatomical features such as organs, and vascular structures, allowing for detailed assessments of changes in tumor growth, and organ development. US imaging is also an integral tool in oncology research, where it helps prescreen tumor models for evaluating study enrollment, monitor tumor growth and evaluate efficacy of anticancer therapies. Non-targeted MBs are used to evaluate blood volume/blood flow for assessing angiogenesis, vascular integrity, and responses to vascular-targeting therapies, while functionalized MBs are used for targeted imaging of specific biomarkers, aiding in the study of various diseases (i.e., angiogenesis), as well as evaluating therapeutic interventions. Additionally, US-image guided cancer cell injections are commonly used for reliably developing orthotopic cancer models while reducing experimental cost by eliminating rework.

A major benefit of US imaging in preclinical to clinical research is in the realm of cardiovascular research where M-mode, Strain and strain rate imaging, and Doppler US modes are routinely employed to assess blood flow, hemodynamics, cardiac function including myocardial deformation, diastolic dysfunction, and post-infarction remodeling, for providing valuable insights into cardiovascular pathophysiology. M-mode imaging is used to evaluate LV function in rodent models to assess heart disease models, such as HFrEF and MI, by measuring parameters like EF and FS. However, its limitations in assessing 2D infarct size can be overcome with advanced techniques like 4D US, which offers real-time, 3D cardiac imaging for a more accurate assessment of myocardial dynamics, LV function, and post-infarction remodeling. Advanced techniques such as strain and strain-rate imaging, quantify myocardial deformation and are sensitive to early dysfunction in conditions such as HFpEF which is difficult to detect by 2D M-mode imaging. Doppler imaging is crucial for assessing diastolic dysfunction, a key feature of HFpEF, by monitoring distinct signatures of PWV and TDI during the cardiac cycle, offering valuable insights into cardiac pathophysiology and improving diagnostic and prognostic evaluations in rodent models.

Although not discussed in the chapter, the US plays a prominent role in assessing the distribution of therapeutic agents, including nanomedicines (assessing scatters from Au nanoparticles). Techniques like US-mediated drug delivery (where the US intense signal generates cavitation-sonoporation) enhances cellular membrane permeability, therefore improving the targeted administration of drugs in preclinical models (Alzaraa et al. 2012). Recently, the use of US in neurological imaging has been explored for brain studies in small animal models. Functional US (fUS) has gained attention as a promising technique for investigating brain activity, blood flow, and the blood-brain barrier in neurovascular research (Brunner et al. 2020; Macé et al. 2018). Overall, the versatility, real-time imaging capabilities, translatability to the clinic, and non-invasive nature make US an indispensable tool in preclinical research, facilitating the understanding of disease mechanisms, the development of new therapeutic strategies, and the optimization of drug delivery systems.

Acknowledgements This project has been funded in whole or in part with Federal funds from the National Cancer Institute, National Institutes of Health, under Contract No. HHSN261201500003I. The content of this publication does not necessarily reflect the views or policies of the Department of Health and Human Services, nor does mention of trade names, commercial products, or organizations imply endorsement by the U.S. Government. The authors thank Ms. Lisa Riffle for her valuable technical expertise in various imaging techniques. Additionally, the authors express their gratitude to Dr. Lino Tessarollo, Dr. Ludmila Szabova, Dr. Christine Heske, and Dr. Serguei Kozlov for providing the animal models used in this book chapter.

Conflict of Interest NLP declares that he joined AstraZeneca after the preparation and submission of this chapter but during the publication process. His affiliation at the time of submission remains listed, but his updated email is provided for correspondence only. NLP also owns or may own AstraZeneca stock options in future. All authors, including NLP, declare that this book chapter was written in the absence of any commercial or financial relationships that could be construed as a potential conflict of interest.

References

Alzaraa A et al (2012) Targeted microbubbles in the experimental and clinical setting. Am J Surg 204:355–366

Balathasan L, Beech JS, Muschel RJ (2013) Ultrasonography-guided intracardiac injection: an improvement for quantitative brain colonization assays. Am J Pathol 183:26–34

Bansal M, Kasliwal RR (2013) How do I do it? Speckle-tracking echocardiography. Indian Heart J 65:117–123

Bauer M et al (2011) Echocardiographic speckle-tracking based strain imaging for rapid cardiovascular phenotyping in mice. Circ Res 108:908–916

Bou About G et al (2019) Ultrasound-guided approaches to improve orthotopic mouse xenograft models for hepatocellular carcinoma. Curr Protoc Mouse Biol 9:e62

Brunner C et al (2020) A platform for brain-wide volumetric functional ultrasound imaging and analysis of circuit dynamics in awake mice. Neuron 108:861–875.e867

Camara Serrano JA (2022) Ultrasound guided surgery as a refinement tool in oncology research. Animals (Basel) 12(23):3445

Chang J et al (2022) Intra-cardiac injection of human prostate cancer cells to create a bone metastasis xenograft mouse model. J Vis Exp 189

Choi SE et al (2023) High-frequency ultrasound-guided intrathecal injections in a young mouse model: targeting the central nervous system in drug delivery. J Neurosci Methods 386:109778

Coatney RW (2001) Ultrasound imaging: principles and applications in rodent research. ILAR J 42:233–247

Cook N, Olive KP, Frese K, Tuveson DA (2008) K-Ras-driven pancreatic cancer mouse model for anticancer inhibitor analyses. Methods Enzymol 439:73–85

Czernuszewicz TJ et al (2022) Development of a robotic shear wave elastography system for noninvasive staging of liver disease in murine models. Hepatol Commun 6:1827–1839

Dandel M, Lehmkuhl H, Knosalla C, Suramelashvili N, Hetzer R (2009) Strain and strain rate imaging by echocardiography – basic concepts and clinical applicability. Curr Cardiol Rev 5:133–148

Daniels LJ et al (2023) Myocardial deformation imaging by 2D speckle tracking echocardiography for assessment of diastolic dysfunction in murine cardiopathology. Sci Rep 13:12344

Dann MM et al (2022) Quantification of murine myocardial infarct size using 2-D and 4-D high-frequency ultrasound. Am J Phys Heart Circ Phys 322:H359–H372

Dawson D et al (2004) Quantitative 3-dimensional echocardiography for accurate and rapid cardiac phenotype characterization in mice. Circulation 110:1632–1637

D'Souza JC et al (2019) B-mode ultrasound for the assessment of hepatic fibrosis: a quantitative multiparametric analysis for a radiomics approach. Sci Rep 9:8708

El Kaffas A et al (2017) Molecular contrast-enhanced ultrasound imaging of radiation-induced p-selectin expression in healthy mice colon. Int J Radiat Oncol Biol Phys 97:581–585

Feng Y, Tamadon A, Hsueh AJW (2018) Imaging the ovary. Reprod Biomed Online 36:584–593

Foster FS, Hossack J, Adamson SL (2011) Micro-ultrasound for preclinical imaging. Interface Focus 1:576–601

Gao S, Ho D, Vatner DE, Vatner SF (2011) Echocardiography in mice. Curr Protoc Mouse Biol 1:71–83

Goetze RG et al (2018) Utilizing high resolution ultrasound to monitor tumor onset and growth in genetically engineered pancreatic cancer models. J Vis Exp 134:56979

Greco A et al (2012) Ultrasound biomicroscopy in small animal research: applications in molecular and preclinical imaging. J Biomed Biotechnol 2012:519238

Hayakawa F et al (2023) Utility of ultrasonography in a mouse model of non-alcoholic steatohepatitis induced by a choline-deficient, high-fat diet and dextran sulfate sodium. Biochem Biophys Rep 36:101575

Jaiswal RS, Singh J, Adams GP (2009) High-resolution ultrasound biomicroscopy for monitoring ovarian structures in mice. Reprod Biol Endocrinol 7:69

Jensen MM, Jorgensen JT, Binderup T, Kjaer A (2008) Tumor volume in subcutaneous mouse xenografts measured by microCT is more accurate and reproducible than determined by 18F-FDG-microPET or external caliper. BMC Med Imaging 8:16

Karlsen S et al (2019) Global longitudinal strain is a more reproducible measure of left ventricular function than ejection fraction regardless of echocardiographic training. Cardiovasc Ultrasound 17:18

Kuroda H et al (2012) Non-invasive determination of hepatic steatosis by acoustic structure quantification from ultrasound echo amplitude. World J Gastroenterol 18:3889–3895

Labora A et al (2023) Generation of liver metastases in a mouse model using ultrasound-guided intravenous injection. STAR Protoc 4:102163

Lessa AS et al (2010) Ultrasound imaging in an experimental model of fatty liver disease and cirrhosis in rats. BMC Vet Res 6:6

Liang HD, Blomley MJ (2003) The role of ultrasound in molecular imaging. Br J Radiol 76(2):S140–S150
Lindsey ML, Kassiri Z, Virag JAI, de Castro Brás LE, Scherrer-Crosbie M (2018) Guidelines for measuring cardiac physiology in mice. Am J Phys Heart Circ Phys 314:H733–H752
Liu H et al (2020) VEGFR1-targeted contrast-enhanced ultrasound imaging quantification of vasculogenic mimicry microcirculation in a mouse model of choroidal melanoma. Transl Vis Sci Technol 9:4
Lyshchik A et al (2007) Ovarian volume measurements in mice with high-resolution ultrasonography. J Ultrasound Med 26:1419–1425
Macé É et al (2018) Whole-brain functional ultrasound imaging reveals brain modules for visuomotor integration. Neuron 100:1241–1251.e1247
McVeigh LE et al (2019) Development of orthotopic tumour models using ultrasound-guided intrahepatic injection. Sci Rep 9:9904
Mitchell JH, Wallace AG, Skinner NS (1963) Intrinsic effects of heart rate on left ventricular performance. Am J Physiol-Legacy Content 205:41–48
Murphy SP, Ibrahim NE, Januzzi JL Jr (2020) Heart failure with reduced ejection fraction: a review. JAMA 324:488–504
Omoto ACM et al (2024) 4D ultrasound-based strain assessment of cardiac dysfunction in male rats with reperfused and nonreperfused myocardial infarction. Physiol Rep 12:e16159
Pachon RE, Scharf BA, Vatner DE, Vatner SF (2015) Best anesthetics for assessing left ventricular systolic function by echocardiography in mice. Am J Physiol Heart Circ Physiol 308:H1525–H1529
Pandit H et al (2019) Utilizing contrast-enhanced ultrasound imaging for evaluating fatty liver disease progression in pre-clinical mouse models. Ultrasound Med Biol 45:549–557
Patten RD, Hall-Porter MR (2009) Small animal models of heart failure. Circ Heart Fail 2:138–144
Price MJ et al (2021) Animal models of metastatic lesions to the spine: a focus on epidural spinal cord compression. World Neurosurg 155:122–134
Ram R, Mickelsen DM, Theodoropoulos C, Blaxall BC (2011) New approaches in small animal echocardiography: imaging the sounds of silence. Am J Physiol Heart Circ Physiol 301:H1765–H1780
Ramirez DG et al (2020) Contrast-enhanced ultrasound with sub-micron sized contrast agents detects insulitis in mouse models of type1 diabetes. Nat Commun 11:2238
Ren M et al (2024) Application of small animal ultrasound imaging technology for identification of polycystic ovary syndrome in a mouse model. Biochem Biophys Res Commun 733:150634
Rojas JD et al (2022) Validation of a combined ultrasound and bioluminescence imaging system with magnetic resonance imaging in orthotopic pancreatic murine tumors. Sci Rep 12:102
Roth D et al (2024) Evaluation of superficial xenograft volume estimation by ultrasound and caliper against MRI in a longitudinal pre-clinical radiotherapeutic setting. PLoS One 19:e0307558
Rutledge C et al (2020) Commercial 4-dimensional echocardiography for murine heart volumetric evaluation after myocardial infarction. Cardiovasc Ultrasound 18:9
Rychak JJ, Lindner JR, Ley K, Klibanov AL (2006) Deformable gas-filled microbubbles targeted to P-selectin. J Control Release 114:288–299
Sastra SA, Olive KP (2013) Quantification of murine pancreatic tumors by high-resolution ultrasound. Methods Mol Biol 980:249–266
Schnelle M et al (2018) Echocardiographic evaluation of diastolic function in mouse models of heart disease. J Mol Cell Cardiol 114:20–28
Sławiński G, Hawryszko M, Liżewska-Springer A, Nabiałek-Trojanowska I, Lewicka E (2023) Global longitudinal strain in cardio-oncology: a review. Cancers (Basel) 15(3):986
Steinl DC et al (2016) Noninvasive contrast-enhanced ultrasound molecular imaging detects myocardial inflammatory response in autoimmune myocarditis. Circ Cardiovasc Imaging 9(8):e004720
Stribbling SM, Beach C, Ryan AJ (2024) Orthotopic and metastatic tumour models in preclinical cancer research. Pharmacol Ther 257:108631
Stypmann J (2007) Doppler ultrasound in mice. Echocardiography 24:97–112

Takalkar AM, Klibanov AL, Rychak JJ, Lindner JR, Ley K (2004) Binding and detachment dynamics of microbubbles targeted to P-selectin under controlled shear flow. J Control Release 96:473–482
Tsujita Y, Kato T, Sussman MA (2005) Evaluation of left ventricular function in cardiomyopathic mice by tissue Doppler and color M-mode Doppler echocardiography. Echocardiography 22:245–253
Viola F, Bustamante M, Bolger A, Engvall J, Ebbers T (2024) Diastolic function assessment with four-dimensional flow cardiovascular magnetic resonance using automatic deep learning E/A ratio analysis. J Cardiovasc Magn Reson 26:101042
Walker SL et al (2023) Ultrasound-guided hepatic portal vein injection is not a reproducible technique for delivery of cell therapies to the liver in mice. Diabet Med 40:e15192
Wang LW, Kesteven SH, Huttner IG, Feneley MP, Fatkin D (2018) High-frequency echocardiography – transformative clinical and research applications in humans, mice, and zebrafish. Circ J 82:620–628
Willmann JK et al (2010) Targeted contrast-enhanced ultrasound imaging of tumor angiogenesis with contrast microbubbles conjugated to integrin-binding knottin peptides. J Nucl Med 51:433–440
Yang XP et al (1999) Echocardiographic assessment of cardiac function in conscious and anesthetized mice. Am J Phys 277:H1967–H1974
Yang F et al (2002) Myocardial infarction and cardiac remodelling in mice. Exp Physiol 87:547–555
Yuan L-J, Wang T, Kahn ML, Ferrari VA (2011) High-resolution echocardiographic assessment of infarct size and cardiac function in mice with myocardial infarction. J Am Soc Echocardiogr 24:219–226
Zhang X, Li K, Cardoso C, Moctezuma-Ramirez A, Elgalad A (2024) Interpreting diastolic dynamics and evaluation through echocardiography. Life (Basel) 14(9):1156
Zhou H, Zhao D (2014) Ultrasound imaging-guided intracardiac injection to develop a mouse model of breast cancer brain metastases followed by longitudinal MRI. J Vis Exp 85:51146

Preclinical Applications of Integration of Magnetic Resonance Imaging and Spectroscopy: Bridging Structural and Metabolic Information

Amit Pratap Singh Chouhan, Ankush Verma, Vandana Singh, and Shagun Agarwal

Abstract

A groundbreaking method in preclinical research, the combination of magnetic resonance imaging (MRI) and Magnetic Resonance Spectroscopy (MRS) allows for a thorough evaluation of metabolic and structural data in biological tissues. The chapter delves into the preclinical uses of combined MRI-MRS, highlighting how it helps to characterize tissues in animal models from many angles. To bridge the gap between structural integrity and metabolic function, MRS supplements MRI's high-resolution anatomical imaging with the ability to quantify and map metabolites. This chapter focusses on the application of MRS and MRI in the study of neurological illnesses, where small biochemical changes may occur before structural alterations become apparent. Characterizing tumor heterogeneity, differentiating malignant from benign tissues, and evaluating therapy response are all made possible in oncology by this dual approach. The integrated imaging approach sheds light on metabolic disease-related changes in tissue metabolism that would otherwise go undetected by standard imaging techniques. The chapter shows how MRI-MRS integration improves the capacity

A. P. S. Chouhan (✉)
Department of Radiology, School of Allied Health Sciences, Galgotias University, Greater Noida, Uttar Pradesh, India
e-mail: amit.chouhan@galgotiasuniversity.edu.in

A. Verma
Department of Radiology, Sharda University, Greater Noida, Uttar Pradesh, India

V. Singh
Department of Microbiology, Sharda University, Greater Noida, Uttar Pradesh, India

S. Agarwal
Department of Physiotherapy, School of Allied Health Sciences, Galgotias University, Greater Noida, Uttar Pradesh, India

I. F. Cengiz et al. (eds.), *Bioimaging Modalities in Bioengineering*, Biomaterials, Bioengineering and Sustainability 9,
https://doi.org/10.1007/978-3-031-99410-4_18

to track disease progression and evaluate treatment efficacy through different case studies. This chapter also discusses the difficulties of putting this combined strategy into practice technically, such as the requirement for sophisticated data analysis tools and the incorporation of multimodal data. We look ahead to a future where interdisciplinary collaboration plays a key role in overcoming present restrictions and where technological breakthroughs may take us.

Keywords

Magnetic resonance spectroscopy · Magnetic resonance imaging · Preclinical research · Tissue metabolism · Multimodal imaging

1 Introduction

Combining magnetic resonance imaging (MRI) with magnetic resonance spectroscopy (MRS) helps one investigate the structural and metabolic properties of biological tissues. MRI provides high-resolution anatomical and functional imaging while MRS allows non-invasive metabolic chemistry in vivo to be examined. This combination is highly helpful in preclinical research as translational medicine depends on a complete knowledge of illness genesis, progression, and pharmacological responses. MRI and MRS increase diagnostic accuracy and give a whole picture of a spectrum of clinical disorders by connecting structural anomalies with metabolic dysfunctions (Ladd et al. 2018).

Preclinical applications of the combination of MRI and MRS abound in many fields, including neurology, cancer, metabolic disorders, and cardiovascular diseases. Neuroscience research on neurodegenerative diseases like multiple sclerosis, Alzheimer's disease, and Parkinson's disease depends totally on these techniques. MRI can detect atrophic changes and assess white matter integrity whereas MRS can reveal biochemical abnormalities include declining levels of N-acetylaspartate (NAA), which signals neuronal death, or elevated myo-inositol levels, which are connected with gliosis. Early diagnosis and sickness progress tracking depend on MRS since it can identify metabolic dysfunctions before structural abnormalities show themselves (Peterson et al. 2021).

Oncology is another area where MRI-MRS integration provides valuable fresh perspectives. Using levels of choline, lactate, and other metabolites connected to tumor metabolism, MRS reveals the metabolic reprogramming that cancer cells undergo to keep their rapid growth. Apart from structural imaging features of MRI, researchers may more precisely assess tumor heterogeneity, assess medication responses, and distinguish benign from malignant cells. MRI-MRS is used in preclinical cancer research to assess therapeutic efficacy, investigate tumor biology, and maximize treatment strategies before they are ever used (Smits 2021).

Integration of MRI-MRS is also helpful for metabolic diseases like obesity and diabetes. MRS allows researchers studying diabetes to analyze hepatic and intramuscular lipid levels, therefore providing vital information on insulin resistance and metabolic dysfunctions. MRI improves organ shape and function studies encompassing pancreatic volume and microvascular integrity. These findings guide knowledge of disease processes and aid to define focused treatments. In same vein, MRI offers structural and functional evaluations of cardiac tissues whereas MRS measures phosphocreatine (PCr)-to-adenosine triphosphate (ATP) ratios to examine myocardial energetics in cardiovascular studies. This combined technique allows us to investigate molecular and macroscopic features of heart failure and other cardiovascular disorders (Weinberg et al. 2021).

2 Fundamentals of Magnetic Resonance Spectroscopy and Imaging

2.1 Principles of Magnetic Resonance Spectroscopy and Imaging

MRI, MRS, and nuclear magnetic resonance are non-invasive imaging modalities offering structural and metabolic information about biological tissues. These approaches are routinely used in preclinical and clinical research to investigate a range of physiological and pathological states.

MRI mostly aims at the spatial distribution of protons, hydrogen nuclei, inside tissues. Strong magnetic field application to a subject result in field alignment for the hydrogen nuclei. These nuclei are excited by a radiofrequency (RF) pulse; when they return to their normal condition, they release detectable signals that are transformed into highly detailed anatomical pictures (Minhas and Oliver 2022).

Depending on their particular magnetic characteristics, the relaxation durations T1 (longitudinal) and T2 (transverse) assist differentiate numerous tissue types. Since MRI can offer high-resolution pictures of soft tissues, it is consequently a really useful tool for identifying anatomical anomalies in the organs, muscles, and brain. Development in imaging technology, pulse sequence creation, and data processing methods has made integration of MRI and MRS in preclinical research possible. Reduced quantity metabolite identification is made possible by high spectral resolution and signal-to-noise ratio (SNR) seen in high-field MRI systems including 7 T and 9.4 T. Two-dimensional correlated spectroscopy and J-difference editing together serve to improve the specificity of metabolite measurement in complex pulse sequence. Moreover, spectrum analysis has been automated, quantification accuracy has risen, and multi-modal data has been merged utilizing artificial intelligence (AI) and machine learning (ML) approaches to comprehensively define a disease (Tampieri et al. 2021).

Despite these developments, preclinical use of MRI-MRS integration still presents various difficulties. MRS's somewhat limited sensitivity in comparison to other molecular imaging modalities, notably positron emission tomography (PET), is one of its main drawbacks. Signal detection has to be raised using optimum acquisition techniques and high-field magnets. Spectral interpretation's intricacy adds even another difficulty as metabolite peaks sometimes overlap and accurate analysis calls for advanced post-processing methods. Moreover, restricting the general use in preclinical laboratories is the great expense and technical knowledge required for MRI-MRS research (Serai 2022).

Preclinical research does, however, hold great promise for MRI-MRS integration. By considerably increasing signal intensity, hyperpolarized MRS and other innovative techniques enable real-time metabolic imaging with until unheard-of sensitivity. Furthermore, the development of multi-nuclear MRS, which extends proton spectroscopy to integrate nuclei such as carbon-13 and phosphorous-31, allows a larger spectrum of metabolic investigation. By means of automated spectrum analysis and AI-driven picture reconstruction, data interpretation is supposed to be simplified, hence improving the usability and efficiency of MRI-MRS integration (Stroman 2024).

An additional intriguing field is the translational potential of MRI-MRS integration. Validation of preclinical results using animal models helps researchers reduce the gap between fundamental knowledge and medicinal applications. Preclinical MRI-MRS studies uncovering biomarkers that may be investigated further in human trials provide tailored treatment regimens and early diagnostic tools. Early metabolic changes observed in animal models might enable researchers studying neurodegenerative disorders to offer non-invasive diagnosis for individuals in early-stage Alzheimer's disease. Customized therapeutic suggestions based on the metabolism of particular tumor can also come from metabolic imaging, therefore supporting precision medicine approaches in preclinical cancer (Singh et al. 2022).

2.2 The Need to Merge MRI and MRS

MRI and MRS greatly affect both preclinical and clinical studies. Combining these two modalities will permit more thorough study of tissue structure and metabolic activity, so increasing treatment assessments, disease surveillance, and diagnostic accuracy (Zhang et al. 2023). Combining MRS and MRI helps mostly with matching anatomical and biochemical data in a single examination. MRI's high-resolution pictures of tissue form help one to spot structural problems even in circumstances when MRS notes on metabolic changes probably preceding clear morphological changes. This is particularly useful in neurodegenerative diseases such as Parkinson's and Alzheimer's where metabolic failure may be seen before structural atrophy as it allows quick treatment and improved patient outcomes by means of diagnosis.

MRS improves MRI in oncology by means of metabolic markers associated with tumor development, aggressiveness, and therapy response. Tumors have different metabolic profiles due to their greater energy needs and development; this is generally observed as higher levels of lactate and choline. Combining MRS with MRI allows doctors and researchers to monitor therapy progress in real time, more precisely distinguish benign from malignant lesions, and assess tumor heterogeneity. Customized treatments based on the metabolism of certain tumors help to improve methods of precision medicine (Küstner et al. 2024).

Apart from neurology and cancer, metabolic illnesses like diabetes and obesity also benefit much from the blend of MRS and MRI. MRS's quantitative data on lipid production in organs like the muscles and liver help one to grasp metabolic malfunction and insulin resistance. Together with the structural and functional imaging features of MRI, this integration enables focused treatment to effectively prevent and cure metabolic diseases (Desale et al. 2024).

Another main advantage of MRI-MRS integration is its non-invasive character; it does not call for radioactive tracers or contrast chemicals, occasionally used in other imaging modalities including PET. When regular evaluations are required to track the development of a disease or the efficacy of treatment over time for long-term study, this is quite useful. The possibility to obtain thorough structural and metabolic data without invasive techniques increases the viability of research and patient safety (Mohammadi et al. 2024).

2.3 MRI-MRS Integration

Technical features combining MRS with MRI demands for modern data processing algorithms, complex gear, and tailored pulse sequences to provide correct and consistent findings. Using high-field MRI systems (usually 3 T and above) is one of the most important technological criteria as greater magnetic fields improve spectral resolution and SNR, therefore allowing more clear differences of metabolites (Zhao et al. 2024). Another crucial component is still the pulse sequence applied in MRI-MRS integration. Point-resolved spectroscopy (PRESS) and stimulated echo acquisition mode (STEAM) are well-known MRS methods producing high-quality spectra. MEGA-PRESS among other J-difference editing methods helps to identify low-concentration metabolites like glutathione and GABA, therefore improving the specificity of metabolic analyses (Kanagasabai et al. 2024). Usually with single- or multi-voxel spectroscopic imaging (MRSI), voxel selection provides spatial localization. While single-voxel MRS gives comprehensive metabolic information from a single, tiny spot, MRSI allows metabolic mapping across a greater tissue volume. Dynamic and automated shimming techniques utilized in modern shimming systems help to lower magnetic field homogeneity and enhance spectral quality (Sheikh-Bahaei et al. 2024). Data processing and analysis demand numerous methods including peak quantification, phase and frequency alignment, and baseline correction. Increasingly automated using AI-driven algorithms, spectrum fitting and interpretation help to lower variability and improve diagnosis accuracy (Chen et al. 2024).

3 Applications of MRS in Neurology, Oncology, and Metabolic Diseases

For the non-invasive evaluation of metabolic alterations in various disorders, MRS is commonly utilized in both clinical and experimental settings. The identification of metabolic syndromes, brain tumors, and neurodegenerative diseases as stated in Table 1 depends on the use of certain biomarkers like lactate, choline, and N-acetylaspartate. In oncology, MRS is also used to evaluate therapeutic responses such that benign and malignant tumors may be separated. Its diagnostic and therapeutic value is further enhanced in psychiatric research where it investigates neurotransmitter imbalances in diseases such as depression and schizophrenia, therefore broadening its relevance (Saleh et al. 2025).

In disorders of the nervous system, MRS offers metabolic insights complementary to structural imaging and finds tremendous use in neurological conditions (as shown in Table 1). Reduced NAA and increased myo-inositol levels in Alzheimer's disease shown by MRS point to glial activation and neuronal shrinkage. Variations in glutamate and GABA levels expose aberrant Parkinson's disease neurotransmitter production. MRS is also fairly important in epilepsy as it helps one discriminate between epileptic and non-epileptic lesions and locates lactate accumulation in seizure sites. MRS also notes differences in NAA and choline levels, which helps assess neurodegeneration in multiple sclerosis. These uses highlight how important MRS is for understanding, diagnosis, and tracking of many neurological illnesses, hence enabling early identification and treatment improvement (Shen et al. 2025).

In the context of oncology, MRS provides metabolic understanding of tumor biology, thereby directing diagnosis, treatment planning, and therapy monitoring in cancer. Higher choline levels resulting from improved cell membrane turnover and lactate buildup resulting from altered glycolysis show different metabolic profiles in malignancies usually identified by these values. Combining MRS with MRI

Table 1 Showing the role of MRS in different disorders

Disorder	Metabolic biomarker	Role of spectroscopy	References
Alzheimer's disease	Decreased NAA, increased Myo-inositol	Early detection and disease progression monitoring	Shen et al. (2025)
Brain tumors	Elevated choline, reduced NAA	Tumor grading and treatment response assessment	Saleh et al. (2025)
Multiple sclerosis	Reduced NAA, increased lipids	Evaluating neurodegeneration and inflammation	Zheng et al. (2025)
Depression	Altered glutamate/GABA levels	Understanding neurotransmitter imbalances	Harastova-Pavlova and Ruda-Kucerova (2025)
Diabetes	Increased hepatic lipids	Assessing metabolic dysfunction	Menon et al. (2025)

allows physicians and researchers to non-invasively assess tumor aggressiveness, separate benign from malignant tumors, and track therapeutic progress over time. MRS can find gliomas from radiation necrosis in brain malignancies based on metabolite patterns. Although increasing choline and reduced N-acetylaspartate (NAA defines high-grade gliomas), necrotic regions reveal more lactate and lipid peaks (Rafanan et al. 2025). This directs the building of more exact treatments and grade malignancies. Besides brain tumors, MRS is helpful for liver tumors, breast cancer, and prostate cancer. Higher citrate and choline ratios help in prostate cancer to differentiate benign from malignant prostate tissue. Compound metabolism, including changes in choline, suggests the aggressiveness of the tumor and the reaction to treatment in breast cancer. Moreover, important for determining the response of a tumor to radiation, chemotherapy, and targeted treatments is MRS. Monitoring metabolic changes after treatment helps doctors to determine the effectiveness of drugs and modify their dosage schemes. This improves patient outcomes and lowers unneeded therapies by way of more customized care strategies (Zheng et al. 2025). As high-field MRS and AI-based spectrum analysis develop, they are expected to enhance non-invasive cancer diagnosis and therapy monitoring even further. MRS is therefore predicted to become more important in cancer.

In metabolic diseases, MRS serves as a necessary tool as it clarifies the biochemical changes related to conditions such as diabetes, obesity, and mitochondrial issues. MRS diagnosis of metabolic dysfunctions and evaluation of therapy effects depend on monitoring metabolites including lipids, glucose, and lactate in organs including the liver, muscles, and brain. Knowing metabolic syndromes, fatty liver disease, and insulin resistance assists one to open the road for tailored treatment programs and early intervention (Menon et al. 2025).

4 Emerging Applications, Challenges and Considerations

MRI-MRS integration marks a rapidly growing discipline with fresh applications outside conventional domains. Among the newly recruited research subjects are psychiatric illnesses, cardiac issues, and precision medicine. Still, there are issues like uniformity, technological complexity, and expensive costs. Solving these challenges and increasing clinical acceptance and translational impact depend on technological improvements, better acquisition strategies, and AI-driven analysis (Selvam et al. 2025).

4.1 Emerging Research Directions in Neurological, Oncological, and Metabolic Research

MRS has evolved into a powerful tool in neuroimaging and cancer when applied with MRI as it offers structural and functional assessments as well as metabolic insights. Particularly in the domains of neurodegenerative disorders, mental diseases, cancer, and metabolic diseases, its uses have lately found new frontiers of research. Early detection and progression tracking in neurodegenerative studies for Parkinson's disease and Alzheimer's disease is under research on MRI-MRS integration. By using neurochemical markers like glutamate, GABA, myo-inositol, and NAA, researchers can identify metabolic alterations happening before structural abnormalities reported with standard MRI. This allows one to identify early symptoms, perhaps pointing to a modification in the direction of treatment and diagnosis. Most importantly in the pathophysiology of mental diseases such schizophrenia, bipolar disorder, and severe depression, MRI-MRS is also being used to look at neurotransmitter imbalances including glutamate and GABA dysfunction. Customized medical treatments are made possible by research showing relationships between the degree of the disease and its response to therapy and changes in metabolite levels in areas of the brain including the hippocampal, prefrontal cortex, and anterior cingulate cortex. Particularly with the unique and defining nature of brain tumors such as gliomas, MRI-MRS integration is also helping cancer research (Sarkar et al. 2025).

MRS offers a standard MRI with a layer of metabolic profile that helps to maximize therapy options, tumor grade categorization, and tumor recurrence diagnosis by determining choline, creatine, and lactate levels. By means of automated tumor type classification and treatment response prediction, the integration of AI and ML into MRI-MRS has even further enhanced diagnostic accuracy. Apart from neurology and cancer, MRI-MRS is allowing non-invasive monitoring of metabolic dysfunctions in the brain and liver, therefore assisting research on metabolic diseases like diabetes and hepatic encephalopathy. MRS has been used, for example, in diabetes studies to investigate changes in brain metabolism linked to insulin resistance and cognitive impairment, therefore offering data on the neurological effects of the condition. Moreover, MRS detects lower myo-inositol and higher glutamine in hepatic encephalopathy, which helps to ascertain the degree of the disease and the effectiveness of the therapy. Under development is also the study of traumatic brain injury, where MRI-MRS is helping to evaluate axonal damage, neuro-inflammation, and post-concussive syndrome. By spotting metabolic changes like lower NAA and higher choline and lactate, MRI-MRS helps in more full evaluation of damage degree and healing capacity.

Moreover, new studies on the gut-brain axis—more especially, in connection with illnesses like inflammatory bowel disease and irritable bowel syndrome are under MRI-MRS. Researchers are looking at the possible impact of changes in gut flora on brain metabolism and function as MRS provides essential information on neurotransmitter imbalances and neuro-inflammation linked with gut dysbiosis. Early diagnosis and treatment programs have great potential from the non-invasive evaluation of metabolites like lactate, NAA, and choline in newborns. Additionally, under study in prenatal and neonatal imaging are newborn metabolic abnormalities, hypoxic-ischemic damage, and brain development using MRI-MRS (Mourão et al. 2025).

Development in ultra-high-field MRI (7 T and beyond) is also helping to enhance MRS spectral and spatial resolution by allowing more precise metabolite measurement and greater separation of overlapping signals. This affects many other fields including metabolic diseases, psychosocial issues, neurodegenerative research, and many more. Notwithstanding these advances, problems still need to be resolved including the necessity of more thorough multicenter studies to validate results throughout many populations, the standardizing of MRS collecting methods, and the challenge of data interpretation. With constant technological developments, AI-driven analysis, and multimodal data integration, MRI-MRS is poised to become an indispensable tool for understanding disease causes, improving early detection, and spreading precision medicine over a spectrum of medical research fields (Veeramachaneni 2025).

4.2 Emerging Research Directions in Cardiovascular, Musculoskeletal, and Systemic Conditions

Through providing structural and metabolic data on many tissues and organs, MRS in association with MRI has revolutionized medical research and diagnosis. Combining these technologies offers considerable potential for research on musculoskeletal disorders, cardiovascular diseases, and other medical illnesses. Through allowing researchers and doctors to investigate metabolic changes in addition to morphological abnormalities, MRI-MRS offers a non-invasive method for early detection, illness monitoring, and tailored therapy planning (Wang et al. 2025).

Cardiac Diseases

Taken all together, heart failure, arrhythmia, coronary artery disease, and cardiomyopathies account for one of the major causes of morbidity and mortality worldwide. MRI-MRS is an approach to consider the structural and metabolic alterations linked to certain diseases. Among the most important applications of MRS in cardiology is analysis of cardiac energy metabolism. The heart has to be precisely balanced in the generation and consumption of energy mostly by oxidative phosphorylation in the mitochondria. Measuring using PCr, ATP, and phosphorous-31 MRS (31P-MRS) provides essentially accurate markers of cardiac energy state. Patients with heart failure have lower PCr/ATP ratios, which indicate poor energy metabolism and mitochondrial malfunction. This data guides therapy initiatives and helps to assess the degree of the illness (Alqudsi and Makaraci 2025).

Apart from energy metabolism, proton MRS (1H-MRS) has been studied on cardiac lipid accumulation, which fuels illnesses such diabetic cardiomyopathy and obesity-related heart disease. All of which cause heart failure include inflammation, oxidative stress, and lipotoxicity suffered by cardiomyocytes with too high lipid content. Through evaluation of myocardial triglyceride content, MRI-MRS can be used to evaluate the efficacy of pharmaceutical or lifestyle interventions aimed to reduce lipid overload and to identify those who run a risk of metabolic cardiomyopathies.

MRI-MRS also enables the assessment of ischemic heart disease and cardiac viability. By evaluating metabolic changes in peri-infarct locations, MRS can complement traditional MRI techniques such late gadolinium enhancement, which offer insights about scar tissue and infarcted areas. MRS measurements of lactate buildup point to anaerobic metabolism and ischemia. This method improves the ability to distinguish between viable and non-viable myocardium, therefore directing treatment choices on revascularization (Mrida et al. 2025).

Musculoskeletal Disorders

Diagnosing and treating musculoskeletal diseases like osteoporosis, rheumatoid arthritis, muscular dystrophies, and osteoarthritis (OA) presents very difficult challenges. MRI-MRS reveals considerable promise for early diagnosis and a more accurate course of illness tracking in assessing bone and muscle metabolism. MRS is extensively used in research on metabolic changes in muscle diseases including mitochondrial myopathies and Duchenne muscular dystrophy. Particularly helpful in evaluating muscle energy metabolism is Phosphorus MRS (31P-MRS), which measures PCr and ATP levels during rest and exercise. Reduced mitochondrial activity and muscle tiredness shown by affected PCr recovery kinetics in muscular dystrophy patients give important information on the degree of the disease and the efficacy of a treatment (Feng et al. 2025).

Research on intramuscular fat accumulation, a unique feature of aging-related sarcopenia, metabolic syndromes, and other neuromuscular illnesses has also benefited much from Proton MRS (1H-MRS). Measuring lipid intrusion in skeletal muscles enables MRI-MRS to guide rehabilitation programs and determine the illness burden; this ties insulin resistance, muscle weakness, and decreased mobility to Measuring lipid intrusion in skeletal muscles. Since MRI-MRS lets one quantify cartilage metabolites, it clearly benefits the research on joint and cartilage diseases like OA. Early-stage OA is well documented to reduce glycosaminoglycans (GAGs), fundamental building blocks of cartilage. Measuring GAG concentration allows salt MRI in conjunction with MRS to identify early on cartilage degeneration before structural damage emerges on conventional MRI tests. This influences their effectiveness as well as disease-modifying OA therapy development.

Using MRI-MRS, studies on bone metabolism have also been conducted; they have evaluated bone marrow adiposity and its correlation to osteoporosis. Reduced bone mineral density and higher risk of fracture have been connected to higher fat components in the bone marrow. By separating the saturated and unsaturated lipid content of bone marrow, MRS offers vital data on the metabolic alterations linked to osteoporosis and other metabolic bone disorders (Din and Yang 2025).

Other Emerging Areas

Apart from musculoskeletal and cardiovascular conditions, MRI-MRS has been used to treat various other disorders including cancer, metabolic diseases, and neurodegenerative diseases.

By means of brain metabolites including NAA, myo-inositol, glutamate, and GABA, MRS provides data on neuronal integrity, neuro-inflammation, and neurotransmitter imbalances, so benefiting research on neurodegenerative diseases including multiple sclerosis, Parkinson's disease, and Alzheimer's disease. For those with neurodegenerative diseases, this data supports early diagnosis, course of treatment efficacy evaluation, and disease monitoring. Using MRS, researchers have investigated how obesity, diabetes, and hepatic encephalopathy affect brain and organ metabolism; although higher glutamine levels in hepatic encephalopathy point to ammonia toxicity and decreased astrocyte function, increased brain lactate and changed glutamate-glutamine cycling have been linked to cognitive impairment in diabetic patients. These metabolic indicators advance knowledge of illness causes and enable the formulation of tailored treatments.

MRI-MRS has especially greatly advanced cancer research in relation to tumor type differentiation, malignancy grade, and therapeutic response monitoring. By monitoring levels of choline, creatine, and lactate, MRS lets brain malignancies like gliomas be distinguished between high-grade and low-grade tumors. Furthermore, used in studies on breast and prostate cancer, MRS has been used to evaluate metabolic markers separating benign from malignant tumors, thereby enhancing therapy planning and diagnostic accuracy (Daimiel Naranjo et al. 2025). Combining AI and ML with MRI-MRS might assist in enabling automatic metabolic pattern analysis and raise diagnostic accuracy. AI-driven models enhance early illness diagnosis and prognosis evaluation by spotting minute metabolic defects undetectable to the human eye.

Furthermore, predicted improvements in spectral resolution and metabolite quantification resulting from developments in ultra-high-field MRI (7 T and higher), hence enhancing the application of MRI-MRS in clinical and experimental environments. Hyperpolarized MRS is one of the new methods for in vivo dynamic metabolic process analysis that could provide real-time metabolic imaging with somewhat improved SNRs (Luo et al. 2025). Although MRI-MRS has enormous promise, broad acceptance of technology still faces various difficulties. Standardizing data processing methods, acquisition procedures, and interpretation guidelines guarantee consistency and dependability throughout several institutions. Particularly in clinical environments, the very long scan periods and motion artifact sensitivity have pragmatic restrictions as well. To solve these difficulties, efforts are in progress to create speedier imaging technologies and motion-correction methods (Luo et al. 2025).

4.3 Technical and Practical Challenges for MRI and MRS Integration

Combining these two approaches helps to improve knowledge of a variety of neurological and neoplastic diseases and raise diagnosis accuracy. However, various logistical and technical obstacles prevent perfect integration, hence software, hardware, and workflow efficiency have to be developed (Feng and Chandarana 2025). One of the technical problems relates to variations in acquisition specifications. Among the acquisition factors varying for MRI and MRS are magnetic field gradients, echo time (TE), and repetition time (TR). MRS usually needs longer acquisition periods to give enough spectral resolution, while MRI benefits from shorter TR and TE for rapid imaging and outstanding spatial resolution. It remains somewhat difficult to reconcile these variations without sacrificing the quality of the data.

Another issue involves SNR restrictions. By nature, MRS has a lower SNR than MRI as biological tissues only contain somewhat low amounts of metabolites. Usually required to improve SNR are longer acquisition durations, more powerful fields, and better RF coils. Still, raising field strength might generate sensitivity artifacts, which would hinder MRI integration. The two modalities also differ in spatial resolution. MRI has exceptional spatial resolution, generally at sub-millimeter levels, even although MRS normally restricts to voxel sizes on the order of millimeters or even centimeters. This disparity makes it more difficult to properly co-register MRS data with MRI images, especially in small areas of interest like the brainstem or hippocampal regions.

Magnetic field inhomogeneity and susceptibility artifacts also pose a challenge. Particularly at sites with notable sensitivity changes (like air-tissue or bone-tissue interfaces), magnetic field inhomogeneity might compromise MRI and MRS quality. Shimming efficiency is affected by field homogeneity; so, enhanced shimming and dynamic field correction techniques are needed to avoid spectrum widening and imprecise metabolite measurement in MRS. Motion-related issues are also relevant. While MRS does not yet have consistent real-time correction solutions, MRI has generated strong motion correction algorithms. MRI and MRS can be affected by motion artifacts which can result from involuntary patient movements, heart pulsations, and breathing changes by producing phase distortions and signal loss. Lastly, standardization and quantification remain major concerns. Unlike MRI, which provides well-known quantitative biomarkers such as diffusion coefficients and T1/T2 relaxation durations, MRS must be properly calibrated using either internal or external standards. Variance in spectrum fitting techniques, baseline modifications, and reference standards complicates metabolite measurement in MRS. Clinical reliability and reproducibility need to be raised by standardized methods for MRS recording and post-processing (Caesar 2025).

4.4 Administrative Difficulties

One issue is that including MRS into a standard MRI protocol lengthens the scan time generally; this might not be feasible in clinical settings with heavy patient load. MRI examinations are especially difficult as MRS sequences frequently need extra setup, shimming, and post-processing methods (Caria 2025). Another challenge relates to the limited number of certified professionals. MRS collecting and interpretation required for particular knowledge of MR physics, spectroscopy, and metabolic profiling. Although many radiologists and MRI technologists have little training in MRS, ordinary clinical practice does not heavily rely on it. If we want to address this discrepancy, training courses must be broadened and automated processing systems made more user-friendly (de Vries et al. 2025).

There are also issues stemming from variations across MRI systems and suppliers. Variations in MRS implementation resulting from hardware, pulse sequences, and data formats between MRI systems made by various vendors. Cross-platform calibration methods and vendor-neutral spectroscopic sequences assist to guarantee compatibility and dependability over numerous scanners in standardizing projects (Ghadimi et al. 2025). Reimbursement and administrative complications also hinder broader use. MRS is still considered at many sites as an advanced or research-based method, even as standard MRI is generally recognized and supported for a variety of therapeutic uses. Clinical MRS applications call for different degrees of regulatory approval, and differing reimbursement regimes inhibit overall adoption. Large-scale validation studies demonstrating the therapeutic value of MRS may help overcome these obstacles (Safari et al. 2025).

Finally, the integration of AI and multimodal data analysis presents both opportunities and hurdles. While conventional MRI is usually accepted and supported for a variety of therapeutic purposes, MRS is still seen as an advanced or research-based technique at many places. Varied degrees of regulatory approval for clinical MRS applications and varied payment schemes cause wide adoption to be impeded. Comprehensive validation studies demonstrating the therapeutic benefits of MRS might help to address these challenges (Sulangi et al. 2025).

5 Conclusions and Future Perspectives

While combining MRI and MRS offers many difficulties, additional developments have significant promise to increase clinical applications and research value. More widespread use demands for addressing logistical issues including staff training in addition to technical ones such SNR, spatial resolution, and motion artifacts. Future studies should largely focus on the development of MRI system cross-platform compatibility, standardizing projects, and AI-driven analytic tools. MRS and MRI used together have the potential to alter the diagnosis and monitoring of neurological and oncological illnesses as high-field MRI technologies and automated processing tools develop. More clinical validation studies and regulatory permits will help MRS to be validated as a required complement to MRI, thereby strengthening MRS and hence improving patient care and the diagnostic capability of medical imaging. Furthermore, encouraging interactions among corporate stakeholders, scientists, and doctors will drive creativity and speed up the application of MRS into approved treatment guidelines. Through collaborative efforts and technology advancements addressing current issues, MRS and MRI together have the ability to revolutionize precision medicine. In many different medical conditions, this would improve therapeutic interventions, treatment monitoring, and early disease identification.

Acknowledgments N/A

Conflict of Interest The authors declare that they have no conflict of interest.

References

Alqudsi Y, Makaraci M (2025) Exploring advancements and emerging trends in robotic swarm coordination and control of swarm flying robots: a review. Proc Inst Mech Eng C J Mech Eng Sci 239(1):180–204

Caesar MK (2025) Recent advances in diagnostic imaging: a literature review on MRI and CT-scan technologies. Global Res Innov Edutech J 1(1):348–353

Caria H (2025) Tech-driven healthcare evolution: navigating through examples of adaptability. In: Leveraging technology for organizational adaptability. IGI Global Scientific Publishing, pp 305–322

Chen X, Li J, Chen D, Zhou Y, Tu Z, Lin M, Kang T, Lin J, Gong T, Zhu L, Zhou J (2024) CloudBrain-MRS: an intelligent cloud computing platform for in vivo magnetic resonance spectroscopy preprocessing, quantification, and analysis. J Magn Reson 358:107601

Daimiel Naranjo I, Bhowmik A, Basukala D, Lo Gullo R, Mazaheri Y, Kapetas P, Eskreis-Winkler S, Pinker K, Thakur SB (2025) Assessment of hypoxia in breast cancer: emerging functional MR imaging and spectroscopy techniques and clinical applications. J Magn Reson Imaging 61(1):83–96

de Vries E, Hagbohm C, Ouellette R, Granberg T (2025) Clinical 7 Tesla magnetic resonance imaging: impact and patient value in neurological disorders. J Intern Med

Desale P, Dhande R, Parihar P, Nimodia D, Bhangale PN, Shinde D, Dhande R, Bhangale PN Jr (2024) Navigating neural landscapes: a comprehensive review of magnetic resonance imaging (MRI) and magnetic resonance spectroscopy (MRS) applications in epilepsy. Cureus 16(3)

Din RU, Yang H (2025) Novel MRI semiquantitative scores for vertebral fat-water assessment in aging and spinal disorders. Chin J Acad Radiol:1–12

Feng L, Chandarana H (2025) Accelerated abdominal MRI: a review of current methods and applications. J Magn Reson Imaging

Feng F, Wang W, Liu S, Liu L, Sun M, Jiang R, Chen M, Luo M (2025) T2 mapping and q-Dixon for assessment of intervertebral disc degeneration in lower back pain. BMC Musculoskelet Disord 26(1):234

Ghadimi DJ, Vahdani AM, Karimi H, Ebrahimi P, Fathi M, Moodi F, Habibzadeh A, Khodadadi Shoushtari F, Valizadeh G, Mobarak Salari H, Saligheh Rad H (2025) Deep learning-based techniques in glioma brain tumor segmentation using multi-parametric MRI: a review on clinical applications and future outlooks. J Magn Reson Imaging 61(3):1094–1109

Harastova-Pavlova I, Ruda-Kucerova J (2025) Translational value of proton MRS in depression: applications to preclinical models. In: Handbook of the biology and pathology of mental disorders. Springer, Cham, pp 1–22

Kanagasabai K, Palaniyappan L, Théberge J (2024) Precision of metabolite-selective MRS measurements of glutamate, GABA and glutathione: a review of human brain studies. NMR Biomed 37(3):e5071

Küstner T, Qin C, Sun C, Ning L, Scannell CM (2024) The intelligent imaging revolution: artificial intelligence in MRI and MRS acquisition and reconstruction. MAGMA 37(3):329–333

Ladd ME, Bachert P, Meyerspeer M, Moser E, Nagel AM, Norris DG, Schmitter S, Speck O, Straub S, Zaiss M (2018) Pros and cons of ultra-high-field MRI/MRS for human application. Prog Nucl Magn Reson Spectrosc 109:1–50

Luo WQ, Liu B, Yang CL, Zhao ZY, Wang GQ, Li XL, Zhang P (2025) MRI shrimp sign and parietooccipital lesion in progressive multifocal leukoencephalopathy. Cerebellum 24(2):1–4

Menon RG, Yepuri G, Martel D, Quadri N, Hasan SN, Manigrasso MB, Shekhtman A, Schmidt AM, Ramasamy R, Regatte RR (2025) Assessment of cardiac and skeletal muscle metabolites using 1H-MRS and chemical-shift encoded magnetic resonance imaging: impact of diabetes, RAGE, and DIAPH1. NMR Biomed 38(1):e5275

Minhas AS, Oliver R (2022) Magnetic resonance imaging basics. In: Electrical properties of tissues: quantitative magnetic resonance mapping, pp 47–82
Mohammadi M, Mohammadi S, Hadizadeh H, Olfati M, Moradi F, Tanzifi G, Ghaderi S (2024) Brain metastases from breast cancer using magnetic resonance imaging: a systematic review. J Med Radiat Sci 71(1):133–141
Mourão TC, Saikali S, Patel E, Dohler M, Patel V, Moschovas MC (2025) Telesurgery applications, current status, and future perspectives in technologies and ethics. In: Handbook of robotic surgery, pp 161–168
Mrida MSH, Rahman MA, Alam MS (2025) AI-driven data analytics and automation: a systematic literature review of industry applications. Strateg Data Manag Innov 2(01):21–40
Peterson P, Trinh L, Månsson S (2021) Quantitative 1H MRI and MRS of fatty acid composition. Magn Reson Med 85(1):49–67
Rafanan J, Ghani N, Kazemeini S, Nadeem-Tariq A, Shih R, Vida TA (2025) Modernizing neuro-oncology: the impact of imaging, liquid biopsies, and AI on diagnosis and treatment. Int J Mol Sci 26(3):917
Safari M, Eidex Z, Chang CW, Qiu RL, Yang X (2025) Advancing MRI reconstruction: a systematic review of deep learning and compressed sensing integration. arXiv preprint arXiv:2501.14158
Saleh SR, Yousif SA, Ahmed IQ (2025) Brain lesion MRI and co-related MRS spectroscopy dataset. Data Brief 59:111288
Sarkar J, Singh R, Chandel S (2025) Understanding LC/MS-based metabolomics: a detailed reference for natural product analysis. Proteomics Clin Appl 19(1):e202400048
Selvam DC, Devarajan Y, Raja T (2025) Exploring the potential of artificial intelligence in nuclear waste management: applications, challenges, and future directions. Nucl Eng Des 431:113719
Serai SD (2022) Basics of magnetic resonance imaging and quantitative parameters T1, T2, T2*, T1rho and diffusion-weighted imaging. Pediatr Radiol 52(2):217–227
Sheikh-Bahaei N, Chen M, Pappas I (2024) Magnetic resonance spectroscopy (MRS) in Alzheimer's disease. In: Biomarkers for Alzheimer's disease drug development. Springer, New York, NY, pp 115–142
Shen Y, Zhang X, Liu S, Xin L, Xuan W, Zhuang C, Chen Y, Chen B, Zheng X, Wu R, Lin Y (2025) CEST imaging combined with 1H-MRS reveal the neuroprotective effects of riluzole by improving neurotransmitter imbalances in Alzheimer's disease mice. Alzheimers Res Ther 17(1):20
Singh DK, Pradhan M, Materny A (eds) (2022) Modern techniques of spectroscopy: basics, instrumentation, and applications, 1st edn. Springer
Smits M (2021) MRI biomarkers in neuro-oncology. Nat Rev Neurol 17(8):486–500
Stroman PW (2024) Essentials of functional MRI: basic concepts to advanced applications. CRC Press
Sulangi AJ, Husain A, Lei H, Okun J (2025) Neuronavigation in glioma resection: current applications, challenges, and clinical. In: Training and education in neurosurgery: strategies and challenges for the next ten years, vol II, p 54
Tampieri A, Szabó M, Medina F, Gulyás H (2021) A brief introduction to the basics of NMR spectroscopy and selected examples of its applications to materials characterization. Phys Sci Rev 6(1):20190086
Veeramachaneni V (2025) Emerging authentication technologies for zero trust in IoT systems. J Adv Re Mobile Comp 7(1):7–21
Wang Q, Zhu H, Deng L, Xu S, Xie W, Li M, Wang R, Tie L, Zhan L, Yu G (2025) Spatial transcriptomics: biotechnologies, computational tools, and neuroscience applications. Methods:2401107
Weinberg BD, Kuruva M, Shim H, Mullins ME (2021). Clinical applications of magnetic resonance spectroscopy (MRS) in of brain tumors: from diagnosis to treatment. Radiologic clinics of North America, 59(3):349
Zhang X, Li J, Li X (2023) Use of magnetic resonance spectroscopy to explore metacognitive ability. Behav Neurosci 137(2):120–126

Zhao S, Gao Y, Wu T, Singh D, Jiang R, Sun H, Sarawata M, Qiu Y, Whittaker W, Higgins I, Du Y (2024) SubT-MRS dataset: pushing SLAM towards all-weather environments. In: Proceedings of the IEEE/CVF conference on computer vision and pattern recognition, pp 22647–22657
Zheng J, Zhang Y, Zhao B, Wang N, Gao T, Zhang L (2025) Metabolic changes of thalamus assessed by 1H-MRS spectroscopy in patients of cervical spondylotic myelopathy following decompression surgery. Front Neurol 15:1513896

Part III

Emerging Technologies and Computational Advances in Bioimaging

Thermoacoustic Imaging: Principles and Applications in Preclinical Research

Tarikul Islam and Kamran Avanaki

Abstract

Thermoacoustic imaging (TAI) is an innovative imaging modality that combines microwave and ultrasound technologies. TAI utilizes the thermoelastic effect to convert microwave energy into acoustic signals, thereby enabling high-contrast imaging of biological tissues with deep penetration. In the electromagnetic spectrum, microwaves (ranging from hundreds of MHz to several GHz) have wavelengths on the order of centimeters to millimeters, which significantly reduces their interaction with cellular and subcellular structures. Hence, they primarily interact with tissue on a macroscopic scale. Tumors, blood clots, and foreign bodies have dielectric properties that differ from healthy tissue and non-coagulated blood, leading to high contrast imaging. TAI's ability to achieve deep tissue penetration and exceptional contrast makes it particularly effective for many medical and biomedical applications. In preclinical research, TAI has shown significant potential for visualizing anatomical structures and tracking disease progression. Applications include breast, prostate, and brain cancer detection and characterization, brain hemorrhage analysis, and joint imaging. TAI has also been utilized for detection of foreign bodies. Despite its advantages, TAI faces technical challenges toward wider implementation, including detection sensitivity optimization, and hardware complexity. For example, there are currently no TAI systems for preclinical research that are commercially available. This chapter explores the fundamental principles of TAI, its core system components, and its diverse preclinical applications. By addressing current challenges and highlighting advancements, this work underscores TAI's potential

T. Islam · K. Avanaki (✉)
Richard and Loan Hill Department of Biomedical Engineering, University of Illinois Chicago, Chicago, IL, USA
e-mail: avanaki@uic.edu

I. F. Cengiz et al. (eds.), *Bioimaging Modalities in Bioengineering*, Biomaterials, Bioengineering and Sustainability 9,
https://doi.org/10.1007/978-3-031-99410-4_19

in biomedical research and diagnostics. Future directions include integrating TAI with machine learning and multimodal imaging systems to enhance its clinical and research utility.

Keywords

Thermoacoustic tomography · Thermoelastic expansion · Preclinical applications · Biomedical imaging · Microwave-induced acoustic imaging

1 Introduction

Biomedical imaging has revolutionized the field of medicine, enabling non-invasive visualization of internal structures and physiological processes, yet each imaging modality exhibits distinct advantages and limitations. X-ray imaging provides high-resolution structural details, but lacks soft tissue contrast and exposes patients to ionizing radiation (Kalender 2011). Computed tomography (CT) enhances X-ray imaging by offering three-dimensional reconstructions, yet it suffers from high radiation exposure and limited soft-tissue contrast (Goh et al. 2025). Magnetic resonance imaging (MRI) offers excellent soft tissue differentiation, but requires long scan times and expensive superconducting magnets (Bruno et al. 2019). Ultrasound imaging is a safe and widely used technique, however, it has limited specificity and struggles with imaging deep-seated structures in the presence of gas or bone (Klibanov and Hossack 2015). Optical imaging techniques such as fluorescence and bioluminescence imaging provide molecular-level sensitivity, but are constrained by strong light scattering in biological tissues, limiting their imaging depth to only a few millimeters (Dhawan et al. 2010). Microwave imaging can also penetrate soft tissue, but has limited spatial resolution due to its longer wavelengths and, because it detects radiation directly, is susceptible to electromagnetic (EM) interference (Zhurbenko 2011).

Given existing imaging modality limitations, there is a growing demand for a highly sensitive and specific imaging modality that can achieve deep tissue imaging while maintaining high contrast and resolution. Thermoacoustic imaging (TAI) emerges as a promising hybrid imaging technique that integrates the advantages of microwave and ultrasound technologies to overcome many of these challenges (Wang et al. 2021a). Unlike purely optical or ultrasound imaging, TAI provides contrast based on tissue dielectric properties, enabling differentiation of tissue types, while the use of ultrasound detection enables high spatial resolution (Wang et al. 2024) compared with microwave imaging. TAI operates on the principle of the thermoelastic effect, where biological tissues absorb very short microwave pulses leading to the generation of acoustic waves that propagate through tissue, are detected using ultrasound transducers, and are used to reconstruct high-resolution images of tissue constituents (Nan et al. 2018). Because microwave energy exhibits lower attenuation in biological tissues compared to optical energy, TAI achieves significantly greater imaging depths than fluorescence, optical coherence tomography or photoacoustic imaging (PAI) while maintaining high imaging contrast (Liu et al. 2022). Simultaneously, the generated acoustic waves ensure spatial resolution comparable to ultrasound imaging.

A major advantage of TAI is its ability to detect soft tissue abnormalities with high contrast, particularly in applications such as breast cancer imaging (Wang et al. 2012a), vascular imaging (Chi et al. 2018), prostate cancer detection (Patch et al. 2015a), brain imaging (Yan et al. 2019; Xu and Wang 2006a; Islam et al. 2024), and foreign body detection (Nie et al. 2009). Unlike X-ray mammography, which struggles with dense breast tissue, or MRI, which requires costly contrast agents, TAI exploits intrinsic tissue dielectric properties to differentiate between normal and pathological tissues. For example, the dielectric properties of tumorous tissues (Martellosio et al. 2016), coagulated blood (Salahuddin et al. 2017), and foreign bodies are distinctly different than healthy tissue and blood. This makes TAI a non-ionizing, cost-effective, and high-contrast imaging tool suitable for preclinical and clinical research (Gao et al. 2016). Despite its numerous advantages, TAI faces challenges such as optimizing the sensitivity of detection systems, managing hardware complexity, and addressing the trade-off between imaging depth and resolution (Liu et al. 2022). At present, work is needed to deliver suitable devices that demonstrate ease of operation and image interpretation for a range of biomedical applications. However, with ongoing advancements in microwave generation, transducer technology, and image processing algorithms, TAI is poised to modernize biomedical imaging by providing high-contrast, deep-penetrating, and non-invasive visualization of biological tissues. This chapter explores the fundamental principles, system components, and preclinical applications of TAI, offering a comprehensive overview of its potential to enhance biomedical imaging and diagnostics.

2 Thermoelastic Expansion and Thermoacoustic (TA) Signal Generation

In TAI, a short-duration microwave pulse is transmitted into biological tissue. This causes the tissue to undergo a rapid thermoelastic expansion within the thermal and stress confinement regime due to the absorption of microwave energy, which leads to heat generation (Mashal et al. 2009; Lin and Lin 2021; Lin 1977, 1980). The generated heat causes local tissue expansion, which induces acoustic pressure waves that are detected by an ultrasound transducer. The detected signals are further processed through an image reconstruction algorithm to generate thermoacoustic (TA) images. The process can be described in three main steps: (i) EM absorption and heat deposition, (ii) thermal expansion and pressure wave generation, and (iii) acoustic wave propagation and detection.

2.1 Electromagnetic (EM) Absorption, Heat Deposition, and Acoustic Wave Generation in TAI

When an EM wave propagates into biological tissue, the tissue absorbs part of the energy, which leads to heating. The power density Q (heat deposition per unit volume) is given by:

$$Q\,(r,t) = \sigma\,(r)|E\,(r,t)|^2 \tag{1}$$

where $Q(r,t)$ is the heat deposition per unit volume at position r and time t, $\sigma(r)$ is the electrical conductivity of the tissue, $E(r,t)$ is the electric field intensity.

For a harmonic EM wave with angular frequency ω:

$$E\,(r,t) = E_0(r)e^{j\omega t}. \tag{2}$$

The time-averaged heat deposition over one cycle is:

$$\langle Q(r)\rangle = \frac{1}{2}\sigma\,(r)|E_0(r)|^2. \tag{3}$$

The specific absorption rate (SAR), which represents the rate of energy absorption per unit mass, is given by:

$$SAR = \frac{\sigma|E|^2}{\rho}. \tag{4}$$

where ρ is the mass density of the tissue.

Following the absorption of EM energy, the resulting heat deposition induces thermoelastic expansion, leading to pressure wave generation and acoustic signal propagation. The absorbed energy causes a rapid temperature rise ΔT, leading to a pressure increase due to thermoelastic expansion. The temperature rise follows the bioheat equation (simplified as heat is not transferred between blood and tissue):

$$\frac{\partial T}{\partial t} = \frac{1}{\rho C_p}Q + D\nabla^2 T - \frac{w_b}{\rho C_p}\,(T - T_b) \tag{5}$$

where C_p is the specific heat capacity of the tissue, D is the thermal diffusivity, w_b is the blood perfusion rate and T_b is the initial temperature of the blood or tissue. The pressure increase due to the thermal expansion is given by:

$$p\,(r,t) = \beta c^2 \int_{-\infty}^{t} Q\left(r,t'\right)e^{-(t-t')/\tau}dt' \tag{6}$$

where β is the thermal expansion coefficient, c is the speed of sound in the tissue, and τ is the thermal relaxation time. For an instantaneous heating source, the pressure can be approximated as:

$$p_0 = \beta c^2 \frac{Q}{\rho C_p}, \tag{7}$$

which shows that the initial pressure p_0 is proportional to the heat deposition Q.

As the thermoelastic pressure wave propagates through the tissue, it undergoes specific acoustic wave dynamics, which are crucial for TA image formation. The generated pressure wave propagates through the tissue as an acoustic wave governed by the TA wave equation (Xu and Wang 2006b):

$$\nabla^2 p(r,t) - \frac{1}{c^2}\frac{\partial^2 p(r,t)}{\partial t^2} = -\frac{\beta}{C_p}\frac{\partial Q(r,t)}{\partial t}. \tag{8}$$

For short-pulsed excitation, where the heating occurs nearly instantaneously, the pressure follows:

$$p(r,t) = \frac{\beta}{C_p}\int_V \frac{1}{4\pi |r-r'|}\frac{\partial Q(r',t')}{\partial t} dV \tag{9}$$

where V is the volume of the heated region. By detecting and reconstructing these acoustic waves using an ultrasound transducer, a TA image can be formed.

2.2 Roles of Tissue Electrical Conductivity and Permittivity in TA Signal Generation

The electrical conductivity in biological tissues mainly comes from free ions present in intracellular and extracellular fluids. The key contributors are:

Water Content: Since water is the primary medium for ion transport, tissues with higher water content (e.g., blood, muscle) tend to have higher conductivity. **Ionic Concentration:** The presence of ions (Na^+, K^+, Cl^-, Ca^{2+}) affects conductivity. Higher concentrations lead to higher conductivity. **Membrane Structures:** The cell membrane acts as a capacitor at low frequencies but contributes to conductivity at higher frequencies. **Frequency Dependence:** Tissue conductivity varies with frequency due to the dispersion effects caused by ionic movement and polarization. Table 1 shows typical electrical conductivities of biological tissues as reported in the literature (Gabriel et al. 1996; Hasgall et al. 2018).

The conductivity of biological tissues increases with frequency due to:

- Increased ionic mobility at higher frequencies.
- Membrane capacitance effects becoming less dominant.
- Relaxation processes like Maxwell-Wagner dispersion (Raicu et al. 1998).

Table 1 Typical electrical conductivities of biological tissues (~1 GHz)

Tissue type	Conductivity (σ) in S/m
Blood	1.5–2.0
Muscle	0.7–1.5
Liver	0.5–1.0
Fat	0.02–0.1
Bone (cortical)	0.01–0.05
Brain (gray matter)	0.6–1.0
Brain (white matter)	0.1–0.5

Based on data from Gabriel et al. (1996) and Hasgall et al. (2018)

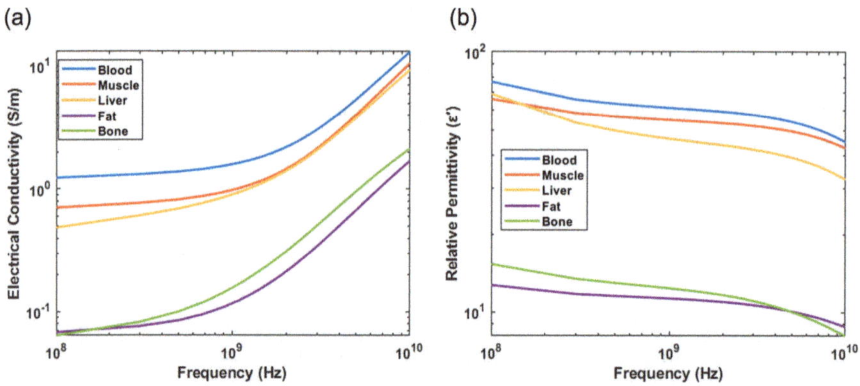

Fig. 1 (**a**) Electrical conductivity vs frequency of different tissue; (**b**) Relative permittivity of different biological tissues vs frequency. Data are plotted using information found in Gabriel et al. (1996)

This means that at microwave frequencies (~1 GHz and above), tissues with high water content (muscle, blood, tumors) show significantly higher conductivity, enhancing TA signal generation. Figure 1a shows a comparison between the electrical conductivity of different biological tissues as reported in (Hasgall et al. 2018).

In addition to electrical conductivity, another critical factor influencing TA signal generation is permittivity, which determines how EM energy is stored and dissipated in tissues. Permittivity (ε) determines how much EM energy is stored in the tissue before dissipation. Complex permittivity is defined as:

$$\varepsilon^* = \varepsilon' - j\varepsilon'' \tag{10}$$

where ε' is the real part (dielectric permittivity), which represents the ability of the tissue to store electrical energy, and ε'' is the imaginary part, which represents energy loss due to conduction and dielectric relaxation. The imaginary part of permittivity is related to the electrical conductivity (σ) by:

$$\varepsilon'' = \frac{\sigma}{\omega} \tag{11}$$

where σ is the electrical conductivity (S/m) and $\omega = 2\pi f$ is the angular frequency of the applied EM field. Thus, the complex permittivity can be rewritten as:

$$\varepsilon^* = \varepsilon' - j\left(\frac{\sigma}{\omega}\right). \tag{12}$$

In TAI, the signal strength is governed by how much EM energy the tissue absorbs and then converts into acoustic waves. This absorption depends on the tissue's complex permittivity, which has both a "dielectric" component and a "conductive" component. In other words, both conductivity and dielectric properties (the real and imaginary parts of permittivity) contribute to the energy absorption—and thus to the amplitude of the resulting TA signal. At lower microwave frequencies (e.g., around 300 MHz up to 3 GHz), electrical conductivity tends to dominate the loss mechanism. At higher microwave frequencies (>3 GHz), dielectric losses (related to polarization effects) become increasingly important.

More specifically, the real part of permittivity (ε') determines how much of the electric field energy is stored in the tissue. A higher contrast in permittivity between tissues leads to stronger reflections and refractions at interfaces (similar to optical scattering in light propagation). Microwaves propagate more efficiently in tissues when there is good impedance matching (low permittivity contrast). If the permittivity difference is large, impedance mismatch occurs, leading to reflection and standing waves rather than deep penetration.

Therefore, similar to optical scattering, permittivity variation leads to changes in microwave propagation (reflection, refraction, and transmission effects), reducing the clarity of deep tissue imaging by preventing uniform wave penetration. The difference is that microwave "scattering" due to permittivity is primarily an impedance mismatch effect rather than physical scattering by sub-wavelength structures. Figure 1b shows the relative permittivity of different biological tissues as reported in (Lin 1977).

As an example, the initial TA pressure generated at 1 cm depth in muscle when a 2.54 GHz, 1 kW continuous-wave (CW) microwave generator is used and modulated to generate 200 ns pulses with a repetition rate of 1 kHz is calculated as follows. Given a short microwave pulse, the initial TA pressure rise p_0 can be determined by how much energy is absorbed locally and converted to pressure. The general relation is (Xu and Wang 2006b):

$$p_0 = \Gamma \mu_a F \tag{13}$$

where, Γ is the Grüneisen parameter (dimensionless thermoelastic conversion efficiency), μ_a is the tissue absorption coefficient (in m^{-1}), and F is the local fluence (energy per area, J/m^2) of the pulse. The microwave generator produces 1 kW pulses of width 200 ns at 1 kHz repetition. The energy per pulse is:

$$E_{pulse} = Power \times pulsewidth = 1000W \times 200 \times 10^{-9} J = 0.2mJ. \quad (14)$$

The microwave is emitted through a rectangular waveguide of 86 mm × 43 mm (area A = 0.086 m × 0.043 m = 3.70 × 10^{-3} m^2). We assume the antenna has gain ~1 (directive), meaning the pulse energy is concentrated over roughly this aperture area (instead of spreading spherically). Thus, at ~2 cm from the aperture (the skin surface), the pulse illuminates approximately the same area.

The energy per unit area (fluence) incident on the skin is:

$$F_0 = \frac{E_{pulse}}{A} = \frac{2 \times 10^{-4} J}{3.70 \times 10^{-3} m^2} \approx 5.4 \times 10^{-2} J/m^2. \quad (15)$$

The absorption coefficient must be calculated or estimated. Muscle is a strong absorber of microwaves in the ~2.5 GHz band. For example, one group calculated 2.45 GHz microwaves have a penetration depth of ~17 mm in muscle (Mohammed and Saber 2014). This corresponds to an absorption coefficient of μ_a roughly:

$$\mu_a = \frac{1}{penetration\_depth} \sim \frac{1}{0.017m} \approx 59m^{-1} \quad (16)$$

For our calculation at 2.54 GHz, we will select μ_a ~ 0.7 cm^{-1} (i.e., 70 m^{-1}) as a representative value. Microwave intensity decays approximately as I(z) = I(0) $e^{-\mu a}$ in tissue. At $z = 1$ cm depth in muscle:

$$I(1cm) = I(0)e^{-\mu_a(1cm)} = I(0)e^{-0.7} \approx 0.50 I(0). \quad (17)$$

In other words, only about 50% of the microwave energy remains at 1 cm, the rest having been absorbed in the overlying tissue. (If μ_a were as high as 1 cm^{-1}, only ~37% would remain; if as low as 0.5 cm^{-1}, about 60% would remain. Either way, a large fraction is lost by 1 cm depth due to absorption). Using our fluence:

$$F(1cm) = F_0 e^{-0.7} \approx 5.4 \times 10^{-2} J/m^2 \times 0.50 \approx 2.7 \times 10^{-2} J/m^2. \quad (18)$$

The energy absorbed within the tissue at ~1 cm depth can be estimated by multiplying the local fluence by μ_a. Essentially, in a thin slice at 1 cm, a fraction μ_a, dz of F(1 ~ cm) is absorbed in thickness dz. In the limit of small dz, the absorbed energy per volume (J/m^3) at that depth is approximately:

$$E_{abs}(1cm) = \mu_a F(1cm) \approx 70m^{-1} \times 2.7 \times 10^{-2} J/m^2 \approx 1.9 J/m^3 \quad (19)$$

since μ_a has units of inverse length. In SI units, we convert μ_a = 0.7 cm^{-1} to 70 m^{-1} and use F(1 cm) in J/m^2.

In order to solve (13), the only parameter left to estimate is the Grüneisen parameter. For soft, fluid-filled tissues at body temperature (≈37 °C), Γ typically ranges 0.1–0.2 depending on tissue composition and temperature; muscle, being water-rich, is usually in the lower end of that range – roughly similar to water's Γ ≈ 0.12–0.15 at 37 °C. We will use Γ ~ 0.15 as a representative value.

Now we multiply the absorbed energy density by Γ:

$$p_0 = \Gamma E_{abs}\,(1cm) \approx 0.15 \times 1.9 J/m^3 \approx 0.285 J/m^3 \tag{20}$$

and recall 1 J/m^3 = 1 Pa (since pressure is energy density in SI). So p_0 ~ 0.29 Pa (approximately 0.3 Pa). Therefore, if the antenna is 2 cm away from the skin, the initial TA pressure generated at 1 cm depth in muscle due to a 2.54 GHz, 1 kW CW microwave generator with 200 ns pulse width and 1 kHz repetition rate is approximately: 0.3 Pa.

3 Main Considerations for Building a TAI System

The key components of a TAI system include the microwave source, the detection system, and the antenna. Additionally, post-processing techniques play a crucial role in the development and optimization of TAI.

3.1 Microwave Sources

The main difference between the more widely known PAI and TAI is the type of excitation source. In PAI, a laser can be used as the excitation source while in TAI, a microwave generator is used for the excitation source. Due to better penetration depth of microwave energy, TAI achieves better imaging depth, ranging from 6–15 cm (Cui et al. 2017). However, because microwave beams are not highly focused due to diffraction, the excitation source needs to be very powerful.

The excitation sources in TAI are primarily categorized into five types: Magnetron, Klystron, High-Voltage Discharge, Solid-State Amplifier (SSA), and Traveling Wave Tube Amplifier (TWTA). A magnetron is a self-oscillating microwave source typically composed of an electron emitter, a power supply, a magnetron cavity, and a waveguide. Electrons emitted from the electron source are accelerated, generating stable electronic oscillations, which are then converted into microwaves within the waveguide. At the waveguide's end, diffraction transforms the microwave energy from a waveguide mode to a radiation pattern. Due to the frequency-selective nature of the waveguide, magnetrons operate at a fixed frequency and are inherently pulsed. However, they have two key limitations: (1) their operating frequency is fixed, preventing frequency tunability, and (2) their wide pulse width (0.2–0.3 μs) restricts imaging resolution to around 0.5 mm. In contrast, a klystron is a vacuum tube-based microwave amplifier that utilizes velocity modulation of an electron beam in a series of resonant cavities to generate and amplify microwaves. Klystrons offer high power stability, frequency tunability, and narrower pulse widths (nanoseconds to microseconds), making them ideal for applications requiring precise microwave control. However, they are bulky, expensive, and require complex cooling systems. Another microwave source, the high-voltage discharge-based generator, operates by rapidly discharging stored electrical energy through a high-speed switch, producing ultra-short microwave pulses (a few nanoseconds). These systems offer high-resolution imaging (micron-scale) due to their narrow pulse width but lack frequency tunability and are often used where short, high-power pulses are needed. The next microwave sources for consideration are solid-state microwave generators, such as SSAs and TWTAs, which provide CW or pulsed microwave output with excellent frequency stability and tunability. SSAs are based on semiconductor technology and are compact, but their power output is lower compared to klystrons and magnetrons. TWTAs, which combine aspects of klystrons and SSAs, provide high power and broadband tunability but are expensive and require cooling. Table 2 reports how these five types of microwave sources compare with each other in regard to these features.

Each microwave source has trade-offs, with magnetrons being cost-effective but lacking tunability, klystrons offering superior stability and control but being bulky and expensive, high-voltage discharge sources achieving the shortest pulses for high-resolution imaging, and solid-state and TWTA sources providing tunable and reliable performance for various applications but lower power output. To achieve sufficient power for imaging, a solid-state radio frequency (RF) amplifier is usually employed in conjunction with these sources (Nan and Arbabian 2014). These amplifiers are often based on gallium nitride (GaN) or gallium arsenide (GaAs) technology (Nan et al. 2018; Liu et al. 2022, 2023a; Wang et al. 2022; Lan et al. 2020; Aliroteh and Arbabian 2017; Singhvi et al. 2019). Recent developments in portable TAI systems have leveraged a combination of a solid-state microwave generator and an amplifier to design compact, high-efficiency imaging devices (Wang et al. 2012a; Sutardja et al. 2022).

Ensuring safety around TAI sources is crucial, as exposure to pulsed microwave radiation can lead to tissue heating and potential health risks. For frequencies between 1 and 6 GHz, safety guidelines focus on limiting the SAR, which measures the rate at which energy is absorbed by body tissues. The Occupational Safety and Health Administration (OSHA) has set an exposure limit of 10 mW/cm^2 (power density exposure) for microwave radiation. For example, average power density for a 2 kW CW, 2.54 GHz microwave source, modulated with 200 ns pulses at a 1 kHz repetition rate, at 10 cm distance (when antenna gain is 1) is 3.18 W/m^2 (0.318 mW/cm^2). As mentioned above, the safety limit for 2.54 GHz is 10 mW/cm^2 (FCC/IEEE standard) therefore, this exposure is considered safe under standard guidelines.

Table 2 Comparison of microwave sources

Microwave source	Power output	Pulse width	Frequency tunability	Continuous-wave (CW) or pulsed	Complexity & cost
Magnetron	High	0.2–0.3 μs	Fixed	Pulsed	Low cost, simple
Klystron	High	ns to μs	Tunable	Pulsed or CW	Expensive, complex
High-Voltage Discharge	High	Few ns	Fixed	Pulsed	Moderate cost, simple
SSA	Low to Moderate	ns to μs	Tunable	CW or Pulsed	Compact, high cost
TWTA	High	ns to μs	Broadband tunable	CW or Pulsed	Expensive, requires cooling

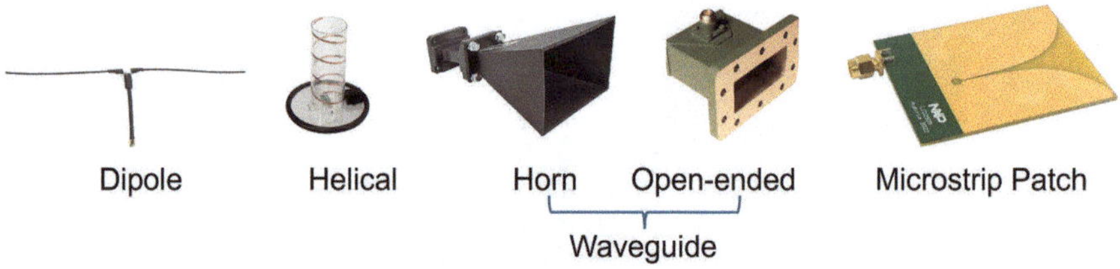

Fig. 2 Different types of antennas used in TAI systems, including dipole, helical, waveguides (horn and open-ended), and microstrip patch. Note: images are not to scale, antenna sizes can vary from the centimeter range (microstrip patch) to ~1 m in length (large horns)

3.2 Types of Antennas

Antennas play a crucial role in delivering microwave energy to the target tissue. The design and type of antenna affects the energy distribution. Previous studies have reported several types of antennas for TAI systems (Liu et al. 2022; Wang et al. 2012a; Tang et al. 2025; Zhao et al. 2021; Huang et al. 2019; Ren et al. 2022), with frequency ranges between 300 MHz to 9 GHz mostly used for imaging soft tissue (Xu et al. 2003) (Fig. 2). Most of the studies have reported the use of waveguide antennas because their resonance frequency is around 3 GHz, and they provide a directional beam path. They can also deliver high power pulsed microwave energy (Ke et al. 2012). There are two types of waveguide antennas, open-ended and horn waveguides. Other types of antennas used in TAI are helical, monopole, microstrip patch, and dipole antennas. The use of multiple antenna elements in a single system has also been reported (Ren et al. 2022; Wu et al. 2021).

Open-ended waveguides are the most widely used antenna in TAI systems (Mashal et al. 2009; Liu et al. 2016; Zhao et al. 2013; Ku et al. 2005; Qin et al. 2014; Jin and Wang 2006; Kruger et al. 2000; Ku and Wang 2000; Xu and Wang 2002a; Nie et al. 2008; Ji et al. 2012; Nan and Arbabian 2017). Systems operating in the 3 GHz – 9 GHz microwave frequency range have predominantly utilized open-ended waveguide antennas for efficient microwave power delivery to the target. For example, the tomography system proposed by Liming Nie et al. in (Nie et al. 2007) used an open ended rectangular waveguide with a cross section of 127×63 mm^2, with which microwave pulses irradiated a sample uniformly with an energy density of 0.4 mJ/cm^2. Or in another example, the spectroscopic thermoacoustic tomography (TAT) system proposed by Bauer, Daniel R., et al. used an open-ended waveguide antenna (WR-187) that carries 2.7–3.1 GHz pulsed microwave signals at 400 ns pulse width (Bauer et al. 2012). The waveguide horn antenna is the second most used antenna in TAI (Tan and Wang 2024; Yan et al. 2018; Chi et al. 2016, 2019; Huang et al. 2013; Wang et al. 2012b). Usually, the horn antennas are bulky and work around the same frequencies as the open-ended waveguide. For example the antenna used in (Song et al. 2013) operates at 2.45 GHz and is capable to delivering 500 ns rectangular pulse with peak power ranging from 0 to 50 kW. Or, the horn antenna used in (Wang et al. 1999) has an aperture of 55 mm length and 74 mm width and operates at 9.4-GHz, transmitting 10 kW of microwave pulses at 2.2 or 0.6 μs pulse widths. Another example is the TAT system proposed by Huang Lin et al. (2013), that uses a horn antenna (114 × 144 mm^2) at 3.0 GHz to deliver a peak power of 70 ~ 100 kW with pulse duration of 0.75 μs. Helical antennas are mostly used for lower frequency bands like 434 MHz and 900 MHz (Gao et al. 2016; Feng et al. 2013) where the size of waveguide or horn antennas would be very large, for example, the helical antenna used by Fei Gao et al. operates at 440 MHz and is able to transmit up to 100 W of microwave power with 3.67 μs pulsed signal (Gao et al. 2013). Other examples of helical antennas are presented in (Kellnberger et al. 2011a; Fu et al. 2014). There are some research groups that have used a monopole antenna. To prevent conducting currents and electric breakdown of the medium, this antenna was covered by an insulating layer of polytetrafluoroethylene (Teflon) material (Razansky et al. 2010). Dipole antennas are also used because of their smaller footprint. The handheld dipole antenna used in (Huang et al. 2019) is one example. The dimension of the antenna was 60 × 60 × 45 mm^3. The maximum average microwave power density at the surface of the antenna was just 8.6 mW/cm^2, when a 10 Hz repetition frequency was used at 3 GHz. The microstrip patch antenna used in (Yan et al. 2025) operates at 1.15–1.7 GHz. The antenna provides a wide axial ratio beamwidth, ensuring uniform energy distribution across the imaging field. The study in (Ren et al. 2022) utilizes two microstrip patch antennas known as miniaturized antipodal Vivaldi antennas, each measuring 51.87 mm × 45 mm × 1.5 mm, with a glass fiber epoxy resin substrate ($\varepsilon_r = 4.4$) and copper foil (0.035 mm thick) coating. These antennas operate at 6.7 GHz, achieving a reflection coefficient (S11) <-10 dB, and are integrated into a coaxial microwave-acoustic probe to ensure uniform microwave illumination and high-sensitivity ultrasound detection for breast tumor screening.

3.3 Detection Systems

TAI detection systems are either single-element ultrasound transducer-based or multi-element transducer-based (including ring array or linear array transducers) (Fig. 3). Each of these detection configurations is designed to optimize imaging performance in terms of signal acquisition speed, spatial resolution, and imaging depth, making them suitable for different biomedical applications such as tumor detection, functional imaging, and deep tissue imaging (Huang and Huang 2023).

Single-element transducer (SET)-based systems utilize a single ultrasonic transducer to sequentially scan the imaging area, capturing TA signals at multiple positions (Islam et al. 2023). These systems are widely used because of their low cost and simplicity. The major limitation of SET-based systems is their slow scanning speed, as they require mechanical movement to acquire signals from different locations before reconstructing a complete image. Wang et al. (2012a) demonstrated that SETs offer excellent depth penetration and signal-to-noise ratio (SNR) but are less suitable for real-time imaging due to the lengthy data acquisition process. Wang et al. (Qin et al. 2014) proposed a computational approach to

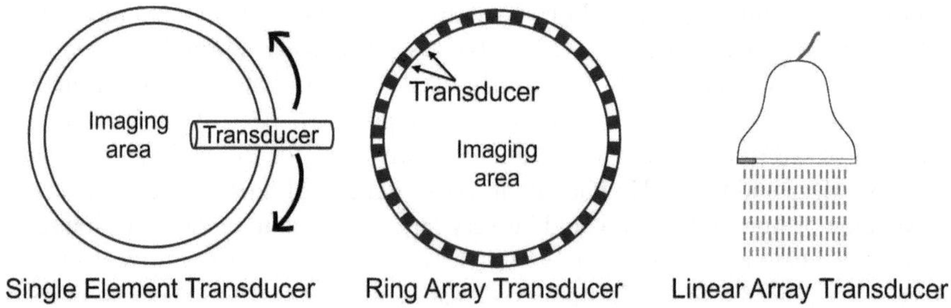

Fig. 3 Different types of detection systems used in TAI systems. Shown are a single element transducer that translates around a circular track (left), a ring array of transducers that surround the imaging area (center) and a linear array of transducers that are aligned in parallel (right)

improve noise reduction and increase scanning efficiency, making SETs more viable for preclinical imaging applications. Similarly, Zhang et al. (2023) explored the impact of frequency optimization on improving image contrast. To overcome the speed limitations of SET systems, multi-element transducer systems have been developed, integrating multiple ultrasonic elements arranged in linear, circular, or 2D array configurations. Unlike SETs that require mechanical movement, multi-element transducer arrays enable simultaneous data acquisition, significantly reducing scanning time and enhancing imaging efficiency.

The primary advantage of multi-element transducer systems is their ability to perform real-time imaging, which is crucial for clinical applications such as cancer detection, functional imaging, and intraoperative monitoring. However, these systems require sophisticated data acquisition and image reconstruction algorithms to manage the large volume of acquired signals. Ding and Zhang (2024) developed an ultrasonic velocity-assisted microwave-induced TAI (US-MTI) system, which enhanced detection accuracy by integrating velocity measurement techniques with array-based transducers. Additionally, Ji et al. (2015) proposed a flexible multi-element transducer array that adapted to the shape of the imaging subject, improving the accuracy of soft tissue imaging. This shape-adaptive technology significantly enhances the sensitivity and resolution of TA images, making it particularly useful for biomedical applications such as brain imaging and vascular imaging. Linear array transducers represent a specialized category of multi-element transducer systems, designed to balance high spatial resolution with fast imaging speed. These systems consist of transducer elements arranged in a linear configuration, allowing for dynamic focusing and electronic scanning, eliminating the need for mechanical movement (Zeng et al. 2009). They allow for beamforming and dynamic focusing, which improves image quality and minimizes artifacts. Recent studies have explored the use of linear array transducers for TAI, highlighting their advantages in achieving high spatial resolution, rapid imaging speed, and efficient signal acquisition. Nie et al. developed a multi-element linear transducer array-based TAI system for detecting foreign bodies in biological tissues, demonstrating high contrast imaging with 1.2 GHz microwave excitation and a 2 MHz ultrasonic detection array, which enabled fast and high-SNR image acquisition (Nie et al. 2007). Ji et al. introduced a portable, handheld TA scanning system based on a linear-array transducer, optimized for breast and soft tissue imaging (Ji et al. 2016). More recently, Wang et al. presented an integrated dual TA and ultrasound imaging system, combining a hollow concave array with a linear transducer array to improve the imaging of vascular structures and soft tissue tumors (Wang et al. 2021a). Additionally, Islam et al. developed a TAI system designed for hemorrhage imaging in the neonatal brain, using ex vivo models to demonstrate the feasibility of linear array-based transducers for neurovascular applications. Their system incorporated a linear array transducer, two waveguides and an advanced image reconstruction algorithm, highlighting the potential of TAI for transcranial applications (Islam et al. 2024). Table 3 provides a review of the range of TA imaging systems reported in the literature.

3.4 Post Processing and Image Reconstruction

Post-processing and image reconstruction play a crucial role in transforming raw TA signals into high-quality images suitable for biomedical diagnostics. The reconstruction process begins with signal preprocessing, which involves noise reduction, baseline correction, and signal amplification to enhance data quality. Effective preprocessing is essential to eliminate background artifacts and enhance the SNR (Xu et al. 2004). Several advanced reconstruction algorithms are employed in TAI, including filtered back-projection (FBP), time-reversal methods, and model-based approaches (Prakash et al. 2024; Zhu et al. 2014; Hristova et al. 2008; Ding et al. 2016). The FBP algorithm is widely used due to its computational efficiency and ability to reconstruct images from multiple angles, making it suitable for circular-scanning TAI systems (Islam et al. 2023). However, FBP tends to introduce streak artifacts and requires additional filtering for optimal image clarity. Time-reversal reconstruction has demonstrated superior performance in heterogeneous tissue by accurately back-propagating the recorded TA signals to reconstruct high-resolution images (Xu and Wang 2002b).

One of the key challenges in TA image reconstruction is accounting for factors such as acoustic wave propagation, tissue heterogeneity, and system geometry. In highly heterogeneous biological tissues, variations in acoustic properties can lead to distortions in the reconstructed image. Deep learning-based reconstruction methods are also emerging, leveraging convolutional neural networks to enhance image quality and suppress artifacts (Xu et al. 2022; Aggrawal et al. 2024). Song et al. (2022) implemented a fully dense U-Net architecture for TAI, demonstrating improved artifact removal and enhanced feature extraction.

Table 3 TA systems implemented by different research groups

Research group	Microwave frequency	Pulse width	Peak power	Microwave source	Amplifier	Acoustic detector	Antenna used	Reference
L. V. Wang	3 GHz 9 GHz 9.4 GHz	0.5 μs 0.5 μs 0.6 μs	2 kW 10 kW 10 kW	Microwave generator	Not reported	Single element	Waveguide	Xu et al. (2003), Feng et al. (2001)
R. A. Kruger	434 MHz	1.0 ms 0.5 μs	25 kW	UHF generator	Not reported	64 element arrays	Waveguide	Kruger et al. (1999, 2000),
D. Xing	6 GHz 1.2 GHz 3 GHz	0.3 μs 0.5 μs 1.2 μs 0.3 μs 10 ns	350 kW 300 kW 4 MW	Ultrashort Pulse generator	Not reported	128 element arrays Linear array Ring array	Waveguide Helical Antenna	Nie et al. (2008, 2009), Ji et al. (2012), Yuan et al. (2019), and Li et al. (2022)
X. Liang	3 GHz	(100–700) ns	35 kW	Microwave generator	Customized	Single element	Dipole Antenna	Liang et al. (2020)
H. Xin	3 GHz 2.7 GHz	0.5 μs 400 ns	4 kW 5.2 kW	Microwave oscillator	Not reported	49 × 49 transducer array Single element	Waveguide	Wang et al. (2012a), Bauer et al. (2012)
V. Ntziachristos	400 MHz 100 MHz	45 ns 10 ns	5.8 MW 70 MW	High power electromagnetic impulse	Not reported	Single element	Helical, Monopole Antenna	Razansky et al. (2010), Omar et al. (2012), and Kellnberger et al. (2011b)
B. Wang	2.45 GHz	0.5 μs	20 kW	Microwave Generator	Not reported	15 elements	Waveguide	Wang et al. (2021b)
L. Huang	3 GHz	750 ns	70 kW 100 kW	Custom-designed microwave generator	Custom designed	Single element	Horn Antenna	Chi et al. (2016), Huang et al. (2013), and Wang et al. (2012b)

Qiang Liu	3 GHz 1.2 GHz	380 ns 300 ns	35 kW 160 kW	In house microwave generator	Custom	Single element	Dipole antenna	Liu et al. (2023b)
Huan Qin	3 GHz	500 ns	70 kW	In house microwave generator	Custom	128-element	Dipole antenna	Huang et al. (2023)
Huabei Jiang	3 GHz	550 ns	65 kW	In house microwave generator	Custom	128-channel	Horn antenna	Du et al. (2024)
Kamran Avanaki	2.4 GHz, 3 GHz	200 ns	100 W, 250 W	Microwave Generator	SSA	Single Element, 64 element linear array	Waveguide	Islam et al. (2023, 2024)

4 Preclinical Applications of TAI

Over the past two decades, researchers have explored various applications of TAI for the development of clinical imaging devices. The main factors that make TAI attractive for clinical use—such as its ability to differentiate tumorous and healthy tissue, clotted and unclotted blood, and foreign bodies—predict that it will be a valuable modality for preclinical research requiring deep penetration and high contrast. This section provides a comprehensive review of preclinical TAI studies focusing on breast cancer, prostate cancer, brain imaging studies (glioma, stroke, and hemorrhage detection), blood vessel imaging and foreign body detection.

4.1 Breast Cancer

One of the most explored applications of TAI is the early detection of breast cancer (Wang et al. 2012a; Zhao et al. 2021; Wu et al. 2021; Kruger et al. 2000; Ji et al. 2012). The breast is primarily composed of fat, which exhibits low microwave absorption. However, cancerous tissue absorbs more microwave energy than healthy tissue due to differences in dielectric properties of tumor tissue and elevated ionic or water content at the tumor margin (Ji et al. 2012). Generally, changes in tissue absorption distributions indicate physiological lesions, and previous findings suggest that malignant breast tissue has higher electrical conductivity than normal tissue (Dessai et al. 2024). Several studies, including those by Kruger et al., Wang et al., and Xing et al., have investigated the applicability of TAI for breast cancer imaging in both small-animal models and clinical settings (Wang et al. 2012a; Nie et al. 2008; Ye et al. 2015; Ji et al. 2012).

Zhao et al. (2021) developed an ultrashort-pulse TAI system for whole-breast tumor screening, addressing challenges such as non-uniform microwave radiation and limited imaging coverage. Their system featured a large-aperture antenna (40 cm × 27 cm) to ensure uniform microwave distribution across the imaging field, enabling comprehensive breast imaging. By delivering ultrashort microwave pulses, the system enhanced spatial resolution and imaging efficiency, covering a 14 cm × 14 cm imaging field. Experimental results demonstrated an imaging depth of 7 cm—sufficient for detecting deep-seated tumors—with a spatial resolution of 290 μm, allowing for clear tumor boundary visualization. Phantom and ex vivo studies using an excised ewe breast embedded with tumors further validated the system's ability to differentiate malignant lesions from healthy tissue, achieving a tumor-to-background contrast ratio of 1:2 (Zhao et al. 2021). Wu et al. (2021) developed a handheld microwave TAI system aimed at overcoming the bulkiness and complexity of conventional TAI scanners. The primary objective was to design a portable, ergonomic imaging system that could be used for real-time breast tumor screening in clinical settings (Wu et al. 2021). Their system integrated a compact impedance-matching microwave-sono probe, which combined a miniaturized microwave antenna, a flexible coaxial cable for microwave excitation, and a linear

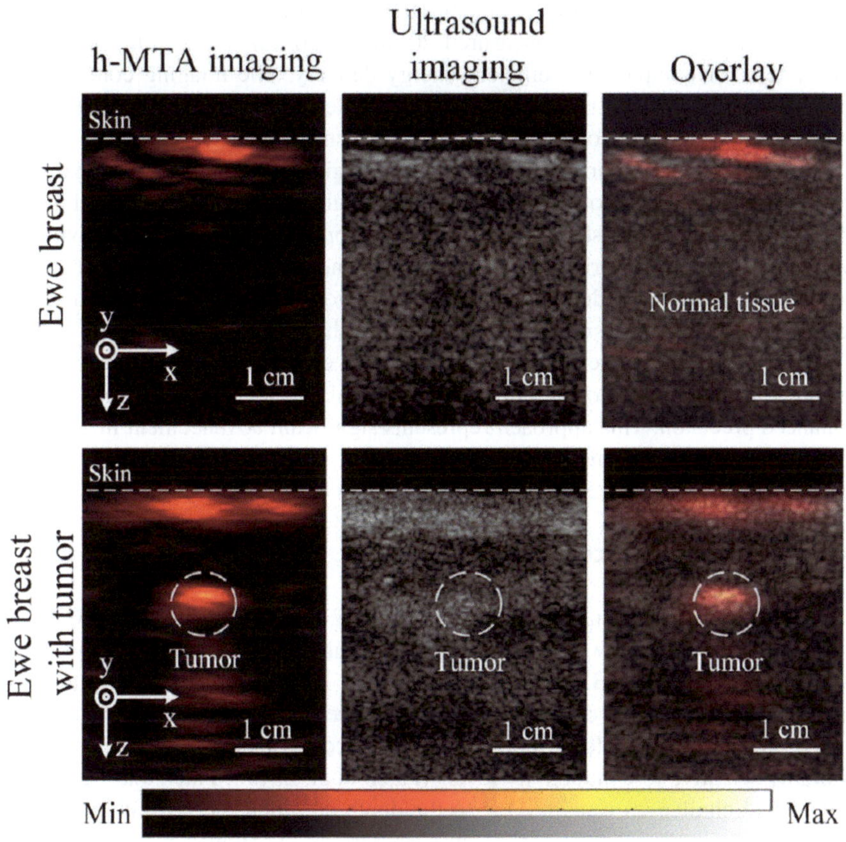

Fig. 4 In vitro TAI imaging of tumor implanted in ewe breast. TA (hot, left), ultrasound (gray, center) and overlaid (right) images of the ewe's breast tissue. Top row: without tumor, bottom row: with tumor. (Reprinted with permission from Fig. 5a of Wu et al. (2021), © 2021 IEEE)

ultrasound transducer array. Additionally, wedge-shaped polystyrene blocks were incorporated to improve acoustic coupling, ensuring efficient signal detection. The handheld system achieved a maximum imaging depth of 5 cm, with a SNR of 8 dB. It also provided high-resolution imaging, with lateral and axial resolutions of 1.5 mm and 0.9 mm, respectively (Fig. 4).

Furthermore, Zhao et al. introduced a flexible detector design aimed at enhancing adaptability and integration for breast cancer detection, providing improved imaging efficiency and patient comfort (Zhao et al. 2018). More recently, Li et al. proposed an integrated TAI technique that incorporates a phased array antenna and compressive sensing, enabling precise microwave focusing at specific depths through phase control. This approach not only enhances imaging accuracy but also significantly reduces measurement time and overall system costs (Li et al. 2021). In their study, Li et al. employed a microwave source operating at 2.45 GHz and a 24-element phased array antenna, with each antenna unit delivering a peak power of 120 W. Each individual antenna unit measured 20 mm × 30 mm × 1.5 mm, optimizing the system for compact and efficient energy delivery. The imaging configuration included a transducer array comprising 13 single elements, each with a central frequency of 1 MHz. To enhance image reconstruction, a compressive sensing method was utilized, demonstrating improved imaging performance while maintaining low computational complexity. This study highlights the clinical feasibility of integrating TAI into breast cancer hyperthermia applications, potentially paving the way for real-time monitoring of thermal therapy and early tumor detection (Li et al. 2021). Huang et al. introduced a finite-element algorithm designed to quantitatively reconstruct conductivity distributions within both tumorous and normal tissues. Their experimental framework involved imaging ex vivo mastectomy specimens, enabling the reconstruction of both morphological and functional parameters with enhanced precision. This approach represents significant advancement in TAI, as it facilitates quantitative imaging (Huang et al. 2013).

4.2 Prostate Cancer

Detecting prostate cancer using conventional imaging modalities such as MRI and ultrasound is a well-known challenge: these modalities have difficulty differentiating benign hyperplasia (BPH) and prostate cancer. Healthy and BPH prostate tissue have high levels of ionic content but prostate cancer reduces the production of ions in the prostate, leading to an opportunity for a contrast mechanism because TAI is highly sensitive to conductivity levels (Patch et al. 2015b; Patch and See 2015). Yet several challenges remain. Due to the depth at which the prostate is located, low-frequency microwave or RF waves are preferred. However, the absorption coefficient of prostate tissue in low-frequency bands is relatively low, necessitating higher peak power for improved contrast. Therefore, the key challenge is to generate ultrashort microwave pulses with an optimal central frequency and sufficient peak power while staying within safety limits. Additionally, an ultrasound transducer with a wide bandwidth is required to detect the TA signals efficiently and achieve adequate SNR. To address these issues, Patch et al. developed a TAI system, utilizing a 108 MHz radio wave with a 900 ns pulse duration and a peak electric field of ~17 kV/m. The system incorporated focused ultrasonic immersion transducers with peak response frequencies ranging from 1 MHz to 3.5 MHz, demonstrating the feasibility of high-contrast TAI for prostate cancer detection (Eckhart et al. 2011). In addition, another ex vivo experiment using VHF-band source (9.4 GHz, 5.66 GHz, 6 GHz, 3 GHz, and 1.2 GHz) with a pulse duration of 700 ns was also performed, where three groups of excised human prostate glands were investigated (Patch et al. 2015a) (Fig. 5). The study in (Ye et al. 2015) utilized 108 MHz EM (VHF) irradiation to generate TA pulses, with a 700-ns pulse width, enabling deeper penetration and quantification of electrical conductivity in biological tissues. The study integrates a clinical ultrasound array with a low-frequency TA channel to improve the sensitivity and specificity of prostate cancer imaging. Their approach

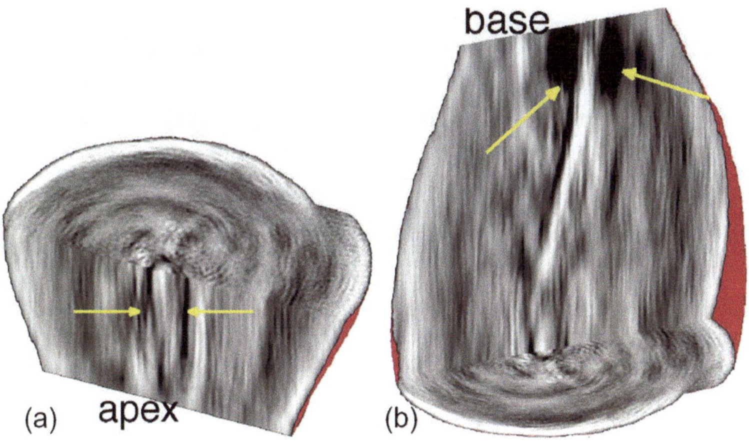

Fig. 5 Visualization of a volumetric TA reconstruction of a prostate (ex vivo) acquired using a 96 channel P4-1 transducer (Verasonics) with receive-only sampling frequency set to 30 MHz. (**a**) A coronal slice (arrow shows the apex); and (**b**) the anterior portion of an axial slice (arrow shows base of the prostate). (Reprinted with permission from Patch et al. (2015b), © 2015 IEEE)

demonstrates the potential of 108 MHz irradiation to enable deep tissue imaging while maintaining high contrast and quantitative accuracy.

4.3 Brain Cancer, Stroke, and Hemorrhage

Recent studies have explored TAI neuroimaging applications, particularly in stroke detection, brain hemorrhage assessment, and transcranial imaging through the skull. One of the major challenges in applying TAI to brain imaging is acoustic heterogeneity due to the skull, which distorts and attenuates acoustic signals. Jin et al. (2008) investigated the effects of skull-induced acoustic aberrations on transcranial brain imaging and proposed compensation methods to correct phase distortions, thereby improving the quality of TAT reconstructions. Similarly, Liu et al. (2023c) developed a full-waveform inversion technique combined with skull phase compensation, demonstrating successful transcranial TAI through a real human skull. These advancements have significantly improved image clarity and accuracy in clinical neuroimaging applications. The feasibility of in vivo brain imaging using TAI was demonstrated by Zhao et al. (2017), who successfully performed TAT on live rat brains. Their findings showed that TAI could resolve key brain structures with high contrast, providing a potential alternative to MRI and CT for functional and structural neuroimaging. Similarly, Xu and Wang (2006a) validated transcranial TAI in rhesus monkeys, proving its ability to penetrate the skull and visualize deep-seated brain tissues, marking an important step toward clinical translation. TAI has also been extensively studied for stroke imaging, particularly hemorrhagic stroke detection. Huang et al. (2017) demonstrated that

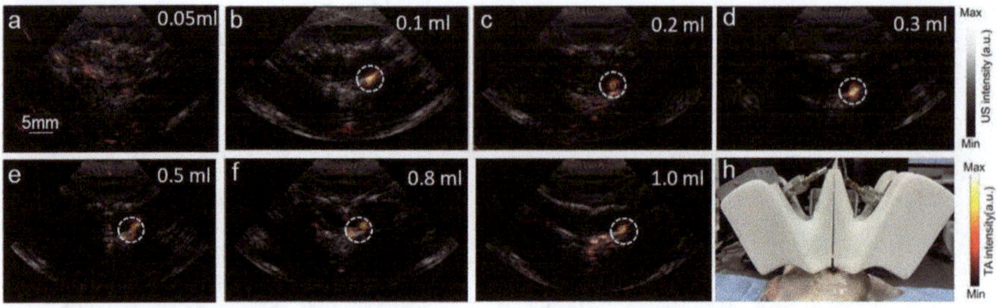

Fig. 6 Ex vivo sheep brain images (ultrasound and TA overlaid; ultrasound is gray, and TA is hot) of artificially induced hemorrhage with different concentration of blood. (**a**) 0.05 mL, (**b**) 0.1 mL, (**c**) 0.2 mL, (**d**) 0.3 mL, (**e**) 0.5 mL, (**f**) 0.8 mL, (**g**) 1.0 mL and (**h**) Ex vivo imaging setup showing angled antennas and ultrasound transducer over the sheep brain. (Reprinted with permission from Islam et al. (2024). Copyright (2024) Society of Photo-Optical Instrumentation Engineers (SPIE))

TAI could effectively image hemorrhagic stroke through a human skull model, highlighting its potential for real-time stroke diagnostics. Expanding on this, Li et al. (Zhang et al. 2022) applied deep learning-based reconstruction techniques (ResAttU-Net) to enhance hemorrhage detection accuracy and image resolution in a TAI system. Meanwhile, Islam et al. (2024) developed a TAI system to detect blood accumulation in the brain, demonstrating its feasibility in ex vivo brain samples (Fig. 6). Further, their simulation studies on transfontanelle TAI (Islam et al. 2022) explored the potential of neonatal brain imaging, suggesting that TAI could be an effective imaging tool for early-stage brain development monitoring. In the context of preclinical research, Liu et al. (2022) applied TAI for intracerebral hemorrhage detection in a canine model, confirming its capability to differentiate between hemorrhagic and normal brain tissues with high sensitivity. Their findings indicate that TAI could be a valuable tool for early stroke detection and monitoring, offering advantages such as real-time imaging, deep penetration, and sensitivity to blood accumulation.

4.4 Joint-Related Diseases

Diagnosing joint-related diseases is challenging due to the complex composition and structural diversity of joint tissues, including tendons, cartilage, and bone. While conventional imaging modalities such as MRI and X-ray are widely used for clinical diagnosis, they have inherent limitations, including high costs, invasiveness, and low sensitivity to soft tissue abnormalities. Given that injured bones exhibit higher dielectric properties than healthy bones, TAI has emerged as a promising alternative for high-contrast, non-invasive joint imaging, offering enhanced detection of pathological changes in musculoskeletal structures (Chi et al. 2016). Huang et al. (Lin et al. 2014) first applied TAI to human finger joint imaging, using a 3 GHz microwave source (80 kW peak power) and a 2.25 MHz transducer, achieving clear visualization of bones and soft tissues with

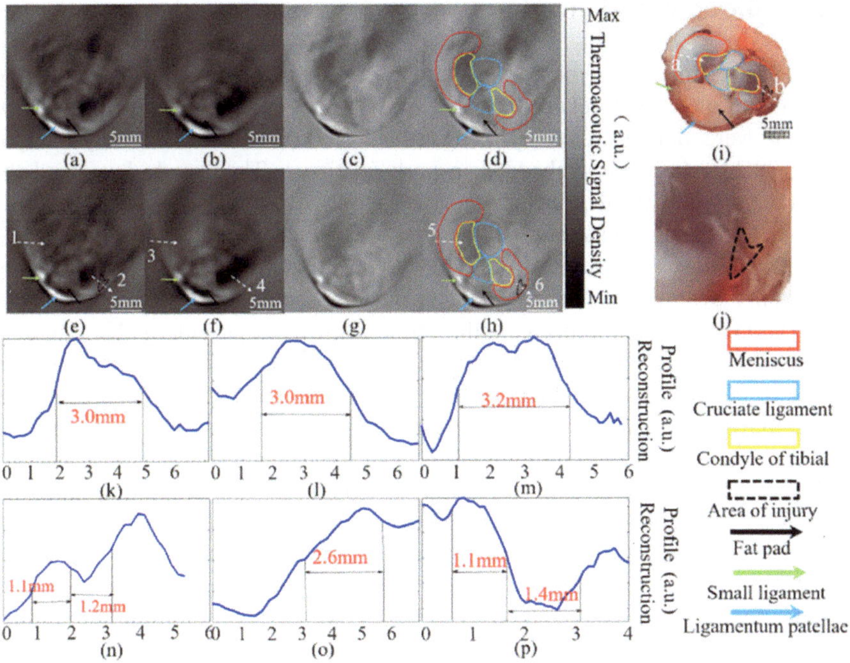

Fig. 7 TAT images of the rabbit knee joint before and after injury. (**a–c**) TAT images of the normal joints corresponding to 1 GHz, 3 GHz, and dual-frequency microwave. (**d**) Image marked with general outlines of the joint tissues, based on (**c**). (**e–g**) Images of the injured joint corresponding to 1 GHz, 3 GHz, and dual-frequency microwave. (**h**) Image marked with general outlines of the joint tissues, based on (**g**). (**i**) The anatomical picture marked with general outlines of tissues; (j) enlarged view of meniscal injury in (**i**); (**k–m**): signal intensity profiles along the dotted lines 1, 3, and 5 in (**e**), (**f**), and (**h**), respectively; (**n–p**): signal intensity profiles along the dotted lines 2, 4, and 6 in (**e**), (**f**), and (**h**), respectively. (Reprinted with permission from Liu et al. (2023d). © 2023 IEEE)

~300 μm resolution (Lin et al. 2014). Later, Chi et al. enhanced the system by incorporating a semi-ring transducer array and a 2.45 GHz microwave source (60 kW peak power), improving imaging speed and spatial resolution (Chi et al. 2018). The method was extended to a dual-frequency TAT system (Liu et al. 2023d) that was able to image a rabbit knee joint and detect meniscal injury (Fig. 7). Further refinements by Chi et al. introduced anti-phase microwave illumination, reducing artifacts and improving joint tissue contrast (Chi et al. 2019). For knee joint imaging, Chi et al. utilized a 3 GHz microwave source (50 kW peak power) and a 2.5 MHz transducer, demonstrating TAI's capability in detecting ligament injuries and cartilage degeneration, making it suitable for osteoarthritis assessment (Chi et al. 2016). More recently, Chi et al. conducted the first human study on RA detection, using a 2.45 GHz pulsed microwave source (80 kW peak power) and a 3.5 MHz transducer array, successfully identifying bone marrow edema and synovitis, highlighting TAI's clinical potential in RA monitoring (Chi et al. 2022).

4.5 Blood Vessel Imaging

Non-invasive imaging of blood vessels plays a crucial role in the screening and diagnosis of blood flow-related diseases. Early efforts in applying TAI for vascular imaging were demonstrated by Zeng et al. (2006), who successfully imaged simulated blood vessels embedded in pork fat tissue. Their study leveraged the high contrast at 1.2 GHz, achieving a high SNR due to the distinct dielectric properties of blood and fat tissue. Further advancements were made by Aliroteh and Arbabian (2017), who developed a TAI system utilizing near-field RF excitation to visualize synthetic phantoms, plant vasculature, and earthworm blood vessels. Using a 2 GHz pulsed microwave with a 167 ns pulse duration and a 5 kHz repetition rate, their system incorporated a radio-frequency dipole applicator with an output power of 42 mW and a 2 MHz SET for detection. The time-reversal algorithm was employed for image reconstruction, achieving high resolution (~300 μm), low peak power (~50 W), and a high SNR (~40 dB), demonstrating the feasibility of high-contrast vascular imaging with TAI. Traditionally, microwave frequencies have been used for angiography imaging, but recent studies have explored RF-induced second harmonic acoustic waves as a promising alternative. Huang et al. (2018) demonstrated that a 3.2 MHz RF source generating a 12 μs RF field could effectively visualize blood vessels in vivo, benefiting from the greater penetration depth of lower-frequency EM waves compared to conventional GHz-range microwave excitation. In addition to ex vivo and animal models, Jiang et al. (Zheng et al. 2018) successfully performed in vivo TAI imaging of human blood vessels, selecting a 3 GHz microwave source and an 8 MHz ultrasound transducer to balance imaging depth and resolution. Their results, including TA and ultrasound images of the forearm blood vessels, demonstrated the feasibility of clinical vascular imaging with TAI.

4.6 Foreign Body Detection

TAI has shown great potential for non-invasive, high-contrast foreign body detection, offering advantages over conventional imaging techniques, particularly for low-density materials such as plastics and rubber, which are difficult to detect using X-rays. By leveraging the conductivity contrast between foreign objects and biological tissues, TAI enables deep tissue penetration and precise localization of embedded materials. Nie et al. (2009) demonstrated that TAT could effectively image low-density foreign bodies in vivo, achieving sub-millimeter resolution at 3 GHz microwave excitation. They imaged a living mouse with a bamboo fragment hidden in its body by X-ray, ultrasound, and TAT. Compared with X-ray and ultrasound images, TAI effectively detects radiolucent foreign objects that remain invisible to X-ray. Their findings were further supported by Nie et al. (2007), who implemented a fast TAI system using a multi-element transducer array, improving imaging speed and accuracy for both metallic and non-metallic foreign bodies.

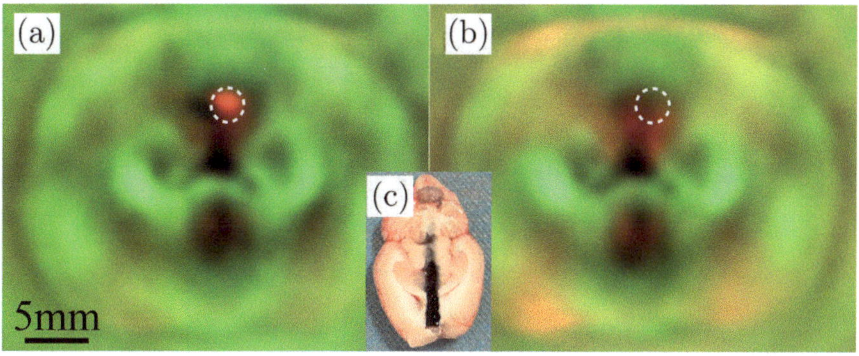

Fig. 8 Images of rat brain. (**a**) In vivo TA image of rat brain (rostral view) containing a foreign object; (**b**) TA image of the same rat brain without the foreign object. Dashed circles in (**a**) represent the edge of foreign object determined by full width half maximum. Dashed circles in (**b**) show the same location (without the foreign object). (**c**) Histological photograph of the rat brain, dorsal view, showing foreign object (plastic tube, 3 mm in diameter, filled with soybean sauce) obtained after imaging. (Reprinted from Zhao et al. (2017) under a Creative Commons Attribution License (CC BY 4.0)

A more recent study by Sun et al. (2021) introduced a low-cost, compressive TAI system capable of detecting hot and cold foreign bodies with improved reconstruction speed and efficiency. Additionally, Zhao et al. (2017) demonstrated TAI's application in neuroimaging, proving its feasibility in detecting foreign bodies within sensitive brain tissues. Figure 8 shows TAT imaging of a rat brain with and without a foreign body embedded. TAI's ability to detect embedded foreign objects with high contrast and deep penetration makes it a promising alternative for trauma diagnostics and surgical guidance. Future advancements in real-time imaging, AI-based reconstruction, and portable TAI systems could further enhance its clinical applicability in emergency medicine.

4.7 Summary of TAI Applications

For TAI, no single configuration of microwave source and ultrasound transducers can cover the range of potential biomedical applications. Table 4 shows the specifications that have been utilized for TAI, along with potential research directions to advance its development.

5 Conclusions and Future Perspectives

TAI has demonstrated significant potential in various preclinical and clinical applications, providing non-invasive, high-contrast, and deep tissue imaging by leveraging the differences in conductivity of biological tissue constituents. This

Table 4 Summary of TAI Applications

Imaging application	System configuration	Disease model/organ	Study type	Sample type	References
Breast Cancer	434 MHz MW source, 64-element ultrasound transducer array, 1 MHz response frequency	Breast cancer detection and tumor screening	Ex vivo (animal studies) and In vivo	Animal (Sheep) Human	Wang et al. (2012a), Zhao et al. (2021), Wu et al. (2021), Kruger et al. (2000), Nie et al. (2008), Ji et al. (2012), Huang et al. (2013), Ye et al. (2015), Zhao et al. (2018), and Li et al. (2021)
Prostate Cancer	108 MHz MW source, 900 ns pulse duration, peak electric field ~17 kV/m	Prostate cancer detection and ablation monitoring	Ex vivo	Animal (canine)	Patch et al. (2015a), Patch et al. (2015b), Patch and See (2015), and Eckhart et al. (2011)
Brain	2.45 GHz MW source, 3.5 MHz transducer array, AI-assisted image reconstruction	Stroke, brain hemorrhage, neurological disorders	In vivo	Human, Animal (rat, monkey)	Xu and Wang (2006a), Islam et al. (2024), Jin et al. (2008), Liu et al. (2023c), Zhao et al. (2017), and Huang et al. (2017)
Joint	2.45 GHz MW source, 3 MHz transducer array, semi-ring transducer array	Arthritis, rheumatoid arthritis, ligament injuries	In vivo	Human, Animal (rabbit)	Chi et al. (2018), Chi et al. (2019), Chi et al. (2016), Lin et al. (2014), and Chi et al. (2022)
Foreign Body Detection	3 GHz MW source, 2.25 MHz single-element transducer	Low-density foreign body detection (plastic, rubber)	In vivo, Ex vivo	Human, Animal	Nie et al. (2009), Nie et al. (2007), and Sun et al. (2021)
Vascular	2 GHz MW source, 2 MHz transducer, RF-induced second harmonic imaging	Angiography, blood vessel imaging, functional vascular imaging	In vivo	Human, Animal	Huang et al. (2018), Zheng et al. (2018)

imaging modality has been successfully applied in breast cancer detection, prostate cancer monitoring, brain imaging, joint disease assessment, foreign body detection, and vascular imaging, each utilizing specialized system configurations tailored for specific diagnostic needs. Compared to conventional imaging techniques such as MRI, X-ray, and ultrasound, TAI offers several advantages, including non-ionizing radiation, high imaging contrast, and functional imaging capabilities. The ability of TAI to visualize deep-seated structures and track disease progression makes it a powerful tool for both research and clinical applications. Moreover, multi-modal integration, such as combining TAI with PAI and ultrasound, could further enhance its diagnostic accuracy and imaging performance. Despite its many advantages, TAI faces challenges, including: the need for expensive, high-power microwave generators; limited microwave penetration; skull-induced acoustic aberrations in brain imaging; and the need to optimize transducer arrays for improved resolution. While addressing these limitations could ultimately lead to clinical adoption, TAI's unique capabilities mean it could also be very beneficial for biomedical research applications. The future of TAI is poised for significant advancements, driven by technological innovations, multi-modal imaging integration, and AI-assisted image reconstruction. One of the primary areas of development lies in enhancing image resolution and signal detection, which requires the optimization of high-frequency ultrasound transducer arrays and improved microwave excitation techniques to achieve better contrast and deeper tissue penetration. Technological innovations could be particularly relevant for applications such as brain and prostate imaging. Additionally, the integration of AI is expected to revolutionize image reconstruction and processing, enabling real-time imaging, automated disease classification, and improved diagnostic accuracy. Another critical focus is the development of portable and low-cost TAI systems, which could enhance their utility in biomedical research including field applications. The combination of TAI with established imaging modalities, such as MRI, CT, and ultrasound, will further expand its clinical applicability by providing multi-contrast and multi-dimensional insights into tissue properties and disease progression.

Acknowledgments Kamran Avanaki acknowledges funding from the National Institutes of Health through grants R01EB027769 and R01EB028661.

Conflict of Interest The authors declare they have no conflict of interest.

References

Aggrawal D, Islam MT, Zafar M, Schonfeld D, Avanaki K (2024) Signal denoising for thermal acoustic imaging. In: Photons plus ultrasound: imaging and sensing 2024, vol 12842, pp 206–212

Aliroteh MS, Arbabian A (2017) Microwave-induced thermoacoustic imaging of subcutaneous vasculature with near-field RF excitation. IEEE Trans Microwave Theory Tech 66(1):577–588

Bauer DR, Wang X, Vollin J, Xin H, Witte RS (2012) Spectroscopic thermoacoustic imaging of water and fat composition. Appl Phys Lett 101(3):033705

Bruno F et al (2019) Advanced magnetic resonance imaging (MRI) of soft tissue tumors: techniques and applications. Radiol Med 124:243–252

Chi Z, Zhao Y, Huang L, Zheng Z, Jiang H (2016) Thermoacoustic imaging of rabbit knee joints. Med Phys 43(12):6226–6233

Chi Z, Zhao Y, Yang J, Li T, Zhang G, Jiang H (2018) Thermoacoustic tomography of in vivo human finger joints. IEEE Trans Biomed Eng 66(6):1598–1608

Chi Z, Huang L, Ge S, Jiang H (2019) Anti-phase microwave illumination-based thermoacoustic tomography of in vivo human finger joints. Med Phys 46(5):2363–2369

Chi Z, Huang L, Wu D, Long X, Xu X, Jiang H (2022) First assessment of thermoacoustic tomography for in vivo detection of rheumatoid arthritis in the finger joints. Med Phys 49(1):84–92

Cui Y, Yuan C, Ji Z (2017) A review of microwave-induced thermoacoustic imaging: excitation source, data acquisition system and biomedical applications. J Innov Opt Health Sci 10(04):1730007

Dessai R et al (2024) A breast tumor monitoring vest with flexible UWB antennas—a proof-of-concept study using realistic breast phantoms. Micromachines 15(9):1153. [Online]. Available: https://www.mdpi.com/2072-666X/15/9/1153

Dhawan AP, D'Alessandro B, Fu X (2010) Optical imaging modalities for biomedical applications. IEEE Rev Biomed Eng 3:69–92

Ding W, Zhang Y (2024) Enhanced thermoacoustic imaging system with parallel ultrasonic velocity measurement for distinguishing types of microwave-absorbing anomalies. Photonics 11(10):916. [Online]. Available: https://www.mdpi.com/2304-6732/11/10/916

Ding L, Deán-Ben XL, Razansky D (2016) Real-time model-based inversion in cross-sectional optoacoustic tomography. IEEE Trans Med Imaging 35(8):1883–1891

Du S, Qiang T, Chi Z, Jiang H (2024) Metal reflector-enhanced thermoacoustic imaging as a guidance for puncture biopsy. J Innov Opt Health Sci 17(6):2450014

Eckhart AT, Balmer RT, See WA, Patch SK (2011) Ex vivo thermoacoustic imaging over large fields of view with 108 MHz irradiation. IEEE Trans Biomed Eng 58(8):2238–2246

Feng D, Xu Y, Ku G, Wang LV (2001) Microwave-induced thermoacoustic tomography: reconstruction by synthetic aperture. Med Phys 28(12):2427–2431

Feng X, Gao F, Zheng Y (2013) Magnetically mediated thermoacoustic imaging toward deeper penetration. Appl Phys Lett 103(8):083704

Fu Y, Ji Z, Ding W, Ye F, Lou C (2014) Thermoacoustic imaging over large field of view for three-dimensional breast tumor localization: a phantom study. Med Phys 41(11):110701

Gabriel S, Lau R, Gabriel C (1996) The dielectric properties of biological tissues: II. Measurements in the frequency range 10 Hz to 20 GHz. Phys Med Biol 41(11):2251

Gao F, Zheng Y, Feng X, Ohl C-D (2013) Thermoacoustic resonance effect and circuit modelling of biological tissue. Appl Phys Lett 102(6):063702

Gao F, Feng X, Zheng Y (2016) Advanced photoacoustic and thermoacoustic sensing and imaging beyond pulsed absorption contrast. J Opt 18(7):074006

Goh CYK, Mohamed Ali PS, Lee KHC, Sim FY, Chong LR (2025) 3D MRI with CT-like bone contrast (3D-BONE): a pictorial review of clinical applications. Acta Radiol 66(2):228–240

Hasgall PA et al (2018) IT'IS database for thermal and electromagnetic parameters of biological tissues. Version 4.0, vol 15, p 2018

Hristova Y, Kuchment P, Nguyen L (2008) Reconstruction and time reversal in thermoacoustic tomography in acoustically homogeneous and inhomogeneous media. Inverse problems 24(5):055006

Huang Y-J, Huang L (2023) Thermoacoustic imaging: from single-element scanning to portable and array-based imaging. J Electr Sci Technol 21(3):100209

Huang L et al (2013) Quantitative thermoacoustic tomography for ex vivo imaging conductivity of breast tissue. Chin Phys Lett 30(12):124301

Huang L, Li T, Jiang H (2017) Thermoacoustic imaging of hemorrhagic stroke: a feasibility study with a human skull. Med Phys 44(4):1494–1499

Huang Y, Kellnberger S, Sergiadis G, Ntziachristos V (2018) Blood vessel imaging using radiofrequency-induced second harmonic acoustic response. Sci Rep 8(1):15522

Huang L, Ge S, Zheng Z, Jiang H (2019) Design of a handheld dipole antenna for a compact thermoacoustic imaging system. Med Phys 46(2):851–856

Huang G, Li Y, Ren M, Zhang H, Qin H (2023) Real-time thermoacoustic imaging for breast tumor biomarker biopsy navigation basing on a semi-ring ultrasonic transducer. Appl Phys Lett 123(18):183702

Islam MT, Manwar R, Lin J, Erricolo D, Avanaki K (2022) Transfontanelle thermoacoustic imaging characterization: simulation study. In: Photons plus ultrasound: imaging and sensing 2022, vol 11960, pp 230–239

Islam MT et al (2023) Development and characterization of thermoacoustic imaging system using rotational single element transducer. In: Photons plus ultrasound: imaging and sensing 2023, vol 12379, pp 181–188

Islam MT et al (2024) Development of a thermoacoustic imaging system to image blood in the brain: preliminary ex-vivo results. In: Photons plus ultrasound: imaging and sensing 2024, vol 12842, pp 166–171

Ji Z, Lou C, Yang S, Xing D (2012) Three-dimensional thermoacoustic imaging for early breast cancer detection. Med Phys 39(11):6738–6744

Ji Z, Ding W, Ye F, Lou C, Xing D (2015) Shape-adapting thermoacoustic imaging system based on flexible multi-element transducer. Appl Phys Lett 107(9):094104

Ji Z, Ding W, Ye F, Lou C (2016) Handheld thermoacoustic scanning system based on a linear-array transducer. Ultrason Imaging 38(4):276–284. https://doi.org/10.1177/0161734615601987

Jin X, Wang LV (2006) Thermoacoustic tomography with correction for acoustic speed variations. Phys Med Biol 51(24):6437

Jin X, Li C, Wang LV (2008) Effects of acoustic heterogeneities on transcranial brain imaging with microwave-induced thermoacoustic tomography. Med Phys 35(7Part1):3205–3214

Kalender WA (2011) Computed tomography: fundamentals, system technology, image quality, applications. Wiley, New York

Ke H, Liu C, Wang LV, Erpelding TN, Jankovic L (2012) Performance characterization of an integrated ultrasound, photoacoustic, and thermoacoustic imaging system. J Biomed Opt 17(5):056010

Kellnberger S, Hajiaboli A, Razansky D, Ntziachristos V (2011a) Near-field thermoacoustic tomography of small animals. Phys Med Biol 56(11):3433

Kellnberger S, Hajiaboli A, Razansky D, Ntziachristos V (2011b) Near-field thermoacoustic tomography of small animals. J Biomed Opt 56(11):3433

Klibanov AL, Hossack JA (2015) Ultrasound in radiology: from anatomic, functional, molecular imaging to drug delivery and image-guided therapy. Investig Radiol 50(9):657–670

Kruger RA, Reinecke DR, Kruger GA (1999) Thermoacoustic computed tomography–technical considerations. Med Phys 26(9):1832–1837

Kruger RA, Miller KD, Reynolds HE, Kiser WL Jr, Reinecke DR, Kruger GA (2000) Breast cancer in vivo: contrast enhancement with thermoacoustic CT at 434 MHz—feasibility study. Radiology 216(1):279–283

Ku G, Wang LV (2000) Scanning thermoacoustic tomography in biological tissue. Med Phys 27(5):1195–1202

Ku G, Fornage BD, Jin X, Xu M, Hunt KK, Wang LV (2005) Thermoacoustic and photoacoustic tomography of thick biological tissues toward breast imaging. Technol Cancer Res Treat 4(5):559–565

Lan L, Li Y, Yang-Tran T, Jiang Y, Cao Y, Cheng J-X (2020. [Online]. Available:) Ultraefficient thermoacoustic conversion through a split ring resonator. Adv Photonics 2(3):036006. https://doi.org/10.1117/1.AP.2.3.036006

Li J et al (2021) A preclinical system prototype for focused microwave breast hyperthermia guided by compressive thermoacoustic tomography. IEEE Trans Biomed Eng 68(7):2289–2300
Li Y et al (2022) Polarization microwave-induced thermoacoustic imaging for quantitative characterization of deep biological tissue microstructures. Photonics Res 10(5):1297–1306
Liang X, Guo H, Liu Q, Wu C, Gong Y, Xi L (2020) Thermoacoustic endoscopy. Appl Phys Lett 116(1):013702
Lin J (1977) Further studies on the microwave auditory effect. IEEE Trans Microwave Theory Tech 25(11):938–943
Lin J (1980) The microwave auditory phenomenon. Proc IEEE 68(1):67–73
Lin JC, Lin JC (2021) The microwave auditory effect. In: Auditory effects of microwave radiation, pp 127–173
Lin H et al (2014) Thermoacoustic imaging of human finger joints and bones. X-Acoustics Imaging Sensing 1(1)
Liu S et al (2016) Analysis of short pulse impacting on microwave induced thermo-acoustic tomography. Prog Electromag Res C 61:37–46
Liu Q, Liang X, Qi W, Gong Y, Jiang H, Xi L (2022) Biomedical microwave-induced thermoacoustic imaging. J Innov Opt Health Sci 15(04):2230007
Liu S-L et al (2023a) Developing microwave-induced thermoacoustic tomography: system, application, and reconstruction. J Electron Sci Technol 21(4):100232. https://doi.org/10.1016/j.jnlest.2023.100232
Liu Q et al (2023b) Split ring resonator topology based microwave induced thermoacoustic imaging (SRR-MTAI). IEEE Trans Med Imaging 42(8):2425–2438
Liu S-L, Shang X, Peng W-T, Wan W-J, Zhang J-B (2023c) Skull phase compensation combined full waveform inversion for transcranial thermoacoustic imaging with a real human skull validated. IEEE J Electromagn RF Microwaves Med Biol 7(4):313–319
Liu Y, Chi Z, Tan Y, Liu X, Jiang H (2023d) Dual-frequency microwave-induced thermoacoustic tomography: an application in joint imaging. In: 2023 IEEE MTT-S International Wireless Symposium (IWS), pp 1–3. https://doi.org/10.1109/IWS58240.2023.10222170
Martellosio A et al (2016) Dielectric properties characterization from 0.5 to 50 GHz of breast cancer tissues. IEEE Trans Microwave Theory Tech 65(3):998–1011
Mashal A, Booske JH, Hagness SC (2009) Toward contrast-enhanced microwave-induced thermoacoustic imaging of breast cancer: an experimental study of the effects of microbubbles on simple thermoacoustic targets. Phys Med Biol 54(3):641
Mohammed YE, Saber AG (2014) Estimation of E-field inside muscle tissue at MICS and ISM frequencies using analytic and numerical methods. J Biomed Engg Technol 2(3):29–33
Nan H, Arbabian A (2014) Stepped-frequency continuous-wave microwave-induced thermoacoustic imaging. Appl Phys Lett 104(22)
Nan H, Arbabian A (2017) Peak-power-limited frequency-domain microwave-induced thermoacoustic imaging for handheld diagnostic and screening tools. IEEE Trans Microwave Theory Tech 65(7):2607–2616
Nan H, Liu S, Buckmaster JG, Arbabian A (2018) Beamforming microwave-induced thermoacoustic imaging for screening applications. IEEE Trans Microwave Theory Tech 67(1):464–474
Nie L, Xing D, Yang D, Zeng L, Zhou Q (2007) Detection of foreign body using fast thermoacoustic tomography with a multielement linear transducer array. Appl Phys Lett 90(17):174109
Nie L, Xing D, Zhou Q, Yang D, Guo H (2008) Microwave-induced thermoacoustic scanning CT for high-contrast and noninvasive breast cancer imaging. Med Phys 35(9):4026–4032
Nie L, Xing D, Yang S (2009) In vivo detection and imaging of low-density foreign body with microwave-induced thermoacoustic tomography. Med Phys 36(8):3429–3437
Omar M, Kellnberger S, Sergiadis G, Razansky D, Ntziachristos V (2012) Near-field thermoacoustic imaging with transmission line pulsers. Med Phys 39(7Part1):4460–4466
Patch S, See W (2015) Broadband & volumetric thermoacoustic imaging of fresh human prostates using a clinical array. In: 2015 IEEE Great Lakes Biomedical Conference (GLBC). IEEE, pp 1–4

Patch S, Hull D, Thomas M, Griep S, Jacobsohn K, See W (2015a) Thermoacoustic contrast of prostate cancer due to heating by very high frequency irradiation. Phys Med Biol 60(2):689
Patch SK, Hull D, See WA, Hanson GW (2015b) Toward quantitative whole organ thermoacoustics with a clinical array plus one very low-frequency channel applied to prostate cancer imaging. IEEE Trans Ultrason Ferroelectr Freq Control 63(2):245–255
Prakash R, Manwar R, Avanaki K (2024) Evaluation of 10 current image reconstruction algorithms for linear array photoacoustic imaging. J Biophotonics 17(3):e202300117
Qin T et al (2014) Experimental validation of a numerical model for thermoacoustic imaging applications. IEEE Antennas Wireless Propagation Lett 14:1235–1238
Raicu V, Saibara T, Enzan H, Irimajiri A (1998) Dielectric properties of rat liver in vivo: analysis by modeling hepatocytes in the tissue architecture. Bioelectrochem Bioenerg 47(2):333–342
Razansky D, Kellnberger S, Ntziachristos V (2010) Near-field radiofrequency thermoacoustic tomography with impulse excitation. Med Phys 37(9):4602–4607
Ren M et al (2022) Portable microwave-acoustic coaxial thermoacoustic probe with miniaturized vivaldi antennas for breast tumor screening. IEEE Trans Biomed Eng 70(1):175–181
Salahuddin S et al (2017) Effects of standard coagulant agents on the dielectric properties of fresh human blood. IEEE Trans Dielectr Electr Insul 24(5):3283–3289. https://doi.org/10.1109/TDEI.2017.006582
Singhvi A, Boyle KC, Fallahpour M, Khuri-Yakub BT, Arbabian A (2019) A microwave-induced thermoacoustic imaging system with non-contact ultrasound detection. IEEE Trans Ultrason Ferroelectr Freq Control 66(10):1587–1599
Song J et al (2013) An integrated simulation approach and experimental research on microwave induced thermo-acoustic tomography system. Prog Electromagn Res 140:385–400
Song J, Shen T, Wang Q (2022) An image post-processing approach based on fully dense u-net for microwave induced thermo-acoustic tomography. IEEE J Electromagn RF Microwaves Med Biol 7(1):59–64
Sun Y, Li C, Wang B, Wang X (2021) A low-cost compressive thermoacoustic tomography system for hot and cold foreign bodies detection. IEEE Sensors J 21(20):23588–23596
Sutardja C, Singhvi A, Fitzpatrick A, Cathelin A, Arbabian A (2022) Multi-watt-level 4.9-GHz silicon power amplifier for portable thermoacoustic imaging. IEEE J Solid State Circuits 57(5):1421–1431
Tan F, Wang H (2024) Design and implementation of an ultra-wideband water immersion antenna for underwater ultrasonic sensing in microwave-induced thermoacoustic tomography. Sensors 24(19):6311
Tang S, Fu Y, Wang Y, Tang X, Zeng L, Qin H (2025) Passive beamforming metasurfaces for microwave-induced thermoacoustic imaging. IEEE Trans Biomed Eng 72(7):2303–2310
Wang LV, Zhao X, Sun H, Ku G (1999) Microwave-induced acoustic imaging of biological tissues. Rev Sci Instrum 70(9):3744–3748
Wang X, Bauer DR, Witte R, Xin H (2012a) Microwave-induced thermoacoustic imaging model for potential breast cancer detection. IEEE Trans Biomed Eng 59(10):2782–2791
Wang X, Bauer DR, Vollin JL, Manzi DG, Witte RS, Xin H (2012b) Impact of microwave pulses on thermoacoustic imaging applications. IEEE Antennas Wireless Propagation Lett 11:1634–1637
Wang X, Huang L, Chi Z, Jiang H (2021a) Integrated thermoacoustic and ultrasound imaging based on the combination of a hollow concave transducer array and a linear transducer array. Phys Med Biol 66(11):115011
Wang B et al (2021b) Microwave-induced thermoacoustic imaging of small animals applying scanning orthogonal polarization excitation. IEEE J RF Microwaves Med Biol 6(1):123–130
Wang W et al (2022) Low frequency terahertz induced thermoacoustic signal characteristics and its application in solid-state terahertz transmitter power detection. IEEE Trans Terahertz Sci Technol 12(6):673–677
Wang Y, Tang X, Qin H (2024) High-spatiotemporal resolution microwave-induced thermoacoustic tomography for imaging biological dynamics in deep tissue. Appl Phys Lett 125(2)

Wu L et al (2021) A handheld microwave thermoacoustic imaging system with an impedance matching microwave-sono probe for breast tumor screening. IEEE Trans Med Imaging 41(5):1080–1086

Xu M, Wang LV (2002a) Pulsed-microwave-induced thermoacoustic tomography: filtered back-projection in a circular measurement configuration. Med Phys 29(8):1661–1669

Xu M, Wang LV (2002b) Time-domain reconstruction for thermoacoustic tomography in a spherical geometry. IEEE Trans Med Imaging 21(7):814–822

Xu Y, Wang LV (2006a) Rhesus monkey brain imaging through intact skull with thermoacoustic tomography. IEEE Trans Ultrason Ferroelectr Freq Control 53(3):542–548

Xu M, Wang LV (2006b) Photoacoustic imaging in biomedicine. Rev Sci Instrum 77(4)

Xu M, Xu Y, Wang LV (2003) Time-domain reconstruction algorithms and numerical simulations for thermoacoustic tomography in various geometries. IEEE Trans Biomed Eng 50(9):1086–1099

Xu Y, Wang LV, Ambartsoumian G, Kuchment P (2004) Reconstructions in limited-view thermoacoustic tomography. Med Phys 31(4):724–733

Xu Q, Zheng Z, Jiang H (2022) Deep learning for image reconstruction in thermoacoustic tomography. Chinese Phys B 31(2):024302

Yan A, Lin L, Na S, Liu C, Wang LV (2018) Large field homogeneous illumination in microwave-induced thermoacoustic tomography based on a quasi-conical spiral antenna. Appl Phys Lett 113(12)

Yan A, Lin L, Liu C, Shi J, Na S, Wang LV (2019) Microwave-induced thermoacoustic tomography through an adult human skull. Med Phys 46(4):1793–1797

Yan A, Zhang Y, Gao C, Ye J, Zhong Z (2025) A novel slot spiral symmetric array antenna with a wide axial ratio beamwidth for microwave-induced thermoacoustic tomography applications. Symmetry 17(2):197. [Online]. Available: https: //www.mdpi.com/2073-8994/17/2/197

Ye F, Ji Z, Ding W, Lou C, Yang S, Xing D (2015) Ultrashort microwave-pumped real-time thermoacoustic breast tumor imaging system. IEEE Trans Med Imaging 35(3):839–844

Yuan C, Qin B, Qin H, Xing D (2019) Increasing dielectric loss of a graphene oxide nanoparticle to enhance the microwave thermoacoustic imaging contrast of breast tumor. Nanoscale 11(46):22222–22229

Zeng L, Xing D, Gu H, Yang D (2006) A fast microwave-induced thermoacoustic tomography system for imaging of biological tissues. In: Fourth international conference on photonics and imaging in biology and medicine, vol 6047, pp 126–132

Zeng L, Liu G, Ji X, Ren Z, Huang Z (2009) High-speed laser photoacoustic imaging system combined with a digital ultrasonic imaging platform. In: International symposium on photo-electronic detection and imaging 2009: laser sensing and imaging, vol 7382, pp 983–989

Zhang J, Li C, Jiang W, Wang Z, Zhang L, Wang X (2022) Deep-learning-enabled microwave-induced thermoacoustic tomography based on sparse data for breast cancer detection. IEEE Trans Antennas Propag 70(8):6336–6348

Zhang H, Ren M, Zhang S, Liu J, Qin H (2023) Microwave-induced thermoacoustic imaging for biomedical applications. Phys Scr 98(3):032001

Zhao Z et al (2013) System development of microwave induced thermo-acoustic tomography and experiments on breast tumor. Prog Electromagn Res 134:323–336

Zhao Y, Chi Z, Huang L, Zheng Z, Yang J, Jiang H (2017) Thermoacoustic tomography of in vivo rat brain. J Innov Opt Health Sci 10(04):1740001

Zhao Y, Ji Z, Qin B, Xing D (2018) A thermoacoustic imaging system with variable curvature and multi-dimensional detection adapted to breast tumor screening. J Appl Phys 124(14):144902

Zhao S et al (2021) Ultrashort-pulse-microwave excited whole-breast thermoacoustic imaging with uniform field of large size aperture antenna for tumor screening. IEEE Trans Biomed Eng 69(2):725–733

Zheng Z, Huang L, Jiang H (2018) Label-free thermoacoustic imaging of human blood vessels in vivo. Appl Phys Lett 113(25):253702

Zhu X, Zhao Z, Wang J, Chen G, Liu QH (2014) Active adjoint modeling method in microwave induced thermoacoustic tomography for breast tumor. IEEE Trans Biomed Eng 61(7):1957–1966

Zhurbenko V (2011) Challenges in the design of microwave imaging systems for breast cancer detection. Adv Electric Comp Engg 11(1):91–96

Applications of Raman Spectroscopy in Preclinical Bioimaging

Maria Mangini, Somayeh Farazpour, Hossein Khadem, Alessandro Esposito, and Anna Chiara DeLuca

Abstract

Raman spectroscopy is a powerful analytical technique with significant potential in preclinical bioimaging due to its non-invasive and label-free molecular insights. This chapter explores its diverse applications in advancing biomedical research and drug development, highlighting unique advantages over traditional imaging methods, such as real-time monitoring, chemical specificity, and compatibility with living tissues. Technological advancements, including confocal Raman microscopy, Raman imaging, Surface-Enhanced Raman Spectroscopy (SERS), and Tip-Enhanced Raman Scattering (TERS), have expanded its applications in mapping biochemical processes and identifying disease-specific biomarkers. The integration of Raman spectroscopy with complementary imaging modalities, such as fluorescence microscopy, atomic force microscopy, and mass spectrometry, underscores its potential to provide comprehensive insights into cellular and tissue-level dynamics. Case studies demonstrate its versatility, with applications ranging from cancer research to neurodegenerative diseases and nanoparticle-based drug delivery, effectively monitoring tumor progression, drug biodistribution, and molecular changes associated with diseases. Furthermore, Raman spectroscopy is progressing toward clinical applications,

M. Mangini (✉) · H. Khadem · A. Esposito · A. C. De Luca (✉)
Institute of Endotypes in Oncology, Metabolism and Immunology 'Gaetano Salvatore' IEOMI-Second Unit, National Research Council CNR, Naples, Italy
e-mail: maria.mangini@cnr.it; annachiara.deluca@cnr.it

S. Farazpour
Institute of Endotypes in Oncology, Metabolism and Immunology 'Gaetano Salvatore' IEOMI-Second Unit, National Research Council CNR, Naples, Italy

Department of Environmental, Biological and Pharmaceutical Sciences and Technologies, University of Campania Luigi Vanvitelli, Caserta, Italy

I. F. Cengiz et al. (eds.), *Bioimaging Modalities in Bioengineering*, Biomaterials, Bioengineering and Sustainability 9,
https://doi.org/10.1007/978-3-031-99410-4_20

driven by technological advancements addressing weak signal intensity and in vivo complexity. Notable developments include endoscopic Raman systems, which combine high chemical specificity with minimally invasive tools for real-time tissue assessment in oncology and other fields. In conclusion, Raman spectroscopy holds great promise for transforming preclinical bioimaging, offering detailed molecular insights that complement existing imaging modalities. Continued technological innovation is expected to bridge the gap between basic research and clinical applications, accelerating the translation of biomedical discoveries into therapeutic innovations.

Keywords

Raman spectroscopy · Preclinical bioimaging · Biomedical applications

1 Introduction

1.1 Overview of Raman Spectroscopies

Understanding and identifying the biochemical dynamics occurring at the microscale is crucial for unraveling the complex mechanisms that govern biological systems. To achieve this, non-invasive, label-free techniques with high molecular sensitivity and specificity are highly sought after as promising tools for bioimaging and diagnostic applications. Among these methods, recent advancements in Raman spectroscopy (RS) have positioned it as a powerful and versatile approach, offering exceptional potential for preclinical research. Its ability to provide detailed molecular signatures of diseases without disrupting biological samples makes it an ideal candidate for exploring new frontiers in preclinical bioimaging.

The Raman scattering phenomenon (Raman and Krishnan, 1929) forms the foundation of Raman spectroscopy. When light interacts with a molecule, most photons are elastically scattered (Rayleigh scattering). However, a small fraction undergoes inelastic scattering, where the scattered photons experience a shift in energy due to molecular vibrations and/or rotations. This shift, known as the Raman scattering, provides a unique spectral fingerprint of the molecule, enabling the identification of its chemical composition and structure. The vibrations of large and small biological molecules, such as amides, polypeptides, and proteins, exhibit strong features in Raman scattering. These characteristics make RS particularly well-suited for preclinical investigations (Shen et al. 2019).

The invention of the laser (Maiman, 1960), the development of Raman microscopy (RM) (Delhaye and Migeon, 1966), and advancements in its instrumentation (Hirschfeld, 1973) have significantly extended the capabilities of RS. A standard Raman microscope uses a conventional optical setup to focus a laser beam of a specific wavelength onto a small spot on the sample, which is positioned on the stage. The scattered light from the sample is collected after passing through a notch or edge filter, which blocks the strong Rayleigh scattered light

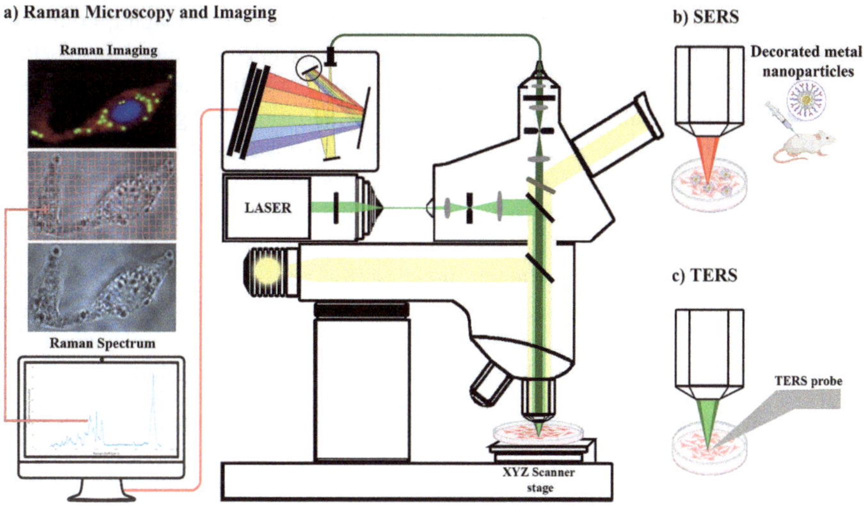

Fig. 1 Schematic of different techniques of Raman spectroscopy: (**a**) Confocal Raman microscopy and Raman imaging: The laser source focuses on the sample to generate a Raman spectrum at each point, enabling high-resolution mapping of chemical composition and distribution in the sample. Raman imaging combines spectral data with spatial mapping for detailed analysis of cellular components. (**b**) Surface-Enhanced Raman Spectroscopy (SERS): Enhanced Raman signals are achieved by using decorated metal nanoparticles, which amplify the Raman scattering of molecules near the surface. This technique is highly sensitive for detecting biomolecules even at trace levels, commonly used in biomedical applications. (**c**) Tip-Enhanced Raman Spectroscopy (TERS): A sharp metallic probe enhances Raman signals at the nanoscale, combining the advantages of atomic force microscopy and Raman spectroscopy for detailed chemical characterization at the nanometer level. (Image in panel A is adapted from (Mangini et al. 2023) under a Creative Commons Attribution License (CC BY 4.0))

(elastic scattering). The remaining scattered light is then separated into different wavelengths, sent to a detector which records the intensity of the Raman signals as a function of wavelength, and processed by a computer to generate a spectrum. This final spectrum reveals the chemical composition of the sample (Turrell and Corset 1996). Confocal RM, an enhanced version, improves spatial resolution and depth focus by adding a confocal optical setup. A pinhole filters out-of-focus light, allowing for sharper imaging and the ability to analyze deeper layers of the sample (Fig. 1a) (Toporski et al. 2018).

1.2 Raman Imaging: High-Resolution Molecular Mapping

Raman microscopes equipped with motorized positioning stages and stepper motors allow for raster-scanning of samples while collecting Raman spectra at specific points. This technique, called Raman imaging (mapping), creates images where each pixel contains a full Raman spectrum, revealing the molecular composition across the sample with micrometer resolution. Hyperspectral imaging combines spatial data with spectral information to generate detailed chemical maps. Preprocessing steps like spectral smoothing, baseline correction, and normalization enhance data quality and reproducibility, reducing variations in signal-to-noise ratio (SNR) due to temperature changes or laser power fluctuations (Vulchi et al. 2024). Tools like the R package *hyperSpec* and web applications, such as the Raman app under development by Politecnico di Milano (available at https://ramapp.io/), enable preprocessing and optimized image generation (Beleites and Sergo 2009).

A common approach in Raman imaging for cell analysis is the classical least squares (CLS) fitting method. This method deconvolutes the cellular Raman spectrum into contributions from reference spectra of known biomolecules (e.g., proteins, lipids, nucleic acids) (Wu et al. 2017). The main steps include:

1. *Data Acquisition:* Collecting Raman spectra from cell regions using a Raman microscope.
2. *Reference Spectra:* Utilizing spectra of individual biomolecules as a fitting library.
3. *CLS Fitting:* Expressing the Raman spectrum as a combination of reference spectra to calculate component proportions.
4. *Component Mapping:* Visualizing biomolecule distribution across a spatially resolved Raman map.

Raman imaging is especially valuable in preclinical studies for detecting early molecular changes linked to disease, often before morphological alterations are visible. This makes it a powerful tool for early diagnostics and monitoring disease progression in biological models (Fig. 1a).

1.3 Surface-Enhanced Raman Scattering (SERS): Boosting Sensitivity

RS is a weak scattering method in which usually fewer than one in a million incoming photons produce a single Raman-scattered photon. Surface-Enhanced Raman Scattering (SERS) was developed to amplify the Raman signal using metallic nanostructures, such as gold or silver (Jain et al. 2008). The SERS mechanism involves the interaction of light with metal nanoparticles, where the collective oscillations of electrons at the surface of the metal create a strong electric field in the vicinity of the nanoparticles. This enhanced field increases the likelihood of Raman scattering from nearby molecules, thereby boosting the overall signal. The effectiveness of SERS is influenced by the size, shape, and spacing of the metal nanoparticles. Smaller particles and specific shapes (like nanostars or nanorods) can create “hot spots”, areas where the electromagnetic field is particularly strong, further enhancing the SERS signal (Fig. 1b) (Cialla-May et al. 2024). Hot spots in SERS are localized regions near nanostructures (e.g., metallic nanoparticles) where the electromagnetic field is significantly amplified, typically at sharp features or gaps between closely spaced nanoparticles. This plasmonic effect dramatically enhances the Raman signal of molecules in these areas.

1.4 Tip-Enhanced Raman Scattering (TERS): Nanoscale Imaging Precision

The integration of RS or SERS with Atomic Force Microscopy (AFM) or Scanning Near-field Optical Microscopy (SNOM) has led to the development of advanced techniques such as Tip-Enhanced Raman Scattering (TERS). In TERS, a sharp metallic tip—often made of or coated with gold or silver—is positioned close to the sample surface. This metallic tip generates a localized electromagnetic field that significantly enhances the Raman signal from the sample, providing an unprecedented level of detail in molecular imaging. Because the enhancement occurs in the immediate vicinity of the tip, TERS can achieve spatial resolutions down to a few nanometers (Bailo and Deckert 2008). This high resolution enables researchers to image fine structural details and study individual molecules or small clusters of molecules. This combination not only provides chemical information through Raman scattering but also topographical data from the microscopy technique (Fig. 1c).

1.5 Integration with Other Imaging Modalities

The integration of RS with other imaging modalities can significantly enhance preclinical bioimaging by offering complementary data that individual techniques alone cannot provide (Kuhar et al. 2018). Specifically, combining Raman spectroscopy with confocal fluorescence microscopy (CFM), atomic force microscopy (AFM), quantitative phase imaging (such as digital holography microscopy, DHM), and mass spectrometry imaging (MSI) has advanced the understanding of molecular and cellular dynamics.

The combination of RS and CFM is a powerful tool in biological imaging, allowing collection of morphological information (via fluorescence) alongside biochemical data (via RS) (Uzunbajakava and Otto 2003). CFM provides high-resolution images of fluorescently labeled molecules within cells but lacks detailed biochemical insights. RS offers label-free molecular information by analyzing vibrational modes of biomolecules. In preclinical bioimaging, this synergy enables researchers to observe both spatial distribution and chemical composition of biomolecules in cells and tissues (Harz et al. 2008). This dual approach has been valuable in cancer research, enabling detailed mapping of cancer cell morphology and molecular signatures, aiding in early cancer detection and therapy monitoring (Lin et al. 2021).

Atomic Force Microscopy (AFM) provides detailed nanoscale insights into the topographical and mechanical properties of cells, such as stiffness and elasticity, which are crucial for understanding cancer progression and metastasis (Cross et al. 2008). By combining AFM with Raman Spectroscopy (RS), researchers can establish correlations between a cell's mechanical characteristics and its biochemical composition (Xiao et al. 2013). AFM excels in producing high-resolution, three-dimensional maps of cellular surfaces, while RS complements this with chemical specificity, identifying biomolecular variations. This correlative approach has proven effective in distinguishing cancerous cells from healthy ones by analyzing differences in biomechanical properties and lipid composition. For example, RS can detect elevated cholesterol levels in cancer cells, while AFM simultaneously measures alterations in cell stiffness, providing a holistic view of cancer-related cellular changes (Cruz et al. 2020; Miletić et al. 2020).

Quantitative phase imaging methods, such as DHM, have been integrated with RS to provide a non-invasive, real-time method for capturing both morphological and biochemical changes in cells (Mangini et al. 2023). DHM reconstructs holographic images by measuring light scattered from cells, enabling observation of cell morphology without staining or labeling (McReynolds et al. 2017). When combined with RS, this technique offers a complete picture of cellular changes. For instance, during apoptosis, RS can detect biochemical shifts like protein degradation, while DHM captures concurrent morphological changes like cell shrinkage (Khmaladze 2017; D'Brant et al. 2019). This powerful combination allows dynamic studies of cell behavior in real time, crucial for preclinical imaging in drug development.

MSI provides detailed molecular maps by analyzing the spatial distribution of biomolecules in tissue sections. When used with RS, MSI extends the range of detectable molecules beyond Raman capabilities, particularly for compounds difficult to ionize (Lanni et al. 2014). RS identifies molecular vibrations, MSI measures molecular masses, allowing detection of a wide variety of biomolecules, such as lipids, proteins, and metabolites within biological tissues (Bocklitz et al. 2015). The combination of MSI and RS has proven useful in cancer research, facilitating the study of spatial heterogeneity within tumors and helping to distinguish between different tissue types and their respective molecular compositions (Yang et al. 2023).

1.6 Importance and Applications in Bioimaging

RS is a powerful molecular technique for the ultrasensitive detection of biomolecules, adaptable for both in vivo and ex vivo studies. Its high chemical specificity enables the non-destructive, label-free differentiation of similar biomolecules. Raman imaging is widely used to study and map tissues, ranging from single cells to complex structures in human and animal samples, offering detailed insights into biological processes. As it relies on light scattering, RS is effective for both transparent and opaque materials, making it ideal for exploring dynamic cellular processes and detecting changes in lipids, proteins, and DNA with high resolution. Moreover, it is particularly well-suited for use in aqueous environments, which are prevalent in biological samples, since water also has a low Raman signal.

Beyond imaging subcellular features, Raman imaging facilitates the tracking and monitoring of drug uptake and the real-time measurement of drug metabolism in the body. The distinct Raman signatures of cancerous and healthy tissues make this technique a valuable tool for tissue analysis in preclinical diagnostics. When combined with metallic nanoparticles, SERS-based imaging can identify biomarkers and, and monitor molecular alterations in specific diseases with exceptional sensitivity. Raman imaging's ability to map protein aggregation, lipid accumulations, and other molecular changes makes it invaluable for early disease detection and understanding disease progression. This solidifies its role as an essential tool for preclinical investigation and biomedical research. The following sections delve into the applications of RS in molecular and cellular imaging for diagnostic purposes, with a focus on its capability to provide molecular-level insights into biological systems in preclinical studies.

2 Molecular and Cellular Imaging for Diagnostic Applications

2.1 Raman Detection of Biomolecules and Cellular Components

RS is a powerful tool for detecting biomolecules such as lipids, proteins, and nucleic acids, offering non-invasive and label-free molecular analysis by capturing the unique vibrational fingerprints of chemical species within a sample. This versatility makes it particularly well-suited for studying cellular dynamics and identifying pathological states. Traditionally, researchers have relied on monitoring specific spectral markers, such as proteins bands associated with aromatic amino acids (e.g., phenylalanine) or specific lipids (e.g., oleic acid), to investigate changes linked to the aggressiveness of certain cancer types (Kirsch et al. 2010).

In recent years, more sophisticated approaches have emerged to harness the full potential of RS. Instead of focusing solely on individual spectral peaks, modern techniques often employ statistical and computational methods to analyze the entire spectral profile. For example, machine learning algorithms (MLA), such as linear discriminant analysis (LDA), are increasingly used to identify patterns and subtle differences across spectral data. These advancements have been particularly impactful in oncology, enabling the classification and subtyping of cancers based on their biochemical signatures, thereby enhancing diagnostic precision and personalized treatment strategies (Suga et al. 2016; Hanna et al. 2022; Spaziani et al. 2024).

Numerous examples of RS applications in oncology can be found in the literature. For instance, several research groups have explored the potential of RS in the classification of breast cancer (BC) samples, allowing for the discrimination of cancerous tissues and cells from normal counterparts by analyzing biochemical variations at the molecular level. For example, Hanna et al. demonstrated the ability of Raman spectroscopy to classify BC molecular subtypes, such as Luminal A, Luminal B, HER2-positive, and triple-negative BC, based on differences in the intensity, position, and shape of certain Raman peaks. By applying LDA on Raman spectra, the study achieved high specificity rates for classifying each subtype, with accuracies ranging from 70% to 100% depending on the subtype (Kallaway et al. 2013). Moreover, the study highlighted those spectral changes, such as increased protein and nucleic acid peaks and reduced lipid-associated peaks, are consistently observed as tissues progress from normal to malignant states. These findings underscore the potential of Raman spectroscopy for precise cancer grading and subtyping, providing valuable information for personalized treatment strategies.

Raman imaging, on the other hand, has also been applied to study tumor heterogeneity and the biochemical composition of metastatic sites, further enhancing its utility in the clinical setting. This imaging methodology is specifically suitable for several biological analysis, such as subcellular mapping (Liu et al. 2017), localizing components like lipids in membranes or proteins in organelles (Notingher and Hench 2006); or in identifying molecular changes in cancer cells or other pathologies (Short et al. 2005). Finally, dynamic studies can be performed also, by observing molecular changes over time, such as during cell division or apoptosis (Rauniyar et al. 2023).

A recent advancement involves the use of Raman imaging in conjunction with digital holographic imaging to identify and differentiate circulating tumor cells (CTCs) from white blood cells (WBCs) in liquid biopsies (Mangini et al. 2023). This novel approach leverages the metabolic peculiarities of cancer cells, specifically their enhanced glucose uptake and subsequent lipid droplet (LD) formation. By employing deuterated glucose as a metabolic tracer, Raman imaging can detect lipid droplets in cancer cells through the unique C-D vibrational bands located in the Raman silent zone (~2120 cm^{-1}), which are absent in WBCs (Fig. 2a and b). Digital holographic imaging further aids in screening large cell populations rapidly by detecting birefringence changes caused by lipid droplets. The combined platform has achieved high sensitivity and specificity, close to 100%, for distinguishing cancer cells from WBCs, highlighting its potential as a label-free, non-destructive diagnostic tool. This innovative method not only improves the isolation and characterization of viable CTCs but also lays the groundwork for enhanced cancer monitoring and personalized treatment strategies.

Beyond cancer, Raman bioimaging applications extend to a wide range of diseases, including neurodegenerative and cardiovascular conditions (Michael et al. 2017).

Despite its strengths, Raman spectroscopy faces challenges such as weak signal intensity, interference from sample preparation (e.g., masking of nucleic acid regions by phosphate vibrational modes in phosphate-buffered saline commonly used in cell biology), and variability in sample handling. These factors can reduce reproducibility and make standardization difficult. Advances in deep learning for spectral analysis and optimization of preparation protocols will improve the reliability of Raman spectroscopy. Combining Raman data with complementary techniques, such as fluorescence, digital holographic imaging, or atomic force microscopy, could provide more comprehensive insights into cellular and molecular dynamics.

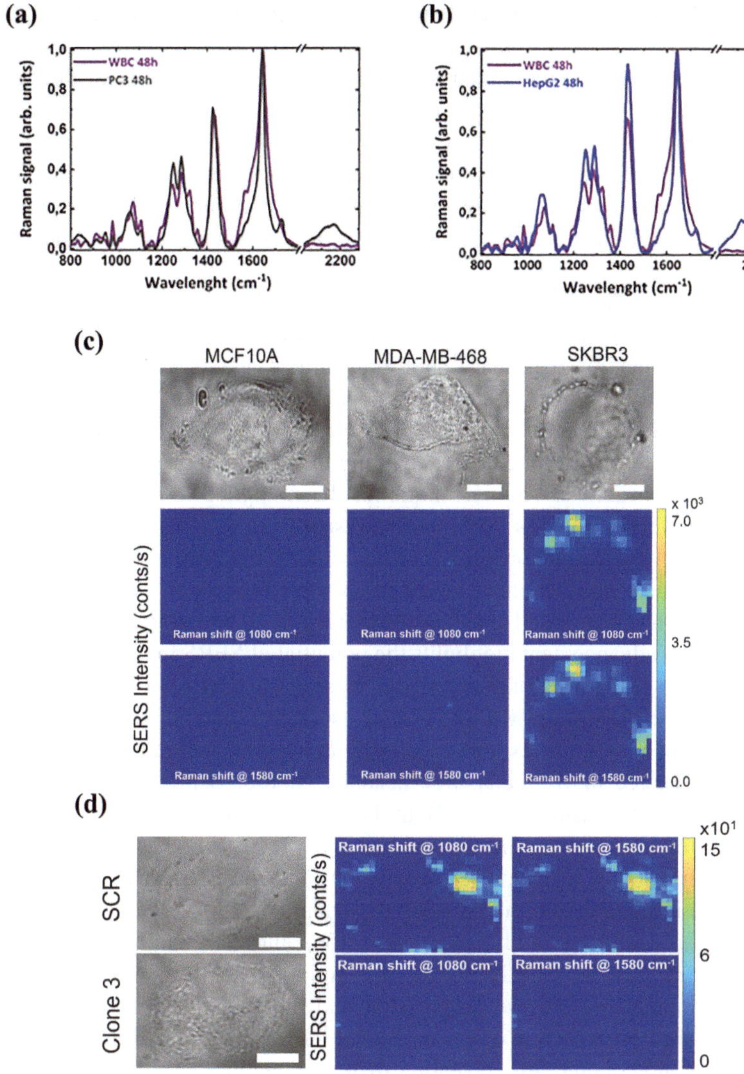

Fig. 2 (**a**) Raman spectra of cancer pancreatic cells (PC3, grey line), co-cultured with White Blood Cells (WBCs, purple line) in presence of deuterated glucose. The deuterium peak (~2120 cm^{-1}) is detectable only in cancer cell Raman spectrum. (**a** and **b**) Raman spectra of hepatocarcinoma cells (HepG2, blue line), co-cultured with WBCs (purple line) in presence of deuterated glucose. The deuterium peak (~2120 cm^{-1}) is detectable only in cancer cell Raman spectrum. (**c**) Brightfield images (upper panel) of breast normal (MCF10A) and cancer (MDA-MB-468 and SKBR3) cell lines expressing different levels of HER2 receptor. Gold nanoparticles were functionalized with an antibody recognizing HER2 and were incubated with the three cells lines. In the lower panels, SERS intensity maps of a selected cell for the Raman peaks at 1080 and 1580 cm^{-1} for each cell line is shown. The functionalized nanoparticles were able to detect only the HER2-expressing cells Scale bar = 10 μm. (**d**) Brightfield images and SERS intensity maps of a selected cell for the peaks at 1080 and 1580 cm^{-1} for scrambled- and HER2-silenced SKBR3 cells incubated with functionalized HER-2 antibody nanoparticles. Scale bar = 10 μm. (Image in panel **a** is adapted from (Mangini et al. 2023), and images in panels **b** and **c** are adapted from (Spaziani et al. 2024), all under a Creative Commons Attribution License (CC BY 4.0))

2.2 SERS Detection of Biomolecules and Cellular Components

Molecular dynamics and biomarker quantitation can also be addressed through RS with the integration of SERS approaches. These advanced imaging and sensing platforms represent transformative tools in precision medicine, offering exceptional sensitivity, molecular specificity, and the ability to perform real-time analyses of complex biological systems. SERS excels in detecting minute biomarker quantities and enabling multiplexed detection, allowing the simultaneous quantification of proteins and nucleic acids in a single assay.

An exciting application of SERS is in the study of neurodegenerative diseases. Recent advancements have demonstrated the use of SERS for the label-free detection and quantification of primary neurotransmitters associated with neurological disorders, including dopamine, serotonin, acetylcholine, and glutamate. Spread spectrum SERS (ss-SERS), a novel approach, enhances signal-to-noise ratio by more than three orders of magnitude, enabling the detection of neurotransmitters at attomolar concentrations. This technique has been applied to monitor biomarkers linked to conditions such as Alzheimer's disease (AD) and Parkinson's disease (PD), where neurotransmitter imbalances are critical indicators of disease progression. For instance, acetylcholine, which plays a pivotal role in cognitive function, is detected with unparalleled sensitivity, offering insights into its role in AD and treatment. This advancement highlights the potential of SERS in the early diagnosis and therapeutic monitoring of neurodegenerative diseases (Lee et al. 2021).

The combined use of SERS and Raman spectroscopy has shown remarkable promise in cancer diagnostics, particularly in identifying and quantifying biomarkers such as HER2 in BC cells. For instance, studies using gold nanoparticles functionalized with trastuzumab (TZ) antibodies have successfully detected HER2 overexpression with high specificity and sensitivity (Fig. 2c). This approach enables precise localization of HER2 on cell membranes and correlates its expression levels with metabolic signatures of malignancy (Spaziani et al. 2024).

SERS also facilitates multiplexed detection, allowing simultaneous analysis of biomarkers such as HER2, EGFR, and lipid metabolites. This capability is pivotal for monitoring disease progression and therapeutic responses. A notable study compared HER2-overexpressing SKBR3 BC cells with triple-negative and non-tumorigenic counterparts, revealing distinct metabolic profiles linked to lipid biosynthesis and protein expression (Talari et al. 2015).

Advances in combining SERS with gene-silencing technologies, such as RNA interference, have further expanded its applications. For example, HER2 silencing in SKBR3 cells, followed by SERS analysis, demonstrated significant reductions in HER2 expression and alterations in associated biochemical pathways, providing valuable insights into the relationship between HER2 expression and lipid metabolism (Fig. 2d) (Spaziani et al. 2024). This methodology supports a tailored approach to HER2-targeted therapies by enhancing the understanding of molecular dynamics in treated cells. Moreover, integrating SERS with machine learning algorithms (MLA) enhances data analysis and interpretation. Techniques such as Principal Component Analysis (PCA) and Linear Discriminant Analysis (LDA) have proven effective in classifying cancer cell subtypes based on Raman and SERS spectral data, achieving accuracies exceeding 90%. These computational tools are essential for refining diagnostic capabilities and optimizing therapeutic strategies. Despite its high sensitivity, SERS faces challenges, including variability in nanoparticle synthesis, inconsistent reproducibility due to irregular hot spot generation, and biocompatibility concerns for in vivo applications. Standardizing nanoparticle production and exploring alternative materials, such as biocompatible nanostructures, could address these issues. Additionally, integrating SERS with advanced computational tools like machine learning and combining it with Raman imaging could enable robust multimodal diagnostics across a wide range of diseases.

2.3 TERS for Cellular Mapping

Tip-Enhanced Raman Spectroscopy (TERS) is a groundbreaking tool for nanoscale chemical imaging, offering spatial resolution as fine as 2.5 nm and high molecular specificity. It addresses key challenges in studying the spatial organization and molecular composition of cellular membranes, which are critical for processes like signal transduction, molecular transport, and cellular communication (Alexander and Schultz 2012; Bonhommeau and Lecomte 2018). TERS has shown remarkable potential in mapping the molecular heterogeneity of cell membranes. For instance, TERS imaging of pancreatic cancer cells (BxPC-3) revealed nanoscale segregation of domains enriched in biomolecules such as phenylalanine, histidine, phosphatidylcholine lipids, proteins, and cholesterol (Mrđenović et al. 2022). These high-resolution maps provide insights into lipid-protein interactions, which are unattainable with diffraction-limited techniques like confocal RM.

TERS is invaluable for studying membrane dynamics, including lipid rafts and protein complexes, in disease progression. It detects subtle Raman shifts linked to functional groups in phospholipids and proteins, revealing critical details of membrane-associated processes. For example, TERS has identified regions where cholesterol and proteins mix, potentially indicating altered fluidity and signaling activity in cancer cells. Despite its promise, TERS faces challenges like instrumental drift and the need for robust, reproducible probes. Advances in probe design, stabilization techniques, and faster imaging protocols are expected to overcome these hurdles, expanding TERS applications in cellular diagnostics. By delivering unprecedented nanoscale insights, TERS is poised to become a cornerstone technique for cell mapping and biomolecular analysis.

3 Raman Microscopy in Preclinical Studies

3.1 Role in Preclinical Imaging

The combination of RS with microscopy (RM) emerged as an essential tool in preclinical imaging due to its capacity to provide detailed molecular information without the need for external labels. Its ability to generate a unique molecular "fingerprint" from the inelastic scattering of photons makes it particularly useful for understanding the biochemical composition of tissues, cells, and biofluids, which is crucial for studying disease mechanisms, drug effects, and pathological conditions. In preclinical studies, where animal models serve to bridge the gap between in vitro research and clinical applications, RS stands out by offering complementary molecular insights alongside traditional imaging techniques such as Magnetic Resonance Imaging (MRI), Computed Tomography (CT), and Positron Emission Tomography (PET). These methods are well-established for structural and functional imaging, but RM adds a layer of specificity by allowing real-time, non-invasive monitoring of biochemical changes within tissues. This has proven invaluable in assessing tumor progression, drug delivery, and therapeutic responses, particularly in fields such as cancer, fibrosis, and neurodegenerative diseases. What sets RM apart from other imaging techniques is its label-free detection, which eliminates the potential issues posed by external markers, such as fluorescence or radiolabels, that can alter biological processes or add artifacts. By preserving the integrity of samples, RM enables more accurate assessments of therapeutic efficacy. Furthermore, it offers the advantage of real-time analysis, providing dynamic insights into biological systems and allowing for continuous observation of molecular changes, which enhances the study of drug effects and disease progression over time.

In comparison to traditional imaging methods, RM offers several key advantages in preclinical research. Unlike fluorescence-based techniques that rely on external labels, RM is non-destructive and can visualize molecular structures in their native state, preserving the sample for future analysis. This makes it particularly useful in situations where the introduction of foreign agents could alter results. Moreover, RM provides molecular specificity, delivering highly detailed insights into the chemical composition of a sample. This specificity is something that structural imaging techniques like MRI and CT cannot achieve, as they are limited to providing anatomical details without molecular information. Functional imaging methods like PET and Single Photon Emission Computed Tomography (SPECT), although informative, require radiolabels that may not be suitable for long-term studies in live animals. In contrast, RM detects even subtle molecular changes, making it ideal for studying complex processes such as drug metabolism, disease progression, and therapeutic responses at a biochemical level. RM also excels in real-time monitoring, eliminating the need for destructive sample preparation often required by histological staining or radiographic methods. This capability is crucial in understanding dynamic processes like tumor growth, drug uptake, and immune responses, all of which can be tracked over time without interrupting the biological system under study. Moreover, the minimally invasive nature of RM, particularly for in vivo applications, reduces the need for repeated invasive procedures in animal models, allowing researchers to monitor disease progression or therapeutic effects with minimal disruption to the subject.

3.2 Case Studies and Applications

Several applications illustrate the strength of RM in preclinical research. Figure 3a–d shows some possible applications of RM in neurodegenerative disease, in cancer and in nanomedicine. One of the most common and extensively studied applications is in cancer research. In particular, RM has been widely applied to monitor tumor progression and evaluate drug delivery, and it has been used to study brain cancer in both animal models and human subjects during surgery procedures. Schackert et al. (Kirsch et al. 2010) presented a pioneering study demonstrating the use of RM to detect intracerebral tumors in vivo through brain surface mapping. This biophotonics method provides a molecular "fingerprint" of tissue composition by probing molecular vibrations without the need for labels. Using a fiber-optic probe, the researchers acquired Raman spectra from the exposed cortex of mice with metastatic brain tumors induced by melanoma cells. The spectral data allowed them to distinguish between normal brain tissue and tumor regions, achieving spatial accuracy of around 250 μm. Their findings underscore the potential of RM as a non-invasive tool for detecting and mapping brain tumors during surgery. The work highlights the technique ability to detect subsurface tumors and offers a framework for future improvements in imaging resolution and acquisition time. Another interesting study was from Li et al. (2021). Particularly, in this study, a hand-held, lab-designed Raman spectroscopic system was developed to enhance the intraoperative detection of pancreatic cancer tissue. Using a convolutional neural network (CNN) model, the system accurately distinguished cancerous from normal pancreatic tissues. This research represents the first reported use of CNN-aided spontaneous RM for pancreatic cancer diagnosis, achieving over 95% sensitivity, specificity, and accuracy. By visualizing CNN features and identifying key Raman peaks, the system highlighted the increased presence of protein (particularly collagen) in cancerous tissue, while normal tissue showed higher lipid and nucleic acid content. These findings suggest that the combination of CNN models and RM could provide real-time, reliable differentiation of pancreatic cancer tissue during surgery, improving resection outcomes. Furthermore, RM is proving to be a valuable tool in drug delivery studies, particularly with nanoparticle-based therapies. Unlike traditional methods such as immunohistochemistry, which are limited in detecting the precise localization and distribution of nanoparticles in tissues, Raman imaging can allow for label-free visualization of nanoparticles penetration within tumor tissues (Fig. 3d). For instance, recently Delle Cave et al. (Delle Cave et al. 2024) successfully detected and visualized the distribution of nanoparticles within tumoral tissue in mice. This is crucial because traditional methods often fail to capture the exact movement and deep tissue infiltration of nanoparticles. Immunohistochem-

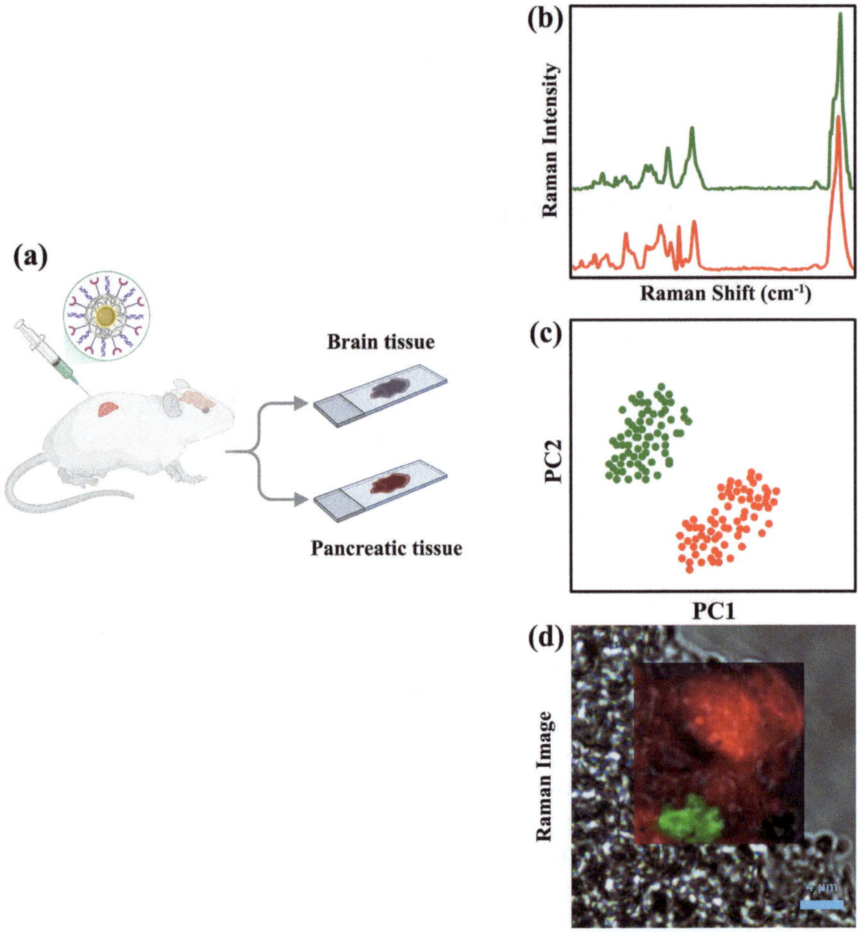

Fig. 3 Overview of the potential applications of Raman microscopy in biomedical research. (**a**) Schematic representation of in vivo studies using Raman-active nanoparticles in a mouse model or for studying other kinds of tissues (e.g., brain or pancreas). (**b** and **c**) Raman spectra acquired by mapping tissues and PCA demonstrate the ability to distinguish between healthy and diseased brain regions in the context of neurodegenerative diseases or cancer cells. (**d**) In nanomedicine, Raman microscopy facilitates label-free imaging and analysis of nanoparticle distribution within tissues, providing insights into their biodistribution and interactions with the microenvironment. Tumoral tissue is shown in red while nanoparticles in green

istry typically relies on antigen-antibody reactions and provides only static, endpoint analysis, whereas RM can continuously monitor the behavior of nanoparticles at the molecular level, offering insights into their therapeutic efficiency and biodistribution that are not possible with conventional techniques. This makes Raman imaging particularly important for advancing nanoparticle-based drug delivery, where the ability to track and quantify nanoparticles' penetration into tumor cells directly correlates with treatment effectiveness. RM has also played a significant role in studying neurodegenerative diseases like AD. For instance, in a recent study of preclinical models, RM has enabled the detection of disease-specific biomarkers, such as lipid alteration in the hippocampus typically associated with AD (Meng et al. 2024). This study utilizes RM and other label-free techniques to analyze the brain of AD mouse models, revealing distinct lipid signatures associated with tau pathology. The findings not only deepen our understanding of tauopathy-associated lipid dysregulation but also highlight the potential of label-free chemical imaging in identifying novel biomarkers and therapeutic targets for AD.

3.3 Challenges and Limitations

Despite its many advantages, RM faces several challenges in preclinical studies. One of the main limitations is the inherently weak signal produced by Raman scattering, which can make it difficult to detect low concentrations of molecules. While techniques like SERS can amplify the signal, they require metal nanoparticles, which complicate in vivo applications. Additionally, RM suffers from limited tissue penetration depth, particularly in the visible spectrum, which restricts its use for deep-tissue imaging. Efforts to extend RM into the near-infrared spectrum have shown promise in overcoming this limitation, though this comes at the cost of reduced spatial resolution. The complexity of biological tissues, which contain a mixture of proteins, lipids, and other biomolecules, also presents a challenge for RM, as overlapping spectral features can make it difficult to deconvolute specific molecular signatures. Advances in computational techniques, such as MLA, are helping to address this issue by improving spectral analysis in complex samples. Finally, the high cost of Raman equipment and the need for specialized expertise limit its accessibility compared to more established imaging techniques like MRI or CT, which remain the gold standard in many preclinical studies.

4 Technological Innovations for Preclinical Bioimaging

4.1 Technological Innovation

Recent advancements in RS have expanded its role in preclinical bioimaging (Kallaway et al. 2013; Pence and Mahadevan-Jansen 2016; Jermyn et al. 2016; Neugebauer et al. 2015; Choo-Smith et al. 2002). Innovations in core components now allow high-resolution, non-invasive chemical information capture, making RS invaluable for biomedical research. Improved sensitivity, signal clarity, and portability have increased its use in applications ranging from early disease detection to therapeutic response monitoring. This progress underscores RS potential to provide critical insights into cellular and tissue-level changes relevant to diagnostic and therapeutic strategies in preclinical studies. Advancements in RS instrumentation have enhanced its utility in preclinical bioimaging. Innovations in lasers, detectors, optical filters, and portable devices have expanded its biomedical applications. The integration of optical fiber probes in endoscopy and the development of spatially offset Raman spectroscopy (SORS) have opened new avenues for non-invasive diagnostics. High-stability, reproducible laser sources ensure consistent excitation, crucial for reliable Raman measurements. Diode lasers operating in the infrared region minimize fluorescence from biological samples, enhancing signal clarity. Narrow frequency band lasers have improved spectral resolution, allowing more precise identification of molecular signatures within complex biological tissues (Praveen et al. 2018; Schwaighofer et al. 2017).

Charge-coupled devices (CCDs) and complementary metal-oxide-semiconductor (CMOS) sensors now offer higher sensitivity and lower noise levels, essential for detecting weak Raman signals from biological specimens (Denson et al. 2007; Quintás et al. 2023; Kuligowski et al. 2013; Kuligowski et al. 2017; Agatonovic-Kustrin and Morton 2023). However, the high sensitivity of detectors can make them susceptible to saturation and damage from strong signals. Advances in optical filters, such as notch and edge filters with superior rejection ratios, have improved the isolation of Raman signals from intense Rayleigh scattered light, further enhancing detection capabilities (Giridhar et al. 2017; Pappas et al. 2000).

The development of portable and handheld Raman spectrometers has made in situ and point-of-care measurements feasible (Quarin and Strobbia 2021; Granger et al. 2016). These compact devices combine miniaturized components with efficient data processing algorithms, enabling rapid, non-invasive analysis of biological samples in clinical and field settings. Their portability facilitates real-time diagnostics and monitoring, particularly beneficial in preclinical studies where swift data acquisition is vital (Ali et al. 2022). Nevertheless, miniaturization often leads to compromises in spectral resolution and sensitivity compared to benchtop systems (Vandenabeele et al. 2014).

Optical fiber Raman probes have revolutionized endoscopic applications (Stevens et al. 2016). By integrating RS with endoscopy, molecular information can be obtained from internal organs without invasive procedures. Different configurations of optical fibers and bundles allow flexible probe designs tailored to specific applications (Fig. 4a). Front-viewing probes enable direct sampling of tissues in the line of sight, while side-viewing probes access lateral tissues (Jayhooni et al. 2019). However, issues such as signal attenuation, background fluorescence from the fibers themselves, and mechanical flexibility can affect the quality of collected data (Motz et al. 2004).

SORS is another significant innovation. SORS allows collection of Raman signals from subsurface layers by spatially separating the illumination and collection zones (Fig. 4b) (Matousek and Stone 2016). This technique enables non-invasive analysis of tissues beneath the surface, useful for detecting buried lesions or monitoring deep-tissue processes (Nicolson et al. 2021). SORS enhances the depth penetration of RS, overcoming traditional limitations associated with superficial tissue analysis. Yet, challenges include the complexity of data interpretation due to overlapping signals from different tissue layers and the need for advanced algorithms to accurately deconvolute subsurface information. Additionally, optimizing

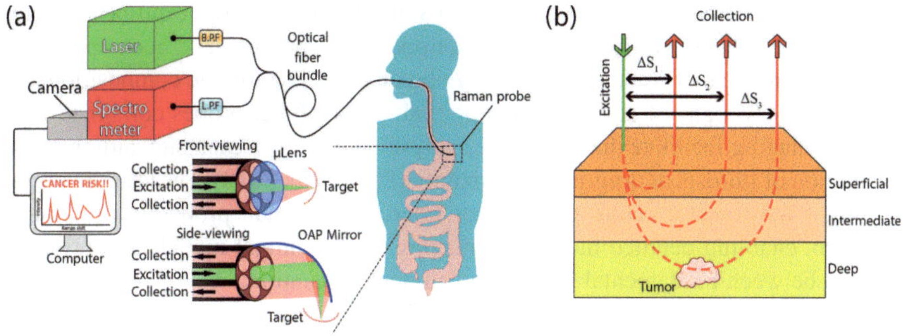

Fig. 4 (**a**) Schematic of the operation of a Raman spectroscopy fiber probe. Laser light (green beam) is coupled into the central core of a fiber optic bundle. Both front-viewing and side-viewing configurations for collection of Raman scattering (red beam) are shown. The Raman scattering is collected through the peripheral cores. BPF: band pass filter, LPF: long pass filter, μLens: micro lens, OAP: off-axis parabolic. (**b**) Schematic of excitation and collection of Raman scattering in spatially offset Raman spectroscopy (SORS). By increasing the transverse distance ΔS between the excitation point and the collection point, Raman spectra from deeper layers can be obtained

the spatial offset for different tissue types and depths remains a technical challenge (Kuhar et al. 2018).

4.2 Applications of Fiber-Optic Raman Probes in Preclinical Studies

The integration of RS with fiber-optic probes has revolutionized its application in preclinical studies, particularly for endoscopic diagnostics. These innovations enable in vivo, non-invasive molecular imaging of internal tissues, offering unprecedented opportunities to study disease progression and therapeutic interventions in animal models. Fiber-optic Raman probes have been employed in preclinical models to study gastrointestinal cancers. Side-viewing fiber-optic probes equipped with SERS capabilities have been used to detect subtle biochemical alterations in early-stage colorectal cancer. These probes provide real-time molecular insights by capturing signals from dysplastic and healthy tissues, enabling precise differentiation of tissue states (Wang et al. 2016). Front-viewing fiber probes integrated with endoscopes allow imaging of pulmonary lesions in animal models, identifying molecular markers associated with malignancy, such as changes in protein or lipid composition (Fernandes et al. 2021). The use of fiber-optic probes for Raman imaging has been extended to cardiovascular research. Raman fiber probes have been used to analyze arterial plaques in preclinical models, distinguishing between stable and unstable plaques based on their lipid and collagen content. This application is critical for studying atherosclerosis and evaluating anti-inflammatory treatments (Jia et al. 2024).

Spatially Offset Raman Spectroscopy (SORS) with fiber-optic probes enables the study of subsurface tissues. SORS-equipped fiber probes have been applied to preclinical models of osteoarthritis, analyzing changes in cartilage composition and detecting early signs of degradation (Raj et al. 2024). Fiber probes combined with Raman imaging facilitate the molecular analysis of liver tissues in situ, aiding in the evaluation of fibrosis or drug-induced toxicity (Pacia et al. 2018). These applications highlight the transformative potential of fiber-optic Raman probes in preclinical research, enabling detailed molecular imaging in live animal models and bridging the gap between fundamental studies and clinical translation.

5 Conclusion and Future Perspectives

RS has proven to be an invaluable tool in preclinical bioimaging, offering unique advantages in the non-invasive, label-free molecular analysis of biological tissues. This chapter highlights the versatile applications of RS and RM across various biomedical fields, including neurodegenerative diseases, cancer research, and nanomedicine. Through its ability to provide high-resolution, real-time information about biochemical compositions, RS has shown potential for monitoring disease progression, evaluating therapeutic efficacy, and studying the distribution of therapeutic nanoparticles. Advancements in Raman techniques, such as SERS, and TERS have further expanded its capabilities. SERS, for example, enhances the Raman signal, enabling more sensitive detection of biomolecules at lower concentrations, which is crucial for early disease diagnosis and monitoring. TERS, on the other hand, allows for molecular-level imaging with nanometer resolution, offering deeper insights into cellular and subcellular structures. These innovations, alongside traditional RS, provide a powerful toolkit for exploring the molecular and structural characteristics of biological tissues with greater precision. Despite the significant promise of RS, several challenges remain. These include limitations in tissue penetration depth, the need for advanced computational tools for data interpretation, and the development of more sensitive and biocompatible Raman-active probes. Addressing these challenges will be crucial for expanding the applicability of Raman-based techniques in clinical settings. Future advancements are expected to further enhance the capabilities of RS. Innovations in instrumentation, such as improved Raman microscopes and the integration of advanced imaging modalities, will allow for deeper tissue penetration and more precise mapping of molecular changes. Additionally, the continued development of more effective analytical methods, such as MLA for spectral analysis, will enable more accurate and faster data processing, facilitating the translation of Raman-based diagnostics from preclinical to clinical applications. In conclusion, RS holds significant promise for revolutionizing the way we study and treat diseases, offering a powerful platform for molecular diagnostics, therapeutic monitoring, and drug development. As technological advancements continue, its role in preclinical and clinical bioimaging will undoubtedly grow, contributing to more personalized and precise medical treatments.

Acknowledgments This work was supported by AIRC IG grant # 21420 and by the NextGenerationEU Project PNRR-SEE LIFE – Strengthening the Italian Infrastructure of Euro-Bioimaging (IR23 – DD MUR 3264/21 CUPB53C22001810006).

Conflict of Interest The authors declare no conflict of interest in this work.

References

Agatonovic-Kustrin S, Morton DW (2023) Advanced spectroscopic detectors for identification and quantification: infrared, Raman, and nuclear magnetic resonance. In: Instrumental thin-layer chromatography, 2nd edn, pp 165–179

Alexander KD, Schultz ZD (2012) Tip-enhanced Raman detection of antibody conjugated nanoparticles on cellular membranes. Anal Chem 84:7408–7414. https://doi.org/10.1021/ac301739k

Ali A, Nettey-Oppong EE, Effah E, Yu CY, Muhammad R, Soomro TA, Byun KM, Choi SH (2022) Miniaturized Raman instruments for SERS-based point-of-care testing on respiratory viruses. Biosensors (Basel) 12. https://doi.org/10.3390/bios12080590

Bailo E, Deckert V (2008) Tip-enhanced Raman scattering. Chem Soc Rev 37:921–930. https://doi.org/10.1039/B705967C

Beleites C, Sergo V (2009) HyperSpec: a package to handle hyperspectral data sets in R

Bocklitz T, Bräutigam K, Urbanek A, Hoffmann F, von Eggeling F, Ernst G, Schmitt M, Schubert U, Guntinas-Lichius O, Popp J (2015) Novel workflow for combining Raman spectroscopy and MALDI-MSI for tissue based studies. Anal Bioanal Chem 407:7865–7873. https://doi.org/10.1007/s00216-015-8987-5

Bonhommeau S, Lecomte S (2018) Tip-enhanced Raman spectroscopy: a tool for nanoscale chemical and structural characterization of biomolecules. ChemPhysChem 19:8–18. https://doi.org/10.1002/cphc.201701067

Choo-Smith L-P, Edwards HGM, Endtz HP, Kros JM, Heule F, Barr H, Robinson JS Jr, Bruining HA, Puppels GJ (2002) Medical applications of Raman spectroscopy: from proof of principle to clinical implementation. Biopolymers 67:1–9. https://doi.org/10.1002/bip.10064

Cialla-May D, Bonifacio A, Bocklitz T, Markin A, Markina N, Fornasaro S, Dwivedi A, Dib T, Farnesi E, Liu C et al (2024) Biomedical SERS—the current state and future trends. Chem Soc Rev 53:8957–8979. https://doi.org/10.1039/D4CS00090K

Cross SE, Jin Y-S, Tondre J, Wong R, Rao J, Gimzewski JK (2008) AFM-based analysis of human metastatic cancer cells. Nanotechnology 19:384003. https://doi.org/10.1088/0957-4484/19/38/384003

Cruz ALS, de A. Barreto E, Fazolini NPB, Viola JPB, Bozza PT (2020) Lipid droplets: platforms with multiple functions in cancer hallmarks. Cell Death Dis 11:105. https://doi.org/10.1038/s41419-020-2297-3

D'Brant LY, Desta H, Khoo TC, Sharikova AV, Mahajan SD, Khmaladze A (2019) Methamphetamine-induced apoptosis in glial cells examined under marker-free imaging modalities. J Biomed Opt 24:046503. https://doi.org/10.1117/1.JBO.24.4.046503

Delle Cave D, Mangini M, Tramontano C, De Stefano L, Corona M, Rea I, De Luca AC, Lonardo E (2024) Hybrid biosilica nanoparticles for in-vivo targeted inhibition of colorectal cancer growth and label-free imaging. Int J Nanomedicine 19:12079–12098. https://doi.org/10.2147/IJN.S480168

Denson SC, Pommier CJS, Denton MB (2007) The impact of array detectors on Raman spectroscopy. J Chem Educ 84:67. https://doi.org/10.1021/ed084p67

Fernandes S, Williams G, Williams E, Ehrlich K, Stone J, Finlayson N, Bradley M, Thomson RR, Akram AR, Dhaliwal K (2021) Solitary pulmonary nodule imaging approaches and the role of optical fibre-based technologies. Eur Respir J 57:2002537. https://doi.org/10.1183/13993003.02537-2020

Giridhar G, Manepalli RRKN, Apparao G (2017) Confocal Raman spectroscopy. In: Spectroscopic methods for nanomaterials characterization, vol 2, pp 141–161

Granger JH, Schlotter NE, Crawford AC, Porter MD (2016) Prospects for point-of-care pathogen diagnostics using surface-enhanced Raman scattering (SERS). Chem Soc Rev 45:3865–3882. https://doi.org/10.1039/c5cs00828j

Hanna K, Krzoska E, Shaaban AM, Muirhead D, Abu-Eid R, Speirs V (2022) Raman spectroscopy: current applications in breast cancer diagnosis, challenges and future prospects. Br J Cancer 126:1125–1139. https://doi.org/10.1038/s41416-021-01659-5

Harz M, Kiehntopf M, Stöckel S, Rösch P, Deufel T, Popp J (2008) Analysis of single blood cells for CSF diagnostics via a combination of fluorescence staining and micro-Raman spectroscopy. Analyst 133:1416–1423. https://doi.org/10.1039/B716132H

Jain PK, Huang X, El-Sayed IH, El-Sayed MA (2008) Noble metals on the nanoscale: optical and photothermal properties and some applications in imaging, sensing, biology, and medicine. Acc Chem Res 41:1578–1586. https://doi.org/10.1021/ar7002804

Jayhooni SMH, Short M, Assadsangabi B, Hohert G, Du C, Zeng H, Takahata K (2019) Side-viewing endoscopic Raman spectroscopy for angle-resolved analysis of luminal organs. Adv Mater Technol 4:1900364. https://doi.org/10.1002/admt.201900364

Jermyn M, Desroches J, Aubertin K, St-Arnaud K, Madore W-J, De Montigny E, Guiot M-C, Trudel D, Wilson BC, Petrecca K et al (2016) A review of Raman spectroscopy advances with an emphasis on clinical translation challenges in oncology. Phys Med Biol 61:R370–R400. https://doi.org/10.1088/0031-9155/61/23/R370

Jia H, Chen X, Shen J, Liu R, Hou P, Yue S (2024) Label-free fiber-optic Raman spectroscopy for intravascular coronary atherosclerosis and plaque detection. ACS Omega 9:27789–27797. https://doi.org/10.1021/acsomega.4c01611

Kallaway C, Almond LM, Barr H, Wood J, Hutchings J, Kendall C, Stone N (2013) Advances in the clinical application of Raman spectroscopy for cancer diagnostics. Photodiagn Photodyn Ther 10:207–219. https://doi.org/10.1016/j.pdpdt.2013.01.008

Khmaladze A (2017) Examining live cell cultures during apoptosis by digital holographic phase imaging and Raman spectroscopy. J Phys Conf Ser 909:012001. https://doi.org/10.1088/1742-6596/909/1/012001

Kirsch M, Schackert G, Salzer R, Krafft C (2010) Raman spectroscopic imaging for in vivo detection of cerebral brain metastases. Anal Bioanal Chem 398:1707–1713. https://doi.org/10.1007/s00216-010-4116-7

Kuhar N, Sil S, Verma T, Umapathy S (2018) Challenges in application of Raman spectroscopy to biology and materials. RSC Adv 8:25888–25908. https://doi.org/10.1039/c8ra04491k

Kuligowski J, Lendl B, Quintás G (2013) Advanced spectroscopic detectors for identification and quantification: FTIR and Raman. In: Liquid chromatography: fundamentals and instrumentation, pp 333–347

Kuligowski J, Lendl B, Quintás G (2017) Advanced IR and Raman detectors for identification and quantification. In: Liquid chromatography: fundamentals and instrumentation, vol 1, 2nd edn, pp 463–477

Lanni EJ, Masyuko RN, Driscoll CM, Dunham SJB, Shrout JD, Bohn PW, Sweedler J (2014) V correlated imaging with C60-SIMS and confocal Raman microscopy: visualization of cell-scale molecular distributions in bacterial biofilms. Anal Chem 86:10885–10891. https://doi.org/10.1021/ac5030914

Lee W, Kang B-H, Yang H, Park M, Kwak JH, Chung T, Jeong Y, Kim BK, Jeong K-H (2021) Spread spectrum SERS allows label-free detection of attomolar neurotransmitters. Nat Commun 12:159. https://doi.org/10.1038/s41467-020-20413-8

Li Z, Li Z, Chen Q, Ramos A, Zhang J, Boudreaux JP, Thiagarajan R, Bren-Mattison Y, Dunham ME, McWhorter AJ et al (2021) Detection of pancreatic cancer by convolutional-neural-network-assisted spontaneous Raman spectroscopy with critical feature visualization. Neural Netw 144:455–464. https://doi.org/10.1016/j.neunet.2021.09.006

Lin J, Graziotto ME, Lay PA, New EJ (2021) A bimodal fluorescence-Raman probe for cellular imaging. Cells 10. https://doi.org/10.3390/cells10071699

Liu W, Wang H, Du J, Jing C (2017) Raman microspectroscopy of nucleus and cytoplasm for human colon cancer diagnosis. Biosens Bioelectron 97:70–74. https://doi.org/10.1016/j.bios.2017.05.045

Mangini M, Ferrara MA, Zito G, Managò S, Luini A, De Luca AC, Coppola G (2023) Cancer metabolic features allow discrimination of tumor from white blood cells by label-free multimodal optical imaging. Front Bioeng Biotechnol 11. https://doi.org/10.3389/fbioe.2023.1057216

Matousek P, Stone N (2016) Development of deep subsurface Raman spectroscopy for medical diagnosis and disease monitoring. Chem Soc Rev 45:1794–1802. https://doi.org/10.1039/c5cs00466g

McReynolds N, Cooke FGM, Chen M, Powis SJ, Dholakia K (2017) Multimodal discrimination of immune cells using a combination of Raman spectroscopy and digital holographic microscopy. Sci Rep 7:43631. https://doi.org/10.1038/srep43631

Meng H, Elliott A, Mansfield J, Bailey M, Frogley M, Cinque G, Moger J, Stone N, Tamagnini F, Palombo F (2024) Identification of tauopathy-associated lipid signatures in Alzheimer's disease mouse brain using label-free chemical imaging. Commun Biol 7:1341. https://doi.org/10.1038/s42003-024-07034-3

Michael R, Lenferink A, Vrensen GFJM, Gelpi E, Barraquer RI, Otto C (2017) Hyperspectral Raman imaging of neuritic plaques and neurofibrillary tangles in brain tissue from Alzheimer's disease patients. Sci Rep 7:15603. https://doi.org/10.1038/s41598-017-16002-3

Miletić M, Aškrabić S, Rüger J, Vasić B, Korićanac L, Mondol AS, Dellith J, Popp J, Schie IW, Dohčević-Mitrović Z (2020) Combined Raman and AFM detection of changes in HeLa cervical cancer cells induced by CeO2 nanoparticles—molecular and morphological perspectives. Analyst 145:3983–3995. https://doi.org/10.1039/C9AN02518A

Motz JT, Hunter M, Galindo LH, Gardecki JA, Kramer JR, Dasari RR, Feld MS (2004) Optical fiber probe for biomedical Raman spectroscopy. Appl Opt 43:542–554. https://doi.org/10.1364/AO.43.000542

Mrđenović D, Ge W, Kumar N, Zenobi R (2022) Nanoscale chemical imaging of human cell membranes using tip-enhanced Raman spectroscopy. Angew Chem Int Ed 61:e202210288. https://doi.org/10.1002/anie.202210288

Neugebauer U, Rösch P, Popp J (2015) Raman spectroscopy towards clinical application: drug monitoring and pathogen identification. Int J Antimicrob Agents 46:S35–S39. https://doi.org/10.1016/j.ijantimicag.2015.10.014

Nicolson F, Kircher MF, Stone N, Matousek P (2021) Spatially offset Raman spectroscopy for biomedical applications. Chem Soc Rev 50:556–568. https://doi.org/10.1039/d0cs00855a

Notingher I, Hench LL (2006) Raman microspectroscopy: a noninvasive tool for studies of individual living cells in vitro. Expert Rev Med Devices 3:215–234. https://doi.org/10.1586/17434440.3.2.215

Pacia MZ, Czamara K, Zebala M, Kus E, Chlopicki S, Kaczor A (2018) Rapid diagnostics of liver steatosis by Raman spectroscopy via fiber optic probe: a pilot study. Analyst 143:4723–4731. https://doi.org/10.1039/C8AN00289D

Pappas D, Smith BW, Winefordner JD (2000) Raman imaging for two-dimensional chemical analysis. Appl Spectrosc Rev 35:1–23. https://doi.org/10.1081/ASR-100101219

Pence I, Mahadevan-Jansen A (2016) Clinical instrumentation and applications of Raman spectroscopy. Chem Soc Rev 45:1958–1979. https://doi.org/10.1039/c5cs00581g

Praveen BB, Ashok PC, Dholakia K (2018) Semiconductor lasers for bio-medical Raman spectroscopy. In: Semiconductor lasers and diode-based light sources for biophotonics, pp 355–376

Quarin S, Strobbia P (2021) Recent advances towards point-of-care applications of surface-enhanced Raman scattering sensing. Front Chem 9. https://doi.org/10.3389/fchem.2021.714113

Quintás G, Lendl B, Kuligowski J (2023) Advanced IR and Raman detectors for identification and quantification. In: Liquid chromatography: fundamentals and instrumentation, vol 1, 3rd edn, pp 727–742

Raj P, Wu L, Almeida C, Conway L, Tanwar S, Middendorf J, Barman I (2024) Shining light on osteoarthritis: spatially offset Raman spectroscopy as a window into cartilage health. ACS Sens 9:3794–3804. https://doi.org/10.1021/acssensors.4c01242

Rauniyar S, Pansare K, Sharda A, Singh SR, Saha P, Chilakapati MK, Gupta S (2023) Raman spectroscopy revealed cell passage-dependent distinct biochemical alterations in radiation-resistant breast cancer cells. ACS Omega 8:5522–5532. https://doi.org/10.1021/acsomega.2c06787

Schwaighofer A, Brandstetter M, Lendl B (2017) Quantum cascade lasers (QCLs) in biomedical spectroscopy. Chem Soc Rev 46:5903–5924. https://doi.org/10.1039/C7CS00403F

Shen Y, Hu F, Min W (2019) Raman imaging of small biomolecules. Annu Rev Biophys 48:347–369. https://doi.org/10.1146/annurev-biophys-052118-115500

Short KW, Carpenter S, Freyer JP, Mourant JR (2005) Raman spectroscopy detects biochemical changes due to proliferation in mammalian cell cultures. Biophys J 88:4274–4288. https://doi.org/10.1529/biophysj.103.038604

Spaziani S, Esposito A, Barisciano G, Quero G, Elumalai S, Leo M, Colantuoni V, Mangini M, Pisco M, Sabatino L et al (2024) Combined SERS-Raman screening of HER2-overexpressing or silenced breast cancer cell lines. J Nanobiotechnol 22:350. https://doi.org/10.1186/s12951-024-02600-7

Stevens O, Iping Petterson IE, Day JCC, Stone N (2016) Developing fibre optic Raman probes for applications in clinical spectroscopy. Chem Soc Rev 45:1919–1934. https://doi.org/10.1039/c5cs00850f

Suga K, Kondo D, Otsuka Y, Okamoto Y, Umakoshi H (2016) Characterization of aqueous oleic acid/oleate dispersions by fluorescent probes and Raman spectroscopy. Langmuir 32:7606–7612. https://doi.org/10.1021/acs.langmuir.6b02257

Talari ACS, Evans CA, Holen I, Coleman RE, Rehman IU (2015) Raman spectroscopic analysis differentiates between breast cancer cell lines. J Raman Spectrosc 46:421–427. https://doi.org/10.1002/jrs.4676

Toporski J, Dieing T, Hollricher O (2018) Confocal Raman microscopy, vol 66. Springer

Turrell G, Corset J (eds) (1996) Raman microscopy developments and applications. Academic Press

Uzunbajakava N, Otto C (2003) Combined Raman and continuous-wave-excited two-photon fluorescence cell imaging. Opt Lett 28:2073–2075. https://doi.org/10.1364/OL.28.002073

Vandenabeele P, Edwards HGM, Jehlička J (2014) The role of mobile instrumentation in novel applications of Raman spectroscopy: archaeometry, geosciences, and forensics. Chem Soc Rev 43:2628–2649. https://doi.org/10.1039/c3cs60263j

Vulchi RT, Morgunov V, Junjuri R, Bocklitz T (2024) Artifacts and anomalies in Raman spectroscopy: a review on origins and correction procedures. Molecules 29. https://doi.org/10.3390/molecules29194748

Wang J, Zheng W, Lin K, Huang Z (2016) Development of a hybrid Raman spectroscopy and optical coherence tomography technique for real-time in vivo tissue measurements. Opt Lett 41:3045–3048. https://doi.org/10.1364/OL.41.003045

Wu Y, Dong Y, Jiang J, Li H, Zhu T, Chen S (2017) Evaluation of the bone-ligament and tendon insertions based on Raman spectrum and its PCA and CLS analysis. Sci Rep 7:38706. https://doi.org/10.1038/srep38706

Xiao L, Tang M, Li Q, Zhou A (2013) Non-invasive detection of biomechanical and biochemical responses of human lung cells to short time chemotherapy exposure using AFM and confocal Raman spectroscopy. Anal Methods 5:874–879. https://doi.org/10.1039/C2AY25951F

Yang E, Kim JH, Tressler CM, Shen XE, Brown DR, Johnson CC, Hahm T-H, Barman I, Glunde K (2023) RaMALDI: enabling simultaneous Raman and MALDI imaging of the same tissue section. Biosens Bioelectron 239:115597. https://doi.org/10.1016/j.bios.2023.115597

Applications of Photon-Counting Detector-Based Micro-Computed Tomography in Preclinical Bioimaging

Aamir Younis Raja, Osama Sikandar Khan, Briya Tariq, Nadine Francis, and Azza Mohamed Ahmed

Abstract

Photon counting detectors (PCDs) have revolutionized micro-computed tomography (micro-CT) imaging technology by directly converting incident X-ray photons into electrical signals, eliminating intermediate scintillation steps. This advancement enables high-resolution, multi-energy CT imaging with reduced electronic noise, enhanced contrast sensitivity, and lower radiation doses compared to conventional systems. By resolving individual photon energies, PCDs allow simultaneous identification and quantification of multiple materials through spectral techniques, providing deeper insights into a sample's composition and spatial distribution. Current applications span biomedical research, preclinical studies, and clinical diagnostics, while emerging innovations in detector efficiency and AI-driven processing promise broader future utility. This chapter examines the fundamental principles of PCDs, evaluates their current and emerging applications in PCD-based micro-CT systems, and addresses the key advancements, challenges, and transformative implications of these technologies for precision medicine and materials science.

A. Y. Raja (✉)
Khalifa University, Department of Physics, Abu Dhabi, United Arab Emirates

Department of Academic Radiology, University of Otago Christchurch, Christchurch, New Zealand
e-mail: aamir.raja@ku.ac.ae

O. S. Khan · B. Tariq · N. Francis · A. M. Ahmed
Khalifa University, Department of Physics, Abu Dhabi, United Arab Emirates

I. F. Cengiz et al. (eds.), *Bioimaging Modalities in Bioengineering*,
Biomaterials, Bioengineering and Sustainability 9,
https://doi.org/10.1007/978-3-031-99410-4_21

Keywords

Photon counting detectors (PCDs) · Energy integrating detectors (EID) · Spectral photon counting computed tomography (SPCCT) · Micro-CT

1 Introduction

Computed tomography (CT) is a fundamental tool in diagnostic radiology, offering highly specific cross-sectional images of anatomical structures such as bones, muscles, and blood vessels. In 1983, Elliot and Dover developed 3D high-resolution CT scanner capable of imaging small animals and rodents with a resolution as fine as 15 μm (Elliott et al. 2008). Over the years, this innovation led to the development and validation of a wide range of high-resolution micro-CT scanners, which have become crucial in understanding the underlying mechanism of human pathology. These micro-CT scanners produce detailed cross-sectional images, facilitating the diagnosis of various health conditions, treatment planning, and monitoring of therapeutic progress. However, conventional micro-CT scanners suffer from loss of photon energy information, reduced geometric dose efficiency and lower detector performance. Moreover, conventional scanners encounter low contrast to soft tissues, limited contrast discrimination, and high susceptibility to artefacts, including metal and beam hardening, which compromise image quality and hinder analytical assessments (Keklikoglou et al. 2021). In 2005, photon counting detectors (PCDs) emerged as a novel technology, addressing many limitations of conventional CT detectors (Tortora et al. 2022). The application and testing of this advanced detector technology in small bore micro scanners was a natural progression. PCDs offer reduced electronic noise, better spatial resolution, and intrinsic spectral sensitivity. Thus, they can effectively preserve high image quality while concurrently reducing radiation dose. In this chapter, we discuss the current and potential future applications of PCDs-based micro-CT. Section 2 explains the two mechanisms underlying radiation detectors in micro-CT technology. Section 3 highlights the capabilities and advanced applications of PCDs, while Sect. 4 examines the challenges associated with using PCDs in spectral photon-counting CT (SPCCT) systems.

2 Radiation Detection Mechanism in Micro-CT System

In CT imaging, detectors employ two primary mechanisms to transform X-ray photons into electronic signals for image reconstruction: indirect and direct conversion. Indirect conversion, used in energy-integrating detectors (EIDs), involves converting X-rays into visible light, which is then transformed into electrical signals. Conversely, PCDs directly convert X-ray photons into electrical signals with an amplitude proportional to the photons' energy. This direct conversion method enables PCDs to detect and process the energy of individual incoming photons, paving the way for advanced multi-energy imaging in SPCCT. This direct conversion technique not only improves spatial resolution and reduces noise but also allows for the differentiation of various materials based on their energy-specific attenuation profiles (Zhou et al. 2018). Understanding the fundamental differences between these two types of detection systems is essential for appreciating the technological advancements in CT imaging.

2.1 Energy-Integrating Detectors

EIDs are the traditional detection technology used in conventional CT systems. These detectors convert X-rays into electrical signals through an indirect process that integrates the total energy of all incoming photons over a specific exposure time as shown in Fig. 1a (So and Nicolaou 2021). The overall mechanism of signal transport in EIDs is as follows:

1. *X-ray Absorption:* When X-rays pass through an object, some photons are absorbed by the material, while others continue to the detector. In an EID, the X-ray photons that reach the detector are absorbed by a scintillator material, such as Gadolinium Oxysulfide, Sodium Iodide (NaI) or Cesium Iodide (CsI).
2. *Light Emission:* Upon absorption, the scintillator material emits visible light proportional to the total energy deposited by all incoming X-ray photons. Reflective septa, thin and opaque material usually made of lead or other high-density metals, are necessary to be used between the individual pixels to reduce crosstalk between pixels and prevent interference caused by the light escaping the detector cell and affecting adjacent detection elements.
3. *Signal Generation:* The photodiode captures the emitted light and converts it into an electrical signal. The strength of this electrical signal corresponds to the total energy of all absorbed photons within a given time frame.
4. *Signal Integration:* The electrical signal is integrated over time and processed to form an image. Since EIDs integrate the energy of all photons and produce a signal that represents the overall exposure, they do not discriminate between different photon energy levels, resulting in the loss of information about the energy spectrum of the X-rays, as shown in Fig. 1b.

EIDs have several limitations that impact image quality and diagnostic accuracy. One major drawback is energy blurring and the loss of spatial resolution. This occurs because of two intermediate processes where the cumulative energy of all incoming X-ray photons is summed, making it impossible to distinguish between photons of different energies. Consequently, spectral information is lost, reducing the ability to differentiate between materials based on their energy-specific attenuation properties. Furthermore, EIDs are more prone to image artefacts, such as beam hardening, blooming, and metal-induced artefacts, which further degrade image quality and compromise analytical precision.

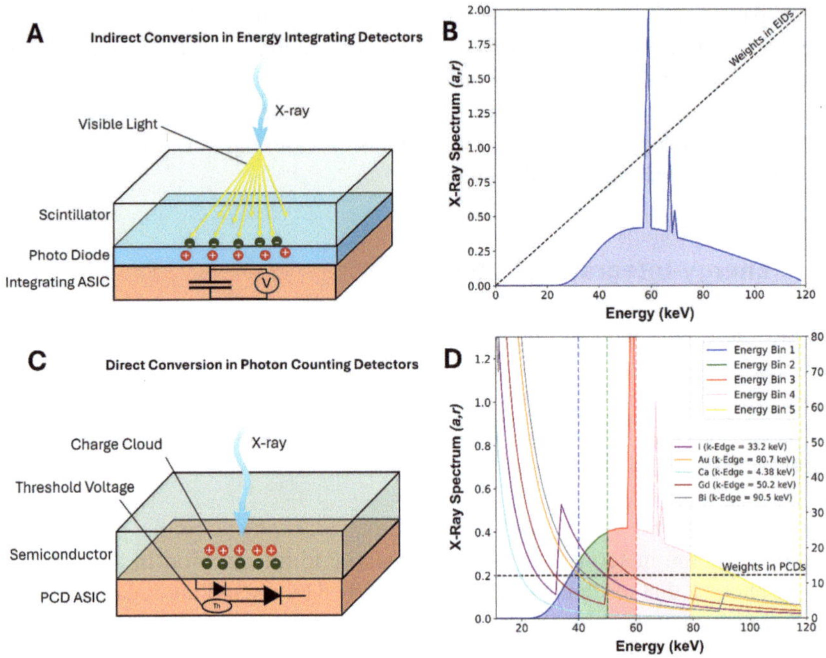

Fig. 1 The figure illustrates the mechanisms of energy-integrating detectors (EIDs) and photon-counting detectors (PCDs). (**a**) Depicts the indirect conversion process and signal generation in an EID using an integrating Application Specific Integrated Circuit (ASIC) that collects charges using a capacitor and then reads the subsequent potential using a voltmeter. (**b**) Uses SpekCalc generated spectrum to highlight the linear weighting applied to photons when using EIDs. (**c**) Demonstrates the direct signal generation in a PCD using a PCD ASIC that counts photons according to energy thresholds. (**d**) Illustrates how energy thresholds can be assigned and detected in PCDs, enabling K-edge imaging for contrast agents like Iodine (I), Gold (Au), Calcium (Ca), Gadolinium (Gd) and Bismuth (Bi)

2.2 Photon-Counting Detectors

In contrast to EIDs, PCDs represent a paradigm shift in X-ray detection technology. PCDs directly convert incoming photons to electrical signals as shown in Fig. 1c. Rather than integrating the energy of all photons over time, PCDs detect, count, and process individual photons as they arrive at the detector. Furthermore, they measure the energy of each photon, allowing for spectral discrimination. This capability is crucial for spectral imaging, enabling the simultaneous acquisition of images at user-defined multiple energy levels or bins (Kreisler 2022). Vertical bar lines in Fig. 1d show the distribution of the energy spectrum into energy bins. The overall mechanism of signal transport in PCDs is as follows:

1. *Direct Photon Detection:* PCDs allow X-ray photons to interact directly with semiconductor materials such as silicon for lower energies (<30 keV) and

cadmium telluride (CdTe), Gallium arsenide (GaAs) or cadmium-zinc-telluride (CZT) for higher energies. As a result of the photoelectric effect, each photon interaction generates electron-hole pairs, creating an electrical signal proportional to the photon's energy. Therefore, the otherwise necessary reflective septa between the individual pixels are also not required since no light photons need to be bundled into a detector cell. This enables precise energy discrimination for each detected photon.

2. *Pulse Processing and Counting:* The electronic system in PCDs processes the electrical pulse generated by each photon interaction. The system measures the pulse height, corresponding to the photon's energy, and registers it as a count within the appropriate energy threshold. Furthermore, PCDs count individual photons irrespective of their energy, assigning equal weighting, as illustrated by the horizontal bar line in Fig. 1d. By giving all detected photons equal contribution, PCDs preserve the intrinsic contrast differences between materials based on their attenuation coefficients. This capability ensures a more accurate representation of low-energy photons, which provide greater contrast information due to their higher interaction probability with soft tissues and other low-density materials.
3. *Formation of Spectral Data:* The spectral data acquired by PCDs is used to create multiple image sets corresponding to the user-defined energy thresholds, which are specific energy ranges that segment detected photons based on their energy levels.

PCDs offer several advantages over traditional EIDs. Their energy discrimination capability facilitates multi-energy imaging, a cornerstone of spectral CT, enabling advanced material characterization and quantitative imaging applications. PCDs also achieve superior spatial resolution due to their smaller detector element size. Another advantage is the use of user-defined multiple energy thresholds, which effectively eliminates electronic noise, enhancing both signal integrity and overall image quality. Furthermore, PCDs mitigate artefacts such as beam hardening by enabling more accurate spectral analysis and correction of the energy spectrum (Hsieh et al. 2021). Most importantly, PCDs allow for significant reductions in radiation dose while maintaining high image quality. This capability is indispensable for clinical applications, where patient safety is paramount, and for preclinical imaging, where preserving the integrity of biological tissues is critical for accurate research. They also allow for more accurate material decomposition (MD), aiding in the differentiation between calcium and iodine in vascular studies and distinguishing lipid-rich tissues and tumors in cancer imaging. Moreover, K-edge imaging, which targets materials' unique high-energy X-ray absorption thresholds (K-edge energies), leverages PCDs' energy-resolving capabilities to distinguish between materials based on the abrupt changes in X-ray attenuation at their characteristic K-edge energies (Wang and Pelc 2021; Jost et al. 2023). Unlike traditional EIDs, PCDs can sort X-ray photons into narrow energy bins, enabling accurate detection of high atomic number (Z) contrast agents like iodine and gadolinium, as shown in Fig. 1d (Schlomka et al. 2008). K-edge imaging is particularly advantageous in applications such as cancer imaging, where high-Z agents enhance tumor visualization, and in vascular studies, where iodine-based agents can be differentiated from calcified plaques. By utilizing K-edge detection, PCDs provide superior diagnostic accuracy while minimizing the need for multiple scans. Overall, with their numerous advantages over conventional EIDs, PCDs have become the cornerstone of SPCCT technology. Table 1 provides a summary of commercially available EID and PCDs based micro-CT systems.

3 Applications of Spectral Photon-Counting Micro-CT

3.1 Material Identification and Quantification

Spectral photon-counting micro-computed tomography (micro-SPCCT) enhances MD using energy-resolved imaging to differentiate materials based on distinct spectral signatures. This enables detailed analysis of complex structures, with applications in biomedical research, materials science, and the imaging of crystal-induced arthropathies. Micro-SPCCT is particularly valuable for analyzing tissue composition, characterizing material properties, and health conditions such as bone micro-cracks and overall bone health.

Conventionally, MD is achieved through basis decomposition, a method that represents a material's effective linear attenuation coefficient as a linear combination of material basis functions, as shown in Eq.3.1. Through this approach, micro-SPCCT can accurately identify and quantify materials, leading to improved accuracy and a deeper understanding of the underlying structures.

$$\mu(E) = a_1 f_1(E) + a_2 f_2(E) + \ldots\cdots + a_m f_m(E) \tag{3.1}$$

which can be simplified as:

$$\mu\,(y, E) = \sum\nolimits_m a_m(y) f_m(E) \tag{3.2}$$

whereas $\mu(E)/\mu(y, E)$ is the composite attenuation coefficient function of multi-energy data, E is the energy bin, y is the pixel position, $a_m/a_m(y)$ is the basis coefficients, and $f_m(E)$ is the basis function. For multi-energy CT imaging, Eq. 3.2 can be further expanded and rewritten in the matrix form as follows:

$$B = AX \tag{3.3}$$

On the right-hand side of Eq. 3.3, two distinct components are present. The first is the material-specific basis vectors (X), which mathematically represent the spectral properties of the materials in the system. The second is the basis coefficients (A), quantifying the relative contribution of each material within a specific voxel. On the left-hand side, B denotes the complete spectral attenuation map of the sample, encapsulating the combined attenuation behavior across all energy bins (Bateman et

Table 1 The table highlights the commercially available micro-CT scanners, comparing energy-integrating detectors (EIDs) and photon-counting detectors (PCDs) used in applications like biomedical research and materials science

Detector type	Scanner manufacturer	Number of energy bins	Field of view (mm)	Scan length (mm)	Voxel size (μm)	Peak tube voltage (kVp)	References
Energy integrating detectors	Bruker SkyScan	01	70–80	70–300	0.4–90	10–100	3D X-ray Microscopy
	Scanco μCT 40	01	36.9	80	3–72	30–70	Scanco MicroCT40 (2018)
	PerkinElmer	01	18–100	200–240	2.3–200	Max 90	PerkinElmer
	MiLabs U-CT	01	Up to 130	Up to 712	Max 2.4	65 & 80	U-CT (2025)
	Inviscan IRIS and IRIS-XL	01	>90	120	<30	–	IRIS
	Molecubes, X-cCube	01	35	63	50	–	Molecubes (2022)
	Sofie Gnext PET-CT	01	104	120	50	25–80	SOFIE
	Mediso nanoScan	01	100	120	30	Up to 80 W	Mediso
	Sedecal SuperArgus Compact PET/CT	01	100	50	50–100	–	SEDECAL (2024)
	MRSolutions MRS*PET/CT 80	01	112	80	Up to 18	40–90	Preclinical PET/CT (2021)
PCD	Mars Bioimaging Preclinical Spectral CT System	08	100	280	30–100	80–120	User (2024)

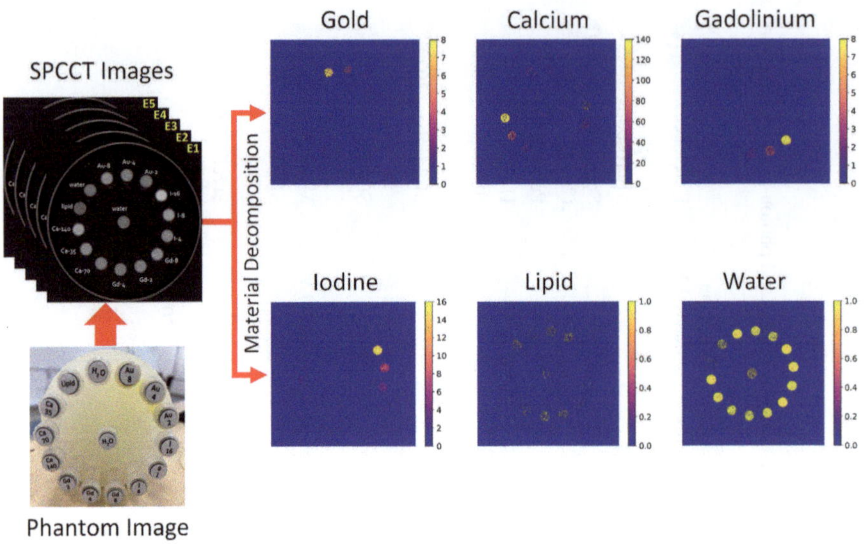

Fig. 2 Illustration of a multi-material phantom scanned using a PCD-based micro SPCCT system (Mars Microlab 5 × 120) at 118 kVp and 80 μA across five energy bins: 7–40 keV, 40–50 keV, 50–60 keV, 60–79 keV, and 79–118 keV. The density images of gold, calcium, gadolinium, iodine, lipid, and water are produced after the application of MD. The concentrations shown in the color bars are in mg/ml. The phantom image is slightly rotated clockwise compared to the SPCCT and material density images, due to differences in its placement within the scanner gantry

al. 2018). Solving this linear system involves employing advanced computational techniques such as the least square method (Ramsey 1969) or various iterative techniques (Chetih and Messali 2015; Chen et al. 2016), depending on the desired accuracy and computational efficiency. Figure 2 demonstrates the scanning of a multi-contrast phantom using an SPCCT system, followed by the application of a MD algorithm to identify and quantify the constituent materials. However, recently, the integration of artificial intelligence in MD techniques has demonstrated the potential to improve its results and eliminate the need for prior knowledge of materials and their concentrations. By leveraging information learned from large databases of CT images, artificial intelligence-driven MD methods can achieve more robust and data-driven results (Zhao et al. 2017; Wu et al. 2020; Mendonça et al. 2010).

3.2 Crystal Imaging

Micro-SPCCT has the potential to advance the assessment of crystal arthropathies and imaging of metallic implants. It is particularly effective at visualizing and differentiating crystal structures based on their spectral characteristics, making it valuable in fields such as geology, mineralogy, and materials science for analyzing crystalline compositions. In medical applications, it has demonstrated the ability to detect and quantify monosodium urate (MSU) crystal deposits in gout-affected fingers ex vivo and to identify and distinguish octacalcium phosphate (OCP) and Calcium pyrophosphate (CPP) crystals in osteoarthritic conditions from calcium hydroxyapatite (CHA) crystal deposits (Stamp et al. 2019; Mourad et al. 2024). Additionally, it shows potential for distinguishing between MSU and CPP, MSU and CHA, CPP and OCP, as well as OCP and CHA in vitro, further broadening its diagnostic applications for crystal arthropathies (Becce et al. 2019; Viry et al. 2018), as shown in Fig. 3a.

3.3 Cardiovascular Diseases

Micro-SPCCT is emerging as a useful imaging tool in cardiovascular research, as shown in Fig. 3b. It offers high spatial resolution with the ability to differentiate multiple soft tissues simultaneously. This technology enables detailed visualization of cardiovascular structures and microvascular pathologies, effectively bridging the gap between preclinical studies and clinical research (Hagen et al. 2023). In coronary artery and plaque assessment, micro-SPCCT provides differentiation of calcified plaques, vessel lumens, and necrotic cores. Its ability to reduce blooming artefacts due to smaller pixel sizes ensures accurate detection and quantification of coronary calcifications, along with detailed visualization of plaque features essential for evaluating stability, lesion severity, and cardiovascular risk (Dahal et al. 2023). Furthermore, its high spatial resolution optimizes coronary stent imaging, enabling improved assessment of stent patency and detection of in-stent restenosis. These advancements are crucial for managing coronary artery disease and monitoring patients' post-percutaneous coronary interventions (Li et al. 2023; Pack et al. 2024; Flohr et al. 2023; Meloni et al. 2023). The advanced material differentiation capabilities of multi-energy CT systems also enable myocardial tissue characterization, facilitating the characterization of fibrosis, infiltration, and contrast agents — key for diagnosing and managing cardiomyopathies. Additionally, it supports the study of myocardial blood perfusion, offering a comprehensive evaluation of cardiac structures and functions previously unattainable with conventional imaging methods (Meloni et al. 2024).

3.4 Molecular Imaging

Micro-SPCCT technology is also instrumental in advancing drug discovery, developing targeted imaging agents, and enabling personalized treatments, as shown in Fig. 3c–e. Its enhanced contrast and material-specific analysis make it a transformative tool for diagnosing and treating conditions such as cancer, infectious diseases, atherosclerosis, and disorders affecting bone, cartilage, and joint health (Moghiseh et al. 2023a; Mutreja et al. 2023). The technology is transforming cancer imaging by using high atomic number (high-Z) nanoparticles (refer to

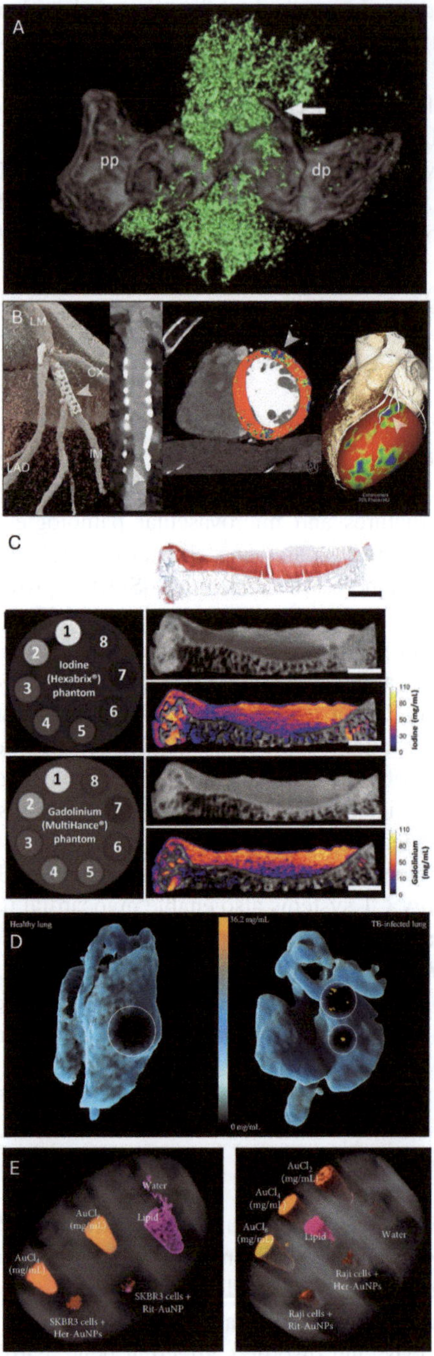

Fig. 3 Different applications of spectral photon-counting CT (SPCCT): (**a**) Crystal Imaging: Detection of crystal deposits and improved spatial resolution in a toe specimen after amputation due to infection and gout. Volume rendering reconstructions with dorsal views (proximal phalanx

Sect. 3.5 for more details), such as gold, hafnium, and bismuth, to enhance soft tissue contrast and enable material differentiation for early tumor detection and characterization. Functionalized nanoparticles with targeting ligands facilitate the imaging of molecular markers, enabling the detection of cellular-level changes and supporting personalized treatment strategies. Dual-contrast imaging with agents like iodine and gadolinium further allows comprehensive assessment of vascular and extracellular tumor spaces, advancing cancer diagnosis and therapy (Moghiseh et al. 2023a; Mutreja et al. 2023; Badea et al. 2019). Furthermore, micro-SPCCT can also assess and track the distribution of therapeutic agents, such as cisplatin-loaded NPs, within tumors. This monitoring ensures the effective delivery and accumulation of therapeutic agents at the target site, optimizing treatment efficacy (Badea et al. 2022).

3.5 Spectral Assessment of New Contrast Agents and Nanoparticles

Contrast agents are essential in enhancing the diagnostic accuracy of CT based imaging techniques by improving the visualization of internal structures for better differentiation. Recent studies have demonstrated that micro-SPCCT systems are well suited for assessing high-atomic number-based contrast agents and nanoparticles. These contrast agents and nanoparticles, typically derived from heavy metal elements ($Z \geq 53$), exploit energy discrimination features, particularly through K-edge imaging as described in Sect. 2.2, enabling multi-material characterization simultaneously. Key candidates for spectral CT-specific contrast agents include ele-

Fig. 3 (continued) (pp), distal phalanx (dp)) show monosodium urate tophi (green). The white arrow indicates juxta-articular erosion with typical overhanging edges; (**b**) Cardiovascular Imaging: Spectral imaging of coronary arteries and the left ventricle in a patient with chronic pulmonary embolism, showcasing: (left) a 3D-cinematic rendering, (middle-left) significant in-stent restenosis causing segmental perfusion delay/defect, (middle-right) a first-pass rest perfusion image with a color-coded overlay, and (right) a 3D volume-rendered coronary anatomy with a color-coded perfusion map highlighting the correspondence between restenosis and the perfusion defect (arrowhead); (**c**) Molecular Imaging in Cartilage Health: SPCCT visualization of a lateral human tibial plateau, (top) A histological image stained with Safranin O (red) to indicate glycosaminoglycan presence, (left) Iodine and gadolinium phantoms with different concentrations for reference, (right-grayscale) Single-energy CT-like images of samples incubated in iodine and gadolinium contrast agents (right-colored) Material-decomposed MARS images showing iodine and gadolinium concentrations in the samples. Scale bars indicate 5 mm; (**d**) Molecular Imaging of Pulmonary Tuberculosis (TB): SPCCT 3D visualization compares a healthy lung (left, no iodine uptake, white circle) with a TB-infected lung (right, iodine uptake in infection lesions, white circle); (**e**) Molecular Imaging in Cancer: SPCCT MD images of phantoms depict gold (yellow), lipid (pink), and water-like material (gray), with color intensity reflecting material concentration. (Images in panels **a–e** are adapted from (Mourad et al. 2024; Meloni et al. 2024; Baer et al. 2021; Lowe et al. 2021; Moghiseh et al. 2018), respectively, all under a Creative Commons Attribution License (CC BY 4.0)) ments with well-defined K-edges, such as iodine, gadolinium, holmium, ytterbium, tantalum, tungsten, gold, and bismuth. A detailed description of some of these contrast agents is given in Table 2. Each element offers distinct imaging advantages and poses specific technical and biological considerations such as biodistribution, pharmacokinetics, and tissue targeting, making them valuable for tailored imaging applications. For instance, gold nanoparticles (AuNP) have been proposed for various biomedical applications, including photoacoustic imaging, photothermal ablation, optical imaging, and cellular tracking. Their biocompatibility and high contrast capability make them promising spectral micro-CT contrast agents. The higher atomic number of gold enhances X-ray attenuation, providing excellent contrast in small animal studies. However, its higher cost may restrict its widespread clinical use (Ibrahim et al. 2024; Cormode et al. 2017). Similarly, gadolinium (Gd), known for its strong paramagnetic properties, has been extensively studied for material differentiation in phantom imaging. Its K-edge energy of 50.2 keV aligns closely with the average photon energy in clinical X-ray spectra, making it an optimal choice for both preclinical and clinical applications (Si-Mohamed et al. 2018). Furthermore, tantalum (K-edge – 67.3 keV) and ytterbium (61.3 keV) exhibit K-edge energies well suited for spectral CT imaging, enabling reduced noise, high contrast-to-noise ratio, and improved MD (Oh et al. 2011; Liu et al. 2012). Hafnium (Hf), with a K-edge of 65 keV offers potential as a contrast agent due to its biocompatibility and targeted delivery capabilities, making it a promising tool for diagnosing conditions such as atherothrombosis (McGinnity et al. 2016). Moreover, bismuth ($Z = 83$) is notable for its high atomic number, chemical stability, and cost-effectiveness, making it a viable candidate for various spectral assessments (Mutreja et al. 2023; Pan et al. 2010; FitzGerald et al. 2016). Different applications of contrast agents are shown in Fig. 4.

3.6 Bone Assessment – Composition and Microarchitecture

Three-dimensional site-specific evaluation of bone microarchitecture, composition, and strength through the energy-discriminating capabilities of Micro-SPCCT is an emerging area of research. The technology has the potential to exceed conventional techniques such as DXA (dual-energy X-ray absorptiometry), which only provides areal density followed by the population-based T-score, and high-resolution peripheral quantitative computed tomography (HR-pQCT), which is limited to only peripheral sites. SPCCT provides volumetric bone mineral density (vBMD) at high spatial resolution to assess the overall bone health and pathology. Bone diseases such as osteoporosis, caused by reduced BMD and structural degradation, can benefit from SPCCT's volumetric measurements and analysis of trabecular and cortical bone, supporting efficient diagnosis and therapy assessment as shown in Fig. 5. Furthermore, the characterization of cartilage health by assessing its thickness, composition, and underlying subchondral bone and evaluating regenerative therapies are also important areas of interest for micro-SPCCT technology for lab assessments (Ostadhossein et al. 2020; Bhattarai et al. 2023).

Table 2 Summary of elements and their properties for SPCCT contrast agents

Element (symbol)	Atomic number (*Z*)	K-edge (keV)	Density (g/cm^3)	Development stage of contrast agents
Iodine (I)	53	33.17	4.9	Iodinated contrast agents are clinically available. Well- studied in human subjects; commonly used as a control for contrast-enhanced CT imaging (Badea et al. 2022; Gaikwad et al. 2018; Panta et al. 2018).
Gadolinium (Gd)	64	50.24	7.9	Gadolinium is clinically approved for MRI imaging and are currently undergoing clinical trials for low-dose CT imaging in CT imaging and SPCCT imaging (Pan et al. 2010; Panta et al. 2018; Feuerlein et al. 2008).
Ytterbium (Yb)	70	61.33	6.9	Several preclinical CT and SPCCT imaging with small animals (Liu et al. 2012; Kim et al. 2018).
Hafnium (Hf)	72	65.35	13.3	Experimental study in Preclinical SPCCT imaging with small animals (McGinnity et al. 2016; Maggiorella et al. 2012).
Tantalum (Ta)	73	67.42	16.4	Preclinical CT and SPCCT imaging with small animals (Kim et al. 2018; Riederer et al. 2019).
Gold (Au)	79	80.72	19.3	Most well-studied and characterized for CT and SPCCT contrast agent in Preclinical CT and SPCCT imaging with small animals (Panta et al. 2018; Ashton et al. 2014).
Bismuth (Bi)	83	90.53	9.8	Numerous studies for CT preclinical CT and SPCCT imaging with small animals (FitzGerald et al. 2016; Kim et al. 2018; Fu et al. 2020).

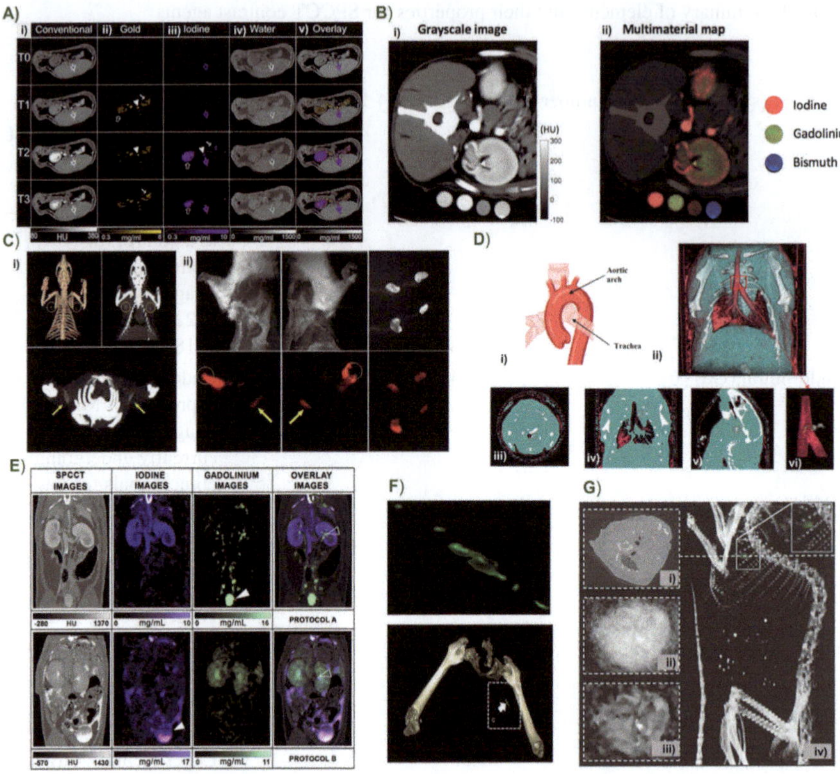

Fig. 4 (**a**) Spectral Photon-Counting CT (SPCCT) images of a rabbit chest injected with gold nanoparticles (AuNP) and an iodinated contrast agent at several time points (i). While conventional images only reveal changes in attenuation, material-specific SPCCT images (ii and iii) differentiate and visualize the distinct distributions of AuNP and iodine within the chest. (Adapted from (Cormode et al. 2017) under a Creative Commons Attribution License (CC BY 3.0)); (**b**) PCD grayscale image reconstructed using all detected photons without energy discrimination at the left pelvis, highlights the dense contrast material within the calibration vials, abdominal aorta, and kidney. In single-energy CT, distinguishing between different contrast agent is not feasible (i). The PCD multimaterial map successfully differentiates contrast agents, with iodine appearing in red, gadolinium in green, and bismuth in blue (ii). (Reproduced with minor reformatting of panels a and b of Fig. 3 from (Symons et al. 2017) with permission from John Wiley and Sons. © 2017 American Association of Physicists in Medicine); (**c**) Sentinel lymph node mapping, (i) CT images 2 h post-injection of PEG-RITC-TaOx (210 mg/mL) show lymph nodes (yellow circles/arrows). (ii) Fluorescence and white light images highlight lymph nodes and injection sites, with dissected nodes shown (right). (Adapted with permission from (Oh et al. 2011). Copyright 2011, American Chemical Society); (**d**) (i) Anatomy of the mouse heart and aortic arch (ii) 3D visualization of fused MD images of the mouse heart and aorta including energy image at E1 (pink), HA channel (white), and soft tissue (blue), (iii–v) axial and coronal slices with AuNPs, (vi) MARS fused MD images of the atherosclerotic plaque showing E1 (pink), HA (white), and AuNPs (yellow). (Adapted from (Moghiseh et al. 2023b) under a Creative Commons Attribution License (CC BY 4.0)); (**e**) Coronal oblique SPCCT images of a rat injected with gadolinium chelate and iodine

3.7 Metal Artefact Assessment

Micro SPCCT is highly effective for assessing and mitigating metal artefacts-related studies. The technology is particularly useful in reducing beam hardening, photon starvation, and scattering effects which are commonly associated with metallic implants. As photons travel through materials of varying densities within the scanned object, a specific attenuation of low-energy photons causes beam hardening. This effect can produce significant visual artefacts, especially near high-attenuation materials such as metals, leading to streaks of dark and bright regions across the image that obscure diagnostic information. The intensity of metal artefacts can fluctuate based on the atomic number, density, and structure of the metal, in addition to the surrounding anatomy and scanned acquisition/reconstruction parameters. Low-energy photons, being more subjected to attenuation, provide detailed data regarding soft tissue contrast. Conversely, high-energy photons penetrate thick materials with greater resistance, diminishing the beam-hardening effects that produce artefacts. By integrating low and high-energy photon data, spectral CT augments image quality and enhances quantification precision for both soft tissues and bone adjacent to metallic structures. High-energy bins improve the precision of assessing tissue densities and material composition adjacent to the scaffold, which is essential for evaluating structural integrity and identifying problems (Tariq et al. 2024; Rajendran et al. 2014; Zhou et al. 2019; Gutjahr et al. 2016; Khan et al. 2024). Figures 6 and 7 show an example of metal artefact reduction utilizing PCD-CT. These novel advancements make spectral technology invaluable for applications such as evaluating orthopedic implants, dental restorations, and metallic nanoparticles in biomedical research, ensuring accurate quantitative and qualitative assessments despite the presence of high-Z materials.

4 Challenges and Limitations

Spectral CT has proven to be a great leap forward in advancing imaging capabilities, but several challenges and limitations hinder its widespread adoption and progression. Addressing these obstacles is essential to unlocking this technology's

Fig. 4 (continued) clearly differentiate their distinct distributions in material-specific images. (Reproduced from (Si-Mohamed et al. 2018), with permission from Wolters Kluwer Heath, Inc. https://journals.lww.com/investigativeradiology/pages/default.aspx); **(f)** SPCCT image of a rabbit showing bismuth signal from an in situ clot (white arrow, left) and enlarged highlighted area (top). (Reproduced with minor reformatting of panels b and c of Fig. 4 from (Pan et al. 2010) with permission from John Wiley and Sons); **(g)** Mouse blood pool imaging following Yb nanocolloid (6 ml/kg) bolus. (i) Heart pseudo-conventional CT slice. (ii–iii) Reconstruction of Yb signal after 1 and 20 repetitions. (iv) Volume-rendered CT with Yb signal (green) demonstrates heart buildup and bone separation. (A adapted with permission from (Pan et al. 2012). Copyright 2011, American Chemical Society)

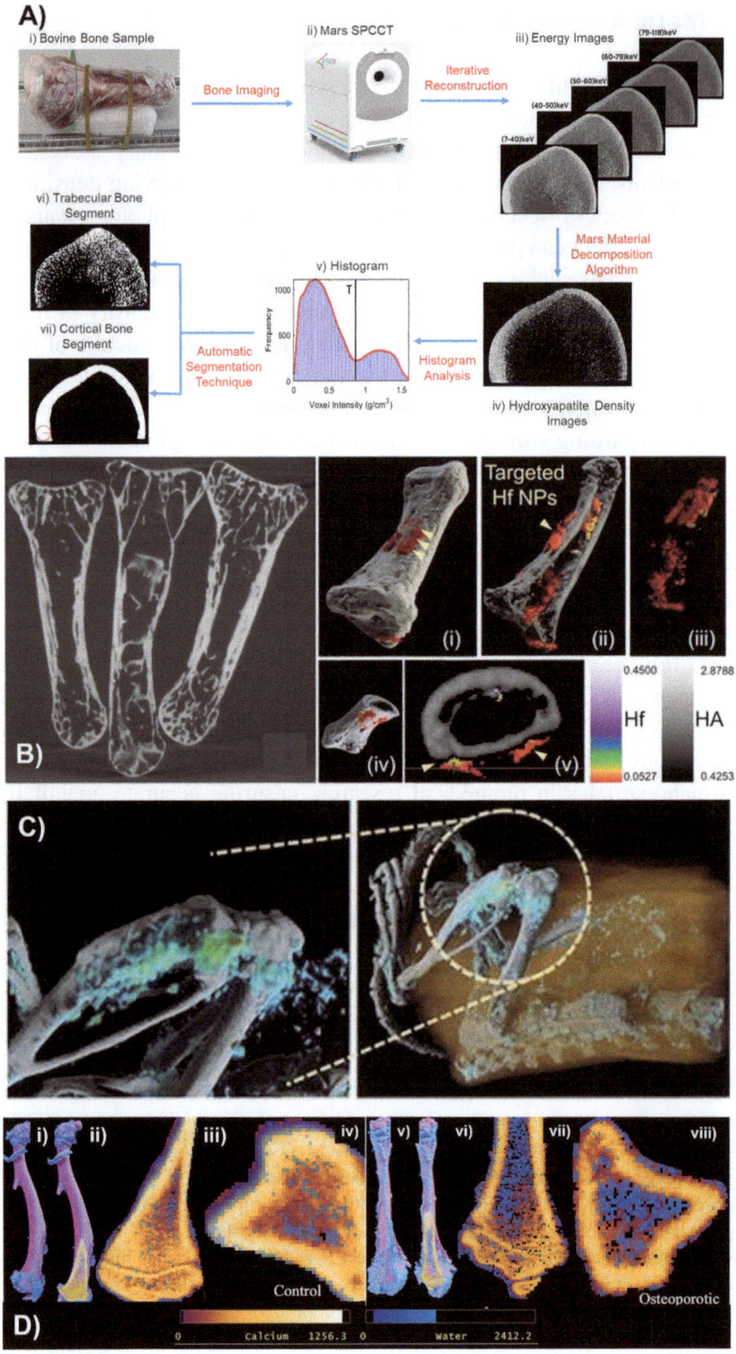

Fig. 5 (**a**) Bone imaging and analysis of a bovine bone sample using MARS SPCCT: The bone sample was scanned with MARS SPCCT and reconstructed across five energy bins. Hydroxyapatite density images were generated using the MARS MD algorithm applied to the multi-energy

full potential and advancing its preclinical, biomedical, and clinical imaging applications.

4.1 Technical Limitations

Despite the benefits, spectral photon-counting-based CT technology is not widely available due to several technical constraints. Compared to conventional CT, which has had 40+ years of development, SPCCT is a relatively new technology which still has many limitations. PCDs rely on high-Z semiconductor materials for their superior energy resolution, but these materials are expensive and often difficult to manufacture and integrate into large detector arrays. The fabrication process requires precise engineering, and only a few companies have managed to produce large-scale PCD arrays with reliable and effective performance (Gallego Manzano et al. 2023; Stephens 2021), further driving up costs and limiting availability. Energy measurements are also distorted by technical issues like charge sharing, which reduces resolution by allowing signals to diffuse across adjacent pixels; Compton scattering, which records secondary photons as distinct events; K-escape effects, which are brought on by fluorescence in the detector material; and pile-up effects, which cause closely arriving photons to be miscounted. The spatial and energy resolution required for precise material differentiation is impacted by these problems, underscoring the financial and technological obstacles to expanding and enhancing PCD-based imaging systems (Greffier et al. 2024).

Fig. 5 (continued) data. Histogram analysis enabled the segmentation of bone tissues, allowing trabecular and cortical bone to be distinguished through specific thresholding techniques. (**b**) Micro-CT imaging of bones in three conditions: clean, soaked in a specific Hf solution, and soaked in an 'as-prepared' Hf nanoparticle solution. (i–v) exhibit SPCCT 3D images of ex vivo human bone microdamage, emphasizing the localization of targeted Hf nanoparticles (red) and bone (white/gray) on a color scale. (Reproduced with minor reformatting (Ostadhossein et al. 2020) with permission from John Wiley and Sons); (**c**) In vivo bone microdamage evaluation and Hf NP differentiation using SPCCT: Bone appears in gray, calcium in white, high-concentration Hf NPs in cyan-green, and low-concentration Hf NPs in blue. SPCCT enables precise localization of microdamage and differentiation of Hf NPs from bone and surrounding tissues. (Reproduced with minor reformatting of panel d of Fig. 8 (Ostadhossein et al. 2020) with permission from John Wiley and Sons); (**d**) Calcium and water distribution maps of the tibia in control and ovariectomized samples. (i and ii) present 3D views of the calcium and water distribution in the control sample, while (iii) and (iv) show sagittal and axial views highlighting the inner trabeculae surrounded by dense cortical bone. (v, vi) provide 3D views of the ovariectomized sample, with sagittal (vii) and axial (viii) views depicting water and calcium distribution in the trabeculae and surrounding dense cortical bone. The ovariectomized sample exhibits lower calcium and higher free water content compared to the control. (Adapted from (Bhattarai et al. 2023) under a Creative Commons Attribution License (CC BY 4.0))

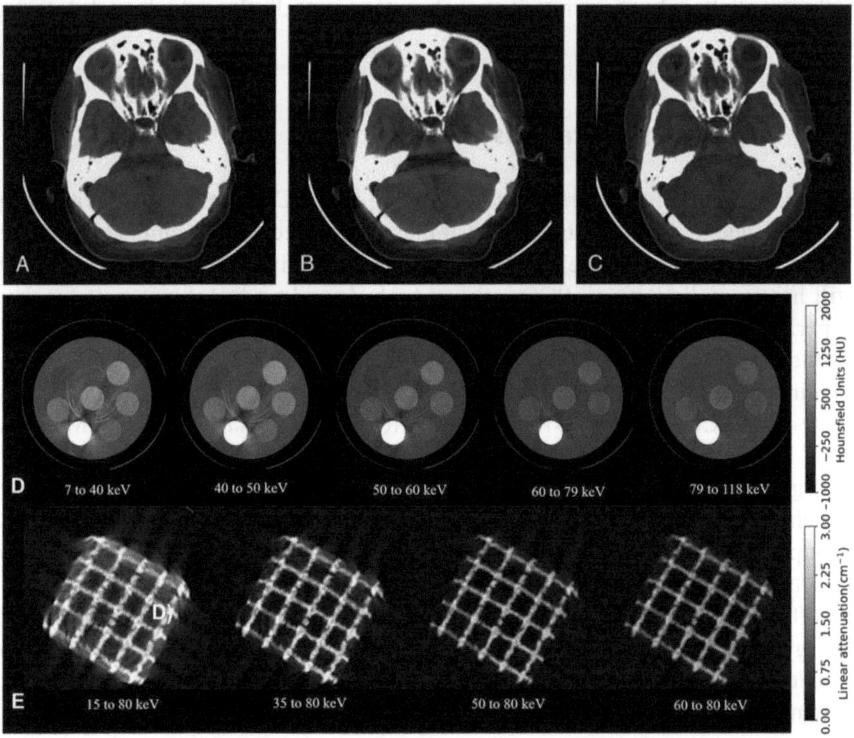

Fig. 6 Images of a cadaver head. Energy-integrating detector images (**a**), PCD images (low-energy bin; **b**), and PCD images (high-energy bin; **c**). The high-energy threshold images from the PCD-CT (**c**) demonstrate reduced blooming and beam-hardening artefacts compared to the low-energy threshold PCD-CT images. (Panel **a**–**c** are reproduced from (Gutjahr et al. 2016), with permission from Wolters Kluwer Heath, Inc. https://journals.lww.com/investigativeradiology/pages/default.aspx); (**d**) The phantom with an aluminum (Al) insert was scanned using a MARS SPCCT scanner. The phantom scan was conducted at 118 kVp and 80 μA across five energy bins: 7–40 keV, 40–50 keV, 50–60 keV, 60–79 keV, and 79–118 keV. (Adapted from (Tariq et al. 2024) under a Creative Commons Attribution License (CC BY 4.0)); (**e**) The titanium scaffold scan was performed at 80 kVp and 60 μA across four energy bins: 15–80 keV, 35–80 keV, 50–80 keV, and 60–80 keV. Low-energy ranges of 7–40 keV in (**d**), and 15–80 keV in (**e**) produced bright streaks affecting non-metal regions, while high-energy ranges of 7–40 keV in (**d**), and 60–80 keV in (**e**), showed reduced artefacts. (Panel **e** is adapted from (Rajendran et al. 2014) under a Creative Commons Attribution License (CC BY 3.0))

4.2 Challenges in Image Processing and Data Analysis

Micro-SPCCT produces high-resolution multi-energy images, which, while useful for in-depth study, also need significant processing time, storage, and computational power. Many research labs or universities struggle to maintain high-performance computing specifications and advanced storage solutions to deal with such large image files. Even basic image processing tasks can consume substantial time and

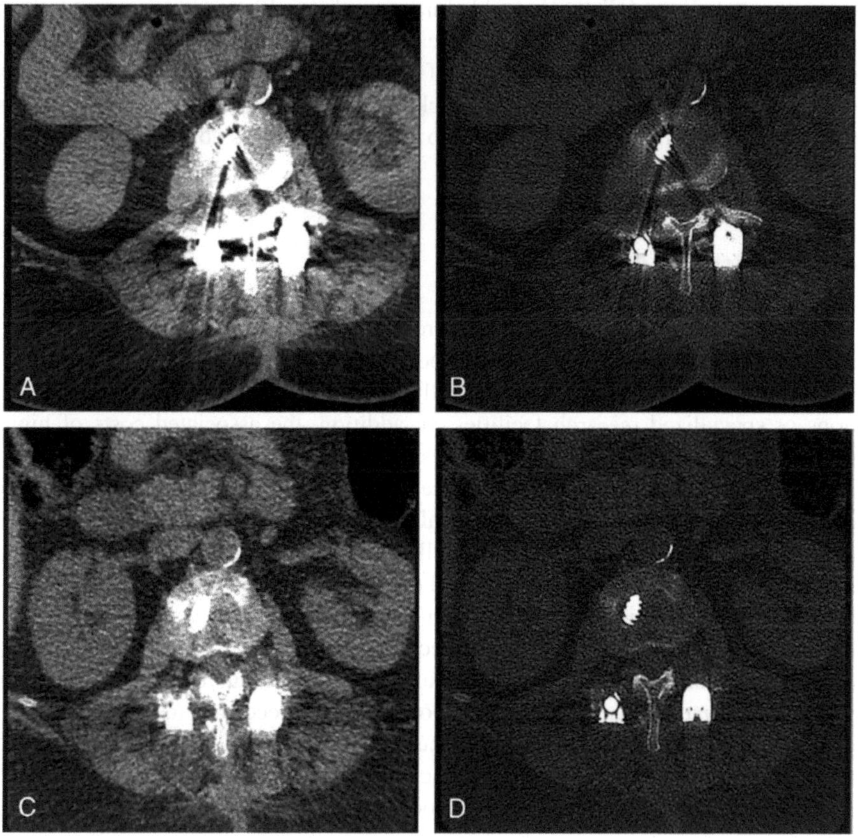

Fig. 7 59-year-old female with posterior rod and screw fixation of L2-L5 and a posterolateral bone graft demonstrating improved visualization of critical anatomic structures of the spine with threshold high (TH) Sn140 kV PCD-CT. Axial images from both EID-CT at 120 kV (**a** and **b**) and TH Sn 140 kV PCD-CT (**c** and **d**) were examined for comparative analysis. The TH Sn 140 kV PCD-CT pictures (**c** and **d**) had diminished metal artefacts relative to the EID-CT images (**a** and **b**), facilitating enhanced visualization of the central canal, neural foramina, and nerve roots, which exhibited the most significant improvement in this investigation. Image settings: (**a** and **c**) Window/Level = 400/40 HU; (**b** and **d**) Window/Level = 1500/450 HU. (Adapted from (Zhou et al. 2019), with permission from Wolters Kluwer Heath, Inc. https://journals.lww.com/investigativeradiology/pages/default.aspx)

resources, slowing work and increasing costs. These computing demands rise even more when spectral CT is combined with deep learning for applications such as automated MD or metal artefact reduction. Deep learning models thrive on comprehensive datasets, but spectral CT's high-resolution imaging increases dataset sizes to an extent that can overwhelm data handling and processing systems. This makes it difficult to develop robust yet lightweight models that run efficiently on existing hardware without compromising performance. Additionally, the absence of open-source spectral CT datasets adds further to the problem. Creating customized datasets for specific deep learning applications is time-intensive, costly, and technically demanding, limiting the ability of researchers to refine existing approaches or experiment with new approaches quickly. Overcoming these obstacles requires the creation of datasets that are easier to access, as well as advancements in data management and hardware efficiency.

4.3 Cost and Accessibility

The high cost of spectral micro CTs remains one of the key obstacles to its accessibility and adoption in the labs. Spectral micro-CT systems and the required hardware for processing images are costly, limiting access to well-funded institutions or specialized research facilities. In addition, the associated costs of high-performance computing resources necessary for spectral CT image processing create an added financial burden. Accessibility is further limited by the lack of publicly accessible datasets and qualified experts in SPCCT, which hinders opportunities for knowledge sharing, collaboration, and training in this small yet impactful area. The limited literature and the need for expensive technology means that advancing expertise and efficiency in SPCCT can be slow, particularly for new researchers entering the field. Reducing equipment and processing costs, improving access to resources and datasets, and cultivating a community of PCCT experts who can innovate and advance the technology are all necessary to overcome these obstacles. These limitations underscore the need for increased funding and resource-sharing initiatives to enable the broader scientific and medical community to utilize the full capabilities of photon-counting spectral CT.

5 Conclusions and Future Perspectives

Spectral photon-counting CT represents a transformative advancement in imaging technology, offering unprecedented capabilities in material differentiation, reduced radiation doses, and enhanced image quality. The integration of PCDs into micro-CT systems has addressed many limitations of conventional EIDs, such as low energy resolution, reduced spatial resolution, and susceptibility to artefacts. By directly converting X-ray photons into electrical signals and measuring their energy, PCDs enable multi-energy imaging, which is crucial for advanced material characterization and quantitative imaging applications. However, several challenges and limitations must be addressed to unlock the full capabilities of PCDs-incorporated SPCCT. Issues such as charge sharing, pile-up effects, and Compton scattering can distort energy measurements and reduce spatial resolution. Future research should focus on developing solutions to mitigate these effects and enhance detector performance.

Looking forward, several key areas hold promise for the continued evolution and application of SPCCT technology. Future research should also focus on improving the efficiency and cost-effectiveness of high-Z semiconductor materials used in PCDs. Innovations in fabrication techniques could lead to more reliable and scalable production of large detector arrays, making SPCCT systems more accessible and affordable. Furthermore, the high-resolution multi-energy images SPCCT produces require significant processing power and storage capacity. Advances in data management, compression techniques, and high-performance computing resources are needed to handle large datasets efficiently. Moreover, the integration of AI and machine learning algorithms with SPCCT data processing can significantly enhance image reconstruction, MD, and artefact reduction. Developing computationally lightweight deep learning models tailored for spectral CT data could streamline complex analyses and democratize access to this technology.

Clinically, SPCCT's potential for precise tissue characterization and early disease detection is exemplified by recent studies highlighting its utility in cardiovascular and oncological imaging. Future clinical studies can further explore its potential in other areas, such as neuroimaging, musculoskeletal imaging, and metabolic disorders, to fully realize its diagnostic capabilities. Emerging preclinical applications, such as dynamic contrast agent tracking and molecular imaging, underscore SPCCT's versatility. Continued research in these areas can lead to new knowledge into disease mechanisms, drug development, and personalized medicine. The high cost of SPCCT systems (whether laboratory or clinical) and the associated hardware for image processing remain significant barriers to widespread adoption. Efforts to reduce equipment and processing costs, improve resource access, and cultivate a community of SPCCT experts are essential for broader utilization. The lack of publicly accessible spectral CT datasets hinders developing and validating new algorithms and applications. Creating and sharing open-source datasets will facilitate collaboration and innovation in the field. By addressing the current challenges and leveraging future advancements, SPCCT can redefine the boundaries of precision medicine, materials science, and beyond.

Acknowledgments This work was supported by the Research and Innovation Grant, Khalifa University of Science and Technology, Abu Dhabi, United Arab Emirates, under Grant 8474000563.

Conflict of Interest Aamir Younis Raja is a shareholder of MARS Bioimaging Ltd. Others have no financial holding in the company nor any other conflict of interest to disclose.

References

3D X-ray Microscopy Ease-of-Use. SKYSCAN 1275. ruker. https://www.bruker.com. Ac-cessed 26 March 2025

Ashton JR, Clark DP, Moding EJ, Ghaghada K, Kirsch DG, West JL, Badea CT (2014) Dual-energy micro-CT functional imaging of primary lung cancer in mice using gold and iodine nanoparticle contrast agents: a validation study. PLoS One 9:e88129. https://doi.org/10.1371/journal.pone.0088129

Badea CT, Clark DP, Holbrook M, Srivastava M, Mowery Y, Ghaghada KB (2019) Functional imaging of tumor vasculature using iodine and gadolinium-based nanoparticle contrast agents:

a comparison of spectral micro-CT using energy integrating and photon counting detectors. Phys Med Biol 64:065007. https://doi.org/10.1088/1361-6560/ab03e2

Badea CT, Clark DP, Allphin A, Ghaghada KB, Mowery YM (2022) Photon counting micro-CT for imaging cisplatin. In: Gimi BS, Krol A (eds) Medical imaging 2022: biomedical applications in molecular, structural, and functional imaging. SPIE, p 55. https://doi.org/10.1117/12.2612574

Baer K, Kieser S, Schon B, Rajendran K, ten Harkel T, Ramyar M, Löbker C, Bateman C, Butler A, Raja A, Hooper G, Anderson N, Woodfield T (2021) Spectral CT imaging of human osteoarthritic cartilage via quantitative assessment of glycosaminoglycan content using multiple contrast agents. APL Bioeng 5. https://doi.org/10.1063/5.0035312

Bateman CJ, Knight D, Brandwacht B, Mahon JM, Healy J, Panta R, Aamir R, Rajendran K, Moghiseh M, Ramyar M, Rundle D, Bennett J, de Ruiter N, Smithies D, Bell ST, Doesburg R, Chernoglazov A, Mandalika VBH, Walsh M, Shamshad M, Anjomrouz M, Atharifard A, Vanden Broeke L, Bheesette S, Kirkbride T, Anderson NG, Gieseg SP, Woodfield T, Renaud PF, Butler APH, Butler PH (2018) MARS-MD: rejection based image domain material decomposition. J Instrumentation 13:P05020. https://doi.org/10.1088/1748-0221/13/05/P05020

Becce F, Viry A, Stamp LK, Pascart T, Budzik J-F, Raja A (2019) Winds of change in imaging of calcium crystal deposition diseases. Joint Bone Spine 86:665–668. https://doi.org/10.1016/j.jbspin.2019.04.005

Bhattarai A, Tanaka R, Yeung AWK, Vardhanabhuti V (2023) Photon-counting CT material decomposition in bone imaging. J Imaging 9:209. https://doi.org/10.3390/jimaging9100209

Chen Y, Zhang Y, Zhang K, Deng Y, Wang S, Zhang F, Sun F (2016) FIRT: filtered iterative reconstruction technique with information restoration. J Struct Biol 195:49–61. https://doi.org/10.1016/j.jsb.2016.04.015

Chetih N, Messali Z (2015) Tomographic image reconstruction using filtered back projection (FBP) and algebraic reconstruction technique (ART). In: 2015 3rd International Conference on Control, Engineering & Information Technology (CEIT). IEEE, pp 1–6. https://doi.org/10.1109/CEIT.2015.7233031

Cormode DP, Si-Mohamed S, Bar-Ness D, Sigovan M, Naha PC, Balegamire J, Lavenne F, Coulon P, Roessl E, Bartels M, Rokni M, Blevis I, Boussel L, Douek P (2017) Multicolor spectral photon-counting computed tomography: in vivo dual contrast imaging with a high count rate scanner. Sci Rep 7:4784. https://doi.org/10.1038/s41598-017-04659-9

Dahal S, Raja AY, Searle E, Colgan FE, Crighton JS, Roake J, Saba L, Gieseg S, Butler APH (2023) Correction to: components of carotid atherosclerotic plaque in spectral photon-counting CT with histopathologic comparison. Eur Radiol 33:9479–9479. https://doi.org/10.1007/s00330-023-09849-w

Elliott JC, Davis GR, Dover SD (2008) In: Stock SR (ed) X-ray microtomography: past and present, p 707803. https://doi.org/10.1117/12.793959

Feuerlein S, Roessl E, Proksa R, Martens G, Klass O, Jeltsch M, Rasche V, Brambs H-J, Hoffmann MHK, Schlomka J-P (2008) Multienergy photon-counting K-edge imaging: potential for improved luminal depiction in vascular imaging. Radiology 249:1010–1016. https://doi.org/10.1148/radiol.2492080560

FitzGerald PF, Colborn RE, Edic PM, Lambert JW, Torres AS, Bonitatibus PJ, Yeh BM (2016) CT image contrast of high- *Z* elements: phantom imaging studies and clinical implications. Radiology 278:723–733. https://doi.org/10.1148/radiol.2015150577

Flohr T, Schmidt B, Ulzheimer S, Alkadhi H (2023) Cardiac imaging with photon counting CT. Br J Radiol 96:20230407

Fu J, Guo J, Qin A, Yu X, Zhang Q, Lei X, Huang Y, Chen M, Li J, Zhang Y, Liu J, Dang Y, Wu D, Zhao X, Lin Z, Lin Y, Li S, Zhang L (2020) Bismuth chelate as a contrast agent for X-ray computed tomography. J Nanobiotechnol 18:110. https://doi.org/10.1186/s12951-020-00669-4

Gaikwad HK, Tsvirkun D, Ben-Nun Y, Merquiol E, Popovtzer R, Blum G (2018) Molecular imaging of cancer using X-ray computed tomography with protease targeted iodinated activity-based probes. Nano Lett 18:1582–1591. https://doi.org/10.1021/acs.nanolett.7b03813

Gallego Manzano L, Monnin P, Sayous Y, Becce F, Damet J, Viry A (2023) Clinical commissioning of the first point-of-care spectral photon-counting CT for the upper extremities. Med Phys 50:2844–2859. https://doi.org/10.1002/mp.16313

Greffier J, Viry A, Robert A, Khorsi M, Si-Mohamed S (2024) Photon-counting CT systems: a technical review of current clinical possibilities. Diagn Interv Imaging 106:53–59. https://doi.org/10.1016/j.diii.2024.09.002

Gutjahr R, Halaweish AF, Yu Z, Leng S, Yu L, Li Z, Jorgensen SM, Ritman EL, Kappler S, McCollough CH (2016) Human imaging with photon counting–based computed tomography at clinical dose levels. Investig Radiol 51:421–429. https://doi.org/10.1097/RLI.0000000000000251

Hagen F, Soschynski M, Weis M, Hagar MT, Krumm P, Ayx I, Taron J, Krauss T, Hein M, Ruile P et al (2023) Photon-counting computed tomography–clinical application in oncological, cardiovascular, and pediatric radiology. In: RöFo-Fortschritte Auf Dem Gebiet Der Röntgenstrahlen Und Der Bildgebenden Verfahren

Hsieh SS, Leng S, Rajendran K, Tao S, McCollough CH (2021) Photon counting CT: clinical applications and future developments. IEEE Trans Radiat Plasma Med Sci 5:441–452. https://doi.org/10.1109/TRPMS.2020.3020212

Ibrahim YO, Maalej N, Pirzada BM, Raja AY, Anjum DH, Jan N, Behouch A, Qurashi AUH (2024) Gold nanoparticles spectral CT imaging and limit of detectability in a new materials contrast-detail phantom. Phys Med 120:103326. https://doi.org/10.1016/j.ejmp.2024.103326

IRIS micro-CT system | Animalab Animalab. https://animalab.eu. Accessed 26 March 2025

Jost G, McDermott M, Gutjahr R, Nowak T, Schmidt B, Pietsch H (2023) New contrast media for K-edge imaging with photon-counting detector CT. Investig Radiol 58:515–522. https://doi.org/10.1097/RLI.0000000000000978

Keklikoglou K, Arvanitidis C, Chatzigeorgiou G, Chatzinikolaou E, Karagiannidis E, Koletsa T, Magoulas A, Makris K, Mavrothalassitis G, Papanagnou E-D, Papazoglou AS, Pavloudi C, Trougakos IP, Vasileiadou K, Vogiatzi A (2021) Micro-CT for biological and biomedical studies: a comparison of imaging techniques. J Imaging 7:172. https://doi.org/10.3390/jimaging7090172

Khan O, Tariq B, Francis N, Maalej N, Behouch A, Kashif A, Waris A, Raja A (2024) Development of residual dense U-net (RDU-net)-based metal artefacts reduction technique using spectral photon counting CT. IEEE Access 12:109735–109749. https://doi.org/10.1109/ACCESS.2024.3439861

Kim J, Bar-Ness D, Si-Mohamed S, Coulon P, Blevis I, Douek P, Cormode DP (2018) Assessment of candidate elements for development of spectral photon-counting CT specific contrast agents. Sci Rep 8:12119. https://doi.org/10.1038/s41598-018-30570-y

Kreisler B (2022) Photon counting detectors: concept, technical challenges, and clinical outlook. Eur J Radiol 149:110229. https://doi.org/10.1016/j.ejrad.2022.110229

Li M, Wu M, Pack J, Wu P, De Man B, Wang A, Nieman K, Wang G (2023) Coronary Atherosclerotic Plaque Characterization with Photon-counting CT: a Simulation-based Feasibility Study. ArXiv Preprint ArXiv:2312.01566

Liu Y, Ai K, Liu J, Yuan Q, He Y, Lu L (2012) A high-performance ytterbium-based nanoparticulate contrast agent for in vivo X-ray computed tomography imaging. Angew Chem Int Ed 51:1437–1442. https://doi.org/10.1002/anie.201106686

Lowe C, Ortega-Gil A, Moghiseh M, Anderson NG, Munoz-Barrutia A, Vaquero JJ, Raja AY, Matanaghi A, Chernoglazov AI, Dapamede T, Adebileje SA, Alexander S, Amma MR, Anjomrouz M, Asghariomabad F, Atharifard A, Atlas J, Baer K, Bell ST, Bheesette S, Butler PH, Carbonez P, Chambers C, Chapagain KM, Clark JA, Colgan F, Crighton JS, Dahal S, Damet J, De Ruiter NJA, Doesburg RMN, Duncan N, Ghodsian N, Gieseg SP, Goulter BP, Gurney S, Healy JL, Kanithi PK, Kirkbride T, Lansley SP, Mandalika VBH, Marfo E, Palmer D, Panta RK, Prebble HM, Renaud P, Sayous Y, Schleich N, Searle E, Sheeja JS, Vanden Broeke L, Vivek VS, Walker EP, Walsh MF, Wijesooriya M, Younger WR, Butler APH (2021) Molecular imaging of pulmonary tuberculosis in an ex-vivo mouse model using spectral photon-counting computed tomography and micro-CT. IEEE Access 9:67201–67208. https://doi.org/10.1109/ACCESS.2021.3076432

Maggiorella L, Barouch G, Devaux C, Pottier A, Deutsch E, Bourhis J, Borghi E, Levy L (2012) Nanoscale radiotherapy with hafnium oxide nanoparticles. Future Oncol 8:1167–1181. https://doi.org/10.2217/fon.12.96

McGinnity TL, Dominguez O, Curtis TE, Nallathamby PD, Hoffman AJ, Roeder RK (2016) Hafnia (HfO_2) nanoparticles as an X-ray contrast agent and mid-infrared biosensor. Nanoscale 8:13627–13637. https://doi.org/10.1039/C6NR03217F

Mediso – NanoScan® PET/CT https://mediso.com. Accessed 26 March 2025

Meloni A, Cademartiri F, Positano V, Celi S, Berti S, Clemente A, La Grutta L, Saba L, Bossone E, Cavaliere C et al (2023) Cardiovascular applications of photon-counting CT technology: a revolutionary new diagnostic step. J Cardiovasc Dev Dis 10:363

Meloni A, Maffei E, Clemente A, De Gori C, Occhipinti M, Positano V, Berti S, La Grutta L, Saba L, Cau R et al (2024) Spectral photon-counting computed tomography: technical principles and applications in the assessment of cardiovascular diseases. J Clin Med 13:2359

Mendonça PRS, Bhotika R, Maddah M, Thomsen B, Dutta S, Licato PE, Joshi MC (2010) Multi-material decomposition of spectral CT images. In: Medical imaging 2010: physics of medical imaging, pp 633–641

Moghiseh M, Lowe C, Lewis JG, Kumar D, Butler A, Anderson N, Raja A (2018) Spectral photon-counting molecular imaging for quantification of monoclonal antibody-conjugated gold nanoparticles targeted to lymphoma and breast cancer: an in vitro study. Contrast Media Mol Imaging 2018:1–9. https://doi.org/10.1155/2018/2136840

Moghiseh M, Clark JA, Amma MR, Chapagain KM, Dixit D, Lowe C, Matanaghi A, Searle E, Sayous Y, Kumar D, Butler APH (2023a) MARS for molecular imaging and preclinical studies, in: photon counting computed tomography. Springer, Cham, pp 63–92. https://doi.org/10.1007/978-3-031-26062-9_4

Moghiseh M, Searle E, Dixit D, Kim J, Dong YC, Cormode DP, Butler A, Gieseg SP (2023b) Spectral photon-counting CT imaging of gold nanoparticle labelled monocytes for detection of atherosclerosis: a preclinical study. Diagnostics 13:499. https://doi.org/10.3390/diagnostics13030499

Molecubes (2022) Systems – molecubes. https://www.molecubes.com. Accessed 26 March 2025

Mourad C, Gallego Manzano L, Viry A, Booij R, Oei EHG, Becce F, Omoumi P (2024) Chances and challenges of photon-counting CT in musculoskeletal imaging. Skeletal Radiol:1–14

Mutreja I, Maalej N, Kaushik A, Kumar D, Raja A (2023) High atomic number nanoparticles to enhance spectral CT imaging aspects. Mater Adv 4:3967–3988. https://doi.org/10.1039/D3MA00231D

Oh MH, Lee N, Kim H, Park SP, Piao Y, Lee J, Jun SW, Moon WK, Choi SH, Hyeon T (2011) Large-scale synthesis of bioinert tantalum oxide nanoparticles for X-ray computed tomography imaging and bimodal image-guided sentinel lymph node mapping. J Am Chem Soc 133:5508–5515. https://doi.org/10.1021/ja200120k

Ostadhossein F, Tripathi I, Benig L, LoBato D, Moghiseh M, Lowe C, Raja A, Butler A, Panta R, Anjomrouz M, Chernoglazov A, Pan D (2020) Multi-"color" delineation of bone microdamages using ligand-directed Sub-5 nm hafnia nanodots and photon counting CT imaging. Adv Funct Mater 30. https://doi.org/10.1002/adfm.201904936

Pack JD, Fitzgerald P, Araujo S, Fan Y, Stevens G, Gerdes J, Wang A, Nieman K, Wang G, De Man B (2024) A coronary artery phantom for task-based CT performance assessment and a comparative study of clinical CT, photon counting CT, and micro CT. ArXiv Preprint ArXiv:2401.04215

Pan D, Roessl E, Schlomka J, Caruthers SD, Senpan A, Scott MJ, Allen JS, Zhang H, Hu G, Gaffney PJ, Choi ET, Rasche V, Wickline SA, Proksa R, Lanza GM (2010) Computed tomography in color: NanoK-enhanced spectral CT molecular imaging. Angew Chem Int Ed 49:9635–9639. https://doi.org/10.1002/anie.201005657

Pan D, Schirra CO, Senpan A, Schmieder AH, Stacy AJ, Roessl E, Thran A, Wickline SA, Proska R, Lanza GM (2012) An early investigation of ytterbium nanocolloids for selective and quantitative "multicolor" spectral CT imaging. ACS Nano 6:3364–3370. https://doi.org/10.1021/nn300392x

Panta RK, Bell ST, Healy JL, Aamir R, Bateman CJ, Moghiseh M, Butler APH, Anderson NG (2018) Element-specific spectral imaging of multiple contrast agents: a phantom study. J Instrum 13:T02001–T02001. https://doi.org/10.1088/1748-0221/13/02/T02001
PerkinElmer Quantum FX: low dose computed tomography. In Quantum FX microCT [Product-note]. https://resources.perkinelmer.com. Accessed 26 March 2025
Preclinical PET/CT for rodents high resolution (2021) MR Solutions. https://www.mrsolutions.com. Accessed 26 March 2025
Rajendran K, Walsh MF, de Ruiter NJA, Chernoglazov AI, Panta RK, Butler APH, Butler PH, Bell ST, Anderson NG, Woodfield TBF, Tredinnick SJ, Healy JL, Bateman CJ, Aamir R, Doesburg RMN, Renaud PF, Gieseg SP, Smithies DJ, Mohr JL, Mandalika VBH, Opie AMT, Cook NJ, Ronaldson JP, Nik SJ, Atharifard A, Clyne M, Bones PJ, Bartneck C, Grasset R, Schleich N, Billinghurst M (2014) Reducing beam hardening effects and metal artefacts in spectral CT using Medipix3RX. J Instrum 9:P03015–P03015. https://doi.org/10.1088/1748-0221/9/03/P03015
Ramsey JB (1969) Tests for specification errors in classical linear least-squares regression analysis. J R Stat Soc Series B Stat Methodol 31:350–371. https://doi.org/10.1111/j.2517-6161.1969.tb00796.x
Riederer I, Bar-Ness D, Kimm MA, Si-Mohamed S, Noël PB, Rummeny EJ, Douek P, Pfeiffer D (2019) Liquid embolic agents in spectral X-ray photon-counting computed tomography using tantalum K-edge imaging. Sci Rep 9:5268. https://doi.org/10.1038/s41598-019-41737-6
Scanco MicroCT40 (2018) Penn state biomechanics laboratory. https://biomechanicslab.psu.edu. Accessed 26 March 2025
Schlomka JP, Roessl E, Dorscheid R, Dill S, Martens G, Istel T, Bäumer C, Herrmann C, Steadman R, Zeitler G, Livne A, Proksa R (2008) Experimental feasibility of multi-energy photon-counting K-edge imaging in pre-clinical computed tomography. Phys Med Biol 53:4031–4047. https://doi.org/10.1088/0031-9155/53/15/002
SEDECAL (2024) SUPER ARGUS PET/CT – SEDECAL. SEDECAL. https://www.sedecal.com. Accessed 26 March 2025
Si-Mohamed S, Thivolet A, Bonnot P-E, Bar-Ness D, Képénékian V, Cormode DP, Douek P, Rousset P (2018) Improved peritoneal cavity and abdominal organ imaging using a biphasic contrast agent protocol and spectral photon counting computed tomography K-edge imaging. Investig Radiol 53:629–639. https://doi.org/10.1097/RLI.0000000000000483
So A, Nicolaou S (2021) Spectral computed tomography: fundamental principles and recent developments. Korean J Radiol 22:86–96. https://doi.org/10.3348/kjr.2020.0144
SOFIE The world's highest sensitivity, highest resolution pet scanner. https://sofie.com. Accessed 26 March 2025
Stamp LK, Anderson NG, Becce F, Rajeswari M, Polson M, Guyen O, Viry A, Choi C, Kirkbride TE, Raja AY (2019) Clinical utility of multi-energy spectral photon-counting computed tomography in crystal arthritis. Arthritis Rheumatol 71:1158–1162
Stephens K (2021) FDA approves Siemens Healthineers' Naeotom alpha CT scanner. AXIS Imaging News
Symons R, Krauss B, Sahbaee P, Cork TE, Lakshmanan MN, Bluemke DA, Pourmorteza A (2017) Photon-counting CT for simultaneous imaging of multiple contrast agents in the abdomen: an in vivo study. Med Phys 44:5120–5127. https://doi.org/10.1002/mp.12301
Tariq B, Sikander O, Francis N, Alkhatib M, Naseer F, Werghi N, Memisoglu E, Maalej N, Raja A (2024) Assessment of material identification and quantification in the presence of metals using spectral photon counting CT. PLoS One 19:e0308658. https://doi.org/10.1371/journal.pone.0308658
Tortora M, Gemini L, D'Iglio I, Ugga L, Spadarella G, Cuocolo R (2022) Spectral photon-counting computed tomography: a review on technical principles and clinical applications. J Imaging 8:112. https://doi.org/10.3390/jimaging8040112
U-CT | MILABS Micro-CT (2025) MILabs. https://www.milabs.com. Accessed 26 March 2025
User MBI (2024) MARS for Researchers—MARS Bioimaging. https://www.marsbioimaging.com. Accessed 26 March 2025

Viry A, Raja AY, Kirkbride TE, Choi C, Stamp LK, Dalbeth N, Combes C, Verdun FR, Anderson NG, Becce F (2018) Multi-energy spectral photon-counting CT in crystal-related arthropathies: initial experience and diagnostic performance in vitro. In: Medical imaging 2018: physics of medical imaging, pp 1280–1286

Wang AS, Pelc NJ (2021) Spectral photon counting CT: imaging algorithms and performance assessment. IEEE Trans Radiat Plasma Med Sci 5:453–464. https://doi.org/10.1109/TRPMS.2020.3007380

Wu W, Yu H, Chen P, Luo F, Liu F, Wang Q, Zhu Y, Zhang Y, Feng J, Yu H (2020) Dictionary learning based image-domain material decomposition for spectral CT. Phys Med Biol 65:245006. https://doi.org/10.1088/1361-6560/aba7ce

Zhao T, Kim K, Wu D, Kalra MK, El Fakhri G, Li Q (2017) Multi-materials decomposition using clinical dualenergy CT. In: 2017 IEEE nuclear science symposium and medical imaging conference (NSS/MIC). IEEE, pp 1–4. https://doi.org/10.1109/NSSMIC.2017.8532936

Zhou W, Lane JI, Carlson ML, Bruesewitz MR, Witte RJ, Koeller KK, Eckel LJ, Carter RE, McCollough CH, Leng S (2018) Comparison of a photon-counting-detector CT with an energy-integrating-detector CT for temporal bone imaging: a cadaveric study. Am J Neuroradiol 39:1733–1738. https://doi.org/10.3174/ajnr.A5768

Zhou W, Bartlett DJ, Diehn FE, Glazebrook KN, Kotsenas AL, Carter RE, Fletcher JG, McCollough CH, Leng S (2019) Reduction of metal artifacts and improvement in dose efficiency using photon-counting detector computed tomography and tin filtration. Investig Radiol 54:204–211. https://doi.org/10.1097/RLI.0000000000000535

Intraoperative Tumor Margin Assessment with Photoacoustic Imaging

Qi Han, Feng He, Keyoumars Ashkan, and Wenfeng Xia

Abstract

Cancers rank among the most life-threatening conditions for humans. Resection surgery remains one of the most prevalent treatments. Intraoperative tumor margin assessment plays a crucial role in the success of surgical resection, influencing recurrence rates and patient prognosis. This importance is exemplified for brain tumors due to the complexities of the anatomical and functional boundaries. Given the inherent limitations of traditional intraoperative guidance techniques, marked by suboptimal imaging resolution, shallow tissue penetration depths and sluggish imaging speeds, there is an urgent imperative for novel and effective imaging methodologies to enhance both accuracy and expediency to shorten surgical time and reduce inadvertent omission of imperceptible tumor cells. Among the emerging guidance techniques, photoacoustic imaging (PAI) stands out for its potential to provide high-speed, high-resolution 3D imaging at large tissue depths. PAI's ability to visualize molecular and structural changes, such as hypoxia, lipid variation, and nuclear morphology, makes it a compelling tool for differentiating malignant tissues from normal tissues during surgery. This chapter reviews recent advances in PAI for intraoperative tumor margin

Authors Qi Han and Feng He have equally contributed to this chapter.

Q. Han · F. He · W. Xia (✉)
School of Biomedical Engineering & Imaging Sciences, King's College London, London, UK
e-mail: wenfeng.xia@kcl.ac.uk

K. Ashkan (✉)
Department of Neurosurgery, Kings College Hospital, Denmark Hill, London, UK
e-mail: k.ashkan@nhs.net

I. F. Cengiz et al. (eds.), *Bioimaging Modalities in Bioengineering*,
Biomaterials, Bioengineering and Sustainability 9,
https://doi.org/10.1007/978-3-031-99410-4_22

assessment, focusing imaging configurations such as clinical array-based PAI systems, ultraviolet photoacoustic microscopy (PAM), and handheld PAM. It also explores the integration of PAI with complementary imaging techniques, such as ultrasound and optical imaging modalities, to enhance comprehensive tumor assessment. Finally, this chapter highlights the potential of PAI to transform surgical oncology by reducing re-excision rates, improving patient outcomes, and shaping the future of intraoperative imaging.

Keywords

Intraoperative tumor margin assessment · Multispectral photoacoustic tomography · Ultraviolet photoacoustic microscopy · Surgical oncology · Hybrid imaging

1 Introduction

Cancer poses a mounting burden on both the economy and social fabric. In 2020, the global incidence of new cancer cases reached 19.3 million, with a staggering ten million succumbing to the disease (Sung et al. 2021). Furthermore, demographic prediction projections show that, with a 77% increase from the 2022 level, the yearly number of new cancer cases would reach 35 million (Sung et al. 2021). According to the most recently released 2025 cancer statistics data released by the American Cancer Society, it is anticipated that there will be 2,041,910 new cancer cases and 618,120 cancer-related deaths in the United States alone in 2025 (Siegel et al. 2025). Consequently, a critical and urgent task lies ahead: the advancement of medical technology for cancer therapy.

Tumors, or neoplasms, represent aberrant tissue masses arising from the uncontrolled proliferation of local cells under the influence of diverse tumorigenic factors (Zhang et al. 2024a). A fundamental classification divides tumors into benign and malignant categories. While benign tumors can pose health risks depending on their location, malignant tumors are of greater concern due to their aggressive nature, rapid growth, invasion of surrounding tissues, and potential for metastasis (Arneth 2019). Malignant tumors exhibit several distinctive features, including abnormal cellular morphology, such as irregularly shaped nuclei and a higher nuclear-to-cytoplasmic ratio, as well as an enhanced proliferative capacity, which contributes to tumor expansion. Additionally, malignant tumors stimulate angiogenesis, forming disorganized and immature blood vessels to support their growth. As tumors grow rapidly, they often outstrip their oxygen supply, leading to hypoxia and even necrosis in some regions, which further promotes malignancy and resistance to treatment (Harris et al. 2022; Chen et al. 2023). These distinct biological characteristics offer valuable opportunities for imaging techniques to distinguish malignant from normal tissues, especially by targeting tumor-specific traits like altered vasculature, increased cellular density, and hypoxic areas (García-Figueiras et al. 2019).

Currently, excisional surgery stands as a cornerstone in many cancer treatments, involving the removal of primary or secondary tumors from surrounding normal tissue (Wyld et al. 2015). However, precise resection is critical, as postoperative complications and tumor recurrence are influenced by the completeness of the tumor removal. Before surgery, imaging techniques like MRI, CT scans, or ultrasound (US) are used to locate the tumor and assess its size, shape, and proximity to vital structures (Lakhani et al. 2023; Bouget et al. 2022; Zeng et al. 2020). A tissue sample may be taken for histological analysis to confirm the diagnosis. Surgeons, oncologists, and radiologists then plan the excision based on imaging and biopsy results, aiming for complete tumor removal with clear margins. Surgeons typically resect the tumor with a margin of healthy tissue to reduce the likelihood of residual cancer cells. Following this, the excised tissue often undergoes frozen section analysis (FSA) by a pathologist to confirm that the margins are free of tumor cells (Peters 2010; Gao et al. 2023a). However, FSA presents certain drawbacks that can compromise surgical outcomes. Notably, the typical duration of FSA, spanning 20–40 min, may prolong surgical procedures and lead to adverse effects like infections and bleeding (Peters 2010). Additionally, its limited resolution may result in the oversight of small tumor cell clusters in the vicinity of cancerous regions, potentially leaving residual tumor tissue post-surgery (Namdar et al. 2021; Demir et al. 2022; Schmidt et al. 2011; De Decker et al. 2021; Kurdi et al. 2022). This scenario could necessitate a subsequent surgical intervention, further jeopardizing patient well-being and increasing healthcare costs. Given the critical importance of achieving thorough resection to prevent tumor recurrence and malignant progression, coupled with the risks posed by extended surgical durations, the imperative to enhance intraoperative analysis techniques becomes apparent. These enhancements should be directed towards swift and precise differentiation of cancerous from non-cancerous tissue within a condensed time frame, thereby ensuring an optimal and safe resection.

In tumor resection surgery, optimal intraoperative imaging methods must meet several essential criteria, including high imaging specificity, sufficient spatial resolution, frame rate, field-of-view, and tissue penetration depth. Several imaging techniques have progressed into clinical applications, while others remain in the laboratory stage. Among these, two imaging modalities have been successfully integrated into clinical navigation: fluorescent imaging and intraoperative magnetic resonance imaging (iMRI). Fluorescent imaging is the most widely used technology for surgical guidance, offering real-time visualization of tumors and surrounding healthy tissue. Clinicians commonly employ contrast agents such as 5-Aminolevulinic acid (5-ALA), Indocyanine green (ICG), and Evan's blue to enhance tumor delineation (Molina et al. 2024; Xiao et al. 2018; Belykh et al. 2023; Wang et al. 2023a; Conq et al. 1957; Lavrador et al. 2024). Additionally, iMRI is employed to provide dynamic feedback during surgical procedures and has been integrated into several neurosurgical centers with promising outcomes in terms of patient survival and local tumor control (Mirzayeva et al. 2024; Altaf et al. 2023; Wu et al. 2023).

Several other imaging methodologies, including Raman spectroscopy, optical coherence tomography (OCT), and positron emission tomography (PET), show potential for tumor delineation but have not been used in patients. While these techniques have demonstrated value in assessing tumor biology, the demand for real-time, highly precise imaging during surgery remains unmet. Recent neurosurgical reports also highlight the potential synergy of iMRI with advanced optical methods for complex brain tumor resection (Lakhani et al. 2023). Despite these advances, more rapid, precise intraoperative approaches are still needed.

Photoacoustic imaging (PAI) is emerging as a promising technology that combines the advantages of optical and US imaging (Xia 2024; Park et al. 2024a; Riksen et al. 2023; Lin and Wang 2022; Beard 2011). By targeting specific tumor features such as abnormal vasculature and hypoxic regions, PAI provides enhanced contrast and spatial resolution, making it a compelling alternative for real-time tumor margin assessment during surgery. One of the key strengths of PAI is its ability to visualize cell nuclei when operated at ultraviolet (UV) wavelengths. This capability offers critical insights into tumor cellularity and malignancy, allowing surgeons to more accurately delineate tumor boundaries and ensure complete resection. Additionally, PAI can monitor vascular changes and assess blood oxygen saturation (sO_2), which is crucial for identifying the chaotic and disorganized blood vessels characteristic of tumor angiogenesis. This technology can also capture the hypoxic regions within tumors, which often correlate with increased malignancy. By visualizing areas of hypoxia and necrosis, PAI allows surgeons to better assess the extent of tumor resection required. Furthermore, the technology can distinguish between varying concentrations of exogenous molecular agents, providing flexibility in tuning wavelengths to target specific molecular markers within the tumor. This adaptability enhances its precision in tumor margin detection.

In this chapter, we first provide an overview of current intraoperative imaging techniques for tumor margin assessment, summarizing their principles, strengths, and limitations. We then discuss the recent advances and Technology Readiness Levels (TRLs) of these imaging modalities, highlighting those that have progressed from laboratory validation to clinical implementation. Next, we focus on PAI methods for tumor margin assessment, covering their configurations, molecular contrast mechanisms, and compatibility with hybrid imaging approaches. Three primary PAI techniques are introduced: clinical array-based PAI, UV photoacoustic microscopy (PAM), and handheld PAM. These techniques enable the visualization of various biological parameters, including DNA/RNA distribution, lipid content, sO_2, and vasculature in the tumor region, which aid in distinguishing tumor tissue from surrounding healthy tissues. Additionally, the integration of PAI with other modalities is also discussed. Finally, we conclude by addressing current challenges, discussing future directions, and positioning PAI within the broader landscape of surgical oncology imaging. This structure aims to provide clinicians, researchers, and developers with a comprehensive roadmap for leveraging advanced imaging technologies in intraoperative tumor margin assessments.

2 Recent Advances in Intraoperative Tumor Margin Assessments

Recent advances in intraoperative tumor margin assessments are reshaping oncologic surgery by providing surgeons with rapid, accurate information to distinguish between cancerous and healthy tissue. These techniques deliver real-time visualization and molecular data, enabling informed decisions with improved speed and accuracy, reducing the need for follow-up surgeries, minimizing recurrence rates, and ultimately leading to better patient outcomes (Kopicky et al. 2024; Dowling et al. 2024; Chen et al. 2024; Carnicelli et al. 2023; Pertzborn et al. 2022; Aaboubout et al. 2021; Kafka and Averch 2019; Shipp et al. 2018; Grootendorst et al. 2017a). In terms of TRL, the progress of these technologies varies, with some methods achieving routine clinical deployment, such as fluorescence guidance and intraoperative US, while others remain in experimental stages, requiring further validation and regulatory approval. This review examines the state-of-the-art in intraoperative tumor margin assessments, categorizing advancements into point-based spectroscopy and imaging techniques, each offering distinct advantages and limitations.

2.1 Point-Based Spectroscopy

Point-based spectroscopy techniques provide localized, high-resolution analysis by leveraging the unique physical, optical, and molecular properties of tissues. These methods excel in differentiating between malignant and benign tissues through precise measurements of their intrinsic characteristics. Unlike imaging modalities that provide wide-field visualizations, point-based techniques focus on obtaining highly detailed information at specific tissue sites, making them valuable for targeted intraoperative decision-making. Despite their potential, these methods often require multiple measurements to cover larger surgical areas, presenting challenges in time-sensitive environments. This section reviews key point-based spectroscopy approaches, emphasizing their mechanisms, applications, and contributions to intraoperative tumor margin assessments.

Electrical-Based Spectroscopy Techniques Electrical-based spectroscopy techniques analyze the electrical properties of tissues, such as conductivity, permittivity, and impedance, to identify malignancy. By exploiting the differences in electrical characteristics between cancerous and normal tissues, these methods offer real-time diagnostic feedback, enabling surgeons to make informed decisions during tumor resections. Techniques like electrical impedance spectroscopy (EIS) and radiofrequency (RF) spectroscopy have shown promise in reducing residual tumor margins, particularly in cancers like breast and cervical. However, as point-based methods, they require precise calibration and careful interpretation to ensure measurement accuracy, especially in heterogeneous tissue environments. This section provides a detailed evaluation of these techniques, discussing their clinical applications, advantages, limitations, and readiness for broader clinical adoption.

EIS measures the impedance of tissues by applying alternating current at varying frequencies and analyzing the resultant voltage response. Cancerous tissues often exhibit distinct impedance profiles due to their higher conductivity and reduced capacitive properties compared to healthy tissues, stemming from differences in cellular density, membrane structure, and water content. EIS has been explored in various cancers, including colorectal and cervical cancers, where it enables differentiation between malignant and normal tissues (Li et al. 2021a; Bergqvist et al. 2023; Oh et al. 2021; Tsampazis et al. 2023). This technique offers several advantages, including non-invasiveness, cost-effectiveness, and real-time feedback, making it well-suited for intraoperative decision-making. However, as a point-based method, EIS lacks spatial resolution and requires multiple measurements to comprehensively assess larger areas. Additionally, tissue heterogeneity and variations in electrode placement can influence measurement accuracy, necessitating careful calibration and advanced interpretation (Cortés et al. 2024; Briz et al. 2023; Carobbio et al. 2023; Abasi et al. 2022; Doležal et al. 2025). EIS currently has a TRL of 5–7 across different cancers, requiring further clinical validation for broader applications.

RF Spectroscopy operates by transmitting RF signals (MHz–GHz range) into tissues and analyzing the reflected or absorbed signals to assess their electrical properties. The FDA-approved MarginProbe represents a prominent clinical application of RF spectroscopy, specifically for breast-conserving surgery. By detecting differences in signal reflection between cancerous and benign tissues, MarginProbe provides real-time feedback to surgeons, significantly reducing re-excision rates (Hoffman and Ashkenazi 2022; Geha et al. 2020; Schnabel et al. 2014; Thill et al. 2014; Armani et al. 2022). RF spectroscopy's strengths include its rapid feedback and seamless integration into surgical workflows. However, its point-based nature limits spatial coverage, necessitating multiple probe placements to evaluate larger areas comprehensively. Additionally, factors such as tissue hydration and probe alignment can impact measurement reliability (Wang et al. 2023b; Qafiti et al. 2022; Grimes 2022). While MarginProbe has achieved a TRL of 9 for breast cancer surgery, RF Spectroscopy's application in other cancers remains at TRL 5–6 due to limited clinical studies and regulatory approvals.

Dielectric Spectroscopy analyses tissues' dielectric properties, such as permittivity and loss factor, over a broad frequency range, offering valuable insights into cellular composition and water content. This technique has been investigated for use in breast and cervical cancers, where malignant tissues exhibit higher dielectric constants due to their increased water content and cellular density (Evans et al. 2024; Kordic and Šarolic 2023; Fahmy et al. 2020; Ghanbarzadeh-Daghian et al. 2020; Hussein et al. 2019). Dielectric spectroscopy provides real-time feedback and is non-invasive, making it a promising tool for intraoperative applications. However, its utility is primarily limited to superficial tissues, as deeper lesions are challenging to detect due to restricted penetration depth. The requirement for advanced instrumentation and complex data analysis further limits its routine clinical adoption. Dielectric Spectroscopy is at TRL 4–5, with most studies confined to preclinical research and early-stage clinical trials.

Optical Spectroscopy Optical spectroscopy techniques utilize the interaction of light with biological tissues to provide molecular and structural information, offering significant advantages in real-time tumor margin assessments. These methods, including fluorescence spectroscopy, diffuse reflectance spectroscopy, and Raman spectroscopy, are non-invasive and capable of detecting subtle biochemical differences between malignant and healthy tissues. Clinically, the use of fluorescent agents such as 5-ALA and ICG has enhanced surgical precision in glioma and liver tumor resections. However, limitations such as shallow tissue penetration and dependence on advanced instrumentation pose challenges for widespread adoption. This section explores the principles, applications, and technical readiness of optical spectroscopy methods for intraoperative oncology.

Fluorescence spectroscopy involves the excitation of tissue fluorophores using specific wavelengths of light, leading to the emission of fluorescence that can be analyzed to distinguish between normal and malignant tissues. Clinically, various fluorescent agents have been utilized: 5-ALA is metabolized into protoporphyrin IX, accumulating in malignant glioma cells and aiding in fluorescence-guided resections (Gautheron et al. 2024; Ma et al. 2024b; Milos et al. 2023; McCracken et al. 2022; Leclerc et al. 2020); ICG is employed for visualizing vascular structures and assessing tissue perfusion during oncological surgeries (Wang et al. 2023a; Achterberg et al. 2024; Leiloglou et al. 2022; Chiba et al. 2022; Farrakhova et al. 2022; Pan et al. 2020); and Evans blue dye has been used in sentinel lymph node mapping to identify metastatic spread (Li et al. 2021b; Lee et al. 2020; Dai et al. 2020). The advantages of fluorescence spectroscopy include real-time feedback, high sensitivity to biochemical changes, and non-invasive application. However, limitations encompass shallow tissue penetration due to light scattering and absorption, dependence on the presence and concentration of fluorophores, and potential photobleaching under prolonged exposure. Fluorescence spectroscopy has achieved a TRL of 8–9 in applications such as fluorescence-guided glioma surgery, where it is integrated into standard clinical practice (Zhao et al. 2024; Zeppa et al. 2022; Maragkos et al. 2021; Sun et al. 2021a; Senders et al. 2017).

Diffuse Reflectance Spectroscopy measures the reflected light from tissues to assess their optical properties, such as absorption and scattering coefficients, which correlate with tissue composition and structure. This technique has been recently explored for intraoperative margin assessment in head and neck cancer surgeries, providing quantitative information to guide resections (Saha et al. 2024; Rickard et al. 2023). DRS offers rapid, non-destructive assessments with high specificity, making it suitable for intraoperative use. However, its shallow penetration depth limits its utility to superficial tissues, and measurement variability caused by probe placement and pressure requires standardized protocols for consistent results. Currently, DRS is at TRL 6–7, with several pilot studies demonstrating feasibility, but broader clinical adoption necessitates further validation (Nogueira et al. 2021;

Gkouzionis et al. 2024; Chaudhry et al. 2023; Nazarian et al. 2024; Zhang et al. 2020a).

Raman spectroscopy detects inelastic scattering of photons, providing a molecular fingerprint of tissues based on the vibrational modes of their constituent molecules. This technique enables precise characterization of biochemical composition, facilitating the identification of malignant transformations (Ma et al. 2024a; Hanna et al. 2022; Ember et al. 2024; Zhang et al. 2024c; Qi et al. 2023; Gaba et al. 2022; Liu et al. 2022; Auner et al. 2018). However, the penetration depth of conventional Raman spectroscopy is limited to a few hundred micrometers beneath the tissue surface, restricting its effectiveness to superficial tissue analysis (Ember et al. 2024; Shaikh et al. 2023). Advancements such as Spatially Offset Raman Spectroscopy have been developed to enhance depth penetration, allowing for subsurface analysis of up to several millimeters (Nicolson et al. 2024; Kotturi et al. 2024; Mordi et al. 2024; Al-Attili et al. 2023; Mosca et al. 2021). Despite its high specificity and label-free detection capabilities, Raman spectroscopy faces challenges, including weak signal intensity, susceptibility to fluorescence interference, and the need for advanced instrumentation. It is currently at TRL 5–6, with ongoing research focused on overcoming these limitations and validating its clinical utility (Ma et al. 2024a; Hanna et al. 2022; Chen et al. 2021; Goff et al. 2024; Li et al. 2023a; Faur et al. 2022).

2.2 Imaging Techniques

Imaging techniques provide real-time visualization of the surgical field, making them valuable for intraoperative tumor margin assessments. Unlike point-based methods, field imaging captures wide-field visualization of tumor margins, offering a broader spatial context and reducing the risk of missed tumor cells. Established modalities like iMRI and US have become routine in many surgical disciplines, while emerging optical imaging methods are advancing towards clinical deployment. This section delves into imaging technologies, categorizing them by their principles, clinical utility, and stages of development, to highlight their transformative potential in oncological surgery.

Intraoperative MRI iMRI, introduced in 1991, has evolved with systems like IMRIS platform that offer movable MRI capabilities (Mislow et al. 2010; Grönemeyer et al. 1996a, b). High-field systems (e.g., those offering functional MRI and diffusion tensor imaging) are primarily used in neurosurgery for glioma and brain tumor resections (Mirzayeva et al. 2024; Li et al. 2024; Leroy et al. 2019). iMRI allows real-time imaging during surgery, helping surgeons visualize tumor location and extent, even when tissue shifts. Its advantages include enhanced resection rates and increased safety through functional and molecular imaging. Some neurosurgical centers have also adopted iMRI in combination with advanced optical or fluorescence guided methods to further refine the resection of gliomas (Mazzucchi et al. 2022; Palmieri et al. 2021; Orillac et al. 2021; Lauwerends et al. 2021; Gupta et al. 2020). In terms of TRL, iMRI is at 9, as it is clinically established, integrated into surgical suites, and widely adopted in neurosurgical centers worldwide (Wu et al. 2023; Barchéus et al. 2024; Rogers et al. 2021). Recent neurosurgical experiences also indicate that iMRI—when used judiciously—can significantly improve local control of high-grade brain tumors (Mirzayeva et al. 2024; Patel and Chavda 2024; Jellema et al. 2023; Fuentes et al. 2022; Wu et al. 2022; Noh et al. 2020). However, it is resource-intensive and can extend surgical times due to setup and imaging processes.

Ionizing Radiation-Based Techniques X-ray-based techniques such as micro-computed tomography (micro-CT) and specimen radiography are valuable for intra-operative tumor margin assessments. Micro-CT provides high-resolution, three-dimensional visualization of tumor margins (Jensen et al. 2024; Dizbay Sak et al. 2023), while specimen radiography highlights density differences between normal and malignant tissues (Lin et al. 2022; van Riet et al. 2021; Kim et al. 2020; Hisada et al. 2016; Laws et al. 2018; Pop et al. 2018). These techniques are at TRL 7–9 depending on the specific implementation. Specimen radiography (e.g., intraoperative breast lesion imaging) is widely used in practice (TRL 9), whereas micro-CT may still be at a lower TRL of around 7–8 due to limited widespread clinical adoption. Despite their benefits, radiation exposure and limited sensitivity can restrict broader application, requiring improved imaging protocols and contrast enhancement.

PET uses radiotracers to visualize active tumor tissues during surgery (Grootendorst et al. 2017a; Darr et al. 2023; Debacker et al. 2021; Marcinkowski et al. 2019; Holland et al. 2011), often in combination with computed tomography (CT) or MRI as PET/CT or PET/MRI. This hybrid technique improves accuracy in locating tumors, though PET's lower spatial resolution and reliance on radioactive tracers limit its widespread use. PET and PET/CT systems are clinically established for diagnostic purposes (TRL 9 preoperatively), but their full intraoperative utilization remains limited (TRL 6–7) due to logistical challenges, radiotracer handling, and integration into the surgical workflow.

Ultrasound (US) High-frequency US plays a vital role in intraoperative tumor margin assessments due to its non-invasive nature and real-time imaging capabilities (Abdelshafy et al. 2018; Alharbi et al. 2016; Dixon et al. 2022; Giammalva et al. 2022; Karanlik et al. 2015; Klein Gunnewiek et al. 2024; Laws et al. 2018; Scaranelo et al. 2016; Sikošek et al. 2014; Tarabichi et al. 2018; Veeratterapillay et al. 2015). By transmitting high-frequency sound waves into tissue and recording the returning echoes, US creates detailed images based on tissue mechanical properties and composition. In terms of TRL, intraoperative US is well-established at level 9, as it is commonly used in surgeries (e.g., breast, liver, and neurosurgery) to guide resection and detect residual tumor.

One major advantage of US is its ability to provide immediate intraoperative feedback without the need for ionizing radiation, allowing surgeons to make real-time decisions. It offers sufficient imaging depth to cover large tissue areas, and repeated scans can be performed safely without exposing patients or staff to radiation, unlike X-ray-based techniques. Additionally, high-frequency US can detect subtle differences in tissue structure, which is crucial for effective tumor margin assessment.

However, US does have limitations. Due to limited soft tissue contrast, it may not always be sufficient to detect very small tumor remnants, and its effectiveness can be impacted by factors such as dense breast tissue or calcifications, which can obscure imaging. Moreover, the interpretation of US images requires significant expertise, and variability between operators can affect diagnostic accuracy, particularly in neurosurgery given the diversity of tumor types. Despite these challenges, the non-destructive, real-time nature of US, combined with its versatility across various types of surgery and relatively low cost, makes it an invaluable tool in intraoperative tumor margin assessments.

Optical Imaging Techniques Optical imaging techniques are increasingly utilized in intraoperative tumor margin assessments due to their ability to provide high-resolution, real-time images without the need for invasive procedures. These techniques offer detailed visualization of tissues, allowing surgeons to better differentiate between malignant and healthy areas during surgery. Confocal microscopy and its variants, such as confocal laser endomicroscopy, probe-based confocal laser endomicroscopy, reflectance confocal microscopy, and fluorescence confocal microscopy, provide real-time, high-resolution imaging of cellular structures (). These methods use laser light to scan tissues and capture subcellular details, making them highly effective for examining tumor margins. While these approaches provide excellent resolution, their imaging depth is limited, and they require specialized interpretation. Typically, their TRL is 6–8, with some confocal endomicroscopy systems undergoing pilot clinical testing.

Structured illumination microscopy (SIM) enhances contrast and delivers optically sectioned images by illuminating tissues with patterned light. SIM is particularly useful for imaging large tissue areas (e.g., breast tumor margins) (Luethy et al. 2020; Johnson and Hagen 2020; Wang et al. 2016, 2018). However, its slower acquisition speeds and sensitivity to out-of-focus light may hamper real-time surgical application. SIM is at TRL 4–5, with research primarily in controlled settings. Microscopy with UV surface excitation (MUSE) uses deep UV light to achieve optical sectioning at the tissue surface, enabling rapid, non-destructive imaging (Lechpammer et al. 2024; Sato et al. 2023; Ye et al. 2023; Wong et al. 2021; Ching-Roa et al. 2021; Xie et al. 2019; Yoshitake et al. 2018; Fereidouni et al. 2017). MUSE allows for imaging of fresh, unfixed tissue without extensive preparation, providing histological images that correspond well with traditional hematoxylin and eosin (H&E) staining. However, its optical section thickness and potential interference from surface contaminants can limit certain applications. MUSE is at TRL 5–6, primarily in pilot clinical studies.

Fluorescence lifetime imaging measures the decay time of fluorescence from tissues, providing detailed information about tissue composition (Fernandes et al. 2024; Weyers et al. 2022; Reichert et al. 2021; Weyers et al. 2019; Wang et al.

2017b). This high-resolution, non-destructive method differentiates between healthy and cancerous tissues without needing exogenous contrast agents, making it suitable for real-time imaging during surgery. Nonetheless, shallow imaging depth and complex interpretation have restricted its widespread adoption (TRL 4–6).

Non-linear microscopy methods, including multiphoton microscopy, enable deeper tissue imaging without external dyes (Cahill et al. 2018; Giacomelli et al. 2018; Parodi et al. 2020; Yan et al. 2014; Yang et al. 2021). Although capable of imaging several hundred micrometers beneath the tissue surface with subcellular resolution, they are slower and require expensive equipment, limiting mainstream clinical adoption (TRL 3–5).

Terahertz imaging provides high-contrast images based on tissue water content and molecular composition, effectively delineating tumor margins (Cassar et al. 2018; Fan et al. 2017; Grootendorst et al. 2017b; Gurjar et al. 2024; Ke et al. 2022; Shi et al. 2022a; Vohra et al. 2020; Yang et al. 2023; Zhan et al. 2023). However, slower acquisition speeds and equipment complexity remain major hurdles, confining it to TRL 3–5 in pilot studies. OCT is another promising 3D label-free imaging modality providing micrometer-scale resolution for real-time tumor margin assessments (Allen et al. 2018; Badhey et al. 2023; Chin et al. 2017; Gubarkova et al. 2023; Ha et al. 2018; Hamdoon et al. 2016; Kuppler et al. 2024; Liu et al. 2023; Ludwig et al. 2018; Park et al. 2024b; Van Manen et al. 2017; Zhang et al. 2024d). Although imaging depth is limited to ~1 mm, OCT has reached TRL 7–8 in certain surgical applications (e.g., ophthalmic surgery) but remains at lower TRLs for broad oncologic usage. Hyperspectral imaging (HSI) captures a wide spectrum of light beyond the visible range, enabling detailed tissue differentiation (Anichini et al. 2024; Campbell et al. 2023; Fei et al. 2017; Parasca et al. 2024; Puustinen et al. 2023; Yoon 2022; Youssef et al. 2024; Zhang et al. 2020b). Although factors like tissue heterogeneity and data interpretation pose challenges, HSI's capacity for wide-field, label-free imaging of large surfaces lends promise for tumor margin assessments (TRL 4–6).

Across the various imaging techniques discussed, recent advances have significantly improved the accuracy and efficiency of intraoperative tumor margin assessments. Point-based spectroscopy excels in providing localized molecular insights, while imaging techniques offer comprehensive spatial resolution for broader surgical contexts. Despite their unique strengths, both modalities face challenges, such as limited penetration depth, measurement variability, and the need for specialized equipment. Addressing these limitations through continued technological innovation and clinical validation will be critical for integrating these methods into routine surgical practice. Collectively, these advances hold the potential to revolutionize oncological surgeries, reducing reoperation rates and enhancing patient outcomes.

3 Photoacoustic Imaging (PAI)

PAI distinguishes itself from other intraoperative imaging modalities by bridging the gap between high-resolution anatomical visualization and functional imaging at clinically relevant depths. Unlike purely optical imaging techniques such as OCT, HSI which rely on optical scattering or a combination of scattering and absorption of tissue for contrast, as a hybrid imaging modality, PAI uniquely leverage optical absorption as its sole contrast source. This fundamental difference enables PAI to provide exceptional specificity, as the optical absorption directly correlates with the presence of biological chromophores, including endogenous molecules like hemoglobin and melanin, and exogenous agents (Xia 2024; Beard 2011), as shown in Fig. 1. By converting absorbed light energy into US waves through the photoacoustic effect, PAI delivers high-resolution images with optical contrast at depths far beyond the capabilities of other purely optical methods (Tsai et al. 2023; Veverka et al. 2023; Gao et al. 2022). By tuning laser wavelengths to match the absorption peaks of specific chromophores, PAI can target a range of biological and molecular features, such as blood oxygenation, lipid content, and even nucleic acid distributions (Zhang et al. 2024b; Vonk et al. 2022; MacCuaig et al. 2020). For example, the high absorption of hemoglobin enables precise mapping of tumor vasculature, while UV PAI highlights nuclear details that correlate with tumor cellularity (Cao et al. 2023a, 2024; Huang et al. 2024; Wong et al. 2017a; Zhao et al. 2022a). This capacity to integrate molecular and structural information in real-time offers a distinct advantage over other modalities, which often rely on indirect or mixed contrast mechanisms (Fig. 1).

PAI also provides a highly adaptable imaging platform. For instance, its resolution and imaging depth can be tailored by switching between PAM for subcellular imaging and photoacoustic tomography (PACT) for deep-tissue imaging (Xia 2024; Lin and Wang 2022). This flexibility makes PAI uniquely suited for the diverse demands of intraoperative tumor resection, where surgeons must navigate varying scales and tissue depths. Moreover, PAI's ability to selectively image key tumor features—such as abnormal vasculature, hypoxic regions, and cell nuclei when paired with UV wavelengths—provides valuable functional insights that complement anatomical imaging (Tsai et al. 2023; Veverka et al. 2023; Gao et al. 2022; Zhao et al. 2022a; Zhao et al. 2019; Kuniyil Ajith Singh and Xia 2021; Xia 2017; Huang et al. 2024; Martell et al. 2023). These capabilities are particularly advantageous in visualizing tumor heterogeneity, aiding in the identification of aggressive regions prone to malignancy.

PAI operates on the principle of the photoacoustic effect. When pulsed or modulated light is directed into tissue, chromophores absorb the light energy and subsequently undergo thermoelastic expansion, generating US waves. The Grüneisen equation captures this relationship between optical absorption and acoustic signal generation (Xu and Wang 2005):

$$p_0 = \Gamma \eta_{th} \mu_a F \tag{1}$$

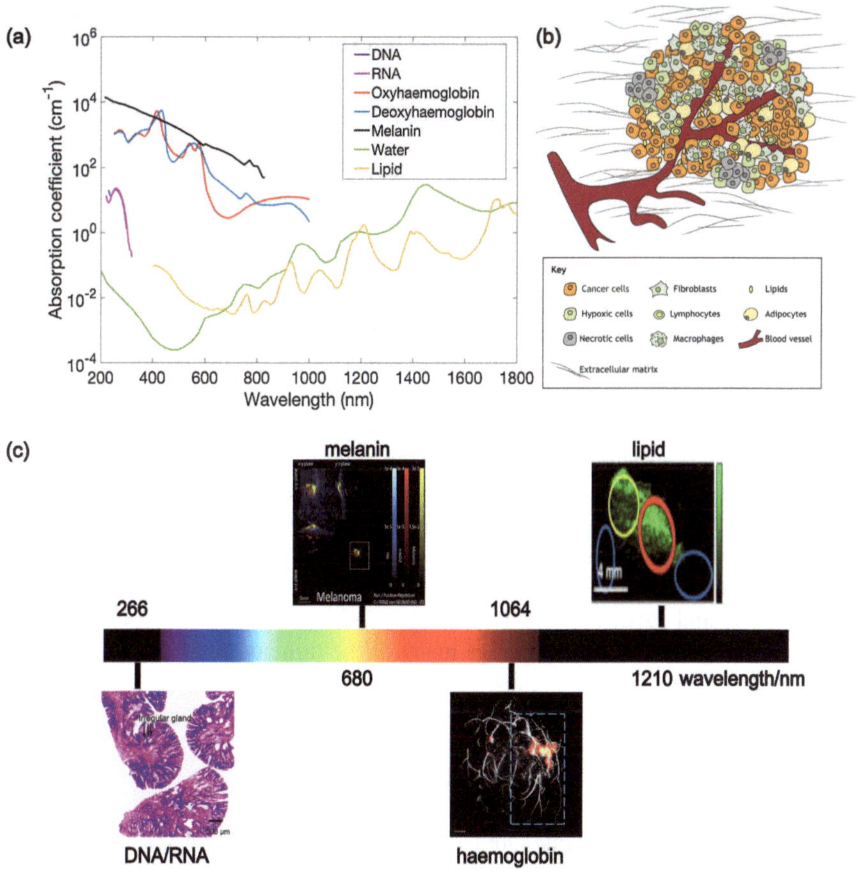

Fig. 1 PAI for tumor detection. (**a**) Optical absorption spectra of various chromophores spanning UV to near-infrared wavelengths, encompassing endogenous molecules such as hemoglobin, lipid, and DNA/RNA. Adapted from (Zhao et al. 2019) under a Creative Commons Attribution License (CC BY 4.0); (**b**) Illustration of the tumor microenvironment, highlighting how these chromophore concentrations (e.g., hemoglobin and lipid) differ between normal and malignant tissues. Reproduced from (Brown et al. 2019) under a Creative Commons Attribution License (CC BY 4.0); (**c**) Photoacoustic images of tumor regions acquired at various wavelengths. By tuning the laser source from deep UV to near-infrared, specific chromophore distributions can be visualized with high contrast, facilitating accurate differentiation of tumor boundaries. The visuals labeled as DNA/RNA, melanin, and hemoglobin were adapted from (Yang et al. 2025; Lin et al. 2018; von Knorring et al. 2025), respectively, all under a Creative Commons Attribution License (CC BY 4.0); while the visual labeled as lipid was adapted with permission from (Li et al. 2015) © 2015 Optical Society of America

where η_{th} represents the fraction of light converted to heat; F represents the local fluence; and the Grüneisen parameter Γ can be expressed as

$$\Gamma = \frac{\beta}{\kappa\rho C_v} \tag{2}$$

Here, β is the thermal coefficient of volume expansion; κ is the isothermal compressibility; and ρ is the density of the tissue (Zhou et al. 2016).

PAI encompasses multiple modalities, each tailored for specific applications with scalable spatial resolution and imaging depths (Xia 2024; Beard 2011). Broadly, these can be categorized into PAM, PACT, and photoacoustic endoscopy (PAE). PAM achieves micrometer-scale spatial resolution, which is ideal for visualizing tumor microvasculature and cellular details. PACT offers greater penetration depths and real-time imaging capabilities, enabling the detection of deep tumor regions. PAE combines flexible probe designs with high resolution for minimally invasive access to internal structures. By choosing the appropriate modality, clinicians and researchers can optimize spatial resolution, penetration depth, and field of view to meet specific needs of surgical or diagnostic procedures.

In subsequent sections, we will delve into each modality, explaining their operating principles, configurations, and comparative advantages. After detailing PAM, PACT, and PAE, we will revisit how PAI, in its various forms, enhances intraoperative guidance and surpasses many of the limitations of existing imaging techniques.

3.1 Photoacoustic Microscopy (PAM)

PAM can be further divided into Optical-Resolution PAM (OR-PAM) and Acoustic-Resolution PAM (AR-PAM), depending on how the focal spot is defined. OR-PAM uses a tightly focused optical beam to achieve lateral resolution on the scale of micrometers or even sub-micrometers, making it suitable for visualizing microvascular networks and cellular structures (Cao et al. 2023b; Shintate et al. 2022; Hu and Wang 2013). However, the imaging depth is limited due to optical attenuation of tissue. In contrast, AR-PAM employs a relatively broad optical beam, with spatial resolution determined by the focal size of the US transducer. This approach allows for deeper imaging penetration (up to a few millimeters), albeit with slightly reduced lateral resolution (Zhang et al. 2023; Moothanchery et al. 2020; Park et al. 2014).

Common PAM configurations employ coaxial alignment of the optical and acoustic beams, which increases detection sensitivity and adaptability for different scanning geometries. These features render PAM highly versatile for applications such as examining superficial tumors, evaluating tumor vasculature, or guiding microsurgeries. Moreover, several research groups have leveraged PAM to visualize the dynamics of tumor growth in animal models (John et al. 2023; Gargiulo et al. 2019), enabling better understanding of the tumor microenvironment. By mapping microvascular networks, researchers can quantify changes in blood flow, observe the development of angiogenesis, and even assess response to anti-angiogenic therapy (Shrestha et al. 2024; Mirg et al. 2022; Zhou et al. 2019). In the context of intraoperative imaging, these detailed vascular and structural insights could help surgeons target regions with active neovascularization or suspected residual disease, improving margin clearance. With the introduction of advanced scanning strategies, including high-speed scanning mirrors or multifocal arrays, PAM's temporal resolution can be further improved, which helps reduce time overheads in the operating theatre (Baik et al. 2021; Cao et al. 2024; Imai et al. 2018; Li et al. 2018, 2022, 2023b; Zhan et al. 2022; Zhu et al. 2024).

3.2 Photoacoustic Computed Tomography (PACT)

PACT extends the imaging depth and field-of-view of PAI by sampling the acoustic field with US detectors. These detectors may include single-element transducers that scan around the object, multi-element transducer arrays, or Fabry-Pérot (FP) sensors (Czaplewski David et al. 2022; Fu et al. 2022; Huynh et al. 2019, 2024; Liang et al. 2020; Manwar et al. 2020; Menozzi and Yao 2024; Yang et al. 2022a; Yao et al. 2011). Instead of point-by-point scanning as in PAM, PACT captures signals from multiple directions to reconstruct cross-sectional or volumetric images of tissue optical absorption. The principle of operation involves illuminating tissue with a short-pulsed laser. The resultant photoacoustic waves emanate from optical absorbers within the tissue and are detected by an array of US transducers arranged in a planar, spherical, or cylindrical configuration. By applying appropriate reconstruction algorithms—such as universal back-projection or time-reversal methods—on the time-resolved signals, a three-dimensional map of tissue absorption can be generated (Wang and Li 2024; Omidi et al. 2022; Sun and Jiang 2022; Dean-Ben and Razansky 2022; Schoonover and Anastasio 2011; Xu and Wang 2005).

Compared to PAM, PACT achieves significantly greater imaging depths, up to several centimeters, making it better suited for visualizing larger or deeper-seated tumors (Menozzi and Yao 2024; Liu et al. 2024a; Kim et al. 2024a). Additionally, its scalable field-of-view accommodates the imaging of entire organs or large lesions. Detector geometries and configurations can be adapted for specific applications. For example, planar-view PACT may utilize 2D matrix piezoelectric transducer arrays or FP interferometers (Huynh et al. 2024; Qiu et al. 2024; Zhu et al. 2023; Chan et al. 2019; Shu et al. 2016; Cox et al. 2007), while spherical or hemispherical arrays are advantageous for more isotropic spatial resolution (Yang et al. 2022a; Kalva et al. 2023; Yip et al. 2022; Deán-Ben et al. 2017a; Deán-Ben et al. 2017b; Xia et al. 2013). By accommodating different architectures and detectors, PACT offers a scalable approach to imaging deeper tissues, facilitating the identification of tumor boundaries in larger lesions or organs.

PACT's depth and volumetric imaging capabilities offer significant advantages for intraoperative tumor assessment. It provides real-time functional information, such as hemoglobin distributions and sO_2 levels, which are critical for assessing tumor hypoxia or necrosis during surgery (Folz et al. 2024; Nasri et al. 2023; Huang et al. 2020; Li et al. 2020; Cao et al. 2017). Research on PACT for intraoperative applications has advanced with the development of bench-top prototypes and pilot clinical systems. Investigators have adapted existing clinical US transducer arrays for dual-modal PAI and US imaging, integrating PACT into sterile surgical environments by tailoring system configurations, including illumination geometry and ergonomic considerations for surgical procedures (Han et al. 2020; Wang et al. 2020; Kuniyil Ajith Singh et al. 2015).

PACT's ability to provide volumetric maps of hemoglobin distribution in real-time has been demonstrated in tissues such as the liver, breast, and prostate, with studies showing that changes in blood vessel density and sO_2 can be monitored intraoperatively (Goh et al. 2024; Tajaldeen et al. 2024; Sun et al. 2023; Gao et al. 2023b; Tong et al. 2023; Rascevska et al. 2022; Sangha et al. 2022; Huang et al. 2022; Wang Lihong et al. 2021; Kosik et al. 2020). This information offers immediate feedback on tumor viability and surgical completeness, aiding surgeons in achieving more precise tumor margin clearance. Advances in high-energy laser diodes and light emitting diodes (Singh et al. 2024; Shi et al. 2022b) and portable laser systems have further improved PACT's accessibility and functionality, bringing it closer to widespread intraoperative use (Seeger et al. 2023; Li et al. 2021c; Kalva and Pramanik 2020; Kuniyil Ajith Singh and Xia 2020; Upputuri and Pramanik 2015).

3.3 Photoacoustic Endoscopy (PAE)

PAE integrates PAI into minimally invasive endoscopic probes, allowing clinicians to access internal tissues directly, circumventing issues related to external illumination and tissue attenuation. In PAE, a pulsed laser beam is delivered through an optical fiber within the endoscope, and the generated photoacoustic signals are collected by a miniaturized US detector at the distal tip. By manipulating the position and orientation of the endoscope, real-time, high-resolution images of vascular and molecular features can be acquired from deep inside the body.

PAE can be adapted to different PAI configurations, including OR-PAM (Kim et al. 2024b; Liang et al. 2023; Sun et al. 2021b; Zhao et al. 2022a, b, c), AR-PAM (Xiao et al. 2022; Pang et al. 2022; Wang et al. 2021a), or PACT-based approaches (Ansari et al. 2018, 2020, 2024; Liu et al. 2024b; Zhang and Beard 2011), offering clinicians the flexibility to choose an imaging strategy tailored to specific clinical requirements. For instance, OR-PAM-based endoscopes (Zhao et al. 2022b, c) could provide cellular-level details in gastrointestinal tissues (Liang et al. 2022; Kim et al. 2022), while AR-PAM or PACT-based endoscopes might be preferred for deeper imaging of vascular structures or tumor infiltration in hollow organs (Liu et al. 2024b; He et al. 2023; Li et al. 2021d; Moore and Jokerst 2019). The endoscopic delivery also facilitates integration with surgical tools, enabling simultaneous imaging and intervention. This integrated, flexible approach parallels the adaptation of other imaging modalities—like US or confocal microscopy—into the surgical workflow, ensuring targeted yet comprehensive guidance for complex procedures.

Recent developments in optical fiber technology (Shabairou et al. 2020; Yang and Hu 2024; Zhao et al. 2022b, c; Zhou and Jokerst 2020) and miniaturized US detectors (Yang et al. 2022b; Zhao et al. 2021, 2022a, b, c) have further boosted PAE's potential. Studies on animal models have demonstrated the feasibility of PAE in detecting early-stage colorectal tumors by mapping microvasculature changes in the intestinal wall (Liang et al. 2022; He et al. 2023; Lin et al. 2024; Ikematsu et al. 2022; Yang et al. 2019). Additionally, for esophageal or bladder cancer, PAE probes have shown promise in locating submucosal lesions that might be missed under conventional white-light endoscopy (Mashimo et al. 2020; Wu et al. 2018; Lerner and Goh 2015; Bergholt et al. 2011). As these prototypes become more robust and user-friendly, the outlook for real-time tumor margin delineation in endoscopic surgeries, such as transurethral resections or endoscopic mucosal resections, become more favorable.

After examining PAM, PACT, and PAE, it becomes clear that PAI offers a unique combination of anatomical and functional information at multiple scales—from microscopic details at shallow depths (PAM) to volumetric imaging of deeper tissues (PACT), and minimally invasive internal assessments (PAE). By complementing other imaging modalities and enhancing real-time feedback, PAI has the potential to streamline intraoperative decision-making, reduce the likelihood of incomplete tumor resections, and improve patient outcomes.

4 PAI for Intraoperative Tumor Margin Assessment

Different from normal organs, tumor tissue typically exhibits specific characteristics such as low sO_2, larger size and density of tumor cells, chaotic and immature blood vessels surrounding the tumor, and a thinner lipid layer compared to normal tissue. As a result, various PAI configurations were proposed to map the distribution of endogenous molecules such as hemoglobin, melanin, and lipid. in order to delineate cancerous tissue for guiding tumor resections (Li et al. 2015). These imaging configurations can be divided into three categories, including clinical US array-based imaging system, UV photoacoustic histology-like imaging, and handheld PAM.

4.1 Clinical US Array-Based Imaging System

Reported clinical US array-based imaging systems for tumor margin assessment can be divided into two categories: multispectral photoacoustic tomography (MS-PAT) and dual-modality photoacoustic and US imaging system. MS-PAT is a multispectral measurement technique that differentiates biomolecules based on their unique optical absorption properties. It uses laser pulses at multiple wavelengths to selectively excite various chromophores, generating photoacoustic signals that correspond to their specific absorption spectra (Yao and Wang 2014; Goh et al. 2018). These signals, produced as ultrasonic waves, are detected, and processed to map the spatial distribution of different biomolecules within tissue. Operating within the 710–1210 nm range, MS-PAT enables high-contrast, label-free visualization of tissue composition, aiding in functional and molecular imaging (Li et al. 2015). Oxyhemoglobin, deoxyhemoglobin, and lipids act as endogenous contrast agents commonly used for tumor margin assessment. Cancerous tissues are often hypoxic, exhibiting lower sO_2, and typically have thinner lipid layers compared to healthy tissue. MS-PAT leverages these differences by quantifying sO_2 levels and detecting lipid variations, making it a powerful tool for intraoperative tumor margin evaluation in resected specimens or exposed tissues.

One critical aspect of tumor characterization is the assessment of sO_2. Solid tumors are often hypoxic due to rapid growth and inadequate vascularization, resulting in regions with low oxygen concentrations that are typically more resistant to therapy. MS-PAT can identify these hypoxic regions by distinguishing oxyhemoglobin from deoxyhemoglobin based on their unique absorption spectra. This information is essential for identifying aggressive tumor areas that may require more extensive resection.

In addition to sO_2, MS-PAT provides valuable information about lipid distribution in tissues. Cancerous tissues frequently display disrupted lipid layers compared to normal tissues. By employing wavelengths around 1210 nm, MS-PAT can effectively visualize lipid variations, offering another critical metric for identifying tumor margins. This capability makes MS-PAT particularly suitable for intraoperative tumor margin assessment because it allows for simultaneous visualization of several key indicators of malignancy, including hypoxia and lipid depletion. Multispectral PAI has already been tested with ex vivo human tumor tissue. In 2015, Li et al. (2015) developed an MS-PAT imaging system extended to the second optical window (1–2 μm) to image both lipid and hemoglobin. The system was illuminated by a Nd:YAG laser with a pulse repetition rate of 10 Hz, and wavelengths switchable from 670 nm to 2300 nm, while photoacoustic signals were collected by a 128-element US array (central frequency 21 MHz). Hemoglobin was visualized at 1064 nm and lipid at 1200 nm; the 3 mm penetration depth and 125 μm axial resolution demonstrated strong potential for assessing tumor margins in excised human breast tissue. In 2018, Goh et al. (2018) applied a commercial multispectral optoacoustic tomography (MSOT) system to capture oxygenated and deoxygenated hemoglobin signals at multiple wavelengths (710, 730, 760, 800, 850, 930, 1050, 1100 nm), and calculated total hemoglobin and sO_2 to delineate malignancy areas. They observed a significant increase in total hemoglobin in tumor regions.

Beyond breast tumors, MS-PAT has also been used to visualize the microvasculature and detect hypoxic regions or lipid changes in other tumor types, including melanoma (Park et al. 2021), oral cancer (Vonk et al. 2022), and more. For instance, Park et al. employed this imaging modality to depict the 3D boundary of human melanomas in vivo (Park 2021). They designed a handheld scanner combining multispectral PAI and US for differentiating melanoma margins in vivo. By tuning

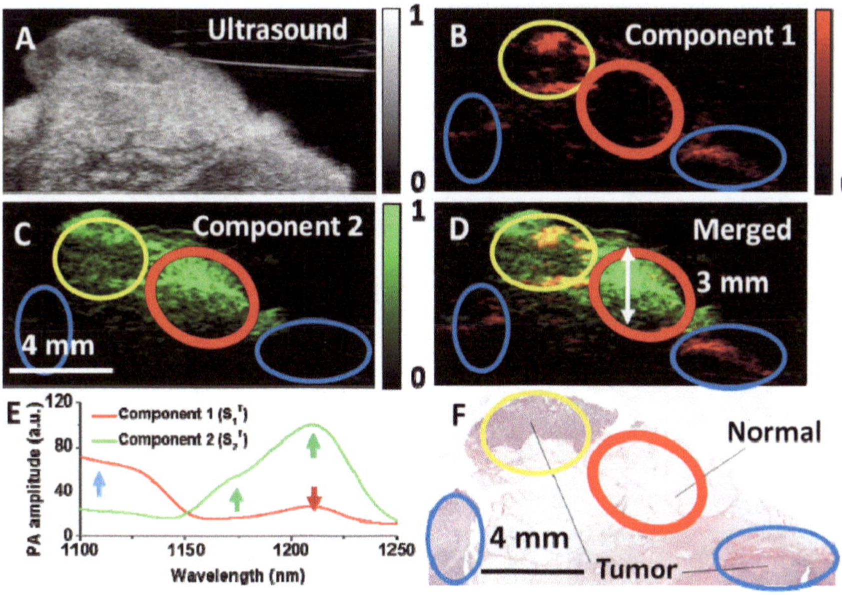

Fig. 2 Multispectral photoacoustic computed tomography imaging system for tumor margin assessment. Representative results from imaging human breast tissue ex vivo: (**a**) US image providing structural information; (**b**) Photoacoustic image of component 1, highlighting regions of high hemoglobin concentration; (**c**) Photoacoustic image of component 2, revealing the distribution of lipid; (**d**) Fused image displaying both hemoglobin and lipid within a ~3 mm imaging depth; (**e**) Overlaid spectra of components 1 and 2, illustrating distinct optical absorption profiles; (**f**) Corresponding H&E histology, validating that the multispectral photoacoustic findings accurately delineate the tumor boundary. (Reprinted with permission from (Li et al. 2015) © 2015 Optical Society of America)

the laser wavelength to 700 nm, they captured strong melanin absorption, clearly identifying tumor boundaries. This approach underscores the potential for rapid, clinically translatable systems that exploit strong optical absorbers such as melanin or hemoglobin to guide tumor resections. Besides, Vonk et al. also integrated MS-PAT with an EGFR-targeted fluorescent tracer to differentiate benign and malignant lymph nodes in oral cancer (Vonk et al. 2022). They adopted the cetuximab-800CW, a type of tumor-specific fluorescent tracer, as the exogeneous contrast agent, aiming at differentiating benign and malignant lymph nodes. These findings underscore the versatility of MS-PAT for mapping various endogenous or exogenous contrast mechanisms. Figure 2 shows an example of the MS-PAT system schematic and its results on excised breast tissue (Li et al. 2015). By combining multiple wavelengths, the tumor region can be distinguished based on hemoglobin content, sO_2, and lipid density, providing comprehensive molecular and functional information vital for tumor margin assessment.

PAI can be seamlessly integrated with conventional US imaging using a clinical US imaging probe (Fig. 3). In this dual-mode approach, the same transducer elements within the probe are employed for both photoacoustic signal detection and

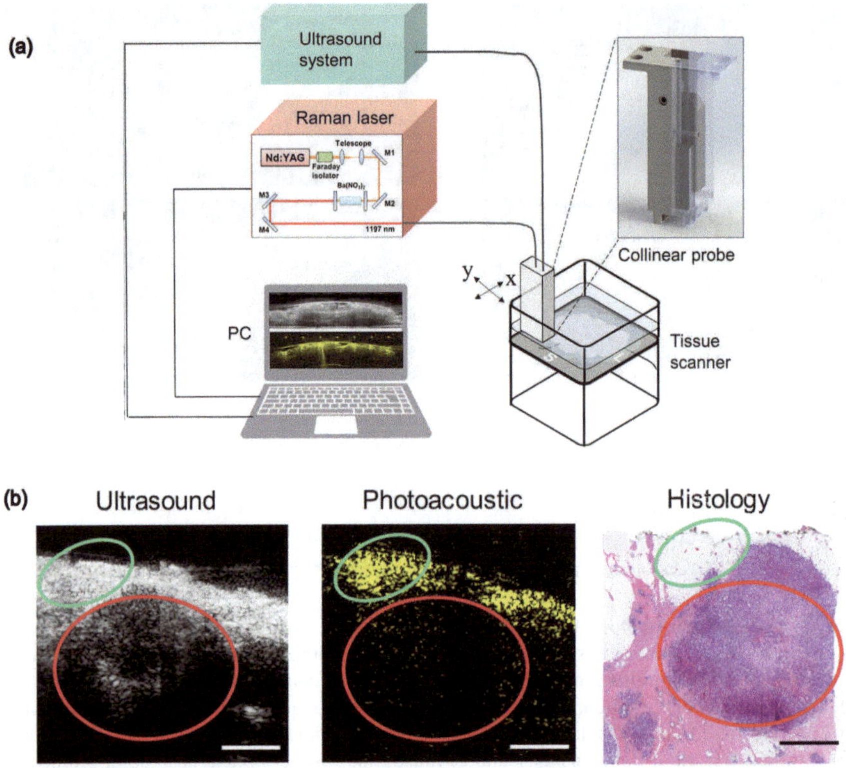

Fig. 3 Dual-modal photoacoustic/US imaging for tumor margin assessment. (**a**) Experimental configuration of the dual-modality system, which couples a pulsed laser source with a 128-element US array. This setup provides both morphological (US-based) and molecular (photoacoustic-based) imaging capabilities in a single scan, enabling comprehensive tissue evaluation. (**b**) Comparative US image, photoacoustic image, and H&E-stained section of excised breast tumor tissue. The green oval delineates adipose tissue, while the red oval highlights the tumor mass. By integrating real-time PAI with conventional US, the system enhances tumor margin detection and complements structural information with functional contrast. (Adapted from (Li et al. 2018) under a Creative Commons Attribution License (CC BY 4.0))

traditional US imaging, ensuring intrinsic spatial co-registration of the two modalities. This integration allows for the simultaneous acquisition of complementary imaging data: US provides high-resolution structural information, while PAI offers molecular and functional contrast based on optical absorption. The resulting PAI/US imaging systems enhance the comprehensive assessment of tissue characteristics while facilitating clinical translation, as clinicians can leverage their familiarity with US imaging to adopt this advanced hybrid modality more readily.

Several research groups have developed dual-modality US and PAI systems for tumor margin assessment. In 2018, Li et al. (2018) introduced a high-speed US/PAI imaging system designed to evaluate excised human breast tumor tissue. Their system utilized a 128-element transducer array to acquire both US and photoacoustic signals, enabling the scanning of a 10 × 10 cm^2 area within approximately 10 min. To further enhance margin detection accuracy, they integrated a convolutional neural network for image analysis. Similarly, Goh et al. (2018) developed a US/PAI dual-modality imaging system by combining a 2D handheld photoacoustic probe with a commercial MSOT platform. This system enabled the visualization of lipid and hemoglobin distributions, providing valuable molecular contrast. Meanwhile, Wang et al. (2021b) applied a comparable approach to assess pancreatic tumor margins in rat models, incorporating an OPO laser to quantify tissue sO_2 levels in hypoxic tumor regions.

4.2 Ultraviolet PAM (UV-PAM)

Ultraviolet PAM (UV-PAM) is a label-free nuclear imaging modality that relies on the strong absorption of DNA and RNA in the cell nucleus over the UV spectrum (Wong et al. 2017b). This capability makes UV-PAM particularly valuable for differentiating between malignant and non-malignant cells, as tumor cells typically have larger, more irregularly shaped nuclei with higher nuclear-to-cytoplasmic ratios compared to healthy cells (Veluponnar et al. 2023). UV-PAM achieves a high lateral resolution of approximately 1 μm (Baik et al. 2021). Notably, it can eliminate the need for tissue processing or staining, making it a potential alternative or supplement to conventional H&E histopathological analysis for tumor margin assessment during surgery.

To date, three types of UV-PAM configurations have been developed for either in vitro or in vivo imaging: transmission mode UV-PAM, reflection mode UV-PAM, and UV photoacoustic remote sensing (UV-PARS). In transmission mode UV-PAM, the sample tissue is positioned between the illuminating laser beam and the US transducer. In 2017, Wong et al. (2017) demonstrated a transmission-mode UV-PAM system (266 nm Nd:YLF laser; 50 MHz ultrasonic transducer) for histology-like imaging of unprocessed human breast tumor tissue, obtaining submicrometric resolution and good concordance with standard H&E histology results. Despite its advancements, transmission-mode systems still face drawbacks, such as limitations imposed by sample thickness constraints.

Reflection mode UV-PAM addresses the thickness limitation by placing both illumination and detection on the same side of the tissue. Baik et al. (2021) introduced a reflection-mode imaging system featuring a micro-electro-mechanical system (MEMS) scanner, achieving a lateral resolution of about 1.2 μm and fast scanning over a 10 × 10 mm^2 area in ~240 seconds. This system can effectively replicate histological analysis for imaging deparaffinized human colon and liver tumor sections. In addition, a recent article on optical-resolution parallel UV-PAM for slide-free histology (Cao et al. 2024) has demonstrated further advancements in high-throughput, high-resolution UV-PAM imaging. The parallel excitation scheme enables faster image acquisition, improving the practical utility of UV-PAM in intraoperative or point-of-care settings. Furthermore, UV-PAM can be integrated with other modalities to enhance its capability in detecting micro-infiltrated tumors. In 2025, Yang et al. (2025) combined Photoacoustic Spectral Analysis (PASA) with reflection-mode UV-PAM, enabling the analysis of intrinsic tissue elastic properties, as shown in Fig. 4. This dual-modality imaging technique was validated using human colorectal tissue samples, achieving both histological visualization and quantitative nuclear property analysis within 20 min. Thus, the joint UV-PAM and in situ PASA imaging system holds promise for evaluating micro-infiltrative colorectal cancer.

A third configuration, UV-PARS, replaces the conventional piezoelectric US sensor with a continuous-wave interrogation beam at 1310 nm. The local optical property changes induced by the photoacoustic signal are detected in a non-contact manner, eliminating the need for an acoustic coupling medium. Abbasi et al. (2020) demonstrated a UV-PARS system with a MEMS scanner for high-speed histology-like imaging (lateral resolution ~1.2 μm, axial resolution ~7.3 μm). Bell et al. (2020) further validated this approach for rapid volumetric imaging of breast tumor margins and healthy human brain tissue, showing excellent agreement with histological analysis.

By measuring nuclear size and density changes—hallmarks of malignancy—UV-PAM can pinpoint cancerous regions with subcellular precision. Consequently, UV-PAM holds promise for real-time margin assessment in surgeries where immediate histological confirmation is essential, potentially reducing reliance on FSA or conventional histopathology. As UV-PAM continues to evolve, its ability to provide

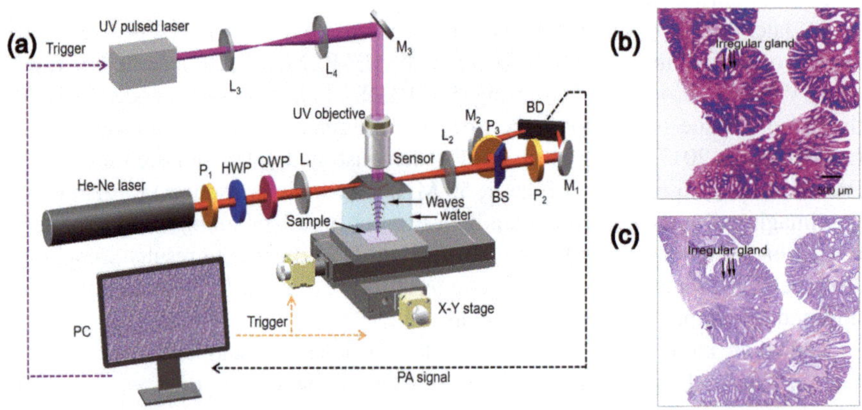

Fig. 4 Joint UV-PAM and in situ PASA imaging system and its acquired UV-PAM image. (**a**) Schematic diagram of the joint PAMI and in situ PASA imaging system, where the sample is positioned at the opposite side of both 266 nm excitation laser beam and an ultrasonic transducer to detect nuclear-level optical absorption from DNA/RNA. (**b**) Representative UV-PAM image of fixed, unprocessed adenoma tissue, highlighting subcellular features associated with malignancy. (**c**) H&E-stained histological image of the same area, obtained after standard processing, sectioning, and staining, confirming that the nuclei-rich regions observed in the UV-PAM image correspond to the malignant tumor. (Adapted from (Yang et al. 2025) under a Creative Commons Attribution License (CC BY 4.0))

label-free contrast based on nuclear morphology positions it as a powerful candidate for supplementing or even replacing FSA in surgical workflows.

4.3 Handheld PAM

Handheld PAM (often single- or dual-wavelength), is normally illuminated by a pulsed laser, which can effectively detect morphological and vascular features of angiogenesis to evaluate tumor margins in real-time. Additionally, melanomas can be readily delineated with single-wavelength photoacoustic systems due to melanin's strong optical absorption. In 2012, Xi et al. (2012) developed an intraoperative system comprising an Nd:YAG 532 nm laser and a custom-designed lead zirconate titanate (PZT) US transducer, with a penetration depth up to 2.3 mm (Fig. 5). The system used a micro-electro-mechanical scanning approach for high-speed imaging in both ex vivo and in vivo mouse tumor models, showing an ~8.5% margin-detection error ex vivo and ~21.5% error in deeper tumor targets when overlaid with chicken breast tissue. Although the penetration depth was modest, the study demonstrated the feasibility of using an intraoperative single-wavelength PAI system to distinguish tumors from surrounding tissues based on vascular distribution.

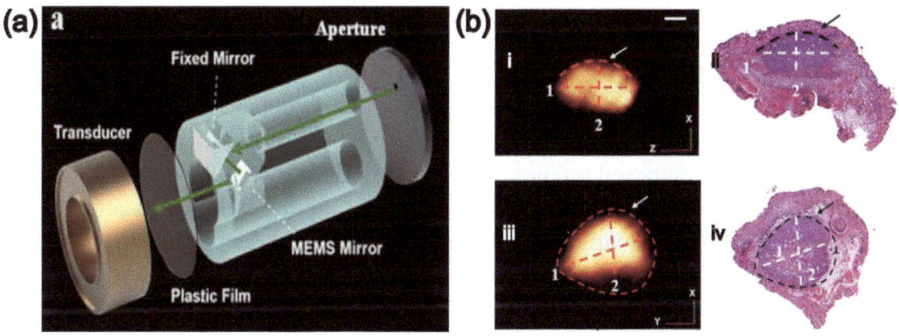

Fig. 5 Handheld PAM system for tumor margin assessment. (**a**) Schematic illustration of the imaging system. The system employs a pulsed Nd:YAG 532 nm laser and a custom-designed PZT US transducer, while an MEMS stage performs high-speed optical scanning to enable real-time mapping of superficial tumors. (**b**) Representative slices from the photoacoustic images compared with H&E histology at varying depths: (i, ii) Photoacoustic slice and H&E section of a tumor located 0.3 mm beneath the mouse skin, demonstrating clear morphological correlation; (iii, iv) Photoacoustic slice and H&E section of a tumor covered by a 1 mm-thick chicken breast layer (white dashed line), showing the system's capability to detect and delineate subsurface tumors through intervening tissue. (Adapted with permission from (Xi et al. 2012) © 2012 Optical Society of America)

5 Conclusions and Future Perspectives

Intraoperative tumor margin assessment is a critical component of cancer surgery. Due to the high aggressiveness and fatality associated with cancer, as well as the need for re-excision procedures when positive margins are detected, there is an urgent need for improved techniques to assess tumor margins and enhance treatment strategies. Currently, FSA is the most widely used method for tumor margin assessment. While accurate, FSA is time-consuming, costly, and can damage tissue structures. PAI addresses these limitations by providing both anatomical and functional information, such as hemoglobin levels, lipid content, and nuclear morphology, offering a comprehensive understanding of the tumor microenvironment. PAI's ability to identify key tumor-specific features—such as chaotic vasculature, hypoxia, and increased nuclear density—makes it a powerful tool for distinguishing malignant from healthy tissue during surgery. This capability ultimately contributes to more precise tumor resections and improved patient outcomes.

Furthermore, the versatility of PAI is enhanced by its compatibility with other imaging modalities, enabling more comprehensive and accurate assessments. Dual-modality imaging systems, such as US-photoacoustic platforms, combine the deep tissue penetration and anatomical detail of US with the molecular and functional insights of PAI. These systems have demonstrated notable success in tumor margin assessments in breast and pancreatic cancer models, where they not only enhanced the detection of malignant tissues but also improved the precision of surgical resections by providing real-time, multimodal feedback (Li et al. 2018; Goh et al. 2018; Wang et al. 2021b).

In addition to US, the integration of PAI with optical imaging techniques offers even greater potential. Raman Spectroscopy, for example, provides detailed chemical information based on molecular vibrations, allowing for the identification of specific biomolecular signatures such as proteins, lipids, and nucleic acids. When combined with PAI, Raman Spectroscopy can provide a deeper understanding of the biochemical environment of the tumor, allowing for the detection of molecular changes that are indicative of malignancy, such as altered lipid content or increased collagen deposition. The high specificity of Raman in detecting these biochemical markers can complement PAI's functional imaging, thereby enabling more precise discrimination between tumor and healthy tissue. OCT, another optical technique, offers high-resolution cross-sectional imaging of tissue structures based on light scattering. OCT provides detailed images of tissue architecture at the micron scale, making it valuable for visualizing cellular morphology and layer structures within tissues. When integrated with PAI, OCT enhances the system's ability to provide both high spatial resolution (from OCT) and functional information (from PAI), offering a more complete picture of tissue composition and tumor boundaries. This combination can be particularly useful in distinguishing malignant tissues from benign ones in critical surgical procedures, such as glioma resections or breast cancer surgeries. Although these dual-modality systems show immense promise, much of the integration of PAI with Raman Spectroscopy and OCT remains in the research phase. However, the potential for these combined imaging platforms to provide both high specificity and functional depth makes them a promising avenue for improving tumor margin assessments in clinical settings. By providing detailed anatomical, functional, and molecular information in real-time, these multimodal imaging systems could lead to more precise tumor resections, reduced recurrence rates, and ultimately, better patient outcomes.

In neurosurgery specifically, combining iMRI with PAI has begun to attract interest, particularly for high-grade gliomas, where accurate delineation is paramount. As multiple research groups continue to refine iMRI usage (Lavrador et al. 2021) and explore synergy with PAI, we may see a paradigm shift in the surgical management of brain tumors. In addition, the development of specialized photoacoustic endoscopic probes may facilitate tumor margin analysis in other challenging anatomical sites such as the gastrointestinal tract, bladder, and thoracic cavity. With continuous advancements in photoacoustic technology, driven by recent comprehensive clinical studies and commercialization efforts (Xia 2024), PAI is poised to become a pivotal tool in the operating room. By bridging the gap between structural and functional imaging, it represents the next generation of intraoperative guidance systems. As these prototypes transition into commercially available devices, clinicians can expect more precise, efficient, and cost-effective procedures, ultimately improving patient outcomes and reducing morbidity and mortality across a wide range of cancers.

Acknowledgments N/A

Conflict of Interest The authors declare that there is no conflict of interest regarding the publication of this chapter.

References

Aaboubout Y, Barroso E, Algoe M, Ewing-Graham P, Ten Hove I, Mast H, Hardillo J, Sewnaik A, Monserez D, Keereweer S, Baatenburg de Jong RJ, Puppels GJ, Koljenovic S et al (2021) Intraoperative assessment of resection margins in Oral cavity cancer: this is the way. J Vis Exp 171:e62446

Abasi S, Aggas JR, Garayar-Leyva GG, Walther BK, Guiseppi-Elie A (2022) Bioelectrical impedance spectroscopy for monitoring mammalian cells and tissues under different frequency domains: a review. ACS Meas Sci Au 2:495–516

Abbasi S, Bell K, Ecclestone B, Reza PH (2020) Real time & 3D photoacoustic remote sensing. In: Proceedings of the optics and the brain. Optica Publishing Group, JTh2A–13

Abdelshafy SZAA, Refaat MM, Saleh GE, Yousef MI (2018) Value of intra-operative ultrasound in localization of palpable or non-palpable breast tumors during breast conserving surgery. Egypt J Radiol Nuclear Med 49:1169–1174

Achterberg FB, Bijlstra OD, Slooter MD, Mulder BGS, Boonstra MC, Bouwense SA, Bosscha K, Coolsen MM, Derksen WJ, Gerhards MF et al (2024) ICG-fluorescence imaging for margin assessment during minimally invasive colorectal liver metastasis resection. JAMA Netw Open 7:e246548

Al-Attili M, Ferreira C, Price C, Faulds K, Chen YC (2023) Development of a spatially offset Raman spectroscopy probe for monitoring pharmaceutical drying. Chem Eng Res Des 192:510–520

Alharbi FM, Chahwan CK, Le Gal SG, Guleryuz KM, Tillou XP, Doerfler AP (2016) Intraoperative ultrasound control of surgical margins during partial nephrectomy. Urol Ann 8:430–433

Allen WM, Foo KY, Zilkens R, Kennedy KM, Fang Q, Chin L, Dessauvagie BF, Latham B, Saunders CM, Kennedy BF (2018) Clinical feasibility of optical coherence micro-elastography for imaging tumor margins in breast-conserving surgery. Biomed Opt Express 9:6331–6349

Altaf A, Shakir M, Malik MJA, Arif A, Islam O, Mubarak F, Knopp E, Siddiqui K, Enam SA (2023) Intraoperative use of low-field magnetic resonance imaging for brain tumors: A systematic review. Surgical. Neurol Int 14:14

Anichini G, Leiloglou M, Hu Z, O'Neill K, Elson D (2024) Hyperspectral and multispectral imaging in neurosurgery: a systematic literature and metanalysis. Eur J Surg Oncol 51:108293

Ansari R, Zhang EZ, Desjardins AE, Beard PC (2018) All-optical forward-viewing photoacoustic probe for high-resolution 3D endoscopy. Light: Sci Appl 7:75

Ansari R, Zhang EZ, Desjardins AE, Beard PC (2020) Miniature all-optical flexible forward-viewing photoacoustic endoscopy probe for surgical guidance. Opt Lett 45:6238–6241

Ansari R, Zhang E, Beard P (2024) Dual-modality rigid endoscope for photoacoustic imaging and white light videoscopy. J Biomed Opt 29:020502–020502

Armani A, Borst J, Douglas S, Goldharber N, Taj R, Blair SL (2022) Intraoperative margin trials in breast cancer. Curr Breast Cancer Rep 14:65–74

Arneth B (2019) Tumor microenvironment. Medicina 56:15

Auner GW, Koya SK, Huang C, Broadbent B, Trexler M, Auner Z, Elias A, Mehne KC, Brusatori MA (2018) Applications of Raman spectroscopy in cancer diagnosis. Cancer Metastasis Rev 37:691–717

Badhey AK, Schwarz JS, Laitman BM, Veremis BM, Westra WH, Yao M, Teng MS, Genden EM, Miles BA (2023) Intraoperative use of wide-field optical coherence tomography to evaluate tissue microstructure in the oral cavity and oropharynx. JAMA Otolaryngol Head Neck Surg 149:71–78

Baik JW, Kim H, Son M, Choi J, Kim KG, Baek JH, Park YH, An J, Choi HY, Ryu SY et al (2021) Intraoperative label-free photoacoustic histopathology of clinical specimens. Laser Photonics Rev 15:2100124

Barchéus H, Peischl C, Björkman-Burtscher IM, Pettersson C, Smits A, Nilsson D, Farahmand D, Eriksson J, Skoglund T, Corell A (2024) Observations from the first 100 cases of intraoperative MRI–experiences, trends and short-term outcomes. BMC Surg 24:268

Beard P (2011) Biomedical photoacoustic imaging. Interface Focus 1:602–631

Bell K, Abbasi S, Dinakaran D, Taher M, Bigras G, van Landeghem FK, Mackey JR, Haji Reza P (2020) Reflection-mode virtual histology using photoacoustic remote sensing microscopy. Sci Rep 10:19121

Belykh E, Bardonova L, Abramov I, Byvaltsev VA, Kerymbayev T, Yu K, Healey DR, Luna-Melendez E, Deneen B, Mehta S et al (2023) 5-aminolevulinic acid, fluorescein sodium, and indocyanine green for glioma margin detection: analysis of operating wide-field and confocal microscopy in glioma models of various grades. Front Oncol 13:1156812

Bergholt M, Zheng W, Lin K, Ho K, Teh M, Yeoh K, So JY, Huang Z (2011) In vivo diagnosis of esophageal cancer using image-guided Raman endoscopy and biomolecular modeling. Technol Cancer Res Treat 10:103–112

Bergqvist L, Heinonen A, Carcopino X, Redman C, Aro K, Kiviharju M, Virtanen S, Omar PL, Kotaniemi-Talonen L, Louvanto K et al (2023) Added value of electrical impedance spectroscopy in adjunction of colposcopy: a prospective cohort study. BMJ Open 13:e074921

Bouget D, Pedersen A, Jakola AS, Kavouridis V, Emblem KE, Eijgelaar RS, Kommers I, Ardon H, Barkhof F, Bello L et al (2022) Preoperative brain tumor imaging: models and software for segmentation and standardized reporting. Front Neurol 13:932219

Briz P, López-Alonso B, Sarnago H, Burdío J, Lucía O (2023) Tumor location on electroporation therapies by means of multi-electrode structures and machine learning. Bioelectrochemistry 154:108510

Brown E, Brunker J, Bohndiek SE (2019) Photoacoustic imaging as a tool to probe the tumour microenvironment. Dis Model Mech 12:dmm039636

Cahill LC, Giacomelli MG, Yoshitake T, Vardeh H, Faulkner-Jones BE, Connolly JL, Sun CK, Fujimoto JG (2018) Rapid virtual hematoxylin and eosin histology of breast tissue specimens using a compact fluorescence nonlinear microscope. Lab Investig 98:150–160

Campbell JM, Habibalahi A, Handley S, Agha A, Mahbub SB, Anwer AG, Goldys EM (2023) Emerging clinical applications in oncology for non-invasive multi-and hyperspectral imaging of cell and tissue autofluorescence. J Biophotonics 16:e202300105

Cao F, Qiu Z, Li H, Lai P (2017) Photoacoustic imaging in oxygen detection. Appl Sci 7:1262

Cao R, Nelson SD, Davis S, Liang Y, Luo Y, Zhang Y, Crawford B, Wang LV (2023a) Label-free intraoperative histology of bone tissue via deep-learning-assisted ultraviolet photoacoustic microscopy. Nat Biomed Eng 7:124–134

Cao R, Zhao J, Li L, Du L, Zhang Y, Luo Y, Jiang L, Davis S, Zhou Q, de la Zerda A et al (2023b) Optical-resolution photoacoustic microscopy with a needle-shaped beam. Nat Photonics 17:89–95

Cao R, Luo Y, Zhao J, Zeng Y, Zhang Y, Zhou Q, le da Zerda A, Wang LV (2024) Optical-resolution parallel ultraviolet photoacoustic microscopy for slide-free histology. Science. Advances 10:eado0518

Carnicelli G, Disconzi L, Cerasuolo M, Casiraghi E, Costa G, De Virgilio A, Esposito AA, Ferreli F, Fici F, Lo Casto A et al (2023) Image-guided intraoperative assessment of surgical margins in oral cavity squamous cell cancer: a diagnostic test accuracy review. Diagnostics 13:1846

Carobbio ALC, Cheng Z, Gianiorio T, Missale F, Africano S, Ascoli A, Fragale M, Filauro M, Marchi F, Guastini L et al (2023) Electric bioimpedance sensing for the detection of head and neck squamous cell carcinoma. Diagnostics 13:2453

Cassar Q, Al-Ibadi A, Mavarani L, Hillger P, Grzyb J, MacGrogan G, Zimmer T, Pfeiffer UR, Guillet JP, Mounaix P (2018) Pilot study of freshly excised breast tissue response in the 300–600 GHz range. Biomed Opt Express 9:2930–2942

Chan J, Zheng Z, Bell K, Le M, Reza PH, Yeow JT (2019) Photoacoustic imaging with capacitive micromachined ultrasound transducers: principles and developments. Sensors 19:3617

Chaudhry N, Albinsson J, Cinthio M, Kröll S, Malmsjö M, Rydén L, Sheikh R, Reistad N, Zackrisson S (2023) Breast cancer diagnosis using extended-wavelength–diffuse reflectance spectroscopy (EW-DRS)—proof of concept in ex vivo breast specimens using machine learning. Diagnostics 13:3076

Chen C, Hao J, Hao X, Xu W, Xiao C, Zhang J, Pu Q, Liu L (2021) The accuracy of Raman spectroscopy in the diagnosis of lung cancer: a systematic review and meta-analysis. Transl Cancer Res 10:3680–3693

Chen Z, Han F, Du Y, Shi H, Zhou W (2023) Hypoxic microenvironment in cancer: molecular mechanisms and therapeutic interventions. Signal Transduct Target Ther 8:70

Chen W, Zhang Y, Zhang L, Luo X, Yang X, Zhu Y, Wang G, Huang W, Zhang D, Zeng Y et al (2024) Intraoperative evaluation of tumor margins using a TROP2 near-infrared imaging probe to enable human breast-conserving surgery. Sci Transl Med 16:eado2461

Chiba R, Ebihara Y, Shiiya H, Ujiie H, Fujiwara-Kuroda A, Kaga K, Li L, Wakasa S, Hirano S, Kato T (2022) A novel system for analyzing indocyanine green (ICG) fluorescence spectra enables deeper lung tumor localization during thoracoscopic surgery. J Thorac Dis 14:2943–2952

Chin L, Latham B, Saunders CM, Sampson DD, Kennedy BF (2017) Simplifying the assessment of human breast cancer by mapping a micro-scale heterogeneity index in optical coherence elastography. J Biophotonics 10:690–700

Ching-Roa VD, Huang CZ, Giacomelli MG (2021) Improved microscopy with ultraviolet surface excitation (MUSE) using high-index immersion illumination. Biomed Opt Express 12:6461–6473

Conq J, Joudiou N, Ucakar B, Vanvarenberg K, Préat V, Gallez B (1957) Assessment of Hyperosmolar Blood–Brain Barrier Opening in Glioblastoma via Histology with Evans Blue and DCE-MRI. Biomedicine 2023:11

Cortés JCG, Carmona JJD, Gutiérrez AIB, Medina JAP, Ramírez AAA, Viscaya JAM, Villegas-Saucillo JJ, Olivarez JP (2024) Modular and portable system design for 3D imaging of breast tumors using electrical impedance tomography. Sensors (Basel, Switzerland) 24:6370

Cox BT, Arridge SR, Beard PC (2007) Photoacoustic tomography with a limited-aperture planar sensor and a reverberant cavity. Inverse Probl 23:S95–S112

Czaplewski David A, Zhang Hao F et al (2022) High-frequency 3D photoacoustic computed tomography using an optical microring resonator. BME Front 2022:9891510

Dai Y, Yu X, Wei J, Zeng F, Li Y, Yang X, Luo Q, Zhang Z (2020) Metastatic status of sentinel lymph nodes in breast cancer determined with photoacoustic microscopy via dual-targeting nanoparticles. Light Sci Appl 9:164

Darr C, Costa PF, Kahl T, Moraitis A, Engel J, Al-Nader M, Reis H, Köllermann J, Kesch C, Krafft U et al (2023) Intraoperative molecular positron emission tomography imaging for intraoperative assessment of radical prostatectomy specimens. Eur Urol Open Sci 54:28–32

De Decker K, Jaroch KH, Edens MA, Bart J, Kooreman LF, Kruitwagen RF, Nijman HW, Kruse AJ (2021) Frozen section diagnosis of borderline ovarian tumors with suspicious features of invasive cancer is a devil's dilemma for the surgeon: A systematic review and meta-analysis. Acta Obstet Gynecol Scand 100:1369–1376

Dean-Ben XL, Razansky D (2022) A practical guide for model-based reconstruction in optoacoustic imaging. Front Phys 10:1028258

Deán-Ben XL, López-Schier H, Razansky D (2017a) Optoacoustic micro-tomography at 100 volumes per second. Sci Rep 7:6850

Deán-Ben XL, Fehm TF, Ford SJ, Gottschalk S, Razansky D (2017b) Spiral volumetric optoacoustic tomography visualizes multi-scale dynamics in mice. Light: Sci Appl 6:e16247–e16247

Debacker JM, Schelfhout V, Brochez L, Creytens D, D'Asseler Y, Deron P, Keereman V, Van de Vijver K, Vanhove C, Huvenne W (2021) High-resolution 18F-FDG PET/CT for assessing three-dimensional intraoperative margins status in malignancies of the head and neck, a proof-of-concept. J Clin Med 10:3737

Demir B, Incaz S, Uckuyulu EI, Oysu C (2022) Accuracy of frozen section examination in oral cavity cancers. Ear Nose Throat J 101:NP354–NP357

Dixon L, Lim A, Grech-Sollars M, Nandi D, Camp S (2022) Intraoperative ultrasound in brain tumor surgery: A review and implementation guide. Neurosurg Rev 45:2503–2515

Dizbay Sak S, Sevim S, Buyuksungur A, Kayı Cangır A, Orhan K (2023) The value of micro-CT in the diagnosis of lung carcinoma: a radio-histopathological perspective. Diagnostics 13:3262

Doležal T, Štumpf M, Hua Q, Franek O (2025) Measurement of electrical conductivity of human tissue: A feasibility study of a novel time-domain approach. Measurement 242:115763

Dowling GP, Hehir CM, Daly GR, Hembrecht S, Keelan S, Giblin K, Alrawashdeh MM, Boland F, Hill AD (2024) Diagnostic accuracy of intraoperative methods for margin assessment in breast cancer surgery: A systematic review & meta-analysis. Breast 76:103749

Ember K, Dallaire F, Plante A, Sheehy G, Guiot MC, Agarwal R, Yadav R, Douet A, Selb J, Tremblay JP et al (2024) In situ brain tumor detection using a Raman spectroscopy system—results of a multicenter study. Sci Rep 14:13309

Evans AL, Buehler D, Schulte JJ, McCarthy DP, Hagness SC (2024) Microwave dielectric properties of normal, fibroelastotic, and malignant human lung tissue. Biomed Phy Eng Express 10:055027

Fahmy HM, Hamad AM, Sayed FAZ, Abdelaziz YS, Abu Serea ES, Mustafa ABED, Mohammed MA, Saadeldin AM (2020) Dielectric spectroscopy signature for cancer diagnosis: A review. Microw Opt Technol Lett 62:3739–3753

Fan B, Neel VA, Yaroslavsky AN (2017) Multimodal imaging for nonmelanoma skin cancer margin delineation. Lasers Surg Med 49:319–326

Farrakhova D, Maklygina Y, Romanishkin I, Yakovlev D, Plyutinskaya A, Bezdetnaya L, Loschenov V (2022) Fluorescence imaging analysis of distribution of indocyanine green in molecular and nanoform in tumor model. Photodiagn Photodyn Ther 37:102636

Faur C, Falamas A, Chirila M, Roman R, Rotaru H, Moldovan M, Albu S, Baciut M, Robu I, Hedesiu M (2022) Raman spectroscopy in oral cavity and oropharyngeal cancer: A systematic review. Int J Oral Maxillofac Surg 51:1373–1381

Fei B, Lu G, Wang X, Zhang H, Little JV, Patel MR, Griffith CC, El-Diery MW, Chen AY (2017) Label-free reflectance hyperspectral imaging for tumor margin assessment: a pilot study on surgical specimens of cancer patients. J Biomed Opt 22:086009

Fereidouni F, Harmany ZT, Tian M, Todd A, Kintner JA, McPherson JD, Borowsky AD, Bishop J, Lechpammer M, Demos SG et al (2017) Microscopy with ultraviolet surface excitation for rapid slide-free histology. Nat Biomed Eng 1:957–966

Fernandes S, Williams E, Finlayson N, Stewart H, Dhaliwal C, Dorward DA, Wallace WA, Akram AR, Stone J, Dhaliwal K et al (2024) Fibre-based fluorescence-lifetime imaging microscopy: a real-time biopsy guidance tool for suspected lung cancer. Transl Lung Cancer Res 13:355–361

Folz J, Jo J, Gonzalez ME, Eido A, Zhai T, Caruso R, Kleer CG, Wang X, Kopelman R (2024) Photoacoustic lifetime oxygen imaging of radiotherapy-induced tumor reoxygenation in vivo. J Photochem Photobiol 21:100241

Fu B, Cheng Y, Shang C, Li J, Wang G, Zhang C, Sun J, Ma J, Ji X, He B (2022) Optical ultrasound sensors for photoacoustic imaging: a narrative review. Quant Imaging Med Surg 12:1608–1631

Fuentes AM, Ansari D, Burch TG, Mehta AI (2022) Use of intraoperative MRI for resection of intracranial tumors: A nationwide analysis of short-term outcomes. J Clin Neurosci 99:152–157

Gaba F, Tipping WJ, Salji M, Faulds K, Graham D, Leung HY (2022) Raman spectroscopy in prostate cancer: techniques, applications and advancements. Cancer 14:1535

Gao X, Chen X, Hu H, Wang X, Yue W, Mu J, Lou Z, Zhang R, Shi K, Chen X et al (2022) A photoacoustic patch for three-dimensional imaging of hemoglobin and core temperature. Nat Commun 13:7757

Gao G, Miyasato D, Barner LA, Serafin R, Bishop KW, Xie W, Glaser AK, Rosenthal EL, True LD, Liu JT (2023a) Comprehensive surface histology of fresh resection margins with rapid open-top light-sheet (otls) microscopy. IEEE Trans Biomed Eng 70:2160–2171

Gao S, Wang Y, Ma X, Zhou H, Jiang Y, Yang K, Lu L, Wang S, Nephew BC, Fichera L et al (2023b) Intraoperative laparoscopic photoacoustic image guidance system in the da Vinci surgical system. Biomed Opt Express 14:4914–4928

García-Figueiras R, Baleato-González S, Padhani AR, Luna-Alcalá A, Vallejo-Casas JA, Sala E, Vilanova JC, Koh DM, Herranz-Carnero M, Vargas HA (2019) How clinical imaging can assess cancer biology. Insights Imaging 10:1–35

Gargiulo S, Albanese S, Mancini M (2019) State-of-the-art preclinical photoacoustic imaging in oncology: recent advances in cancer theranostics. Contrast Media Mol Imaging 2019:5080267–5080224

Gautheron A, Bernstock J, Picart T, Guyotat J, Valdés P, Montcel B (2024) 5-ALA induced PpIX fluorescence spectroscopy in neurosurgery: a review. Front Neurosci 18:1310282

Geha RC, Taback B, Cadena L, Borden B, Feldman S (2020) A single institution's randomized double-armed prospective study of lumpectomy margins with adjunctive use of the MarginProbe in nonpalpable breast cancers. Breast J 26:2157–2162

Ghanbarzadeh-Daghian A, Ahmadian MT, Ghanbarzadeh-Dagheyan A (2020) Quick, single-frequency dielectric characterization of blood samples of pediatric cancer patients by a cylindrical capacitor: pilot study. Electronics 9:95

Giacomelli MG, Yoshitake T, Cahill LC, Vardeh H, Quintana LM, Faulkner-Jones BE, Brooker J, Connolly JL, Fujimoto JG (2018) Multiscale nonlinear microscopy and widefield white light imaging enables rapid histological imaging of surgical specimen margins. Biomed Opt Express 9:2457–2475

Giammalva GR, Ferini G, Musso S, Salvaggio G, Pino MA, Gerardi RM, Brunasso L, Costanzo R, Paolini F, Di Bonaventura R et al (2022) Intraoperative ultrasound: emerging technology and novel applications in brain tumor surgery. Front Oncol 12:818446

Gkouzionis I, Zhong Y, Nazarian S, Darzi A, Patel N, Peters CJ, Elson DS (2024) A YOLOv5-based network for the detection of a diffuse reflectance spectroscopy probe to aid surgical guidance in gastrointestinal cancer surgery. Int J Comput Assist Radiol Surg 19:11–14

Goff NK, Ashby L, Ashour R (2024) Meta-analysis of the efficacy of Raman spectroscopy and machine-learning-based identification of glioma tissue. World Neurosurg 189:26–32

Goh Y, Balasundaram G, Moothanchery M, Attia A, Li X, Lim HQ, Burton N, Qiu Y, Putti TC, Chan CW et al (2018) Multispectral optoacoustic tomography in assessment of breast tumor margins during breast-conserving surgery: a first-in-human case study. Clin Breast Cancer 18:e1247–e1250

Goh Y, Balasundaram G, Tan HM, Putti TC, Renzhe B, Hartman M, Buhari SA, Ng CWQ, Lui SA, Fang E et al (2024) Ultrasound-guided photoacoustic (US-PA) tomography of the breast: biochemical differentiation using intrinsic tissue markers—lipids, collagen and hemoglobin with histopathologic correlation. Sci Rep 14:18054

Grimes DR (2022) Radiofrequency radiation and cancer: a review. JAMA Oncol 8:456–461

Grönemeyer D, Seibel R, Erbel R, Schmidt A, Melzer A, Plaßmann J, Deli M, Schmermund A, Baumgart D, Welsch R et al (1996a) Equipment configuration and procedures: preferences for interventional micro therapy. J Digit Imaging 9:81–96

Grönemeyer DH, Seibel RM, Schmidt A, Melzer A, Deli M (1996b) Two-and three-dimensional imaging for interventional MRI and CT guidance. In: Proceedings of the medicine meets virtual reality. IOS Press, pp 62–76

Grootendorst MR, Cariati M, Pinder SE, Kothari A, Douek M, Kovacs T, Hamed H, Pawa A, Nimmo F, Owen J et al (2017a) Intraoperative assessment of tumor resection margins in breast-conserving surgery using 18F-FDG Cerenkov luminescence imaging: a first-in-human feasibility study. J Nucl Med 58:891–898

Grootendorst MR, Fitzgerald AJ, De Koning SGB, Santaolalla A, Portieri A, Van Hemelrijck M, Young MR, Owen J, Cariati M, Pepper M et al (2017b) Use of a handheld terahertz pulsed imaging device to differentiate benign and malignant breast tissue. Biomed Opt Express 8:2932–2945

Gubarkova E, Kiseleva E, Moiseev A, Vorontsov D, Kuznetsov S, Plekhanov A, Karabut M, Sirotkina M, Gelikonov G, Gamayunov S et al (2023) Intraoperative assessment of breast cancer tissues after breast-conserving surgery based on mapping the attenuation coefficients in 3D cross-polarization optical coherence tomography. Cancer 15:2663

Gupta S, Guha D, Taslimi S, Priola SM, Waggass G, Heyn C, Graham S, Foster S, Kongkham P, Sinclair J et al (2020) Multi-modality imaging assisted fluorescence-guided resection of glioblastoma: case report. Interdiscip Neurosurg 19:100593

Gurjar N, Bailey K, El-Shenawee M (2024) Polarimetry terahertz imaging of human breast cancer surgical specimens. J Med Imaging 11:065503–065503

Ha R, Friedlander LC, Hibshoosh H, Hendon C, Feldman S, Ahn S, Schmidt H, Akens MK, Fitzmaurice M, Wilson BC et al (2018) Optical coherence tomography: a novel imaging method for post-lumpectomy breast margin assessment—a multi-reader study. Acad Radiol 25:279–287

Hamdoon Z, Jerjes W, McKenzie G, Jay A, Hopper C (2016) Optical coherence tomography in the assessment of oral squamous cell carcinoma resection margins. Photodiagn Photodyn Ther 13:211–217

Han M, Choi W, Ahn J, Ryu H, Seo Y, Kim C (2020) In vivo dual-modal photoacoustic and ultrasound imaging of sentinel lymph nodes using a solid-state dye laser system. Sensors 20:3714

Hanna K, Krzoska E, Shaaban AM, Muirhead D, Abu-Eid R, Speirs V (2022) Raman spectroscopy: current applications in breast cancer diagnosis, challenges and future prospects. Br J Cancer 126:1125–1139

Harris B, Saleem S, Cook N, Searle E (2022) Targeting hypoxia in solid and haematological malignancies. J Exp Clin Cancer Res 41:318

He H, Englert L, Ntziachristos V (2023) Optoacoustic endoscopy of the gastrointestinal tract. ACS Photon 10:559–570

Hisada T, Sawaki M, Ishiguro J, Adachi Y, Kotani H, Yoshimura A, Hattori M, Yatabe Y, Iwata H (2016) Impact of intraoperative specimen mammography on margins in breast-conserving surgery. Mol Clin Oncol 5:269–272

Hoffman A, Ashkenazi I (2022) The efficiency of MarginProbe in detecting positive resection margins in epithelial breast cancer following breast conserving surgery. Eur J Surg Oncol 48:1498–1502

Holland JP, Normand G, Ruggiero A, Lewis JS, Grimm J (2011) Intraoperative imaging of positron emission tomographic radiotracers using Cerenkov luminescence emissions. Mol Imaging 10:7290–2010

Hu S, Wang LV (2013) Optical-resolution photoacoustic microscopy: auscultation of biological systems at the cellular level. Biophys J 105:841–847

Huang W, Wang K, An Y, Meng H, Gao Y, Xiong Z, Yan H, Wang Q, Cai X, Yang X et al (2020) In vivo three-dimensional evaluation of tumour hypoxia in nasopharyngeal carcinomas using FMT-CT and MSOT. Eur J Nucl Med Mol Imaging 47:1027–1038

Huang G, Lv J, He Y, Yang J, Zeng L, Nie L (2022) In vivo quantitative photoacoustic evaluation of the liver and kidney pathology in tyrosinemia. Photoacoustics 28:100410

Huang B, Wong IH, Zhang Y, Wong TT (2024) Ultraviolet photoacoustic microscopy for histopathology. In: Biomedical photoacoustics: technology and applications. Springer, pp 499–532

Hussein M, Awwad F, Jithin D, El Hasasna H, Athamneh K, Iratni R (2019) Breast cancer cells exhibits specific dielectric signature in vitro using the open-ended coaxial probe technique from 200 MHz to 13.6 GHz. Sci Rep 9:4681

Huynh N, Lucka F, Zhang E, Betcke M, Arridge SR, Beard PC, Cox BT (2019) Single-pixel camera photoacoustic tomography. J Biomed Opt 24:121907–121907

Huynh N, Zhang E, Francies O, Kuklis F, Allen T, Zhu J, Abeyakoon O, Lucka F, Betcke M, Jaros J et al (2024) A fast all-optical 3D photoacoustic scanner for clinical vascular imaging. Nat Biomed Eng 9:1–18

Ikematsu H, Ishihara M, Okawa S, Minamide T, Mitsui T, Kuwata T, Ito M, Kinoshita T, Fujita T, Yano T et al (2022) Photoacoustic imaging of fresh human surgically and endoscopically resected gastrointestinal specimens. DEN Open 2:e28

Imai T, Shi J, Wong TT, Li L, Zhu L, Wang LV (2018) High-throughput ultraviolet photoacoustic microscopy with multifocal excitation. J Biomed Opt 23:036007–036007

Jellema PE, Wijnen JP, De Luca A, Mutsaerts HJ, Obdeijn IV, van Baarsen KM, Lequin MH, Hoving EW (2023) Advanced intraoperative MRI in pediatric brain tumor surgery. Front Physiol 14:1098959

Jensen M, Clemmensen A, Hansen JG, van Krimpen Mortensen J, Christensen EN, Kjaer A, Ripa RS (2024) 3D whole body preclinical micro-CT database of subcutaneous tumors in mice with annotations from 3 annotators. Sci Data 11:1021

John S, Hester S, Basij M, Paul A, Xavierselvan M, Mehrmohammadi M, Mallidi S (2023) Niche preclinical and clinical applications of photoacoustic imaging with endogenous contrast. Photoacoustics 32:100533

Johnson KA, Hagen GM (2020) Artifact-free whole-slide imaging with structured illumination microscopy and Bayesian image reconstruction. Gigascience 9:giaa035

Kafka IZ, Averch TD (2019) Intraoperative assessment of tumor resection margins. In: Smith's textbook of endourology. Wiley, pp 1097–1100

Kalva SK, Pramanik M (2020) Photoacoustic tomography with high-energy pulsed laser diodes. SPIE

Kalva SK, Deán-Ben XL, Reiss M, Razansky D (2023) Spiral volumetric optoacoustic tomography for imaging whole-body biodynamics in small animals. Nat Protoc 18:2124–2142

Karanlik H, Ozgur I, Sahin D, Fayda M, Onder S, Yavuz E (2015) Intraoperative ultrasound reduces the need for re-excision in breast-conserving surgery. World J Surg Oncol 13:1–8

Ke J, Jia L, Hu Y, Jiang X, Mo H, An X, Yuan W (2022) Clinical and experimental study of a terahertz time-domain system for the determination of the pathological margins of laryngeal carcinoma. World J Surg Oncol 20:339

Kim JY, Jung YS, Han SH, Sung JH, Hur MH (2020) Indications and methods of intraoperative specimen radiography in breast-conserving surgery. Transl Cancer Res 9:6625–6628

Kim M, Lee KW, Kim K, Gulenko O, Lee C, Keum B, Chun HJ, Choi HS, Kim CU, Yang JM (2022) Intra-instrument channel workable, optical-resolution photoacoustic and ultrasonic mini-probe system for gastrointestinal endoscopy. Photoacoustics 26:100346

Kim J, Lee J, Choi S, Lee H, Yang J, Jeon H, Sung M, Kim WJ, Kim C (2024a) 3d multiparametric photoacoustic computed tomography of primary and metastatic tumors in living mice. ACS Nano 18:18176–18190

Kim J, Heo D, Cho S, Ha M, Park J, Ahn J, Kim M, Kim D, Jung DH, Kim HH et al (2024b) Enhanced dual-mode imaging: superior photoacoustic and ultrasound endoscopy in live pigs using a transparent ultrasound transducer. Sci Adv 10:eadq9960

Klein Gunnewiek K, van Baarsen KM, Graus EH, Brink WM, Lequin MH, Hoving EW (2024) Navigated intraoperative ultrasound in pediatric brain tumors. Child's Nervous Syst 41:1–9

Kopicky L, Fan B, Valente SA (2024) Intraoperative evaluation of surgical margins in breast cancer. In: Proceedings of the seminars in diagnostic pathology, vol 41. Elsevier, pp 293–300

Kordic A, Šarolic A (2023) Dielectric spectroscopy shows a permittivity contrast between meningioma tissue and brain white and gray matter—A potential physical biomarker for meningioma discrimination. Cancer 15:4153

Kosik I, Brackstone M, Kornecki A, Chamson-Reig A, Wong P, Carson JJ (2020) Lipid-weighted intraoperative photoacoustic tomography of breast tumors: volumetric comparison to preoperative MRI. Photoacoustics 18:100165

Kotturi D, Paterson S, McShane M (2024) Surface-enhanced spatially offset Raman spectroscopy in tissue. Biosensors 14:81

Kuniyil Ajith Singh M, Xia W (2020) Portable and affordable light source-based photoacoustic tomography. Sensors 20:6173

Kuniyil Ajith Singh M, Xia W (2021) Biomedical photoacoustic imaging and sensing using affordable resources. Sensors 21(7):2572. https:/\penalty\z@/\penalty\z@doi.org/\penalty\z@10.3390/\penalty\z@s21072572

Kuniyil Ajith Singh M, Steenbergen W, Manohar S (2015) Handheld probe-based dual mode ultrasound/photoacoustics for biomedical imaging. In: Frontiers in biophotonics for translational medicine: in the celebration of year of light. Springer, pp 209–247

Kuppler P, Strenge P, Lange B, Spahr-Hess S, Draxinger W, Hagel C, Theisen-Kunde D, Brinkmann R, Huber R, Tronnier V et al (2024) Microscope-integrated optical coherence tomography for in vivo human brain tumor detection with artificial intelligence. J Neurosurg 1:1–9

Kurdi M, Baeesa S, Maghrabi Y, Bardeesi A, Saeedi R, Al-Sinani T, Samkari A, Lary A, Hakamy S (2022) Diagnostic discrepancies between intraoperative frozen section and permanent histopathological diagnosis of brain tumors. Turk J Pathol 38:34

Lakhani DA, Sabsevitz DS, Chaichana KL, Quiñones-Hinojosa A, Middlebrooks EH (2023) Current state of functional MRI in the presurgical planning of brain tumors. Radiol Imaging Cancer 5:e230078

Lauwerends LJ, van Driel PB, de Jong RJB, Hardillo JA, Koljenovic S, Puppels G, Mezzanotte L, Löwik CW, Rosenthal EL, Vahrmeijer AL et al (2021) Real-time fluorescence imaging in intraoperative decision making for cancer surgery. Lancet Oncol 22:e186–e195

Lavrador JP, Ghimire P, Brogna C, Furlanetti L, Patel S, Gullan R, Ashkan K, Bhangoo R, Vergani F (2021) Pre-and intraoperative mapping for tumors in the primary motor cortex: decision-making process in surgical resection. J Neurol Surg Part A: Cent Eur Neurosurg 82:333–343

Lavrador JP, Marchi F, Elhag A, Kalyal N, Mthunzi E, Awan M, Wroe-Wright O, Díaz-Baamonde A, Mirallave-Pescador A, Reisz Z et al (2024) In situ light-source delivery during 5-Aminulevulinic acid-guided high-grade glioma resection: spatial, functional and oncological informed surgery. Biomedicines 12:2748

Laws A, Brar MS, Bouchard-Fortier A, Leong B, Quan ML (2018) Does intra-operative margin assessment improve margin status and re-excision rates? A population-based analysis of outcomes in breast-conserving surgery for ductal carcinoma in situ. J Surg Oncol 118:1205–1211

Lechpammer M, Todd A, Tang V, Morningstar T, Borowsky A, Shahlaie K, Kintner JA, McPherson JD, Bishop JW, Fereidouni F et al (2024) Neuropathological applications of microscopy with ultraviolet surface excitation (MUSE): a concordance study of human primary and metastatic brain tumors. Brain Sci 14:108

Leclerc P, Ray C, Mahieu-Williame L, Alston L, Frindel C, Brevet PF, Meyronet D, Guyotat J, Montcel B, Rousseau D (2020) Machine learning-based prediction of glioma margin from 5-ALA induced PpIX fluorescence spectroscopy. Sci Rep 10:1462

Lee JY, Kim HY, Lee YS, Seo HY, Park JY, Jeong JM (2020) Preparation of a multi-modal agent for sentinel lymph node mapping using Evans blue and 99mTc-labeled mannosylated human serum albumin conjugate. Nucl Med Biol 84:38–45

Leiloglou M, Kedrzycki MS, Chalau V, Chiarini N, Thiruchelvam PT, Hadjiminas DJ, Hogben KR, Rashid F, Ramakrishnan R, Darzi AW et al (2022) Indocyanine green fluorescence image processing techniques for breast cancer macroscopic demarcation. Sci Rep 12:8607

Lerner SP, Goh A (2015) Novel endoscopic diagnosis for bladder cancer. Cancer 121:169–178

Leroy HA, Delmaire C, Le Rhun E, Drumez E, Lejeune JP, Reyns N (2019) High-field intraoperative MRI and glioma surgery: results after the first 100 consecutive patients. Acta Neurochir 161:1467–1474

Li R, Wang P, Lan L, Lloyd FP Jr, Goergen CJ, Chen S, Cheng JX (2015) Assessing breast tumor margin by multispectral photoacoustic tomography. Biomed Opt Express 6:1273–1281

Li R, Lan L, Xia Y, Wang P, Han LK, Dunnington GL, Obeng-Gyasi S, Sandusky GE, Medley JA, Crook ST et al (2018) High-speed intra-operative assessment of breast tumour margins by multimodal ultrasound and photoacoustic tomography. Med Devices Sens 1:e10018

Li J, Chekkoury A, Prakash J, Glasl S, Vetschera P, Koberstein-Schwarz B, Olefir I, Gujrati V, Omar M, Ntziachristos V (2020) Spatial heterogeneity of oxygenation and haemodynamics in breast cancer resolved in vivo by conical multispectral optoacoustic mesoscopy. Light: Sci Appl 9:57

Li P, Highfield PE, Lang ZQ, Kell D (2021a) Cervical cancer prognosis and diagnosis using electrical impedance spectroscopy. J Electr Bioimpedance 12:153–162

Li L, Gao N, Yang AQ, Xu WH, Ding Y, Chu J, Lin XN, Liu JQ (2021b) Application of fluorescein combined with methylene blue in sentinel lymph node biopsy of breast cancer. Sci Rep 11:12119

Li X, Tsang VT, Kang L, Zhang Y, Wong TT (2021c) High-speed high-resolution laser diode-based photoacoustic microscopy for in vivo microvasculature imaging. Visual Comput Ind Biomed Art 4:1–6

Li Y, Lu G, Zhou Q, Chen Z (2021d) Advances in endoscopic photoacoustic imaging. Proc Photon MDPI 8:281

Li X, Kang L, Lo CT, Tsang VT, Wong TT (2022) High-speed ultraviolet photoacoustic microscopy for histological imaging with virtual-staining assisted by deep learning. J Vis Exp:63649

Li C, Feng C, Xu R, Jiang B, Li L, He Y, Tu C, Li Z (2023a) The emerging applications and advancements of Raman spectroscopy in pediatric cancers. Front Oncol 13:1044177

Li L, Qin W, Li T, Zhang J, Li B, Xi L (2023b) High-speed adaptive photoacoustic microscopy. Photon Res 11:2084–2092

Li Z, Song Y, Hameed NF, Yuan S, Wu S, Gong X, Zhuang D, Lu J, Zhu F, Qiu T et al (2024) Effect of high-field iMRI guided resection in cerebral glioma surgery: a randomized clinical trial. Eur J Cancer 199:113528

Liang Y, Li L, Li Q, Liang H, Jin L, Wang L, Guan BO (2020) Photoacoustic computed tomography by using a multi-angle scanning fiber-laser ultrasound sensor. Opt Express 28:8744–8752

Liang Y, Fu W, Li Q, Chen X, Sun H, Wang L, Jin L, Huang W, Guan BO (2022) Optical-resolution functional gastrointestinal photoacoustic endoscopy based on optical heterodyne detection of ultrasound. Nat Commun 13:7604

Liang S, Zhou J, Guo Z, He D, Yang W, Ye Z, Shao W, Jing L, Chen SL (2023) Miniature probe for optomechanical focus-adjustable optical-resolution photoacoustic endoscopy. IEEE Trans Med Imaging 42:2400–2413

Lin L, Wang LV (2022) The emerging role of photoacoustic imaging in clinical oncology. Nat Rev Clin Oncol 19:365–384
Lin L, Peng H, Shi J, Appleton CM, Maslov K, Li L, Zhang R, Wang LV (2018) Single-breath-hold photoacoustic computed tomography of the breast. Nat Commun 9(1):2352
Lin C, Wang KY, Chen HL, Xu YH, Pan T, Chen YD (2022) Specimen mammography for intraoperative margin assessment in breast conserving surgery: a meta-analysis. Sci Rep 12:18440
Lin R, Lv S, Lou W, Wang X, Xie Z, Zeng S, Chen R, Gao W, Jiang T, Cheng KWE et al (2024) In-vivo assessment of a rat rectal tumor using optical-resolution photoacoustic endoscopy. Biomed Opt Express 15:2251–2261
Liu K, Zhao Q, Li B, Zhao X (2022) Raman spectroscopy: a novel technology for gastric cancer diagnosis. Front Bioeng Biotechnol 10:856591
Liu HC, Lin MH, Ting CH, Wang YM, Sun CW (2023) Intraoperative application of optical coherence tomography for lung tumor. J Biophotonics 16:e202200344
Liu S, Wang T, Zheng X, Zhu Y, Tian C (2024a) On the imaging depth limit of photoacoustic tomography in the visible and first near-infrared windows. Opt Express 32:5460–5480
Liu L, Li A, Zhao Y, Zhu L, Zhao Y, Gao F (2024b) An umbrella-inspired snap-on robotic 3D photoacoustic endoscopic probe for augmented intragastric sensing: proof of concept study. Photoacoustics 35:100568
Ludwig WW, Wobker SE, Ball MW, Zysk AM, Yemul KS, Pierorazio PM, Gorin MA, Allaf ME (2018) Margin assessment in renal surgery using a handheld optical coherence tomography probe. Urology 113:241–245
Luethy S, Tulman DB, Brown JQ (2020) Automated gigapixel circumferential surface microscopy of the prostate. Sci Rep 10:131
Ma M, Zhang J, Liu Y, Wang X, Han B (2024a) Advances in the clinical application of Raman spectroscopy in breast cancer. Appl Spectrosc Rev:1–35
Ma R, Livermore LJ, Taylor L, Laycock J, Williams S, Ansorge O, Vallance C, Plaha P (2024b) Endoscopic 5-Aminolevulinic Acid–Induced Fluorescence-Guided Intraparenchymal Brain Tumor Resection—Can the Endoscope Detect More Fluorescence Than the Microscope? World Neurosurg 185:e1268–e1279
MacCuaig WM, Jones MA, Abeyakoon O, McNally LR (2020) Development of multispectral optoacoustic tomography as a clinically translatable modality for cancer imaging. Radiology: imaging. Cancer 2:e200066
Manwar R, Kratkiewicz K, Avanaki K (2020) Overview of ultrasound detection technologies for photoacoustic imaging. Micromachines 11:692
Maragkos GA, Schüpper AJ, Lakomkin N, Sideras P, Price G, Baron R, Hamilton T, Haider S, Lee IY, Hadjipanayis CG et al (2021) Fluorescence-guided high-grade glioma surgery more than four hours after 5-aminolevulinic acid administration. Front Neurol 12:644804
Marcinkowski R, Keereman V, Van Holen R, Vandenberghe S, Van Bockstal M, Van Dorpe J, Brans B, Depypere H, Van den Broecke R (2019) Intraoperative margin assessment for breast conserving surgery using 18F-FDG micro-PET/CT: a proof-of-concept study. Eur J Surg Oncol 45:e31–e32
Martell MT, Haven NJ, Cikaluk BD, Restall BS, McAlister EA, Mittal R, Adam BA, Giannakopoulos N, Peiris L, Silverman S et al (2023) Deep learning-enabled realistic virtual histology with ultraviolet photoacoustic remote sensing microscopy. Nat Commun 14:5967
Mashimo H, Gordon SR, Singh SK (2020) Advanced endoscopic imaging for detecting and guiding therapy of early neoplasias of the esophagus. Ann N Y Acad Sci 1482:61–76
Mazzucchi E, La Rocca G, Hiepe P, Pignotti F, Galieri G, Policicchio D, Boccaletti R, Rinaldi P, Gaudino S, Ius T et al (2022) Intraoperative integration of multimodal imaging to improve neuronavigation: a technical note. World Neurosurg 164:330–340
McCracken DJ, Schupper AJ, Lakomkin N, Malcolm J, Painton Bray D, Hadjipanayis CG (2022) Turning on the light for brain tumor surgery: a 5-aminolevulinic acid story. Neuro-Oncology 24:S52–S61

Menozzi L, Yao J (2024) Deep tissue photoacoustic imaging with light and sound. npj Imaging 2:44
Milos P, Haj-Hosseini N, Hillman J, Wårdell K (2023) 5-ALA fluorescence in randomly selected pediatric brain tumors assessed by spectroscopy and surgical microscope. Acta Neurochir 165:71–81
Mirg S, Turner KL, Chen H, Drew PJ, Kothapalli SR (2022) Photoacoustic imaging for microcirculation. Microcirculation 29:e12776
Mirzayeva L, Uçar M, Kaymaz AM, Temel E (2024) Intraoperative magnetic resonance imaging in glioma surgery: a single-center experience. J Neuro-Oncol 168:249–257
Mislow JM, Golby AJ, Black PM (2010) Origins of intraoperative MRI. Magn Reson Imaging Clin 18:1–10
Molina ES, Bruneau M, Reuter G, Shahein M, Cavallo LM, Daniel RT, Kasper EM, Froelich S, Jouanneau E, Manet R et al (2024) Fluorescence-guidance in Skull Base surgery: applications and limitations–a systematic review. Brain Spine 4:103328
Moore C, Jokerst JV (2019) Strategies for image-guided therapy, surgery, and drug delivery using photoacoustic imaging. Theranostics 9:1550–1571
Moothanchery M, Dev K, Balasundaram G, Bi R, Olivo M (2020) Acoustic resolution photoacoustic microscopy based on microelectromechanical systems scanner. J Biophotonics 13:e201960127
Mordi A, Karunakaran V, Marium Mim U, Marple E, Rajaram N (2024) Design and validation of a multimodal diffuse reflectance and spatially offset Raman spectroscopy system for in vivo applications. J Biophotonics 18:e202400333
Mosca S, Dey P, Salimi M, Gardner B, Palombo F, Stone N, Matousek P (2021) Spatially offset Raman spectroscopy—how deep? Anal Chem 93:6755–6762
Namdar ZM, Omidifar N, Arasteh P, Akrami M, Tahmasebi S, Nobandegani AS, Sedighi S, Zangouri V, Talei A (2021) How accurate is frozen section pathology compared to permanent pathology in detecting involved margins and lymph nodes in breast cancer? World J Surg Oncol 19:1–7
Nasri D, Manwar R, Kaushik A, Er EE, Avanaki K (2023) Photoacoustic imaging for investigating tumor hypoxia: a strategic assessment. Theranostics 13:3346–3367
Nazarian S, Gkouzionis I, Murphy J, Darzi A, Patel N, Peters CJ, Elson DS (2024) Real-time classification of tumour and non-tumour tissue in colorectal cancer using diffuse reflectance spectroscopy and neural networks to aid margin assessment. Int J Surg 110:1983–1991
Nicolson F, Andreiuk B, Lee E, O'Donnell B, Whitley A, Riepl N, Burkhart DL, Cameron A, Protti A, Rudder S et al (2024) In vivo imaging using surface enhanced spatially offset raman spectroscopy (SESORS): balancing sampling frequency to improve overall image acquisition. npj Imaging 2:7
Nogueira MS, Maryam S, Amissah M, Lu H, Lynch N, Killeen S, O'Riordain M, Andersson-Engels S (2021) Evaluation of wavelength ranges and tissue depth probed by diffuse reflectance spectroscopy for colorectal cancer detection. Sci Rep 11:798
Noh T, Mustroph M, Golby AJ (2020) Intraoperative imaging for high-grade glioma surgery. Neurosurg Clin N Am 32:47–54
Oh TI, Kang MJ, Jeong YJ, Zhang T, Yeo SG, Park DC (2021) Tissue characterization using an electrical bioimpedance spectroscopy-based multi-electrode probe to screen for cervical intraepithelial neoplasia. Diagnostics 11:2354
Omidi P, Yip L, Rascevska E, Diop M, Carson J (2022) PATLAB: A graphical computational software package for photoacoustic computed tomography research. Photoacoustics 28:100404
Orillac C, Stummer W, Orringer DA (2021) Fluorescence guidance and intraoperative adjuvants to maximize extent of resection. Neurosurgery 89:727–736
Palmieri G, Cofano F, Salvati LF, Monticelli M, Zeppa P, Perna GD, Melcarne A, Altieri R, La Rocca G, Sabatino G et al (2021) Fluorescence-guided surgery for high-grade gliomas: state of the art and new perspectives. Technol Cancer Res Treat 20:15330338211021605

Pan J, Deng H, Hu S, Xia C, Chen Y, Wang J, Wang Y (2020) Real-time surveillance of surgical margins via ICG-based near-infrared fluorescence imaging in patients with OSCC. World J Surg Oncol 18:1–8

Pang W, Wang Y, Guo L, Wang B, Lai P, Xiao J (2022) Two-dimensional photoacoustic/ultrasonic endoscopic imaging based on a line-focused transducer. Front Bioeng Biotechnol 9:807633

Parasca SV, Calin MA, Manea D, Radvan R (2024) Hyperspectral imaging with machine learning for in vivo skin carcinoma margin assessment: a preliminary study. Phys Eng Sci Med 47:1–12

Park S, Lee C, Kim J, Kim C (2014) Acoustic resolution photoacoustic microscopy. Biomed Eng Lett 4:213–222

Park B, Bang C, Lee C, Han J, Choi W, Kim J, Park G, Rhie J, Lee J, Kim C (2021) 3D wide-field multispectral photoacoustic imaging of human melanomas in vivo: a pilot study. J Eur Acad Dermatol Venereol 35:669–676

Park J, Choi S, Knieling F, Clingman B, Bohndiek S, Wang LV, Kim C (2024a) Clinical translation of photoacoustic imaging. Nat Rev Bioeng 3:1–20

Park JS, Yoon T, Park SA, Lee BH, Jeun SS, Eom TJ (2024b) Delineation of three-dimensional tumor margins based on normalized absolute difference mapping via volumetric optical coherence tomography. Sci Rep 14:7984

Parodi V, Jacchetti E, Osellame R, Cerullo G, Polli D, Raimondi MT (2020) Nonlinear optical microscopy: from fundamentals to applications in live bioimaging. Front Bioeng Biotechnol 8:585363

Patel V, Chavda V (2024) Intraoperative glioblastoma surgery-current challenges and clinical trials: An update. Cancer Pathogen Ther 2:256–267

Pertzborn D, Nguyen HN, Hüttmann K, Prengel J, Ernst G, Guntinas-Lichius O, von Eggeling F, Hoffmann F (2022) Intraoperative assessment of tumor margins in tissue sections with hyperspectral imaging and machine learning. Cancer 15:213

Peters SR (2010) A practical guide to frozen section technique. Springer

Pop MM, Cristian S, Hanko-Bauer O, Ghiga DV, Georgescu R (2018) Obtaining adequate surgical margin status in breast-conservation therapy: intraoperative ultrasound-guided resection versus specimen mammography. Clujul Med 91:197–202

Puustinen S, Vrzáková H, Hyttinen J, Rauramaa T, Fält P, Hauta-Kasari M, Bednarik R, Koivisto T, Rantala S, von und zu Fraunberg M et al (2023) Hyperspectral imaging in brain tumor surgery—evidence of machine learning-based performance. World Neurosurg 175:e614–e635

Qafiti F, Layton C, McDonald KA (2022) Radiofrequency spectroscopy with intraoperative pathological assessment for breast carcinoma: synergistic or redundant? Expert Rev Med Devices 19:369–373

Qi Y, Liu Y, Luo J (2023) Recent application of Raman spectroscopy in tumor diagnosis: from conventional methods to artificial intelligence fusion. PhotoniX 4:22

Qiu C, Zhang Z, Xu Z, Qiao L, Ning L, Zhang S, Su M, Wu W, Song K, Xu Z et al (2024) Transparent ultrasonic transducers based on relaxor ferroelectric crystals for advanced photoacoustic imaging. Nat Commun 15:1–14

Rascevska E, Yip L, Omidi P, Brackstone M, Carson J (2022) Investigating the feasibility of a hand-held photoacoustic imaging probe for margin assessment during breast conserving surgery. Photoacoustics 28:100424

Reichert D, Erkkilae MT, Gesperger J, Wadiura LI, Lang A, Roetzer T, Woehrer A, Andreana M, Unterhuber A, Wilzbach M et al (2021) Fluorescence lifetime imaging and spectroscopic co-validation for protoporphyrin IX-guided tumor visualization in neurosurgery. Front Oncol 11:741303

Rickard AG, Mikati H, Mansourati A, Stevenson D, Krieger M, Rocke D, Esclamado R, Dewhirst MW, Ramanujam N, Lee WT et al (2023) A clinical study to assess diffuse reflectance spectroscopy with an auto-calibrated, pressure-sensing optical probe in head and neck cancer. Curr Oncol 30:2751–2760

Riksen JJ, Nikolaev AV, van Soest G (2023) Photoacoustic imaging on its way toward clinical utility: a tutorial review focusing on practical application in medicine. J Biomed Opt 28:121205–121205

Rogers CM, Jones PS, Weinberg JS (2021) Intraoperative MRI for brain tumors. J Neuro-Oncol 151:479–490

Saha PS, Yan J, Zhu C (2024) Diffuse reflectance spectroscopy for optical characterizations of orthotopic head and neck cancer models in vivo. Biomed Opt Express 15:4176–4189

Sangha GS, Hu B, Li G, Fox SE, Sholl AB, Brown JQ, Goergen CJ (2022) Assessment of photoacoustic tomography contrast for breast tissue imaging using 3D correlative virtual histology. Sci Rep 12:2532

Sato J, Matsumoto T, Nakao R, Tanaka H, Nagahara H, Niioka H, Takamatsu T (2023) Deep UV-excited fluorescence microscopy installed with CycleGAN-assisted image translation enhances precise detection of lymph node metastasis towards rapid intraoperative diagnosis. Sci Rep 13:21363

Scaranelo AM, Moshonov H, Escallon J (2016) A prospective pilot study of analysis of surgical margins of breast cancers using high-resolution sonography. Springerplus 5:251

Schmidt RL, Hunt JP, Hall BJ, Wilson AR, Layfield LJ (2011) A systematic review and meta-analysis of the diagnostic accuracy of frozen section for parotid gland lesions. Am J Clin Pathol 136:729–738

Schnabel F, Boolbol SK, Gittleman M, Karni T, Tafra L, Feldman S, Police A, Friedman NB, Karlan S, Holmes D et al (2014) A randomized prospective study of lumpectomy margin assessment with use of MarginProbe in patients with nonpalpable breast malignancies. Ann Surg Oncol 21:1589–1595

Schoonover RW, Anastasio MA (2011) Image reconstruction in photoacoustic tomography involving layered acoustic media. JOSA A 28:1114–1120

Seeger M, Stylogiannis A, Prade L, Glasl S, Ntziachristos V (2023) Overdriven laser diode optoacoustic microscopy. Sci Rep 13:19542

Senders JT, Muskens IS, Schnoor R, Karhade AV, Cote DJ, Smith TR, Broekman ML (2017) Agents for fluorescence-guided glioma surgery: a systematic review of preclinical and clinical results. Acta Neurochir 159:151–167

Shabairou N, Lengenfelder B, Hohmann M, Klämpfl F, Schmidt M, Zalevsky Z (2020) All-optical, an ultra-thin endoscopic photoacoustic sensor using multi-mode fiber. Sci Rep 10:9142

Shaikh R, Daniel A, Lyng FM (2023) Raman spectroscopy for early detection of cervical cancer, a global women's health issue—a review. Molecules 28:2502

Shi H, Li T, Liu Z, Zhao J, Qi F (2022a) Early detection of gastric cancer via high-resolution terahertz imaging system. Front Bioeng Biotechnol 10:1052069

Shi M, Vercauteren T, Xia W (2022b) Spatiotemporal singular value decomposition for denoising in photoacoustic imaging with a low-energy excitation light source. Biomed Opt Express 13:6416–6430

Shintate R, Ishii T, Ahn J, Kim JY, Kim C, Saijo Y (2022) High-speed optical resolution photoacoustic microscopy with MEMS scanner using a novel and simple distortion correction method. Sci Rep 12:9221

Shipp DW, Rakha EA, Koloydenko AA, Macmillan RD, Ellis IO, Notingher I (2018) Intra-operative spectroscopic assessment of surgical margins during breast conserving surgery. Breast Cancer Res 20:1–14

Shrestha B, Stern NB, Zhou A, Dunn A, Porter T (2024) Current trends in the characterization and monitoring of vascular response to cancer therapy. Cancer Imaging 24:143

Shu W, Ai M, Salcudean T, Rohling R, Abolmaesumi P, Tang S (2016) Broadening the detection view of 2D photoacoustic tomography using two linear array transducers. Opt Express 24:12755–12768

Siegel RL, Kratzer TB, Giaquinto AN, Sung H, Jemal A (2025) Cancer statistics, 2025. CA: Cancer J Clin 75:10–45

Sikošek NC, Dovnik A, Arko D, Takacˇ I (2014) The role of intraoperative ultrasound in breast-conserving surgery of nonpalpable breast cancer. Wien Klin Wochenschr 126:90–94

Singh MKA, Sato N, Ichihashi F, Xia W, Sankai Y (2024) The evolution of led-based photoacoustic imaging: from labs to clinics. In: Biomedical photoacoustics: technology and applications. Springer, pp 573–608

Sun Y, Jiang H (2022) Enhancing finite element-based photoacoustic tomography by localized reconstruction method. Proc Photon MDPI 9:337

Sun R, Cuthbert H, Watts C (2021a) Fluorescence-guided surgery in the surgical treatment of gliomas: past, present and future. Cancer 13:3508

Sun H, Wang W, Zhang Z, Wang L, Zhang W, Xiong K, Yang S (2021b) Real-time optical-resolution photoacoustic endoscope. Appl Phys Express 14:042012

Sun T, Lv J, Zhao X, Li W, Zhang Z, Nie L (2023) In vivo liver function reserve assessments in alcoholic liver disease by scalable photoacoustic imaging. Photoacoustics 34:100569

Sung H, Ferlay J, Siegel RL, Laversanne M, Soerjomataram I, Jemal A, Bray F (2021) Global cancer statistics 2020: GLOBOCAN estimates of incidence and mortality worldwide for 36 cancers in 185 countries. CA Cancer J Clin 71:209–249

Tajaldeen A, Alrashidi M, Alsaadi MJ, Alghamdi SS, Alshammari H, Alsleem H, Jafer M, Aljondi R, Alqahtani S, Alotaibi A et al (2024) Photoacoustic imaging in prostate cancer: A new paradigm for diagnosis and management. Photodiagn Photodyn Ther 47:104225

Tarabichi O, Kanumuri V, Juliano AF, Faquin WC, Cunnane ME, Varvares MA (2018) Intra-operative ultrasound in oral tongue cancer resection: feasibility study and early outcomes. Otolaryngol Head Neck Surg 158:645–648

Thill M, Dittmer C, Baumann K, Friedrichs K, Blohmer JU (2014) MarginProbe®–final results of the German post-market study in breast conserving surgery of ductal carcinoma in situ. Breast 23:94–96

Tong X, Lin L, Hu P, Cao R, Zhang Y, Olick-Gibson J, Wang LV (2023) Non-invasive 3D photoacoustic tomography of angiographic anatomy and hemodynamics of fatty livers in rats. Adv Sci 10:2205759

Tsai WY, Breimann S, Shen TW, Frishman D (2023) Photoacoustic and absorption spectroscopy imaging analysis of human blood. PLoS One 18:e0289704

Tsampazis N, Vavoulidis E, Siarkou CM, Siarkou GM, Pratilas GC, Symeonidou M, Intzes S, Petousis S, Papanikolaou A, Dinas K (2023) Diagnostic comparison of electrical impedance spectroscopy with colposcopy and HPV mRNA-testing in the prediction of CIN2+ women in Greece. J Obstet Gynaecol Res 49:1222–1229

Upputuri PK, Pramanik M (2015) Performance characterization of low-cost, high-speed, portable pulsed laser diode photoacoustic tomography (PLD-PAT) system. Biomed Opt Express 6:4118–4129

Van Manen L, Stegehuis PL, Farina-Sarasqueta A, De Haan LM, Eggermont J, Bonsing BA, Morreau H, Lelieveldt BP, Van De Velde CJ, Vahrmeijer AL et al (2017) Validation of full-field optical coherence tomography in distinguishing malignant and benign tissue in resected pancreatic cancer specimens. PLoS One 12:e0175862

van Riet YE, Schipper RJ, van Merrienboer F, Orsini RG, Bloemen JG, Jansen FH, Nieuwenhuijzen GA (2021) Is specimen radiography still necessary in patients with non-palpable breast cancer undergoing breast-conserving surgery using radioactive I-125 seed localization? Clin Imaging 69:311–317

Veeratterapillay R, Bromby A, Patel A, Sakthivel A, Abdelbakhy A, Gowda B, Mutton A, Nagarajan S, Naisby G, Bhatti A (2015) Intraoperative and surgical specimen (ex vivo) ultrasound in the assessment of margins at partial nephrectomy. Int Urol Nephrol 47:1665–1669

Veluponnar D, Dashtbozorg B, Jong LJS, Geldof F, Guimaraes MDS, Peeters MJTV, van Duijnhoven F, Sterenborg HJ, Ruers TJ, de Boer LL (2023) Diffuse reflectance spectroscopy for accurate margin assessment in breast-conserving surgeries: importance of an optimal number of fibers. Biomed Opt Express 14:4017–4036

Veverka M, Menozzi L, Yao J (2023) The sound of blood: photoacoustic imaging in blood analysis. Med Novel Technol Devices 18:100219

Vohra N, Bowman T, Bailey K, El-Shenawee M (2020) Terahertz imaging and characterization protocol for freshly excised breast cancer tumors. J Vis Exp:10–3791. https:/\penalty\z@/\penalty\z@doi.org/\penalty\z@10.3791/\penalty\z@61007 von Knorring T, Ihlemann TB, Blanche P, Reichl C, Israelsen NM, Olesen CM, Yuksel YT, Mogensen M (2025) Normal and melanoma skin visualized, quantified and compared by in vivo photoacoustic imaging. Photoacoustics 42:100693

Vonk J, Kukačka J, Steinkamp P, De Wit J, Voskuil F, Hooghiemstra W, Bader M, Jüstel D, Ntziachristos V, Van Dam G et al (2022) Multispectral optoacoustic tomography for in vivo detection of lymph node metastases in oral cancer patients using an EGFR-targeted contrast agent and intrinsic tissue contrast: a proof-of-concept study. Photoacoustics 26:100362

Wang Y, Li C (2024) Comprehensive framework of GPU-accelerated image reconstruction for photoacoustic computed tomography. J Biomed Opt 29:066006

Wang Lihong V et al (2021) Recent advances in photoacoustic tomography. BME Front 2021:9823268

Wang M, Tulman DB, Sholl AB, Kimbrell HZ, Mandava SH, Elfer KN, Luethy S, Maddox MM, Lai W, Lee BR et al (2016) Gigapixel surface imaging of radical prostatectomy specimens for comprehensive detection of cancer-positive surgical margins using structured illumination microscopy. Sci Rep 6:27419

Wang M, Tang F, Pan X, Yao L, Wang X, Jing Y, Ma J, Wang G, Mi L (2017) Rapid diagnosis and intraoperative margin assessment of human lung cancer with fluorescence lifetime imaging microscopy. BBA Clin 8:7–13

Wang M, Tulman DB, Sholl AB, Mandava SH, Maddox MM, Lee BR, Quincy Brown J (2018) Partial nephrectomy margin imaging using structured illumination microscopy. J Biophotonics 11:e201600328

Wang S, Zhao Y, Xu Y (2020) Recent advances in applications of multimodal ultrasound-guided photoacoustic imaging technology. Visual Comput Ind Biomed Art 3:24

Wang B, Wang C, Zhong F, Pang W, Guo L, Peng K, Xiao J (2021a) 3D acoustic resolution-based photoacoustic endoscopy with dynamic focusing. Quant Imaging Med Surg 11:685–696

Wang Y, Jhang DF, Tsai CH, Chiang NJ, Tsao CH, Chuang CC, Chen LT, Chang WSW, Liao LD (2021b) In vivo assessment of hypoxia levels in pancreatic tumors using a dual-modality ultrasound/photoacoustic imaging system. Micromachines 12:668

Wang H, Ji T, Qu H, Yan T, Li D, Yang R, Tang X, Guo W (2023a) Indocyanine green fluorescence imaging may detect tumour residuals during surgery for bone and soft-tissue tumours. Bone Joint J 105:551–558

Wang J, Zhang L, Pan Z (2023b) Evaluating the impact of radiofrequency spectroscopy on reducing reoperations after breast conserving surgery: a meta-analysis. Thoracic Cancer 14:1413–1419

Weyers BW, Marsden M, Sun T, Bec J, Bewley AF, Gandour-Edwards RF, Moore MG, Farwell DG, Marcu L (2019) Fluorescence lifetime imaging for intraoperative cancer delineation in transoral robotic surgery. Transl Biophoton 1:e201900017

Weyers BW, Birkeland AC, Marsden MA, Tam A, Bec J, Frusciante RP, Gui D, Bewley AF, Abouyared M, Marcu L et al (2022) Intraoperative delineation of p16+ oropharyngeal carcinoma of unknown primary origin with fluorescence lifetime imaging: preliminary report. Head Neck 44:1765–1776

Wong TT, Zhang R, Zhang C, Hsu HC, Maslov KI, Wang L, Shi J, Chen R, Shung KK, Zhou Q et al (2017a) Label-free automated three-dimensional imaging of whole organs by microtomy-assisted photoacoustic microscopy. Nat Commun 8:1386

Wong TT, Zhang R, Hai P, Zhang C, Pleitez MA, Aft RL, Novack DV, Wang LV (2017b) Fast label-free multi-layered histology-like imaging of human breast cancer by photoacoustic microscopy. Sci Adv 3:e1602168

Wong IH, Zhang Y, Chen Z, Kang L, Wong TT (2021) Slide-free histological imaging by microscopy with ultraviolet surface excitation using speckle illumination. Photon Res 10:120–125

Wu IC, Syu HY, Jen CP, Lu MY, Chen YT, Wu MT, Kuo CT, Tsai YY, Wang HC (2018) Early identification of esophageal squamous neoplasm by hyperspectral endoscopic imaging. Sci Rep 8:13797

Wu A, Wu JY, Lim M (2022) Updates in intraoperative strategies for enhancing intra-axial brain tumor control. Neuro-Oncology 24:S33–S41

Wu O, Clift GW, Hilliard S, Ip M (2023) Evaluating the use of intraoperative magnetic resonance imaging in paediatric brain tumour resection surgeries: a literature review. J Med Radiat Sci 70:479–490

Wyld L, Audisio RA, Poston GJ (2015) The evolution of cancer surgery and future perspectives. Nat Rev Clin Oncol 12:115–124

Xi L, Grobmyer SR, Wu L, Chen R, Zhou G, Gutwein LG, Sun J, Liao W, Zhou Q, Xie H et al (2012) Evaluation of breast tumor margins in vivo with intraoperative photoacoustic imaging. Opt Express 20:8726–8731

Xia W (2017) Emerging clinical applications of photoacoustic imaging. J Biomed Imaging Bioeng 1:13

Xia W (2024) Biomedical Photoacoustics: technology and applications. Springer Nature

Xia W, Piras D, Van Hespen JC, Van Veldhoven S, Prins C, Van Leeuwen TG, Steenbergen W, Manohar S (2013) An optimized ultrasound detector for photoacoustic breast tomography. Med Phys 40:032901

Xiao Q, Chen T, Chen S (2018) Fluorescent contrast agents for tumor surgery. Exp Ther Med 16:1577–1585

Xiao J, Jiang J, Zhang J, Wang Y, Wang B (2022) Acoustic-resolution-based spectroscopic photoacoustic endoscopy towards molecular imaging in deep tissues. Opt Express 30:35014–35028

Xie W, Chen Y, Wang Y, Wei L, Yin C, Glaser AK, Fauver ME, Seibel EJ, Dintzis SM, Vaughan JC et al (2019) Microscopy with ultraviolet surface excitation for wide-area pathology of breast surgical margins. J Biomed Opt 24:026501

Xu M, Wang LV (2005) Universal back-projection algorithm for photoacoustic computed tomography. Phys Rev E Stat Nonlinear Soft Matter Phys 71:016706

Yan J, Zhuo S, Chen G, Milsom JW, Zhang H, Lu J, Zhu W, Xie S, Chen J, Ying M (2014) Real-time optical diagnosis for surgical margin in low rectal cancer using multiphoton microscopy. Surg Endosc 28:36–41

Yang S, Hu S (2024) Perspectives on endoscopic functional photoacoustic microscopy. Appl Phys Lett 125:030502

Yang G, Amidi E, Chapman WC Jr, Nandy S, Mostafa A, Abdelal H, Alipour Z, Chatterjee D, Mutch M, Zhu Q (2019) Co-registered photoacoustic and ultrasound imaging of human colorectal cancer. J Biomed Opt 24:121913–121913

Yang L, Park J, Marjanovic M, Chaney EJ, Spillman DR, Phillips H, Boppart SA (2021) Intraoperative label-free multimodal nonlinear optical imaging for point-of-procedure cancer diagnostics. IEEE J Select Top Quantum Electron 27:1–12

Yang J, Choi S, Kim C (2022a) Practical review on photoacoustic computed tomography using curved ultrasound array transducer. Biomed Eng Lett 12:1–17

Yang L, Xu D, Chen G, Wang A, Li L, Sun Q (2022b) Miniaturized fiber optic ultrasound sensor with multiplexing for photoacoustic imaging. Photoacoustics 28:100421

Yang S, Ding L, Wang S, Du C, Feng L, Qiu H, Zhang C, Wu J, Fan K, Jin B et al (2023) Studying oral tissue via real-time high-resolution terahertz spectroscopic imaging. Phys Rev Appl 19:034033

Yang F, Yang Z, Zhu Z, Zhu S, Song W, Yang Y, Yuan X (2025) A joint photoacoustic imaging and broadband spectral analysis for early-stage intraoperative pathology assessment: a case study with colorectal cancer. Photoacoustics 1(43):100712

Yao J, Wang LV (2014) Sensitivity of photoacoustic microscopy. Photoacoustics 2:87–101

Yao J, Maslov KI, Zhang Y, Xia Y, Wang LV (2011) Label-free oxygen-metabolic photoacoustic microscopy in vivo. J Biomed Opt 16:076003–076003

Ye S, Zou J, Huang C, Xiang F, Wen Z, Wang N, Yu J, He Y, Liu P, Mei X et al (2023) Rapid and label-free histological imaging of unprocessed surgical tissues via dark-field reflectance ultraviolet microscopy. iScience 26:105849

Yip LC, Omidi P, Rascevska E, Carson JJ (2022) Approaching closed spherical, full-view detection for photoacoustic tomography. J Biomed Opt 27:086004

Yoon J (2022) Hyperspectral imaging for clinical applications. Biochip J 16:1–12

Yoshitake T, Giacomelli MG, Quintana LM, Vardeh H, Cahill LC, Faulkner-Jones BE, Connolly JL, Do D, Fujimoto JG (2018) Rapid histopathological imaging of skin and breast cancer surgical specimens using immersion microscopy with ultraviolet surface excitation. Sci Rep 8:4476

Youssef A, Moa B, El-Sharkawy YH (2024) A novel visible and near-infrared hyperspectral imaging platform for automated breast-cancer detection. Photodiagn Photodyn Ther 46:104048

Zeng Z, Amin A, Roy A, Pulliam NE, Karavites LC, Espino S, Helenowski I, Li X, Luo Y, Khan SA (2020) Preoperative magnetic resonance imaging use and oncologic outcomes in premenopausal breast cancer patients. NPJ Breast Cancer 6:49

Zeppa P, De Marco R, Monticelli M, Massara A, Bianconi A, Di Perna G, Greco Crasto S, Cofano F, Melcarne A, Lanotte MM et al (2022) Fluorescence-guided surgery in glioblastoma: 5-ALA, SF or both? Differences between fluorescent dyes in 99 consecutive cases. Brain Sci 12:555

Zhan H, Sun C, Xu M, Luo T, Wang G, Xi G, Liu Z, Zhuo S (2022) Analysis of intraoperative microscopy imaging techniques and their future applications. Front Phys 10:991279

Zhan X, Liu Y, Chen Z, Luo J, Yang S, Yang X (2023) Revolutionary approaches for cancer diagnosis by terahertz-based spectroscopy and imaging. Talanta 259:124483

Zhang EZ, Beard PC (2011) A miniature all-optical photoacoustic imaging probe. In: Proceedings of the photons plus ultrasound: imaging and sensing 2011, vol 7899. SPIE, pp 291–296

Zhang Y, Moy AJ, Feng X, Nguyen HT, Sebastian KR, Reichenberg JS, Markey MK, Tunnell JW (2020a) Diffuse reflectance spectroscopy as a potential method for nonmelanoma skin cancer margin assessment. Transl Biophoton 2:e202000001

Zhang Y, Wu X, He L, Meng C, Du S, Bao J, Zheng Y (2020b) Applications of hyperspectral imaging in the detection and diagnosis of solid tumors. Transl Cancer Res 9:1265–1277

Zhang Z, Jin H, Zhang W, Lu W, Zheng Z, Sharma A, Pramanik M, Zheng Y (2023) Adaptive enhancement of acoustic resolution photoacoustic microscopy imaging via deep CNN prior. Photoacoustics 30:100484

Zhang S, Xiao X, Yi Y, Wang X, Zhu L, Shen Y, Lin D, Wu C (2024a) Tumor initiation and early tumorigenesis: molecular mechanisms and interventional targets. Signal Transduct Target Ther 9:149

Zhang J, Arroyo J, Bell MAL (2024b) Multispectral photoacoustic imaging of breast cancer tissue. In: Proceedings of the 2024 IEEE Ultrasonics, Ferroelectrics, and Frequency Control Joint Symposium (UFFC-JS). IEEE, pp 1–4

Zhang Y, Li Z, Li Z, Wang H, Regmi D, Zhang J, Feng J, Yao S, Xu J (2024c) Employing Raman spectroscopy and machine learning for the identification of breast cancer. Biol Proced Online 26:28

Zhang S, Yang B, Yang H, Zhao J, Zhang Y, Gao Y, Monteiro O, Zhang K, Liu B, Wang S (2024d) Potential rapid intraoperative cancer diagnosis using dynamic full-field optical coherence tomography and deep learning: a prospective cohort study in breast cancer patients. Sci Bull 69:1748–1756

Zhao T, Desjardins AE, Ourselin S, Vercauteren T, Xia W (2019) Minimally invasive photoacoustic imaging: current status and future perspectives. Photoacoustics 16:100146

Zhao T, Ma MT, Ourselin S, Vercauteren T, Xia W (2021) Towards ultrathin fiber-optic probe for simultaneous photoacoustic and fluorescence endoscopy. In: Proceedings of the 2021 IEEE International Ultrasonics Symposium (IUS). IEEE, pp 1–4

Zhao T, Ma MT, Ourselin S, Vercauteren T, Xia W (2022a) Video-rate dual-modal photoacoustic and fluorescence imaging through a multimode fibre towards forward-viewing endomicroscopy. Photoacoustics 25:100323

Zhao T, Pham TT, Baker C, Ma MT, Ourselin S, Vercauteren T, Zhang E, Beard PC, Xia W (2022b) Ultrathin, high-speed, all-optical photoacoustic endomicroscopy probe for guiding minimally invasive surgery. Biomed Opt Express 13:4414–4428

Zhao T, Zhang M, Ourselin S, Xia W (2022c) Wavefront shaping-assisted forward-viewing photoacoustic endomicroscopy based on a transparent ultrasound sensor. Appl Sci 12:12619

Zhao H, Li C, Shi X, Zhang J, Jia X, Hu Z, Gao Y, Tian J (2024) Near-infrared II fluorescence-guided glioblastoma surgery targeting monocarboxylate transporter 4 combined with photothermal therapy. EBioMedicine 106:105243

Zhou J, Jokerst JV (2020) Photoacoustic imaging with fiber optic technology: a review. Photoacoustics 20:100211

Zhou Y, Yao J, Wang LV (2016) Tutorial on photoacoustic tomography. J Biomed Opt 21:061007–061007

Zhou HC, Chen N, Zhao H, Yin T, Zhang J, Zheng W, Song L, Liu C, Zheng R (2019) Optical-resolution photoacoustic microscopy for monitoring vascular normalization during anti-angiogenic therapy. Photoacoustics 15:100143

Zhu J, Huynh N, Ogunlade O, Ansari R, Lucka F, Cox B, Beard P (2023) Mitigating the limited view problem in photoacoustic tomography for a planar detection geometry by regularized iterative reconstruction. IEEE Trans Med Imaging 42:2603–2615

Zhu X, Menozzi L, Cho SW, Yao J (2024) High speed innovations in photoacoustic microscopy. npj Imaging 2:46

Applications of Secondary Ion Mass Spectrometry Imaging in Preclinical Research

Inci Barut Helm, Simon Uzoni, and John S. Fletcher

Abstract

Secondary ion mass spectrometry (SIMS) is a versatile analytical technique that uses an energetic beam of ions to eject atoms and molecules from a sample surface. The ability to finely focus the impinging ion beam allows chemical maps of a sample surface to be produced while the gradual erosion of a sample under an ion beam bombardment provides access to a third dimension of chemical information. Both hard and soft materials are accessible for analysis by SIMS with the caveat that the technique operates at high vacuum and hence samples may need to be prepared or preserved (e.g., cryogenically) for optimum results. This chapter provides an overview of the fundamentals of the technique, recent progress in instrument development including the introduction of gas cluster ion beams (GCIBs) for SIMS analysis and the resulting development of new mass analyzer configurations to maximize the potential of these beams for mass spectrometry. and examples of applications in bioengineering and biomaterial analysis highlighting pre-clinical application, cancer research and neuroscience applications along with biomaterial development.

Keywords

SIMS · Imaging mass spectrometry · Preclinical research · Biomaterials · Tissue engineering

I. B. Helm (✉)
Department of Pharmacy, Basic Pharmaceutical Sciences, Gazi University, Ankara, Turkiye

Department of Chemistry and Molecular Biology, University of Gothenburg, Gothenburg, Sweden
e-mail: incibarut@gazi.edu.tr

S. Uzoni · J. S. Fletcher
Department of Chemistry and Molecular Biology, University of Gothenburg, Gothenburg, Sweden

I. F. Cengiz et al. (eds.), *Bioimaging Modalities in Bioengineering*,
Biomaterials, Bioengineering and Sustainability 9,
https://doi.org/10.1007/978-3-031-99410-4_23

1 Introduction

1.1 What Is Secondary Ion Mass Spectrometry (SIMS)?

Mass spectrometry is a unique analytical method that can provide broad analyte coverage, high chemical specificity, and high sensitivity. Mass spectrometry is most commonly applied to liquid or gaseous samples, typically coupled to a suitable chromatography system, however there are also several approaches that can be applied to solid samples. For the analysis of solid samples, a step is required to liberate the analyte from the solid material to eventually produce gas phase ions for detection. Approaches include laser ablation that can be used to eject atoms from a sample surface (as used in laser ablation inductively couple mass spectrometry, LA-ICP-MS), matrix assisted laser desorption ionization (MALDI) can be used to eject intact molecules from a surface once the sample is coated with a suitable matrix, electro-sprayed liquids can also be used to desorb molecules from a surface (desorption electrospray ionization). Secondary ion mass spectrometry (SIMS) uses a beam of energetic ions to eject a mixture of atoms, molecules, and fragmented molecules from a sample surface. Hence, SIMS bridges the gap between elemental analysis and molecular detection and can be applied to both hard and soft materials. The ions fired at the sample are referred to as primary ions and those ejected (sputtered) are termed secondary ions and these are extracted into the mass spectrometer as depicted in Fig. 1a. A unique feature of SIMS is the inherent surface sensitivity associated with ion beam induced sputtering where the emitted secondary ions originate from the top few atomic layers (Lockyer et al. 2024).

By discreetly probing the sample point by point a mass spectral image can be produced where each pixel point contains a full mass spectrum. The SIMS image data can be interrogated retrospectively where regions of interest in the MS image can be selected and the local mass spectrum for that region outputted or, alternatively, ions of interest in the mass spectrum can be selected and their distribution over the image visualized (Lockyer et al. 2024). As complex hyper spectral data sets are produced multivariate analysis methods, recently expanding to machine learning approaches, are often applied for spectral classification, image segmentation/classification, contrast enhancement or general data reduction (Yokoyama et al. 2015; Matsuda and Aoyagi 2022; Fransson et al. 2024; Aoyagi et al. 2021; Gardner et al. 2022). While the secondary ions emitted from a single primary ion impact are characteristic of the surface of the sample, continued ion bombardment will gradually erode the material. Hence, repeated analysis can be used to investigate chemical changes as a function of depth into the sample that can be displayed as a plot of signal intensity (chemical composition) versus depth (a depth profile plot) or 3-dimensional ion images.

SIMS instruments operate at high to ultrahigh vacuum and there are a range of different combinations of primary ion source and mass analyzers that lend themselves to different application areas (e.g., semiconductor analysis, geo-science research, life science/biomaterials applications). In this chapter we will focus on the use of SIMS for the general analysis of complex materials. For such analysis

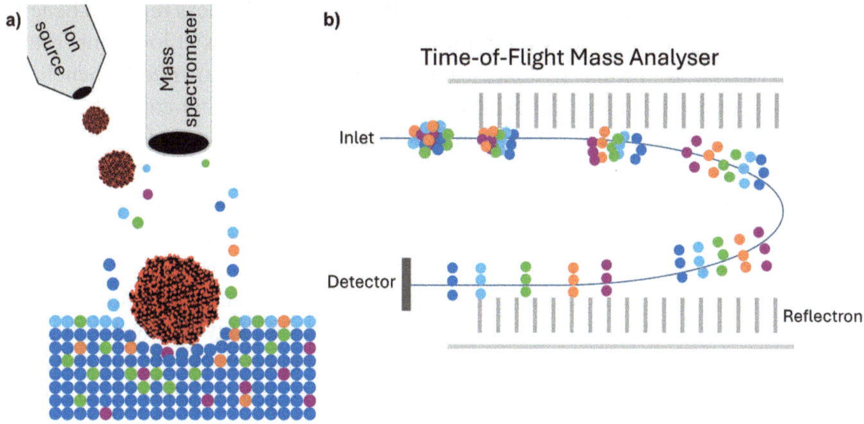

Fig. 1 (**a**) The process of sputtering and extracting secondary ions into a mass spectrometer. From the ion source, a primary ion beam of gas cluster ions is fired onto the sample, the impact ejects material that originated from the impact site as secondary ions which are extracted into the mass spectrometer to be analyzed. The light blue balls represent the top layer of the surface, the dark blue represents subsurface material, and the other color balls represent different molecules; (**b**) Separation of secondary ions with varying mass-to-charge ratio (*m/z*) using a ToF mass analyzer equipped with a reflectron to measure ions *m/z* based on flight time to the detector, secondary ions with different *m/z* represented by balls with different colors. The ions enter the ToF mass analyzer unorganized and as the ions travel through the flight tube and reflected towards the detector, lighter ions will travel with higher velocity and reach the detector before heavier ions, separating the ions based on *m/z*

variations of time-of-flight (ToF) mass analyzers, depicted in Fig. 1b, are most common due to their high sensitivity, parallel detection capability and broad mass range, although the use of Fourier transform ion trap based analyzers such as the Orbitrap™ that provide very high mass resolution at the cost of speed and sensitivity is increasing.

1.2 Sample Preparation

One advantage of SIMS is that the technique is applicable to a broad range of sample types with minimal sample preparation. However, that is not to say that the optimum result is obtained without some specific sample preparation. The vacuum environment required for SIMS analysis presents a challenge for the preservation of hydrated samples, for example biological specimens or hydrogels. The gold standard for the preservation and analysis of these samples is to use cryogenic preservation with the sample analyzed at cryogenic temperatures (ideally below 150 K) to prevent water sublimation (Lockyer et al. 2024). There are several benefits associated with hydrated sample analysis. First the microstructure and distribution of molecules in biological samples is preserved and second there have been multiple observations of secondary ion signal improvements in frozen samples (Angerer et al. 2016a; Sämfors et al. 2019; Conlan et al. 2006). A third, but less reported benefit, has been the improvement in depth profiling of organic samples at low temperature (Sjovall et al. 2010). Frozen hydrated sample analysis while known to be beneficial is however generally more technically challenging and in some cases there may be a trade-off between the more complex sample preparation and experimental reproducibility (Schaepe et al. 2015).

SIMS analysis also benefits from the use of flat samples as topographical variation can result in shadowing of different parts of the ample from the incoming ion beam and a variation in emitted secondary ion energies as the originate from different parts of the electrostatic extraction field. This energy spread can have implications for ion transmission and mass resolution on many SIMS instruments. Tissue samples are typically sectioned to a few micrometers thickness and mounted onto a conductive substrate prior to analysis while cultured cells are analyzed directly on a suitable flat support (Barut and Fletcher 2023). Sample modification using organic matrices, similar to those used in MALDI MS, have been used in SIMS (matrix enhanced, ME-SIMS) (Wu and Odom 1996) as has with metal addition (metal assisted (MA-SIMS) (Heile et al. 2008), and in some cases exposure to reactive vapor (e.g., trifluoroacetic acid, TFA, or ammonia) (Angerer et al. 2015, 2016a) has also been employed.

Inside the SIMS instruments ion beams may also be used to prepare the sample either by pre-etching the sample to remove the outer surface layers or focused ion beam milling to expose buried material.

1.3 SIMS for Pre-clinical Research

The flexibility of SIMS means that it can be utilized in multiple areas of pre-clinical research. SIMS can be used to perform molecular pathology producing chemical maps of tissue sections. The data in these maps can be subsequently analyzed and used for sample classification, biomarker discovery or to specifically study chemical changes in the samples in order to improve our understanding of, for example, disease progression or treatment target interactions. The high spatial resolution of SIMS makes it applicable to cultured cell analysis including the detection of drugs within single cells, for example the distribution of the antiarrhythmic agent amiodarone in macrophages (Passarelli et al. 2015) and on a smaller scale the imaging of ampicillin and tetracycline in E. coli bacteria (Tian et al. 2017). As SIMS is not limited to soft materials or materials where analyte can be extracted, the method can be employed to investigate samples such as pharmaceutical formulations where heterogeneity and crystallinity can be probed (Paladino et al. 2022). Harder tissues including bone can be analyzed using SIMS where soft tissue and mineral data can be collected together (Kern et al. 2023). Further, the surface sensitive nature of the technique is especially powerful for investigating material fouling and chemical modifications to prevent fouling, for example in medical implants (Malmberg et al. 2021), catheters (Tyler et al. 2017) and drug eluting stents (Fisher et al. 2009). Chemical modifications that prevent biofilm formation can be characterized (Wong et al. 2023) and recently biofilms themselves have been assessed in 3D (Akbari et al. 2023).

2 Fundamentals of SIMS

When a focused primary ion beam is fired onto a sample surface, the impact causes particles to be ejected from the surface in a process called sputtering, most of these partials are neutral molecules but a small part are electrons, and positive and negative secondary ions. The secondary ions are extracted into a mass spectrometer that measures and records the mass-to-charge-ration (*m/z*) of the ions, producing a mass spectrum that describes the chemical composition of the surface. Images of the surface can be generated by scanning the ion beam over the sample surface and generating a mass spectrum for each spot that represents a pixel in an image, known as imaging mass spectrometry (IMS) or mass spectrometry imaging (MSI). This is commonly used to study how the intensity and composition of chemicals varies over a surface, for example to understand how lipids composition depends on the type of cell and tissue in an organ (Lockyer et al. 2024; Vickerman 2009).

The information provided by the SIMS analysis can depend greatly on the nature of the primary ion used for analysis. Signal levels in SIMS are influenced by the mass and composition of the primary ion beam along with the energy of the impacting ion and the angle of incidence of the ion beam. The sputtering event can be described in two different ways depending on the type of primary ion beam that impacts the surface. Primary ion beams composed of atomic (e.g., Cs^+ or O^-) or small cluster (e.g., $Au_n{}^+$, $Bi_n{}^+$) ions penetrate the sample upon impact, transferring the kinetic energy into the sample, some of that energy is transferred to the surface by a chain reaction of colliding atoms and molecules, known as collision cascade, and causing material from the very surface to be sputtered. The energy which does not reach the surface is absorbed by the molecules beneath the surface, resulting in molecular fragmentation which can have negative effects when analyzing organic compounds by complicating the mass spectrum or limiting the analysis to the sample surface. This has created two types of SIMS. The first is static SIMS for which the analysis is kept to the surface layer by limiting the primary ion dose such that only 1% of the sample surface is impacted by the primary ions. This typically corresponds to an ion dose of 1E+12 ions/cm^2 for organic samples and ensures that the mass spectrum is representative of the pristine surface chemistry. It should be noted that the dose limit for static SIMS will vary depending on the type of primary ion used and the type of sample. The second mode of SIMS is termed dynamic SIMS. Here high ion beam flux is used to erode the sample during the experiment exposing the material underneath the surface. Dynamic SIMS analysis is common in inorganic material analysis but, historically, was not possible for organic material due to the accumulation of sub-surface ion beam induced damage.

Increasingly, SIMS analysis of organic and biological samples uses polyatomic ion beams, such as $C_{60}{}^+$, or gas cluster ion beams (GCIBs) that are larger than atomic cluster ions, especially GCIBs, such as $Ar_{2000}+$, which produce clusters

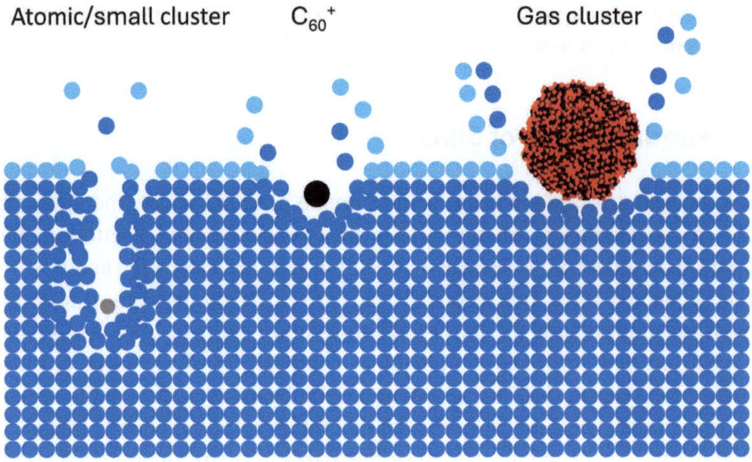

Fig. 2 Illustrative comparison how the sputtering process and the sample is affected by atomic/small cluster (left), C_{60}^{+} (middle), and GCIB (right) primary ions impacting a sample surface. Atomic/small cluster primary ion penetrates significantly deeper into the sample material and disturbs subsurface material compared to C_{60}^{+} and GCIB. C_{60}^{+} and GCIB forms craters on the surface and limits the effect of the impact closer to the impact site. The top layer is represented by the light blue balls while the subsurface material is represented by the dark blue balls

composed of several thousand atoms or molecules, such as Ar, CO_2, and H_2O as the ion beam. These ions have lower energy density relative to their mass compared to atomic and small cluster ions, resulting in shallower penetration and subsurface damage accumulation closer to the sample surface, reduced molecular fragmentation, and increased ejection of larger intact molecules. All of which are improved as the energy density is reduced by increasing the gas cluster size. The reduction of sub-surface damage accumulation in organic samples when using these beams makes it possible to depth-profile intact molecular ions in organic and biological materials and to generate 3D molecular images from organic samples. Figure 2 illustrates how the impact of three different primary ion beams, atomic/small cluster, C_{60}^{+}, and GCIBs affect the sample during the sputtering process. The introduction of GCIBs for biological analysis has made the detection of intact molecular ions with *m/z* over 2000, commonplace compared to atomic ion beam analysis of organic samples where the mass range is typically limited to a few hundred Da. However, the GCIBs come with some disadvantages. The beams are harder to focus with most of the reported work using spot sizes of several micron in diameter compared to 200 nm for C_{60}^{+} ion beam (Fletcher et al. 2006) and 50 nm with atomic and atomic cluster ions (Lockyer et al. 2024; Vickerman 2009). Table 1 details the spatial resolution, use cases, advantages, and disadvantages with atomic and small cluster-, C_{60}^{+}-, and GCIBs.

Table 1 Comparison between atomic and small cluster-, C_{60}^+-, and GCIBs

Ion beam	Spatial resolution	Use case	Advantage	Disadvantage
Atomic/small cluster Au_n^+, Ga_n^+, Bi_n^+, O^-, Cs^+	50 nm	Material analysis (metals, alloys, semiconductors, minerals) Elemental depth profiling Subcellular isotope imaging	High spatial resolution High depth resolution (inorganic material)	High subsurface damage accumulation High fragmentation
Polyatomic C_{60}^+	200 nm	Molecular depth profiling Material analysis (inorganics to organics)	High spatial resolution Wide material compatibility	Limited high mass ionization
Gas cluster Ar_n^+, $(CO_2)_n^+$, $(H_2O)_n^+$	1 μm	Organic material analysis (polymers) Bio sample analysis (tissues and cells) Molecular analysis Molecular depth profiling	High mass ionization Low fragmentation Low subsurface damage accumulation	Low spatial resolution Limited compatibility with inorganic material

Organic molecules are ionized by protonation $[M+H]^+$, deprotonation $[M-H]^-$, cationization by sodium or potassium $[M+Na/K]^+$, or by fragmentation $[M-X]^{\pm}$. The preferred ion form depends on the analytes functional groups and molecular structure, basic functional groups tend to protonate while deprotonation is more common for acidic functional groups (Vickerman and Briggs 2013). While gas cluster beams have aided the detection of intact molecules with SIMS molecular fragmentation is still common and as such SIMS is considered a relatively hard ionization technique. Fragment peaks in the mass spectrum can be used to improve confidence in peak assignment/identification; lipids have predictable fragmentation pathways that depend on the lipid type and can be used interpret a mass spectrum containing intact and fragment ions. However, chemically complex samples can be more complicated to interpret due to overlapping fragment ions from different lipid types, e.g., phosphatidylserine lipids can fragment and be detected as phosphatidic acid lipids as a negative ion by neutral loss of 87 from the headgroup, and all lipids can produce negative fatty acids (FA) fragment ions (Adams et al. 2016; Pi et al. 2016; Vandergrift et al. 2023; Angerer et al. 2019). On the other hand, combined with IMS, fragmentation can be useful for localizing and identifying lipids without the need for MS/MS (Angerer et al. 2019).

3 The Role of SIMS Imaging in Drug Development

3.1 SIMS as an Imaging Tool to Map Drug Distribution in Biological Model Systems

In drug development research, an important application of SIMS imaging is to map the distribution of drugs in biological model systems. Researchers can image drug penetration, accumulation and metabolism in various tissues and organs with high spatial resolution (Lagator et al. 2024; Belsey et al. 2023). Molecular imaging is very helpful in increasing the effectiveness and decreasing the potential side effects of a drug, hence optimizing its formulation. Imaging with SIMS does not require labels which enables analyzing and locating drugs and their metabolites inside organic or inorganic matrices directly without label interruption (Spruill et al. 2022; Topolski et al. 2024). Additionally, it is possible to image drugs in 3D, which enables researchers to visualize drug penetration and distribution in different layers of tissues (Belsey et al. 2023; Johnson et al. 2021). For example, SIMS has been used to image penetration of the depth and accumulation of a topical drug in different skin layers (Belsey et al. 2023). SIMS imaging is also very well suited to the detection of the distribution of metal-labelled drugs in cell and tissues. In a recent study, ToF-SIMS imaging has been used to show the distribution of a trastuzumab complex, a cancer drug, conjugated with zirconium in xenograft ovarian and breast tumor sections. In the study, heterogenic drug distribution was visualized using a bismuth ion gun and a 12 μm lateral resolution achieved preceding the resolution of the conventional methods such as positron emission tomography (approximately 5 mm) (Penen et al. 2022).

3.2 Imaging Applications of SIMS in Drug Delivery Systems

Recently the usage of SIMS imaging increased in the characterization and optimization of drug delivery systems due to its advantages. Delivery systems like microparticles, nanoparticles, liposomes and hydrogels aim to improve the solubility and release of the drugs. SIMS can be used to understand drug capsulation efficiency, release kinetics and spatial distribution of these systems (Johnson et al. 2021; Kret et al. 2023; Vennemann et al. 2022; Robinson et al. 2021).

Additionally, combining SIMS imaging with other MS imaging modalities provides extensive monitoring of drugs and their release systems. In one study, ToF-SIMS with 3D depth profiling and MALDI MS imaging and histopathological techniques were combined to track down and compare a respiratory drug and its excipients (microparticle controlled release system) with different formulations over time in rat lung tissue. Using consecutive sections for different analyze modalities, firstly MALDI MSI provided the localization information of the drug inside the tissue while 3D ToF-SIMS performed on the locations picked by MALDI MSI revealing drug and excipient colocalizations. With the help of ToF-SIMS depth profiling, vertical distribution of drug mapped inside the microparticles and drug traces detected after dosing in timely manner (Robinson et al. 2021).

4 Imaging Applications of SIMS in Bioengineering and Biomaterials Research

4.1 SIMS Imaging in Biomaterials Research

Biomaterials research is an evolving area that involves crucial applications in preclinical research such as drug delivery and controlled release, tissue engineering and regenerative medicine (Ratner and Zhang 2020). It is very important to assess the function, biocompatibility, and location of different biomaterials especially during the discovery of effective treatment technologies. Often incorporating complex components, these materials have to be tested both for their quality and their interactions with biological surroundings. The decomposition rate of biodegradable materials is very critical in terms of sustaining the mechanical and structural stability of the material while compensating the degradation rate in a timely manner. Besides, the biomaterial of interest should meet the criteria ensuring the unique necessities of the tissue will be aimed at and potential complications that the biomaterial can trigger, such as inflammation and infection, should be evaluated (Kayadurmus et al. 2024). At that point, imaging mass spectrometry provides a versatile analytical tool with its ability to measure chemical composition and the location of both biological tissue and the biomaterials. Hence, MSI is being used for quality control of new biomaterials such as implants, nanofibers, and different surface modification items (Kret et al. 2023). As an important research area for aging populations, bone research includes the usage of biomaterials aiming to decrease the prevalence of osteoporosis which is a very important public health problem. MSI has been successfully used to assess the efficiency of different biomaterials in different bone sections. For instance, in one recent study different modes of ToF-SIMS and several SIMS instruments have been used to visualize mineralized and nonmineralized tissue with or without different biomaterial applications and bone/implant interphases (Fig. 3) (Kern et al. 2023). In another study SIMS using a bismuth primary ion source has been used to monitor the released components from biomaterials implanted to the rat bones. Strontium-modified calcium phosphate cement paste which also contains mesoporous bioactive glass particles has been tested. In the defect area mineralized bone and collagen, together with release and incorporation of strontium and silicon, were imaged. Figure 4 shows imaging results after ToF-SIMS analysis of defected rat bone tissue compartments with different biomaterials (Fig. 4a). Decrease in Sr^{+} mass signal distribution from biomaterial implant to the surrounding tissue are shown in Fig. 4b. Chemical imaging of these different biomaterials such as hydroxyapatite (HAP), collagen, strontium (Sr), and silicon (Si, for bioactive glass) is shown in Fig. 4c–f respectively (Kern et al. 2020).

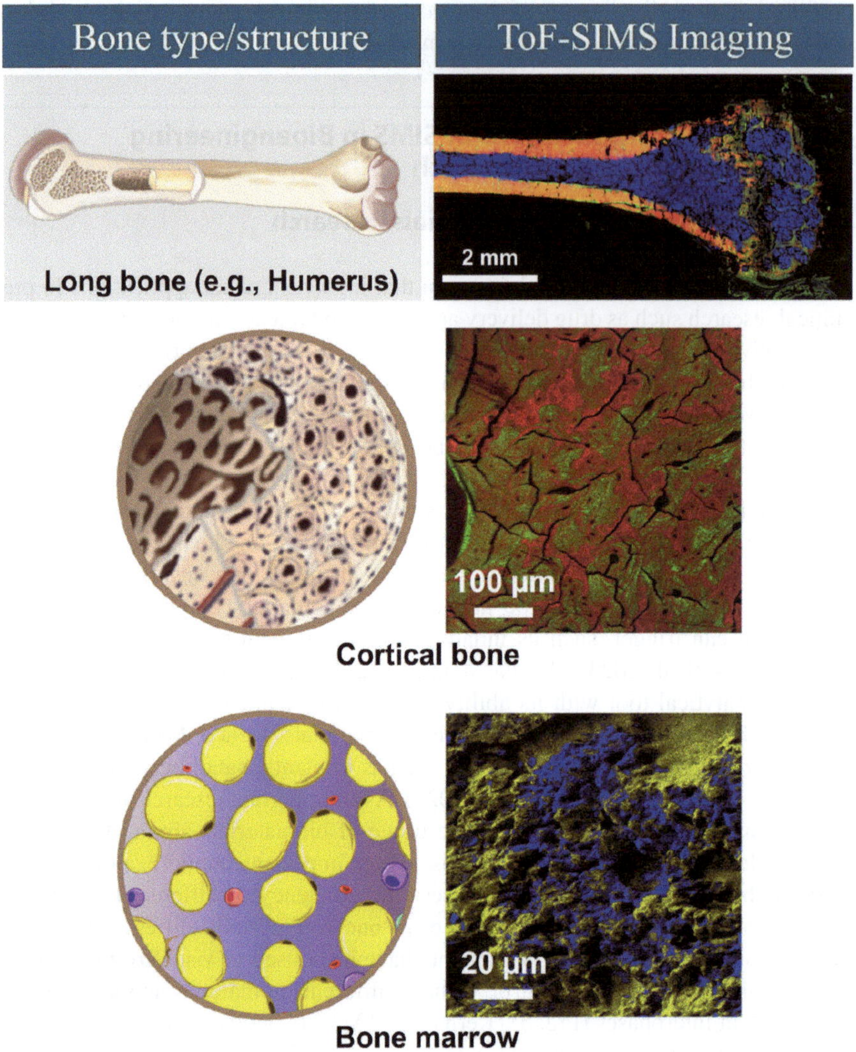

Fig. 3 ToF-SIMS imaging allows imaging of different types and structures of bones (e.g., humerus, cortical bone, bone marrow), discriminates mineralized bone (red) and non-mineralized bone (green) as well as bone marrow (blue) and lipids (orange) in whole bone sections. Adapted from (Kern et al. 2023) under a Creative Commons Attribution License (CC BY 4.0)

4.2 Applications of SIMS Imaging in Tissue Engineering

Tissue engineering is a highly multidisciplinary area merging recent advancements mainly from materials science, engineering, and medicine; hence allowing the healing of wounded tissues, tissue replacements in the case of tissue or organ loss and general tissue restoration and regeneration. Various types of scaffolds

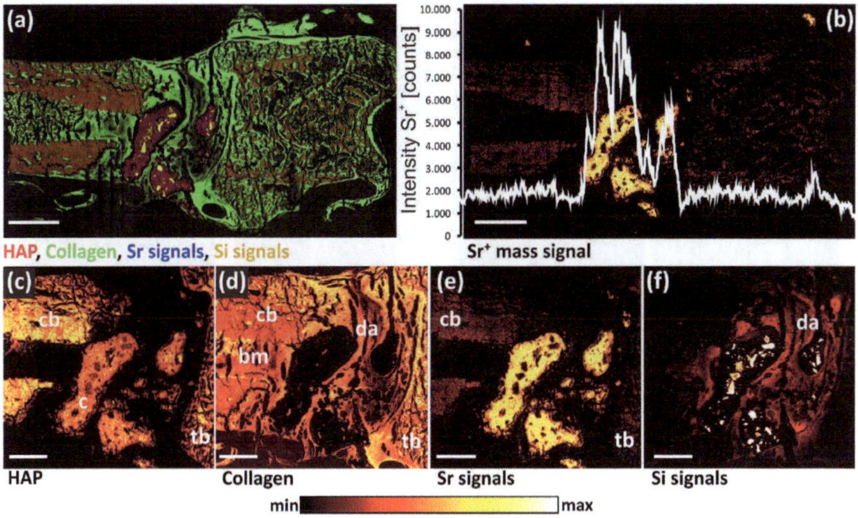

Fig. 4 (**a**) Mass fragment distribution of hydroxyapatite (red), collagen (green), mass signals of strontium (blue) and silicon (yellow); (**b**) line scan of strontium showing the Sr^{+} intensity in the biomaterial implant cement increased and lessen in the surrounding bone tissue; (**c**) hydroxyapatite (HAP) signal distribution in the bone cement; (**d**) chemical image of collagen structure in the defected bone area; (**e**) strontium distribution in the remaining bone cement fragments while showing lower intensity in different bones, cortical (cb) and trabecular bone (tb); (**f**) silicon signal mass distribution reveals glass particles of cement remains, and incorporation of silicon incorporated in collagenous tissue around the former defect area (da). The scale bars indicate 2 mm in panels a and b, and 1 mm in panels c-f. (Reprinted with permission from (Kern et al. 2020). Copyright 2020, American Vacuum Society)

have been used in the area to aid these treatments made of synthetic materials or biomaterials. These scaffolds allow cell growth and tissue formation which need to be biocompatible. Developing compatible tissue engineering methods requires dynamic monitoring between scaffolds and cellular behaviors (Khan and Tanaka 2018). SIMS imaging helps provide a detailed understanding of the interactions between how synthetic or biomaterials interact with biological structures and monitorization of possible biomolecular events (Ilin and Kraft 2015). Ruiter et al. used ToF-SIMS to confirm scaffold functionalization showing thiol and peptide groups´ attachment to be used for human corneal stromal cells (Ruiter et al. 2020). Cometta et al. used melamine modified scaffolds to prevent biomaterial infections caused by biofilm formation and used ToF-SIMS for melamine characterization and monitoring on the scaffold surface at submicron spatial resolution (Cometta et al. 2022). ToF-SIMS imaging was also used to identify degradation products and patterns (Fig. 5) of newly designed ultrathin stents made of zinc scaffolds using microalloying by adding lithium (Yang et al. 2023).

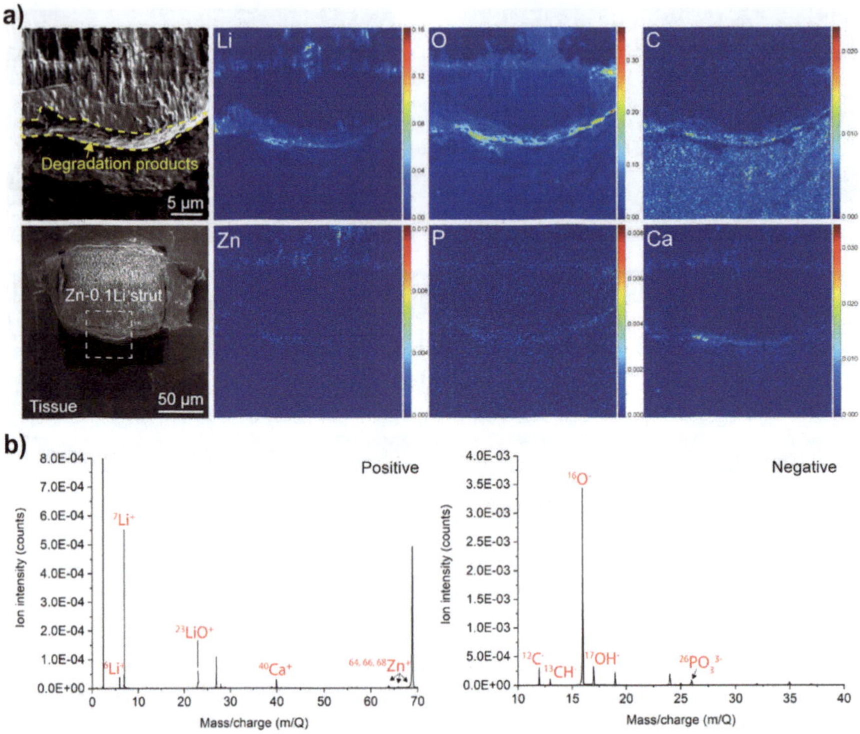

Fig. 5 Identification of degradation products of ultrathin stents. (**a**) Mapping of elements with the regions of interest (white rectangle) from a 12-month sample; (**b**) ToF-SIMS able to map and compare the ions abundant in degradation products. Detected negative and positive ion spectrums (Li^+, O^-, C^-, Zn^+, $P5^+$ and Ca^+) analyzed by ToF-SIMS showing the abundance of the ions coming from the sample proportional to peak intensities. (Adapted from (Yang et al. 2023), with permission from John Wiley and Sons)

5 Applications of SIMS Imaging in Disease Models

5.1 SIMS Imaging in Cancer Research

ToF-SIMS has been used to correlate chemical signal from light lipids and fragment ions with pathological diagnosed cancer tissues and cells, such as breast cancer (Bluestein et al. 2016; Robinson et al. 2016), pancreatic cancer (Bluestein et al. 2018), prostate cancer (Gazi et al. 2004), colon cancer (Park et al. 2008), bladder cancer (Gostek et al. 2015), and thyroid cancer (Hansson et al. 2008), with high spatial resolution using small cluster and C_{60}^+ ion beams. The improved ability to detect intact lipids with GCIBs has been beneficial for cancer research as it allows the lipid composition to be correlated with pathological diagnosis and gives insights how lipid pathways are affected by breast cancer (Tian et al. 2021; Angerer et al.

2016b) and skin cancers, such melanoma (Neittaanmäki et al. 2024) and basal cell carcinoma (Nilsson et al. 2020).

Angerer et al. showed the possibility to use GCIBs to distinguish between cancerous and inflammatory areas based on FA saturation in a breast cancer biopsy tissue using single-ion images of intact lipids and FA. Phosphatidylinositol (PI) 38:4 and FA 20:4 were seen to be most intense in the inflamed tissue surround the tumor while PI 38:3, PI 38:2, FA 20:3, and FA 20:2 was more intense in the tumor (Angerer et al. 2016b). Tian et al. demonstrated a multi-omics approach to obtain more information from breast cancer tissue than possible with lipid analysis alone. By first analyzing the samples using H_2O-GCIB under cryogenic conditions for lipidomics and metabolomics, and subsequently staining the tissue with lanthanide-labelled antibodies and re-analyzing the tissue using a C_{60}^+ ion beam, it was possible to correlate lipids, metabolites and proteins to different types of cells in the tissue (Tian et al. 2021). Neittaanmäki et al. used GCIB-SIMS in a pilot study to compare the lipid composition in healthy skin, and melanoma tissues that were primary, in-transit, and distant metastasis tumors. Multivariate data analysis using principal components canonical variate analysis, indicated an increase in phosphatidylethanolamine relative to PI lipids in primary tumors compared to metastatic tumors, and that metastatic tumors had increased GM3 gangliosides signal, longer chain GM3 correlated with distant metastasis and shorter chains with in-transit metastasis (Neittaanmäki et al. 2024). The localization of three GM3 gangliosides to metastasis melanoma compared to non-cancerous tissue is also shown in Fig. 6.

5.2 SIMS Imaging in Neurodegenerative Disease Models

Compared to cancer research applications, where surgically accessible tumors provide an abundant source of human samples, the application of SIMS to neuroscience

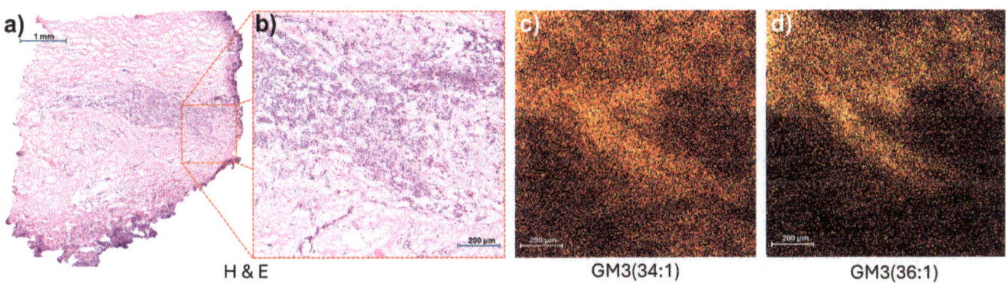

Fig. 6 (**a**) Hematoxylin-eosin stained histological image of metastatic melanoma tissue section with cancerous (purple) and non-cancerous (pink) tissue; (**b**) magnified view of a selected region from panel a; and (**c** and **d**) single-ion SIMS image of GM3 ganglioside 34:1 and 36:1 (**d**), respectively, from a consecutive tissue section. The images show that GM3 gangliosides are involved in metastatic melanoma, as the ion images show that the gangliosides are localized to the tumor. Scale bars indicate 1 mm in panel a, and 200 μm panels b–d

has predominantly involved analysis of cell or animal model samples. Fly brains, specifically the fruit fly drosophila melanogaster, is a popular model organism due to its relatively simple genome and the availability of multiple mutants (Phan et al. 2014; Le et al. 2018). Several studies have used SIMS to study the changes in the lipid composition of the fly brain in response to drug treatment either with the attention deficit hyperactivity disorder drug methylphenidate or the narcotic cocaine observing different changes in the lipidome caused by these drugs that both effect dopamine neurotransmission (Phan et al. 2015; Philipsen et al. 2018). Despite the drugs having overall similar actions on the dopamine transport system, albeit with different rates of dopamine accumulation/increase major lipids showed opposite changes upon oral administration to the fly.

Cell Models

Vitamin E, more specifically α-tocopherol has been analyzed in a range of samples using ToF-SIMS. Including sea slug neurons (Aplysia californica) (Tucker et al. 2012), selected due to their large size and relatively easy sample handling but also in induced pluripotent stem cells (Bruinen et al. 2018). SIMS imaging of cultured cells has been combined with electrochemical measurement of neurotransmitter release during exocytosis to study the dose dependent effects of the proteosome inhibitor lactacystin (Barut et al. 2023).

Rodent Brain Tissue

Rodent brain tissue has been well studied using SIMS both for development of sample preparation methodology, where cholesterol migration has been shown to be a particular problem in brain samples (Angerer et al. 2016a), and also in applied studies. For example, amyloid plaque exhibiting transgenic mouse brains have been studied with initial reports showing changes in the distribution of cholesterol and sulfatides (Sole-Domenech et al. 2013) and more recent examples, using the softer desorption available from GCIBs, mapping ganglioside accumulation at plaque sites (Barut and Fletcher 2023).

ToF-SIMS has been applied in studies of glioblastoma where the SIMS imaging was used to investigate inter- and intratumoral heterogeneity and revealed differences in the chemical signatures of tumors from young versus old patients. Similar to the use of lanthanide labelling mentioned previously, in this study Au nanobody labelling of the protein Caveolin-1 was employed for correlated metabolite/protein mapping (Gularyan et al. 2020). In a SIMS imaging study of neonatal rat pups the effects of inhaled nitric oxide were assessed. Changes to the lipid composition of the white matter regions of the brain were observed, specifically an increase in sulfatide lipid signals upon the nitric oxide treatment indicating an increase in myelination following inhalation (Kadar et al. 2014).

6 Conclusions and Future Perspectives

The ability of SIMS to provide unique information from a wide range of materials probing both the specific surface chemistry, and chemical changes as a function of depth, has made it very useful for obtaining as much chemical information as possible from tissues, cells, and other biological samples. SIMS has been used for preclinical studies on the delivery and release of drugs, the incorporation and development of biomaterials in different research areas including tissue engineering and biotechnologies for improved healthcare, lipid changes in tissues and cells caused by cancer, and in model brains from drug treatment and amyloid plaque, and effects on neurotransmitters from protein inhibitors. Recent advancements in SIMS technology have shifted the focus to analyzing larger intact molecules and improved molecular depth profiling in biological samples from the introduction of GCIBs while Orbitrap™ has brought increased mass resolving power to reduce uncertainties in molecular identification. This allows for new opportunities to revisit samples to gain new molecular information with subcellular spatial resolution and higher mass resolving power, but also to study samples with SIMS that were previously not possible or challenging due to limited spatial resolution and detectable mass range, such single-cell analysis of intact biomolecules. Furthermore, as analysis under cryogenic conditions becomes more common, the quality of the analysis and the information about the sample will increase as chemical composition in biological samples will be kept as life-like as possible. There is also hope for technological improvements that would result in high-throughput SIMS instruments that could analyze samples much faster than current instruments. These instruments could allow for SIMS to be incorporated into clinical studies for studying large sample sets but potentially as a medical diagnostic tool as well. All these improvements will result in larger and more complex datasets that detail the chemical composition in cells and tissues. This is an evolving challenge and will require more advanced and powerful data processing methods, such as machine learning and artificial intelligence, to fully interpret the data but may give new chemical and molecular insights into biological processes and diseases progression that can be useful to improve and solve health related problems.

Acknowledgements The authors gratefully acknowledge financial support from the Swedish Research Council (VR, Grant: 2022-04498).

Conflict of Interest The authors declare no conflict of interest.

References

Adams KJ, DeBord JD, Fernandez-Lima F (2016) Lipid specific molecular ion emission as a function of the primary ion characteristics in TOF-SIMS. J Vac Sci Technol B 34(5):051804. https://doi.org/10.1116/1.4961461

Akbari A, Galstyan A, Peterson RE, Arlinghaus HF, Tyler BJ (2023) Label-free sub-micrometer 3D imaging of ciprofloxacin in native-state biofilms with cryo-time-of-flight secondary ion mass spectrometry. Anal Bioanal Chem 415(5):991–999. https://doi.org/10.1007/s00216-022-04496-4

Angerer TB, Pour MD, Malmberg P, Fletcher JS (2015) Improved molecular imaging in rodent brain with time-of-flight-secondary ion mass spectrometry using gas cluster ion beams and reactive vapor exposure. Anal Chem 87(8):4305–4313. https://doi.org/10.1021/ac504774y

Angerer TB, Mohammadi AS, Fletcher JS (2016a) Optimizing sample preparation for anatomical determination in the hippocampus of rodent brain by ToF-SIMS analysis. Biointerphases 11(2):02A319. https://doi.org/10.1116/1.4941064

Angerer TB, Magnusson Y, Landberg G, Fletcher JS (2016b) Lipid heterogeneity resulting from fatty acid processing in the human breast cancer microenvironment identified by GCIB-ToF-SIMS imaging. Anal Chem 88(23):11946–11954. https://doi.org/10.1021/acs.analchem.6b03884

Angerer TB, Velickovic D, Nicora CD, Kyle JE, Graham DJ, Anderton C et al (2019) Exploiting the Semidestructive nature of gas cluster ion beam time-of-flight secondary ion mass spectrometry imaging for simultaneous localization and confident lipid annotations. Anal Chem 91(23):15073–15080. https://doi.org/10.1021/acs.analchem.9b03763

Aoyagi S, Fujiwara Y, Takano A, Vorng JL, Gilmore IS, Wang YC et al (2021) Evaluation of time-of-flight secondary ion mass spectrometry spectra of peptides by random Forest with amino acid labels: results from a Versailles project on advanced materials and standards Interlaboratory study. Anal Chem 93(9):4191–4197. https://doi.org/10.1021/acs.analchem.0c04577

Barut I, Fletcher JS (2023) Cell and tissue imaging by secondary ion mass spectrometry. Biointerphases 18(6). https://doi.org/10.1116/6.0003140

Barut I, He XL, Sener E, Samfors S, Ewing AG, Fletcher JS (2023) Correlative cellular mass spectrometry imaging and amperometry show dose dependent changes in lipid composition and exocytosis. Angew Chem Int Edit 62(15):e202217993. https://doi.org/10.1002/anie.202217993

Belsey NA, Dexter A, Vorng J-L, Tsikritsis D, Nikula CJ, Murta T et al (2023) Visualisation of drug distribution in skin using correlative optical spectroscopy and mass spectrometry imaging. J Control Release 364:79–89. https://doi.org/10.1016/j.jconrel.2023.10.026

Bluestein BM, Morrish F, Graham DJ, Guenthoer J, Hockenbery D, Porter PL et al (2016) An unsupervised MVA method to compare specific regions in human breast tumor tissue samples using ToF-SIMS. Analyst 141(6):1947–1957. https://doi.org/10.1039/c5an02406d

Bluestein BM, Morrish F, Graham DJ, Huang L, Hockenbery D, Gamble LJ (2018) Analysis of the Myc-induced pancreatic β cell islet tumor microenvironment using imaging ToF-SIMS. Biointerphases 13(6):06D402. https://doi.org/10.1116/1.5038574

Bruinen AL, Fisher GL, Balez R, van der Sar AM, Ooi L, Heeren RMA (2018) Identification and high-resolution imaging of α-tocopherol from human cells to whole animals by TOF-SIMS tandem mass spectrometry. J Am Soc Mass Spectrom 29(8):1571–1581. https://doi.org/10.1007/s13361-018-1979-x

Cometta S, Jones RT, Juárez-Saldivar A, Donose BC, Yasir M, Bock N et al (2022) Melimine-modified 3D-printed Polycaprolactone scaffolds for the prevention of biofilm-related biomaterial infections. ACS Nano 16(10):16497–16512. https://doi.org/10.1021/acsnano.2c05812

Conlan XA, Lockyer NP, Vickerman JC (2006) Is proton cationization promoted by polyatomic primary ion bombardment during time-of-flight secondary ion mass spectrometry analysis of frozen aqueous solutions? Rapid Commun Mass Spectrom 20(8):1327–1334. https://doi.org/10.1002/rcm.2446

Fisher GL, Belu AM, Mahoney CM, Wormuth K, Sanada N (2009) Three-dimensional time-of-flight secondary ion mass spectrometry imaging of a Pharmaceutical in a Coronary Stent Coating as a function of elution time. Anal Chem 81(24):9930–9940. https://doi.org/10.1021/ac901587k

Fletcher JS, Lockyer NP, Vickerman JC (2006) C60, buckminsterfullerene: its impact on biological ToF-SIMS analysis. Surf Interface Anal 38(11):1393–1400. https://doi.org/10.1002/sia.2461

Fransson A, Dimovska Nilsson K, Henderson A, Farewell A, Fletcher JS (2024) PCA, PC-CVA, and random Forest of GCIB-SIMS data for the elucidation of bacterial envelope differences in antibiotic resistance research. Anal Chem 96(35):14168–14177. https://doi.org/10.1021/acs.analchem.4c02093

Gardner W, Winkler DA, Maliki R, Cutts SM, Ellis S, Anderson RL et al (2022) Fusing ToF-SIMS images for spatial-spectral resolution enhancement using a convolutional neural network. Advanced materials. Interfaces 9(34). https://doi.org/10.1002/admi.202201464

Gazi E, Dwyer J, Lockyer N, Gardner P, Vickerman JC, Miyan J et al (2004) The combined application of FTIR microspectroscopy and ToF-SIMS imaging in the study of prostate cancer. Faraday Discuss 126:41–59. https://doi.org/10.1039/b304883g

Gostek J, Awsiuk K, Pabijan J, Rysz J, Budkowski A, Lekka M (2015) Differentiation between single bladder cancer cells using principal component analysis of time-of-flight secondary ion mass spectrometry. Anal Chem 87(6):3195–3201. https://doi.org/10.1021/ac504684n

Gularyan SK, Gulin AA, Anufrieva KS, Shender VO, Shakhparonov MI, Bastola S et al (2020) Investigation of inter- and Intratumoral heterogeneity of glioblastoma using TOF-SIMS*. Mol Cell Proteomics 19(6):960–970. https://doi.org/10.1074/mcp.RA120.001986

Hansson M, Grunditz T, Isaksson M, Jansson S, Lausmaa J, Mölne J et al (2008) Iodine content and distribution in Extratumoral and tumor thyroid tissue analyzed with X-Ray fluorescence and time-of-flight secondary ion mass spectrometry. Thyroid 18(11):1215–1220. https://doi.org/10.1089/thy.2008.0020

Heile A, Lipinsky D, Wehbe N, Delcorte A, Bertrand P, Felten A et al (2008) Metal-assisted SIMS and cluster ion bombardment for ion yield enhancement. Appl Surf Sci 255(4):941–943. https://doi.org/10.1016/j.apsusc.2008.05.007

Ilin Y, Kraft ML (2015) Secondary ion mass spectrometry and Raman spectroscopy for tissue engineering applications. Curr Opin Biotechnol 31:108–116. https://doi.org/10.1016/j.copbio.2014.10.011

Johnson ME, Bennett J, Montoro Bustos AR, Hanna SK, Kolmakov A, Sharp N et al (2021) Combining secondary ion mass spectrometry image depth profiling and single particle inductively coupled plasma mass spectrometry to investigate the uptake and biodistribution of gold nanoparticles in Caenorhabditis elegans. Anal Chim Acta 1175:338671. https://doi.org/10.1016/j.aca.2021.338671

Kadar H, Pham H, Touboul D, Brunelle A, Baud O (2014) Impact of inhaled nitric oxide on the Sulfatide profile of neonatal rat brain studied by TOF-SIMS imaging. Int J Mol Sci 15(4):5233–5245

Kayadurmus HM, Ayran M, Goktug S, Gunduz O, Dogan C (2024) Spectroscopic characterization of biomaterials for tissue engineering applications. Biomed Mater Devices 3:153–169. https://doi.org/10.1007/s44174-024-00177-1

Kern C, Ray S, Gelinsky M, Bellew AT, Pirkl A, Rohnke M (2020) New insights into ToF-SIMS imaging in osteoporotic bone research. Biointerphases 15(3):031005. https://doi.org/10.1116/6.0000051

Kern C, Kern S, Henss A, Rohnke M (2023) Secondary ion mass spectrometry for bone research. Biointerphases 18(4). https://doi.org/10.1116/6.0002820

Khan F, Tanaka M (2018) Designing smart biomaterials for tissue engineering. Int J Mol Sci 19(1):17

Kret P, Bodzon-Kulakowska A, Drabik A, Ner-Kluza J, Suder P, Smoluch M (2023) Mass spectrometry imaging of biomaterials. Materials 16(18):6343

Lagator M, Patel B, Sheraz S, Lockyer N (2024) Reactive gas cluster ion beams for enhanced drug analysis by secondary ion mass spectrometry. Anal Chem 96(38):15185–15193. https://doi.org/10.1021/acs.analchem.4c02144

Le MUT, Son JG, Shon HK, Park JH, Lee SB, Lee TG (2018) Comparison between thaw-mounting and use of conductive tape for sample preparation in ToF-SIMS imaging of lipids in Drosophila microRNA-14 model. Biointerphases 13(3):03B414. https://doi.org/10.1116/1.5019597

Lockyer NP, Aoyagi S, Fletcher JS, Gilmore IS, van der Heide PAW, Moore KL et al (2024) Secondary ion mass spectrometry. Nat Rev Methods Primers 4(1):32. https://doi.org/10.1038/s43586-024-00311-9

Malmberg P, Lopes VR, Billström GH, Gallinetti S, Illies C, Linder LKB et al (2021) Targeted ToF-SIMS analysis of macrophage content from a human cranial Triphasic calcium phosphate implant. ACS Appl Bio Mater 4(9):6791–6798. https://doi.org/10.1021/acsabm.1c00513

Matsuda K, Aoyagi S (2022) Sparse autoencoder-based feature extraction from TOF-SIMS image data of human skin structures. Anal Bioanal Chem 414(2):1177–1186. https://doi.org/10.1007/s00216-021-03744-3

Neittaanmäki N, Zaar O, Cehajic KS, Nilsson KD, Katsarelias D, Bagge RO et al (2024) ToF-SIMS imaging reveals changes in tumor cell lipids during metastatic progression of melanoma. Pigm Cell Melanoma R 37(6):793–800. https://doi.org/10.1111/pcmr.13182

Nilsson KD, Neittaanmäki N, Zaar O, Angerer TB, Paoli J, Fletcher JS (2020) TOF-SIMS imaging reveals tumor heterogeneity and inflammatory response markers in the microenvironment of basal cell carcinoma. Biointerphases 15(4):041012. https://doi.org/10.1116/6.0000340

Paladino E, Doerr FJS, Bordos E, Onyemelukwe II, Lamprou DA, Florence AJ et al (2022) High spatial resolution ToF-SIMS imaging and image analysis strategies to monitor and quantify early phase separation in amorphous solid dispersions. Int J Pharm 628:122191. https://doi.org/10.1016/j.ijpharm.2022.122191

Park JW, Shon HK, Yoo BC, Kim IH, Moon DW, Lee TG (2008) Differentiation between human normal colon mucosa and colon cancer tissue using ToF-SIMS imaging technique and principal component analysis. Appl Surf Sci 255(4):1119–1122. https://doi.org/10.1016/j.apsusc.2008.05.102

Passarelli MK, Newman CF, Marshall PS, West A, Gilmore IS, Bunch J et al (2015) Single-cell analysis: visualizing pharmaceutical and metabolite uptake in cells with label-free 3D mass spectrometry imaging. Anal Chem 87(13):6696–6702. https://doi.org/10.1021/acs.analchem.5b00842

Penen F, Raavé R, Kip A, Heskamp S, Malmberg P (2022) Time of flight secondary ion mass spectrometry imaging for precise localization of zirconium-labelled trastuzumab in xenograft cancer tumour tissues. Microchem J 181:107860. https://doi.org/10.1016/j.microc.2022.107860

Phan NT, Fletcher JS, Sjovall P, Ewing AG (2014) ToF-SIMS imaging of lipids and lipid related compounds in Drosophila brain. Surf Interface Anal 46(Suppl 1):123–126. https://doi.org/10.1002/sia.5547

Phan NTN, Fletcher JS, Ewing AG (2015) Lipid structural effects of Oral Administration of Methylphenidate in drosophila brain by secondary ion mass spectrometry imaging. Anal Chem 87(8):4063–4071. https://doi.org/10.1021/acs.analchem.5b00555

Philipsen MH, Phan NTN, Fletcher JS, Malmberg P, Ewing AG (2018) Mass spectrometry imaging shows cocaine and methylphenidate have opposite effects on major lipids in drosophila brain. ACS Chem Neurosci 9(6):1462–1468. https://doi.org/10.1021/acschemneuro.8b00046

Pi JJ, Wu X, Feng YF (2016) Fragmentation patterns of five types of phospholipids by ultra-high-performance liquid chromatography electrospray ionization quadrupole time-of-flight tandem mass spectrometry. Anal Methods 8(6):1319–1332. https://doi.org/10.1039/c5ay00776c

Ratner BD, Zhang G (2020) 1.1.2 – A history of biomaterials. In: Wagner WR, Sakiyama-Elbert SE, Zhang G, Yaszemski MJ (eds) Biomaterials science, 4th edn. Academic, pp 21–34

Robinson MA, Graham DJ, Morrish F, Hockenbery D, Gamble LJ (2016) Lipid analysis of eight human breast cancer cell lines with ToF-SIMS. Biointerphases 11(2):02A303. https://doi.org/10.1116/1.4929633

Robinson E, Giffen P, Hassall D, Ball D, Reid H, Coe D et al (2021) Multimodal imaging of drug and excipients in rat lungs following an inhaled administration of controlled-release drug laden PLGA microparticles. Analyst 146(10):3378–3390. https://doi.org/10.1039/d0an02333g

Ruiter FAA, Sidney LE, Kiick KL, Segal JI, Alexander C, Rose FRAJ (2020) The electrospinning of a thermo-responsive polymer with peptide conjugates for phenotype support and extracellular matrix production of therapeutically relevant mammalian cells. Biomater Sci 8(9):2611–2626. https://doi.org/10.1039/C9BM01965K

Sämfors S, Ståhlman M, Klevstig M, Borén J, Fletcher JS (2019) Localised lipid accumulation detected in infarcted mouse heart tissue using ToF-SIMS. Int J Mass Spectrom 437:77–86

Schaepe K, Kokesch-Himmelreich J, Rohnke M, Wagner AS, Schaaf T, Wenisch S et al (2015) Assessment of different sample preparation routes for mass spectrometric monitoring and imaging of lipids in bone cells via ToF-SIMS. Biointerphases 10(1):019016. https://doi.org/10.1116/1.4915263

Sjovall P, Rading D, Ray S, Yang L, Shard AG (2010) Sample cooling or rotation improves C60 organic depth profiles of multilayered reference samples: results from a VAMAS interlaboratory study. J Phys Chem B 114(2):769–774. https://doi.org/10.1021/jp9095216

Sole-Domenech S, Sjovall P, Vukojevic V, Fernando R, Codita A, Salve S et al (2013) Localization of cholesterol, amyloid and glia in Alzheimer's disease transgenic mouse brain tissue using time-of-flight secondary ion mass spectrometry (ToF-SIMS) and immunofluorescence imaging. Acta Neuropathol 125(1):145–157. https://doi.org/10.1007/s00401-012-1046-9

Spruill ML, Maletic-Savatic M, Martin H, Li F, Liu X (2022) Spatial analysis of drug absorption, distribution, metabolism, and toxicology using mass spectrometry imaging. Biochem Pharmacol 201:115080. https://doi.org/10.1016/j.bcp.2022.115080

Tian H, Six DA, Krucker T, Leeds JA, Winograd N (2017) Subcellular chemical imaging of antibiotics in single bacteria using C60-secondary ion mass spectrometry. Anal Chem 89(9):5050–5057. https://doi.org/10.1021/acs.analchem.7b00466

Tian H, Sparvero LJ, Anthonymuthu TS, Sun W-Y, Amoscato AA, He R-R et al (2021) Successive high-resolution (H2O)n-GCIB and C60-SIMS imaging integrates multi-omics in different cell types in breast cancer tissue. Anal Chem 93(23):8143–8151. https://doi.org/10.1021/acs.analchem.0c05311

Topolski JE, Sheng H, Moore C, Rading D, Moyer TB, Punia A et al (2024) Leveraging ToF-SIMS imaging to investigate tenofovir disoproxil fumarate degradation at excipient interfaces in oral compressed tablets. J Pharm Biomed Anal 239:115863. https://doi.org/10.1016/j.jpba.2023.115863

Tucker KR, Li Z, Rubakhin SS, Sweedler JV (2012) Secondary ion mass spectrometry imaging of molecular distributions in cultured neurons and their processes: comparative analysis of sample preparation. J Am Soc Mass Spectrom 23(11):1931–1938. https://doi.org/10.1007/s13361-012-0472-1

Tyler BJ, Hook A, Pelster A, Williams P, Alexander M, Arlinghaus HF (2017) Development and characterization of a stable adhesive bond between a poly(dimethylsiloxane) catheter material and a bacterial biofilm resistant acrylate polymer coating. Biointerphases 12(2):02C412. https://doi.org/10.1116/1.4984011

Vandergrift GW, Lukowski JK, Taylor MJ, Zemaitis KJ, Alexandrov T, Eder JG, et al. Are phosphatidic acids ubiquitous in mammalian tissues or overemphasized in mass spectrometry imaging applications? Anal Sens 2023;3(5). https://doi.org/10.1002/anse.202200112

Vennemann A, Breitenstein D, Tallarek E, Mørch Ý, Schmid R, Wiemann M (2022) Subcellular detection of PEBCA particles in macrophages: combining darkfield microscopy, confocal Raman microscopy, and ToF–SIMS analysis. Drug Deliv Transl Res 12(9):2075–2088. https://doi.org/10.1007/s13346-022-01128-3

Vickerman JC (2009) Molecular surface mass spectrometry by SIMS. In: Surface analysis – the principal techniques, pp 113–205

Vickerman JC, Briggs D (2013) ToF-SIMS: materials analysis by mass spectrometry. SurfaceSpectra

Wong SY, Hook AL, Gardner W, Chang C-Y, Mei Y, Davies MC et al (2023) Exploring the relationship between polymer surface chemistry and bacterial attachment using ToF-SIMS and self-organizing maps. Adv Mater Interfaces 10(9):2202334. https://doi.org/10.1002/admi.202202334

Wu KJ, Odom RW (1996) Matrix-enhanced secondary ion mass spectrometry: a method for molecular analysis of solid surfaces. Anal Chem 68(5):873–882. https://doi.org/10.1021/ac950717i

Yang H, Jin D, Rao J, Shi J, Li G, Wang C et al (2023) Lithium-induced optimization mechanism for an ultrathin-strut biodegradable Zn-based vascular scaffold. Adv Mater 35(19):2301074. https://doi.org/10.1002/adma.202301074

Yokoyama Y, Kawashima T, Ohkawa M, Iwai H, Aoyagi S (2015) Extraction of hidden information of ToF-SIMS data using different multivariate analyses. Surf Interface Anal 47(4):439–446. https://doi.org/10.1002/sia.5731

Second Harmonic Generation Imaging in Preclinical Research

Tong Wu, Xiao-Hua Hu, Michael J. Wilhelm, Yujie Li, Jianqiang Ma, and Hai-Lung Dai

Abstract

Second Harmonic Generation (SHG) is a nonlinear optical technique that enables high-resolution, label-free imaging of non-centrosymmetric structures such as lipid membranes, collagen fibers, microtubules, and tissues. By applying a near-infrared laser, SHG generates signals at half the wavelength, facilitating specific interface imaging with minimal photodamage. This chapter begins with an explanation of the physical principles underlying SHG, highlighting why it is particularly effective for surface applications in biological samples. It then delves into the applications of SHG imaging, including pharmacokinetic studies of cell membranes, imaging of collagen fibers, and distinguishing tumor tissues from healthy tissues. The chapter concludes with a critical evaluation of SHG's strengths—such as high specificity, non-invasive nature, and minimal sample preparation—and its limitations, including its restricted applicability to centrosymmetric molecules and potential signal interference. SHG imaging not only provides detailed insights into tissue structures independently but also, when

T. Wu (✉)
Division of Infectious Disease, Perelman School of Medicine, University of Pennsylvania, Philadelphia, PA, USA

Department of Chemistry, Temple University, Philadelphia, PA, USA

Department of Pathology and Laboratory Medicine, Children's Hospital of Philadelphia, Philadelphia, PA, USA

Institute for Membranes and Interfaces, Temple University, Philadelphia, PA, USA
e-mail: WuTong@pennmedicine.upenn.edu; tong.wu@temple.edu

X.-H. Hu · M. J. Wilhelm · Y. Li · J. Ma · H.-L. Dai
Department of Chemistry, Temple University, Philadelphia, PA, USA

Institute for Membranes and Interfaces, Temple University, Philadelphia, PA, USA

I. F. Cengiz et al. (eds.), *Bioimaging Modalities in Bioengineering*, Biomaterials, Bioengineering and Sustainability 9,
https://doi.org/10.1007/978-3-031-99410-4_24

combined with other imaging techniques, creates a comprehensive and detailed image of the tissue being studied. The advancement toward minimally invasive, tissue-preserving microscopy opens up expansive possibilities in diagnosing a wide range of diseases, potentially influencing therapeutic decisions. The capabilities of nonlinear optics in medicine have been demonstrated in recent years and are anticipated to be further harnessed as clinical application areas continue to evolve and expand.

Keywords

Second harmonic generation · Nonlinear optical microscopy · Biomedical imaging · Cellular membrane · Collagen fibers

1 Introduction

Over the past two decades, second harmonic generation (SHG) imaging has emerged as a powerful and versatile tool for studying biological materials, owing to its label-free, non-invasive nature, high specificity, and exceptional spatial resolution (Aghigh et al. 2023; Campagnola and Loew 2003; Chen et al. 2012; Wilhelm and Dai 2020b). Unlike fluorescence imaging, SHG generates signals directly from non-centrosymmetric structures—materials or biological assemblies lacking inversion symmetry—within the sample, eliminating the need for dyes or fluorophores. This unique characteristic makes SHG free from photobleaching, ensuring long-term signal stability and suitability for extended imaging sessions.

SHG signals are inherently tied to the physical properties of the sample, persisting indefinitely as long as laser illumination is sustained. This feature makes SHG particularly advantageous for monitoring dynamic biological processes over extended periods. Moreover, SHG signal generation is instantaneous, bypassing the delays associated with molecular excitation and emission in fluorescence imaging. This instantaneous nature enables SHG to achieve extremely high temporal resolution, capturing rapid dynamic changes with exceptional precision. Additionally, the ability to operate seamlessly with high-repetition-rate lasers further enhances imaging speed, making SHG ideal for high-speed scanning and real-time monitoring of fast biological dynamics. Together, these features establish SHG as a crucial technique for exploring delicate biological systems in their native states with unparalleled precision and efficiency.

These properties have driven the widespread use of SHG in preclinical research on biological materials, including pharmacokinetic studies of cellular membranes (Hu et al. 2024; Li et al. 2024b; Mizuguchi et al. 2018; Sharifian Gh et al. 2019; Wu et al. 2022a, b; Zeng et al. 2013), visualization of collagen (Avila et al. 2015; Bueno et al. 2019; Bueno et al. 2011; Cicchi et al. 2013; Hanson and Bardeen 2009), and distinguishing between tumor and healthy tissues (Dudenkova et al. 2019; Guimaraes et al. 2024). SHG imaging not only offers detailed insights into various tissue structures on its own but also, when integrated with other imaging techniques, enables the creation of a comprehensive and detailed image of the tissue being studied (Nejim et al. 2022; Nikolaev et al. 2024). This chapter begins by outlining the physical principles underlying SHG and emphasizing its effectiveness for surface applications in biological samples. Subsequently, it explores the preclinical applications of SHG imaging. Finally, the chapter concludes with an evaluation of SHG's strengths and limitations.

2 Physical Principles of SHG Imaging

SHG, also known as frequency doubling, is a nonlinear optical process in which two photons of the same frequency interact with a nonlinear medium to produce a new photon with twice the energy of the original photons (resulting in doubled frequency and halved wavelength) (Boyd 2020). SHG was the first nonlinear optical phenomenon to be discovered, made possible by the advent of lasers as high-intensity coherent light sources. Franken and colleagues reported the discovery of SHG in 1961 (Franken et al. 1961). They employed a 694 nm ruby laser focused onto a quartz sample, and the spectral record captured on photographic paper revealed an output wavelength of 347 nm.

Traditional single-photon fluorescence emission and nonlinear optical two-photon fluorescence processes are accompanied by energy absorption. In contrast, SHG can be described as the annihilation of two fundamental incident photons within a nonlinear medium, simultaneously producing a new photon with doubled frequency, as shown in Fig. 1a. Unlike fluorescence-based processes, SHG does not involve an energy transition; the energy of the two incident photons with frequency ω equals the energy of the single emitted photon with frequency 2ω. The optical response of a nonlinear molecule can be determined by the polarization P induced by the incident radiation field E. The total polarization can be expressed in Eq. 1:

$$
\begin{aligned}
P &= P^{(0)} + P^{(1)} + P^{(2)} + P^{(3)} + \cdots\cdots \\
&= P^{(0)} + \varepsilon_0\chi^{(1)} \cdot E + \varepsilon_0\chi^{(2)} : EE + \varepsilon_0\chi^{(3)} \vdots EEE + \cdots\cdots
\end{aligned}
\tag{1}
$$

Where, $P^{(0)}$ is the permanent dipole of the molecule, ε_0 is the vacuum permittivity (8.85×10^{-12} F/m), $\chi^{(n)}$ represents the n-th order nonlinear susceptibility tensor, $\chi^{(1)}$ is the linear polarization susceptibility, $\chi^{(2)}$ is the second-order polarization susceptibility, and $\chi^{(3)}$ is the third-order polarization susceptibility. The second term $P^{(1)} = \varepsilon_0\chi^{(1)} \bullet E$ corresponds to the linear optical response of the medium, which includes phenomena such as light absorption, scattering, and reflection. The third term, $P^{(2)} = \varepsilon_0\chi^{(2)} : EE$, represents second-order nonlinear optical effects, such as SHG and sum-frequency generation (SFG), which occur in materials without inversion symmetry (Hu et al. 2019, 2020). The fourth term, $P^{(3)} = \varepsilon_0\chi^{(3)} \vdots EEE$, describes third-order nonlinear effects such as two-photon and three-photon absorption, third harmonic generation (THG), stimulated Raman scattering (SRS), and coherent anti-Stokes Raman scattering (CARS). The susceptibility of a material is

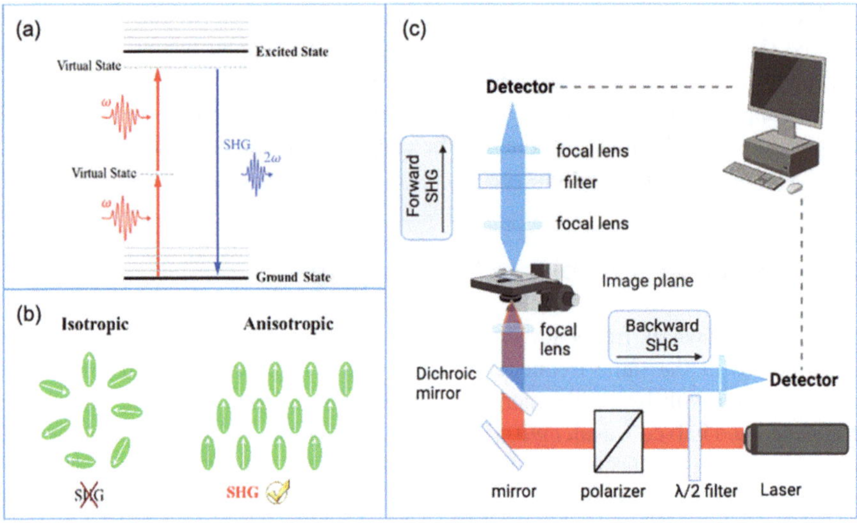

Fig. 1 (**a**) Energy level diagram of SHG. SHG is a nonlinear optical process where two photons of the same frequency interact in a nonlinear medium, combining to form a single photon with twice the frequency (2ω). (**b**) The schematic illustrates the molecular arrangement required for SHG generation. In isotropic ordering, opposite polarization cancels out SHG due to centrosymmetric. In contrast, anisotropic ordering lacks inversion symmetry, allowing aligned molecular dipoles to generate an SHG signal. (**c**) Schematic of a typical SHG microscope with forward and backward detection channels. A laser beam (red) is directed through a λ/2 filter, which adjusts polarization to optimize SHG signal generation. The polarized light then passes through a polarizer to ensure the correct polarization state before being reflected by a mirror toward a dichroic mirror. The dichroic mirror selectively transmits the excitation wavelength while reflecting the generated SHG signal (blue). At the image plane, the SHG signal propagates in both forward and backward directions. The forward SHG is collected by a high numerical aperture objective and passes through a set of focal lenses and a filter to remove residual excitation light before being detected. The backward SHG is collected and directed toward a second detector via another optical path. The system enables comprehensive imaging of SHG-active structures, with forward detection typically providing stronger signals for thick samples, while backward detection is useful for thin structures or interfaces

a key parameter characterizing the interaction between light and the medium. It reflects properties such as the refractive index and spatial symmetry of the material. As a result, nonlinear optical phenomena can be used to probe and reveal structural features and other characteristics of the medium.

To produce SHG signals, three criteria must be met simultaneously. First, the medium must lack a center of inversion symmetry (i.e., anisotropic orientation), as in the electric dipole approximation, the induced polarization in centrosymmetric materials (i.e., isotropic orientation) cancels out due to equal magnitude but opposite direction, preventing SHG, as depicted in Fig. 1b (Hu and Dai 2024; Hu et al. 2024; Wu et al. 2022a, b). Second, the medium must have a high second-order susceptibility $\chi^{(2)}$, since SHG intensity (I_{SHG}) is proportional to the square of $\chi^{(2)}$, i.e., $I_{SHG} \propto |\chi^{(2)}|^2$, making a large essential for observable SHG. Materials with inversion symmetry have $\chi^{(2)} = 0$, resulting in no SHG. Third, phase-matching conditions, which obey the conservation of momentum law, are essential for ensuring strong coupling between the fundamental and harmonic waves, enabling efficient conversion of fundamental power into SHG power. Together, these criteria determine the feasibility and efficiency of SHG in nonlinear media.

The setup for SHG imaging involves a scanning confocal microscope paired with a femtosecond laser as the excitation light source. Currently, numerous commercial multiphoton confocal microscopes are available for SHG imaging, alongside custom-built experimental devices. These setups typically utilize a configuration similar to the one illustrated in Fig. 1c. Traditionally, lasers used for SHG imaging consist of a solid-state Nd:YAG laser pumping a tunable Ti:sapphire oscillator; however, fiber-based femtosecond lasers are quickly rising in popularity due to their high power, adjustable repetition rates, and long-term stability. These lasers generally produce pulses with a duration of approximately 100 fs at a repetition rate of several tens of MHz, covering excitation wavelengths between 700 and 1000 nm.(Aghigh et al. 2023; Chen et al. 2012) Such wavelengths allow imaging to depths of several hundred microns within tissue. The half-wave plate (λ/2) or quarter-wave plate (λ/4) is used to accurately control the laser's polarization. Finally, the SHG signals are collected through forward detection and backward detection, as shown in Fig. 1c. Notably, SHG signals primarily originate from biological materials themselves and exogenous probes.

2.1 SHG Signal Comes from Biological Materials Themselves

Biological structural proteins often assemble into large, ordered, and non-centrosymmetric structures, such as collagen, actin, and tubulin, which are highly polarized and capable of generating SHG signals (Aghigh et al. 2023; Chen et al. 2012). These SHG signals, originating from nanostructures or other spatially irregular systems, provide valuable structural insights through their angular distribution. The angular intensity pattern of scattered SHG signals depends on factors such as the size, shape, symmetry of the scatterers, and the polarization state of the incident light. For instance, anisotropic structures like elongated fibers produce distinct directional patterns. SHG is particularly effective for analyzing systems with complex geometries or heterogeneous compositions. By systematically studying angular scattering patterns under various polarization configurations, researchers can extract detailed information about molecular orientation, structural anisotropy, and dynamic changes in the system.

2.2 SHG Signal from Exogenous Probes

While certain tissues and structures in living organisms can produce endogenous SHG signals, the majority of tissues and cells lack the ability to generate sufficiently intense SHG signals for reliable detection. To overcome this limitation, exogenous probes are commonly employed to label cells, enabling strong SHG signal generation for imaging studies. Surface SHG is a specialized application that occurs at the interface between two media, such as a solid and a liquid, where the bulk inversion symmetry is disrupted. In centrosymmetric bulk media, the electric dipole contribution to $\chi^{(2)}$ is null due to symmetric constraints, making SHG forbidden in the bulk (Hu and Dai 2024; Sharifian Gh et al. 2019; Wu et al. 2022a, b). However, at surfaces or interfaces where inversion symmetry is inherently broken, $\chi^{(2)}$ becomes nonzero, enabling SHG to occur. This interfacial specificity makes SHG a powerful technique for isolating information about thin interfacial layers even in the presence of overwhelming bulk media. Its non-invasive nature and high sensitivity have rendered SHG invaluable in studying biological systems, such as cell membranes, lipid bilayers, and protein adsorption.

3 SHG Microscopy for Quantitative Analysis the Pharmacokinetics

Understanding how molecules interact with plasma membranes in living cells is essential for both fundamental and applied biology, particularly in predicting and optimizing pharmacological interactions. Traditional experimental approaches have predominantly focused on selective ion transport through membrane-embedded protein channels, often neglecting the broader context of passive molecular transport. Many of these methods rely on reporter labels, which can inadvertently alter the interactions under study. In contrast, SHG microscopy provides a label-free optical approach capable of imaging molecular transport across plasma membranes in living cells. By enabling real-time visualization of these processes, SHG reveals regional differences in molecule-membrane interactions with high spatial resolution (Wilhelm and Dai 2020b). These advancements establish SHG microscopy as a critical tool for investigating molecular interactions and transport mechanisms in living cells, paving the way for deeper insights into cellular dynamics and pharmacological processes.

3.1 Label-Free and Dye-Assisted Imaging of Lipid Membrane Dynamics

SHG microscopy takes advantage of the intrinsic non-centrosymmetric of neuronal structures, such as microtubules and other ordered cytoskeletal elements, to generate SHG signals. These non-centrosymmetric arrangements allow for high-resolution, label-free imaging of living mammalian neurons (Macias-Romero et al. 2014), capturing structural and functional details without requiring external dyes. In addition to neurons, SHG signals are generated by hydrated water molecules organized by the charged membrane surface, enabling label-free imaging of artificial lipid membranes (Tarun et al. 2018). Membrane hydration dynamics and electrostatic potentials are mapped in real time, revealing lipid domain mobility, membrane potential changes, and phase separation. However, weak SHG signals from disordered or sparsely organized membranes can result in incomplete or biased representations of the cellular environment. Additionally, its reliance on highly controlled experimental conditions restricts broader applicability to complex or less-defined systems.

The use of a non-invasive, photostable dye, like AP3 or FM4–64, has enabled the study of membrane potential in living cortical neurons (Dombeck et al. 2005; Nuriya et al. 2016) and high-resolution, selective imaging of the plasma membrane in living Chinese hamster ovary (CHO) cells (Mizuguchi et al. 2018). AP3, an impermeable dye, generates a robust SHG signal with minimal two-photon fluorescence (TPF), effectively circumventing the high background signals typically associated with intracellular organelle fluorescence in traditional methods (Mizuguchi and Nuriya 2020). The SHG signal from AP3 increased during depolarization and decreased during hyperpolarization, providing direct visualization of localized voltage dynamics across different cellular regions. This revealed significant spatial heterogeneity in membrane potential. With millisecond temporal resolution, AP3's SHG signal proved invaluable for investigating neuronal excitability and synaptic transmission in live cells.

3.2 Imaging of Adsorption and Uptake of Antibiotics in Individual Living Cells

Malachite green (MG), an amphiphilic cationic dye with a strong SHG signal at 400 nm, has been extensively utilized to study molecular adsorption and transport dynamics in artificial lipid bilayers and living cell membranes (Sharifian Gh et al. 2016; Wilhelm and Dai 2020a; Wilhelm et al. 2014, 2015b, 2019; Wu et al. 2022a, b). It should be noted that MG is also an organic compound that is used as an antimicrobial on microorganisms like *Staphylococcus*, *Enterobacteriaceae*, and *Candida* (Junqueira et al. 2010). Its positive charge naturally interacts with negatively charged membrane surfaces, enabling the quantification of molecular adsorption density (Zeng et al. 2013, 2015), surface charge (Wilhelm et al. 2021), and phase separation (Wu et al. 2022a, b) through modified Langmuir adsorption isotherms by calculating maximum molecular adsorption densities and equilibrium constants. Additionally, MG, with its membrane-permeable properties, has been applied in pharmacokinetics to investigate drug transport rates (Li et al. 2024b; Wilhelm et al. 2015a; Wu et al. 2022a, b). For drugs lacking intrinsic SHG activity, a competitive method leveraging MG adsorption can deduce the transport rate, as the displacement of MG by competing molecules correlates directly with their uptake

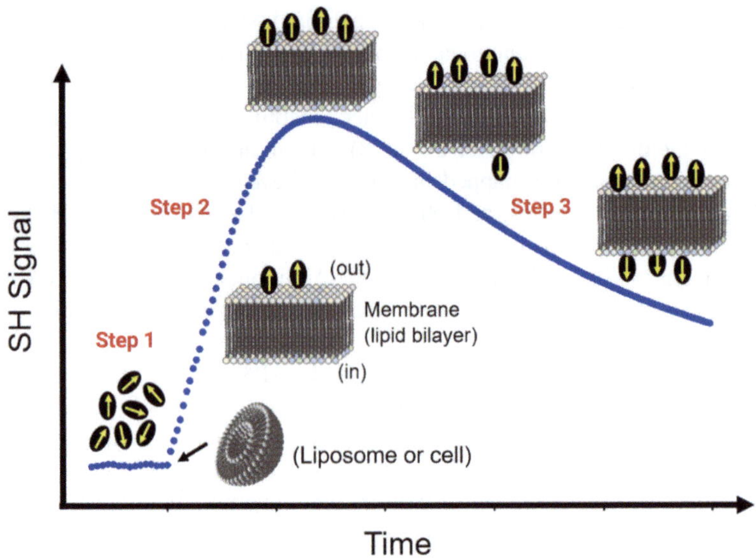

Fig. 2 Characteristic time-resolved SHS response illustrating the transport dynamics of SHG-active molecules across a lipid bilayer. Step 1: SHG-active molecules are initially dispersed randomly in solution, generating a minimal SHG signal. Step 2: The molecules adsorb onto the outer leaflet of the lipid bilayer, leading to a rapid SHG signal increase due to their non-centrosymmetric alignment. Step 3: Some molecules translocate to the inner leaflet or undergo reorientation, resulting in a gradual SHG signal decrease over time. This characteristic SHG signal evolution provides valuable insights into membrane transport mechanisms and molecular interactions with lipid bilayers

kinetics (Sharifian Gh et al. 2018; Wu et al. 2022a, b). These transport dynamics are depicted in the SHG molecular membrane interaction diagram (Fig. 2). When SH-active molecules are randomly distributed in a solution, the SH light is undetectable due to destructive interference of the SH fields emitted from neighboring molecules (step 1). Upon adsorption onto the outer surface of a membrane, the molecules align with each other. This alignment leads to constructive interference of the SH fields from neighboring molecules, generating a detectable SH signal that scales quadratically with surface concentration (step 2). As the molecules diffuse across the bilayer and adsorb onto the interior surface, their orientation reverses relative to the exterior surface. This results in destructive interference between SH fields from opposing sides of the bilayer, attenuating the SH signal (step 3). The time-dependent rise and fall of the SH signal provide insights into adsorption and transport rates.

Given the heterogeneity of cell membranes, different locations exhibit intrinsically varied permeation rates. Unlike the ensemble-averaged response from SHG spectroscopy measurement, SHG imaging permits precise measurements of surface adsorption and transport of non-endogenous molecules across a plasma membrane in individual living cells with sub-cellular spatial resolution (Sharifian Gh et al. 2019). Quantification of region-specific adsorption and transport rates was achieved

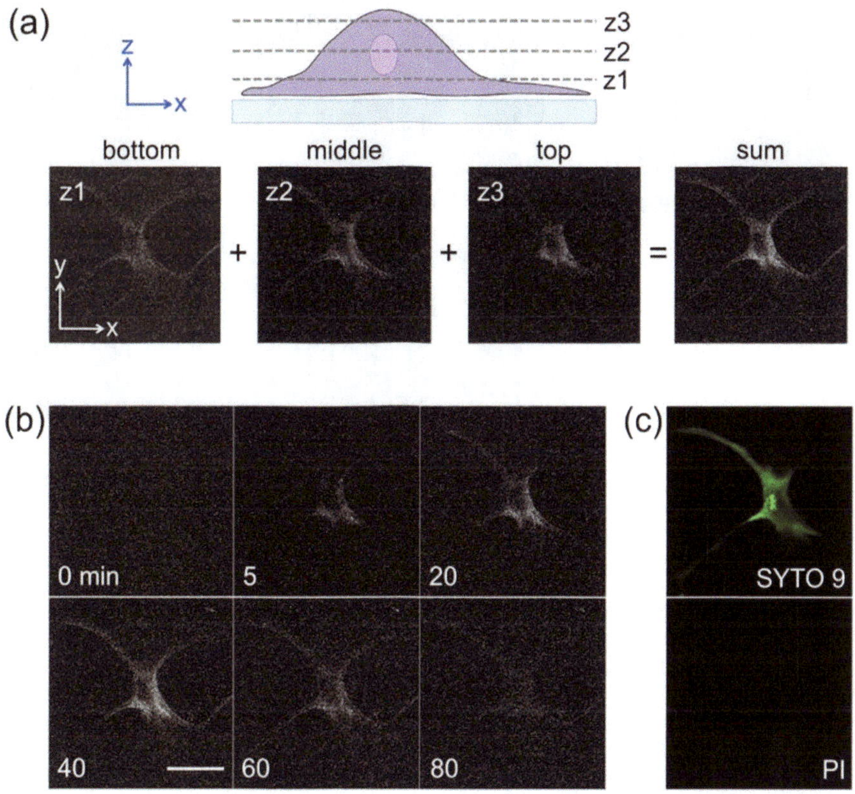

Fig. 3 Time-resolved SHS images. (**a**) Cartoon schematic of a single HDF cell showing a height profile and typical z-stack image sampling planes along the z-axis. For each time-point, a set of three images at distinct heights along the z-axis (z1, z2, and z3) are collected and summed into a single representative image. (**b**) Representative time-resolved SHS images of MG interacting with a single HDF. The scale bar represents 100 μm. Over time, as MG molecules adsorbed onto the HDF membrane, the cell image gradually became visible, reaching maximum intensity after 30 to 40 minutes. Subsequently, as MG molecules crossed the membrane and were reabsorbed onto the inner leaflet, the SHS signal began to decrease. (**c**) shows the results of viability staining assays conducted after SHS imaging using SYTO 9 (top) and PI (bottom) staining. The cell was rapidly stained by SYTO9, but not PI, indicating the membrane was still intact following the 80 minutes imaging session. (Reprinted with permission from (Sharifian Gh et al. 2019). Copyright 2019 American Chemical Society)

through time-resolved SHG imaging of membranes in individual living human dermal fibroblast (HDF) cells cultured in vitro (Fig. 3) (Sharifian Gh et al. 2019). To capture signals from the entire exposed membrane surface, a z-stack of three images, separated by 3 microns along the z-axis, was acquired every minute. To minimize sample heating, scanning was restricted to the first 2.6 seconds of each minute, allowing the sample to cool down afterward. The scanning parameters, including laser power, pixel dwell time, and dye concentration, were carefully optimized

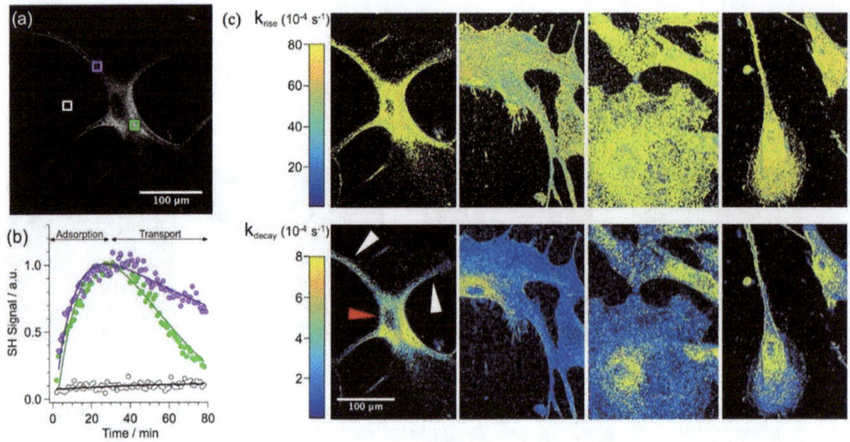

Fig. 4 Kinetics of adsorption and transport for MG interacting with a single HDF cell. (**a**) Representative MG + HDF cell SHG image (t = 40 min) with three distinct ROI's (green square on the main cell body, violet square on a filopodia, and white square on background noise). (**b**) Normalized time-dependent kinetic traces (colored circles) and best fits (lines) for the total signal measured in each of the three ROI's shown in (**a**) using Eq. 2. (**c**) Regional specific rate heat maps corresponding to (top) surface adsorption and (bottom) molecular transport for MG interacting with individual HDF cells. The deduced rate constants were derived from pixel-wise image analysis using Eq. 2 to construct heat map images. Transport rates show distinctions related to the location of the nucleus (red arrow) and between the main cell body and the filopodia (white arrows). These characteristics were faithfully repeated across different experiments. In general, membranes covering the nucleus and filopodia exhibit the slowest transport rates (blue regions), whereas for membranes surrounding the nuclear bulge and the main cell body, transport rates are considerably faster (yellow regions). (The images in panel (a) and (b) are adapted, and the image in panel (c) is reprinted from (Sharifian Gh et al. 2019). Copyright 2019 American Chemical Society)

to achieve sufficient signal intensity while avoiding cell damage. Post-processing involved summing the three z-stack images at each time point into a single three-dimensional (3D) projection image (Fig. 3a), providing a comprehensive representation of the membrane dynamics over time.

The spatial dependence of transport rates was revealed in Fig. 4, highlighting three representative regions: (1) the main cell body (green), (2) filopodia (violet), and (3) background (no cell) (white). Distinct spatial differences were clearly observed, with each region exhibiting unique dynamic SH signals indicative of adsorption and transport processes. Specifically, the kinetic trace of the measured SHS signal at each pixel was extracted and fit to deduce the rates describing the signal rise (adsorption) and signal decay (transport), as described by Eq. 2.

$$\sqrt{I_{SHG}(t)} = \varphi_0 + \varphi_1 \left[1 - \exp\left(-k_{rise}t\right)\right] \times \exp\left(-k_{decay}t\right) \qquad (2)$$

Pixel-wise fit analysis normalized the SH signal relative to the global maximum, producing rate constants that were subsequently used to construct heat maps for surface adsorption and passive transport (Fig. 4c).

The heat map images revealed uniform adsorption rates across the plasma membrane, attributed to the homogeneous distribution of anionic adsorption sites. This observation aligns with the understanding that adsorption is primarily driven by electrostatic interactions between cationic MG and individual anionic sites spread over the membrane surface. In contrast, passive diffusion across the HDF plasma membrane occurs with regional specificity. Areas over the nucleus and filopodia exhibited the slowest transport rates, represented by blue regions on the map, while faster rates were observed around the nuclear bulge and the main cell body (yellow regions). These variations correlated with membrane strain, where increased curvature or localized stress reduced molecular transport rates. This phenomenon is likely due to compositional modifications of the membrane, possibly as part of the cell's self-preservation mechanism. For example, HDF cells, being adherent, bind to both neighboring cells and the substrate through anchoring proteins. These proteins are stabilized by localized enrichment of the membrane with rigidifying compounds such as cholesterol and unsaturated phospholipids. Such enrichment has been shown to decrease the passive transport rates across membranes, as demonstrated in earlier studies by Eisenthal and colleagues (Eisenthal and Eisenthal 2000). While the precise origin of these regional variations remains unclear, the ability of SHG imaging to correlate membrane structural changes with kinetic transport properties provides valuable insights into the interplay between cellular architecture and functional dynamics.

4 SHG Microscopy for Quantitative Analysis of Collagen Fibers

In animals, collagen, a major component of the extracellular matrix, is the most abundant protein type. Structural changes in collagen are early indicators of various diseases such as fibrosis, cancer, and connective tissue disorders (Karsdal et al. 2017). Understanding its hierarchical organization, from nanoscale collagen molecules to microscale collagen fiber, provides essential insight into how tissues withstand mechanical stress and resist strain. Additionally, the spatial arrangement of collagen plays a pivotal role in guiding the development of synthetic scaffolds and engineered tissues for regenerative medicine. Therefore, obtaining detailed, non-invasive, 3D structural information on collagen is crucial for both fundamental research and clinical diagnostics.

Unlike the lipid bilayer membrane as we discussed in Sect. 3, collagen fibers and fibrils exhibit strong SHG signals due to their intrinsically non-centrosymmetric structure (Fig. 5a). Collagen fibrils, especially types I and II, are composed of repeating triple-helix units formed by three polypeptide chains. These chains twist around each other in a specific manner that lacks a center of symmetry (Campagnola and Loew 2003). The SHG signals generated by these structures provide critical insights into their arrangement. The polarization sensitivity of SHG further enhances its utility. For example, when the light polarization aligns parallel to the fiber structure, the SHG signal is strongest, while it is weakest when polarized perpendicular to the fiber structure. This sensitivity allows for precise characterization of the structural organization of proteins like collagen at the molecular level. Type I collagen, which comprises approximately 90% of the body's collagen, has the most highly ordered structure and produces the strongest SHG signal compared to other collagen types due to its well-organized molecular arrangement (Dudenkova et al. 2019). Additionally, the staggered, overlapping molecular packing of collagen molecules within fibrils creates a periodic, directionally ordered arrangement known as the 67 nm D-periodic banding, which further amplifies SHG signal generation. The aggregation of fibrils into larger fibers, stabilized by hydrogen bonds and covalent crosslinks, reinforces this ordered structure, ensuring consistent and strong SHG emissions. These structural features collectively make collagen highly efficient for SHG-based imaging and analysis.

The first application of SHG imaging to study collagen fibers was reported in 1986, focusing on the orientation of collagen fibers in rat tail tendon. This pioneering work laid the foundation for subsequent research investigating collagen-rich tissues such as corneas (Batista et al. 2022; Bueno et al. 2019; Bueno et al. 2011; Park et al. 2015), skin (Bancelin et al. 2015; Guimaraes et al. 2024; Hanson and Bardeen 2009; Lynch et al. 2017), muscles (Both et al. 2004; Campagnola and Loew 2003; Fedorovskiy et al. 2024; Liu et al. 2013) and bones (Houle et al. 2015; Pendleton et al. 2020; Thomas et al. 2017), enabling the study of structural organization changes in various physiological and pathological processes. Entropy (S) and orientation index (OI) are standard quantitative metrics in SHG image processing used to evaluate fiber alignment, disorder, and overall tissue structure. The OI quantifies the degree of fibers aligned along the dominant orientation, while S, derived from statistical entropy, quantifies the randomness of fiber orientations independently of any dominant orientation (Bancelin et al. 2015; Lynch et al. 2017). These metrics are mathematically defined as shown in Eqs. 3–5 below:

$$S = -\sum_{\theta=-90^\circ}^{90^\circ} p(\theta) \bullet \ln[p(\theta)] \tag{3}$$

$$OI = \frac{\max(p(\theta))}{\sum p(\theta)} \tag{4}$$

where

$$p(\theta) = \frac{I(\theta)}{\sum_{\theta=-90^\circ}^{90^\circ} I(\theta)} \tag{5}$$

$p(\theta)$ is the normalized probability distribution of fiber orientations, derived from the intensity $I(\theta)$ of the SHG signal at a given orientation angle θ. Higher OI

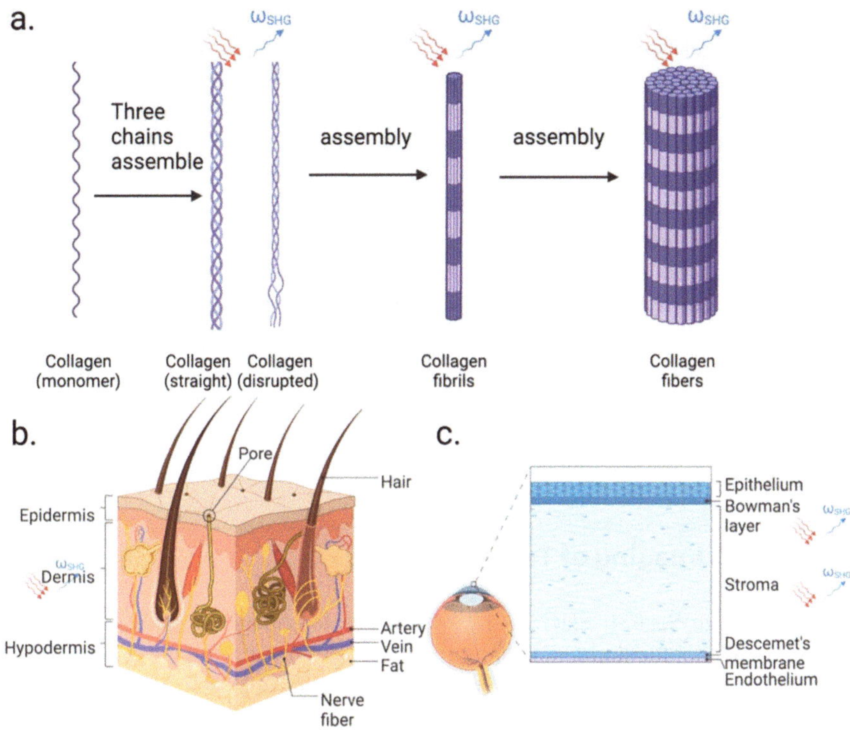

Fig. 5 Structural organization of collagen and SHG signal generation in biological tissues. (**a**) Collagen assembles hierarchically, beginning as monomeric chains that form triple helices, which then organize into fibrils and fibers. SHG signals are primarily generated from collagen fibrils and fibers due to their non-centrosymmetric structure, enabling efficient second harmonic generation. (**b**) Cross-sectional view of the cornea, illustrating SHG signal production predominantly from collagen fibrils within the **Bowman's layer** and **stroma**, where collagen is highly organized. (**c**) Cross-sectional view of the skin, highlighting SHG signal generation in the **collagen-rich dermis layer**, which provides structural integrity. SHG imaging is particularly useful for visualizing collagen distribution in these tissues due to its label-free nature and specificity for non-centrosymmetric structures. (Created with BioRender)

values indicate a more alignment of fibers along a dominant orientation, while lower values suggest more random orientations. Conversely, higher S values reflect a more disordered collagen network, while lower S implies greater alignment within collagen fibers.

Utilizing near-infrared (NIR) light, SHG imaging penetrates up to 200–300 μm, making it highly effective for examining the collagen-dense dermis (Fig. 5b) (Chen et al. 2012). This layer generates strong SHG signals and is extensively studied for its role in skin biomechanics (Bancelin et al. 2015), including damage (Nuriya et al. 2016), aging (Lynch et al. 2017), and wound healing (Gitanjal et al. 2012). SHG imaging is also widely applied in cornea research (Fig. 5c). The cornea's high transparency allows light to penetrate deeply, making SHG microscopy an effective tool for imaging the Bowman's layer and stroma, which is valuable for studying corneal conditions like cross-linking (CXL) and keratoconus (Bueno et al. 2019; Bueno et al. 2011). Polarization-resolved SHG (P-SHG) microscopy has been used to visualize and measure the orientation of collagen fibrils in the cornea (Avila et al. 2015; Cicchi et al. 2013). Advanced image analysis methods, including Fourier Transform (FT), Structure Tensor (ST), Gray-Level Co-occurrence Matrix (GLCM), and Log-Gabor filter bank analysis, enable quantitative evaluation of collagen organization (Guimaraes et al. 2024). P-SHG studies have revealed significant collagen disorganization in keratoconic corneas, characterized by increased entropy (indicating more random fibril orientation) and decreased orientation index (reflecting reduced alignment of fibrils) (Raoux et al. 2021). Among the tested metrics, STSTD and Log-Gabor were the most reliable across various imaging conditions, while FT-based metrics were less effective in noisy or rotated environments, emphasizing the need for robust analytical methods in corneal structure assessment (Guimaraes et al. 2024).

5 SHG Imaging of Tumor Tissue

As discussed in Sect. 4, SHG is highly sensitive to collagen structure, making it an invaluable tool for studying the tumor microenvironment, which is enriched with collagen in conditions such as cancer, fibrosis, and atherosclerosis (Brett et al. 2020; Keikhosravi et al. 2014; Tilbury and Campagnola 2015). SHG imaging precisely detects changes in collagen properties, including density, alignment, bundling, and crosslinking (Chakraborty et al. 2024). When combined with other nonlinear optical imaging techniques, such as CARS, THG, and TPF, researchers can obtain complementary insights into the structural and biochemical composition of tumors.

CARS is particularly sensitive to lipids due to the strong vibrational "fingerprint" of C-H bonds, which allows it to provide biochemical data, such as identifying lipid-rich environments. When combined, SHG and CARS enable a comprehensive analysis by merging structural data on collagen with biochemical insights into lipid metabolism (Bocklitz et al. 2016). THG, on the other hand, is sensitive to changes in refractive index, enabling the distinction between collagenous ECM, non-collagenous ECM, and intracellular structures (Gavgiotaki et al. 2020; Weigelin et al. 2016). Finally, TPF provides detailed information on cellular structures and endogenous fluorophores, linking collagen remodeling with cellular activities (Javier et al. 2011; Li et al. 2024a; Zoumi et al. 2002). This multimodal approach greatly enhances our understanding of the tumor microenvironment by integrating structural, biochemical, and cellular dynamics.

While collagen remains the primary target for SHG imaging, other non-centrosymmetric structures, such as myosin, microtubules, and lipid interfaces, also generate SHG signals, broadening its applications in tumor tissue studies. Myosin filaments, with their highly ordered coiled rod domains, provide insight into sarcomeric organization within tumor-associated muscle tissue without the need for dyes (Both et al. 2004; Karunendiran et al. 2022; Liu et al. 2013; Odin et al. 2008;

Plotnikov et al. 2006; Zhao et al. 2019). Microtubules, as part of the cytoskeleton, produce weaker yet detectable SHG signals, offering valuable information about intracellular architecture and alignment, which are often altered in tumor cells (Dombeck et al. 2003; Yu et al. 2014). Additionally, lipid interfaces, such as those in cell membranes and lipid droplets, contribute to SHG signals due to the asymmetry of their molecular arrangements (Macias-Romero et al. 2014; Tarun et al. 2018; Wu et al. 2022a;, b). These SHG sources enable the detailed examination of tumor microenvironments, lipid metabolism, and cytoskeletal dynamics, providing a comprehensive understanding of structural and functional changes within tumor tissues.

6 Conclusions and Future Perspectives

SHG imaging has emerged as a powerful tool in biomedical preclinical research due to its unique advantages. As a label-free and non-invasive technique, SHG offers high specificity for non-centrosymmetric structures, enabling detailed visualization of tissue architecture in its native state. Its high spatial and temporal resolution, along with the absence of photobleaching, makes SHG particularly well-suited for long-term imaging and real-time monitoring of dynamic biological processes. Furthermore, SHG generates intrinsic contrast without external dyes, minimizes sample preparation and eliminates potential artifacts, allowing for a more accurate assessment of structural and functional changes in tissues.

Despite its many strengths, SHG imaging faces limitations that hinder its broader clinical adoption. Its reliance on highly ordered, non-centrosymmetric structures restrict its applicability to specific biological targets. Additionally, the relatively weak SHG signals require advanced and costly laser systems, limiting accessibility and scalability. Furthermore, many current SHG imaging systems are custom-built for specific labs, presenting challenges in terms of reproducibility and portability. Standardization and unification of SHG imaging processes are essential to address these issues and advance its clinical utility.

The transition of SHG imaging from preclinical research to clinical practice faces several significant challenges. A key barrier is the need for user-friendly, compact, and cost-effective imaging systems that can seamlessly integrate into clinical workflows. Additionally, robust signal analysis algorithms are required to handle the complexity of human tissues, ensuring consistent and reproducible results across patients. A further challenge is the relatively limited understanding and training among clinicians, which may slow the adoption of SHG-based diagnostic tools. To overcome these obstacles, future research should focus on the development of portable and cost-efficient SHG imaging systems, coupled with advancements in laser technology to enhance signal strength and imaging depth. Improved image processing algorithms, powered by machine learning, could enable automated analysis and facilitate the interpretation of SHG data in clinical settings. Furthermore, interdisciplinary collaboration between researchers, engineers, and clinicians is vital to designing SHG systems that are tailored to specific clinical applications, such as cancer diagnostics, tissue fibrosis monitoring, and wound healing assessment. In conclusion, SHG imaging holds immense potential to revolutionize biomedical research and clinical diagnostics by providing high-resolution, label-free visualization of tissue structure and dynamics. However, overcoming its current limitations through technological advancements and enhanced clinical integration is crucial for unlocking its full potential as a minimally invasive diagnostic and monitoring tool in healthcare.

Acknowledgments We would like to express our sincere gratitude to the members of the Dai Group and our collaborators for their invaluable contributions to the research presented in this book chapter.

Conflict of Interest The authors declare no competing interests.

References

Aghigh A, Bancelin S, Rivard M, Pinsard M, Ibrahim H, Légaré F (2023) Second harmonic generation microscopy: a powerful tool for bio-imaging. Biophys Rev 15(1):43–70. https://doi.org/10.1007/s12551-022-01041-6

Avila FJ, Del Barco O, Bueno JM (2015) Polarization dependence of aligned collagen tissues imaged with second harmonic generation microscopy. J Biomed Opt 20(8):86001. https://doi.org/10.1117/1.JBO.20.8.086001

Bancelin S, Lynch B, Bonod-Bidaud C, Ducourthial G, Psilodimitrakopoulos S, Dokladal P, Allain JM, Schanne-Klein MC, Ruggiero F (2015) Ex vivo multiscale quantitation of skin biomechanics in wild-type and genetically-modified mice using multiphoton microscopy. Sci Rep 5:17635. https://doi.org/10.1038/srep17635

Batista A, Guimaraes P, Domingues JP, Quadrado MJ, Morgado AM (2022) Two-photon imaging for non-invasive corneal examination. Sensors (Basel) 22(24). https://doi.org/10.3390/s22249699

Bocklitz TW, Salah FS, Vogler N, Heuke S, Chernavskaia O, Schmidt C, Waldner MJ, Greten FR, Brauer R, Schmitt M, Stallmach A, Petersen I, Popp J (2016) Pseudo-HE images derived from CARS/TPEF/SHG multimodal imaging in combination with Raman-spectroscopy as a pathological screening tool. BMC Cancer 16:534. https://doi.org/10.1186/s12885-016-2520-x

Both M, Vogel M, Friedrich O, von Wegner F, Kunsting T, Fink RH, Uttenweiler D (2004) Second harmonic imaging of intrinsic signals in muscle fibers in situ. J Biomed Opt 9(5):882–892. https://doi.org/10.1117/1.1783354

Boyd RW (2020) Chapter 1 – The nonlinear optical susceptibility. In: Boyd RW (ed) Nonlinear optics (Fourth Edition). Academic Press, pp 1–64. https://doi.org/10.1016/B978-0-12-811002-7.00010-2

Brett EA, Sauter MA, Machens HG, Duscher D (2020) Tumor-associated collagen signatures: pushing tumor boundaries. Cancer Metab 8:14. https://doi.org/10.1186/s40170-020-00221-w

Bueno JM, Avila FJ, Martinez-Garcia MC (2019) Quantitative analysis of the corneal collagen distribution after in vivo cross-linking with second harmonic microscopy. Biomed Res Int 2019:3860498. https://doi.org/10.1155/2019/3860498

Bueno JM, Gualda EJ, Giakoumaki A, Perez-Merino P, Marcos S, Artal P (2011) Multiphoton microscopy of ex vivo corneas after collagen cross-linking. Invest Ophthalmol Vis Sci 52(8):5325–5331. https://doi.org/10.1167/iovs.11-7184

Campagnola PJ, Loew LM (2003) Second-harmonic imaging microscopy for visualizing biomolecular arrays in cells, tissues and organisms. Nat Biotechnol 21(11):1356–1360. https://doi.org/10.1038/nbt894

Chakraborty I, Mazumder N, Gogoi A, Chen MC, Zhuo GY (2024) Recent advances in label-free imaging techniques based on nonlinear optical microscopy to reveal the heterogeneity of the tumor microenvironment. Biophys Rev 16(5):581–590. https://doi.org/10.1007/s12551-024-01229-y

Chen X, Nadiarynkh O, Plotnikov S, Campagnola PJ (2012) Second harmonic generation microscopy for quantitative analysis of collagen fibrillar structure. Nat Protoc 7(4):654–669. https://doi.org/10.1038/nprot.2012.009

Cicchi R, Vogler N, Kapsokalyvas D, Dietzek B, Popp J, Pavone FS (2013) From molecular structure to tissue architecture: collagen organization probed by SHG microscopy. J Biophotonics 6(2):129–142. https://doi.org/10.1002/jbio.201200092

Dombeck DA, Kasischke KA, Vishwasrao HD, Ingelsson M, Hyman BT, Webb WW (2003) Uniform polarity microtubule assemblies imaged in native brain tissue by second-harmonic generation microscopy. Proc Natl Acad Sci 100(12):7081–7086. https://doi.org/10.1073/pnas.0731953100

Dombeck DA, Sacconi L, Blanchard-Desce M, Webb WW (2005) Optical recording of fast neuronal membrane potential transients in acute mammalian brain slices by second-harmonic generation microscopy. J Neurophysiol 94(5):3628–3636. https://doi.org/10.1152/jn.00416.2005

Dudenkova VV, Shirmanova MV, Lukina MM, Feldshtein FI, Virkin A, Zagainova EV (2019) Examination of collagen structure and state by the second harmonic generation microscopy. Biochemistry (Mosc) 84:S89–S107. https://doi.org/10.1134/S0006297919140062

Eisenthal ECYY, Eisenthal KB (2000) Effect of cholesterol on molecular transport of organic cations across liposome bilayers probed by second harmonic generation. Biophys J 79:898–903

Fedorovskiy AG, Antropov DN, Dome AS, Puchkov PA, Makarova DM, Konopleva MV, Matveeva AM, Panova EA, Shmendel EV, Maslov MA, Dmitriev SE, Stepanov GA, Markov OV (2024) Novel efficient lipid-based delivery systems enable a delayed uptake and sustained expression of mRNA in human cells and mouse tissues. Pharmaceutics 16(5). https://doi.org/10.3390/pharmaceutics16050684

Franken PA, Hill AE, Peters CW, Weinreich G (1961) Generation of optical harmonics. Phys Rev Lett 7(4):118–119. https://doi.org/10.1103/PhysRevLett.7.118

Gavgiotaki E, Filippidis G, Tsafas V, Bovasianos S, Kenanakis G, Georgoulias V, Tzardi M, Agelaki S, Athanassakis I (2020) Third Harmonic Generation microscopy distinguishes malignant cell grade in human breast tissue biopsies. Sci Rep 10(1):11055. https://doi.org/10.1038/s41598-020-67857-y

Gitanjal D, Wei-Wen W, Fu-Jen K (2012) In vivo wound healing diagnosis with second harmonic and fluorescence lifetime imaging. J Biomed Opt 18(6):061222. https://doi.org/10.1117/1.JBO.18.6.061222

Guimaraes P, Morgado M, Batista A (2024) On the quantitative analysis of lamellar collagen arrangement with second-harmonic generation imaging. Biomed Opt Express 15(4):2666–2680. https://doi.org/10.1364/BOE.516817

Hanson KM, Bardeen CJ (2009) Application of nonlinear optical microscopy for imaging skin. Photochem Photobiol 85(1):33–44. https://doi.org/10.1111/j.1751-1097.2008.00508.x

Houle MA, Couture CA, Bancelin S, Van der Kolk J, Auger E, Brown C, Popov K, Ramunno L, Legare F (2015) Analysis of forward and backward Second Harmonic Generation images to probe the nanoscale structure of collagen within bone and cartilage. J Biophotonics 8(11–12):993–1001. https://doi.org/10.1002/jbio.201500150

Hu X-H, Dai H-L (2024) Effect of molecular structure on membrane diffusion: triphenylmethanes across Escherichia coli studied by second harmonic light scattering. J Chem Phys 161(12):124701. https://doi.org/10.1063/5.0232591

Hu X-H, Fu L, Hou J, Zhang Y-N, Zhang Z, Wang H-F (2020) N–H Chirality in folded peptide LK7β is governed by the Cα–H Chirality. J Phys Chem Lett 11(4):1282–1290. https://doi.org/10.1021/acs.jpclett.9b03470

Hu X-H, Li Y, Wu T, Ma J, Wu Y, Wilhelm MJ, Dai H-L (2024) Effect of exogenous glucose on molecular transport through bacterial membranes studied by second harmonic light scattering. J Phys Chem B 128(50):12379–12388. https://doi.org/10.1021/acs.jpcb.4c06134

Hu X-H, Wei F, Wang H, Wang H-F (2019) α-quartz crystal as absolute intensity and phase standard in sum-frequency generation vibrational spectroscopy. J Phys Chem C 123(24):15071–15086. https://doi.org/10.1021/acs.jpcc.9b03202

Javier FAMD, Vitor BP, Leverson FLC, André Ad T, Diogo BA, Carlos LC, Luciana P, Liliana AA, Fatima B-L (2011) Recognition of serous ovarian tumors in human samples by multimodal nonlinear optical microscopy. J Biomed Opt 16(9):096017. https://doi.org/10.1117/1.3626575

Junqueira JC, Ribeiro MA, Rossoni RD, Barbosa JO, Querido SMR, Jorge AOC (2010) Antimicrobial photodynamic therapy: photodynamic antimicrobial effects of Malachite Green on Staphylococcus, Enterobacteriaceae, and Candida. Photomed Laser Surg 28(S1):S-67-S-72. https://doi.org/10.1089/pho.2009.2526

Karsdal MA, Nielsen SH, Leeming DJ, Langholm LL, Nielsen MJ, Manon-Jensen T, Siebuhr A, Gudmann NS, Ronnow S, Sand JM, Daniels SJ, Mortensen JH, Schuppan D (2017) The good and the bad collagens of fibrosis – Their role in signaling and organ function. Adv Drug Deliv Rev 121:43–56. https://doi.org/10.1016/j.addr.2017.07.014

Karunendiran A, Mirsanaye K, Stewart BA, Barzda V (2022) Second Harmonic Generation properties in Chiral Sarcomeres of Drosophila Larval Muscles. Front Phys 10. https://doi.org/10.3389/fphy.2022.758709

Keikhosravi A, Bredfeldt JS, Sagar AK, Eliceiri KW (2014) Second-harmonic generation imaging of cancer. Methods Cell Biol 123:531–546. https://doi.org/10.1016/B978-0-12-420138-5.00028-8

Li B, Li J, Chen S, Yuan Q, Fang C, Gan W (2024a) Monitoring the response of a model protocell to dye and surfactant molecules through second harmonic generation and fluorescence imaging. Phys Chem Chem Phys 26(10):8148–8157. https://doi.org/10.1039/D4CP00009A

Li Y, Wilhelm Michael J, Wu T, Hu X-H, Ruiz Oscar N, Dai H-L (2024b) Quantifying bacterial efflux within subcellular domains of Pseudomonas aeruginosa. Appl Environ Microbiol 90(11):e01447–e01424. https://doi.org/10.1128/aem.01447-24

Liu W, Raben N, Ralston E (2013) Quantitative evaluation of skeletal muscle defects in second harmonic generation images. J Biomed Opt 18(2):26005. https://doi.org/10.1117/1.JBO.18.2.026005

Lynch B, Bonod-Bidaud C, Ducourthial G, Affagard JS, Bancelin S, Psilodimitrakopoulos S, Ruggiero F, Allain JM, Schanne-Klein MC (2017) How aging impacts skin biomechanics: a multiscale study in mice. Sci Rep 7(1):13750. https://doi.org/10.1038/s41598-017-13150-4

Macias-Romero C, Didier ME, Jourdain P, Marquet P, Magistretti P, Tarun OB, Zubkovs V, Radenovic A, Roke S (2014) High throughput second harmonic imaging for label-free biological applications. Opt Express 22(25):31102–31112. https://doi.org/10.1364/OE.22.031102

Mizuguchi T, Nuriya M (2020) Applications of second harmonic generation (SHG)/sum-frequency generation (SFG) imaging for biophysical characterization of the plasma membrane. Biophys Rev 12(6):1321–1329. https://doi.org/10.1007/s12551-020-00768-4

Mizuguchi T, Yasui M, Nuriya M (2018) High-resolution plasma membrane-selective imaging by second harmonic generation. iScience 9:359–366. https://doi.org/10.1016/j.isci.2018.11.008

Nejim Z, Navarro L, Morin C, Badel P (2022) Quantitative analysis of second harmonic generated images of collagen fibers: a review. Res Biomed Eng 39(1):273–295. https://doi.org/10.1007/s42600-022-00250-y

Nikolaev VV, Kistenev YV, Kroger M, Zuhayri H, Darvin ME (2024) Review of optical methods for noninvasive imaging of skin fibroblasts-From in vitro to ex vivo and in vivo visualization. J Biophotonics 17(1):e202300223. https://doi.org/10.1002/jbio.202300223

Nuriya M, Fukushima S, Momotake A, Shinotsuka T, Yasui M, Arai T (2016) Multimodal two-photon imaging using a second harmonic generation-specific dye. Nat Commun 7:11557. https://doi.org/10.1038/ncomms11557

Odin C, Guilbert T, Alkilani A, Boryskina OP, Fleury V, Grand YL (2008) Collagen and myosin characterization by orientation field second harmonic microscopy. Opt Express 16(20):16151–16165. https://doi.org/10.1364/OE.16.016151

Park CY, Lee JK, Chuck RS (2015) Second harmonic generation imaging analysis of collagen arrangement in human cornea. Invest Ophthalmol Vis Sci 56(9):5622–5629. https://doi.org/10.1167/iovs.15-17129

Pendleton EG, Tehrani KF, Barrow RP, Mortensen LJ (2020) Second harmonic generation characterization of collagen in whole bone. Biomed Opt Express 11(8):4379–4396. https://doi.org/10.1364/BOE.391866

Plotnikov SV, Millard AC, Campagnola PJ, Mohler WA (2006) Characterization of the myosin-based source for second-harmonic generation from muscle sarcomeres. Biophys J 90(2):693–703. https://doi.org/10.1529/biophysj.105.071555

Raoux C, Schmeltz M, Bied M, Alnawaiseh M, Hansen U, Latour G, Schanne-Klein MC (2021) Quantitative structural imaging of keratoconic corneas using polarization-resolved SHG microscopy. Biomed Opt Express 12(7):4163–4178. https://doi.org/10.1364/BOE.426145

Sharifian Gh M, Wilhelm MJ, Dai H-L (2016) Label-free optical method for quantifying molecular transport across cellular membranes in vitro. J Phys Chem Lett 7(17):3406–3411. https://doi.org/10.1021/acs.jpclett.6b01483

Sharifian Gh M, Wilhelm MJ, Dai H-L (2018) Azithromycin-induced changes to Bacterial membrane properties monitored in vitro by second-harmonic light scattering. ACS Med Chem Lett 9(6):569–574. https://doi.org/10.1021/acsmedchemlett.7b00499

Sharifian Gh M, Wilhelm MJ, Moore M, Dai H-L (2019) Spatially resolved membrane transport in a single cell imaged by second harmonic light scattering. Biochemistry 58(14):1841–1844. https://doi.org/10.1021/acs.biochem.9b00110

Tarun OB, Hannesschlager C, Pohl P, Roke S (2018) Label-free and charge-sensitive dynamic imaging of lipid membrane hydration on millisecond time scales. Proc Natl Acad Sci USA 115(16):4081–4086. https://doi.org/10.1073/pnas.1719347115

Thomas B, McIntosh D, Fildes T, Smith L, Hargrave F, Islam M, Thompson T, Layfield R, Scott D, Shaw B, Burrell CL, Gonzalez S, Taylor S (2017) Second-harmonic generation imaging of collagen in ancient bone. Bone Rep 7:137–144. https://doi.org/10.1016/j.bonr.2017.10.005

Tilbury K, Campagnola PJ (2015) Applications of second-harmonic generation imaging microscopy in ovarian and breast cancer. Perspect Med Chem 7:21–32. https://doi.org/10.4137/PMC.S13214

Weigelin B, Bakker GJ, Friedl P (2016) Third harmonic generation microscopy of cells and tissue organization. J Cell Sci 129(2):245–255. https://doi.org/10.1242/jcs.152272

Wilhelm MJ, Dai H-L (2020a) Molecule-membrane interactions in biological cells studied with second harmonic light scattering. Chemistry Asian J 15(2):200–213. https://doi.org/10.1002/asia.201901406

Wilhelm MJ, Dai HL (2020b) Molecule-membrane interactions in biological cells studied with second harmonic light scattering. Chem Asian J 15(2):200–213. https://doi.org/10.1002/asia.201901406

Wilhelm MJ, Sharifian Gh M, Dai H-L (2015a) Chemically induced changes to membrane permeability in living cells probed with nonlinear light scattering. Biochemistry 54(29):4427–4430. https://doi.org/10.1021/acs.biochem.5b00600

Wilhelm MJ, Sharifian Gh M, Wu T, Li Y, Chang C-M, Ma J, Dai H-L (2021) Determination of bacterial surface charge density via saturation of adsorbed ions. Biophys J 120(12):2461–2470. https://doi.org/10.1016/j.bpj.2021.04.018

Wilhelm MJ, Sharifian Gh M, Dai H-L (2019) Influence of molecular structure on passive membrane transport: a case study by second harmonic light scattering. J Chem Phys 150(10). https://doi.org/10.1063/1.5081720

Wilhelm MJ, Sheffield JB, Gonella G, Wu Y, Spahr C, Zeng J, Xu B, Dai H-L (2014) Real-time molecular uptake and membrane-specific transport in living cells by optical microscopy and nonlinear light scattering. Chem Phys Lett 605–606:158–163. https://doi.org/10.1016/j.cplett.2014.05.009

Wilhelm MJ, Sheffield JB, Sharifian Gh M, Wu Y, Spahr C, Gonella G, Xu B, Dai H-L (2015b) Gram's stain does not cross the bacterial cytoplasmic membrane. ACS Chem Biol 10(7):1711–1717. https://doi.org/10.1021/acschembio.5b00042

Wu T, Wilhelm MJ, Li Y, Ma J, Dai HL (2022a) Indole facilitates antimicrobial uptake in bacteria. ACS Infect Dis 8(6):1124–1133. https://doi.org/10.1021/acsinfecdis.1c00618

Wu T, Wilhelm MJ, Ma J, Li Y, Wu Y, Dai H-L (2022b) Influence of phase transitions on diffusive molecular transport across biological membranes. Angew Chem Int Ed 61(42):e202205608. https://doi.org/10.1002/anie.202205608

Yu CH, Langowitz N, Wu HY, Farhadifar R, Brugues J, Yoo TY, Needleman D (2014) Measuring microtubule polarity in spindles with second-harmonic generation. Biophys J 106(8):1578–1587. https://doi.org/10.1016/j.bpj.2014.03.009

Zeng J, Eckenrode HM, Dai H-L, Wilhelm MJ (2015) Adsorption and transport of charged vs. neutral hydrophobic molecules at the membrane of murine erythroleukemia (MEL) cells. Colloids Surf B: Biointerfaces 127:122–129. https://doi.org/10.1016/j.colsurfb.2015.01.014

Zeng J, Eckenrode HM, Dounce SM, Dai H-L (2013) Time-resolved molecular transport across living cell membranes. Biophys J 104(1):139–145. https://doi.org/10.1016/j.bpj.2012.11.3814

Zhao H, Cisek R, Karunendiran A, Tokarz D, Stewart BA, Barzda V (2019) Live imaging of contracting muscles with wide-field second harmonic generation microscopy using a high power laser. Biomed Opt Express 10(10):5130–5135. https://doi.org/10.1364/BOE.10.005130

Zoumi A, Yeh A, Tromberg BJ (2002) Imaging cells and extracellular matrix in vivo by using second-harmonic generation and two-photon excited fluorescence. Proc Natl Acad Sci 99(17):11014–11019. https://doi.org/10.1073/pnas.172368799

Emerging Approaches in Metabolic Imaging Using Photoacoustics

Anandhu Anil Sindu, Swathi Padmanabhan, and Jaya Prakash

Abstract

Photoacoustic imaging (PAI) can provide non-invasive, label-free visualization of biological tissue chromophores deep tissue. This chapter explores the field of metabolic imaging using PAI, highlighting its ability to quantify various metabolic chromophores, including lipids, glucose, and reactive oxygen species, providing a comprehensive metabolic profile. Furthermore, the diverse applications of these emerging PAI techniques in preclinical and clinical settings, including the study of cancer metabolism, diabetes, and cardiovascular diseases were discussed. A highlight of the latest breakthroughs in PAI-based metabolic imaging, as well as future directions, is also presented. This chapter aims to provide a comprehensive overview of the transformative potential of PAI in advancing our understanding and management of metabolic diseases.

Keywords

Photoacoustic imaging · Metabolic imaging · Glucose · Fatty acids · Oxygen saturation

1 Introduction

Metabolism is a transformation of energy with a combination of enzyme mediated anabolic and catabolic reactions, generating the energy required for proper functioning of the body. These reactions are almost the same in all organisms and are mostly

A. A. Sindu · S. Padmanabhan · J. Prakash (✉)
Department of Instrumentation and Applied Physics, Indian Institute of Science, Bengaluru, India
e-mail: jayap@iisc.ac.in

I. F. Cengiz et al. (eds.), *Bioimaging Modalities in Bioengineering*,
Biomaterials, Bioengineering and Sustainability 9,
https://doi.org/10.1007/978-3-031-99410-4_25

mediated by enzymes (Sánchez López de Nava and Raja 2025). Basal metabolic rate refers to the rate of energy production and is affected by numerous factors like fat-free mass (FFM), fat mass (FM), age, sex, diet, and presence of diseases (Lazzer et al. 2010). Therefore, measuring the metabolites can act as an indirect marker to understand the cause of underlying condition and hence can be used for disease diagnosis.

Various imaging modalities are being developed to visualize and quantify these metabolic activities in vivo using positron emission tomography (PET), nuclear magnetic resonance methods like magnetic resonance imaging (MRI) (Apostolova et al. 2016), fluorescent imaging, and bioluminescence imaging. Even though PET has high sensitivity and specificity, it utilizes radioactive tracers having small half-life (thereby reducing the signal to noise ratio) alongside being carcinogenic in the long term (Di Gialleonardo et al. 2016). MRI is a non-ionizing, non-invasive alternative but suffers from low sensitivity, and slow acquisition times. Fluorescence and bioluminescent imaging-based probes have potential; however, these technologies suffer due to their poor spatial resolution. The penetration depth of fluorescence molecular tomography (FMT) depends on the optical attenuation of the organ being examined and the strong background autofluorescence (Nievelstein et al. 2012).

Photoacoustic imaging (PAI) (also known as optoacoustic imaging) is an established technique that offers high resolution metabolic imaging of various biological processes. PAI can image both endogenous and exogeneous contrast agents (Stuker et al. 2011; Hu 2016). In label free (endogenous) method, the chromophores that are naturally present inside the body absorb at a specific wavelength of the nanosecond pulsed laser light to produce photoacoustic signals due to the thermoelastic expansion of the tissue components (Fig. 1). Endogenous chromophores like water, hemoglobin, lipids, and melanin naturally absorb light, enabling label-free imaging as indicated in Fig. 1. Additionally, exogenous (extrinsic) contrast agents such as indocyanine green (ICG), Rose Bengal, methylene blue, quantum dots, and nanoparticles can be used to enhance image contrast and specificity (Stuker et al. 2011; Hu 2016).

Clinical PAI systems operate at the near-infrared (NIR) spectral range, enabling deeper tissue penetration. For instance, oxygenated and deoxygenated hemoglobin can be imaged by illuminating the tissue at 650–900 nm wavelength range. Tissue oxygenation saturation maps can be derived by computing the ratio of oxygenated hemoglobin to the total blood hemoglobin. The rate of oxygen saturation change is linked to oxygen utilization (reflecting oxidative metabolism), which is relevant for various endocrine conditions. PAI can be used for imaging lipids, by illuminating the tissue with 930 nm or 1210 nm (lipid peak) light source. PA measurements at 930 nm provide insights into parameters associated with white adipose tissue (WAT), brown adipose tissue (BAT), and lipid concentrations in the bloodstream and specific tissue compartments (Gujrati et al. 2017).

Label-free PAI methods have been developed to image and detect subcellular metabolic molecules within the tissue with high signal-to-noise and signal-to-contrast ratios. Mid-infrared optoacoustic microscopy (MiROM) offers the poten-

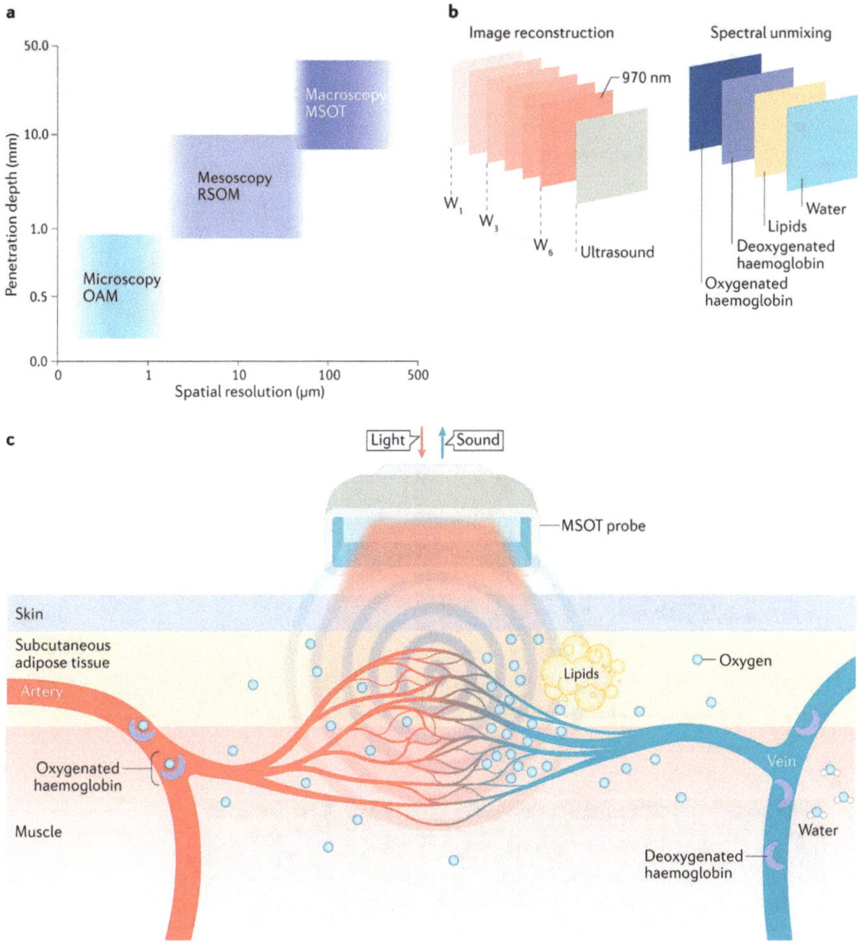

Fig. 1 PAI for metabolic imaging. (**a**) Scalable PAI technologies classified according to the penetration depth and resolution. (**b**) Data acquisition and image reconstruction (**c**) The PA scanner sends light pulses (red arrow) at near infrared (NIR) wavelengths into the tissue. These light pulses are absorbed, and the generated ultrasound waves are then detected via a transducer and then reconstructed into a photoacoustic frame. (Reproduced from Karlas et al. 2021, with permission from Springer Nature)

tial to monitor levels of carbohydrates, proteins, and lipids. Miguel et al. utilized MiROM to image carbohydrates (with absorption peaks at approximately 1081 cm^{-1} and 1084 cm^{-1}) and lipids (with a vibrational absorption peak at approximately 2853 cm^{-1}) (Fig. 2). MiROM was further applied to investigate adipogenesis and lipolysis (Karlas et al. 2021), as shown in Fig. 2. Multispectral optoacoustic tomography (MSOT) is an emerging biological imaging technique that offers high spatial and temporal resolution at imaging ICG marked sentinel

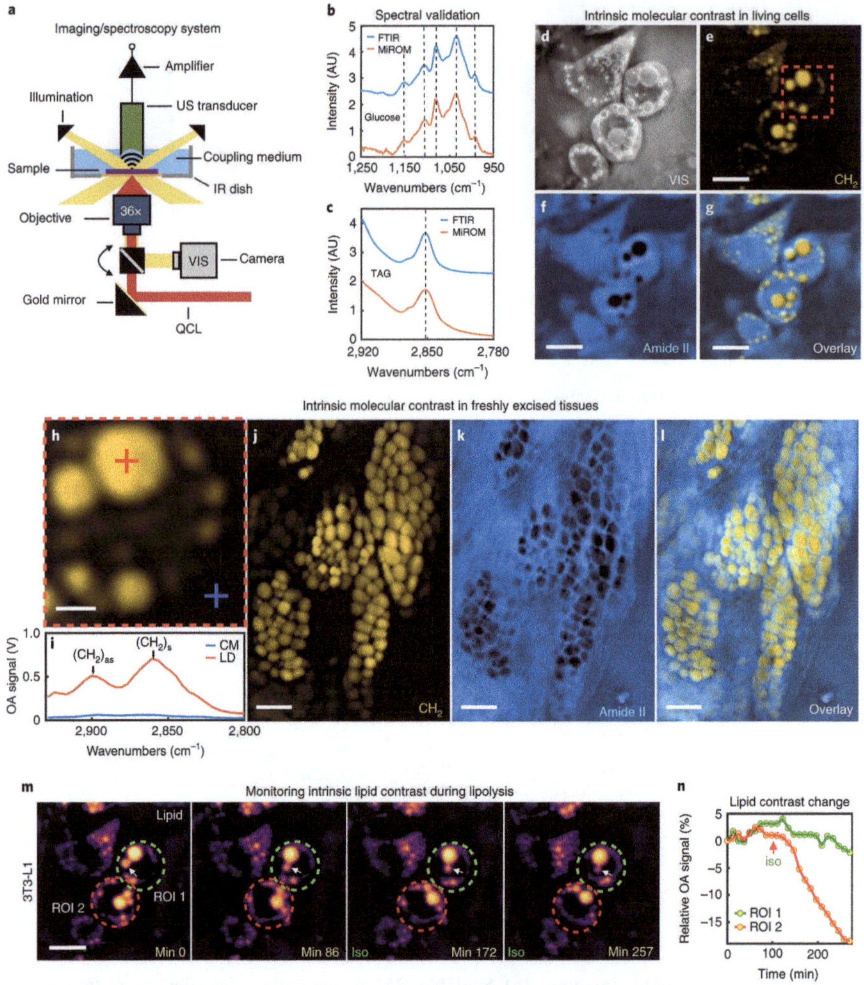

Fig. 2 Imaging of cellular metabolites using MiROM (**a**) MiROM system schematic. (**b, c**) MiROM and FTIR spectrum of glucose and 1,2- 1,2-dioleoyl-3-palmitoyl-rac-glycerol (TAG). (**d**) Bright field images of adipocyte cells derived from 3 T3-L1 cells. (**e**) Endogenous lipid contrast obtained using MiROM at 2857 cm^{-1} wavenumber. (**f**) Protein contrast obtained using label-free MiROM at 1550 cm^{-1} wavenumber. (**g**) Overlay of e and f. Scale bars represent 40 μm. (**h**) Zoomed in region from the red dashed square in e, Scale bar, 10 μm. (**j–l**) MiROM images of freshly excised 4 mm thick pancreatic mouse tissue. Scale bar, 100 μm showing (m) Monitoring intrinsic lipid at 2857 cm^{-1} during lipolysis. (**n**) Relative lipid contrast change for region of interest (ROI)-1 and 2 in m; the red arrow indicates when isoproterenol (ISO) was added. (Reproduced from Pleitez et al. 2020, with permission from Springer Nature)

lymph nodes at a depth of more than 5 cm of tissue (Pleitez et al. 2020). There are many reports where MSOT has been successfully applied for imaging metabolic activity. A few key metabolites and their role in metabolism will be elaborated in the following sections. Moreover, how photoacoustic imaging (PAI) is used to image these metabolites will be discussed and its advantages will be highlighted.

2 Metabolic Imaging

Metabolic imaging refers to the visualization or quantification of metabolic activities that occur in the cells or the whole body. Steady-state levels of metabolic parameters, such as pH, oxygen, and glucose, are intricately linked to functional parameters like metabolic turnover rates and blood flow (Stoffels et al. 2015). The local oxygen levels and most metabolites remain stable in the organs of a healthy human. Deviations from these normal levels in any organ may signify metabolic disorders. Metabolic imaging focuses on evaluating different metabolic pathways, which can be later used for disease diagnosis, monitoring the disease progression, to identify potential drug targets and response to a particular treatment. By examining key metabolites in different pathways such as glycolysis, the tricarboxylic acid (TCA) cycle, choline metabolism, and amino acid metabolism, researchers and clinicians have gained insights into various conditions like cancer, brain metabolic imaging (neurological disorders), cardiac metabolic imaging, inflammation, and infection studies (Walenta et al. 2002).

2.1 Oxygen: The Fuel for Cellular Respiration

Oxygen plays a key role in cellular respiration, especially in oxidative phosphorylation that generates Adenosine Triphosphate (ATP). The generated ATP is used in many of the metabolic pathways (Gialleonardo et al. 2016). Any disruption in homeostasis of the oxygen microcirculation throughout the body can lead to abnormal metabolism. Oxygen metabolism can be used as a marker for abnormal conditions like diabetes, cancer, stroke, Alzheimer's disease, and aging (Brahimi-Horn and Pouysségur 2007). The gold standard for oxygen sensing is via needle-based probes, but these probes cannot be used to probe large surface areas. Notably there can be significant variations in the oxygen concentration within a single organ or cellular structure making needle-based probes ineffective. To fully understand the influence of oxygen on physiology and disease, it is crucial to quantify molecular oxygen at tissue, cellular, and subcellular levels in vivo. Other modalities that can be used for detection of intracellular and tissue oxygen levels are MRI (Yoshihara et al. 2017), electron paramagnetic resonance oximetry (EPRO) (Krishna et al. 2013), PET (Gallez et al. 1999), and via luminescent-based probes (Shishido et al. 1996). These modalities have a lot of drawbacks making their clinical applicability limited. Blood oxygen level dependent functional MRI (BOLD fMRI) is sensitive to blood oxygenation (Hyder et al. 2001); however, fMRI is known to have high false positive rates (Eklund et al. 2016). PET can allow for indirect imaging of oxygen content in the tissue by correlating with cell metabolic activity but requires intravenous and intraarterial injection as well as significant exposure to radiation. On the other hand, luminescent and fluorescent probes-based imaging are widely used in vitro, but they have less depth to resolution ratio in vivo, making them unsuitable for in vivo metabolic imaging applications.

PAI can be used for studying label free single cell metabolism, Hai, et al. were able to trap single cells in an oxygen diffusion limited high density microwell array (Hai et al. 2019). The system utilized 532 nm and 559 nm laser wavelengths for photoacoustic excitation, with a laser spot size of 15 μm and a scanning step size of 80 μm. PAI was then used to measure the oxygen consumption rate in these cells based on hemoglobin oxygen changes (Hai et al. 2019). This technology can identify metabolic vulnerabilities during cancer therapy and enable the development of targeted metabolic treatments at the single-cell level. Photoacoustic lifetime imaging (PALI) is another novel method derived from the concept of phosphorescence imaging and can directly show the spatio-temporal distribution of oxygen in tissue samples. PALI uses a series of laser pulses (laser output power of less than 100 mW and a low repetition rate of 10 Hz) to measure oxygen levels (Shao and Ashkenazi 2015).

2.2 Glucose: The Body's Primary Energy Source

Glucose is the primary energy source for most living organisms. Glucose is mainly metabolized via oxidative phosphorylation to produce the energy currency known as ATP. Cancerous cells have uncontrollable proliferation behavior leading to increased uptake of glucose, and glucose is metabolized via less efficient glycolysis pathway. The glucose uptake is up to 200 times more in cancer cells compared to normal cells and is known as the Warburg effect (Warburg et al. 1927). Currently 2-deoxy-2-[18F] fluoro-glucose (18F-FDG)-based PET/computed tomography (PET/CT) is used for glucose metabolism imaging. Alternatively, other radiolabeled tracers used for PET/CT imaging include 18F, 68Ga, and 99mTc (Feng et al. 2019). Despite its high sensitivity, (18F-FDG)-based PET faces limitations for longitudinal studies due to its high cost, the short half-life of radioisotopes, and the potential carcinogenic effects of accumulated radiation exposure. Therefore, this technique has been applied widely for cancer staging. Despite extensive development in bioluminescent agents for imaging they are constrained to two-dimensional surface-weighted signals, thereby limiting their use in in vivo metabolic imaging. FMT can enable three-dimensional imaging but suffers from low resolution due to photon diffusion, thus restricting the use of FMT to superficial tumor (8 mm below skin) metabolism imaging (Han et al. 2023).

2-Deoxyglucose (2-DG) or analogues of 2-DG is commonly used in molecular imaging to study glucose metabolism in tissues. 2DG or its analogues are typically conjugated with a near infrared dye to make them optically detectable using PAI. The conjugated compound can absorb light in the near infrared range which enables deep tissue penetration and low background noise. PAI can be used to study the relationship between glucose metabolism and vascular properties in tumor (Han et al. 2023). Ring-shaped confocal photoacoustic computed tomography (RC-PACT) was developed to simultaneous visualization anatomical details and tumor glucose metabolism at high spatial resolution. RC-PACT system utilized a tunable Ti-sapphire laser (690–990 nm) as the excitation source. For imaging, 776 nm, 796 nm, and 820 nm were selected (with a maximum light intensity of ~15 mJ/cm^2). The custom-built transducer array was operated at a central frequency of 5 MHz, with a bandwidth of 80% at −6 dB and had a diameter of 5 cm. By utilizing IRDye800-2DG as a glucose tracer, researchers were able to study the metabolic activity of tumors with a single imaging modality (Chatni et al. 2012).

2.3 The Impact of Fatty Acids on Metabolic Functions

Fat metabolism involves the process of storing and later breaking fat down to obtain energy. There are mainly two types of adipose tissue, WAT (stores energy in the form of triglycerides within the fat droplets) and BAT (breaks carbohydrates and fat to generate heat). BAT physiology can be associated with abnormalities like obesity, and diabetes. 18F-fluorodeoxyglucose (FDG)-based PET/CT is considered as the gold standard in studying BAT activity (Reber et al. 2018). However, PET/CT cannot be used for detection of inactive BAT as the method relies on glucose uptake. MSOT is a non-invasive tool that enables label-free and non-ionizing detection of BAT metabolism (Reber et al. 2018). MSOT can also distinguish between BAT and WAT as well as give information about the lipid, water content in BAT and WAT. The ability of MSOT to detect the changes in BAT morphology and vascularization is useful to test out the efficacy of drugs that target BAT as well as for personalized medicine (Reber et al. 2018).

2.4 Imaging of Another Endogenous Metabolism

In addition to oxygen and glucose, cancer cells exhibit abnormalities in amino acid metabolism, which can serve as targets for detecting and monitoring cancer progression and drug response. Several amino acid transporter families are upregulated in breast cancer cells (Cha et al. 2018). Traditional detection methods, such as PET (Najjar et al. 2018), Magnetic Resonance Spectroscopy (MRS) (Łmigielska-Kuzia et al. 2010), Hyperpolarized MRI (hMRI) (Ross and Bluml 2001) and Chemical Exchange Saturation Transfer (CEST) (Goldenberg and Pagel 2019) can be used to study amino acid metabolism.

Other nitrogen-based compounds have also been explored to provide information about metabolic state of the body like amino acids (Dedon and Tannenbaum 2004). For example, a nitrogen derivative, nitric oxide (NO) (synthesized by nitric oxide synthase (NOS)) plays a major role in many biological processes including nerve signaling, immune response, and blood vessel regulation (Förstermann and Sessa

2012). Therefore, imaging NO can help in understanding the role of NO in health and disease. However, most of the current methods are limited to image NO in ex vivo samples and fail to provide NO distribution in complex in vivo situations. Methods like MRI and electron paramagnetic resonance (EPR) technology have been utilized for the detection of NO in vivo *but* have poor sensitivity (Sharma et al. 2014).

Reinhardt et al. utilized PAI to noninvasively detect NO metabolism in vivo by developing a class of probes called Acoustogenic Probes for Nitric Oxide (APNO) (Reinhardt et al. 2018). Specifically, they employed the probe (APNO-5) in a murine lipopolysaccharide (LPS)-mediated inflammation model (Reinhardt et al. 2018). By administering APNO-5 to mice with LPS-induced inflammation and irradiating the tissue at specific wavelengths (770 nm for APNO-5 and 680 nm for its product tAPNO-5), they were able to measure changes in the PA signal, indicating the presence of NO. This approach offers a significant advantage over traditional imaging methods due to real-time in vivo imaging of NO dynamics (Reinhardt et al. 2018).

2.5 Pharmacokinetic Imaging

The efficacy of most drugs is influenced by factors like local pH and oxygen levels. Consequently, both the pharmacological effects and the pharmacokinetics of therapeutic agents are impacted by the metabolic milieu and its influence on microcirculation (Lewis 2011). Compared to normal blood vessels, tumor vasculature differs significantly in terms of exhibiting abnormal characteristics such as dilation, tortuosity, distortion, and disorganization (Jain 2005). These abnormal characteristics were targeted initially for therapy, thereby destroying existing tumor vessels, and inhibiting angiogenesis, resulting in starvation of cancer cells to death. However, this approach led to increased tumor hypoxia and impaired drug delivery, prompting a reduction in the effectiveness of chemotherapy and radiotherapy.

A more recent strategy involved normalizing tumor vasculature to induce anti-angiogenic effects without compromising drug delivery (Jain 2005). To monitor the impact of such therapies and assess the balance between vascular normalization and hypoxia, non-invasive imaging techniques are crucial. While CT perfusion imaging can visualize tumor angiogenesis, [18F] FDG-PET was used for metabolism measurements. PAI offers a relatively low-cost, label-free approach to study vascular changes induced by anti-angiogenic therapies.

Zheng et al. employed near-infrared-dye-labeled 2-Deoxyglucose (2-DG-800) in an orthotopic breast tumor mouse model to assess intratumoral pharmacokinetics using Molecularly Targeted Optical Imaging (Han et al. 2023). They achieved precise segmentation of arteries and detailed quantification of intratumoral hemodynamic parameters. The rapid acquisition time of MSOT significantly reduced motion artifacts, ensuring accurate data collection. The voxel-wise analysis allowed for detailed characterization of intratumoral heterogeneity in metabolic, hemodynamic, and oxygenation parameters.

In the study conducted by Lv et al., they developed a spatiotemporally resolved clearance pathway tracking (SRCPT) method using photoacoustic tomography (PAT) to monitor drug clearance dynamics in the liver and kidneys. PAT was used to study the clearance of mitoxantrone, a clinical drug, and immunoglobulin G. The results demonstrated reduced liver clearance of mitoxantrone in cases of impaired hepatic function and revealed distinct clearance pathways for immunoglobulin G (Lv et al. 2024). Neuschmelting et al. were able to monitor the effects of vascular targeted photodynamic therapy (VTP) on tumor tissues, specifically in a renal cell cancer mouse model, they were able to achieve this by non-invasively capturing the oxygen saturation within the tumors after the administration of a photosensitizer called WST11 (Neuschmelting et al. 2018).

3 Applications

The applications of metabolic imaging, spanning academic research and clinical standpoints will be discussed in this section.

3.1 Translational Gene Expression Studies

Traditional methods for studying gene expression, such as Northern blotting, real-time PCR, and microarrays, require cell lysis and in vitro analysis. Emerging techniques employing fluorescent, luminescent, or dual reporter probes allow for in vivo imaging, however these techniques suffer due to poor depth to resolution ratio. A notable advancement in study of gene expression involves combining PAI with a tyrosinase-based reporter system, enabling the study of gene expression, including microRNA regulation in living organisms (Jathoul et al. 2015). Tyrosinase (Tyr) catalyzes the formation of pigment melanin from cellular tyrosine (Yun et al. 2021) which has high photostability as well as strong broadband optical absorption which make melanin an excellent candidate for PAI. Jathoul et al. used a novel retroviral vector to ensure stable and long-term expression of Tyrosinase in the cell of interest, with good contrast between Tyr expressing cells and background tissue (Jathoul et al. 2015). Jathoul et al. injected Tyr-expressing K562 and 293 T cells subcutaneously into the flanks of nude mice and performed longitudinal imaging of the xenograft mice at multiple wavelengths and time points, achieving high-resolution three-dimensional images at depths approaching 1 cm. The Fabry-Pérot (FP) sensor used as an all-optical ultrasound detector, which enhanced the performance of PAI. FP consists of a thin-film structure formed by a pair of highly reflective dichroic mirrors separated by a polymer spacer, which creates an etalon. This design allows the sensor to be highly reflective at 1500 nm to 1600 nm wavelength range, which is used to detect photoacoustic waves. Notably, the sensor is transparent to the excitation laser wavelengths (600–900 nm) used for imaging (Jathoul et al. 2015). Therefore, co-expressing genes like tyrosine, which enhance photoacoustic contrast, alongside genes of interest enables in vivo study of gene expression, like the approach demonstrated by Jathoul et al. On the other hand, Yun et al. used a genetically engineered strain of E. coli that expresses Tyr (E-Tyr), making bacterial detection possible using PAI. E-Tyr was further then injected into mice bearing CT26 tumors to visualize the tumors (Yun et al. 2021). Future development of multispectral optoacoustic endoscopes holds the potential to extend the clinical applicability of these theranostic approaches to organs beyond the penetration depth of transcutaneous MSOT.

3.2 Cancer Research and Diagnosis

As previously discussed, the metabolic activity of cancer cells is different compared to healthy cells. This metabolic difference can be utilized for diagnosing, imaging, and monitoring cancer progression. Key areas of focus include the Warburg effect and hemoglobin oxygen saturation, reflecting elevated glucose consumption and hypoxic conditions within the tumor microenvironment. Qian Yu et al. were able to detect microscopic melanoma of 400 μm size at a depth up to 7 mm in vivo (Yu et al. 2020). This resolution surpassed the capabilities of ultrasound and MRI at similar imaging depths. Additionally, PAI showed a significantly higher tumor-to-liver contrast ratio compared to PET/CT, enabling more effective differentiation between tumors and surrounding tissues (Yu et al. 2020). On the other hand, Li et al. were able to demonstrate that mesoscopic optoacoustic microscopy (MSOM) was capable of imaging entire deep tissue tumor morphology and physiology to quantifying tumor heterogeneity, including oxygenation (sO2) and vascularization patterns, at the whole-tumor level and within specific regions such as the core and rim, providing insights into tumor metabolism (Li et al. 2020). With this ability, MSOM has significant potential in drug screening, therapeutic evaluation, and advancing precision oncology.

Tumor targeting bacteria serves as a promising theranostic platform by specifically localizing tumors, delivering therapy, and enabling diagnostic imaging (Yun et al. 2021). However, current imaging methods, including optical and radiological techniques, face limitations in resolution, sensitivity, and cost, restricting detailed analysis of bacterial cancer targeting and treatment efficacy. Lin et al. were able to develop a novel imaging technique called Single-Breath-Hold Photoacoustic Computed Tomography (SBH-PACT) for breast cancer detection. SBH-PACT involves a donut-shaped laser beam and a full-ring ultrasonic transducer array to create high-resolution images of breast tissue without using ionizing radiation or contrast agents. The system detects tumors by analyzing blood vessel density and includes elastography to assess tissue compliance, enhancing tumor detection capabilities (as shown in Fig. 3) (Lin et al. 2018). Diot et al. conducted a pilot study using handheld MSOT that employed a unique single-frame-per-pulse (SFPP) image acquisition method for imaging breast cancer (Diot et al. 2017).

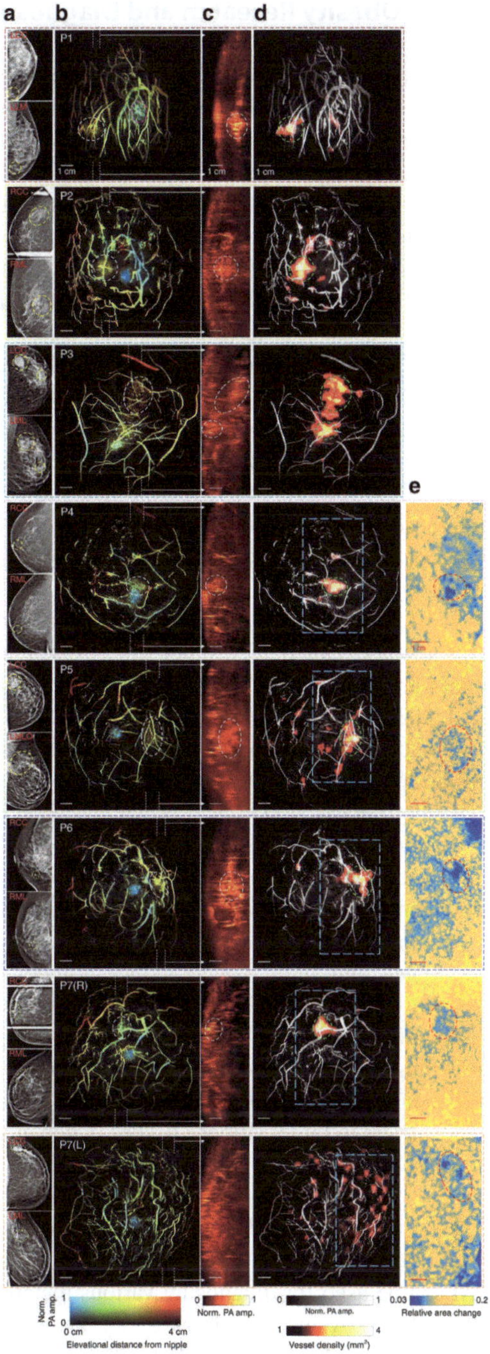

Fig. 3 SBH-PACT images of cancerous breast. (**a**) X-ray mammograms of 7 breast cancer patients. (**b**) Angiograms obtained from SBH-PACT (tumors marked by white circles) (**c**) Maximum amplitude projection in sagittal planes. (**d**) Automatic tumor detection on vessel density maps. Tumors are identified by green circles. (Reprinted from Lin et al. 2018 under a Creative Commons Attribution License (CC BY 4.0))

3.3 Obesity Research and Diagnosis

Obesity is a rising concern worldwide that significantly raises the risk of cardiovascular disease and premature death. Adipose tissue, beyond its role in storing energy, releases various bioactive substances that impact multiple physiological processes. These bioactive substances influence not only body weight regulation but also contribute to the development of insulin resistance (a hallmark of type-2 diabetes) as well as alterations in lipid levels, blood pressure, blood clotting, and inflammation, resulting in endothelial dysfunction and atherosclerosis (Karlas et al. 2017). This prompts us to investigate various treatments that can safely and effectively combat obesity.

MSOT has proven to be a valuable tool in assessing various aspects of vascular health. Particularly in the context of diabetes, MSOT can be used to image the peripheral vasculature and assess endothelial macrovascular function in the radial artery, an early biomarker of atherosclerosis, which is accelerated in diabetes (Karlas et al. 2017). Additionally, MSOT has shown promise in imaging the intestines and muscles, which can be affected in diabetes, leading to conditions like diabetic enteropathy and myopathy (Karlas et al. 2020). Raster-scanning optoacoustic mesoscopy (RSOM) is another powerful technique that allows for detailed imaging of the skin's microvasculature, a key indicator of diabetes-related microangiopathy. This method can detect subtle microvascular impairments even before clinical symptoms manifest, offering a valuable early diagnostic tool (He et al. 2023; Uluç et al. 2024).

Notably BAT and WAT can be studied by using photoacoustic, the effectiveness of various obesity treatments can also be studied. One of the studies used photodynamic therapy (PDT) as a browning agent (rosiglitazone) to target specifically WAT leading to adipose tissue reduction. PAI was used for real time monitoring of treatment induced changes, including complex activity like lipid catabolism and angiogenesis (Chen et al. 2021). PDT involved a photosensitizer (zinc phthalocyanine tetrasulfonate) that produces reactive oxygen species (ROS) in the presence of oxygen and light, which induce adipocyte cell death. Adipose browning is the process where WAT is converted into thermogenic BAT promoting energy dissipation as heat. Note that PDT is effective against fats at shallow depths whereas browning affects deeper fats thereby providing an effective and complementary treatment. Hepatitis B core (HBc) protein virus-like particles (VLPs) were used as nanocarriers to target adipose tissues (Chen et al. 2021).

3.4 Metabolic Disease Diagnosis

PAI can be used for diagnosing various metabolic diseases or abnormal conditions in the body. A few examples of the same include (but not limited to) imaging of hydroxyl radicals, metabolic acidosis, and brain metabolism study/monitoring. Hydroxyl radicals are highly reactive molecules that contribute to the development and progression of cardiovascular disease by initiating lipid peroxidation, leading to cellular damage and inflammation. Due to their short lifespan and high reactivity, detecting hydroxyl radicals is challenging. Cy-OH-LP is an optical probe with NIR absorption and fluorescence properties, which has been used to effectively detect hydroxyl radicals in lipid-rich environments. By utilizing PAI, Yang et al. identified atherosclerotic lesions and the presence of hydroxyl radicals within them. They confirmed their findings through the histological staining of the aortas (Yang et al. 2024).

Metabolic acidosis is a condition where there is an excessive accumulation of acids in bodily fluids. Traditional diagnostic methods like digital subtraction angiography (DSA) and computed tomography angiograph (CTA) are invasive and time-consuming (Kik et al. 2022; Grossberg et al. 2019) thereby limiting their use in vivo. Monitoring cerebral vessel diameter can help in assessing metabolic acidosis (Zhu et al. 2023).

Cerebral blood flow is dynamically regulated for meeting the metabolic demands of neurons. Local changes in blood vessel diameter, influenced by neurotransmitters and vasoactive factors, reflect these metabolic needs. Oxygen, as a natural contrast agent, can be utilized in photoacoustic microscopy (PAM) to visualize and quantify these changes in real-time in vivo (resolution was in the range of a few micrometers, which is crucial for detailed imaging of cerebral vessels). Zhu et al. used dual-wavelength PAM (532 nm and 559 nm), to visualize and quantify changes in cerebral vessel diameter. They established a diagnostic nomogram based on these vascular changes to assess the severity of acidosis and its effects on oxygen metabolism in the brain. This method allowed insights into the real-time monitoring of acid-base imbalances and cerebral oxygenation (Zhu et al. 2023).

4 Conclusions and Future Perspectives

PAI is a non-invasive technique that offers high spatial resolution and the ability to image functional parameters such as blood oxygenation, blood flow, and metabolic activity. Its versatility is further enhanced by its compatibility with other imaging modalities, like ultrasound imaging which provides complementary structural information. Recent advancements in contrast agents and novel instrumentation have expanded the potential of PAI for metabolic imaging. For example, Padmanabhan et al. have developed a novel deep tissue chiral molecule sensing technique using polarization enhanced photoacoustics in the NIR II window. Photoacoustic-based Polarization Enhanced Optical Rotation Sensing (PAPEORS) system was developed to estimate the optical rotation parameter from the measured photoacoustic signal. This system can be further developed for enhancing metabolic imaging as there are quite a few chiral metabolites, for example, glucose studied in this work is a chiral metabolite that takes part in many metabolic reactions (Padmanabhan and Prakash 2024).

PAI can also be combined with other imaging modalities, to enhance the overall imaging capabilities and provide a cross-validation mechanism for the obtained information. Chen et al. combined MR with PAI and was able to provide a comprehensive view of the brain by integrating detailed anatomical information from MR with functional data like blood oxygenation from PAI (Chen et al. 2022). This cross-validated, multi-modal imaging platform opens up possibilities to get deeper insights into neurological disorders like Alzheimer's and Parkinson's.

Since image reconstruction is a crucial component of all imaging modalities, numerous studies have been conducted to enhance the photoacoustic image reconstruction component. One way to achieve significantly improved contrast-to-noise ratio (CNR) and structural similarity (SSIM) was by introducing a regional Laplacian method to integrate structural information from ultrasound images as soft priors for photoacoustic reconstruction. This multimodal approach enabled better visualization and reduced limited-view artifacts. The method demonstrates potential in vascular imaging, further enhancements, such as light fluence correction and addressing morphologic mismatches between ultrasound and photoacoustic slice thicknesses, could expand its clinical applicability in metabolic imaging (Yang et al. 2020). In contrast, Munjal et al. were able to enhance PAI by rectifying transducer-specific spatial impulse response (SIR) artifacts through the use of SALSA and FD-UNet deconvolution techniques. This demonstrated how deep learning models could be used to correct SIR artifacts and made it possible to use the same method to address several physical artifacts in PAI (Munjal and Prakash 2024). In conclusion, PAI is a promising tool for metabolic research, offering the potential to revolutionize our understanding of biological processes and disease mechanisms. Future advancements in technology will likely lead to even greater applications of PAI in this field.

Acknowledgments The authors would like to acknowledge the Department of Biotechnology (DBT)—Innovative Young Biotechnologist Award (IYBA) Program, and Science and Engineering Research Board (SERB)—Startup Research Grant (SRG).

Conflict of Interest The authors declare no conflict of interest in this study.

References

Apostolova I, Wedel F, Brenner W (2016) Imaging of tumor metabolism using positron emission tomography (PET). Recent Results Cancer Res 207:177–205

Brahimi-Horn MC, Pouysségur J (2007) Oxygen, a source of life and stress. FEBS Lett 581(19):3582–3591

Cha YJ, Kim E-S, Koo JS (2018) Amino acid transporters and glutamine metabolism in breast cancer. Int J Mol Sci 19(3):907. https://doi.org/10.3390/ijms19030907

Chatni MR, Xia J, Maslov KI, Guo Z, Wang K, Anastasio MA, Wang LV, Sohn R, Arbeit JM, Zhang Y, Xia Y (2012) Tumor glucose metabolism imaged in vivo in small animals with whole-body photoacoustic computed tomography. J Biomed Opt 17(7):076012. https://doi.org/10.1117/1.JBO.17.7.076012

Chen R, Huang S, Lin T et al (2021) Photoacoustic molecular imaging-escorted adipose photodynamic–browning synergy for fighting obesity with virus-like complexes. Nat Nanotechnol 16:455–465. https://doi.org/10.1038/s41565-020-00844-6

Chen Z, Gezginer I, Augath MA, Ren W, Liu YH, Ni R et al (2022) Hybrid magnetic resonance and optoacoustic tomography (MROT) for preclinical neuroimaging. Light Sci Appl 11(1):332

Dedon PC, Tannenbaum SR (2004) Reactive nitrogen species in the chemical biology of inflammation. Arch Biochem Biophys 423(1):12–22

Di Gialleonardo V, Wilson DM, Keshari KR (2016) The potential of metabolic imaging. In: Seminars in nuclear medicine, vol 46, No. 1. WB Saunders, pp 28–39

Diot G, Metz S, Noske A, Liapis E, Schroeder B, Ovsepian SV et al (2017) Multispectral optoacoustic tomography (MSOT) of human breast cancer. Clin Cancer Res 23(22):6912–6922

Eklund A, Nichols TE, Knutsson H (2016) Cluster failure: why fMRI inferences for spatial extent have inflated false-positive rates. Proc Natl Acad Sci 113(28):7900–7905

Feng H, Wang X, Chen J, Cui J, Gao T, Gao Y, Zeng W (2019) Nuclear imaging of glucose metabolism: beyond 18F-FDG, Contrast Media. Mol Imaging 2019:7954854, 12 pages. https://doi.org/10.1155/2019/7954854

Förstermann U, Sessa WC (2012) Nitric oxide synthases: regulation and function. Eur Heart J 33(7):829–837. https://doi.org/10.1093/eurheartj/ehr304

Gallez B, Jordan BF, Baudelet C, Misson PD (1999) Pharmacological modifications of the partial pressure of oxygen in murine tumors: evaluation using in vivo EPR oximetry. Magn Reson Med 42(4):627–630

Gialleonardo VD, Wilson DM, Keshari KR (2016) The potential of metabolic imaging. Semin Nucl Med 46(1):28–39. https://doi.org/10.1053/j.semnuclmed.2015.09.004

Goldenberg JM, Pagel MD (2019) Assessments of tumor metabolism with CEST MRI. NMR Biomed 32(10):e3943

Grossberg JA, Howard BM, Saindane AM (2019) The use of contrast-enhanced, time-resolved magnetic resonance angiography in cerebrovascular pathology. Neurosurg Focus 47:E3

Gujrati V, Mishra A, Ntziachristos V (2017) Molecular imaging probes for multi-spectral optoacoustic tomography. Chem Commun 53(34):4653–4672. https://doi.org/10.1039/C6CC09421J

Hai P, Imai T, Xu S et al (2019) High-throughput, label-free, single-cell photoacoustic microscopy of intratumoral metabolic heterogeneity. Nat Biomed Eng 3:381–391. https://doi.org/10.1038/s41551-019-0376-5

Han Z, MacCuaig WM, Gurcan MN, Claros-Sorto J, Garwe T, Henson C, Holter-Chakrabarty J, Hannafon B, Chandra V, Wellberg E, McNally LR (2023) Dynamic 2-deoxy-D-glucose-enhanced multispectral optoacoustic tomography for assessing metabolism and vascular hemodynamics of breast cancer. Photoacoustics 32:100531

He H, Fasoula NA, Karlas A, Omar M, Aguirre J, Lutz J et al (2023) Opening a window to skin biomarkers for diabetes stage with optoacoustic mesoscopy. Light: Sci Appl 12(1):231

Hu S (2016) Emerging concepts in functional and molecular photoacoustic imaging. Curr Opin Chem Biol 33:25–31. https://doi.org/10.1016/j.cbpa.2016.04.003

Hyder F, Kida I, Behar KL, Kennan RP, Maciejewski PK, Rothman DL (2001) Quantitative functional imaging of the brain: towards mapping neuronal activity by BOLD fMRI. NMR Biomed 14(7–8):413–431

Jain RK (2005) Normalization of tumor vasculature: an emerging concept in antiangiogenic therapy. Science 307:58–62. https://doi.org/10.1126/science.1104819

Jathoul A, Laufer J, Ogunlade O et al (2015) Deep in vivo photoacoustic imaging of mammalian tissues using a tyrosinase-based genetic reporter. Nat Photon 9:239–246. https://doi.org/10.1038/nphoton.2015.22

Karlas A, Reber J, Diot G, Bozhko D, Anastasopoulou M, Ibrahim T, Schwaiger M, Hyafil F, Ntziachristos V (2017) Flow-mediated dilatation test using optoacoustic imaging: a proof-of-concept. Biomed Opt Express 8:3395–3403

Karlas A, Kallmayer M, Fasoula N-A et al (2020) Multispectral optoacoustic tomography of muscle perfusion and oxygenation under arterial and venous occlusion: a human pilot study. J Biophotonics 13:e201960169. https://doi.org/10.1002/jbio.201960169

Karlas A, Pleitez MA, Aguirre J et al (2021) Optoacoustic imaging in endocrinology and metabolism. Nat Rev Endocrinol 17:323–335. https://doi.org/10.1038/s41574-021-00482-5

Kik CC, Slooff WM, Moayeri N, de Jong PA, Muijs SPJ, Öner FC (2022) Diagnostic accuracy of computed tomography angiography (CTA) for diagnosing blunt cerebrovascular injury in trauma patients: a systematic review and meta-analysis. Eur Radiol 32:2727–2738

Krishna MC, Matsumoto S, Saito K, Matsuo M, Mitchell JB, Ardenkjaer-Larsen JH (2013) Magnetic resonance imaging of tumor oxygenation and metabolic profile. Acta Oncol 52(7):1248–1256. https://doi.org/10.3109/0284186X.2013.819118

Lazzer S, Bedogni G, Lafortuna CL, Marazzi N, Busti C, Galli R, de Col A, Agosti F, Sartorio A (2010) Relationship between basal metabolic rate, gender, age, and body composition in 8,780 white obese subjects. Obesity 18:71–78. https://doi.org/10.1038/oby.2009.162

Lewis MR (2011) Looking through the vascular normalization window: timing antiangiogenic treatment and chemotherapy with 99mTc-Annexin A5. J Nucl Med 52(11):1670–1672. https://doi.org/10.2967/jnumed.111.097725

Li J, Chekkoury A, Prakash J et al (2020) Spatial heterogeneity of oxygenation and haemodynamics in breast cancer resolved in vivo by conical multispectral optoacoustic mesoscopy. Light Sci Appl 9:57. https://doi.org/10.1038/s41377-020-0295-y

Lin L, Hu P, Shi J, Appleton CM, Maslov K, Li L et al (2018) Single-breath-hold photoacoustic computed tomography of the breast. Nat Commun 9(1):2352

Łmigielska-Kuzia J, Boćkowski L, Sobaniec W, Kułak W, Sendrowski K (2010) Amino acid metabolic processes in the temporal lobes assessed by proton magnetic resonance spectroscopy (1H MRS) in children with down syndrome. Pharmacol Rep 62(6):1070–1077

Lv J, Lan H, Qin A et al (2024) Dynamic synthetic-scanning photoacoustic tracking monitors hepatic and renal clearance pathway of exogeneous probes in vivo. Light Sci Appl 13:304. https://doi.org/10.1038/s41377-024-01644-6

Munjal I, Prakash J (2024) Deep-learning based deconvolution for correcting spatial impulse response of transducer in optoacoustic tomography. IEEE Trans Instrum Meas 73:1–12. https://doi.org/10.1109/TIM.2024.3417542

Najjar AM, Johnson JM, Schellingerhout D (2018) The emerging role of amino acid PET in neuro-oncology. Bioengineering 5(4):104

Neuschmelting V, Kim K, Malekzadeh-Najafabadi J, Jebiwott S, Prakash J, Scherz A et al (2018) WST11 vascular targeted photodynamic therapy effect monitoring by multispectral optoacoustic tomography (MSOT) in mice. Theranostics 8(3):723–734

Nievelstein RA, Quarles van Ufford HM, Kwee TC, Bierings MB, Ludwig I, Beek FJ, de Klerk JM, Mali WP, de Bruin PW, Geleijns J (2012) Radiation exposure and mortality risk from CT and PET imaging of patients with malignant lymphoma. Eur Radiol 22(9):1946–1954. https://doi.org/10.1007/s00330-012-2447-9

Padmanabhan S, Prakash J (2024) Deep tissue sensing of chiral molecules using polarization enhanced Photoacoustics. arXiv [Physics.Med-Ph]. Retrieved from http://arxiv.org/abs/2401.10812

Pleitez MA, Khan AA, Soldà A et al (2020) Label-free metabolic imaging by mid-infrared optoacoustic microscopy in living cells. Nat Biotechnol 38:293–296. https://doi.org/10.1038/s41587-019-0359-9

Reber J, Willershäuser M, Karlas A, Paul-Yuan K, Diot G, Franz D et al (2018) Non-invasive measurement of Brown fat metabolism based on optoacoustic imaging of hemoglobin gradients. Cell Metab 27(3):689–701.e4. https://doi.org/10.1016/j.cmet.2018.02.002

Reinhardt CJ, Zhou EY, Jorgensen MD, Partipilo G, Chan J (2018) A Ratiometric Acoustogenic probe for in vivo imaging of endogenous nitric oxide. J Am Chem Soc 140(3):1011–1018. https://doi.org/10.1021/jacs.7b10783

Ross B, Bluml S (2001) Magnetic resonance spectroscopy of the human brain. Anat Rec 265(2):54–84

Sánchez López de Nava A, Raja A (2025) Physiology, metabolism [Updated 2022 Sep 12]. In: StatPearls [Internet]. StatPearls Publishing, Treasure Island (FL). Available from: https://www.ncbi.nlm.nih.gov/books/NBK546690/

Shao Q, Ashkenazi S (2015) Photoacoustic lifetime imaging for direct in vivo tissue oxygen monitoring. J Biomed Opt 20(3):036004. https://doi.org/10.1117/1.JBO.20.3.036004

Sharma R, Seo JW, Kwon S (2014) In vivo imaging of nitric oxide by magnetic resonance imaging techniques. J Nanomater 2014(1):523646

Shishido F, Uemura K, Inugami A et al (1996) Cerebral oxygen and glucose metabolism and blood flow in mitochondrial encephalomyopathy: a PET study. Neuroradiology 38:102–107. https://doi.org/10.1007/BF00604789

Stoffels I et al (2015) Metastatic status of sentinel lymph nodes in melanoma determined noninvasively with multispectral optoacoustic imaging. Sci Transl Med 7:317ra199. https://doi.org/10.1126/scitranslmed.aad1278

Stuker F, Ripoll J, Rudin M (2011) Fluorescence molecular tomography: principles and potential for pharmaceutical research. Pharmaceutics 3(2):229–274. https://doi.org/10.3390/pharmaceutics3020229

Uluç N, Glasl S, Gasparin F, Yuan T, He H, Jüstel D et al (2024) Non-invasive measurements of blood glucose levels by time-gating mid-infrared optoacoustic signals. Nat Metab 6(4):678–686

Walenta S, Schroeder T, Mueller-Klieser W (2002) Metabolic mapping with bioluminescence: basic and clinical relevance. Biomol Eng 18(6):249–262

Warburg O, Wind F, Negelein E (1927) The metabolism of tumors in the body. J Gen Physiol 8:519–530. https://doi.org/10.1085/jgp.8.6.519

Yang H, Jüstel D, Prakash J, Karlas A, Helfen A, Masthoff M et al (2020) Soft ultrasound priors in optoacoustic reconstruction: improving clinical vascular imaging. Photoacoustics 19:100172. https://doi.org/10.1016/j.pacs.2020.100172

Yang F-K, Cao J, Zhang T, Jiang H-X, Cui H-B, Wang K (2024) Dual-activated photoacoustic probe for reliably detecting hydroxyl radical in ischemic cardiovascular disease in mouse and human samples. ACS Sensors 9(10):5445–5453. https://doi.org/10.1021/acssensors.4c01665

Yoshihara T, Hirakawa Y, Hosaka M, Nangaku M, Tobita S (2017) Oxygen imaging of living cells and tissues using luminescent molecular probes. J Photochem Photobiol C: Photochem Rev 30:71–95

Yu Q, Huang S, Wu Z, Zheng J, Chen X, Nie L (2020) Label-free visualization of early cancer hepatic micrometastasis and intraoperative image-guided surgery by photoacoustic imaging. J Nucl Med 61(7):1079–1085. https://doi.org/10.2967/jnumed.119.233155

Yun M, You SH, Nguyen VH, Prakash J, Glasl S, Gujrati V et al (2021) Reporter gene-based optoacoustic imaging of E. Coli targeted colon cancer in vivo. Sci Rep 11(1):24430

Zhu B, Li H, Xie C et al (2023) Photoacoustic microscopic imaging of cerebral vessels for intensive monitoring of metabolic acidosis. Mol Imaging Biol 25:659–670. https://doi.org/10.1007/s11307-023-01815-8

Laser Speckle Contrast Imaging in Preclinical Research: Recent Advances and Emerging Applications

Anastasiia Neganova and Dmitry D. Postnov

Abstract

Laser Speckle Contrast Imaging (LSCI) is a high-speed, wide-field optical imaging technique that enables real-time characterization of microvascular blood flow. This chapter provides a concise overview of LSCI's core principles, tracing its development from simplified single-exposure approaches to more sophisticated models accounting for non-ergodic scattering. We highlight recent methodological progress—from refined theoretical models and specialized hardware configurations to novel computational strategies—all aimed at enhancing measurement accuracy, precision, and depth penetration. In addition, we examine emerging applications across multiple organ systems, including the brain, retina, kidney, and liver. These advancements underscore LSCI's growing prominence in preclinical research and signal its potential for broader adoption in microvascular research.

Keywords

Laser speckle contrast imaging · Blood flow · Accuracy · Precision · Pre-clinical research

A. Neganova · D. D. Postnov (✉)
Center of Functionally Integrative Neuroscience, Department of Clinical Medicine, Aarhus University, Aarhus, Denmark
e-mail: aneganova@cfin.au.dk; dpostnov@cfin.au.dk

I. F. Cengiz et al. (eds.), *Bioimaging Modalities in Bioengineering*,
Biomaterials, Bioengineering and Sustainability 9,
https://doi.org/10.1007/978-3-031-99410-4_26

1 Introduction

Laser Speckle Contrast Imaging (LSCI) is an optical imaging technique that enables rapid, wide-field visualization of light-scattering particles' dynamics, most commonly flowing red blood cells (RBCs) in microvasculature. By illuminating a region of interest with coherent laser light and recording the resulting interference (speckle) pattern on a camera sensor, LSCI exploits the temporal blurring of speckles to deduce the relative motion of scattering particles. The origins of LSCI trace back to early findings showing that speckle patterns, initially deemed noise, could be quantified to measure flow, laying the groundwork for its broad range of applications. This chapter presents an overview of LSCI's fundamental principles and recent methodological and technological advances, followed by a discussion of its emerging applications in pre-clinical research.

The theoretical foundations of laser speckle statistics were first established in the early 1970s (Goodman 1975), followed by a proposal to use speckle patterns for noninvasive assessment of microcirculatory perfusion (Stern 1975). In 1981, Fercher and Briers demonstrated that time-averaged speckle patterns, captured with sufficiently long exposure, could be analyzed to map blood flow (Fercher and Briers 1981). By high-pass filtering a photograph of a laser-illuminated human retina, they produced the first in vivo speckle contrast image (Fig. 1a), thus conceiving what is now referred to as LSCI. Improved computing and sensor technologies in the 1990s made real-time flow measurements feasible (Briers and Webster 1996). By

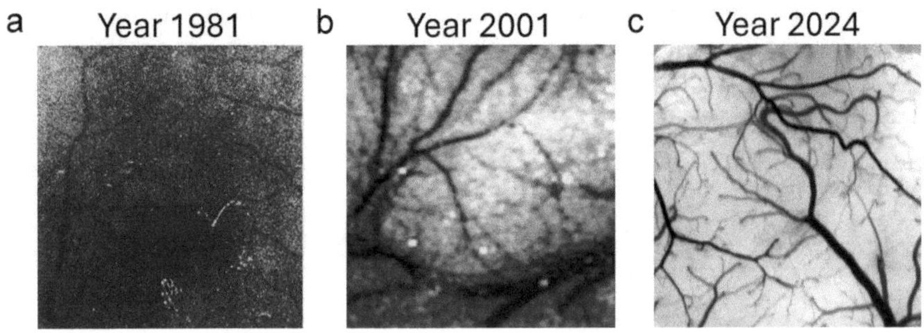

Fig. 1 Examples of laser speckle contrast images made in different years. (**a**) the first in vivo laser speckle contrast image published by in Fercher and Briers (1981). (**b**) the first contrast image of mouse cortex published in Dunn et al. (2001). (**c**) typical contrast image of mouse cortex obtained by our group, 2024. Although the LSCI progress is evident, it is essential to note that the difference between images does not entirely reflect improvement in the LSCI data quality as the processing steps, such as frame averaging, may vary. The image in panel A is adapted from Optics Communications, 37(5), Fercher AF, Briers JD. Flow visualization by means of single-exposure speckle photography, 326–330, copyright ©1981 Elsevier, with permission of Elsevier. The image in panel (**b**) is adapted from the Journal of Cerebral Blood Flow & Metabolism, 21(3), Dunn AK, Bolay H, Moskowitz MA, Boas DA. Dynamic imaging of cerebral blood flow using laser speckle, copyright ©2001 The International Society for Cerebral Blood Flow and Metabolism, with permission of Sage Publications Journal of Cerebral Blood Flow and Metabolism

the early 2000s, theoretical insights clarified how speckle contrast reflects flow dynamics (Zakharov et al. 2006; Bandyopadhyay et al. 2005), while practical implementations approached quantitative assessments (Yuan et al. 2005; Völker et al. 2005; Parthasarathy et al. 2008).

The working principle of LSCI can briefly be described as follows - coherent light scattered by RBCs produces an electric field, which can be characterized using the normalized field autocorrelation function $g_1(\tau)$:

$$g_1(\tau) = \frac{\langle E(t) \cdot E^*(t+\tau) \rangle}{\langle E(t) \cdot E^*(t) \rangle}, \tag{1}$$

where E is the electric field, t is time, and τ is the time lag. A common assumption is that $g_1(\tau)$ decays exponentially:

$$g_1(\tau) = e^{-\frac{\tau}{\tau_c}}, \tag{2}$$

where τ_c is the characteristic decorrelation time, which is inversely related to particle velocity in the case of linear motion. In practice, photodetectors and cameras record intensity rather than the electric field, so the intensity autocorrelation function $g_2(\tau)$ has to be used instead of $g_1(\tau)$:

$$g_2(\tau) = \frac{\langle I(t)I(t+\tau) \rangle}{\langle I(t) \rangle^2}, \tag{3}$$

where I is the recorded light intensity. Under the assumption of ergodicity, the Siegert relation connects $g_1(\tau)$ and $g_2(\tau)$:

$$g_2(\tau) = 1 + \beta |g_1(\tau)|^2, \tag{4}$$

where β accounts for the correlation loss due to speckle averaging, polarization, and laser coherence. Although measuring $g_2(\tau)$ directly would allow one to compute τ_c, in most practical cases, it will not be possible due to temporal averaging of the speckle intensity caused by relatively long exposure times. This drawback is elegantly converted into an advantage by using the estimate of spatial variance σ_s^2 in speckle patterns averaged over the exposure time T (Goodman 1975; Bandyopadhyay et al. 2005):

$$\sigma_s^2(T) = \frac{2}{T} \int_0^T (1 - \frac{\tau}{T}) C_t(\tau) d\tau, \tag{5}$$

where $C_t(\tau)$ is the intensity autocovarience. For stationary processes it can be defined as:

$$C_t(\tau) = \langle I(t)I(t+\tau) \rangle - \langle I(t) \rangle^2. \tag{6}$$

Combining (3), (4), and (6):

$$C_t(\tau) = (g_2(\tau) - 1) < I >_t^2 = \beta |g_1(\tau)|^2 < I >_t^2, \tag{7}$$

and substituting it into (8) we arrive to

$$\sigma_s^2(T) = \frac{2\beta < I >_t^2}{T} \int_0^T (1 - \frac{\tau}{T}) e^{-\frac{\tau}{\tau_c}} d\tau. \tag{8}$$

Finally, under the ergodicity assumption, it can be simplified to

$$K = \frac{\sigma_s(T)}{< I >} = \beta^{0.5} \left\{ \frac{\tau_c}{T} + \frac{\tau_c^2}{2T^2} \left[exp\left(- \frac{2T}{\tau_c} \right) - 1 \right] \right\}^{0.5}. \tag{9}$$

where K is speckle contrast. Several studies have challenged the ergodicity assumption, stating that, in biomedical applications, static scattering is non-negliblie (Boas and Yodh 1997; Zakharov et al. 2006; Parthasarathy et al. 2008), making it non-ergodic. They have suggested a modified version of the Eq. (9) that accounts for the electric field comprising both fluctuating and static components:

$$K = \beta^{0.5} \left\{ \rho^2 \frac{e^{-2x} - 1 + 2x}{2x^2} + 4\rho(1 - \rho) \frac{e^{-x} - 1 + x}{x^2} + (1 - \rho)^2 \right\}^{0.5}, \tag{10}$$

where $x = T/\tau_c$ and ρ is the fraction of dynamically scattered light. Although one can numerically solve the Eqs. (9) or (10) for τ_c, such an approach requires substantial computation and estimates of β and ρ. A simplified form of the equation is generally used to approximate perfusion measurements via the blood flow index (BFI) and expedite real-time imaging, assuming $T \gg \tau_c$ and $\rho = 1$ (Cheng and Duong 2007):

$$K \approx \left(\beta \frac{\tau_c}{T} \right)^{0.5}, \tag{11}$$

$$BFI = 1/K^2 \approx \frac{1}{\beta} \frac{T}{\tau_c}. \tag{12}$$

Conveniently, since β depends on the system configuration and does not change within a single experiment, Eq. (12) allows β-independent estimation of the relative changes in blood flow index, which became the most widely used outcome of LSCI measurements.

Such simplicity facilitated the widespread adoption of LSCI. Building on the early pioneering studies (Dunn et al. 2001) (Fig. 1b), LSCI soon became indispensable for monitoring cerebral blood flow during functional activation and under pathological conditions such as stroke, traumatic brain injury, or migraine. These early demonstrations highlighted LSCI's potential, and to date, the technique has been applied to virtually every organ for which blood perfusion is of research interest.

2 Recent Advances in Laser Speckle Contrast Imaging

In reviewing recent LSCI advancements, we focus on three principle objectives mentioned above: accuracy of contrast interpretation, measurement precision, and depth-resolved imaging. The first relates to more comprehensive models connecting contrast to decorrelation time, including accounting for non-ergodic conditions and alternative approaches for characterizing speckle blurring. The second addresses how system configuration and key components affect the signal-to-noise ratio when tracking perfusion changes and how to optimize them. Finally, we highlight recent strategies for imaging deeper structures, usually involving hardware or experimental modifications.

2.1 Accuracy and Interpretation

The debate on the appropriate model for interpreting laser speckle fluctuations has persisted for decades. Factors such as static scattering (Parthasarathy et al. 2008, 2010; Zakharov 2017; Zakharov et al. 2006, 2009) and correlation loss (Postnov et al. 2019; Kirkpatrick et al. 2008) alter the relation between $g_2(\tau)$ and $g_1(\tau)$. The form of $g_1(\tau)$ further depends on whether the light experiences single or multiple scattering and whether particle motion is ordered or unordered (Bonner and Nossal 1981; Duncan and Kirkpatrick 2008; Davis et al. 2016; Postnov et al. 2020). Although $g_1(\tau) = \exp(-\tau/\tau_c)$ is commonly assumed, recent studies have shown that it applies primarily to vessels of certain sizes, while in most other cases, $g_1(\tau) = \exp(-\sqrt{\tau/\tau_c})$, which corresponds to multiple scattering for unordered motion, is more accurate (Davis et al. 2016; Postnov et al. 2020; Zheng and Mertz 2022a; Yi et al. 2022). Liu et al. (2021a) have derived an updated contrast model and quantified the errors resulting from an incorrect choice of $g_1(\tau)$ or the model parameters:

$$K = \beta^{0.5}\left\{\rho^2 \frac{e^{-2\sqrt{x}}(4x + 6\sqrt{x} + 3) + 2x - 3}{2x^2} + 8\rho(1-\rho)\frac{e^{-\sqrt{x}}(2x + 6\sqrt{x} + 6) + x - 6}{x^2} + (1-\rho)^2\right\}^{0.5}. \tag{13}$$

The proposed model improved the accuracy of flow estimates by up to 100% for small vessels while introducing relatively smaller error regions, which are better described by conventional $g_1(\tau)$ form (Liu et al. 2021a). Furthermore, it has improved accuracy for relative changes in blood flow index in the presence of unknown static scattering. It is important to note that just as the original model (Fercher and Briers 1981), Eq. (13) was derived for contrast calculated over space. If $\rho = 1$, meaning that the system is ergodic and there is no static scattering, temporal and spatial statistics would naturally converge to the same equations (Cheng et al. 2008). For non-ergodic cases, however, one should account for the difference in how static scattering affects temporal and spatial statistics (Liu et al. 2024).

Although these refined models can theoretically yield the most accurate results, they demand considerable computation and prior knowledge of which model best fits each pixel. Recently, an alternative, model-agnostic approach, was proposed (Zheng and Mertz 2022a), using generalized definitions of decorrelation time and contrast:

$$K(\tau) = \sqrt{\frac{\langle I(t+\tau)I(t)\rangle - \langle I\rangle^2}{\langle I\rangle^2}}, \tag{14}$$

$$\tau_c = \frac{2}{\beta}\int_0^{\infty} K^2(\tau)d\tau \approx \frac{T}{\beta}\sum_{m=0}^{M-1}(K_m^2 + K_{m+1}^2). \tag{15}$$

This method does not require computationally intensive numerical fitting and completely circumvents the problem of model selection in the case of ergodic measurements but becomes less robust, depending more strongly on $g_1(\tau)$ as static scattering increases (Zheng and Mertz 2022a):

$$\tau_c \approx \frac{1}{\beta\left(\rho^2 + 2\rho(1-\rho)\eta\right)}\int_{-\infty}^{\infty} K^2(\tau)d\tau, \tag{16}$$

where $\eta = \frac{\int_{-\infty}^{\infty} \mathrm{Re}[g_1(\tau)]d\tau}{\int_{-\infty}^{\infty} |g_1(\tau)|^2 d\tau}$ is a parameter defining the sharpness of field autocorrelation function. Unlike conventional LSCI analysis, the model-agnostic approach is most optimal when $T \lesssim \tau_c$, while for $T \gg \tau_c$, it converges to the original simplified model described in the Eq. (11).

Most recently, the same authors have modified their method, now utilizing spatial statistics (Zheng et al. 2024). Unlike conventional contrast estimation, it does not suffer from noise bias, producing more accurate results in noisy and low-light recordings. A similar improvement was demonstrated utilizing a two-dimensional entropy algorithm to obtain better estimates of contrast, which are less biased by the measurements noise and insufficient speckle statistics (Liu et al. 2021b).

Equations (10), (13), and (16) offer advanced theoretical descriptions, yet they rely on accurately measuring static scattering to ensure precise contrast interpretation. Multi-Exposure Laser Speckle Contrast Imaging (MESI), introduced in Parthasarathy et al. (2008, 2010), tackles this requirement by collecting speckle patterns at multiple exposure times and numerically fitting Eq. (10) to estimate τ_c,ρ

and β. Although MESI substantially enhances quantitative accuracy, it demands more complex instrumentation and reduces temporal resolution. A more recent version of the technique (Dragojević et al. 2015) uses a high-speed camera to capture frames and short exposure time, which are then averaged to mimic varying laser pulse lengths used in the original MESI. Most recently, machine learning analysis of MESI data (Fredriksson et al. 2019; Hultman et al. 2020; Stebakov et al. 2021; Hao et al. 2023), has abolished the need for numerical fitting while showing improved speed and accuracy in calculating flow dynamics. Altogether, these approaches mark an essential leap toward a truly quantitative LSCI, closing the gap between theoretical potential and experimental feasibility.

2.2 Precision and Signal-to-noise Ratio

Accurately characterizing rapid changes in blood flow using Laser Speckle Contrast Imaging demands precision that often surpasses the importance of absolute accuracy. In particular, achieving high precision and signal-to-noise ratio is crucial for detecting the subtle and transient hemodynamic events associated with functional activation, vasomotion, cardiac pulsatility, or pharmacological interventions. In the following paragraphs, we discuss how the parameters of a typical LSCI system (Fig. 2a)—speckle size, polarization, exposure time, illumination characteristics, and camera noise—can be optimized to enhance measurement precision.

Several studies have highlighted the significance of speckle size relative to the pixel dimensions for enhancing the signal-to-noise ratio. Kirkpatrick et al. (2008) argued that speckles should be at least twice the pixel size to maximize image contrast and thereby improve SNR. This notion was later contested, as smaller speckles were observed to yield better measurement precision (Qiu et al. 2013). A recent study González Olmos et al. (2023) has confirmed this observation and pointed to a nuanced relationship: while larger speckles may enhance the accuracy of the contrast estimation, they also reduce the number of independent speckle samples per pixel, ultimately degrading precision due to insufficient speckle statistics. Therefore, for spatial laser speckle contrast analysis, it would be recommended to use a speckle-to-pixel size ratio of 1 and smaller to achieve the best precision, in combination with either β-calibration (Liu et al. 2021a; González Olmos et al. 2023; Boas and Dunn 2010) or a bias correcting approach (Zheng and Mertz 2022b) to preserve overall accuracy.

Alongside the speckle size, the polarization of the detected light has been known as a crucial factor for image quality, depth penetration, and possibly the signal-to-noise ratio since the development of Laser Speckle Contrast Imaging. Early studies noted that non-polarized detection leads to correlation loss (Kirkpatrick et al. 2008; Briers et al. 2013), effectively lowering β in Eq. (9) and thereby reducing speckle contrast range and heighting susceptibility to noise. Recent experimental and theoretical work (Colin et al. 2022; Amini et al. 2024; González Olmos et al. 2023; Akther et al. 2023) have reinforced these observations, confirming that introducing a polarizer in the system (Fig. 2a) and cross-polarizing the detected

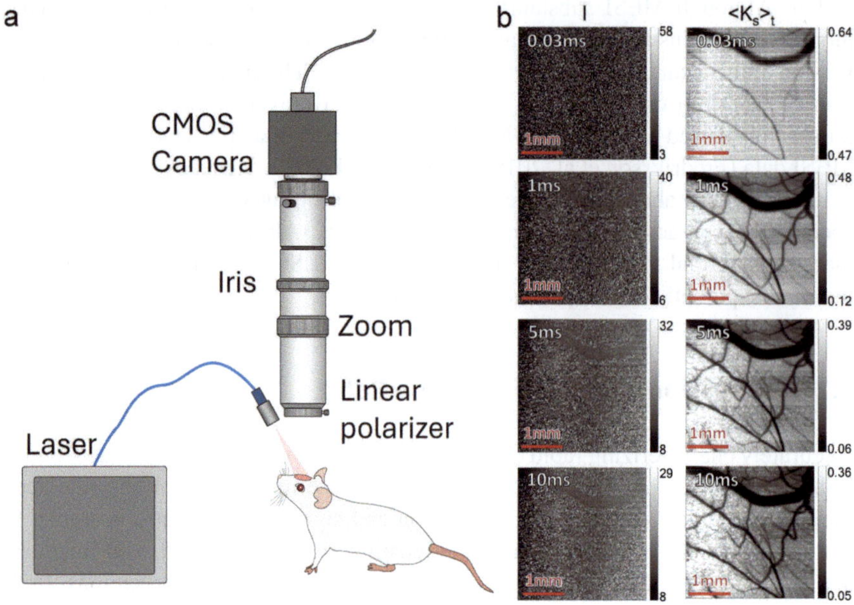

Fig. 2 Key components and parameters of a typical LSCI system and effect of different exposure times on intensity and contrast images. (**a**) a typical LSCI system consists of a laser, camera, and optics, in this case - a machine vision lens with integrated zoom, iris, and a linear polarizer mounted in front of the objective. (**b**) examples of intensity and contrast images of the mouse brain cortex captured at an exposure time $T = 0.03$, 1, 5, and 10 milliseconds. One can see that longer exposure times make measurements more sensitive to flow in small vessels (slow dynamics) while reducing the intensity dynamic range

light yields higher β, deeper penetration, fewer specular reflections, and ultimately a better SNR.

In addition to polarization and speckle size, exposure time is another parameter that the end-user can easily adjust and is decisive for achieving high contrast sensitivity to a wide range of flow velocities (Fig. 2b). An exposure time of about 5 ms was suggested as optimal for brain perfusion imaging (Yuan et al. 2005). Yet, subsequent theoretical developments and practical insights revealed that shorter exposures may be preferable to increase sensitivity to rapid hemodynamic events (Liu et al. 2021a; González Olmos et al. 2023) and even enable more quantitative and accurate analysis approaches (Dragojević et al. 2015; Zheng and Mertz 2022a). Conversely, prolonged exposure times can diminish absolute and relative sensitivity to rapid flow dynamics, especially when substantial static scattering is present (González Olmos et al. 2023).

Although not adjustable during the experiments, the choice of the light source and the camera can be crucial to achieving high-precision measurements with LSCI. A recent study has demonstrated that the spectral width of the laser illumination plays a vital role in defining measurement precision (Postnov et al. 2019). The conventional semiconductor diodes have a larger than desired spectral width, effectively lowering β parameter and strongly reducing the speckle contrast range. The study suggests that using stabilized laser diodes with spectral width below 1 nm would result in up to fourfold improvement in SNR in biomedical applications. Finally, although various sources of camera-related noise, such as readout and shot noise, were once considered a significant limitation, these effects are largely mitigated in most LSCI applications using modern complementary metal-oxide semiconductor (CMOS) sensors (González Olmos et al. 2023; Cheng et al. 2024; Zilpelwar et al. 2022). Camera noise becomes more relevant only when illumination intensity is weak—either due to low-power lasers or very short exposures. Hence, investing in more expensive or higher-quality camera systems may yield limited improvements in LSCI precision.

It is worth noting that various analysis approaches, such as those mentioned in the previous section (Zheng et al. 2024; Liu et al. 2021b; Stebakov et al. 2021; Fredriksson et al. 2019; Hultman et al. 2020; Hao et al. 2023; Guilbert and Desjardins 2022), can further improve measurement precision by mitigating noise in speckle contrast estimates. These methodological refinements and advancements in hardware—spanning laser illumination, camera sensors, and computational algorithms—now enable higher pixel densities, faster frame rates, and substantially reduced noise. As a result, LSCI can now reliably capture rapid and subtle flow changes, underscoring its expanding role in quantitative biomedical imaging.

2.3 Deep Imaging

Depth resolution and deep imaging remain among the most desired capabilities for Laser Speckle Contrast Imaging. These attributes are where LSCI currently lags behind other imaging modalities, such as optical coherence tomography or two-photon microscopy, which offer superior penetration and depth-specific resolution (Davis et al. 2014). Nonetheless, substantial progress has been made to narrow this gap, driven by innovations in transmissive imaging, multifocal methods, and endoscopic techniques.

Transmissive imaging in LSCI implies directing coherent light through the tissue and capturing the transmitted light on the opposite side, enhancing sensitivity to deep regions. Unlike the backscattering-based LSCI, transmissive setups enable higher signal-to-background ratios by minimizing superficial light interactions and isolating signals from deeper tissue layers (Li et al. 2021; Chen et al. 2023a,b). Li et al. (2021) have employed Monte Carlo simulations and in vivo experiments to validate the superior performance of transmissive-detected LSCI (TR-LSCI) in imaging thick tissues, highlighting its robustness for subcutaneous and muscle layers. Despite these advancements, transmissive imaging faces challenges, such as resolution degradation due to multiple scattering and out-of-focus blurring. These issues were recently addressed by combining transmissive LSCI with multifocal imaging and deep-learning algorithms for 3D reconstruction of vasculature and blood flow, achieving accurate depth-specific blood flow measurements with improved spatiotemporal resolution (Chen et al. 2023a,b). Notably, a multifocal approach does not necessarily require a transmissive illumination and was recently used to enhance focus accuracy and quantify flow at depths up to 300 μm (Zheng et al. 2021).

Endoscopic LSCI systems and special surgical preparations have been proposed as an alternative solution for applications demanding access to specific regions deep in the tissue. These methods physically bypass superficial layers, allowing the imaging system to focus on the target tissue. Microendoscopes, for example, have been used in tandem with LSCI to visualize deep brain regions (Chen et al. 2018). At the same time, laparoscopic setups offer real-time perfusion imaging from within the body cavity, such as during surgical procedures (Wildeboer et al. 2022). The surgical approach utilizes deep optical windows, which enable longitudinal monitoring of, for example, hippocampal blood flow using a conventional LSCI system (Mikkelsen et al. 2024).

Most recently, short-separation speckle contrast optical spectroscopy (ss-SCOS) has emerged as a modification of traditional LSCI to address the limitations of probing deeper tissue. Unlike wide-field LSCI, ss-SCOS uses point illumination and spatially separated detectors to increase sensitivity to deeper structures by minimizing the confounding effects of superficial scattering. Despite these modifications, it retains the essence of Laser Speckle Contrast Imaging since the recorded speckle pattern is still analyzed for contrast. It was demonstrated that ss-SCOS can capture both rapid cerebral blood flow and slower cellular dynamics in in vivo mouse model (Liu et al. 2023) by leveraging a fiber-based array with controllable source-detector separation distances to provide depth-specific measurements. This approach enhances the detection of deeper-layer signals and opens avenues for investigating metabolic events and cellular responses to injury—particularly relevant in conditions such as stroke or other pathologies involving cellular motility and metabolism.

Altogether, these advances in transmissive imaging, multifocal methods, endoscopic techniques, and ss-SCOS significantly improve LSCI's ability for deep tissue imaging. While LSCI does not yet match the full-depth capabilities of OCT or two-photon microscopy, the ongoing integration of machine learning and advanced optical setups aims to expand LSCI's depth-specific imaging potential further. This progress positions LSCI as an increasingly competitive modality for deep tissue studies in biomedical research.

3 Emerging Applications of LSCI in Preclinical Research

Building on its high spatiotemporal resolution and straightforward implementation, Laser Speckle Contrast Imaging has demonstrated broad utility in preclinical studies, spanning diverse organs, animal models, and disease states. In the following sections, we highlight recent investigations that harness LSCI's capabilities and new applications, underscoring its growing impact on biomedical research.

3.1 Brain

Preclinical research of cerebral hemodynamics is essential for unraveling the fundamental mechanisms of neurovascular interactions and brain-related disease progression, from stroke to neurodegenerative conditions. Over the past few decades, the need for accurate assessment of brain blood flow and the associated challenges have been driving the progress in multiple imaging modalities, including Laser Speckle Contrast Imaging (Fig. 1). Since the first applications in Dunn et al. (2001), LSCI has proven to be a versatile tool for investigating cerebral perfusion, particularly in murine models. By offering high spatiotemporal resolution of blood flow dynamics, LSCI has opened new avenues in stroke research, neurovascular coupling studies, and investigations of vasomotion and vasoreactivity. Recent improvements in LSCI's accuracy and precision have positioned it as a strong alternative—if not a complete replacement—to the historically more established laser Doppler flowmetry.

LSCI has become indispensable in stroke research, facilitating dynamic visualization of ischemia severity over the cortex area. For instance, real-time LSCI was employed to develop a chronic awake mouse model of targeted pial vessel occlusion via photothrombosis, enabling precise tracking of localized ischemic events and subsequent recovery (Sunil et al. 2020). Most recently, continuous cerebral blood flow monitoring with LSCI became key in studies involving therapeutic interventions (El-Awaisi et al. 2024; Huang et al. 2022; Erdener et al. 2021), and establishing the association between ischemia degree and tissue damage (Yin et al. 2022). Finally, as noted in the previous section, short-separation speckle contrast optical spectroscopy was demonstrated capable of detecting cellular motility changes during stroke (Liu et al. 2023), highlighting future opportunities for applying laser speckle contrast analysis in stroke research and diagnostics.

LSCI's high spatiotemporal resolution is likewise critical for studying functional hyperemia, the localized increase in blood flow triggered by neuronal activity (Boas and Dunn 2010). Among the recent advances, LSCI was used to show that endothelial Piezo1 channel activation acts as a mechano-feedback system to regulate blood flow recovery, advancing our understanding of neurovascular coupling mechanisms (Lim et al. 2024). It has also played a crucial role in revealing exaggerated neurovascular coupling in a familial hemiplegic migraine model, suggesting potential therapeutic targets for correcting vascular dysfunction (Staehr et al. 2024). Another type of vasoreactivity - spontaneous rhythmic oscillations in blood vessel diameters, called vasomotion, has recently attracted attention as a possible mechanism regulating cerebral perfusion (Broggini et al. 2024). LSCI is one of the main tools to study vasomotion on par with Two-Photon Microscopy because it provides full-field perfusion monitoring with high temporal resolution (Golubova et al. 2023). Most recently, it was used to examine long-term alterations in vasomotion dynamics, revealing how these oscillations contribute to maintaining cerebral perfusion under various conditions (Broggini et al. 2024).

Beyond the examples above, LSCI has been applied in various other contexts to expand our understanding of cerebral blood flow, such as sex-associated perfusion differences in disease models (Salinero et al. 2020), or cerebral blood flow changes in response to global ischemic events (He et al. 2019). Finally, aside from being used for basic research, ongoing clinical and preclinical studies investigate how it can be translated into a practical surgical tool (Mangraviti et al. 2020; Goldberg et al. 2024).

3.2 Retina

Retinal microvasculature carries early information about eye diseases and is considered the most accessible outpost of the central nervous system and systemic vascular conditions. Variations in retinal blood flow can signal the onset of diseases ranging from diabetic retinopathy and glaucoma to systemic disorders like hypertension and diabetes (Neganova et al. 2016, 2017; Patel and Lipinski 2020; Patel et al. 2021).

Retinal blood flow imaging was the first in vivo demonstration of LSCI (Fercher and Briers 1981; Briers and Fercher 1982), and it remains one of the technique's most promising clinical targets, as it offers a non-invasive assessment of ocular perfusion in humans. Over the past two decades, notable strides have been made to adapt LSCI for preclinical retinal studies. In particular, methods to mitigate signal-quality reductions associated with unevenly bright illumination in animal retinal imaging were developed (Cheng et al. 2008; Cheng et al. 2009). It was followed by studies demonstrating that LSCI can capture light-evoked and pressure-evoked changes in retinal blood flow in strong agreement with more invasive confocal microscopy and microsphere-based results (Srienc et al. 2010; Wang et al. 2012).

Development of an endoscope-based LSCI system for retinal imaging (Ponticorvo et al. 2013) has made retinal blood flow imaging in rodents more accessible. Several studies employed this methodology to investigate how systemic stimuli such as vasoconstrictors, vasodilators, and anesthesia affect microvascular dynamics in the retina (Neganova et al. 2016, 2017). Most recently, (Patel and Lipinski 2020; Patel et al. 2021) have introduced retinal LSCI systems featuring improved illumination and collection optics, enhancing signal and image quality. Using these refinements, they detected early hemodynamic changes in an Alzheimer's disease mouse model—characterized by elevated pulsatility and resistance indices—thus identifying possible retinal biomarkers for disease progression (Tate et al. 2024). Despite its importance to basic and clinical research, accurately assessing retinal perfusion, especially in mouse animal models, remains a challenging task due to the constraints introduced by the optical properties of the eye and rapid ocular movements. Nonetheless, continuous advancements in LSCI data analysis, hardware, and experimental approaches are steadily improving the feasibility of retinal perfusion assessments in preclinical models, opening new opportunities for completely non-invasive translational research.

3.3 Kidney

Renal blood flow is uniquely governed by coupled autoregulation mechanisms (myogenic response and tubuloglomerular feedback) that operate over different timescales, ensuring efficient filtration, solute reabsorption, and homeostasis in a wide range of systemic conditions (Bezemer et al. 2010; Holstein-Rathlou et al. 2011; Scully et al. 2013; Postnov and Marsh 2022). Studying the complex perfusion dynamics arising from these regulating processes is vital for understanding renal function as well as its role in hypertension or chronic kidney disease. Laser speckle contrast imaging has recently emerged as a tool to address this need by providing noninvasive, high-resolution mapping of blood flow oscillations across broad regions of the renal cortex.

LSCI was first applied in renal research to characterize microcirculatory changes following ischemia-reperfusion (Bezemer et al. 2010). The same study has demonstrated that real-time speckle contrast measurements could detect rapid alterations in perfusion that conventional methods often miss. Subsequent studies used LSCI to unravel the interplay of vasomotion and tubuloglomerular feedback, contributing to the complex autoregulation of pressure and blood flow. In particular, it allowed visualizing large-scale perfusion dynamics in the renal cortex (Holstein-Rathlou et al. 2011), showing how TGF-mediated blood flow oscillations emerge and propagate among nephron populations. These findings were refined by using higher temporal resolution LSCI to map TGF-associated pressure oscillations more precisely, revealing how synchronization patterns shift under various physiological and pharmacological conditions (Scully et al. 2013). Further research has confirmed that these oscillations could synchronize, strengthening the hypothesis that collective behavior in the renal vasculature is critical for maintaining stable filtration rates (Brazhe et al. 2014). Most recently, multi-scale and high-resolution LSCI approaches have been introduced (Lee et al. 2022; Postnov and Marsh 2022), enabling researchers to visualize tubuloglomerular feedback oscillations and local pinpoint local flow disturbances at the level of individual star-vessels (Fig. 3a and b). Beyond rodent models, blood flow imaging using LSCI was validated in porcine kidney preparations (Heeman et al. 2021; Nwaiwu et al. 2023; Arildsen et al. 2024), underlining its translational potential. These large-animal studies highlight how LSCI may ultimately aid in real-time decision-making during kidney transplantation—providing noninvasive, dynamic assessments of reperfusion quality to improve post-surgical outcomes.

3.4 Liver

The liver's dual blood supply from the hepatic artery and portal vein creates a distinctive vascular architecture, where oxygen-rich arterial flow merges with nutrient-rich portal flow before passing through the sinusoids (Sturesson et al. 2013; Zhukov et al. 2024). This intricate arrangement underpins vital metabolic

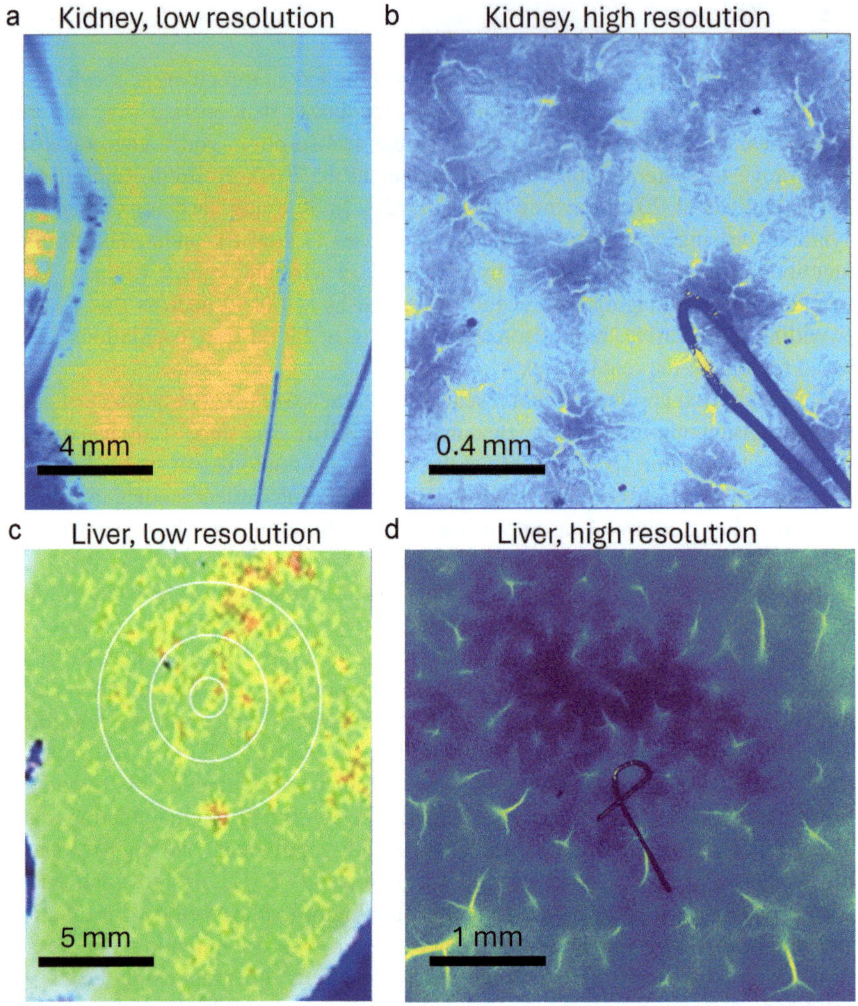

Fig. 3 High-resolution LSCI image of renal and hepatic perfusion. (**a**), (**b**) conventional and high-resolution images of perfusion in the kidney. (**c**), (**d**) conventional (Sturesson et al. 2013), and most recent, high-resolution blood flow index images of liver perfusion. Image in panel C is adapted from Microvascular Research 87, Sturesson C, Milstein DM, Post IC, Maas AM, van Gulik TM, Laser speckle contrast imaging for assessment of liver microcirculation, copyright ©2013 Elsevier, with permission of Elsevier

and detoxification processes and introduces complex hemodynamic regulation via mechanisms like the hepatic arterial buffer response, which ensures steady perfusion when portal flow fluctuates. The portal venous system, uniquely contractile and sensitive to vasoactive factors, further modulates flow distribution, resulting in a heterogeneous sinusoidal environment. Deciphering these intravital hemodynamics is pivotal for both fundamental liver biology and for refining diagnostic or therapeutic approaches that target hepatic microcirculation.

After its successful introduction to kidney perfusion studies, LSCI gained traction as a versatile method for assessing liver hemodynamics. The first study validating the technique by correlating speckle-derived flow maps with conventional measures of liver function (Sturesson et al. 2013) (Fig. 3c) was followed by studies of liver perfusion during ischemia, hepatectomy, and reperfusion (Li et al. 2014, 2017). Most recently, a high-resolution LSCI approach was introduced to characterize vasomotion within the liver microcirculation, revealing distinct spatiotemporal patterns of sinusoidal contractions (Zhukov et al. 2024) (Fig. 3d). Such detailed insights into minute oscillatory behavior deepen our understanding of how hepatic vasoregulation responds to injury, metabolic demands, or pharmacological interventions, reinforcing LSCI as a powerful imaging modality for basic liver research and potential clinical translation.

3.5 Skin

As the body's largest organ, the skin houses a versatile network of blood vessels that balances local demands like temperature regulation and barrier function against systemic needs. Moreover, it is one of the most attractive subjects for translational research and testing pharmacological interventions, as it allows direct and minimally invasive study design relevant to basic physiology, disease diagnostics, and the beauty industry alike.

Over the past decade, interest in preclinical applications of LSCI for monitoring skin microcirculation has steadily increased, even though the technique has traditionally been more widespread in clinical contexts for burn or wound assessment. The first feasibility study was conducted in Choi et al. (2004), using a rodent dorsal skin model to show that LSCI could capture dynamic changes in perfusion during physiological or pharmacological challenges. Subsequent studies leveraged these capabilities to visualize vascular responses to specific irritants, such as methyl salicylate, quantifying inflammatory reactions and local blood flow alterations (Kalchenko et al. 2014; Ponticorvo et al. 2017; Im et al. 2020). Most recently, methodological advances were made aiming to refine the quality and depth penetration of skin LSCI. For instance, topical optical clearing agents have been used to reduce light scattering within superficial layers of the skin, thereby improving image contrast (Kalchenko et al. 2019; Couturier et al. 2022), while transmissive illumination setups provide an alternative configuration that can enhance signal detection by capturing forward-scattered light (Li et al. 2021).

4 Conclusions and Future Perspectives

Laser Speckle Contrast Imaging has evolved into a highly versatile and accessible modality for preclinical perfusion studies, bridging fundamental physiology and translational research. Recent advances in the theory and the introduction of new image-processing approaches have notably strengthened LSCI's capacity for quantifying hemodynamic changes. In parallel, hardware innovations have improved measurement precision and deep-tissue imaging abilities.

These developments and the unique capacity for rapid, wide-field imaging assessment of tissue perfusion make LSCI increasingly popular. Its applications now extend beyond the organs discussed in this chapter, encompassing the heart (El-Awaisi et al. 2024), mesentery (Nyvad et al. 2017), intestine, reproductive organs, and specialized animal models such as the chicken embryo's chorioallantoic membrane (Chen et al. 2021). By enabling real-time, high-resolution assessments of microvascular flow, LSCI not only deepens our understanding of physiological regulation but also offers new avenues for monitoring disease progression and guiding therapeutic interventions. Despite the advances mentioned above, LSCI is still far from reaching its full potential and will continue to evolve. We envision three crucial directions that may shape future LSCI and its applications:

- **Multimodal imaging.** Synergic co-implementation with other imaging modalities is a strong side of LSCI on par with the high spatiotemporal resolution. Combining it with fluorescent techniques, spectroscopy, optical coherence tomography, and even ultrasound imaging will further enrich the acquired data, providing more comprehensive information on physiological processes. While it is already being done by some of the research groups (Sunil et al. 2020; Amki et al. 2020; El-Awaisi et al. 2024; Skøtt et al. 2024), we expect this approach to become more widespread and the data from multiple modalities to be analyzed in a more co-operative way, rather than independently. Examples of such co-operative analysis would include utilizing fluorescent imaging for selecting regions of interest in LSCI data, comparing blood flow and oxygenation dynamics to determine how efficient oxygen extraction is, or combining flow and molecular fluorescence signals could detail the interplay between tissue features (e.g., amyloid plaques deposition) and vascular perfusion.
- **AI-enhanced image analysis.** Machine learning tools—especially diffusion-based and generative models—are poised to transform LSCI data processing. By learning from high-quality multi-exposure or validated ground-truth data, AI can address the long-standing challenge of extracting the absolute velocity of red blood cells. In future implementations, neural networks could ingest raw speckle frames and output velocity maps, bridging the gap between relative perfusion indices and physiologically relevant absolute flow measurements. Alternatively, AI-driven algorithms can target improving the statistical characterization of speckle intensity distribution, providing more accurate and precise contrast

estimation without increasing the required number of speckles (or pixels) and further enhancing LSCI resolution in both time and space.

- **Novel vascular biomarkers.** Additional microvascular features and biomarkers can be extracted as the resolution of LSCI systems and data-analysis pipelines advance. These include slow tissue dynamics (Liu et al. 2023), microvascular pulsatility (Skøtt et al. 2024), local capillary velocity heterogeneity, and the spatio-temporal propagation speeds of neurovascular coupling responses. LSCI-based identification of distributed oscillatory phenomena—such as synchronized vasomotion in the kidney cortex or propagating dilation waves in the brain—could yield new indicators of systemic or localized vascular health. Furthermore, once accurately estimated, the static scattering, which is often viewed as a nuisance due to the increased complexity of contrast interpretation, might become a marker of edema or other slow tissue processes.

The outlined strategies underscore how LSCI might transcend its current capabilities. As these developments come to fruition, they will make it an even more potent and comprehensive tool, allowing new applications to emerge in preclinical and clinical research and transforming how we study microvascular function and perfusion.

Acknowledgments Dmitry Postnov was supported by the Lundbeck Foundation (grant R345-2020-1782) and the Independent Research Fund Denmark (grant 3122-00012B).

Conflict of Interest The authors have no conflicts of interest to declare that are relevant to the content of this chapter.

Images presented in this chapter are either adopted from respective articles with permissions or previously unpublished data from our lab. In the latter case, relevant experimental protocols were approved by the Danish National Animals Experiments Inspectorate and conducted according to the respective guidelines and the Directive 2010/63/EU of the European Parliament on the protection of animals used for scientific purposes.

References

Akther S, Mikkelsen MB, Postnov DD (2023) Biomed Opt Express 15(1):336
Amini N, Esteki A, Ahmadi M, Sasanpour P (2024) Sci Rep 14(1):26652
El Amki M, Glück C, Binder N, Middleham W, Wyss MT, Weiss T, Meister H, Luft A, Weller M, Weber B et al (2020) Cell Rep 33(2)
Arildsen MM, Thrane L, Staulund J, Eijken M, Jespersen B, Postnov D (2024) Al-Mashhadi RH, Pedersen M, Artif Organs 48(4):347
Bandyopadhyay R, Gittings A, Suh S, Dixon P, Durian DJ (2005) Rev Sci Instrum 76(9):093110
Bezemer R, Legrand M, Klijn E, Heger M, Post IC, van Gulik TM, Payen D, Ince C (2010) Opt Express 18(14):15054
Boas DA, Dunn AK (2010) J Biomed Opt 15(1):011109
Boas DA, Yodh AG (1997) JOSA A 14(1):192
Bonner R, Nossal R (1981) Appl Opt 20(12):2097
Brazhe AR, Marsh DJ, Holstein-Rathlou NH, Sosnovtseva O (2014) PloS One 9(9):e105879
Briers J, Fercher A, Invest Ophthalmol Vis Sci 22(2):255

Briers JD, Webster S (1996) J Biomed Opt 1(2):174
Briers D, Duncan DD, Hirst ER, Kirkpatrick SJ, Larsson M, Steenbergen W, Stromberg T, Thompson OB (2013) J Biomed Opt 18(6):066018
Broggini T, Duckworth J, Ji X, Liu R, Xia X, Mächler P, Shaked I, Munting LP, Iyengar S, Kotlikoff M, van Veluw SJ (2024) Long-wavelength traveling waves of vasomotion modulate the perfusion of cortex. Neuron. 112(14):2349–2367
Chen M, Wen D, Huang S, Gui S, Zhang Z, Lu J, Li P (2018) Opt Lett 43(22):5627
Chen L, Wang S, Feng Y, Zhang J, Du Y, Zhang J, Van Ongeval C, Ni Y, Li Y, Cells 10(2):463
Chen R, Tong S, Miao P (2023a) Opt Lett 48(11):2913
Chen R, Miao P, Tong S (2023b) J Innovative Opt Health Sci 16(05):2350005
Cheng H, Duong TQ (2007) Opt Lett 32(15):2188
Cheng H, Yan Y, Duong TQ (2009) Ophthalmic technologies XIX, vol. 7163. SPIE, pp. 15–21
Cheng H, Yan Y, Duong TQ (2008) Opt Express 16(14):10214
Cheng TY, Kim B, Zimmermann BB, Robinson MB, Renna M, Carp SA, Franceschini MA, Boas DA, Cheng X (2024) Sci Rep 14(1):11915
Choi B, Kang NM, Nelson JS (2004) Microvasc Res 68(2):143
Colin E, Plyer A, Golzio M, Meyer N, Favre G, Orlik X (2022) J Biomed Opt 27(4):046003
Couturier A, Bouvet R, Cracowski JL, Roustit M (2022) Microvasc Res 141:104319
Davis MA, Kazmi SS, Dunn AK (2014) J Biomed Opt 19(8): 086001
Davis MA, Gagnon L, Boas DA, Dunn AK (2016) Biomed Opt Express 7(3):759
Dragojević T, Bronzi D, Varma HM, Valdes CP, Castellvi C, Villa F, Tosi A, Justicia C, Zappa F, Durduran T (2015) Biomed Opt Express 6(8):2865
Duncan DD, Kirkpatrick SJ (2008) JOSA A 25(8):2088
Dunn AK, Bolay H, Moskowitz MA, Boas DA (2001) J Cerebral Blood Flow Metab 21(3):195
El-Awaisi J, Kavanagh DP, Kalia N (2024) Front Cardiovasc Med 11:1358472
Erdener ŞE, Tang J, Kılıç K, Postnov D, Giblin JT, Kura S, Chen ICA, Vayisoğlu T, Sakadžić S, Schaffer CB et al (2021) J Cerebral Blood Flow Metab 41(2):236
Fercher A, Briers JD (1981) Opt Commun 37(5):326
Fredriksson I, Hultman M, Strömberg T, Larsson M (2019) J Biomed Opt 24(1):016001
Goldberg J, Miller DR, Dimanche A, Kissling C, Müller T, Müller MD, Jesse CM, Murek M, Bervini D, Dunn AK et al (2024) Neurosurgery 94(5):983
Golubova N, Potapova E, Seryogina E, Dremin V (2023) Biomed Signal Process Control 85:104969
González Olmos A, Zilpelwar S, Sunil S, Boas DA, Postnov DD (2023) Sci Rep 13(1):17970
Goodman JW (1975) Laser speckle and related phenomena. Springer, pp. 9–75
Guilbert J, Desjardins M (2022) J Biophoton 15(1):e202100218
Hao X, Wu S, Lin L, Chen Y, Morgan SP, Sun S (2023) Opt Lasers Eng 166:107587
He J, Lu H, Young L, Deng R, Callow D, Tong S, Jia X (2019) J Cerebral Blood Flow Metab 39(6):1161
Heeman W, Maassen H, Calon J, Van Goor H, Leuvenink H, van Dam GM, Boerma EC (2021) J Biomed Opt 26(5):056004
Holstein-Rathlou NH, Sosnovtseva OV, Pavlov AN, Cupples WA, Sorensen CM, Marsh DJ (2011) Am J Physiol-Renal Physiol 300(2):F319
Huang Y, Zhang X, Zhang C, Xu W, Li W, Feng Z, Zhang X, Zhao K (2022) Mediators Inflammation 2022(1):3855698
Hultman M, Larsson M, Strömberg T, Fredriksson I (2020) J Biomed Opt 25(11):116007
Im J, Kong TH, Choi JS, Seo YJ, Choi EC, Jung B, Kim JH (2020) Microvasc Res 132:104050
Kalchenko V, Kuznetsov Y, Preise D, Meglinski I, Harmelin A (2014) J Biomed Opt 19(6):060502
Kalchenko V, Meglinski I, Sdobnov A, Kuznetsov Y, Harmelin A (2019) J Biomed Opt 24(6):060501
Kirkpatrick SJ, Duncan DD, Wells-Gray (2008) Opt Lett 33(24):2886
Lee B, Sosnovtseva O, Sørensen CM, Postnov DD (2022) Biomed Opt Express 13(4):2312
Li CH, Wang HD, Hu JJ, Ge XL, Pan K, Zhang AQ, Dong JH (2014) Microvasc Res 94:28
Li CH, Ge XL, Pan K, Wang PF, Su YN, Zhang AQ (2017) Microvasc Res 110:14

Li DY, Xia Q, Yu TT, Zhu JT, Zhu D (2021) Light Sci Appl 10(1):241
Lim XR, Abd-Alhaseeb MM, Ippolito M, Koide M, Senatore AJ, Plante C, Hariharan A, Weir N, Longden TA, Laprade KA et al (2024) Nature Commun 15(1):8686
Liu C, Kılıç K, Erdener SE, Boas DA, Postnov DD (2021a) Biomed Opt Express 12(6):3571
Liu X, Yang H, Li R (2021b) IEEE Access 9:148925
Liu B, Shah S, Küreli G, Devor A, Boas DA, Cheng X (2023) Biomed Opt Express 14(9):4790
Liu B, Postnov D, Boas DA, Cheng X (2024) Biomed Opt Express 15(2):579
Mangraviti A, Volpin F, Cha J, Cunningham SI, Raje K, Brooke MJ, Brem H, Olivi A, Huang J, Tyler BM et al (2020) Sci Rep 10(1):7614
Mikkelsen SH, Skøtt MV, Gutierrez E, Postnov DD (2024) Biomed Opt Express 15(2):1268
Neganova AY, Postnov DD, Jacobsen CB, Sosnovtseva O (2016) Biomed Opt Express 7(4):1375
Neganova AY, Postnov DD, Sosnovtseva O, Jacobsen JCB (2017) PloS One 12(3):e0173805
Nwaiwu CA, Buharin VE, Mach A, Grandl R, King ML, Dechert AF, O'Shea L, Schwaitzberg SD, Kim PC (2023) Surg Endosc 37(2):1086
Nyvad J, Mazur A, Postnov DD, Straarup MS, Soendergaard AM, Staehr C, Brøndum E, Aalkjaer C, Matchkov VV (2017) J Physiol 595(15):5037
Parthasarathy AB, Tom WJ, Gopal A, Zhang X, Dunn AK (2008) Opt Express 16(3):1975
Parthasarathy AB, Kazmi SS, Dunn AK (2010) Biomed Opt Express 1(1):246
Patel DD, Lipinski DM (2020) Sci Rep 10(1):1
Patel DD, Dhalla AH, Viehland C, Connor TB, Lipinski DM (2021) Transl Vis Sci Technol 10(9):19
Ponticorvo A, Cardenas D, Dunn AK, Ts'o D, Duong TQ (2013) J Biomed Opt 18(9):090501
Ponticorvo A, Burmeister DM, Rowland R, Baldado M, Kennedy GT, Saager R, Bernal N, Choi B, Durkin AJ (2017) Lasers Surg Med 49(3):293
Postnov D, Marsh DJ, Cupples WA, Holstein-Rathlou NH, Sosnovtseva O (2022) Elife 11:e75284
Postnov DD, Cheng X, Erdener SE, Boas DA (2019) Sci Rep 9(1):2542
Postnov DD, Tang J, Erdener SE, Kılıç K, Boas DA (2020) Sci Adv 6(45):eabc4628
Qiu J, Li Y, Huang Q, Wang Y, Li P (2013) Opt Express 21(23):28902
Salinero AE, Robison LS, Gannon OJ, Riccio D, Mansour F, Abi-Ghanem C, Zuloaga KL (2020) FASEB J Official Publication Federation Am Soc Exp Biol 34(11):15108 (2020)
Scully CG, Mitrou N, Braam B, Cupples WA, Chon KH (2013) Am J Physiol-Regulatory Integrative Comparative Physiol 304(11):R929
Skøtt MV, Melnikova E, Gutiérrez-Jiménez E, Matchkov VV, Østergaard L, Postnov DD (2024) bioRxiv
Srienc AI, Kurth-Nelson ZL, Newman E (2010) Front Neuroenergetics 2:128
Staehr C, Guldbrandsen HØ, Homilius C, Johnsen LØ, Postnov D, Pedersen TM, Pierre S, Sandow SL, Matchkov VV (2024) J Cerebral Blood Flow Metab 0271678X241305562
Stebakov I, Kornaeva E, Stavtsev D, Potapova E, Dremin V (2021) Vibroeng Proc 38:50
Stern M (1975) Nature 254(5495):56
Sturesson C, Milstein DM, Post IC, Maas AM, van Gulik TM (2013) Microvasc Res 87:34
Sunil S, Erdener SE, Lee BS, Postnov D, Tang J, Kura S, Cheng X, Chen IA, Boas DA, Kilic K (2020) Neurophotonics 7(1):015005
Tate AJ, Lipinski DM (2024) Invest Ophthalmol Visual Sci 65(7):4931
Völker AC, Zakharov P, Weber B, Buck F, Scheffold F (2005) Opt Express 13(24):9782
Wang L, Cull GA, Piper C, Burgoyne CF, Fortune B (2012) Invest Ophthalmol Visual Sci 53(13):8303
Wildeboer A, Heeman W, van der Bilt A, Hoff C, Calon J, Boerma EC, Al-Taher M, Bouvy N (2022) Life 12(8):1251
Yi C, Jung J, Im J, Lee KC, Chung E, Lee SA (2022) Optica 9(11):1227
Yin L, Yu T, Cheng L, Liu X, Zhang W, Zhang H, Du L, He W (2022) BMC Neurosci 23(1):80
Yuan S, Devor A, Boas DA, Dunn AK (2005) Appl Opt 44(10):1823
Zakharov P (2017) Opt Lett 42(12):2299
Zakharov P, Völker A, Buck A, Weber B, Scheffold F (2006), Opt Lett 31(23):3465

Zakharov P, Völker A, Wyss M, Haiss F, Calcinaghi N, Zunzunegui C, Buck A, Scheffold F, Weber B (2009) Opt Express 17(16):13904
Zheng S, Mertz J (2022a) Biomed Opt Express 13(8):4118
Zheng S, Mertz J (2022b) Opt Lett 47(24):6333
Zheng S, Xiao S, Kretsge L, Cruz-Martín A, Mertz J (2021) Opt Lett 46(19):5059
Zheng S, Davison I, Garrett A, Lin X, Chitkushev N, Roblyer D, Mertz J (2024) Optica 11(12):1733
Zhukov O, Postnov DD, Hejn KH, Ravnskjaer K, Sosnovtseva O (2024) bioRxiv
Zilpelwar S, Sie EJ, Postnov D, Chen AI, Zimmermann B, Marsili F, Boas DA, Cheng X (2022) Biomed Opt Express 13(12):6533

Integrating Surface-Enhanced Raman Scattering (SERS) Imaging and Artificial Intelligence (AI) for Advanced Bioimaging Modalities in Antimicrobial Resistance

Zakarya Al-Shaebi, Munevver Akdeniz, and Omer Aydin

Abstract

The global rise of antimicrobial resistance (AMR) presents a critical challenge to public health, necessitating the development of advanced diagnostic tools for rapid and accurate detection. Imaging technologies facilitate examination of biological processes at molecular and cellular levels, explanation of resistance mechanisms, and monitoring resistant strains. The integration of artificial intelligence (AI) with surface-enhanced Raman spectroscopy (SERS) imaging has emerged as an innovative approach to tackle this challenge. SERS imaging offers high sensitivity and molecular specificity, enabling the identification of resistant bacterial strains and their mechanisms at a single-cell level. AI-driven data analysis further enhances diagnostic potential by automating pattern identification, revealing minor spectrum properties, and improving the speed and accuracy of AMR detection. This combination allows for real-time, label-

Z. Al-Shaebi · M. Akdeniz
Department of Biomedical Engineering, Erciyes University, Kayseri, Türkiye

Nanothera Lab, Drug Application and Research Center (ERFARMA), Erciyes University, Kayseri, Türkiye

O. Aydin
Department of Biomedical Engineering, Erciyes University, Kayseri, Türkiye

Nanothera Lab, Drug Application and Research Center (ERFARMA), Erciyes University, Kayseri, Türkiye

Clinical Engineering Research and Implementation Center (ERKAM), Erciyes University, Kayseri, Türkiye

Nanotechnology Research and Application Center (ERNAM), Erciyes University, Kayseri, Türkiye
e-mail: omeraydin@erciyes.edu.tr; biomer@umich.edu

I. F. Cengiz et al. (eds.), *Bioimaging Modalities in Bioengineering*,
Biomaterials, Bioengineering and Sustainability 9,
https://doi.org/10.1007/978-3-031-99410-4_27

free Raman imaging of microbial populations, providing insights into resistance mechanisms, and the interactions between pathogens and antimicrobial agents. By combining the complementary capabilities of AI and SERS imaging, this unique approach offers reliable AMR diagnosis and facilitates the development of treatment solutions to address the global AMR crisis.

Keywords

Antimicrobial resistance · SERS · Imaging · Artificial intelligence

1 Introduction

Advances in medical diagnostics and bioimaging have made significant strides in combating the global threat posed by antimicrobial resistance (AMR), one of the most pressing challenges of modern medicine (Ahmed et al. 2024; Tang et al. 2023). According to the 2022 Global Research on Antimicrobial Resistance (GRAM) report published in The Lancet, AMR was directly associated with approximately 1.27 million deaths worldwide in 2019, and indirectly associated with nearly 4.95 million deaths. Economically, AMR could result in a global cost of up to $100 trillion by 2050, according to a 2016 review commissioned by the UK government (Murray et al. 2022) As the overuse and misuse of antibiotics continue to accelerate the emergence of drug-resistant pathogens, early and accurate diagnosis of infections is critical to controlling the spread of AMR and ensuring appropriate treatment decisions (Salam et al. 2023). In this context, bioimaging technologies have emerged as powerful tools, allowing clinicians and researchers to visualize and study biological processes at molecular and cellular levels. These advances are pivotal not only for diagnostic purposes but also for understanding the mechanisms of resistance, tracking the evolution of resistant strains, and developing targeted antimicrobial therapies.

Among the leading technologies revolutionizing the bioimaging landscape is Raman spectroscopy, a vibrational spectroscopic technique that enables the acquisition of molecular fingerprints from biological samples (Ferraro 2003). Unlike conventional imaging techniques that rely on external labels or dyes, Raman spectroscopy offers a label-free and non-invasive approach, allowing for the direct analysis of molecular structures within cells, tissues, or biological fluids. This feature is particularly advantageous in clinical diagnostics, where minimizing sample preparation and preserving the integrity of the biological specimen are of utmost importance.

However, one of the primary challenges with the traditional Raman spectroscopy is its low sensitivity. The weak intensity of the Raman signal often limits its application, especially in detecting trace amounts of biological markers or pathogens that are present in very low concentrations. This limitation is particularly critical in the context of early diagnosis of infections, where detecting low levels of bacterial pathogens or specific molecular indicators of resistance is essential. To address this challenge, researchers have developed Surface-Enhanced Raman Spectroscopy (SERS), an advanced variant of Raman spectroscopy that significantly enhances the Raman signal.

SERS amplifies the Raman scattering of molecules by several orders of magnitude through the use of nanostructured metallic surfaces, such as gold or silver nanoparticles (Fleischmann et al. 1974). These nanoparticles or noble metallic surfaces create localized surface plasmons. Localized surface plasmons refer to the collective oscillations of electrons in metallic nanoparticles that occur when they interact with incident light. These oscillations are confined to the surface of the nanoparticle, creating intense electromagnetic fields in the immediate vicinity, which is crucial for the enhancement of the Raman signal in SERS (Willets and Van Duyne 2007; Langer et al. 2020). The resulting electromagnetic field enhancement around the nanoparticles increases the intensity of the Raman signal, enabling the detection of even single molecules. This ultra-sensitive detection capability makes SERS an invaluable tool for identifying minute quantities of bacterial biomarkers or antibiotic-resistant pathogens in clinical samples. Additionally, SERS allows for multiplexed detection, wherein multiple molecular targets can be identified simultaneously within a complex biological sample, making it ideal for studying polymicrobial infections or detecting a variety of biomarkers associated with AMR (Al-Shaebi et al. 2022, 2023; Uysal Ciloglu et al. 2020).

SERS imaging takes this step further by providing spatially resolved molecular information within a biological sample, enabling researchers to visualize the distribution of specific molecular species across tissues, cells, or other biological matrices (Krafft et al. 2017; Plou et al. 2022; Liu et al. 2022; Lin et al. 2021; Aydin et al. 2009a, b). In SERS imaging, nanostructures that enhance the Raman signal are applied to the sample, and the resulting highly sensitive Raman spectra are collected from different locations across the sample (Zhang et al. 2011). By mapping the intensity of specific Raman peaks across the sample, SERS imaging creates detailed chemical images that highlight the spatial distribution of key biomolecules. This capability is especially important in studying heterogeneous biological samples, such as bacterial colonies, biofilms, or tissue sections, where understanding the spatial arrangement of molecular markers is essential for diagnosing infections or tracking disease progression.

One of the unique advantages of SERS is its ability to perform both label-free and label-based imaging. In the label-free approach, SERS leverages the intrinsic Raman signatures of biomolecules, allowing for the detection of molecular species without the need for external labels or probes. This is particularly beneficial in complex biological systems, where label-free Raman imaging can provide a more accurate reflection of the native molecular environment. On the other hand, label-based SERS imaging employs Raman-active tags that bind selectively to target molecules, enhancing the specificity of detection (Fig. 1). This dual capability allows SERS to be adapted to a wide range of bioimaging applications, from the direct detection of biomolecules to targeted imaging of specific pathogens or biomarkers.

In the context of AMR, SERS imaging offers a unique advantage in the visualization of bacterial communities, including the ability to detect resistant and

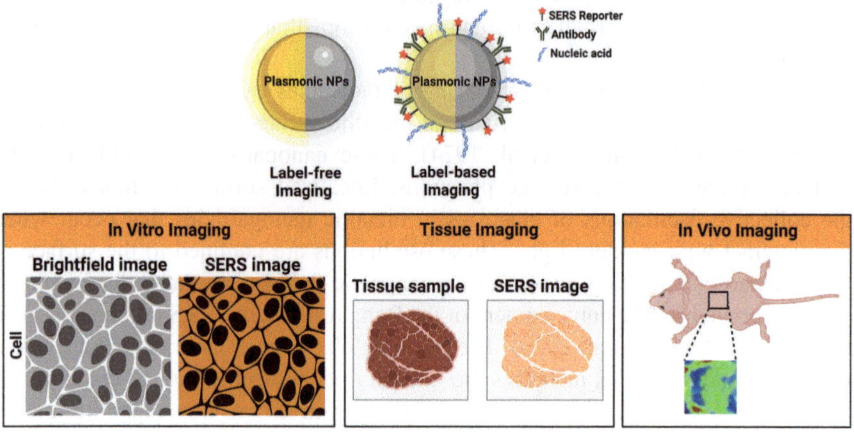

Fig. 1 Schematic illustration of nanoparticles for both label-free and label-based SERS imaging. Label-free Raman imaging utilizes inherent Raman signals, whereas label-based imaging incorporates Raman-active tags for the targeted detection of molecules. (Created with BioRender.com)

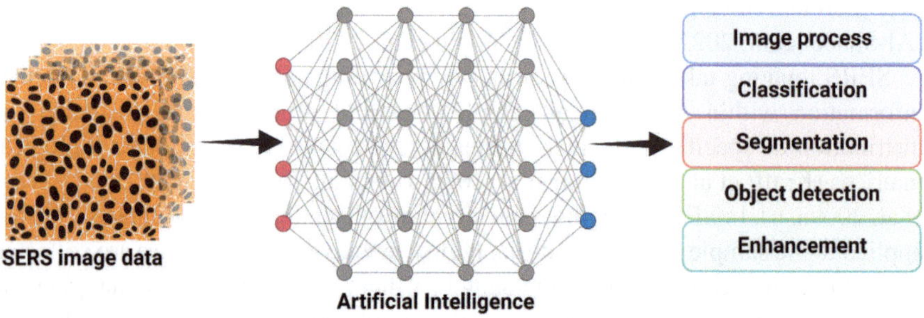

Fig. 2 Schematic representation of AI-enhanced SERS imaging workflow. SERS-generated spectral or imaging data are processed through an AI-driven model, enabling tasks such as image processing, classification, segmentation, object detection, and enhancement. (Created with BioRender.com)

susceptible strains within the same sample. For example, SERS imaging can be used to detect specific resistance-associated biomarkers and visualize their distribution in bacterial colonies or infected tissues (Bodelón et al. 2018). This spatial information is invaluable for understanding how resistance mechanisms are organized within a population and for identifying potential weak points in bacterial defense strategies that could be targeted by new antibiotics or treatments.

When combined with the power of AI, SERS-based imaging introduces new possibilities for improving the detection, monitoring, and characterization of pathogens, including drug-resistant bacteria (Ho et al. 2019; Taha et al. 2024). As illustrated in Fig. 2, AI-driven algorithms process SERS-based imaging data through various stages, including image processing, classification, segmentation, object detection, and enhancement. The integration of AI with SERS-based bioimaging introduces a level of automation and precision that is crucial in the analysis of large and complex datasets typically generated by SERS experiments. AI-driven algorithms, especially those based on machine learning (ML), can efficiently process and analyze vast amounts of spectral data, identifying subtle patterns and relationships that might be difficult for human analysts to detect. This makes AI-enhanced SERS imaging a powerful tool for distinguishing between resistant and susceptible bacterial strains and providing personalized insights into infection dynamics.

The spatial resolution of SERS imaging, combined with AI, adds a new dimension to the study of AMR. By generating detailed molecular maps of bacterial populations or infected tissues, researchers can use AI algorithms to identify spatial patterns of resistance or pinpoint molecular signatures that are correlated with antibiotic resistance. This capability is particularly important for studying biofilms, where resistant bacteria may be organized in specific regions, making it harder for antibiotics to penetrate and eliminate the infection. AI can analyze these complex SERS imaging datasets to identify regions of resistance and predict the overall susceptibility of the biofilm, offering new avenues for precision medicine in infections caused by AMR pathogens.

In routine clinical practice, where time is often a critical factor, AI can streamline the interpretation of SERS imaging data, reducing the time to diagnosis and enabling real-time decision-making. For instance, AI can quickly classify bacterial species based on their Raman spectra, identify specific resistance phenotypes, and predict the likelihood of treatment failure, allowing clinicians to make informed choices about the most appropriate antimicrobial therapies. By integrating SERS with AI, diagnostics can move from being reactive to predictive and personalized, helping to combat AMR by ensuring that patients receive the right treatment at the right time.

Moreover, the potential of AI-enhanced SERS imaging extends beyond diagnostics. The detailed molecular information provided by SERS, coupled with AI's ability to handle big data, holds promise for understanding the molecular mechanisms underlying AMR. By analyzing vast datasets of SERS spectra from different bacterial strains and imaging their spatial distribution, AI can help uncover how resistance evolves, identify new resistance genes, and track the spread of resistant pathogens across populations. This knowledge is crucial for developing new strategies to combat AMR, such as designing more effective antibiotics or devising alternative therapeutic approaches.

The combination of SERS imaging and AI represents a transformative approach to bioimaging and diagnostics in the fight against AMR. SERS provides the molecular sensitivity required for detecting and characterizing resistant pathogens, while AI adds a layer of intelligence that enhances the accuracy, speed, and predictive power of diagnostic analyses. Together, these technologies offer a powerful solution for addressing the urgent need for early detection, precise monitoring, and effective treatment of infections caused by antibiotic-resistant bacteria.

2 Bioimaging Applications of SERS and Its Relevance to Antimicrobial Resistance

In the ever-evolving landscape of bioimaging, SERS stands out for its exceptional sensitivity in detecting molecular signatures in biological samples. The ability of SERS to provide spatially resolved molecular information offers unique insights into the detection and characterization of bacterial pathogens, especially those resistant to antibiotics. While Sect. 1 highlighted the fundamental advantages of SERS in diagnostics, this section delves into specific bioimaging applications and explores the role SERS plays in addressing AMR.

2.1 Applications of SERS in Bioimaging

2.1.1 Label-Free Based Applications

The primary strength of SERS lies in its ability to generate highly sensitive and specific molecular fingerprints from biological samples. In bioimaging, SERS has expanded the possibilities of molecular visualization by mapping the distribution of key biomolecular alteration across tissues (Nicolson et al. 2019; Wang et al. 2015, 2016; Li et al. 2023a; Verdin et al. 2020), cells (Verdin et al. 2020; Lee et al. 2014; Dong et al. 2022; Qiu et al. 2023; Jimenez de Aberasturi et al. 2020), or biofilms (Ivleva et al. 2010a; Ravindranath et al. 2012; Guo et al. 2018). Unlike traditional imaging methods that rely on dyes or external labels (Lv et al. 2022; Li et al. 2023b; Austin et al. 2013; Han et al. 2023a; Gao et al. 2020; Heiner et al. 2017; Kumar et al. 2017; Zheng et al. 2019; Lee et al. 2021; Köker et al. 2018; Zavaleta et al. 2009; Oseledchyk et al. 2017; Gao et al. 2017; Wang et al. 2019; Wen et al. 2021; Wei et al. 2021), SERS provides a label-free alternative (El-Mashtoly et al. 2014; Lima et al. 2021; Spadea et al. 2021; Gvazava et al. 2023; De Marchi et al. 2019; Barzan et al. 2021), making it highly suitable for clinical applications where sample integrity must be preserved. De Marchi et al. explores the use of SERS spectroscopy as a tool for imaging bioactive metabolites in mixed bacterial populations. Bacteria often exist in complex communities where they secrete metabolites that influence their environment and interactions. Understanding these chemical exchanges is crucial for comprehending microbial community dynamics. Traditional methods have focused on separate co-cultures, but this study demonstrates the application of SERS for non-invasive detection and visualization of metabolites directly in mixed bacterial cultures, specifically *Escherichia coli* and *Pseudomonas aeruginosa*. The research highlights the ability of SERS to detect and visualize indole and pyocyanin, revealing that indole signaling can down-regulate pyocyanin expression. This capability underscores the potential of SERS in studying the chemical interactions within multispecies microbial communities, offering insights into the complex chemistry that governs these ecosystems (De Marchi et al. 2019). The study by Barzan et al. explores the use of hyperspectral chemical imaging to analyze the structure of single bacterial cells, specifically *Escherichia coli*, using Raman spectroscopy combined with machine learning techniques. This approach allows for the identification and spatial mapping of biomolecules such as membrane proteins, lipids, and DNA within a single cell. Raman spectroscopy is highlighted for its advantages in being a direct, nondestructive, and cost-effective method that requires minimal sample preparation and no chemical labeling or staining (Barzan et al. 2021).

In addition to label-free SERS imaging in bacteria, numerous studies in literature focus on its application in other biological areas. Xu et al. presents a rapid, label-free SERS method for identifying living cancer cells within 10 min using iodide-modified Ag nanoparticles (Ag IMNPs) and ultra-fast Raman mapping. Ag IMNPs offer a superior balance of sensitivity, reproducibility, and biocompatibility compared to traditional nanoparticles. Statistical analysis of Raman spectra provides detailed intracellular chemical information for distinguishing normal and cancer cells across three cell lines (7701/7703/Caski). This approach can identify cancer cells, monitor chemical changes, and detect reactive oxygen species, particularly in liver cancer cells (7703), offering potential for early cancer detection and monitoring (Xu et al. 2024). Milewska et al. uses noninvasive surface-enhanced Raman scattering (SERS) to study mesenchymal stromal cell (MSC) osteogenic differentiation. By employing biocompatible gold nanoisland SERS substrates, they track molecular species and chemical events on the cell membrane, detecting bone tissue formation from early stages to mature osteoblasts without labeling. The results closely correlate with classical differentiation assays, demonstrating that SERS imaging can effectively monitor osteogenic differentiation, enabling continuous label-free screening of live cells (Milewska et al. 2021). Olson et al. demonstrates a label-free SERS imaging technique combining SERS and STORM algorithms for super-resolved chemical imaging. Using ultra-thin silver substrates, it achieves <50 nm spatial resolution and spectral content for identifying chemical signatures in bacteria. The simple setup, requiring only silver film deposition, allows imaging in turbid environments and at cell/substrate interfaces, making it suitable for diverse biological studies (Olson et al. 2017). In addition, the study by El-Mashtoly et al. explores the use of label-free Raman microscopy to investigate the distribution and metabolism of the molecular targeted drug erlotinib in colon cancer cells. Targeted cancer therapies, such as those inhibiting the epidermal growth factor receptor (EGFR), are crucial in blocking cancer growth. The research highlights the advantages of using label-free Raman imaging techniques, which avoid the potential toxicity and invasiveness of dyes or fluorescent labels (El-Mashtoly et al. 2014). In another study, Horgan et al. reported Raman spectroscopic imaging extracellular vesicle (EV) using three deuterium-labelled metabolites, deuterium oxide (D_2O), deuterated choline chloride (d-Chol), and deuterated D-glucose (d-Gluc) in 2D and 3D. They demonstrated how deuterium is metabolically incorporated into EVs to allow for both thorough molecular characterization of EVs in solution and in vitro (Horgan et al. 2020).

Exosomes carry specific proteins and nucleic acids that reflect the state of their cells of origin, making them promising biomarkers for early cancer detection. SERS offers a powerful method for analyzing exosomes with high sensitivity, enabling the detection of cancer-related molecular signatures in body fluids. By targeting specific exosomal markers, SERS imaging facilitates early cancer diagnostics and provides insights into tumor progression and metastasis. Similarly, SERS is also used in the detection of circulating tumor cells, allowing for minimally invasive "liquid biopsy" approaches to cancer monitoring (Akdeniz et al. 2024).

2.1.2 Label-Based Applications

Label-based SERS applications have emerged as powerful tools in biological and medical research, enabling the detection and visualization of molecular interactions with high sensitivity and specificity. These approaches rely on functionalized nanostructures, such as core-shell nanoparticles or specialized nanoprobes, designed to bind to specific molecular targets through conjugation with antibodies, aptamers, or other biorecognition elements. By incorporating Raman reporter molecules and enhancing the Raman signal, label-based SERS facilitates precise molecular identification and spatial mapping. Regarding imaging bacteria using this method, a SERS sandwich assay leveraging 3-mercaptophenylboronic acid (3-MPBA) as a capturer and reporter molecule was developed for rapid bacterial detection through SERS imaging. The assay demonstrated high sensitivity, detecting *Salmonella enterica* at concentrations as low as 10^2 CFU mL^{-1} with quantification capabilities, using unique bacterial SERS signals generated by interactions among bacteria, 3-MPBA, and silver nanoparticles (Pearson et al. 2017). Further advancements expanded its application to environmental and food matrices, where SERS imaging outperformed the conventional aerobic plate count (APC) method in sensitivity for detecting and quantifying bacteria in pond water, spinach leaves, and high-sugar juice matrices. These developments underscore the potential of SERS imaging as a powerful tool for rapid and reliable bacterial detection in real-world food safety and environmental monitoring applications (Pearson et al. 2018). Furthermore, a surface cell imprinted substrate (SCIS) integrated with SERS mapping enabled specific and quantitative detection of *E. coli* in real samples. Using SERS tags for labeling, the platform achieved high specificity, a wide detection range (10^2–10^8 CFU/mL), and a low detection limit (~1.35 CFU/mL). It was successfully applied to complex food matrices, demonstrating the potential of label-based SERS imaging for bacterial detection (Yang et al. 2022).

In addition to its use in bacteria, label-based SERS imaging can be applied in various other fields. For example, Qui et al. reported SERS probes, are made of Au@Ag core–shell nanoparticles, decorated with Raman reporter double-layer on the surface of Au core and Ag shell with functional polyethylene glycol (HS-PEG-NHS) layer for the antibody conjugation, capable of achieving precise diagnosis and accurate therapeutic evaluation of breast cancer through the detection of growth factor reporters on tumor cell membranes (Qiu et al. 2023). A 3D nano-popcorn substrate for the detection of influenza A/H1N1 virus was reported to have been created by Chen et al. To produce a strong Raman signal, thiolated capture DNAs anchored on the 3D nano-popcorn substrate were hybridized with Cy3-labeled aptamer probes (Chen et al. 2020). Lee et al. reported SERS nano tags consisting of silica-encapsulated hollow gold nanospheres (SEHGNs) conjugated with specific antibodies to detect and quantify breast cancer phenotypic markers expressed on cell surfaces (Lee et al. 2014). Deng et al. developed nuclear-targeted

SERS nanoprobes (NLS-RGD-PEG-AuNRs) for real-time monitoring of specific diagnostic and therapeutic processes in HepG2 cells using functionalized aptamer complex (TLS11a-GC-Dox) of a targeted antitumor agent (doxorubicin) (Deng et al. 2017).

SERS imaging enables researchers to not only detect but also visualize the spatial arrangement of molecules, creating detailed chemical maps. This spatial information is invaluable in understanding complex biological systems where molecular heterogeneity exists, such as bacterial biofilms, which are often characterized by varying levels of resistance within different regions. By using nanostructures that enhance the Raman signal, SERS imaging can reveal these molecular distributions with high resolution, allowing for in-depth analysis of how pathogens behave within host tissues or within their protective biofilm environments. A notable feature of SERS is its multiplexing capability. Researchers can use SERS to detect multiple biomarkers simultaneously within a single sample, enabling comprehensive analysis of polymicrobial infections or complex biochemical processes (Berry et al. 2021; Guo et al. 2019). This is particularly beneficial in cases where bacterial resistance mechanisms are multifaceted, involving various molecular targets or pathways. The ability to capture this complexity in a single imaging session offers a significant advantage in both research and clinical diagnostics.

2.2 Importance of Antimicrobial Resistance and SERS Applications in Bioimaging

The global rise of antimicrobial resistance has underscored the need for advanced diagnostic technologies capable of detecting resistant pathogens with high precision. SERS-based bioimaging offers critical advantages in this regard, providing a highly sensitive, spatially resolved, and non-invasive means of detecting bacterial pathogens and resistance markers. The label-free nature of SERS further adds to its appeal, allowing for the analysis of pathogens in their native state without the risk of altering their molecular environment.

Regarding AMR, one of the most impactful applications of SERS is its ability to differentiate between antibiotic-resistant and susceptible bacterial strains. By analyzing the Raman spectra of bacterial samples, SERS can identify specific biochemical signatures associated with resistance, such as altered cell wall structures, metabolic shifts, or the presence of resistance genes (Fig. 3) (Uysal Ciloglu et al. 2020; Al-Shaebi et al. 2022, 2024; Ciloglu et al. 2021, 2022; Galvan and Yu 2018; Zhang et al. 2022; Skvortsova et al. 2022; Wang and Vikesland 2023; Lu et al. 2022; Wang et al. 2022). This high level of molecular detail provides a deeper understanding of the mechanisms that drive resistance, offering valuable insights for both diagnostics and therapeutic development.

Moreover, SERS imaging's ability to deliver spatially resolved data is particularly relevant when studying bacterial colonies or biofilms, where resistant strains may be unevenly distributed. Biofilms are notorious for harboring bacteria with varying degrees of resistance, often making infections more difficult to treat.

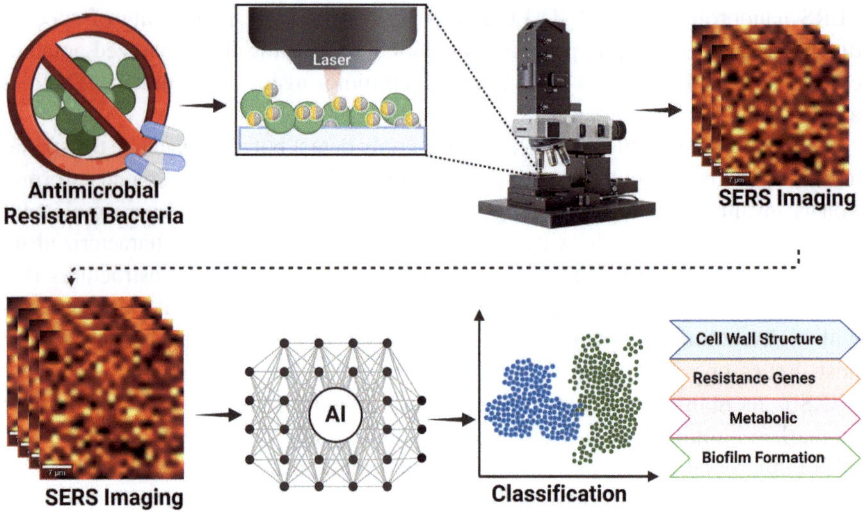

Fig. 3 Role of SERS imaging in understanding antimicrobial resistance. SERS imaging detects specific biochemical signatures linked to antimicrobial resistance, including changes in cell wall structure, metabolic alterations, biofilm formation, and the presence of resistance genes. The schematic illustrates the integration of SERS with AI, enabling the analysis of complex datasets to uncover patterns and mechanisms associated with resistance. (Created with BioRender.com)

"Pockets of resistance" refers to localized areas within a biofilm where bacteria exhibit higher levels of resistance compared to surrounding regions. These pockets may result from variations in nutrient availability, oxygen levels, or the presence of protective molecules in the extracellular matrix. SERS imaging can detect these pockets of resistance within the biofilm, allowing researchers to visualize how resistance is organized within bacterial communities and molecules released into extracellular matrix by resistant bacteria (Ivleva et al. 2010a, b; Bodelón et al. 2016; Zheng et al. 2022; Kim and Chin 2023). Kim et al. monitoring MRSA biofilm development on solid culture media from 0 to 48 h by measuring DNA/RNA-associated Raman peaks and protein/lipid-associated peaks (Kim and Chin 2023). Bodelón et al. reported imaging of pyocyanin produced by *P. aeruginosa* biofilms grown on mesostructured Au@TiO_2 thin films (Bodelón et al. 2016). These studies are crucial for developing strategies to disrupt these microenvironments, ultimately improving treatment outcomes.

Another critical application of SERS in the fight against AMR is its use in real-time monitoring of bacterial responses to antibiotics. SERS can track biochemical changes in bacterial samples as they are exposed to different antimicrobial agents, allowing for rapid assessment of treatment efficacy (Han et al. 2023b; Gukowsky and He 2022). Han et al. reported antibiotic susceptibility testing (AST) protocol based on SERS-AST for bacteremia caused by eight clinically relevant Gram-positive and Gram-negative pathogens treated with seven commonly administered antibiotics (Han et al. 2023b). Gukowsky et al. investigated spectra from bacterial samples, including Escherichia coli, Bacillus cereus, and Salmonella enterica, collected on a filter membrane after incubation with antibiotics by revealing characteristic antibiotic response patterns with SERS spectra according to the peak at 738 cm^{-1} (Gukowsky and He 2022). This ability to monitor bacterial behavior in response to therapy is especially important in clinical settings, where timely adjustments to treatment can be the difference between success and failure. By providing this real-time feedback, SERS offers clinicians a powerful tool for personalized medicine, ensuring that treatments are tailored to the specific resistance profiles of the pathogens involved.

SERS is also gaining traction as a method for studying the molecular evolution of resistance. By collecting and analyzing SERS spectra from various bacterial strains over time, researchers can track the emergence of resistance markers and gain insights into how resistance evolves under selective pressure. This application has far-reaching implications for public health, as understanding the pathways through which resistance develops can guide the design of new antibiotics or alternative therapies that are less prone to resistance (Liu et al. 2016). Lui et al. showed that when a susceptible *Staphylococcus aureus* or *Escherichia coli* strain was exposed to the antibiotics oxacillin and imipenem, respectively, the intensity of specific biomarkers in SERS spectra decreased significantly within 2 h (Liu et al. 2016).

The application of SERS in bioimaging has revolutionized the study of antimicrobial resistance. By offering highly sensitive, spatially resolved, and real-time molecular insights, SERS provides an unparalleled platform for detecting resistant pathogens, understanding their molecular underpinnings, and informing both diagnostics and therapeutic interventions. As AMR continues to pose significant challenges to global health, the integration of SERS into clinical microbiology and infectious disease research will be essential in curbing the spread of resistant pathogens and improving patient outcomes.

3 Role of Artificial Intelligence (AI) in Enhancing Bioimaging Modalities

AI algorithms are strong image processing and data interpretation in SERS bioimaging by providing faster, more accurate, and detailed insights. AI algorithms, especially those based on deep learning, can preprocess SERS bioimaging data to reduce noise and improve the signal-to-noise ratio. Techniques such as denoising autoencoders and convolutional neural networks (CNNs) can enhance the clarity of SERS spectra and images, making it easier to identify subtle spectral features. He et al. demonstrated that training a deep convolutional neural network to statistically convert low-resolution, high-speed input images into high-resolution images significantly enhances the spatial resolution and imaging speed of line-scanning Raman microscopy (He et al. 2019). In another study, Roman et al. used three vibrational spectroscopy techniques (high-resolution Fourier transform infrared (FT-IR), Raman, and atomic force microscopy infrared (AFM-IR)) for chemical imaging of cholesterol derivatives in PC-3 prostate cancer cells, and these three methods were compared in terms of image quality, spectral pattern, and chemical information (Roman et al. 2020).

Using machine learning algorithms or neural networks, AI can classify SERS spectra into categories, such as distinguishing between healthy and diseased tissues. These models can be trained to recognize complex spectral patterns, reducing the reliance on manual interpretation. He et al. reported imaging and differentiation of melanoma cancer cells and non-cancer cells by revealing molecular difference on the cells using machine learning assisted Raman spectroscopic imaging (He et al. 2022). Moreover, AI supports real-time data processing, quantitative molecular mapping, and the integration of multi-dimensional datasets, providing a comprehensive perspective of biological samples (Barzan et al. 2021). Wang et al. employed Raman spectroscopic imaging to identify the spectral-pathological characteristics of ductal carcinoma in situ and lobular hyperplasia in contrast to healthy breast tissue. The linear discrimination analysis (LDA) technique was utilized in conjunction with principal component analysis (PCA) to categorize the spectral characteristics of tissues at various stages of breast cancer progression (Wang et al. 2021). PCA is a statistical method used to reduce the dimensionality of large datasets while retaining the most important information. In this context, PCA identifies patterns in the spectral data by transforming them into a set of orthogonal components (principal components), which capture the most variance within the dataset, making it easier to distinguish between tissue types. Hollon et al. investigated histological samples using stimulated Raman histology (SRH) with deep convolutional neural networks (CNNs) to facilitate automatic, near real-time diagnostic predictions. The CNN, trained on more than 2.5 million SRH images, demonstrated the capability to predict brain tumor diagnoses in the operating room in under 150 s, significantly outpacing conventional methods (Hollon et al. 2020). Hossain et al. developed a Raman microspectroscopy-based method combined with multivariate curve resolution analysis to identify, distinguish and visualize structurally similar glucan polysaccharides. With this method, they successfully separated Raman spectra of α- and β-glucans from mixtures with high specificity and visualized heterogeneous molecular distributions in a label-free manner at the single cell level during sporulation of yeast (Hossain et al. 2023). By automating and streamlining these procedures, AI not only accelerates discoveries in biomedical research but also provides the framework for breakthrough applications in diagnostics and customized therapy.

4 Applications of AI-Integrated SERS Imaging in Antimicrobial Resistance Research

The combination of SERS imaging with AI technologies possesses the ability to change bioimaging in the fight against AMR. SERS imaging offers remarkable sensitivity and specificity in revealing bacterial structures and their biochemical contexts, facilitating the identification of resistant strains and their underlying processes. AI systems effectively analyze complex, high-dimensional SERS data, detecting minor image and spectrum differences that distinguish resistant bacteria from non-resistant strains. This synergy enables rapid classification of pathogens and real-time monitoring of their responses to antimicrobial treatments. Moreover, AI-powered SERS imaging can detect molecular changes in microbial populations, delivering crucial insights into the evolution of resistance at both the cellular and molecular levels. By combining these technologies, researchers may develop new diagnostic tools that facilitate early identification of AMR and enable personalized treatments to combat resistance more effectively. This new technique not only accelerates the knowledge of AMR mechanisms but also helps the formulation of personalized treatments, which are essential in interacting with this rising global health concern. One of our studies reviews the integration of SERS with AI as a transformative approach to combat AMR (Al-Shaebi et al. 2023). The article highlights advancements in SERS-AI methodologies, focusing on their ability to deliver rapid, accurate, and cost-effective detection of drug-resistant bacteria. The authors discuss the synergistic role of machine learning and deep learning in enhancing the sensitivity and specificity of SERS, emphasizing its potential for clinical applications. The study underscores the critical need for continued research and innovation in SERS-AI systems to address the global AMR crisis effectively (Al-Shaebi et al. 2023).

However, research into the application of SERS imaging in AMR remains limited (Zhang et al. 2023; Hong et al. 2018; Chen et al. 2023). Hong et al. demonstrated that the metabolic uptake of deuterated glucose in a single living bacterium can be quantitatively monitored via hyperspectral stimulated Raman spectroscopy imaging using vancomycin-susceptible and -resistant enterococci *E. faecalis*. The study demonstrated that the metabolic activity of susceptible bacteria reacts differently to antibiotics compared to the resistant strain within just 0.5 h of antibiotic administration (Hong et al. 2018). Chen et al. investigated molecular changes in *C.Albicans* in the presence of antifungal drugs, amphotericin B and fluconazole, using stimulated Raman spectroscopy imaging and deuterium probing (Chen et al. 2023). This research highlights the exciting possibilities of integrating SERS imaging with AI, facilitating significant breakthroughs in the diagnosis and treatment of AMR.

5 Conclusions and Future Perspectives

This chapter highlights the transformative potential of integrating SERS imaging with AI to address the pressing global challenge of AMR. The molecular sensitivity and spatial resolution offered by SERS imaging enable unparalleled detection and visualization of resistant pathogens, while AI adds a layer of intelligence by automating data interpretation, identifying complex patterns, and delivering actionable insights. Together, these technologies redefine the possibilities for early detection, personalized treatment, and a comprehensive understanding of resistance mechanisms.

Looking ahead, several key areas demand attention to fully harness this potential. First, advances in nanomaterial design and synthesis can further enhance the sensitivity, specificity, and multiplexing capabilities of SERS, enabling the detection of even trace levels of resistance biomarkers. Second, the development of more robust and scalable AI algorithms tailored for clinical applications will be crucial for translating laboratory successes into real-world settings. These algorithms must integrate seamlessly with existing clinical workflows while providing interpretable and reliable outputs.

Moreover, expanding the scope of SERS-AI applications beyond diagnostics to include real-time monitoring of treatment efficacy and understanding resistance evolution offers exciting prospects. This requires the standardization of experimental protocols and the creation of comprehensive spectral databases, representing diverse bacterial species and resistance mechanisms. Collaborative efforts between researchers, clinicians, and policymakers will also be essential to overcome regulatory, technical, and adoption barriers.

Ultimately, the integration of SERS imaging and AI has the potential to revolutionize infectious disease diagnostics, enabling a shift from reactive to predictive and personalized healthcare. As the global community continues to confront the challenges posed by AMR, these technologies provide a beacon of hope, promising faster, more accurate, and actionable solutions to safeguard public health.

Acknowledgments This work has been supported by the Erciyes University Scientific Research Projects Coordination Unit under grant number FBAÜ-2023-12265. Zakarya Al-Shaebi has been funded by the National Early-Stage Researchers Program (The Scientific and Technological Research Council of Türkiye (TUBITAK) 121C221). Munevver Akdeniz has been funded by the TUBITAK 2211-A National PhD Scholarship Program.

Conflict of Interest The authors declare no conflicts of interest related to the content of this chapter.

References

Ahmed SK et al (2024) Antimicrobial resistance: impacts, challenges, and future prospects. J Med Surg Public Health 2:100081

Akdeniz M et al (2024) Label-free acute leukemia diagnosis from exosome with SERS and artificial intelligence. In: Proceedings of the 9th world congress on recent advances in nanotechnology (RAN 2024), London, United Kingdom

Al-Shaebi Z et al (2022) Highly accurate identification of bacteria's antibiotic resistance based on Raman spectroscopy and U-net deep learning algorithms. ACS Omega 7(33):29443–29451

Al-Shaebi Z et al (2023) Breakthrough solution for antimicrobial resistance detection: surface-enhanced Raman spectroscopy-based on artificial intelligence. Adv Mater Interfaces:2300664

Al-Shaebi Z et al (2024) Staphylococcus aureus-related antibiotic resistance detection using synergy of surface-enhanced Raman spectroscopy and deep learning. Biomed Signal Process Control 91:105933

Austin LA, Kang B, El-Sayed MA (2013) A new nanotechnology technique for determining drug efficacy using targeted plasmonically enhanced single cell imaging spectroscopy. J Am Chem Soc 135(12):4688–4691

Aydin Ö et al (2009a) Differentiation of healthy brain tissue and tumors using surface-enhanced Raman scattering. Appl Spectrosc 63(10):1095–1100

Aydin Ö et al (2009b) Surface-enhanced Raman scattering of rat tissues. Appl Spectrosc 63(6):662–668

Barzan G et al (2021) Hyperspectral chemical imaging of single bacterial cell structure by Raman spectroscopy and machine learning. Appl Sci 11. https://doi.org/10.3390/app11083409

Berry ME et al (2021) Surface enhanced Raman scattering for the multiplexed detection of pathogenic microorganisms: towards point-of-use applications. Analyst 146(20):6084–6101

Bodelón G et al (2016) Detection and imaging of quorum sensing in Pseudomonas aeruginosa biofilm communities by surface-enhanced resonance Raman scattering. Nat Mater 15(11):1203–1211

Bodelón G et al (2018) Surface-enhanced Raman scattering spectroscopy for label-free analysis of P. aeruginosa Quorum Sensing. Front Cell Infect Microbiol 8:143

Chen H et al (2020) SERS imaging-based aptasensor for ultrasensitive and reproducible detection of influenza virus A. Biosens Bioelectron 167:112496

Chen C et al (2023) Rapid antifungal susceptibility testing based on single-cell metabolism analysis using stimulated Raman scattering imaging. Anal Chem 95(42):15556–15565

Ciloglu FU et al (2021) Drug-resistant Staphylococcus aureus bacteria detection by combining surface-enhanced Raman spectroscopy (SERS) and deep learning techniques. Sci Rep 11(1):18444

Ciloglu FU et al (2022) SERS-based sensor with a machine learning based effective feature extraction technique for fast detection of colistin-resistant Klebsiella pneumoniae. Anal Chim Acta 1221:340094

De Marchi S et al (2019) Surface-enhanced Raman scattering (SERS) imaging of bioactive metabolites in mixed bacterial populations. Appl Mater Today 14:207–215

Deng R et al (2017) Tracing the therapeutic process of targeted aptamer/drug conjugate on cancer cells by surface-enhanced Raman scattering spectroscopy. Anal Chem 89(5):2844–2851

Dong C et al (2022) Simultaneous visualization of dual intercellular signal transductions via SERS imaging of membrane proteins dimerization on single cells. ACS Nano 16(9):14055–14065

El-Mashtoly SF et al (2014) Label-free imaging of drug distribution and metabolism in colon cancer cells by Raman microscopy. Analyst 139(5):1155–1161

Ferraro JR (2003) Introductory raman spectroscopy. Elsevier

Fleischmann M, Hendra PJ, McQuillan AJ (1974) Raman spectra of pyridine adsorbed at a silver electrode. Chem Phys Lett 26(2):163–166

Galvan DD, Yu Q (2018) Surface-enhanced Raman scattering for rapid detection and characterization of antibiotic-resistant bacteria. Adv Healthc Mater 7(13):1701335

Gao X et al (2017) Guiding brain-tumor surgery via blood–brain-barrier-permeable gold nanoprobes with acid-triggered MRI/SERRS signals. Adv Mater 29(21):1603917

Gao Y et al (2020) Bifunctional peptide-conjugated gold nanoparticles for precise and efficient nucleus-targeting bioimaging in live cells. Anal Chem 92(19):13595–13603

Gukowsky JC, He L (2022) Development of a portable SERS method for testing the antibiotic sensitivity of foodborne bacteria. J Microbiol Methods 198:106496

Guo J et al (2018) A multifunctional SERS sticky note for real-time quorum sensing tracing and inactivation of bacterial biofilms. Chem Sci 9(27):5906–5911

Guo J et al (2019) "Three-in-one" SERS adhesive tape for rapid sampling, release, and detection of wound infectious pathogens. ACS Appl Mater Interfaces 11(40):36399–36408

Gvazava N et al (2023) Label-free high-resolution photothermal optical infrared spectroscopy for spatiotemporal chemical analysis in fresh, hydrated living tissues and embryos. J Am Chem Soc 145(45):24796–24808

Han L et al (2023a) Confined target-triggered hot spots for in situ SERS analysis of intranuclear genotoxic markers. Anal Chem 95(15):6312–6322

Han Y-Y et al (2023b) SERS-based rapid susceptibility testing of commonly administered antibiotics on clinically important bacteria species directly from blood culture of bacteremia patients. World J Microbiol Biotechnol 39(10):282

He H et al (2019) Speeding up the line-scan Raman imaging of living cells by deep convolutional neural network. Anal Chem 91(11):7070–7077
He Q et al (2022) Label-free differentiation of cancer and non-cancer cells based on machine-learning-algorithm-assisted fast Raman imaging. Biosensors 12. https://doi.org/10.3390/bios12040250
Heiner Z et al (2017) Surface-enhanced hyper Raman hyperspectral imaging and probing in animal cells. Nanoscale 9(23):8024–8032
Ho C-S et al (2019) Rapid identification of pathogenic bacteria using Raman spectroscopy and deep learning. Nat Commun 10(1):4927
Hollon TC et al (2020) Near real-time intraoperative brain tumor diagnosis using stimulated Raman histology and deep neural networks. Nat Med 26(1):52–58
Hong W et al (2018) Antibiotic susceptibility determination within one cell cycle at single-bacterium level by stimulated Raman metabolic imaging. Anal Chem 90(6):3737–3743
Horgan CC et al (2020) Molecular imaging of extracellular vesicles in vitro via Raman metabolic labelling. J Mater Chem B 8(20):4447–4459
Hossain IM et al (2023) Direct visualization of structurally similar polysaccharides in single yeast cells in vivo by multivariate analysis assisted Raman microspectroscopy. J Phys Chem B 127(23):5249–5256
Ivleva NP et al (2010a) Label-free in situ SERS imaging of biofilms. J Phys Chem B 114(31):10184–10194
Ivleva NP et al (2010b) Raman microscopy and surface-enhanced Raman scattering (SERS) for in situ analysis of biofilms. J Biophotonics 3(8–9):548–556
Jimenez de Aberasturi D et al (2020) Using SERS tags to image the three-dimensional structure of complex cell models. Adv Funct Mater 30(14):1909655
Kim J, Chin YW (2023) Antimicrobial agent against methicillin-resistant Staphylococcus aureus biofilm monitored using Raman spectroscopy. Pharmaceutics 15(7)
Köker T et al (2018) Cellular imaging by targeted assembly of hot-spot SERS and photoacoustic nanoprobes using split-fluorescent protein scaffolds. Nat Commun 9(1):607
Krafft C et al (2017) Label-free molecular imaging of biological cells and tissues by linear and nonlinear Raman spectroscopic approaches. Angew Chem Int Ed 56(16):4392–4430
Kumar S et al (2017) Myoglobin and polydopamine-engineered Raman nanoprobes for detecting, imaging, and monitoring reactive oxygen species in biological samples and living cells. Small 13(43):1701584
Langer J et al (2020) Present and future of surface-enhanced Raman scattering. ACS Nano 14(1):28–117
Lee S et al (2014) Rapid and sensitive phenotypic marker detection on breast cancer cells using surface-enhanced Raman scattering (SERS) imaging. Biosens Bioelectron 51:238–243
Lee JH et al (2021) Real-time surface-enhanced Raman scattering-based live cell monitoring of the changes in mitochondrial membrane potential. Nanoscale Adv 3(12):3470–3480
Li Q et al (2023a) Design and synthesis of SERS materials for in vivo molecular imaging and biosensing. Adv Sci 10(8):2202051
Li J et al (2023b) Imaging immune checkpoint networks in cancer tissues with supermultiplexed SERS nanoprobes. Biomaterials 302:122327
Lima C et al (2021) Imaging isotopically labeled bacteria at the single-cell level using high-resolution optical infrared photothermal spectroscopy. Anal Chem 93(6):3082–3088
Lin L et al (2021) Surface-enhanced Raman scattering nanotags for bioimaging. J Appl Phys 129(19):191101
Liu CY et al (2016) Rapid bacterial antibiotic susceptibility test based on simple surface-enhanced Raman spectroscopic biomarkers. Sci Rep 6:23375
Liu H et al (2022) SERS tags for biomedical detection and bioimaging. Theranostics 12(4):1870–1903
Lu S et al (2022) On-site detection of multiple extracellular antibiotic resistance genes using SERS. Sensors Actuators B Chem 369:132262

Lv J et al (2022) Live-cell profiling of membrane sialic acids by fluorescence imaging combined with SERS labelling. Sensors Actuators B Chem 351:130877

Milewska A, Sigurjonsson OE, Leosson K (2021) SERS imaging of mesenchymal stromal cell differentiation. ACS Appl Bio Mater 4(6):4999–5007

Murray CJL et al (2022) Global burden of bacterial antimicrobial resistance in 2019: a systematic analysis. Lancet 399(10325):629–655

Nicolson F et al (2019) Non-invasive in vivo imaging of cancer using surface-enhanced spatially offset Raman spectroscopy (SESORS). Theranostics 9(20):5899–5913

Olson AP et al (2017) Chemically imaging bacteria with super-resolution SERS on ultra-thin silver substrates. Sci Rep 7(1):9135

Oseledchyk A et al (2017) Folate-targeted surface-enhanced resonance Raman scattering nanoprobe ratiometry for detection of microscopic ovarian cancer. ACS Nano 11(2):1488–1497

Pearson B et al (2017) Innovative sandwich assay with dual optical and SERS sensing mechanisms for bacterial detection. Anal Methods 9(32):4732–4739

Pearson B et al (2018) Rationalizing and advancing the 3-MPBA SERS sandwich assay for rapid detection of bacteria in environmental and food matrices. Food Microbiol 72:89–97

Plou J et al (2022) Prospects of surface-enhanced Raman spectroscopy for biomarker monitoring toward precision medicine. ACS Photonics 9(2):333–350

Qiu C et al (2023) Highly sensitive surface-enhanced Raman scattering (SERS) imaging for phenotypic diagnosis and therapeutic evaluation of breast cancer. Chem Eng J 459:141502

Ravindranath SP et al (2012) Intracellularly grown gold nanoislands as SERS substrates for monitoring chromate, sulfate and nitrate localization sites in remediating bacteria biofilms by Raman chemical imaging. Anal Chim Acta 745:1–9

Roman M et al (2020) Comparison between high definition FT-IR, Raman and AFM-IR for subcellular chemical imaging of cholesteryl esters in prostate cancer cells. J Biophotonics 13(5):e201960094

Salam MA et al (2023) Antimicrobial resistance: a growing serious threat for global public health. Healthcare 11:1946. https://doi.org/10.3390/healthcare11131946

Skvortsova A et al (2022) SERS and advanced chemometrics – utilization of Siamese neural network for picomolar identification of beta-lactam antibiotics resistance gene fragment. Anal Chim Acta 1192:339373

Spadea A et al (2021) Analysis of fixed and live single cells using optical photothermal infrared with concomitant Raman spectroscopy. Anal Chem 93(8):3938–3950

Taha BA et al (2024) Synergizing nanomaterials and artificial intelligence in advanced optical biosensors for precision antimicrobial resistance diagnosis. ACS Synth Biol 13(6):1600–1620

Tang KWK, Millar BC, Moore JE (2023) Antimicrobial Resistance (AMR). 80

Uysal Ciloglu F et al (2020) Identification of methicillin-resistant Staphylococcus aureus bacteria using surface-enhanced Raman spectroscopy and machine learning techniques. Analyst 145(23):7559–7570

Verdin A et al (2020) Multiplex micro-SERS imaging of cancer-related markers in cells and tissues using poly(allylamine)-coated Au@Ag nanoprobes. Anal Bioanal Chem 412(28):7739–7755

Wang W, Vikesland PJ (2023) Metabolite-mediated bacterial antibiotic resistance revealed by surface-enhanced Raman spectroscopy. Environ Sci Technol 57(36):13375–13383

Wang YW et al (2015) In vivo multiplexed molecular imaging of esophageal cancer via spectral endoscopy of topically applied SERS nanoparticles. Biomed Opt Express 6(10):3714–3723

Wang Y et al (2016) Surgical guidance via multiplexed molecular imaging of fresh tissues labeled with SERS-coded nanoparticles. IEEE J Sel Top Quantum Electron 22(4):154–164

Wang W et al (2019) A SERS optophysiological probe for the real-time mapping and simultaneous determination of the carbonate concentration and pH value in a live mouse brain. Angew Chem Int Ed 58(16):5256–5260

Wang S et al (2021) Studying the pathological and biochemical features in breast cancer progression by confocal Raman microspectral imaging of excised tissue samples. J Photochem Photobiol B Biol 222:112280

Wang L et al (2022) Machine learning analysis of SERS fingerprinting for the rapid determination of Mycobacterium tuberculosis infection and drug resistance. Comput Struct Biotechnol J 20:5364–5377

Wei Q et al (2021) Intraoperative assessment and photothermal ablation of the tumor margins using gold nanoparticles. Adv Sci (Weinh) 8(5):2002788

Wen Y, Truong VX, Li M (2021) Real-time intraoperative surface-enhanced Raman spectroscopy-guided thermosurgical eradication of residual microtumors in orthotopic breast cancer. Nano Lett 21(7):3066–3074

Willets KA, Van Duyne RP (2007) Localized Surface Plasmon Resonance Spectroscopy and Sensing. Annu Rev Phys Chem 58:267–297

Xu L et al (2024) Rapid identification of living cancer cells based on label-free surface-enhanced Raman spectroscopy. Opt Commun 569:130806

Yang Y et al (2022) Specific and quantitative detection of bacteria based on surface cell imprinted SERS mapping platform. Biosens Bioelectron 215:114524

Zavaleta CL et al (2009) Multiplexed imaging of surface enhanced Raman scattering nanotags in living mice using noninvasive Raman spectroscopy. Proc Natl Acad Sci USA 106(32):13511–13516

Zhang Y et al (2011) Molecular imaging with SERS-active nanoparticles. Small 7(23):3261–3269

Zhang P et al (2022) Dynamic insights into increasing antibiotic resistance in Staphylococcus aureus by label-free SERS using a portable Raman spectrometer. Spectrochim Acta A Mol Biomol Spectrosc 273:121070

Zhang M et al (2023) Fingerprint stimulated Raman scattering imaging unveils Ergosteryl Ester as a metabolic signature of azole-resistant Candida albicans. Anal Chem 95(26):9901–9913

Zheng X-S et al (2019) Cell-penetrating peptide conjugated SERS nanosensor for in situ intracellular pH imaging of single living cells during cell cycle. Anal Chem 91(13):8383–8389

Zheng Z et al (2022) 2D scanning SERS probe for early biofilm boundary determination. Sensors Actuators B Chem 363:131822

Applications of Computer Vision and Machine Learning in Bioimaging

Ali Özgür Argunşah, Ertunc Erdil, and Devrim Ünay

Abstract

Accurate and reproducible bioimage analysis is crucial for understanding cellular structures and functions. Over the years, the field has progressed from relying heavily on manual analysis and classical image processing techniques to incorporate advanced machine learning and deep learning algorithms. These modern approaches have greatly improved key aspects such as preprocessing, segmentation, feature extraction, object tracking, and classification, enabling automated and reproducible analysis while minimizing subjectivity. The integration of classical and modern computational methods has significantly expanded the capabilities of bioimage analysis. User-friendly software tools and annotation platforms have further made these advanced techniques accessible to researchers lacking extensive computational expertise. This democratization has accelerated the adoption of machine learning in bioimage analysis across various biological disciplines. Real-world applications range from detecting subcellular structures to single-cell segmentation and tissue classification in clinical diagnostics. The ongoing synergy between traditional computer vision and deep learning continues to enhance bioimage analysis and interpretation, paving the way for

A. Ö. Argunşah (✉)
Brain Research Institute, University of Zurich, Zürich, Switzerland
e-mail: argunsah@hifo.uzh.ch

E. Erdil
ETH Zürich, Zürich, Switzerland
e-mail: erdile@ethz.ch

D. Ünay
Department of Electrical-Electronics Engineering, Izmir Democracy University, Faculty of Engineering, Üçkuyular Mahallesi, Gürsel Aksel Bulvarı, İzmir, Türkiye
e-mail: unaydevrim@gmail.com

I. F. Cengiz et al. (eds.), *Bioimaging Modalities in Bioengineering*,
Biomaterials, Bioengineering and Sustainability 9,
https://doi.org/10.1007/978-3-031-99410-4_28

new discoveries in biology and medicine. Consequently, this chapter explores the fundamentals, advances, and challenges in bioimage analysis, highlighting the integration of traditional methods and modern machine learning approaches to revolutionize the field.

Keywords

Bioimage analysis · Computer vision · Machine learning · Fluorescence microscopy · Annotation

1 Introduction

1.1 Historical Perspective and Fundamentals of Bioimage Analysis

Bioimage analysis has advanced significantly in recent years, driven by innovations in imaging technologies and experimental designs which resulted in the collection of vast datasets that require advanced analysis methods (Moen et al. 2019; Duncan et al. 2019; Jan et al. 2024). Through the integration of computer vision (CV) and machine learning (ML), researchers achieve greater accuracy, efficiency, and reproducibility, advancing biomedical research and diagnostics. This chapter covers the fundamentals of bioimage analysis, challenges, CV and ML techniques, image annotation tools, open datasets, and available analysis solutions, culminating in examples where integrated CV-ML approaches yield promising results. Historically, bioimage analysis relied heavily on traditional image processing techniques and subsequent subjective/expert interpretation of the results. Classic image processing focuses on enhancing or modifying images through tasks like noise reduction and sharpening without interpreting their content. In contrast, CV uses data to extract meaningful insights that go beyond apparent visual enhancements. CV has shifted the bioimage analysis paradigm, enabling automated, streamlined, more precise, and objective analysis. This section provides an overview of the bioimage types, the fundamental principles of bioimage analysis, and the key challenges therein.

1.2 Types of Bioimages

Bioimages encompass a wide range of imaging modalities, spanning various spatiotemporal scales in biological research, from microscopic cellular details to whole-organism levels. Figure 1 illustrates examples of bioimaging techniques across these scales. Microscopy stands at the forefront of bioimaging, offering diverse modalities ranging from bright-field imaging to advanced techniques such as confocal microscopy (Fig. 1a and b), multiphoton fluorescence imaging (Fig. 1c and d), and electron microscopy. These techniques enable high-resolution imaging of cellular structures and their function. Specialized approaches like light-sheet

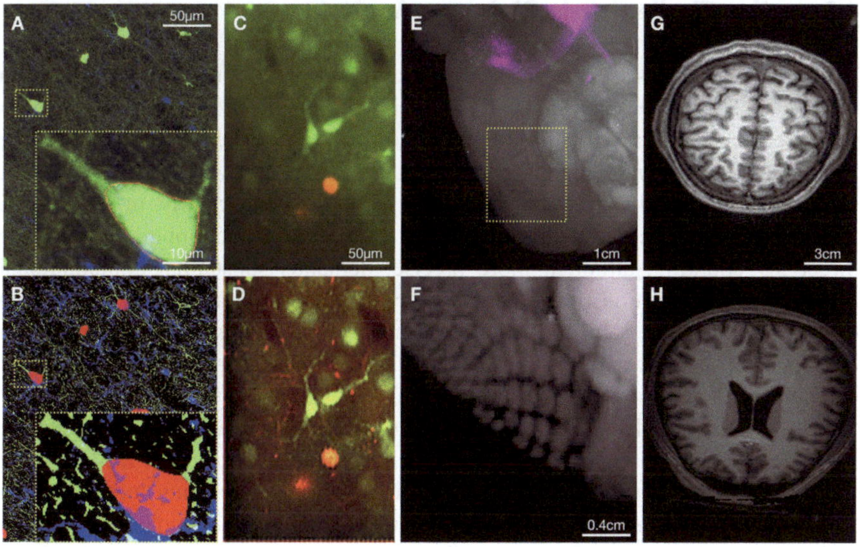

Fig. 1 Bioimages at different scales. (**a**) Maximum intensity projection (MIP) of a confocal image-stack of mouse cortical tissue, sagittal section. (**b**) Segmented version of the same image in (**a**), soma in red, dendrites in green, microglia in blue. (**c**) In vivo two-photon image of mouse cortical tissue, pyramidal neurons in green, interneurons in red. (**d**) Super-resolution enhancement of the image in (**c**). (**e**) 2D projection of a 3D light-sheet image-stack of a mouse brain injected with a neuronal tracer (magenta). (**f**) Contrast adjusted version of the yellow box in E showing barrels of barrel cortex. (**g**) and (**h**) Axial slices of a brain MRI at different depths. (Images in panels (**g**) and (**h**) were obtained from the OpenfMRI database, accession number ds000002 Poldrack and Gorgolewski 2017) and are reprinted under the Open Data Commons Public Domain Dedication and License (PDDL) v1.0

microscopy (Fig. 1e and f) have revolutionized the imaging of large, optically cleared samples such as entire organs or embryos (Huisken et al. 2009; Voigt et al. 2019; Vladimirov et al. 2024). By enabling rapid 3D data acquisition with minimal photodamage, this technique supports long-term imaging studies that were previously challenging to conduct. Time-series imaging further extends these capabilities by capturing dynamic biological processes over time, including cell movement, tissue remodeling, and organism development. Researchers can now observe key cellular mechanisms, such as mitosis, migration, and intracellular transport, in unprecedented detail. At larger scales, techniques like magnetic resonance imaging (MRI) (Fig. 1g and h) and computed tomography (CT) provide three-dimensional views of larger biological structures, offering high-resolution images of organs and tissues that reveal anatomical details and pathological changes in vivo. By leveraging this diverse array of imaging modalities, researchers can explore biological systems at multiple scales, enabling deeper insights into their structure and function.

1.3 Key Challenges in Bioimage Analysis

Analyzing bioimages is inherently complex, as scientists must study phenomena across different scales (Fig. 1), from molecular movements within cells to changes in entire tissues during growth or response to treatments. The diversity of imaging techniques further complicates the development of standardized analysis methods. One significant challenge is accurate cell counting, particularly in 3D stacks. Flattening 3D images into 2D for analysis can obscure overlapping cells, while direct 3D counting, although more precise, demands substantial computational resources. Segmentation, the process of delineating structures, presents its own difficulties. In tissue samples, cellular boundaries often blur due to overlapping components, as shown in Fig. 1a and b. The complexity increases with rapidly changing cell shapes, such as immune cells, or intricate structures like neuron branches. Traditional methods frequently struggle with variability in brightness, indistinct edges, and complex textures, as illustrated in Fig. 1e and f. Identifying irregularly shaped structures also remains a challenge, as existing methods often fail to perform consistently across varying image conditions, sample preparations, and specimen differences. Tracking cellular dynamics over time adds another layer of complexity. Tasks such as monitoring cell division, molecular diffusion, or particle movement in dense environments are hindered by occlusions and the rapid pace of cellular processes, which are often constrained by temporal limitations. Microscopy images also face issues with low signal-to-noise ratios (Ucar et al. 2024; Meiniel et al. 2018; Micó et al. 2019; Haan et al. 2019). While image restoration methods can enhance quality, it is critical to avoid introducing artificial features or compromising essential biological information. Figure 1c and d demonstrate an approach to improve the resolution of two-photon projections by leveraging time-series data (Aydeniz et al. 2022). Standardization across laboratories remains a persistent obstacle. Differences in imaging settings, sample preparation, and analysis workflows make cross-study comparisons difficult. Processing speed and memory usage are additional challenges, especially with large datasets like whole-brain imaging or high-throughput screening. For instance, light-sheet data (Fig. 1e) can exceed 2 TB in size, requiring significant computational resources. Finally, training machine learning systems requires labeled datasets. However, manual annotation of large image sets is tedious, time-consuming, and error-prone. While semi-automated labeling and synthetic training data offer potential solutions, they come with their own challenges, as discussed in Sect. 3.4.

2 Computer Vision Techniques in Bioimage Analysis

Today, CV encompasses a broad range of techniques and applications, from feature extraction and object detection to segmentation and motion analysis (registration and tracking). These advancements are now being integrated into bioimage analysis to tackle the analysis of complex biological datasets, enabling new insights and discoveries in life sciences.

2.1 Feature Extraction and Representation

Feature extraction and representation are essential in bioimage analysis, akin to how the human visual system identifies distinguishing traits. Computational methods analyze key features in images, transforming raw pixel data into structured information. In bioimage analysis, features span a spectrum from basic attributes to complex patterns, analogous to how humans might progress from noticing simple brightness variations to understanding intricate cellular morphology (Meyer-Baese and Schmid 2014). These features can include low-level characteristics such as local intensity distributions and textural patterns, to more sophisticated elements like morphological shapes and distinctive keypoints that define cellular structures (Al-Thelaya et al. 2023). Just as facial recognition relies on key distinguishing features, these biological markers serve as crucial identifiers for computational analysis, allowing automated systems to differentiate between cell types, track developmental changes, or identify pathological conditions (Rittscher 2010).

2.1.1 Texture Analysis

Texture, defined by the arrangement of intensity variations or patterns within an image, serves as a crucial feature in bioimage analysis. It helps differentiate regions that may appear similar in shape or size but vary structurally. Biological tissues and cellular structures often exhibit repetitive patterns, making texture analysis invaluable. For instance, different tissue types in histological images can be identified based on their textural characteristics (Cataldo and Ficarra 2017). Methods like the gray level co-occurrence matrix (GLCM) quantify texture by examining spatial relationships between pixel intensities (Haralick et al. 1973; Sebastian et al. 2012). GLCM has been successfully applied to identify fine-grained structures, such as distinguishing healthy tissue from fibrotic tissue (Chaddad et al. 2015). Wavelet transforms, which capture both frequency and spatial information, are especially useful for analyzing multiscale structures, such as those found in electron microscopy images (Workman et al. 2015). Local Binary Patterns (LBP) encode the local structure around each pixel by comparing its intensity to neighboring pixels and generating a binary code (Ojala et al. 1996). LBP is computationally efficient and has been applied to distinguish cell types (Nanni et al. 2010), identify abnormal tissue patterns (Wan et al. 2017) or classify histopathological images (Kather et al. 2020). Combining LBP with other methods, such as GLCM or wavelets, can improve accuracy in texture-based classification and segmentation (Hamouchene et al. 2014; Ghalati et al. 2021).

2.1.2 Shape Analysis

Shape analysis is essential for understanding the morphology of biological structures in bioimages. Entities such as cells, nuclei, and organelles are often differentiated by shape, reflecting their functional or pathological states (Young 2007). For example, cancer cells frequently exhibit irregular shapes, and quantifying these deviations aids in disease diagnosis (Yevick and Martin 2018). Geometric features

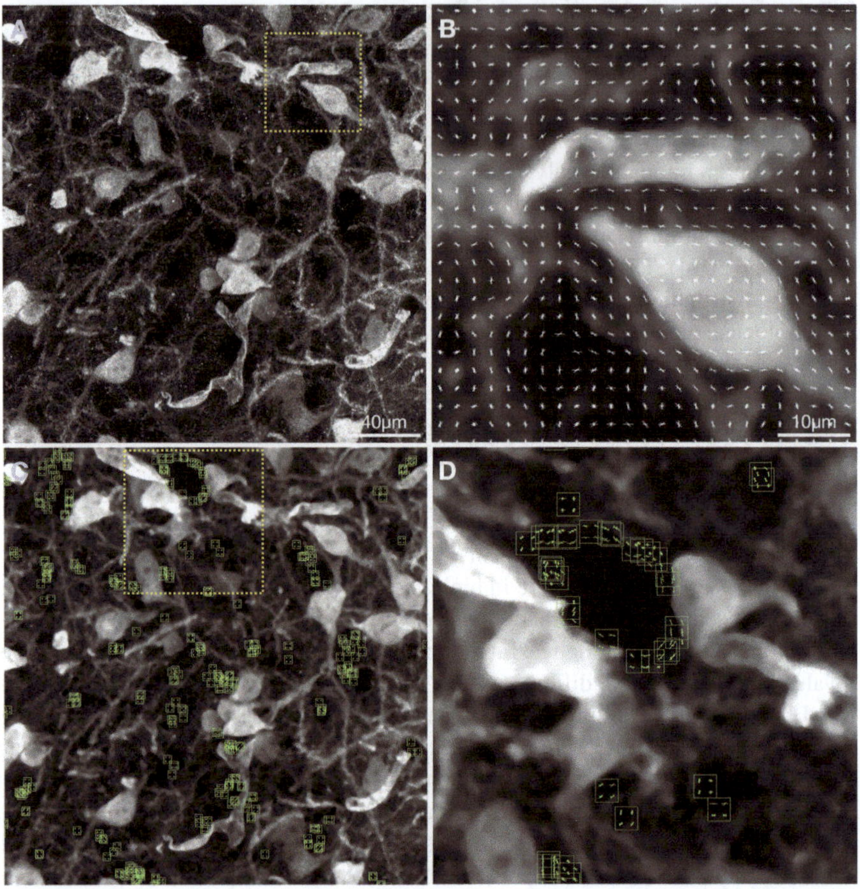

Fig. 2 An example confocal image showing different cells such as neuronal soma, dendrites and blood vessels. (**a**) Two-dimensional maximum intensity projection (MIP) of 3D image stack of mouse brain tissue. (**b**) Histogram of Oriented Gradients (HOG) features of the region in the *yellow box* in (**a**). (**c**) SURF Features of the image in (**a**). (**d**) Zoomed version of the SURF feature image in (**c**)

describing the contours or boundaries of objects can be extracted using contour tracing and region-based methods. Metrics such as eccentricity, circularity, and aspect ratio quantify the form of objects, helping classify cell types or detect abnormal shapes in diagnostic imaging. Histogram of Oriented Gradients (HOG) (Dalal and Triggs 2005) encodes structural and contour information by analyzing gradient orientation distributions (Fig. 2a and b). When paired with classifiers like Support Vector Machines (SVMs) (Vapnik 1997), HOG descriptors are effective for shape-based object detection in bioimages (Liu et al. 2014). For more intricate structures, Fourier descriptors or shape context methods provide detailed, scale-invariant shape information. These approaches have been used to analyze cellular morphology during apoptosis and track developmental shape changes in embryos (Schott et al. 2018; Boquet-Pujadas et al. 2021). Shape descriptors also assist in identifying mitotic figures in histopathology images, where dividing cells exhibit characteristic shape alterations (Gurcan et al. 2009). Additionally, methods such as disjunctive normal shape models are employed to classify dendritic spines based on their morphological features (Ghani et al. 2017).

2.1.3 Keypoint Detection and Description

Keypoint detection and description focus on identifying distinct points or regions in an image that serve as landmarks for subsequent analysis. In bioimages, keypoints often correspond to biologically significant features, such as cell centroids, neuronal branch points, or dendritic tree junctions. These landmarks are crucial for tasks like alignment, tracking, or matching across images. The Scale-Invariant Feature Transform (SIFT) (Lowe 2004) is a widely used method for keypoint detection, robustly identifying features at varying scales and orientations. Similarly, Speeded-Up Robust Features (SURF) (Bay et al. 2008) offers an efficient alternative for detecting keypoints in large bioimage datasets (Fig. 2c and d). These keypoints are described using feature vectors that capture the local image structure around each point, enabling reliable matching across images (Ünay and Stanciu 2018). Keypoint detection is particularly useful in time-lapse microscopy for tracking cell migration or monitoring dynamic cellular interactions (Yang et al. 2023). It is also employed in developmental biology, where the shapes of organisms at different stages are compared by analyzing the positions of predefined anatomical landmarks (Bookstein 1997).

2.2 Object Detection and Segmentation

Detection refers to determining whether there is a statistical distinction, either spatially or temporally. It can be seen as a one-class classification problem (Wei et al. 2018). Classical object detection methods aim to identify specific objects by locating relevant regions in an image and matching them with predefined templates or feature descriptors. Template matching is effective when the objects to be detected have stereotypical shapes. This method involves searching for occurrences of a template image within a larger target image by sliding the template across the image and calculating similarity scores at each position, typically using 2D cross-correlations in either the intensity or Fourier domains. Template matching often requires little to no image preprocessing or annotation. For heterogeneous objects, using multiple template images can improve detection by accounting for varying appearances or orientations (Thomas and Gehrig 2020). However, this approach may face challenges when detecting highly deformable objects or dealing with significant variations in appearance (Uchida 2013).

Segmentation plays a critical role in bioimage analysis, where the goal is to partition images into meaningful structures or regions that correspond to specific biological features, such as cells, tissues, or organs. This step is essential for sub-

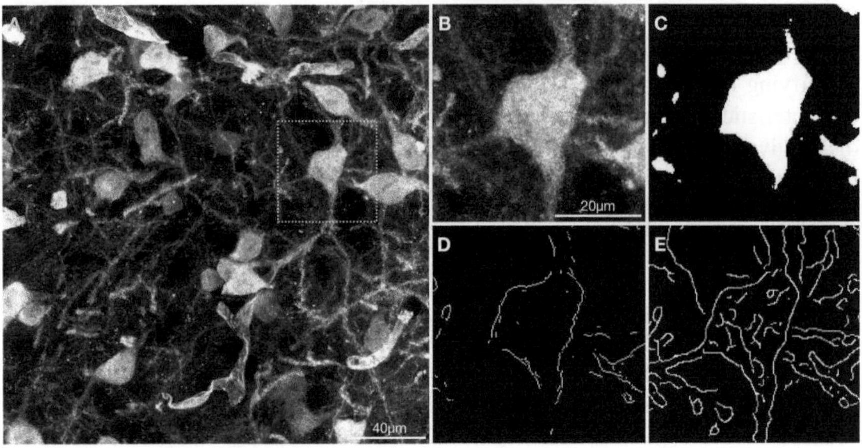

Fig. 3 (**a**) Same MIP of confocal stack as in Fig. 2a. (**b**) Zoom of the soma shown in yellow box in (**a**). (**c**) Otsu-thresholded version of the image in (**b**). (**d**) Edge detection performed on image in (**b**) using Prewitt algorithm. (**e**) Edge detection performed on image in (**b**) using Canny algorithm

sequent tasks such as quantification, classification, and disease diagnosis. Several CV techniques are used for bioimage segmentation, including thresholding, edge detection, and region-based methods. Each of these approaches offers unique ways to extract relevant structures from bioimages.

2.2.1 Thresholding

Thresholding often refers to Otsu's method of segmentation (Otsu et al. 1975), a non-parametric, unsupervised technique for converting grayscale images into binary ones (see Fig. 3a–c). It determines a threshold value that minimizes intra-class variance, assuming a bimodal histogram (foreground and background). However, global methods like Otsu may fail with non-uniform intensity variations (see Fig. 3c). Local thresholding methods, which segment images within localized neighborhoods, are more effective in handling intensity variations or uneven lighting (Luo et al. 2018). Combining local thresholding with Otsu's global approach improves segmentation performance, especially under challenging lighting or contrast conditions (Li et al. 2020).

2.2.2 Edge Detection

Edge detection is a fundamental computer vision technique (see Fig. 3d and e), particularly useful for images with well-defined, non-overlapping objects possessing strong morphological features. It identifies sharp intensity or color contrasts to detect boundaries, aiding in segmentation for tasks like object recognition, tracking, or measurement (see Fig. 3b–e). Common techniques include Sobel, Canny, Prewitt, and Difference or Laplacian of Gaussian filters. The Difference of Gaussians (DoG) filter highlights regions of rapid intensity change by subtracting one blurred version of an image from another (Marr and Poggio 1979; Man and Vision 1982). Similarly, the Laplacian of Gaussian (LoG) filter detects edges through zero-crossings after Gaussian smoothing. In medical imaging, edge detection helps in delineating organ contours (Zhang et al. 2008), identifying lesion boundaries, and detecting tissue abnormalities (Pathak et al. 2000; Somkantha et al. 2010), supporting diagnostic assessment of size, shape, and location.

2.2.3 Intensity and Shape Prior-Based Methods

Before deep learning became dominant, classical region-based methods in CV relied on hand-crafted features, statistical models, and mathematical techniques for tasks like object detection, segmentation, and tracking (Adiga et al. 2000; Wienert et al. 2012). These methods often struggled with complex scenes, object variations, lighting changes, and occlusions. In bioimage analysis, classical region-based segmentation leveraged intrinsic image properties such as fluorescence intensity, spectral characteristics, or texture to group pixels into meaningful regions. Methods like region-growing start from seed points (e.g., organelle intensity or nuclear texture) and expand to neighboring pixels with similar properties. To address the need for global context in complex samples, graph-based algorithms were introduced (Ta et al. 2009). These treat bioimages as graphs, with pixels or regions as nodes and edges representing similarities. Techniques such as normalized cuts enhance segmentation by balancing local and global features, improving accuracy in cellular and tissue analysis across diverse contexts (Shi and Malik 1997).

Level set methods have been pivotal in segmentation by representing shapes as implicit functions (Osher and Fedkiw 2001). Level sets are highly effective shape representations for segmenting complex shapes thanks to their ability to handle various topologies. Earlier methods find optimal level sets that minimize an objective function by only using intensity-based features (Chan and Vese 2001). Then, shape prior information was also used along with the intensity information to deal with ambiguous intensities, which are common in biomedical imaging (Kim et al. 2007; Uzunbas et al. 2010; Erdil et al. 2017, 2016). Similar objectives have also been optimized by using different shape representations such as Disjunctive Normal Shape Models (DNSM) (Mesadi et al. 2017) and snakes (Guo et al. 2021; Delgado-Gonzalo et al. 2014).

2.3 Object Classification

Object classification in bioimage analysis involves categorizing detected and/or segmented biological structures into predefined classes based on their characteristics. This process enables automated analysis of large-scale biological datasets by identifying specific cell types, organelles, tissue structures, or pathological conditions. Traditional methods rely on hand-crafted features extracted during the feature extraction phase (Sect. 2.1). Traditional classifiers such as support vector machines (SVMs), random forests, and k-Nearest Neighbors (k-NN) use features derived from texture analysis, shape measurements (Sect. 2.1.2), and keypoint descriptions (Sect. 2.1.3). These methods are particularly effective when working with limited training data and when the distinguishing characteristics between classes are well-understood. For instance, in blood cell classification (Tai et al. 2011; Lamberti 2021), traditional methods can effectively differentiate between cell types based on size, shape, and texture parameters.

Biological objects often require classification into multiple categories or hierarchical organization. For instance, a cell may need to be classified not only by its type but also by its cell cycle phase (Holtzer et al. 1972), or a tissue section may need to be categorized by both its organ of origin and pathological state. Such cases demand more sophisticated classification strategies, such as assigning multiple non-exclusive labels to a single object (Wang et al. 2011), following a tree-like classification structure from general to specific, or using ensemble methods (Rokach 2010) that combine multiple classifiers to improve accuracy and robustness.

2.4 Registration and Tracking

Registration and tracking are critical components of bioimage analysis, enabling the alignment and temporal analysis of biological structures across images or time frames. Registration involves aligning images from the same or different modalities to a common reference, facilitating comparisons and integration of complementary information. Tracking, on the other hand, focuses on following objects, such as cells or particles, over time to study their dynamics. These techniques are essential for applications like studying cellular behavior, monitoring disease progression, and integrating multimodal datasets. This section explores methods for unimodal and multimodal registration, as well as approaches for effective tracking in bioimaging.

2.4.1 Unimodal Image Registration or Tracking

Unimodal registration aligns images of the same structure captured with the same imaging modality at consecutive time points. An example is dendritic spine imaging, where registration tracks neuronal spine morphological changes over time using two-photon microscopy (Rada et al. 2018). This enables monitoring dynamic processes such as spine shape alterations, volume changes, structural remodeling during plasticity, and spine formation or elimination. For robust alignment advanced algorithms are employed utilizing pixel intensity matching, geometric transformations, and computational methods like mutual information and normalized cross-correlation.

Biological image tracking monitors the movement, deformation, and evolution of structures over time (Meijering et al. 2006). For dendritic spines (Argunşah et al. 2022), tracking enables temporal sequence analysis, quantification of dynamics, trajectory reconstruction, and feature extraction (Ghani et al. 2016b) (see Fig. 4a and b). Techniques range from manual annotation to advanced computational methods, including point-based, contour-based, and region-based tracking.

Classical region-based tracking methods first identify an initial object region and then estimate its evolution over time. For instance, the Mean-Shift algorithm (Schnell 1964; Fukunaga and Hostetler 1975) iteratively shifts a region to maximize

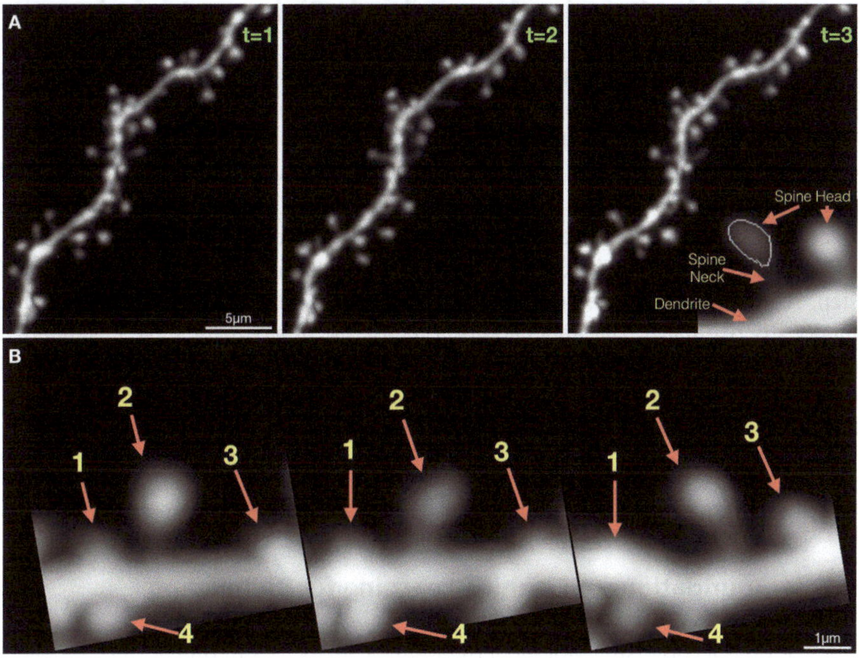

Fig. 4 (**a**) Maximum intensity projection (MIP) of a two-photon 3D image stacks of a dendritic segment of a CA1 hippocampal neuron at three different time points. Inset at the 3rd time point shows the names of different substructures including one segmented spine head. (**b**) Tracking problem concerns with the repeated detection of the same structure and their identity over time. Here there are 4 different dendritic spines as their shapes change after a naturalistic plasticity inducing stimulus

similarity (e.g., based on color histograms). Optical flow estimates pixel motion between frames, while Kalman Filters (Kalman 1960) predict object positions based on prior observations, refining detection through prediction.

2.4.2 Multimodal Image Registration

Multimodal image registration integrates data from complementary imaging techniques to provide a more comprehensive view of biological structures and functions. Examples include registering CT scans with MRI, aligning T1- and T2-weighted MR images, or combining two-photon and confocal microscopy data. In dendritic spine analysis, multimodal registration is used to correlate functional activity from two-photon calcium imaging with structural details captured by confocal microscopy, as well as ultra-high-resolution morphology obtained from electron microscopy. This approach enhances the understanding of the structure-function relationships in neurons (Eastman et al. 2011).

3 Machine Learning Approaches in Bioimage Analysis

The integration of ML into bioimage analysis is fundamentally transforming biological research, offering efficient and accurate interpretations of complex bioimaging data (Moen et al. 2019). Deep learning (DL), as a sophisticated subset of ML, emerges as exceptionally well-suited for analyzing the high-dimensional and heterogeneous datasets characteristic of biological imaging (Meijering 2020). The successful implementation of DL applications relies on three steps. First, researchers must undertake the tedious task of annotating training data. The second step, selection and training of neural network architectures, directly influence the model's performance and applicability. Finally, the deployment stage ensures that these sophisticated trained models can be seamlessly integrated into experimental workflows, bridging the gap between computational innovation and practical scientific investigation. The growing availability of curated datasets and open-source tools is democratizing access to deep learning, enabling automated analyses and novel experimental designs.

3.1 Foundations of Machine Learning for Bioimaging

ML serves as a cornerstone for modern bioimage analysis, offering powerful tools to extract meaningful information from complex and high-dimensional biological data. Unlike traditional rule-based approaches, ML algorithms learn patterns directly from data, enabling automated detection, classification, and segmentation tasks. In bioimaging, the diversity of imaging modalities, demands robust and flexible ML approaches. Key techniques range from supervised learning, where labeled datasets drive the learning process, to unsupervised learning, which uncovers hidden patterns in unlabeled data.

3.1.1 Unsupervised Learning

Unsupervised learning involves finding patterns in unlabeled data, identifying inherent structures without predefined labels. Common tasks include clustering and dimensionality reduction (Nassif et al. 2021). In bioimage analysis, clustering is used for segmenting cellular or subcellular structures in microscopy images, classifying tissue regions in histopathology, and grouping cells based on morphological features (Irshad et al. 2013). Dimensionality reduction techniques, such as principal component analysis (PCA) and t-distributed stochastic neighbor embedding (t-SNE), simplify multi-channel bioimages or extract key features from high-dimensional datasets (Kutsuna et al. 2012; Ejdrup et al. 2022; Wan et al. 2023; Nieves et al. 2023).

3.1.2 Clustering Methods

Clustering is an unsupervised learning technique that automatically assigns data points into clusters such that points within the same cluster are more similar to each other than to points in other clusters. Several methods exist, each suitable for different tasks. K-means is a widely used and simple clustering algorithm that partitions data into k clusters, assigning each data point to the nearest centroid (Lloyd 1982; Macqueen 1967; Forgy 1965). While k is typically user-defined, Akaike or Bayesian Information Criterion (AIC or BIC) can help determining the optimal number of clusters (Akaike 2025; Vrieze 2012). K-means is extensively used in bioimage analysis for tasks like segmenting microscopy images into biologically relevant regions or classifying cell phenotypes based on morphological (Pchitskaya and Bezprozvanny 2020) and intensity features (Safdar et al. 2022; Song et al. 2016; Ghani et al. 2016a). Hierarchical clustering creates a hierarchy of clusters through either agglomerative (bottom-up) or divisive (top-down) methods, producing a dendrogram that illustrates relationships between clusters at various levels of granularity. Unlike K-means, it does not require pre-specifying the number of clusters, but it is computationally expensive for large datasets. In bioimage analysis, it is used to segment tissue regions in histopathological images and classify subcellular structures in fluorescence microscopy based on intensity and spatial patterns, revealing cellular organization (Mittal et al. 2022; Caicedo et al. 2017). Density-Based Spatial Clustering of Applications with Noise (DBSCAN) identifies clusters as regions of high density, making it effective for detecting clusters of arbitrary shape and handling noise (Ester et al. 1996). Unlike K-means, DBSCAN does not require the number of clusters to be specified and can handle non-spherical clusters and outliers. In bioimage analysis, DBSCAN has been used to detect irregularly shaped cell colonies in stem cell research (Hartsock et al. 2024) and group molecules in super-resolution microscopy to reveal nanoscale cellular structures (Hyun and Kim 2023). Gaussian Mixture Models (GMM) is a probabilistic clustering method that represents data as a mixture of multiple Gaussian distributions, each corresponding to a cluster. Unlike K-means, GMM supports soft clustering, where points can belong to multiple clusters with different probabilities, offering flexibility for Gaussian-distributed data. In bioimage analysis, GMM has been applied to segment microscopy images, such as distinguishing cellular from background regions in fluorescence images. Its probabilistic approach is particularly useful for overlapping data points, enabling nuanced biological interpretations (Elisa Gelasca et al. 2009; Hallou et al. 2021).

3.1.3 Dimensionality Reduction

In the context of bioimage analysis and unsupervised learning, dimensionality reduction techniques like PCA, t-SNE and uniform manifold approximation and projection (UMAP) play crucial roles in simplifying complex high-dimensional data while retaining important patterns (Jan et al. 2024). PCA transforms high-dimensional data into fewer dimensions, summarizing key features and reducing computational expense. It is particularly helpful for handling numerous features, such as pixel intensity values, and visualizing complex bioimage data in 2D or 3D to uncover hidden patterns. However, PCA assumes linear relationships and may not capture highly correlated patterns. Additionally, scale matters; variables with larger magnitudes can dominate the results, necessitating standardization for features on different scales (Shamir et al. 2010; Song et al. 2016; Nanni et al. 2018; Chen et al. 2018b). t-SNE is a successful dimensionality reduction method for visualizing high-dimensional bioimage data in 2D or 3D, often revealing clusters missed by PCA (Hinton and Roweis 2002; Van der Maaten and Hinton 2008). It has been applied to single-cell imaging, identifying immune cell subpopulations based on fluorescence profiles in flow cytometry (Saeys et al. 2016), and high-content screening assays to analyze phenotypic diversity under drug treatments (Chandrasekaran et al. 2021). UMAP, a more recent addition to the dimensionality reduction toolkit, has gained popularity in bioimage analysis due to its ability to preserve both local and global structure in the data (McInnes et al. 2018). UMAP often provides faster computation times compared to t-SNE and can handle larger datasets more efficiently. These techniques are unsupervised methods which make them valuable tools for exploratory data analysis in bioimage research.

3.1.4 Supervised Learning

Supervised learning is one of the most widely used approaches in bioimage analysis, where models are trained on labeled data to perform tasks such as classification, segmentation, or object detection. The success of supervised learning relies on the ability to learn from example input-output pairs, meaning that each input image must be accompanied by a corresponding label or annotation that describes the relevant information in the image. Deep learning, particularly through convolutional neural networks (CNNs), has significantly advanced the performance of supervised learning in bioimage analysis. However, these models are highly data-dependent, requiring large volumes of labeled data to achieve high accuracy and robustness. Bioimage data, especially from modalities like microscopy or medical imaging, can be high-dimensional, with millions of pixels needing precise annotations, and biological datasets often exhibit class imbalance, where certain structures like abnormal cells may be rarer compared to others, complicating model training. To overcome these challenges, ensuring access to properly labeled datasets is critical, and innovative annotation tools and workflows have been developed to aid this process. Effective annotation not only enhances the model's ability to generalize to unseen data but also reduces the likelihood of introducing biases or errors. In the following section, we will discuss the importance of annotations and training data, as well as specific tools and methods that facilitate the creation of labeled datasets for supervised learning in bioimage analysis.

3.1.5 Annotations and Training Data

Supervised learning relies on annotated data provided by domain experts. ML models usually need a vast amount of labeled data which is very time consuming to produce. It is very important to have the right data annotations tools in hand to create as many useful annotations as possible. There are multiple alternatives that are free to be used for such purposes each of which has different advantages and disadvantages. An example segmentation mask annotation created using LabelStudio is shown in Fig. 5a and b. Figure 5c shows results of a deep network after training with dendritic spine segmentation and classification dataset (Ghani et al. 2017). In

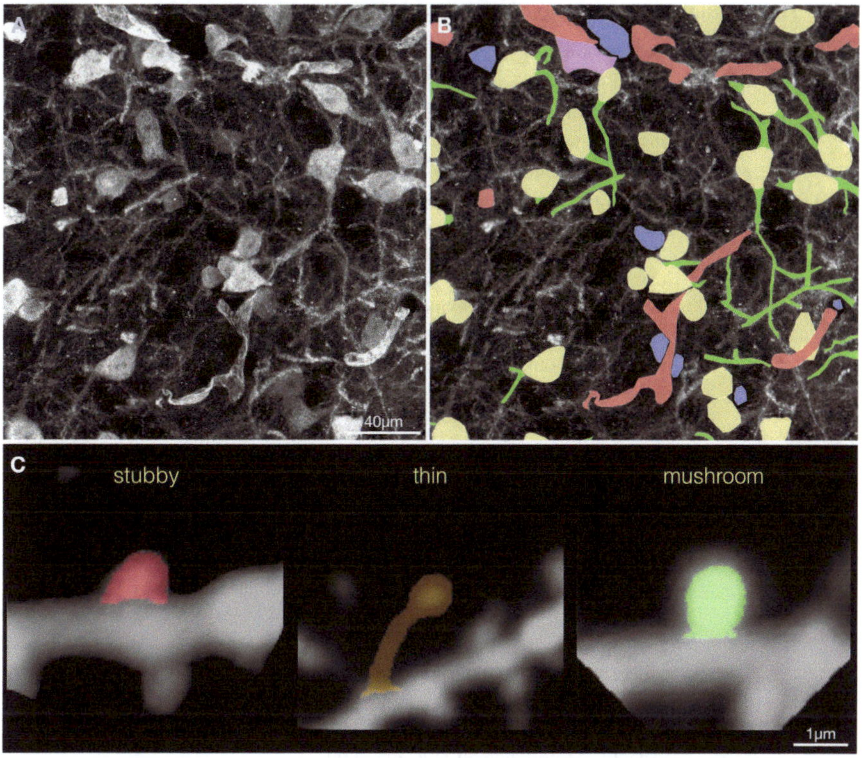

Fig. 5 (**a**) A MIP of a confocal image stack of a mouse brain tissue. (**b**) Manual segmentations of different structures using LabelStudio. *Yellow*: Cell body, *Red*: Blood vessel, *Green*: Dendrite, *Purple*: Negative cell body, *Pink*: Compound structure of multiple objects occluding each other. While cell bodies have more streotypical shapes, dendrites and blood vessels can be more heterogeneous. (**c**) Mask-RCNN segmentation and classification of three different dendritic spine class in a different dataset. Manual annotations for this task were made using a simple annotator GUI in Matlab

Table 1, we will provide a non-exhaustive list of open and/or free annotation tools that are mostly free for individuals or small teams.

3.2 Deep Learning in Bioimage Analysis

DL has completely changed bioimage analysis by automatically extracting and learning hierarchical features from data. Unlike traditional ML, DL eliminates the need for manual feature engineering, making it highly effective for complex tasks.

3.2.1 Detection Models

Modern DL methods have revolutionized object detection, mimicking how our brains quickly identify objects in a room. Instead of eyes and neural pathways,

Table 1 A brief list of open source and/or free tools for data annotation

Tool	Description	Platform
COCO annotator	A tool designed for COCO-style dataset creation, supporting multiple categories and instances with bounding boxes and masks. https://github.com/jsbroks/coco-annotator	Web-based
CVAT	An advanced tool developed by Intel, supporting various annotation types, collaborative annotation, tracking, and automatic annotation using models. https://github.com/openvinotoolkit/cvat	Web-based
Labelbox	A platform providing both free and paid plans, supporting bounding boxes, polygons, and segmentation masks with collaborative features. https://labelbox.com/	Web-based
LabelImg	A graphical image annotation tool supporting bounding box annotations for object detection tasks. https://github.com/tzutalin/labelImg	Windows, MacOS, Linux
LabelMe	An MIT-developed tool with a web interface for polygonal annotation, supporting collaborative annotation. https://github.com/wkentaro/labelme	Web-based
LabelStudio	A flexible data labeling tool for all data types. Image, natural language processing, speech, voice, and video models. https://labelstud.io	Web-based
Napari-Cellseg3D	A plugin for 3D cell segmentation and annotation in bioimages. https://github.com/AdaptiveMotorControlLab/CellSeg3D	Napari
Napari-Labelling	A plugin for interactive image labeling and annotation. https://github.com/haesleinhuepf/napari-labeling	Napari
RectLabel	A simple and powerful tool for bounding box annotation, supporting YOLO, Keras, and TensorFlow formats.https://rectlabel.com/	MacOS
Segmentation editor	A plugin for manual and semi-automated segmentation of images. https://imagej.net/Segmentation_Editor	ImageJ/Fiji
Simple neurite tracer	A plugin for semi-automated tracing and annotation of neurites in 3D images. https://imagej.net/plugins/snt/	ImageJ/Fiji
SVG editor	A generic vector graphics editor for image annotation, supporting various formats with high customization. https://github.com/SVG-Edit/svgedit	Web-based

(continued)

Table 1 (continued)

Tool	Description	Platform
TotalSegmentator annotate	This platform enables global collaboration by splitting the segmentation work over thousands of people to annotate large datasets. https://annotate.totalsegmentator.com/	Web-based
Trainable Weka segmentation	A ML-based plugin for pixel classification and segmentation. https://imagej.net/Trainable_Weka_Segmentation	ImageJ/Fiji
VGG image annotator	A lightweight, standalone software supporting image, video, and audio annotation. Highly customizable. https://github.com/ox-vgg/via	Web-based

they use sophisticated neural network algorithms. One notable detection method, YOLO (You Only Look Once), mimics the way we observe a scene to recognize multiple objects simultaneously (Redmon et al. 2016). It divides an image into a grid, analyzes each section once, and determines what objects are present and their locations. This grid-based approach ensures both speed and accuracy. In contrast, Faster R-CNN employs a two-stage process for more precise detection. It first generates region proposals with a region proposal network (RPN) and then classifies these regions and refines their bounding boxes through a convolutional neural network (Ren et al. 2015). This method offers higher precision than YOLO, making it ideal for accuracy-critical applications like bioimaging. DL solutions are increasingly applied to detect cell nuclei (Xie et al. 2018), mitochondria (Li et al. 2019), and cellular events like mitosis (Albarqouni et al. 2016; Phan et al. 2018). However, extensive evaluations are needed to fully assess their capabilities (Meijering 2020).

3.2.2 Segmentation Models

U-Net is a DL model designed for biological image segmentation. It uses an encoder-decoder structure to capture context and detail, enabling precise separation of individual cells or structures in complex microscopy images. The system operates like a two-part process: first, it "zooms out" to understand the bigger picture (the encoder part), then "zooms in" to draw precise boundaries around each structure (the decoder part). This approach is particularly effective for biological images where cells might overlap or cluster together, much like trying to count and outline each fish in a crowded school. The result is remarkably accurate identification of cell boundaries, shapes, and locations, enabling scientists to count cells, analyze tissue structures, and track how cellular components move and change over time with unprecedented precision. By combining low-level spatial features with high-level semantic information through skip connections, U-Net preserves both fine details and broader context, making it well-suited for handling complex bioimaging data (Ronneberger et al. 2015). Variants of U-Net, such as 3D U-Net (Çiçek et al. 2016) and attention U-Net (Oktay et al. 2018), further refine segmentation by adapting to volumetric data and improving focus on salient features, respectively. These models are pivotal in applications like tumor boundary detection and neuron mapping, where precise and automated segmentation is essential for advancing biological research and clinical diagnostics (Abdusalomov et al. 2023). Accordingly, deep learning is increasingly penetrating the field of bioimage analysis with individual segmentation solutions (Araújo et al. 2019; Long 2020; Allehaibi et al. 2019; Mamunur Rahaman et al. 2020; Sait Erdem et al. 2023) as well as international competitions (Caicedo et al. 2019; Uhlmann et al. 2022; Arganda-Carreras et al. 2015; Sirinukunwattana et al. 2017).

3.2.3 Classification Models

CNNs and RNNs have become integral to bioimaging for learning hierarchical and temporal features from complex biological data. CNNs, originally for image recognition, use convolutional layers to extract spatial features, making them ideal for bioimaging (Xing et al. 2017). RNNs, especially Long Short-Term Memory (LSTM) networks, excel at capturing sequential information crucial for temporal bioimaging data (Yu-Ting et al. 2017). Tasks like tracking cell movement and analyzing time-lapse microscopy use RNNs to model temporal changes, offering insights into dynamic cellular processes. By learning dependencies across frames, RNNs enable automated analysis of cell migration, development, and protein dynamics. Analyzing cell videos requires capturing spatial features like shapes and temporal patterns like movement. Hybrid models combining CNNs for spatial details and RNNs for temporal dynamics excel in tasks like cell tracking and nucleus segmentation, enabling robust spatial-temporal analysis (Wang et al. 2018; Xing et al. 2018).

3.3 Deep Learning with Limited Labels

Advanced learning paradigms address limitations of conventional approaches by leveraging unlabeled data, reducing dependency on extensive annotations, and adapting models to new domains. These techniques, including self-supervised, semi-supervised, and transfer learning, have become increasingly valuable in bioimage analysis, where labeled datasets are often scarce or challenging to acquire.

3.3.1 Self- and Semi-Supervised Learning

Supervised DL methods have achieved state-of-the-art performance across numerous applications, thanks to their ability to learn from large labeled datasets. A pivotal moment came with the release of the ImageNet dataset (Deng et al. 2009), which enabled Krizhevsky et al. (Krizhevsky et al. 2012) to train a deep network that reduced the error rate of prior top-performing models by nearly 50%, showcasing the transformative potential of DL in CV. This success spurred advancements in medical image analysis, where significant accuracy gains have been achieved despite the scarcity of large labeled datasets through architectures like U-Net (Ronneberger et al. 2015), V-Net (Milletari et al. 2016), and DeepMedic (Kamnitsas et al. 2016). However, the performance of supervised methods still heavily depends on access to extensive labeled data, which remains a challenge to obtain due to the labor-intensive, costly, and expertise-driven annotation process.

Various strategies have been developed to tackle the challenge of limited labeled datasets. Data augmentation is one of the simplest and most cost-effective approaches, expanding the labeled training set by applying random transformations to existing data, significantly improving the performance of deep networks (Zhang et al. 2020). Common techniques include random affine transformations (Cireşan et al. 2011), contrast adjustments (Ronneberger et al. 2015), and elastic deformations (Perez et al. 2018; Hong et al. 2017). Advanced methods like mix-up (Zhang 2017; Eaton-Rosen et al. 2018) create new training samples by alpha-blending two randomly selected images and their labels. Synthetic data generation using deep generative models, such as GANs (Goodfellow et al. 2020) and diffusion models (Sohl-Dickstein et al. 2015; Ho et al. 2020; Kingma et al. 2021), have also proven effective (Trabucco et al. 2023; Chaitanya et al. 2021; Hoo-Chang et al. 2018).

A common approach to addressing the limited labeled dataset problem is leveraging the abundance of unlabeled data, which is often readily available. Semi-supervised learning is a family of techniques designed to utilize both labeled and unlabeled data (Chapelle et al. 2009; Grandvalet and Bengio 2004; Kingma et al. 2014; Dong-Hyun et al. 2013; Laine and Aila 2016; Miyato et al. 2018; Rasmus et al. 2015; Sajjadi et al. 2016; Tarvainen and Valpola 2017). Among these techniques, self-training is the most widely used and has been applied to numerous problems including medical image analysis (Bai et al. 2017). Self-training begins by training a neural network on the limited labeled dataset. This network is then used to predict labels for the unlabeled samples. Predictions with high confidence, referred to as pseudo-labels, are incorporated into the training set. This process is repeated iteratively throughout the training phase. The method assumes that the high-confidence predictions are generally correct. However, deep neural networks are known to make overconfident yet incorrect predictions (Guo et al. 2017). As a result, relying solely on confidence to select pseudo-labeled samples can lead to the inclusion of incorrect labels in the training set, which may significantly degrade the model's performance. Another effective group of semi-supervised learning methods is based on adversarial training. A pioneering approach introduced by Zhang et al. (2017) for medical image segmentation employs two interconnected network architectures: a segmentation network and an evaluation network. The segmentation network generates segmentation predictions for both labeled and unlabeled images, while the evaluation network assesses the quality of these predictions. The two networks are trained jointly using an adversarial loss. The evaluation network is trained to differentiate between segmentation predictions from the unlabeled set and those from the labeled set. Simultaneously, the segmentation network is optimized to produce segmentations for unlabeled images that the evaluation network cannot distinguish from the segmentations generated for labeled samples.

Self-supervised learning (SSL) is another way of exploiting unlabeled datasets when training deep neural networks. SSL methods have been extremely powerful and to go approach especially when it comes to exploiting from massive amount of unlabeled datasets. They are successfully used in many applications including natural language processing (Devlin et al. 2018), computer vision (Oquab et al. 2023), and biomedical image analysis (Pérez-García et al. 2024). One group of SSL methods are called as pretext task-based methods which focus on pretext tasks, where labels are automatically and easily derived from unlabeled data. These tasks are used to train the network on the unlabeled data, enabling it to learn a strong initialization. The network is then fine-tuned on a labeled training dataset with limited annotations using supervised learning. The assumption is that solving the pretext task helps the network capture useful features that benefit downstream tasks like classification or segmentation. This robust initialization allows the model to train on limited labeled data without quickly overfitting. Some commonly used pretext tasks, particularly in earlier works, include predicting the rotation of the input image (Gidaris et al. 2018), filling in missing pixel values through inpainting (Pathak et al. 2016), restoring spatial context (Chen et al. 2019), and various others (Doersch et al. 2015; Noroozi and Favaro 2016; Zhang et al. 2016; Dosovitskiy et al. 2014).

More recently, with the success of transformer architectures (Vaswani et al. 2017) in computer vision (Dosovitskiy 2020), methods like masked image modeling (MIM) (Xie et al. 2022) and masked autoencoders (MAE) (He et al. 2022) have gained significant popularity due to their ability to learn rich feature representations. These approaches work by masking portions of an image and using the context from the visible parts to reconstruct the missing areas. Specifically, images are divided into patches, with a large proportion masked out. The remaining visible patches are processed by a transformer-based encoder, such as ViT (Dosovitskiy 2020) or Swin (Liu et al. 2021), while a lightweight decoder is used to reconstruct the masked regions. After self-supervised pretraining, the encoder is fine-tuned for downstream tasks. MIMs and MAEs are highly effective and widely adopted, with numerous variations (Chen et al. 2022; Gandelsman et al. 2022; Yi et al. 2022; Lee et al. 2022; Huang et al. 2024; Hondru et al. 2024), and have been successfully applied in biomedical image analysis as well (Zhou et al. 2023; Xiao et al. 2023; Quan et al. 2024; Buess et al. 2024). Another widely used and highly successful group of SSL methods is based on contrastive learning. Although contrastive learning concepts were explored earlier (Bachman et al. 2024), its popularity surged after the seminal work by Chen et al. (2020), which introduced the SimCLR method. SimCLR's architecture includes an encoder E and a projection head h. Latent representations are generated for two differently augmented versions, x' and $\hat{x}$, of an image x as $z' = E(h(x'))$ and $\hat{z} = E(h(\hat{x}))$. The network is trained using a contrastive loss objective, which minimizes the cosine distance between these two representations as positive pairs while maximizing the distance between these representations and those of other samples in the mini-batch as negative pairs. A limitation of SimCLR is that negative pairs within the mini-batch may include samples with similar contextual information to the positive samples. Forcing their representations to diverge in such cases can be suboptimal. To address this issue in volumetric bioimage segmentation, Chaitanya et al. (2020, 2023) leverage domain knowledge by selecting slices that are likely to be anatomically similar as positive pairs, while treating the remaining slices as negative pairs. Other methods bypass the need for negative pairs but risk trivial solutions, mapping all samples to the same representation. To prevent this, they employ regularization techniques. Notable and highly successful approaches in this category include BYOL (Grill et al. 2020), SiamSiam (Chen and He 2021), Barlow Twins (Zbontar et al. 2021), and VicReg (Bardes et al. 2021). Contrastive learning-based methods have been successfully applied to many biomedical image analysis problems (Perakis et al. 2021; Li et al. 2021; Wu et al. 2024; Ciga et al. 2022; Wang et al. 2022).

3.3.2 Transfer Learning

Transfer learning, a popular approach since early DL, uses a pretrained network on a large dataset and fine-tunes its parameters with limited labeled data. A common strategy is to use pretrained network architectures (e.g., ResNet He et al. 2016, VGG Simonyan and Zisserman 2014) trained on the ImageNet dataset as the starting point and adapt them to the target dataset. Despite the differences between ImageNet and medical images, transferring models from ImageNet to medical datasets has become a standard practice (Matsoukas et al. 2022). More recently, with the emergence of so-called *foundational models* (e.g., DinoV2 Oquab et al. 2023, SAM Kirillov et al. 2023) trained on datasets much larger than ImageNet, the use of these models as a starting point for fine-tuning has gained increasing popularity. However, the domain gap between popular foundational models and medical images remains a challenge, as these models are primarily trained on natural images. To address this issue while leveraging the capabilities of foundational models, domain-specific foundational models are becoming increasingly common. Some examples are MedSAM (Ma et al. 2024), RadDino (Pérez-García et al. 2024), TotalSegmentator (D'Antonoli et al. 2024), and SAM2-PATH (Zhang et al. 2024), which are specifically designed for biomedical image analysis.

Parameter-efficient fine-tuning (PEFT) addresses selecting parameters to update when fine-tuning foundational models, avoiding catastrophic forgetting. It complements traditional methods like updating last layers or batch normalization parameters. LoRA (Hu et al. 2021), initially designed for language models, introduces learnable low-rank matrices to approximate pretrained parameters. During fine-tuning, only these matrices are updated, preserving the original network's knowledge. Similarly, Ladder fine-tuning (Chai et al. 2023) employs a CNN-based encoder-decoder architecture alongside the foundational model, training only the CNN parameters. Many PEFT methods, originally from NLP, are being adapted for image-based models including biomedical image analysis (Cekmeceli et al. 2024).

3.4 Challenges of Using Deep Learning Methods

DL has achieved state-of-the-art results across numerous applications, including biomedical image analysis. However, its deployment in real-world scenarios still faces significant challenges. In this section, we explore some of the common obstacles in biomedical image analysis.

3.4.1 Domain Shift

Supervised ML, particularly DL, has achieved expert-level accuracy in biomedical image analysis but faces challenges in real-world deployment, notably due to distribution shifts between training and testing data. Castro et al. (2020) categorize distribution shifts in biomedical images into population, annotation, prevalence, manifestation, and acquisition shifts. Acquisition shift, caused by factors like imaging modality, contrast materials, device vendor, and operator variability, is the most common and has been the focus of much research in the field. Addressing acquisition (covariate) shift in biomedical image analysis is challenging. Ideally, a new model would be trained for each domain, but this requires large, labeled datasets, which are costly and time-intensive to create, as noted in Sect. 3.3.1. Supervised transfer learning can mitigate the need for extensive labeled data from the target domain but may still be impractical in certain scenarios since it still requires labeling effort. Unsupervised domain adaptation (UDA) approaches aim to minimize the dependence on labeled data in the target domain by utilizing labeled data from the source domain alongside unlabeled data from the target domain for domain adaptation (Tsai et al. 2018; Ganin and Lempitsky 2015; Kamnitsas et al. 2017; Hoyer et al. 2022). However, a significant challenge in applying UDA methods lies in their reliance on access to labeled data from the source domain. In many cases, access to the training dataset is no longer available after the model has been trained. Even when access is maintained, sharing the dataset is typically restricted due to privacy regulations, quite common in medical imaging. For example, consider a scenario where a model is trained in one hospital using its proprietary labeled dataset. To deploy this model in another hospital, both the trained model and the labeled training data would need to be transferred to the target hospital to perform UDA and adapt the model to the new environment. However, such data transfers would breach patient privacy and violate ethical and legal standards. Domain generalization methods address this issue by training neural networks on the source domain in a way that enables the model to generalize across multiple domains. Earlier approaches, such as data augmentation (Zhang et al. 2020) and meta-learning (Li et al. 2018; Khandelwal and Yushkevich 2020), have proven to be highly effective. More recently, Cekmeceli et al. (2024) have shown that foundational models, even those trained on natural images, exhibit remarkable generalization capabilities in medical imaging tasks. Test-time or source-free domain adaptation addresses distribution shifts without requiring source domain data or labeled target domain samples, adapting using only test images. Key questions include which parameters to adapt and the strategy for adaptation. Wang et al. (2020) suggest updating batch normalization parameters by minimizing prediction entropy to boost confidence. Karani et al. (2021) propose a test-time segmentation method using shape priors, employing a normalizer network and a pretrained denoising autoencoder (DAE) to correct segmentations. Adaptation is guided by the DAE's reconstruction loss, updating the normalizer's parameters. Bateson et al. (2022) extend this with shape moments as priors, while Prabhudesai et al. (2023) pretrain a class-conditional diffusion model, adapting it at test time by minimizing its objective conditioned on predicted classes.

3.4.2 Explainability

Deep learning methods are often regarded as "black box" models, making their decision-making processes difficult to interpret and the rationale behind their predictions challenging to understand. This lack of transparency is particularly critical in safety-sensitive domains such as biomedical image analysis, where trustworthiness is very important. Explaining the reasoning behind the decisions of these models not only builds trust but also aids in debugging these systems, identifying potential failure points, and driving improvements. The integration of explainable AI (XAI) into clinical workflows has become essential for gaining acceptance among clinicians and biologists, particularly in high-stakes fields like histopathology and neuroimaging. Explainable AI (XAI) aims to improve the interpretability and transparency of machine learning models, especially deep learning systems, which are increasingly applied in bioimage analysis. In biomedical applications, XAI addresses critical concerns related to trustworthiness, accountability, and regulatory compliance. Techniques such as saliency maps, Grad-CAM (Gao et al. 2023), and SHAP (SHapley Additive exPlanations) (Gevaert et al. 2024) allow researchers to visualize which image features influence model predictions. These methods help ensure that models focus on biologically relevant patterns, while feature attribution techniques provide deeper insights into how specific characteristics—such as texture, shape, and intensity—contribute to decision-making. Recent advancements in XAI have extended to vision-language models (VLMs), where the reasoning of neural networks is explained using another neural network based on large language models (LLMs) (Rehman Hashmi et al. 2024). Leveraging LLMs to communicate the decisions of deep neural networks is a promising research direction, as it aligns more closely with human communication methods.

3.4.3 Other Challenges

Data imbalance is a significant challenge in applying machine learning to biomedical data, where datasets often have highly skewed class distributions. Models trained on such data tend to favor the majority class, leading to poor performance on the minority class, which is often the most clinically significant. Errors in minority class predictions can have severe consequences, such as misdiagnosing critical conditions (Sangalli et al. 2021). Addressing this issue commonly involves resampling techniques, such as oversampling the minority class, undersampling the majority class, or using synthetic data generation methods. Additionally, assigning higher weights to the loss contributions of the minority class or implementing algorithms designed to account for class imbalance can improve performance. Communicating uncertainty is another critical challenge, as deep neural networks are prone to making overconfident predictions even when they are wrong (Guo et al. 2017). Reliable uncertainty estimation is crucial in systems where humans and AI collaborate, allowing uncertain predictions to be referred to experts and reducing overall errors (Sangalli et al. 2024). Methods for uncertainty quantification include Bayesian neural networks (Blundell et al. 2015), deep ensembles (Fort et al. 2019), MC dropout (Gal and Ghahramani 2016), and specialized tools like Probabilistic UNet (Kohl et al. 2018) and PhiSeg (Baumgartner et al. 2019; Gantenbein et al. 2020). Conformal prediction (Angelopoulos and Bates 2021) has also gained attention for refining uncertainty estimates with guarantees based on initial measurements.

3.5 Datasets and Tools

Robust datasets and accessible tools are fundamental to advancing bioimage analysis. High-quality datasets provide the foundation for training and validating ML models. Furthermore, a wide array of software tools, frameworks, and workflows has been developed to streamline data annotation, analysis, and model deployment. This section highlights key datasets in bioimaging and reviews popular tools and platforms.

3.5.1 Key Datasets for Bioimaging

There are several excellent resources available for finding free datasets of bioimages:

1. **BioImage Archive:** A free, publicly available resource that stores and distributes bioimages (Hartley et al. 2022). It accepts submissions from any modality, as long as the data are associated with a peer-reviewed publication or have value beyond a single experiment. The archive also provides AI-ready datasets with relevant tags and visualizations.
2. **AI-ready datasets in the BioImage Archive:** A collection of AI/Machine Learning datasets, many of which have been converted to OME-Zarr format (Hartley et al. 2022; Zulueta-Coarasa et al. 2023). These datasets cover various biological imaging applications, including:

   - Zebrafish embryo development
   - 3D cell segmentation
   - Multi-channel fluorescence images of tissues and cells

3. **BIII (Bio-Image Information Index):** BIII provides a curated list of bioimage datasets (Zhang et al. 2023), including:

   - SNEMI3D: 3D Segmentation of neurites in EM images
   - Nuclei segmentation in histopathology images
   - MoNuSeg - Multi-organ nuclei segmentation challenge
   - FAIRsharing Euro-BioImaging collection

4. **Broad Bioimage Benchmark Collection (BBBC):** Contains various image sets for benchmarking bioimage analysis algorithms (Zhang et al. 2023).

5. **Other resources:**

- **CREMI challenge:** Provides serial section Transmission Electron Microscopy (ssTEM) data (Zhang et al. 2023)
- **Diadem challenge:** Offers manually traced digital neural reconstructions (Zhang et al. 2023)
- **NCI Imaging Data Commons (IDC):** Contains histopathology images, including 2D brightfield whole slide images and multiplex IF datasets (Fedorov et al. 2021)
- **CAMAD and WHAD:** A collection of manually annotated phase contrast microscopy time-lapse images from wound healing (WHAD) and cell adhesion and motility (CAMAD) assays (Iheme et al. 2024)
- **Open Microscopy Environment (OME) Database:** A platform offering access to standardized bioimaging data (Goldberg et al. 2005)
- **The Cancer Imaging Archive (TCIA):** A repository of medical images, particularly focused on cancer research (Clark et al. 2013)
- **Allen Institute for Cell Science:** Provides a comprehensive collection of cell biology images (Chen et al. 2018a)

3.5.2 Software Tools, Frameworks, and Workflows

A non-exhaustive list of software tools, frameworks, and workflows is presented in Table 2. Interested readers may consult the European BioImage Analysis Symposium webpage for a more comprehensive list or the works (Miura and Sladoje 2020, 2022) for collections of bioimage analysis workflows.

3.6 Integrating Computer Vision and Machine Learning for Specific Bioimaging Applications

The integration of CV and ML has transformed bioimaging, enabling tasks such as cell segmentation, counting, and tissue classification with greater accuracy and automation. Traditional CV techniques, while effective for simpler images, struggled with complex bioimages featuring densely packed or irregularly shaped cells. Machine learning, especially deep learning models like U-Net, has addressed these challenges by learning hierarchical features directly from annotated data, making segmentation and counting tasks more robust and adaptable. For example, in pathology, deep learning-based models quantify tumor cell populations, aiding in assessing tumor margins and heterogeneity. Similarly, in live-cell imaging, segmentation tools track cell migration and division, essential in cancer research and developmental biology. In neuroscience, segmentation of individual neurons from dense brain tissue images helps study neural circuit functionality. These automated approaches process large datasets efficiently, minimizing manual annotation and improving diagnostic workflows. In cases of overlapping or occluded cells, density-based models, such as Fully Convolutional Networks (FCNs), estimate cell counts

Table 2 A non-exhaustive list of tools, workflows, and frameworks designed in different platforms for bioimage analysis. See https://eubias.org/demo/linked_data_demos.html for a more comprehensive list

Tool/Category	Description/Workflow	Platform
ImageJ workflows	(1) Electrophoresis gel image analysis (Natale et al. 2011), (2) High-throughput cell contour extraction (Erguvan et al. 2019), (3) Quantification of neurodegeneration in Drosophila (Brazill et al. 2018), (4) Segmentation and analysis of retinal microglia (Ash et al. 2021)	ImageJ/Fiji
Generic workflows	(1) Analysis of wound healing and cell motility assays from phase-contrast optical microscopy time-lapse images (Sait Erdem et al. 2023)	Web-based
Matlab workflows	(1) Spatiotemporal quantification of cell migration (Zabary and Zaritsky 2022), (2) Resolving clathrin-mediated endocytosis using correlative light and electron microscopy (Schorb and Paul-Gilloteaux 2020)	Matlab
ImageJ macros/plugins	(1) Neurohistological image analysis suite (Timothy and Forlano 2019), (2) Myopic Choroidal neovascularization analysis (Deshpande et al. 2023), (3) Whole slide analysis (Vincenzo Della et al. 2017), (4) Fiber diameter quantification (Huling et al. 2022), (5) Mitochondrial network morphology analysis (Valente et al. 2017), (6) Axon regeneration quantification (Patel et al. 2018), (7) Fluorescence lifetime image analysis (Gao et al. 2020), (8) Electron microscopy image denoising (Roels et al. 2020), (9) 3D segmentation of nuclear objects (Dubos et al. 2022), (10) Visualizing and measuring colocalization in cells (Stauffer et al. 2018).	ImageJ/Fiji
Softwares	(1) ImageJ/Fiji: A popular software for bioimage analysis tool with numerous plugins (2) CellProfiler: A software for automated analysis of bioimages. Used for tracking, segmentation, and quantification (3) ilastik: A machine learning-based software for image segmentation and classification (4) Cytomine: A platform for collaborative and scalable analysis of large histopathology images (5) BioImage Suite: A suite of bioimage analysis tools for 3D visualization, segmentation, and modeling of biological structures (6) KNIME: An open-source platform for data analytics and image processing workflows, used for biological image analysis	Stand-alone
Frameworks	(1) BIAFLOWS: A collaborative framework to reproducibly deploy and benchmark bioimage analysis workflows (Rubens et al. 2020)	Stand-alone

using density maps. This method is particularly useful in scenarios like stem cell research or hematology, where traditional segmentation struggles. For instance, counting white blood cells in blood smears or clustering stem cell colonies can now be performed with higher accuracy using these models.

ML has also revolutionized tissue classification and grading, essential for understanding disease severity and prognosis. Models like convolutional neural networks (CNNs) and Vision Transformers (ViTs) excel at identifying tissue types and classifying pathological characteristics. In cancer diagnosis, for instance, ML models trained on histopathological images can classify tumor types and grades with high accuracy, providing consistent, objective support for clinical decisions. Automated grading based on cellular differentiation and tissue structures improves the speed and precision of assessments critical for treatment planning. Figure 6 illustrates how integrating CV and ML enhances segmentation. When neural networks are trained on limited data, segmentation may be suboptimal (Fig. 6a–d). Preprocessing with structure-enhancing filters significantly improves results (Fig. 6e–h). Neuroimaging has also benefited from ML, improving brain region segmentation, disease diagnosis, and studying brain connectivity. Traditional methods like thresholding struggled with variability and noise, but deep learning models such as 3D U-Nets have improved segmentation accuracy for structures like the hippocampus and cortex. These tools enable longitudinal studies by ensuring consistent segmentation across time points, aiding in tracking disease progression. In diagnosing neurodegenerative diseases such as Alzheimer's or multiple sclerosis

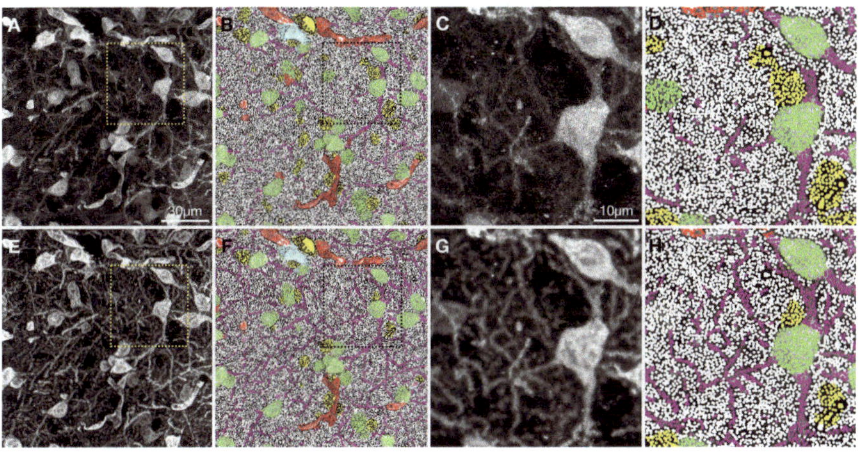

Fig. 6 (**a**) MIP of a confocal image stack of a cortical mouse brain tissue. (**b**) Automatic segmentations of different structures using a deep neural network after preparing segmentation mask annotations using LabelStudio. (**c**) Zoom of the region shown in yellow box in (**a**). (**d**) Zoom of the segmented region shown in black box in (**b**). (**e**) Enhancement of tubular structures in A using a multiscale vessel enhancement filter. (**f**) Using the same annotations combined with the tubular enhancement improves the segmentation of the tubular structures. (**g**) Zoom of the region shown in yellow box in (**e**). (**h**) Zoom of the segmented region shown in black box in (**f**)

(MS), ML models detect subtle imaging patterns indicative of early disease stages (Yalciner et al. 2019; Cash et al. 2015; Carass et al. 2017). For example, CNNs can identify atrophy, plaques, and lesions in MRI and PET scans. These models also provide prognostic information, predicting disease progression and treatment efficacy (Unay 2012). ML techniques enhance precision medicine by tailoring treatment plans based on imaging profiles, enabling earlier intervention and improved outcomes. This integration of CV and ML has fundamentally advanced bioimaging by automating complex tasks, improving diagnostic accuracy, and enabling novel experimental approaches across biological research and clinical applications.

4 Conclusions and Future Perspectives

As bioimaging technologies advance, integrating more sophisticated machine learning models will continue to enhance cell segmentation and counting. One key area of ongoing research is improving the generalization capabilities of machine learning models across different imaging platforms (e.g., brightfield vs. fluorescence microscopy) and biological conditions (e.g., healthy vs. pathological tissue). Another promising direction is the application of self-supervised learning and transfer learning, which can reduce the dependency on large amounts of annotated data, further accelerating the adoption of machine learning in bioimaging applications. Moreover, the integration of these methods into clinical workflows is already proving to enhance diagnostic accuracy and efficiency, as well as reducing inter-/intra-observer variability in radiology and pathology. As computational power and algorithmic sophistication continue to evolve, these technologies will become even more integral to both research and clinical applications, paving the way for more automated, reproducible, and scalable bioimage analysis.

Advancements in bioimaging are evolving hand-in-hand with developments in AI and imaging technologies, with emerging trends focused on improving analysis accuracy, expanding applications, and addressing key limitations across the field. Self-supervised learning is reducing dependency on labeled data by leveraging unlabeled datasets for pretraining models (Azizi et al. 2023), while multimodal imaging analysis integrates data from various sources like MRI, fluorescence, and electron microscopy to provide a more comprehensive understanding of biological systems (Bischof et al. 2024). The field is also making strides in spatio-temporal analysis to capture dynamic processes such as cell migration and tissue development, while Graph Neural Networks (GNNs) are enabling more nuanced analysis of complex biological structures like cellular interactions and brain networks (Cui et al. 2022). Additionally, AI-driven hypothesis generation through unsupervised models is opening new avenues for scientific discovery (Xianyu et al. 2024), while federated learning is facilitating collaborative model training across institutions while preserving data privacy (Rehman et al. 2023). The increasing emphasis on explainability and ethics ensures the development of interpretable and unbiased AI systems, and real-time analysis capabilities are enabling live bioimage analysis during experiments, particularly for applications like cellular behavior tracking.

Acknowledgments This work is partially supported by the Scientific and Technological Research Council of Türkiye (TUBITAK) under grants 119E578, 113E603 and 111E083; and by the EU COST Action CA15124. Authors would like to thank Theofanis Karayannis and his lab members, especially George Kanatouris, Linbi Cai, Asim Iqbal, Martha Gjikolaj and Lorenzo Gesuita for providing bioimages and corresponding segmentation masks.

Conflict of Interest The authors have no conflicts of interest to declare that are relevant to the content of this chapter.

References

Abdusalomov AB, Mukhiddinov M, Whangbo TK (2023) Brain tumor detection based on deep learning approaches and magnetic resonance imaging. Cancers 15(16):4172

Adiga PSU, Chaudhuri BB (2000) Region based techniques for segmentation of volumetric histopathological images. Comput Methods Programs Biomed 61(1):23–47

Akaike H (2025) Akaike's Information Criterion. In: Lovric M (ed) International encyclopedia of statistical science. Springer, Berlin/Heidelberg. https://doi.org/10.1007/978-3-662-69359-9_14

Al-Thelaya K, Gilal NU, Alzubaidi M, Majeed F, Agus M, Schneider J, Househ M (2023) Applications of discriminative and deep learning feature extraction methods for whole slide image analysis: a survey. J Pathol Inf 14:100335

Albarqouni S, Baur C, Achilles F, Belagiannis V, Demirci S, Navab N (2016) Aggnet: deep learning from crowds for mitosis detection in breast cancer histology images. IEEE Trans Med Imaging 35(5):1313–1321

Allehaibi KHS, Nugroho LE, Lazuardi L, Prabuwono AS, Mantoro T et al (2019) Segmentation and classification of cervical cells using deep learning. IEEE Access 7:116925–116941

Angelopoulos AN, Bates S (2021) A gentle introduction to conformal prediction and distribution-free uncertainty quantification. Preprint, arXiv:2107.07511

Araújo FHD, Silva RRV, Ushizima DM, Rezende MT, Carneiro CM, Bianchi AGC, Medeiros FNS (2019) Deep learning for cell image segmentation and ranking. Comput Med Imaging Graph 72:13–21

Arganda-Carreras I, Turaga SC, Berger DR, Cireşan D, Giusti A, Gambardella LM, Schmidhuber J, Laptev D, Dwivedi S, Buhmann JM et al (2015) Crowdsourcing the creation of image segmentation algorithms for connectomics. Front Neuroanat 9:152591

Argunşah AÖ, Erdil E, Ghani MU, Ramiro-Cortés Y, Hobbiss AF, Karayannis T, Çetin M, Israely I, Ünay D (2022) An interactive time series image analysis software for dendritic spines. Sci Rep 12(1):12405

Ash NF, Massengill MT, Harmer L, Jafri A, Lewin AS (2021) Automated segmentation and analysis of retinal microglia within imagej. Exp Eye Res 203:108416

Aydeniz B, Metin SC, Turkan M, Unay D, Karayannis T, Argunsah AÖ (2022) Enhancing two-photon images for anatomical visualisation using super-resolution. In 2022 medical technologies congress (TIPTEKNO). IEEE, pp. 1–4

Azizi S, Culp L, Freyberg J, Mustafa B, Baur S, Kornblith S, Chen T, Tomasev N, Mitrović J, Strachan P et al (2023) Robust and data-efficient generalization of self-supervised machine learning for diagnostic imaging. Nature Biomed Eng 7(6):756–779

Bachman P, Hjelm RD, Buchwalter W (2024) Learning representations by maximizing mutual information across views. Adv Neural Inf Process Syst 32

Bai W, Oktay O, Sinclair M, Suzuki H, Rajchl M, Tarroni G, Glocker B, King A, Matthews PM, Rueckert D (2017) Semi-supervised learning for network-based cardiac mr image segmentation. In Medical image computing and computer-assisted intervention- MICCAI 2017: 20th international conference, Quebec City, QC, Canada, September 11–13, 2017, Proceedings, Part II 20. Springer, pp. 253–260

Bardes A, Ponce J, LeCun Y (2021) Vicreg: variance-invariance-covariance regularization for self-supervised learning. Preprint, arXiv:2105.04906

Bateson M, Lombaert H, Ayed IB (2022) Test-time adaptation with shape moments for image segmentation. In International conference on medical image computing and computer-assisted intervention. Springer, pp. 736–745

Baumgartner CF, Tezcan KC, Chaitanya K, Hötker AM, Muehlematter UJ, Schawkat K, Becker AS, Donati O, Konukoglu E (2019) Phiseg: capturing uncertainty in medical image segmentation. In Medical image computing and computer assisted intervention–MICCAI 2019: 22nd international conference, Shenzhen, China, October 13–17, 2019, Proceedings, Part II 22. Springer, pp. 119–127

Bay H, Ess A, Tuytelaars T, Van Gool L (2008) Speeded-up robust features (surf). Comput Vis Image Understanding 110(3):346–359

Bischof J, Fletcher G, Verkade P, Kuntner C, Fernandez-Rodriguez J, Chaabane L, Rose LA, Walter A, Vandenbosch M, van Zandvoort MAMJ et al (2024) Multimodal bioimaging across disciplines and scales: challenges, opportunities and breaking down barriers. npj Imaging 2(1):5

Blundell C, Cornebise J, Kavukcuoglu K, Wierstra D (2015) Weight uncertainty in neural network. In International conference on machine learning (PMLR), pp. 1613–1622

Bookstein FL (1997) Landmark methods for forms without landmarks: morphometrics of group differences in outline shape. Med Image Anal 1(3):225–243

Boquet-Pujadas A, Olivo-Marin J-C, Guillén N (2021) Bioimage analysis and cell motility. Patterns 2(1)

Brazill JM, Zhu Y, Li C, Zhai RG (2018) Quantitative cell biology of neurodegeneration in drosophila through unbiased analysis of fluorescently tagged proteins using imagej. J Vis Exp Jove (138)

Buess L, Stollenga MF, Schinz D, Wiestler B, Kirschke J, Maier A, Navab N, Keicher M (2024) Video-CT MAE: self-supervised video-CT domain adaptation for vertebral fracture diagnosis. Proceedings of the 7nd international conference on medical imaging with deep learning, in proceedings of machine learning research, vol 250, pp 151–167. Available from https://proceedings.mlr.press/v250/buess24a.html

Caicedo JC, Cooper S, Heigwer F, Warchal S, Qiu P, Molnar C, Vasilevich AS, Barry JD, Bansal HS, Kraus R et al (2017) Data-analysis strategies for image-based cell profiling. Nature Methods 14(9):849–863

Caicedo JC, Goodman A, Karhohs KW, Cimini BA, Ackerman J, Haghighi M, Heng CK, Becker T, Doan M, McQuin C et al (2019) Nucleus segmentation across imaging experiments: the 2018 data science bowl. Nature Methods 16(12):1247–1253

Carass A, Roy S, Jog A, Cuzzocreo JL, Magrath E, Gherman A, Button J, Nguyen J, Prados F, Sudre CH, Cardoso MJ, Cawley N, Ciccarelli O, Wheeler-Kingshott CAM, Ourselin S, Catanese L, Deshpande H, Maurel P, Commowick O, Barillot C, Tomas-Fernandez X, Warfield SK, Vaidya S, Chunduru A, Muthuganapathy R, Krishnamurthi G, Jesson A, Arbel T, Maier O, Handels H, Iheme LO, Unay D, Jain S, Sima DM, Smeets D, Ghafoorian M, Platel B, Birenbaum A, Greenspan H, Bazin P-L, Calabresi PA, Crainiceanu CM, Ellingsen LM, Reich DS, Prince JL, Pham DL (2017) Longitudinal multiple sclerosis lesion segmentation: resource and challenge. NeuroImage 148:77–102

Cash DM, Frost C, Iheme LO, Ünay D, Kandemir M, Fripp J, Salvado O, Bourgeat P, Reuter M, Fischl B, Lorenzi M, Frisoni GB, Pennec X, Pierson RK, Gunter JL, Senjem ML, Jack CR, Guizard N, Fonov VS, Louis Collins D, Modat M, Jorge Cardoso M, Leung KK, Wang H, Das SR, Yushkevich PA, Malone IB, Fox NC, Schott JM, Ourselin S (2015) Assessing atrophy measurement techniques in dementia: results from the miriad atrophy challenge. NeuroImage 123:149–164

Castro DC, Walker I, Glocker B (2020) Causality matters in medical imaging. Nature Commun 11(1):3673

Cekmeceli K, Himmetoglu M, Tombak GI, Susmelj A, Erdil E, Konukoglu E (2024) Do vision foundation models enhance domain generalization in medical image segmentation? Preprint, arXiv:2409.07960

Chaddad A, Zinn PO, Colen RR (2015) Radiomics texture feature extraction for characterizing gbm phenotypes using glcm. In 2015 IEEE 12th international symposium on biomedical imaging (ISBI). IEEE, pp. 84–87

Chai S, Jain RK, Teng S, Liu J, Li Y, Tateyama T, Chen Y-W (2023) Ladder fine-tuning approach for sam integrating complementary network. Preprint, arXiv:2306.12737

Chaitanya K, Erdil E, Karani N, Konukoglu E (2020) Contrastive learning of global and local features for medical image segmentation with limited annotations. Adv Neural Inf Process Syst 33:12546–12558

Chaitanya K, Karani N, Baumgartner CF, Erdil E, Becker A, Donati O, Konukoglu E (2021) Semi-supervised task-driven data augmentation for medical image segmentation. Med Image Anal 68:101934

Chaitanya K, Erdil E, Karani N, Konukoglu E (2023) Local contrastive loss with pseudo-label based self-training for semi-supervised medical image segmentation. Med Image Anal 87:102792

Chan TF, Vese LA (2001) Active contours without edges. IEEE Trans Image Process 10(2):266–277

Chandrasekaran SN, Ceulemans H, Boyd JD, Carpenter AE (2021) Image-based profiling for drug discovery: due for a machine-learning upgrade? Nature Rev Drug Discov 20(2):145–159

Chapelle O, Scholkopf B, Zien A (2009) Semi-supervised learning (Chapelle, O. et al., eds.; 2006) [book reviews]. IEEE Trans Neural Netw 20(3):542–542

Chen X, He K (2021) Exploring simple siamese representation learning. In Proceedings of the IEEE/CVF conference on computer vision and pattern recognition, pp. 15750–15758

Chen J, Ding L, Viana MP, Lee HW, Sluezwski MF, Morris B, Hendershott MC, Yang R, Mueller IA, Rafelski SM (2018a) The allen cell and structure segmenter: a new open source toolkit for segmenting 3d intracellular structures in fluorescence microscopy images. BioRxiv, p. 491035

Chen W, Li W, Dong X, Pei J (2018b) A review of biological image analysis. Curr Bioinf 13(4):337–343

Chen L, Bentley P, Mori K, Misawa K, Fujiwara M, Rueckert D (2019) Self-supervised learning for medical image analysis using image context restoration. Med Image Anal 58:101539

Chen T, Kornblith S, Norouzi M, Hinton G (2020) A simple framework for contrastive learning of visual representations. In International conference on machine learning (PMLR), pp. 1597–1607

Chen Y, Liu Y, Jiang D, Zhang X, Dai W, Xiong H, Tian Q (2022) Sdae: self-distillated masked autoencoder. In European conference on computer vision. Springer, pp. 108–124

Çiçek Ö, Abdulkadir A, Lienkamp SS, Brox T, Ronneberger O (2016) 3d u-net: Learning dense volumetric segmentation from sparse annotation. In International conference on medical image computing and computer-assisted intervention (MICCAI). Springer, pp. 424–432

Ciga O, Xu T, Martel AL (2022) Self supervised contrastive learning for digital histopathology. Mach Learn Appl 7:100198

Cireşan DC, Meier U, Masci J, Gambardella LM, Schmidhuber J (2011) High-performance neural networks for visual object classification. Preprint, arXiv:1102.0183

Clark K, Vendt B, Smith K, Freymann J, Kirby J, Koppel P, Moore S, Phillips S, Maffitt D, Pringle M et al (2013) The cancer imaging archive (tcia): maintaining and operating a public information repository. J Digital Imaging 26:1045–1057

Cui H, Dai W, Zhu Y, Kan X, Gu AAC, Lukemire J, Zhan L, He L, Guo Y, Yang C (2022) Braingb: a benchmark for brain network analysis with graph neural networks. IEEE Trans Med Imaging 42(2):493–506

D'Antonoli TA, Berger LK, Indrakanti AK, Vishwanathan N, Weiß J, Jung M, Berkarda Z, Rau A, Reisert M, Küstner T et al (2024) Totalsegmentator mri: Sequence-independent segmentation of 59 anatomical structures in mr images. Preprint, arXiv:2405.19492

Dalal N, Triggs B (2005) Histograms of oriented gradients for human detection. In 2005 IEEE computer society conference on computer vision and pattern recognition (CVPR'05), vol. 1. IEEE, pp. 886–893 de Haan K, Rivenson Y, Wu Y, Ozcan A (2019) Deep-learning-based image reconstruction and enhancement in optical microscopy. Proc IEEE 108(1):30–50

Delgado-Gonzalo R, Uhlmann V, Schmitter D, Unser M (2014) Snakes on a plane: a perfect snap for bioimage analysis. IEEE Signal Process Mag 32(1):41–48

Deng J, Dong W, Socher R, Li L-J, Li K, Fei-Fei L (2009) Imagenet: a large-scale hierarchical image database. In 2009 IEEE conference on computer vision and pattern recognition. IEEE, pp. 248–255

Deshpande A, Raman S, Dubey A, Susvar P, Raman R (2023) An imagej macro tool for octa-based quantitative analysis of myopic choroidal neovascularization. PLOS ONE 18(4):1–11

Devlin J (2018) Bert: pre-training of deep bidirectional transformers for language understanding. Preprint, arXiv:1810.04805

Di Cataldo S, Ficarra E (2017) Mining textural knowledge in biological images: applications, methods and trends. Comput Struct Biotechnol J 15:56–67

Doersch C, Gupta A, Efros AA (2015) Unsupervised visual representation learning by context prediction. In Proceedings of the IEEE international conference on computer vision, pp. 1422–1430

Dong-Hyun L et al (2013) Pseudo-label: the simple and efficient semi-supervised learning method for deep neural networks. In Workshop on challenges in representation learning, ICML, vol. 3, p. 896

Dosovitskiy A (2020) An image is worth 16x16 words: transformers for image recognition at scale. Preprint, arXiv:2010.11929

Dosovitskiy A, Springenberg JT, Riedmiller M, Brox T (2014) Discriminative unsupervised feature learning with convolutional neural networks. Adv Neural Inf Proc Syst 27

Dubos T, Poulet A, Thomson G, Péry E, Chausse F, Tatout C, Desset S, van Wolfswinkel JC, Jacob Y (2022) Nodej: an imagej plugin for 3d segmentation of nuclear objects. BMC Bioinf 23(1):216

Duncan JS, Insana MF, Ayache N (2019) Biomedical imaging and analysis in the age of big data and deep learning [scanning the issue]. Proc IEEE 108(1):3–10

Eastman RD, Netanyahu NS, Le Moigne J (2011) Survey of image registration methods. Image Registration Remote Sens 21:35–76

Eaton-Rosen Z, Bragman F, Ourselin S, Cardoso MJ (2018) Improving data augmentation for medical image segmentation. In: international conference on medical imaging with deep learning

Ejdrup AL, Lycas MD, Lorenzen N, Konomi A, Herborg F, Madsen KL, Gether U (2022) A density-based enrichment measure for assessing colocalization in single-molecule localization microscopy data. Nature Commun 13(1):4388

Elisa Gelasca D, Obara B, Fedorov D, Kvilekval K, Manjunath BS (2009) A biosegmentation benchmark for evaluation of bioimage analysis methods. BMC Bioinf 10:1–12

Erdil E, Yildirim S, Cetin M, Tasdizen T (2016) Mcmc shape sampling for image segmentation with nonparametric shape priors. In Proceedings of the IEEE conference on computer vision and pattern recognition, pp. 411–419

Erdil E, Ghani MU, Rada L, Argunsah AO, Unay D, Tasdizen T, Cetin M (2017) Nonparametric joint shape and feature priors for image segmentation. IEEE Trans Image Process 26(11):5312–5323

Erguvan Ö, Louveaux M, Hamant O, Verger S (2019) Imagej surfcut: a user-friendly pipeline for high-throughput extraction of cell contours from 3d image stacks. BMC Biol 17:1–12

Ester M, Kriegel H-P, Sander J, Xu X et al (1996) A density-based algorithm for discovering clusters in large spatial databases with noise. In kdd, vol. 96, pp. 226–231

Fedorov A, Longabaugh WJR, Pot D, Clunie DA, Pieper S, Aerts HJWL, Homeyer A, Lewis R, Akbarzadeh A, Bontempi D et al (2021) Nci imaging data commons. Cancer Res 81(16):4188–4193

Forgy EW (1965) Cluster analysis of multivariate data: efficiency versus interpretability of classifications. Biometrics 21:768–769

Fort S, Hu H, Lakshminarayanan B (2019) Deep ensembles: a loss landscape perspective. Preprint, arXiv:1912.02757

Fukunaga K, Hostetler L (1975) The estimation of the gradient of a density function, with applications in pattern recognition. IEEE Trans Inf Theory 21(1):32–40

Gal Y, Ghahramani Z (2016) Dropout as a bayesian approximation: representing model uncertainty in deep learning. In International conference on machine learning (PMLR), pp. 1050–1059

Gandelsman Y, Sun Y, Chen X, Efros A (2022) Test-time training with masked autoencoders. Adv Neural Inf Process Syst 35:29374–29385

Ganin Y, Lempitsky V (2015) Unsupervised domain adaptation by backpropagation. In International conference on machine learning (PMLR), pp. 1180–1189

Gantenbein M, Erdil E, Konukoglu E (2020) Revphiseg: a memory-efficient neural network for uncertainty quantification in medical image segmentation. In Uncertainty for safe utilization of machine learning in medical imaging, and graphs in biomedical image analysis: second international workshop, UNSURE 2020, and third international workshop, GRAIL 2020, held in conjunction with MICCAI 2020, Lima, Peru, October 8, 2020, Proceedings 2. Springer, pp. 13–22

Gao D, Barber PR, Chacko JV, Sagar MAK, Rueden CT, Grislis AR, Hiner MC, Eliceiri KW (2020) Flimj: an open-source imagej toolkit for fluorescence lifetime image data analysis. PloS ONE 15(12):e0238327

Gao Y, Liu J, Li W, Hou M, Li Y, Zhao H (2023) Augmented grad-cam++: super-resolution saliency maps for visual interpretation of deep neural network. Electronics 12(23):4846

Gevaert A, Rousseau A-J, Becker T, Valkenborg D, De Bie T, Saeys Y (2024) Evaluating feature attribution methods in the image domain. Mach Learn 1–46

Ghalati MK, Nunes A, Ferreira H, Serranho P, Bernardes R (2021) Texture analysis and its applications in biomedical imaging: a survey. IEEE Rev Biomed Eng 15:222–246

Ghani MU, Erdil E, Kanık SD, Argunşah AÖ, Hobbiss AF, Israely I, Ünay D, Taşdizen T, Cetin M (2016a) Dendritic spine shape analysis: a clustering perspective. In Computer vision–ECCV 2016 workshops: Amsterdam, The Netherlands, October 8–10 and 15–16, 2016, Proceedings, Part I 14. Springer, pp. 256–273

Ghani MU, Mesadi F, Kanık SD, Argunşah AÖ, Israely I, Ünay D, Taşdizen T, Çetin M (2016b) Dendritic spine shape analysis using disjunctive normal shape models. In 2016 IEEE 13th international symposium on biomedical imaging (ISBI). IEEE, pp. 347–350

Ghani MU, Mesadi F, Kanık SD, Argunşah AÖ, Hobbiss AF, Israely I, Ünay D, Taşdizen T, Çetin M (2017) Dendritic spine classification using shape and appearance features based on two-photon microscopy. J Neurosci Methods 279:13–21

Gidaris S, Singh P, Komodakis N (2018) Unsupervised representation learning by predicting image rotations. Preprint, arXiv:1803.07728

Goldberg IG, Allan C, Burel J-M, Creager D, Falconi A, Hochheiser H, Johnston J, Mellen J, Sorger PK, Swedlow JR (2005) The open microscopy environment (ome) data model and xml file: open tools for informatics and quantitative analysis in biological imaging. Genome Biol 6:1–13

Goodfellow I, Pouget-Abadie J, Mirza M, Xu B, Warde-Farley D, Ozair S, Courville A, Bengio Y (2020) Generative adversarial networks. Commun ACM 63(11):139–144

Grandvalet Y, Bengio Y (2004) Semi-supervised learning by entropy minimization. Adv Neural Inf Process Syst 17

Grill J-B, Strub F, Altché F, Tallec C, Richemond P, Buchatskaya E, Doersch C, Pires BA, Guo Z, Azar MG et al (2020) Bootstrap your own latent-a new approach to self-supervised learning. Adv Neural Inf Process Syst 33:21271–21284

Guo C, Pleiss G, Sun Y, Weinberger KQ (2017) On calibration of modern neural networks. In International conference on machine learning (PMLR), pp. 1321–1330

Guo L, Liu Y, Wang Y, Duan Y, Tai X-C (2021) Learned snakes for 3d image segmentation. Signal Process 183:108013

Gurcan MN, Boucheron LE, Can A, Madabhushi A, Rajpoot NM, Yener B (2009) Histopathological image analysis: a review. IEEE Rev Biomed Eng 2:147–171

Hallou A, Yevick H, Dumitrascu B, Uhlmann V (2021) Deep learning for bioimage analysis. Preprint, arXiv:2107.02584
Hamouchene I, Aouat S, Lacheheb H (2014) Texture segmentation and matching using lbp operator and glcm matrix. In Intelligent systems for science and information: extended and selected results from the science and information conference 2013. Springer, pp. 389–407
Haralick RM, Shanmugam K, Dinstein IH (1973) Textural features for image classification. IEEE Trans Syst Man Cybern (6):610–621
Hartley M, Kleywegt GJ, Patwardhan A, Sarkans U, Swedlow JR, Brazma A (2022) The bioimage archive–building a home for life-sciences microscopy data. J Mol Biol 434(11):167505
Hartsock I, Park E, Toppen J, Bubenik P, Dimitrova ES, Kemp ML, Cruz DA (2024) Topological data analysis of pattern formation of human induced pluripotent stem cell colonies. bioRxiv
He K, Zhang X, Ren S, Sun J (2016) Deep residual learning for image recognition. In Proceedings of the IEEE conference on computer vision and pattern recognition, pp. 770–778
He K, Chen X, Xie S, Li Y, Dollár P, Girshick R (2022) Masked autoencoders are scalable vision learners. In Proceedings of the IEEE/CVF conference on computer vision and pattern recognition, pp. 16000–16009
Hinton GE, Roweis S (2002) Stochastic neighbor embedding. Adv Neural Inf Process Syst 15
Ho J, Jain A, Abbeel P (2020) Denoising diffusion probabilistic models. Adv Neural Inf Process Syst 33:6840–6851
Holtzer H, Weintraub H, Mayne R, Mochan B (1972) The cell cycle, cell lineages, and cell differentiation. Curr Top Dev Biol 7:229–256
Hondru V, Croitoru FA, Minaee S, Ionescu RT, Sebe N (2024) Masked image modeling: a survey. Preprint, arXiv:2408.06687
Hong J, Park B-Y, Park H (2017) Convolutional neural network classifier for distinguishing barrett's esophagus and neoplasia endomicroscopy images. In 2017 39th annual international conference of the IEEE engineering in medicine and biology society (EMBC). IEEE, pp. 2892–2895
Hoo-Chang S, Tenenholtz NA, Rogers JK, Schwarz CG, Senjem ML, Gunter JL, Andriole KP, Michalski M (2018) Medical image synthesis for data augmentation and anonymization using generative adversarial networks. In Simulation and synthesis in medical imaging: third international workshop, SASHIMI 2018, held in conjunction with MICCAI 2018, Granada, Spain, September 16, 2018, Proceedings 3. Springer, pp. 1–11
Hoyer L, Dai D, Van Gool L (2022) Daformer: improving network architectures and training strategies for domain-adaptive semantic segmentation. In Proceedings of the IEEE/CVF conference on computer vision and pattern recognition, pp. 9924–9935
Hu EJ, Shen Y, Wallis P, Allen-Zhu Z, Li Y, Wang S, Wang L, Chen W (2021) Lora: low-rank adaptation of large language models. Preprint, arXiv:2106.09685
Huang Z et al (2024) Contrastive masked auto encoders are stronger vision learners. IEEE Trans Pattern Anal Mach Intell 46(4):2506–2517. https://doi.org/10.1109/TPAMI.2023.3336525
Huisken J, Didier Y, Stainier R (2009) Selective plane illumination microscopy techniques in developmental biology. Development 136(12):1963–1975. https://doi.org/10.1242/dev.022426
Huisken J, Stainier DYR (2009) Selective plane illumination microscopy techniques in developmental biology. Development 136(12):1963–1975. https://doi.org/10.1242/dev.022426
Huling J, Götz A, Grabow N, Illner S (2022) Gift: an imagej macro for automated fiber diameter quantification. PLOS ONE 17(10):1–13
Hyun Y, Kim D (2023) Recent development of computational cluster analysis methods for single-molecule localization microscopy images. Comput Struct Biotechnol J 21:879–888
Iheme LO, Onal S, Erdem YS, Ucar M, Yalcin-Ozuysal O, Pesen-Okvur D, Toreyin BU, Unay D (2024) Collection: wound healing assay dataset (whad) and cell adhesion and motility assay dataset (camad). IEEE Data Descr 1:95–102
Irshad H, Veillard A, Roux L, Racoceanu D (2013) Methods for nuclei detection, segmentation, and classification in digital histopathology: a review–current status and future potential. IEEE Rev Biomed Eng 7:97–114

Jan M, Spangaro A, Lenartowicz M, Usaj MM (2024) From pixels to insights: machine learning and deep learning for bioimage analysis. BioEssays 46(2):2300114

Kalman RE (1960) A new approach to linear filtering and prediction problems. ASME J Basic Eng 82(1):35–45. https://doi.org/10.1115/1.3662552

Kamnitsas K, Ferrante E, Parisot S, Ledig C, Nori AV, Criminisi A, Rueckert D, Glocker B (2016) Deepmedic for brain tumor segmentation. In Brainlesion: glioma, multiple sclerosis, stroke and traumatic brain injuries: second international workshop, BrainLes 2016, with the challenges on BRATS, ISLES and mTOP 2016, held in conjunction with MICCAI 2016, Athens, Greece, October 17, 2016, Revised Selected Papers 2. Springer, pp. 138–149

Kamnitsas K, Baumgartner C, Ledig C, Newcombe V, Simpson J, Kane A, Menon D, Nori A, Criminisi A, Rueckert D et al (2017) Unsupervised domain adaptation in brain lesion segmentation with adversarial networks. In Information processing in medical imaging: 25th international conference, IPMI 2017, Boone, NC, USA, June 25–30, 2017, Proceedings 25. Springer, pp. 597–609

Karani N, Erdil E, Chaitanya K, Konukoglu E (2021) Test-time adaptable neural networks for robust medical image segmentation. Med Image Anal 68:101907

Kather JN et al (2020) Classification of tissue regions in histopathological images: comparison between pre-trained convolutional neural networks and local binary patterns variants. In: Nanni L, Brahnam S, Brattin R, Ghidoni S, Jain L (eds) Deep learners and deep learner descriptors for medical applications. Intelligent systems reference library, vol 186. Springer, Cham. https://doi.org/10.1007/978-3-030-42750-4_3

Khandelwal P, Yushkevich P (2020) Domain generalizer: a few-shot meta learning framework for domain generalization in medical imaging. In Domain adaptation and representation transfer, and distributed and collaborative learning: second MICCAI workshop, DART 2020, and first MICCAI workshop, DCL 2020, held in conjunction with MICCAI 2020, Lima, Peru, October 4–8, 2020, Proceedings 2. Springer, pp. 73–84

Kim J, Çetin M, Willsky AS (2007) Nonparametric shape priors for active contour-based image segmentation. Signal Process 87(12):3021–3044

Kingma DP, Mohamed S, Rezende DJ, Welling M (2014) Semi-supervised learning with deep generative models. Adv Neural Inf Process Syst 27

Kingma D, Salimans T, Poole B, Ho J (2021) Variational diffusion models. Adv Neural Inf Process Syst 34:21696–21707

Kirillov A, Mintun E, Ravi N, Mao H, Rolland C, Gustafson L, Xiao T, Whitehead S, Berg AC, Lo W-Y et al (2023) Segment anything. In Proceedings of the IEEE/CVF international conference on computer vision, pp. 4015–4026

Kohl S, Romera-Paredes B, Meyer C, De Fauw J, Ledsam JR, Maier-Hein K, Eslami SM, Rezende DJ, Ronneberger O (2018), A probabilistic u-net for segmentation of ambiguous images. Adv Neural Inf Process Syst 31

Krizhevsky A, Sutskever I, Hinton GE (2012) Imagenet classification with deep convolutional neural networks. Adv Neural Inf Process Syst 25

Kutsuna N, Higaki T, Matsunaga S, Otsuki T, Yamaguchi M, Fujii H, Hasezawa S (2012) Active learning framework with iterative clustering for bioimage classification. Nature Commun 3(1):1032

Laine S, Aila T (2016) Temporal ensembling for semi-supervised learning. Preprint, arXiv:1610.02242

Lamberti WF (2021) Blood cell classification using interpretable shape features: a comparative study of svm models and cnn-based approaches. Comput Methods Program Biomed Update 1:100023

Lee Y, Willette J, Kim J, Lee J, Hwang SJ (2022) Exploring the role of mean teachers in self-supervised masked auto-encoders. Preprint, arXiv:2210.02077

Li D, Yang Y, Song Y-Z, Hospedales T (2018) Learning to generalize: meta-learning for domain generalization. In Proceedings of the AAAI conference on artificial intelligence, vol. 32

Li R, Zeng X, Sigmund SE, Lin R, Zhou B, Liu C, Wang K, Jiang R, Freyberg Z, Lv H et al (2019) Automatic localization and identification of mitochondria in cellular electron cryo-tomography using faster-rcnn. BMC Bioinf 20:75–85
Li Z, Guo R, Li M, Chen Y, Li G (2020) A review of computer vision technologies for plant phenotyping. Comput Electron Agric 176:105672
Li H, Xue F-F, Chaitanya K, Luo S, Ezhov I, Wiestler B, Zhang J, Menze B (2021) Imbalance-aware self-supervised learning for 3d radiomic representations. In Medical image computing and computer assisted intervention–MICCAI 2021: 24th international conference, Strasbourg, France, September 27–October 1, 2021, Proceedings, Part II 24. Springer, pp. 36–46
Liu K, Skibbe H, Schmidt T, Blein T, Palme K, Brox T, Ronneberger O (2014) Rotation-invariant hog descriptors using fourier analysis in polar and spherical coordinates. Int J Comput Vision 106:342–364
Liu Z, Lin Y, Cao Y, Hu H, Wei Y, Zhang Z, Lin S, Guo B (2021) Swin transformer: Hierarchical vision transformer using shifted windows. In Proceedings of the IEEE/CVF international conference on computer vision, pp. 10012–10022
Lloyd S (1982) Least squares quantization in pcm. IEEE Trans Inf Theory 28(2):129–137
Long F (2020) Microscopy cell nuclei segmentation with enhanced u-net. BMC Bioinf 21(1):8
Lowe DG (2004) Distinctive image features from scale-invariant keypoints. Int J Comput Vis 60:91–110
Luo TL, Eisenberg MC, Hayashi MAL, Gonzalez-Cabezas C, Foxman B, Marrs CF, Rickard AH (2018) A sensitive thresholding method for confocal laser scanning microscope image stacks of microbial biofilms. Sci Rep 8(1):13013
Ma J, He Y, Li F, Han L, You C, Wang B (2024) Segment anything in medical images. Nature Commun 15(1):654
Macqueen J (1967) Some methods for classification and analysis of multivariate observations. In Proceedings of 5-th Berkeley symposium on mathematical statistics and probability. University of California Press, California
Mamunur Rahaman M, Li C, Wu X, Yao Y, Hu Z, Jiang T, Li X, Qi S (2020) A survey for cervical cytopathology image analysis using deep learning. IEEE Access 8:61687–61710
Man D, Vision A (1982) A computational investigation into the human representation and processing of visual information, vol. 1:1. Freeman and Company, WH San Francisco, San Francisco
Marr D, Poggio T (1979) A computational theory of human stereo vision. Proc R Soc Lond Series B Biol Sci 204(1156):301–328
Matsoukas C, Haslum JF, Sorkhei M, Söderberg M, Smith K (2022) What makes transfer learning work for medical images: feature reuse & other factors. In Proceedings of the IEEE/CVF conference on computer vision and pattern recognition, pp. 9225–9234
McInnes L, Healy J, Melville J (2018) Umap: uniform manifold approximation and projection for dimension reduction. Preprint, arXiv:1802.03426
Meijering E (2020) A bird's-eye view of deep learning in bioimage analysis. Comput Struct Biotechnol J 18:2312–2325
Meijering E, Smal I, Danuser G (2006) Tracking in molecular bioimaging. IEEE Signal Process Mag 23(3):46–53
Meiniel W, Olivo-Marin J-C, Angelini ED (2018) Denoising of microscopy images: a review of the state-of-the-art, and a new sparsity-based method. IEEE Trans Image Process 27(8):3842–3856
Mesadi F, Erdil E, Cetin M, Tasdizen T (2017) Image segmentation using disjunctive normal bayesian shape and appearance models. IEEE Trans medical Imaging 37(1):293–305
Meyer-Baese A, Schmid VJ (2014) Pattern recognition and signal analysis in medical imaging. Elsevier, Amsterdam
Micó V, Zheng J, Garcia J, Zalevsky Z, Gao P (2019) Resolution enhancement in quantitative phase microscopy. Adv Opt Photon 11(1):135–214
Milletari F, Navab N, Ahmadi S-A (2016) V-net: fully convolutional neural networks for volumetric medical image segmentation. In 2016 fourth international conference on 3D vision (3DV). IEEE, pp. 565–571

Mittal H, Pandey AC, Saraswat M, Kumar S, Pal R, Modwel G (2022) A comprehensive survey of image segmentation: clustering methods, performance parameters, and benchmark datasets. Multimedia Tools Appl 1–26

Miura K, Sladoje N (2020) Bioimage data analysis workflows. Springer Nature, Berlin

Miura K, Sladoje N (2022) Bioimage data analysis workflows–advanced components and methods. Springer Nature, Berlin

Miyato T, Maeda S, Koyama M, Ishii S (2018) Virtual adversarial training: a regularization method for supervised and semi-supervised learning. IEEE Trans Pattern Anal Mach Intell 41(8):1979–1993

Moen E, Bannon D, Kudo T, Graf W, Covert M, Van Valen D (2019) Deep learning for cellular image analysis. Nature Methods 16(12):1233–1246

Nanni L, Lumini A, Brahnam S (2010) Local binary patterns variants as texture descriptors for medical image analysis. Artif Intell Med 49(2):117–125

Nanni L, Brahnam S, Ghidoni S, Lumini A (2018) Bioimage classification with handcrafted and learned features. IEEE/ACM Trans Comput Biol Bioinf 16(3):874–885

Nassif AB, Talib MA, Nasir Q, Dakalbab FM (2021) Machine learning for anomaly detection: a systematic review. IEEE Access 9:78658–78700

Natale M, Maresca B, Abrescia P, Bucci EM (2011) Image analysis workflow for 2-d electrophoresis gels based on imagej. Proteomics Insights 4:PRI–S7971

Nieves DJ, Pike JA, Levet F, Williamson DJ, Baragilly M, Oloketuyi S, De Marco A, Griffié J, Sage D, Cohen EAK et al (2023) A framework for evaluating the performance of smlm cluster analysis algorithms. Nature Methods 20(2):259–267

Noroozi M, Favaro P (2016) Unsupervised learning of visual representations by solving jigsaw puzzles. In European conference on computer vision. Springer, pp. 69–84

Ojala T, Pietikäinen M, Harwood D (1996) A comparative study of texture measures with classification based on featured distributions. Pattern Recogn 29(1):51–59

Oktay O, Schlemper J, Folgoc LL, Lee M, Heinrich M, Misawa K, Mori K, McDonagh S, Hammerla NY, Kainz B et al (2018) Attention u-net: learning where to look for the pancreas. In International conference on medical image computing and computer-assisted intervention (MICCAI). Springer, pp. 206–214

Oquab M, Darcet T, Moutakanni T, Vo H, Szafraniec M, Khalidov V, Fernandez P, Haziza D, Massa F, El-Nouby A et al (2023) Dinov2: learning robust visual features without supervision. Preprint, arXiv:2304.07193

Osher S, Fedkiw RP (2001) Level set methods: an overview and some recent results. J Comput Phys 169(2):463–502

Otsu N et al (1975) A threshold selection method from gray-level histograms. Automatica 11(285–296):23–27

Patel A, Li Z, Canete P, Strobl H, Dulin J, Kadoya K, Gibbs D, Poplawski GHD (2018) Axontracer: a novel imagej plugin for automated quantification of axon regeneration in spinal cord tissue. BMC Neurosci 19:1–9

Pathak SD, Haynor DR, Kim Y (2000) Edge-guided boundary delineation in prostate ultrasound images. IEEE Trans Med Imaging 19(12):1211–1219

Pathak D, Krahenbuhl P, Donahue J, Darrell T, Efros AA (2016) Context encoders: feature learning by inpainting. In Proceedings of the IEEE conference on computer vision and pattern recognition, pp. 2536–2544

Pchitskaya E, Bezprozvanny I (2020) Dendritic spines shape analysis–classification or clusterization? perspective. Front Synaptic Neurosci 12:31

Perakis A, Gorji A, Jain S, Chaitanya K, Rizza S, Konukoglu E (2021) Contrastive learning of single-cell phenotypic representations for treatment classification. In Machine learning in medical imaging: 12th international workshop, MLMI 2021, held in conjunction with MICCAI 2021, Strasbourg, France, September 27, 2021, Proceedings 12. Springer, pp. 565–575

Perez F, Vasconcelos C, Avila S, Valle E (2018) Data augmentation for skin lesion analysis. In International workshop on computer-assisted and robotic endoscopy. Springer International Publishing, Cham, pp 303–311

Pérez-García F, Sharma H, Bond-Taylor S, Bouzid K, Salvatelli V, Ilse M, Bannur S, Castro DC, Schwaighofer A, Lungren MP et al (2024) Rad-dino: exploring scalable medical image encoders beyond text supervision. Preprint, arXiv:2401.10815

Phan HTH, Kumar A, Feng D, Fulham M, Kim J (2018) Unsupervised two-path neural network for cell event detection and classification using spatiotemporal patterns. IEEE Trans Med Imaging 38(6):1477–1487

Poldrack RA, Gorgolewski KJ (2017) Openfmri: open sharing of task fmri data. Neuroimage 144:259–261

Prabhudesai M, Ke T-W, Li AC, Pathak D, Fragkiadaki K (2023) Diffusion-tta: test-time adaptation of discriminative models via generative feedback. In Thirty-seventh conference on neural information processing systems

Quan H, Li X, Chen W, Bai Q, Zou M, Yang R, Zheng T, Qi R, Gao X, Cui X (2024) Global contrast-masked autoencoders are powerful pathological representation learners. Pattern Recogn 156:110745

Rada L, Kilic B, Erdil E, Ramiro-Cortés Y, Inbal Israely, Unay D, Cetin M, Argunsah AÖ (2018) Tracking-assisted detection of dendritic spines in time-lapse microscopic images. Neuroscience 394:189–205

Rasmus A, Berglund M, Honkala M, Valpola H, Raiko T (2015) Semi-supervised learning with ladder networks. Adv Neural Inf Process Syst 28

Redmon J, Divvala S, Girshick R, Farhadi A (2016) You only look once: Unified, real-time object detection. In Proceedings of the IEEE conference on computer vision and pattern recognition (CVPR), pp. 779–788

Rehman MHU, Hugo Lopez Pinaya W, Nachev P, Teo JT, Ourselin S, Cardoso MJ (2023) Federated learning for medical imaging radiology. Br J Radiol 96(1150):20220890

Rehman Hashmi AU, Mahapatra D, Yaqub M (2024) Envisioning medclip: a deep dive into explainability for medical vision-language models. Preprint, arXiv:2403.18996

Ren S, He K, Girshick R, Sun J (2015) Faster r-cnn: towards real-time object detection with region proposal networks. In Advances in neural information processing systems (NeurIPS), vol 28, pp. 91–99

Rittscher J (2010) Characterization of biological processes through automated image analysis. Ann Rev Biomed Eng 12(1):315–344

Roels J, Vernaillen F, Kremer A, Gonçalves A, Aelterman J, Luong HQ, Goossens B, Philips W, Lippens S, Saeys Y (2020) An interactive imagej plugin for semi-automated image denoising in electron microscopy. Nature Commun 11(1):771

Rokach L (2010) Pattern classification using ensemble methods, vol. 75. World Scientific, Singapore

Ronneberger O, Fischer P, Brox T (2015) U-net: convolutional networks for biomedical image segmentation. In International conference on medical image computing and computer-assisted intervention (MICCAI). Springer, pp. 234–241

Rubens U, Mormont R, Paavolainen L, Bäcker V, Pavie B, Scholz LA, Michiels G, Maška M, Ünay D, Ball G et al (2020) Biaflows: a collaborative framework to reproducibly deploy and benchmark bioimage analysis workflows. Patterns 1(3)

Saeys Y, Van Gassen S, Lambrecht BN (2016) Computational flow cytometry: helping to make sense of high-dimensional immunology data. Nat Rev Immunol 16(7):449–462

Safdar S, Rizwan M, Gadekallu TR, Javed AR, Rahmani MKI, Jawad K, Bhatia S (2022) Bio-imaging-based machine learning algorithm for breast cancer detection. Diagnostics 12(5):1134

Sait Erdem Y, Ayanzadeh A, Mayalı B, Balıkçi M, Belli ÖN, Uçar M, Özyusal ÖY, Okvur DP, Önal S, Morani K et al (2023) Automated analysis of phase-contrast optical microscopy time-lapse images: application to wound healing and cell motility assays of breast cancer. In Diagnostic biomedical signal and image processing applications with deep learning methods. Elsevier, pp. 137–154

Sajjadi M, Javanmardi M, Tasdizen T (2016) Regularization with stochastic transformations and perturbations for deep semi-supervised learning. Adv Neural Inf Process Syst 29

Sangalli S, Erdil E, Hötker A, Donati O, Konukoglu E (2021) Constrained optimization to train neural networks on critical and under-represented classes. Adv Neural Inf Process Syst 34:25400–25411
Sangalli S, Erdil E, Konukoglu E (2024) Expert load matters: operating networks at high accuracy and low manual effort. Adv Neural Inf Process Syst 36
Schnell P (1964) Eine methode zur auffindung von gruppen. Biom Z 6(1):47–48
Schorb M, Paul-Gilloteaux P (2020) Resolving the process of clathrin mediated endocytosis using correlative light and electron microscopy (clem). Bioimage Data Anal Workflows 143
Schott B, Traub M, Schlagenhauf C, Takamiya M, Antritter T, Bartschat A, Löffler K, Blessing D, Otte JC, Kobitski AY et al (2018) Embryominer: a new framework for interactive knowledge discovery in large-scale cell tracking data of developing embryos. PLOS Comput Biol 14(4):e1006128
Sebastian BV, Unnikrishnan A, Balakrishnan K (2012) Gray level co-occurrence matrices: generalisation and some new features. Preprint, arXiv:1205.4831
Shamir L, Delaney JD, Orlov N, Eckley DM, Goldberg IG (2010) Pattern recognition software and techniques for biological image analysis. PLoS Comput Biol 6(11):e1000974
Shi J, Malik J (1997) Normalized cuts and image segmentation. In Proceedings of IEEE computer society conference on computer vision and pattern recognition. IEEE, pp. 731–737
Simonyan K, Zisserman A (2014) Very deep convolutional networks for large-scale image recognition. Preprint, arXiv:1409.1556
Sirinukunwattana K, Pluim JPW, Chen H, Qi X, Heng P-A, Guo YB, Wang LY, Matuszewski BJ, Bruni E, Sanchez U et al (2017) Gland segmentation in colon histology images: the glas challenge contest. Med Image Anal 35:489–502
Sohl-Dickstein J, Weiss E, Maheswaranathan N, Ganguli S (2015) Deep unsupervised learning using nonequilibrium thermodynamics. In International conference on machine learning (PMLR), pp. 2256–2265
Somkantha K, Theera-Umpon N, Auephanwiriyakul S (2010) Boundary detection in medical images using edge following algorithm based on intensity gradient and texture gradient features. IEEE Trans Biomed Eng 58(3):567–573
Song Y, Cai W, Huang H, Feng D, Wang Y, Chen M (2016) Bioimage classification with subcategory discriminant transform of high dimensional visual descriptors. BMC Bioinf 17:1–15
Stauffer W, Sheng H, Lim HN (2018) Ezcolocalization: an imagej plugin for visualizing and measuring colocalization in cells and organisms. Sci Rep 8(1):15764
Ta V-T, Lézoray O, Elmoataz A, Schüpp S (2009) Graph-based tools for microscopic cellular image segmentation. Pattern Recogn 42(6):1113–1125
Tai W-L, Hu R-M, Hsiao HCW, Chen R-M, Tsai JJP (2011) Blood cell image classification based on hierarchical svm. In 2011 IEEE international symposium on multimedia. IEEE, pp. 129–136
Tarvainen A, Valpola H (2017) Mean teachers are better role models: weight-averaged consistency targets improve semi-supervised deep learning results. Adv Neural Inf Process Syst 30
Thomas LSV, Gehrig J (2020) Multi-template matching: a versatile tool for object-localization in microscopy images. BMC Bioinf 21:1–8
Timothy M, Forlano PM (2019) A versatile macro-based neurohistological image analysis suite for imagej focused on automated and standardized user interaction and reproducible data output. J Neurosci Methods 324:108286
Trabucco B, Doherty K, Gurinas M, Salakhutdinov R (2023) Effective data augmentation with diffusion models. Preprint, arXiv:2302.07944
Tsai Y-H, Hung W-C, Schulter S, Sohn K, Yang M-H, Chandraker M (2018) Learning to adapt structured output space for semantic segmentation. In Proceedings of the IEEE conference on computer vision and pattern recognition, pp. 7472–7481
Ucar M, Iheme LO, Onal S, Pesen-Okvur D, Yalcin-Ozuysal O, Toreyin BU, Unay D (2024) Blank frame and intensity variation distortion detection and restoration pipeline for phase-contrast microscopy time-lapse images. Electrica 24(1)

Uchida S (2013) Image processing and recognition for biological images. Dev Growth Differ 55(4):523–549
Uhlmann V, Donati L, Sage D (2022) A practical guide to supervised deep learning for bioimage analysis: challenges and good practices. IEEE Signal Process Mag 39(2):73–86
Unay D (2012) Local and global volume changes of subcortical brain structures from longitudinally varying neuroimaging data for dementia identification. Comput Med Imaging Graph 36(6):464–473
Ünay D, Stanciu SG (2018) An evaluation on the robustness of five popular keypoint descriptors to image modifications specific to laser scanning microscopy. IEEE Access 6:40154–40164
Uzunbas MG, Soldea O, Ünay D, Cetin M, Unal G, Erçil A, Ekin A (2010) Coupled nonparametric shape and moment-based intershape pose priors for multiple basal ganglia structure segmentation. IEEE Trans Med Imaging 29(12):1959–1978
Valente AJ, Maddalena LA, Robb EL, Moradi F, Stuart JA (2017) A simple imagej macro tool for analyzing mitochondrial network morphology in mammalian cell culture. Acta Histochemica 119(3):315–326
Van der Maaten L, Hinton G (2008) Visualizing data using t-sne. J Mach Learn Res 9(11)
Vapnik VN (1997) The support vector method. In International conference on artificial neural networks. Springer, pp. 261–271
Vaswani A, Noam S, Niki P, Jakob U, Llion J, Aidan NG, Łukasz K, Illia P (2017) Attention is all you need. Adv Neural Inf Process Syst 30
Vincenzo Della M, Baroni GL, Pilutti D, Di Loreto C (2017) Slidej: an imagej plugin for automated processing of whole slide images. PLOS ONE 12(7):1–9
Vladimirov N, Voigt FF, Naert T, Araujo GR, Cai R, Reuss AM, Zhao S, Schmid P, Hildebrand S, Schaettin M et al (2024) Benchtop mesospim: a next-generation open-source light-sheet microscope for cleared samples. Nature Commun 15(1):2679
Voigt FF, Kirschenbaum D, Platonova E, Pagès S, Campbell RAA, Kastli R, Schaettin M, Egolf L, Van Der Bourg A, Bethge P et al (2019) The mesospim initiative: open-source light-sheet microscopes for imaging cleared tissue. Nature Methods 16(11):1105–1108
Vrieze SI (2012) Model selection and psychological theory: a discussion of the differences between the akaike information criterion (aic) and the bayesian information criterion (bic). Psychol Methods 17(2):228
Wan S, Lee H-C, Huang X, Xu T, Xu T, Zeng X, Zhang Z, Sheikine Y, Connolly JL, Fujimoto JG et al (2017) Integrated local binary pattern texture features for classification of breast tissue imaged by optical coherence microscopy. Med Image Anal 38:104–116
Wan J, Dong Y, Yao Y, Xiao W, Huang R, Xue J-H, Peng R, Pei H, Tian X, Liao R et al (2023) Unsupervised learning of pixel clustering in mueller matrix images for mapping microstructural features in pathological tissues. Commun Eng 2(1):88
Wang H, Shen X, Pan W (2011) Large margin hierarchical classification with mutually exclusive class membership. J Mach Learn Res 12(9), 2011.
Wang S, Shen Y, Zhang C, Wong HC, Ren D (2018) Deep learning for automatic cell tracking: a case study in cyst nematode and root-knot nematode. Plant Methods 14(1):58
Wang D, Shelhamer E, Liu S, Olshausen B, Darrell T (2020) Tent: fully test-time adaptation by entropy minimization. Preprint, arXiv:2006.10726
Wang X, Yang S, Zhang J, Wang M, Zhang J, Yang W, Huang J, Han X (2022) Transformer-based unsupervised contrastive learning for histopathological image classification. Med Image Anal 81:102559
Wei Q, Ren Y, Hou R, Shi B, Lo JY, Carin L (2018) Anomaly detection for medical images based on a one-class classification. In Medical imaging 2018: computer-aided diagnosis, vol. 10575. SPIE, pp. 375–380
Wienert S, Heim D, Saeger K, Stenzinger A, Beil M, Hufnagl P, Dietel M, Denkert C, Klauschen F (2012) Detection and segmentation of cell nuclei in virtual microscopy images: a minimum-model approach. Sci Rep 2(1):503

Workman MJ, Serov A, Halevi B, Atanassov P, Artyushkova K (2015) Application of the discrete wavelet transform to sem and afm micrographs for quantitative analysis of complex surfaces. Langmuir 31(17):4924–4933

Wu L, Zhuang J, Chen H (2024) Voco: a simple-yet-effective volume contrastive learning framework for 3d medical image analysis. In Proceedings of the IEEE/CVF conference on computer vision and pattern recognition, pp. 22873–22882

Xianyu Z, Correia C, Ung CY, Zhu S, Billadeau DD, Li H (2024) The rise of hypothesis-driven artificial intelligence in oncology. Cancers 16(4):822

Xiao J, Bai Y, Yuille A, Zhou Z (2023) Delving into masked autoencoders for multi-label thorax disease classification. In Proceedings of the IEEE/CVF winter conference on applications of computer vision, pp. 3588–3600

Xie Y, Xing F, Shi X, Kong X, Su H, Yang L (2018) Efficient and robust cell detection: a structured regression approach. Med Image Anal 44:245–254

Xie Z, Zhang Z, Cao Y, Lin Y, Bao J, Yao Z, Dai Q, Hu H (2022) Simmim: a simple framework for masked image modeling. In Proceedings of the IEEE/CVF conference on computer vision and pattern recognition, pp. 9653–9663

Xing F, Xie Y, Su H, Liu F, Yang L (2017) Deep learning in microscopy image analysis: a survey. IEEE Trans Neural Netw Learn Syst 29(10):4550–4568

Xing F, Xie Y, Yang L (2018) An automatic learning-based framework for robust nucleus segmentation. IEEE Trans Med Imaging 35(2):550–566

Yalciner BZ, Kandemir M, Taskale S, Tepe SM, Unay D (2019) Modified visual magnetic resonance rating scale for evaluation of patients with forgetfulness. Can J Neurol Sci J Can Sci Neurol 46(1):71–78

Yang X, Chen D, Sun Q, Wang Y, Xia Y, Yang J, Lin C, Dang X, Cen Z, Liang D et al (2023) A live-cell image-based machine learning strategy for reducing variability in psc differentiation systems. Cell Discov 9(1):53

Yevick HG, Martin AC (2018) Quantitative analysis of cell shape and the cytoskeleton in developmental biology. Wiley Interdiscip Rev Dev Bio 7(6):e333

Yi K, Ge Y, Li X, Yang S, Li D, Wu J, Shan Y, Qie X (2022) Masked image modeling with denoising contrast. Preprint, arXiv:2205.09616

Young KD (2007) Bacterial morphology: why have different shapes? Curr Opin Microbiol 10(6):596–600

Yu-Ting S, Lu Y, Chen M, Liu A-A (2017) Spatiotemporal joint mitosis detection using cnn-lstm network in time-lapse phase contrast microscopy images. IEEE Access 5:18033–18041

Zabary Y, Zaritsky A (2022) A MATLAB pipeline for spatiotemporal quantification of monolayer cell migration. In: Miura K, Sladoje N (eds) Bioimage data analysis workflows – advanced components and methods. Learning materials in biosciences. Springer, Cham. https://doi.org/10.1007/978-3-030-76394-7_8

Zbontar J, Jing L, Misra I, LeCun Y, Deny S (2021) Barlow twins: self-supervised learning via redundancy reduction. In International conference on machine learning (PMLR), pp. 12310–12320

Zhang H (2017) mixup: beyond empirical risk minimization. Preprint, arXiv:1710.09412

Zhang X, Fujita H, Qin T, Zhao J, Qin Y, Gao C, Long L, Zhang Z (2008) A novel method for extraction of spleen by using thin-plate splines (tps) deformation and edge detection from abdominal ct images. In 2008 international conference on biomedical engineering and informatics, vol. 1. IEEE, pp. 830–834

Zhang R, Isola P, Efros AA (2016) Colorful image colorization. In Computer vision–ECCV 2016: 14th European conference, Amsterdam, The Netherlands, October 11–14, 2016, Proceedings, Part III 14. Springer, pp. 649–666

Zhang Y, Yang L, Chen J, Fredericksen M, Hughes DP, Chen DZ (2017) Deep adversarial networks for biomedical image segmentation utilizing unannotated images. In Medical image computing and computer assisted intervention (MICCAI 2017): 20th international conference, Quebec City, QC, Canada, September 11–13, 2017, Proceedings, Part III 20. Springer, pp. 408–416

Zhang L, Wang X, Yang D, Sanford T, Harmon S, Turkbey B, Wood BJ, Roth H, Myronenko A, Xu D et al (2020) Generalizing deep learning for medical image segmentation to unseen domains via deep stacked transformation. IEEE Trans Med Imaging 39(7):2531–2540

Zhang C, Gaignard A, Kalas M, Levet F, Delestro F, Lindblad J, Sladoje N, Plantard L, Latour A, Haase R et al (2023) Bio-image informatics index biii: a unique database of image analysis tools and workflows for and by the bioimaging community. Preprint, arXiv:2312.11256

Zhang M, Wang L, Gu L, Li Z, Wang Y, Ling T, Tao X (2024) Sam2-path: a better segment anything model for semantic segmentation in digital pathology. arXiv e-prints, arXiv–2408

Zhou L, Liu H, Bae J, He J, Samaras D, Prasanna P (2023) Self pre-training with masked autoencoders for medical image classification and segmentation. In 2023 IEEE 20th international symposium on biomedical imaging (ISBI). IEEE, pp. 1–6

Zulueta-Coarasa T, Jug F, Mathur A, Moore J, Muñoz-Barrutia A, Anita L, Babalola K, Bankhead P, Gilloteaux P, Gogoberidze N et al (2023) Mifa: metadata, incentives, formats, and accessibility guidelines to improve the reuse of ai datasets for bioimage analysis. Preprint, arXiv:2311.10443

Quantifying Biological Processes in Motion and Across Scales

Diego Ulisse Pizzagalli, Raffaella Fiamma Cabini, and Inge M. N. Wortel

Abstract

A fundamental characteristic of living organisms is their capacity to change over time. These changes manifest across various scales: from molecular dynamics to modifications in the shape or arrangement of cells and, ultimately, changes at the level of tissues and organs. Imaging techniques to record biomedical videos have become integral to studying these processes in motion, but are notoriously difficult to analyze. While exciting advances in computer vision technology hold the promise to solve open challenges in quantifying these videos, their application to the field of biomedical imaging is far from trivial. This book chapter summarizes the state-of-the-art in analyzing motion from

The original version of this chapter was previously published without open access. A correction to this chapter is available at https://doi.org/10.1007/978-3-031-99410-4_31

D. U. Pizzagalli (✉)
Euler Institute, USI, Lugano, Switzerland

International Center for Advanced Computing in Medicine, University of Pavia, Pavia, Italy

Theodor Kocker Institute, University of Bern, Bern, Switzerland
e-mail: pizzad@usi.ch

R. F. Cabini
Euler Institute, USI, Lugano, Switzerland

International Center for Advanced Computing in Medicine, University of Pavia, Pavia, Italy
e-mail: raffaella.fiamma.cabini@usi.ch

Inge M. N. Wortel (✉)
Data Science, Institute for Computing and Information Sciences, Radboud University, Nijmegen, Netherlands
e-mail: inge.wortel@ru.nl

I. F. Cengiz et al. (eds.), *Bioimaging Modalities in Bioengineering*,
Biomaterials, Bioengineering and Sustainability 9,
https://doi.org/10.1007/978-3-031-99410-4_29

biomedical videos, and highlights opportunities and challenges for the translation of computer vision techniques to this field.

Keywords

Movement analysis · Tracking · Motility metrics · Video analysis

1 Introduction

Historically, our understanding of the dynamics of biological processes has gone hand in hand with the development of the microscopes used to visualize them—from Antoni van Leeuwenhoek's observations of 'wee animalcules' in the seventeenth century and Metchnikoff's observations of phagocytosis and theories on cell-mediated immunity in the late 1800s (Masters 2008), to the recordings of immune cells in lymph nodes facilitated by the first two-photon microscopes in the early 2000s. Likewise, recent advances in computer vision are revolutionizing our ability to *analyze* the motion in these videos. Video analysis enables the precise quantification of dynamic processes such as cell migration and interaction, intracellular transport, tissue, and organ remodeling—but very similar techniques can be applied to analyze human behavior, movement disorders, and crowd dynamics.

This chapter outlines classical and more recent methodologies for quantifying motion across different resolutions. While the primary focus is on cells observed through time-lapse microscopy, many of the principles discussed can also be applied to other imaging systems recording motile objects at various scales—ranging from molecular diffusion and vesicle trafficking to organ-level motility and blood flow. We discuss what is needed to successfully translate advances in computer vision to the field of bioimaging.

2 Track-based Quantification of Motion

Classical approaches to quantify motion start with **tracking**, the process of reconstructing the position of an object over time in a given space (i.e., one-dimensional (1D) along one given direction, 2D movement in a plane, or 3D movement in space). They are largely applied to time-lapse microscopy to track individual cells (Fig. 1a), but can also be applied to study the motility of molecules or intracellular vesicles. This section illustrates how to obtain tracks and how to describe them using quantitative metrics.

2.1 Tracking Algorithms

The first step in obtaining tracks from time-lapse videos is to identify objects, typically via **segmentation** or centroid **detection**, followed by linking positions of each individual cell across frames (tracking, Fig. 1a). Facilitated by software tools such as the Trackmate plugin in FIJI (Tinevez et al. 2017), manual or semi-

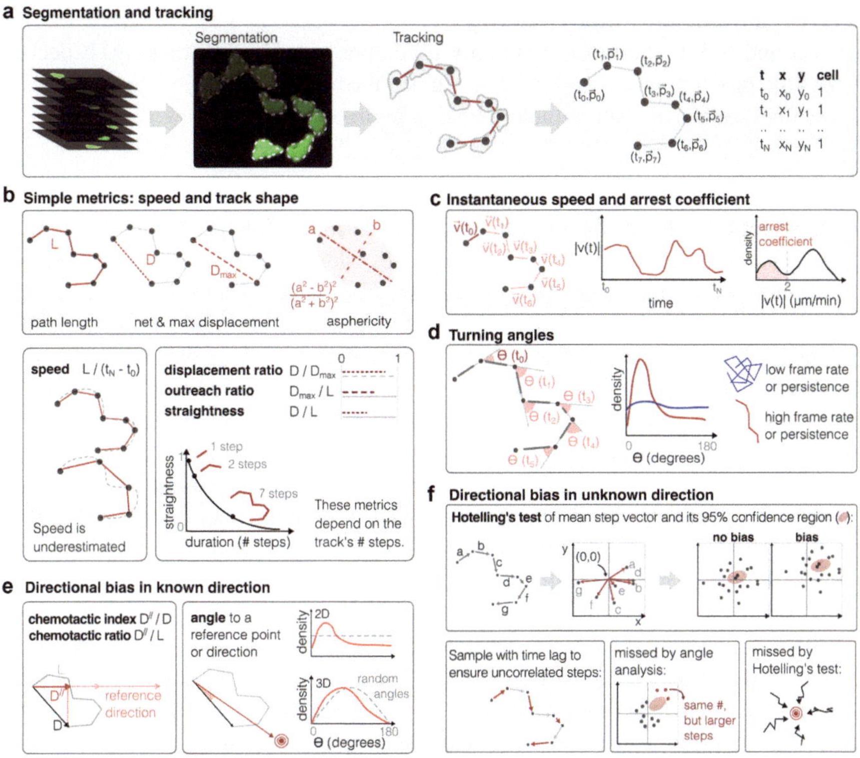

Fig. 1 Extracting track-based motility metrics. (**a**) Segmentation and tracking yields a table of cell positions over time. (**b**) Simple metrics describing tracks. (**c**) Deriving the arrest coefficient from instantaneous speeds. (**d**) Lower turning angles can indicate higher persistence or higher imaging frame rates. (**e**) Detecting bias in a known direction using metrics or angle analysis. Expected random angle distributions differ between 2D and 3D. (**f**) Hotelling's test can detect bias in an unknown direction by evaluating if the 95% confidence region of the mean step vector includes **0**. Since it assumes that steps are independent, a time lag is required for persistent cells. Compared to angle analysis, Hotelling's test is more sensitive to orthotaxis (where steps in the reference direction are faster, but not more frequent), but it can miss symmetrical biases with a mean step vector of **0**

automated tracking has long been the standard and still remains helpful in complex scenarios.

Advances in computer vision are now shifting the focus towards automated tracking methods (Hirsch et al. 2022). Despite significant algorithmic advancements, automated tracking of motile objects in biomedical imaging remains challenging due to issues like large deformations, unclear boundaries, cell interactions, and a limited field of view. Current state-of-the-art methods for automated tracking can be divided into two main approaches: **detect-and-link** and **space-time-connected**, discussed further below.

2.1.1 Detect-and-link Tracking

Detect-and-link is the most commonly used approach for cell tracking. It decomposes the problem into two steps that can be tuned independently: detecting cells in individual frames and subsequently linking them over time.

Cell Detection

Cell **detection** is the problem of finding cell centroids. This process often includes segmentation of the cell shape, which provides additional morphodynamic features but is not strictly necessary for cell tracking. In images where the (fluorescence) intensity is roughly uniform within the cell and across the field of view, a global intensity threshold can segment cells. In practice, however, the threshold often requires local adaptation, for example using background subtraction techniques. In transmitted light microscopy, higher variations of intensity are expected on the cell boundaries, and intensity thresholding may lead to hollow objects that are difficult to track. An effective technique to facilitate cell detection is **brushing**, where users annotate the foreground (cells) and background by drawing lines on the image (Berg et al. 2019; Arganda et al. 2017; Pizzagalli et al. 2022a). Points on these lines are then used to train pixel classifiers.

Once foreground pixels are detected, they are grouped into objects based on intensity (Li et al. 2010), texture features (Al et al. 2006), or size filtering using clustering, connected components, or region growing/merging methods. Challenges arise with non-globular shapes and with overlapping cells, requiring specific rules or advanced clustering algorithms (Pizzagalli et al. 2019a). When the expected size of the objects is known, methods based on intensity gradients such as the Laplacian-of-Gaussian or shape-based models such as Hough Transform, can effectively differentiate even overlapping objects.

More recently, state-of-the-art deep learning (DL) based methods for cell segmentation such as Cellpose (Stringer et al. 2021) and StarDist (Schmidt et al. 2018) have been developed to segment cells of diverse shapes and sizes, offering pre-trained and customizable models for integration into downstream tools like TrackMate. General-purpose models like YOLO have significantly improved multiple object detection and subsequent tracking in other fields, e.g., vehicles or soccer players (Markappa et al. 2024; Zuraimi et al. 2021; Krishna et al. 2021). Promising results are also expected from general-purpose models such as SAM (segment anything model), which can distinct objects within the same image. These have been recently extended to solve the Segment and Tracking Anything problem (Cheng et al. 2023; Rajič et al. 2023). For an overview of these and other cell segmentation methods, see (Schienstock et al. 2022; Moen et al. 2019).

Cell Linking

Once centroids are detected for each frame, **linking** involves finding the most likely correspondence of objects across frames. Simple heuristics, like maximal overlap or nearest neighbor, can be sufficient in simple settings, but often fail when objects are crowded and/or displace far between frames, such that the object with the minimal displacement is no longer always the correct choice. Graph-based methods address this issue by modeling tracks as paths on a graph, where nodes represent detected objects and edges indicate possible transitions. Tracking is achieved by numerical optimization methods that minimize path-cost functions either for individual tracks or globally across all tracks (Jiang et al. 2017; Turetken et al. 2016; Magnusson et al. 2012; Chenouard et al. 2013). A key problem of detect-and-link tracking is that linking algorithms are sensitive to errors in the detection step, causing interrupted tracks or erroneous links (Berclaz et al. 2011). Graph-based approaches mitigate this issue by incorporating special nodes for object (dis)appearance. Alternatively, detection and linking can be performed simultaneously, discussed below.

2.1.2 Space-time Connected Tracking

Instead of treating detection and linking as separate problems, **space-time connected** approaches perform both simultaneously by reconstructing tracks as structures in space-time, essentially segmenting cells in both space and time at once. Contour evolution methods fit an initial contour (or 3D mesh) to object shapes by minimizing an energy functional. Usually, a data term and a regularization term are included to drive the contour towards desired properties, enabling the reconstruction of object shapes even without clear edges (Chan et al. 2001). An advantage of space-time approaches is the ability to identify special structures; for example, so-called "T-junctions" were shown to help recover the position of two interacting objects—one of the most challenging problems in tracking (Apostoloff et al. 2005, 2006). However, joint segmentation and tracking is challenging, and most methods focus on single objects or use contours from one frame as a starting points for the next. DL methods, using convolutional or recurrent neural networks, enable precise segmentation of 3D structures and show promise for reconstructing multiple space-time tracks (Milan et al. 2017).

2.2 Track-based Motility Descriptors

Once obtained, tracks can be described quantitatively using a simple yet rich set of metrics.

2.2.1 Speed and Stopping

The most commonly used descriptor of a track is its **speed**, defined as the total distance traveled divided by track duration (Fig. 1b). While seemingly straightforward, this definition will always underestimate the true speed as cells are observed at discrete intervals (Beltman et al. 2009b). This bias can be problematic when cells turn relatively far between frames (Fig. 1b); thus, reliable estimates of cell speed may require an imaging frame rate tuned to the cell's turning behavior.

Aside from average speed, "instantaneous" speeds between subsequent frames can vary by cell type and condition (Fig. 1c). For example, early in vivo microscopy studies observed immune cells in lymph nodes alternating between rapid motion (up to 25 μm/min) and a non-motile "stop" state (Miller et al. 2002, 2003; Mempel et al. 2004; Bousso et al. 2003). Stopping is quantified using the **arrest coefficient**, the fraction of time where the speed is below some threshold (Fig. 1c; often 2 μm/min, but the ideal threshold can vary). As this fraction is computed by counting discrete steps, a more smooth, adapted version is preferable for downstream applications such as machine learning (ML) (Pizzagalli et al. 2019b).

2.2.2 Turning and Persistence Time

Another key track feature is the degree and frequency of turns in direction, which can vary substantially across cell types (Maiuri et al. 2015; Miller et al. 2003; Krummel et al. 2016) but can also reflect factors like confinement, obstacles, or cell-cell interactions (Beltman et al. 2007). A simple way to quantify turning is by analyzing **turning angles** between subsequent "steps" in the track, where lower angles indicate straighter motion (Fig. 1d). However, turning angle distributions can be difficult to interpret as they depend heavily on the imaging frame rate, limiting comparability across studies. Alternative metrics to describe turning behavior include the **displacement ratio**, **outreach ratio**, and **straightness** (Mokhtari et al. 2013) (Fig. 1b; where the latter is also referred to as "directness" Ibidi 2025, "confinement ratio" Beltman et al. 2009b; Mokhtari et al. 2013, or "directionality ratio" Gorelik et al. 2014). These metrics range between 0 and 1; however, their value depends strongly on the total duration of the track (Fig. 1b), and should be either compared between tracks of similar duration, or corrected for duration dependency (Beltman et al. 2009b; Gorelik et al. 2014). Track **asphericity** is also bounded by 0 and 1, but comes with the caveat that it is invariant to the order in which coordinates were visited (Mokhtari et al. 2013) (Fig. 1b). Another key metric is the **persistence time** representing how long motion remains relatively straight; however, this can be defined and measured in different ways as discussed in Sect. 3.1.

2.2.3 Directional Bias

Several metrics exist to detect directional bias arising when cells follow environmental structures or chemotactic gradients. The **chemotactic index** (also called "forward migration index" Ibidi 2025) is obtained by dividing the parallel component of the track's displacement ($D^{//}$) by the overall displacement (Fig. 1e); the **McCutcheon index** ("chemotactic ratio") divides $D^{//}$ by the total path length L instead (Fig. 1e) (McCutcheon et al. 1946). Another indicator of directionality is the angle of the cell's displacement vector to a reference point, direction, or plane. Without directional bias, the expected angle is 90° (although its distribution differs between 2D and 3D data; Beltman et al. 2009b, Fig. 1e). Enrichment of lower angles then indicates directed motion. This approach has proven useful to detect subtle directionality in various studies (Beltman et al. 2011; Ariotti et al. 2015).

All these approaches assume that the direction of bias is known. When it is not, directional bias can be detected using the **Rayleigh test** or **Hotelling's T^2 test**. The former is a non-parametric test against the null hypothesis that displacement directions are uniformly distributed (Ibidi 2025; Moore 1980); the latter instead takes the average step-wise displacement across all tracks, and tests if it differs significantly from the null vector (Hotelling 1931; Oja et al. 2004; Textor et al. 2011) (Fig. 1f). A caveat of Hotelling's test is that it assumes independence of data points—an assumption that is violated when we extract subsequent steps from a track with directional persistence. This problem can be circumvented by downsampling the data, allowing sufficient spacing between steps that they are uncorrelated (Fig. 1f). It has been shown that Hotelling's test can detect directionality with similar sensitivity as angle analysis, even when the reference direction is unknown (Textor et al. 2011). Moreover, unlike angle analysis, it could detect "orthotaxis" (a mechanism where movement in the direction of the chemokine is not necessarily more frequent, but occurs at higher speed) (Fig. 1f). There are also cases where angle analysis is preferable; radial attraction towards a point may not be visible using Hotelling's (or Rayleigh's) test as displacement directions cancel out (Fig. 1f).

3 Analyzing Cell Dynamics Across Populations and Over Time

3.1 Applying Motility Metrics

3.1.1 Track-based, Tracklet-based, and Step-based Analyses

After defining a set of metrics, tracks can be described and compared between cell types or experimental conditions. These comparisons can be performed in multiple ways (Fig. 2a) (Beltman et al. 2009b; Mokhtari et al. 2013). The most straightforward approach is to compute metrics in a **track-based** fashion, with one value per cell track. However, this approach is highly misleading when tracks differ in duration. Some metrics are highly sensitive to these differences (Fig. 1b)—even if the cells do not truly differ in their underlying motion. Instead, analysis can be performed on **tracklets**, track segments of a fixed duration (also called "subtracks", Fig. 2a). This approach solves the issue of confounding by track duration. However, it discards tracks with durations below the chosen tracklet size, which can strengthen selection bias effects (see Sect. 3.3.1). A third method is to perform **step-based** analyses, computing metrics on pooled, single steps extracted from the data (Fig. 2a). This approach does not discard shorter tracks and therefore lessens selection bias, but it also has disadvantages. For example, it is sensitive to measurement noise in cell positions, less intuitive than track-based analysis, unsuitable for identifying subpopulations (Beltman et al. 2009b), and inapplicable to some metrics (e.g., "straightness" of a single step is meaningless). Statistical methods that combine strengths of both approaches have also been proposed (Letendre et al. 2015). Generally, whether an analysis should be track-, tracklet-, or step-based depends both on the research question at hand and the variation in duration among the tracks in the data.

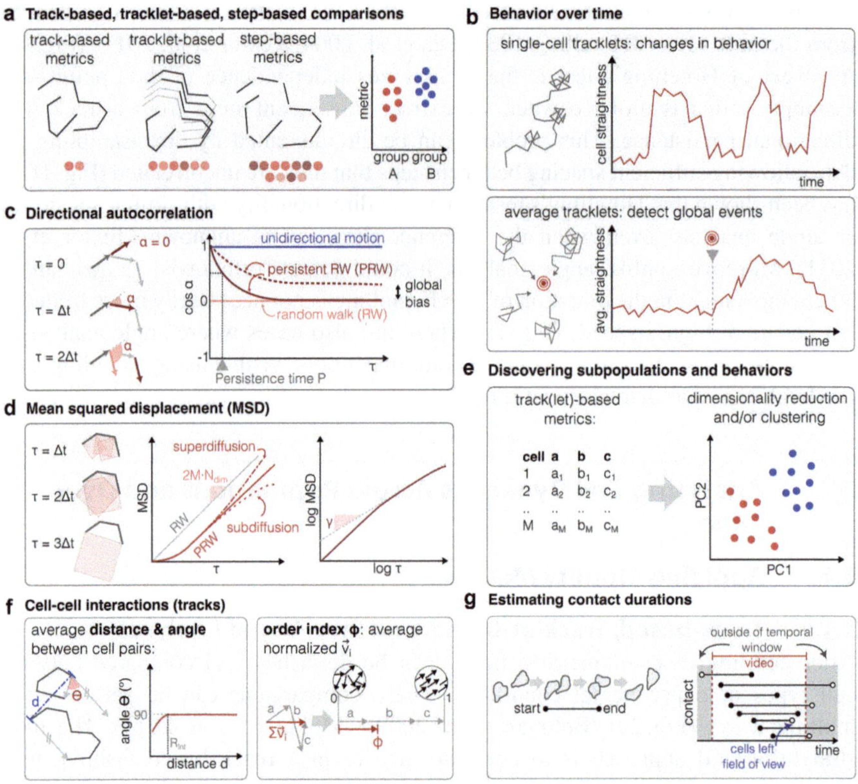

Fig. 2 Approaches to quantifying cell motility. (**a**) Metrics can be computed per track, per "tracklet" (track segment of fixed duration), or per individual "step". (**b**) Tracklet-based metrics can be analyzed over time, either for individual tracks (top, detecting changes in behavior), or for population averages (bottom, detecting global changes such as an increase in directed motion after an event of interest occurs). (**c**) The directional autocorrelation compares a cell's direction at time t to that at time $t + \tau$ to infer the persistence time. Reference curves are shown for an uncorrelated random walk (RW), persistent random walk (PRW) with and without global directionality, and unidirectional motion. (**d**) Like autocorrelation, the mean squared displacement (MSD) is computed over all possible tracklet durations τ. Expected MSD curves are shown for (P)RW motion and super-/subdiffusive motion. (**e**) Track features can be used for clustering and/or dimensionality reduction, allowing for exploration of subpopulations. (**f**) Distance-angle plots of cell pairs and the system's order index can help identify correlated, collective motion. (**g**) Physical contact can be detected from cell segmentations, but estimates for their duration are biased since we often do not observe the beginning and/or end. For unbiased estimates, contact durations should therefore be treated as time-to-event data

3.1.2 Single Cell Behavior Over Time

One advantage of step- or tracklet-based metrics is that they can be analyzed as a time series (Figs. 1c and 2b). When applied to single tracks, this approach can help identify changes in behavior over time, such as transitions between different migration modes (Crainiciuc et al. 2022). The same method can be applied to averages of tracklets from different cells in order to detect global events, for example an overall increase in directionality in response to tissue damage or sudden release of chemokines (Fig. 2b). An open question is to what extent general methodology from the field of time series can be applied to cell motility data to infer changes in behavior automatically. Simple methods such as hidden Markov models may be of use in this context (Schienstock et al. 2022), but applying them is non-trivial given the short tracks typically available in practical settings.

3.1.3 Expectations Over Time: Autocorrelation and Mean Squared Displacement

Since most metrics strongly depend on track duration (Fig. 1), several approaches involve computing these metrics on tracklets of varying duration τ in order to derive general, duration-independent motility parameters. One example is the directional **autocorrelation** function, defined as the cosine of the angle between the cell's direction at time t and $t + \tau$ (Fig. 2c). To construct the autocorrelation curve, this value is computed and averaged over all tracklets of the given duration τ. For an uncorrelated random walk (RW), the autocorrelation is zero for all $\tau > 0$, whereas for perfectly ballistic motion it always equals one.

For most cells, the autocorrelation is a function of τ, with high correlations at low τ indicating directional correlation in the short term. For such a **persistent random walk (PRW)**, the correlation decays approximately exponentially over longer time scales (Banigan et al. 2015) and becomes zero for $\tau >> P$—unless some other directional bias ensures further correlation on longer time scales. The half-life of this decay is the **persistence time** P and no longer depends on imaging frame rate or track duration. Alternative to the directional autocorrelation, the velocity autocorrelation (which includes not only the angle between vectors but also their magnitudes) is also frequently used (Banigan et al. 2015).

Similar to the autocorrelation, the **mean squared displacement (MSD)** is computed over all possible tracklet durations τ (Fig. 2d). Analytical derivations of the MSD are available for various motility models (Codling et al. 2008; Textor et al. 2013; Rosen et al. 2021). For example, the expected MSD for RW-motility is a linear function:

$$\mathrm{MSD}(\tau) = 2MN_{\mathrm{dim}}\tau \tag{1}$$

with N_{dim} the dimensionality, and M the **motility coefficient**, a parameter analogous to a diffusion coefficient. For PRW-motion, the MSD depends on both the motility coefficient M and the persistence time P according to Fürth's equation (Fürth 1920):

$$\mathrm{MSD}(\tau) = 2MN_{\mathrm{dim}}\left[\tau - P(1 - e^{-\tau/P})\right] \tag{2}$$

which starts non-linear but becomes approximately linear for $\tau >> P$ (Fig. 2d). Fitting Fürth's equation to the empirical MSD, one can derive estimates for M and P that are mostly independent of track duration and imaging frame rate. However, for cells imaged at high frame rates, fitting a modified version of Fürth's equation may be more appropriate as cell positions (at high frame rates) are ill-defined (Thomas et al. 2020).

The linear MSD at large τ observed for (P)RW-motion is characteristic of diffusive behavior. MSD curves that grow faster or slower than linearly with τ can also be observed, indicating "superdiffusive" and "subdiffusive" behavior, respectively. These behaviors are also reflected by the slope γ of the log-log MSD plot (Fig. 2d); where $\gamma = 1$ indicates diffusive behavior, $\gamma = 2$ for ballistic motion, $1 < \gamma < 2$ is superdiffusive motion, and $\gamma < 1$ indicates subdiffusion.

While autocorrelation and MSD reduce dependency on track duration and imaging frame rate, they have some limitations. First, for larger τ, fewer tracklets are available for time averaging (Fig. 2c and d), increasing noise in both autocorrelation and MSD; exclusion of high τ values can therefore be required for robust fitting of P and M. Second, although deviations in MSD from regular (P)RW behavior can be of biological interest, they may also arise from selection bias imposed by the imaging window (see Sect. 3.3.1). Third, as fitting autocorrelation and MSD curves is less robust for individual tracks, P and M are typically derived for the entire population—potentially masking heterogeneity within the population (Mokhtari et al. 2013).

3.1.4 Exploring Subpopulations Using Clustering and Dimensionality Reduction

Finally, an interesting application of motility metrics is to discover subpopulations and behaviors in the data. After computing a set of metrics on each track (or tracklet), popular dimensionality reduction methods such as principal component analysis and UMAP (McInnes et al. 2018) can be used for data exploration, and clustering methods can help identify subpopulations (Fig. 2e). This approach can be applied both to entire tracks, revealing subpopulations of cells (McInnes et al. 2024), or to shorter tracklets, also showing cell switching between behavioral modes over time (Crainiciuc et al. 2022). However, as many metrics depend on imaging frame rate and track duration, clusters must be interpreted cautiously, especially when combining data from different experiments.

3.2 Quantifying Cell-Cell Interaction

The analyses discussed so far have mostly treated cells as independently moving objects—but a frequently asked question is to what extent cells affect the motion of their neighbors through cell-cell interaction.

3.2.1 Correlated Motion

Measurements of correlated motion, commonly studied in animal flocking and collective behavior (Vicsek et al. 2012), can also be applied to cell migration. In some settings such as wound healing and tissue development (Camley and Rappel 2017), cell-cell interaction gives rise to collective motion with cells aligning on a global scale. The resulting directional bias can then be detected as discussed in Fig. 1e and f. Another frequently used metric to detect alignment in a collective is the **order index**, or order parameter, ϕ, which ranges from 0 to 1 (Fig. 2f) (Vicsek et al. 2012) and has also been applied to cell migration (Banigan et al. 2015; Niculescu et al. 2015; Szabó et al. 2006):

$$\phi = \frac{\left|\sum_i^N \mathbf{v}_i\right|}{\sum_i^N |\mathbf{v}_i|} = \frac{1}{N v_{\text{avg}}} \left|\sum_i^N \mathbf{v}_i\right| \tag{3}$$

with N the number of tracks, tracklets, or steps and $\mathbf{v}_i$ their velocity vectors.

However, many cells do not align their directions globally, and only align with their local neighbors—at the very least, some interaction is expected for distances close to the average cell radius due to physical exclusion effects. One way to take this spatial scale into account is to consider the angles and distances between pairs of track(let)s in the data (Fig. 2f) (Beltman et al. 2009b). Interactions result in average angles below 90° within an interaction distance R_{int}. If R_{int} is substantially larger than the average cell radius, this may indicate that cells are aligning over larger distances than expected from physical exclusion alone.

3.2.2 Cell-Cell Contact Dynamics

The dynamics of cell-cell contact are key for biological processes such as immune responses (Mempel et al. 2004; Davis 2009). Unfortunately, physical contact is hard to infer directly from tracks. While cells within a minimal threshold distance (R_{min}) can be marked as "contacting", this is often inaccurate as distance distributions for contacting and non-contacting cells can overlap substantially due to variability in cell sizes and shapes (Wortel et al. 2024). When available, segmentation masks (Fig. 1a) can help detect contact more reliably. Yet even when contacts are detected with complete accuracy in single frames, estimating contact duration remains difficult. We often do not observe their beginning and/or end—for example when the recording ends or when cells leave the field of view before the contact is resolved (Fig. 2g) (Beltman et al. 2009b). This makes cell-cell contact durations a form **time-to-event data** that is both "truncated" (as we do not observe contacts at all when they fall outside the imaged time window) and "right-censored" (as we underestimate the duration when either the beginning or end of the interaction is unobserved). These issues can cause substantial bias in the estimation of contact durations and should be corrected for; for details, see Beltman et al. (2009a).

3.3 Robustly Interpreting Track-based Analyses

Many methods exist to quantify cell motion and interactions, but they can introduce bias if not applied carefully. Here, we describe potential pitfalls in the classical track analysis pipeline, ways to mitigate them, and how simulations can facilitate robust data interpretation.

3.3.1 Artefacts from imaging, detection and tracking

Errors during image acquisition, cell detection, and tracking can introduce artifacts in downstream analysis. Common imaging errors include **tissue drift**—where movement of the tissue specimen during imaging adds a global drift velocity on top of actual cell movement, and **calibration errors** in the axial (z) dimension, causing false anisotropy. Depending on the algorithms used, common errors from cell detection and tracking include: **track gaps**—when cells are not detected in intermediate frames, resulting in tracks with either inconsistent time step (causing outliers in step-based metrics) or **track splitting** (causing an abundance of short tracks). **Track switching** occurs when cell paths cross each other, and the algorithm mistakenly switches the cell identity at the crossover point; frequent switches can bias track metrics towards preferences of the linking algorithm (e.g., a bias towards straighter tracks). Finally **border tracking** is an artefact causing artificially straight tracks at the image border. These errors have been extensively discussed elsewhere (Beltman et al. 2009b; Banigan et al. 2015), and we refer the reader to these resources for strategies to diagnose and mitigate them.

Another issue in track-based analysis is **detection noise**: as cells are deformable, their position is ill-defined (Thomas et al. 2020), and positions estimated during tracking can fluctuate even for static cells. This noise can distort motility metrics, especially for short tracklets, leading to overestimated instantaneous speeds and directional changes, an underestimated arrest coefficient, a faster decay of directional autocorrelation (Fig. 2c) and an underestimated persistence time. Importantly, as detection noise may differ between cell types with different (nuclear) shapes and sizes, this effect can create the illusion of motility differences. To diagnose this, it is useful to assess tracks of static cells, and to manually visualize detections on top of the original image. Noise can be reduced by smoothing raw tracks before track analysis using methods such as Kalman smoothing (Kalman 1960) (for an example, see Wortel et al. 2024). However, these methods also come with parameters that should be carefully tuned to avoid removing true variation.

3.3.2 Bias from Imaging Window

A key issue in time-lapse microscopy is the limited size of the imaging window, especially in the axial dimension for in vivo imaging, which introduces **selection**

bias: faster, straighter cells leave the imaging window sooner, and thus have shorter tracks. This effect has been shown to cause substantial artefacts in some analyses—especially for cell- and tracklet-based analyses where tracks below a minimum duration are removed (Beltman et al. 2009b). For example, in MSD curves it can give a false impression of subdiffusive behavior (Fig. 2d) as faster cells contribute progressively less to the average as τ increases (Beltman et al. 2009b). Similarly, directional bias may go unnoticed in MSD analysis (Textor et al. 2011). Plotting track metrics (e.g., speed) against track duration can help diagnose this bias.

3.3.3 Dealing with Short Tracks

A common problem in tracking data—especially after automated tracking—is the occurrence of short-duration tracks. Short tracks are often removed before further analysis, but this approach strongly exacerbates the effects of selection bias described above. If selection bias is suspected, step-based analyses are recommended instead (Sect. 3.1) (Beltman et al. 2009b). Short tracks can also result from undetected cells in certain frames, splitting tracks into multiple shorter segments. For analyses that require tracks of longer durations, such as MSD and directional autocorrelation, it may therefore be preferable to keep tracks intact and interpolate missing positions, to include as many cells as possible (see e.g., supplementary materials of Wortel et al. (2021) for details).

3.3.4 Video Heterogeneity

Finally, a common issue in motility analysis is the substantial variation in motility between video replicates, due to biological or technical factors like timing or temperature differences. This variation can obscure differences between populations across videos. In settings where heterogeneity is substantial, it is useful to include an internal control population that occurs in all videos and for which the motility can be assumed to be consistent (see e.g., the supplements of Wortel et al. 2024; McInnes et al. 2024).

3.3.5 Improving Interpretations Using Simulation

Cell motility analysis can reveal interesting biological mechanisms, but their sensitivity to measurement errors and biases means they should be interpreted carefully. An interesting approach to assess the robustness of track-based analysis is through **simulation**. For example, common issues such as track gaps, track splitting, track switching, or a limited imaging window can be artificially introduced in either real or simulated data to check sensitivity (see, for example, Textor et al. 2011). Similarly, detection noise can be introduced artificially to real or simulated data to evaluate its impact on analysis, helping determine how accurate the tracking needs to be. Simulations of simple models, such as random walks (Codling et al. 2008), are easy to implement, and some are available in open-source tools (e.g., MotilityLab and CelltrackR Textor et al. 2015; Wortel et al. 2021).

Simulations can also help design robust analyses to detect mechanisms of interest. For example, the stop-and-go behavior of lymphocytes in lymph nodes was originally attributed to an "internal clock" after Fourier analysis seemed to indicate periodicity in speeds; however, later simulations with and without an internal clock (akin to a positive and negative experimental control) showed that the data were more consistent with the baseline scenario (Beltman et al. 2007).

4 Optical Flow

The quantification approaches discussed so far have all relied on object tracking. Alternatively, it is possible to bypass object tracking entirely using **Optical Flow** (OF). OF is similar to tracking, but instead of tracking object centroids, it tracks every pixel across adjacent images (Fig. 3a). It does not require object detection, making it useful when objects cannot be defined or detected, and provides additional information on morphological changes over time. OF is applied in various fields, such as studying human behavior (Perš et al. 2010), cell transportation in blood vessels (Guo 2013), and heart motion (Gröger et al. 2006), and can be combined with motion models to enhance tracking (Rodrigues et al. 2014).

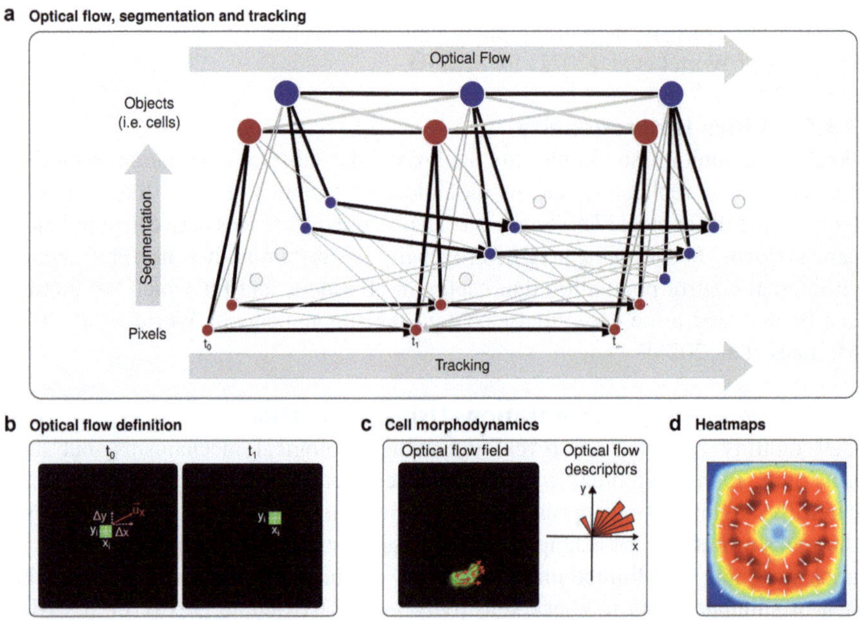

Fig. 3 Schematic overview of the OF calculation process. (**a**) OF is a pixel tracking process. (**b**) For each pixel i, the displacement vector $\mathbf{u}_x$ is calculated, mapping its position (x_i, y_i, t_0) to (x_i, y_i, t_1). (**c**) Global descriptors are derived from the OF vectors. (**d**) OF is applied to generate tissue-level motility maps

4.1 Definition and Considerations for Optical Flow in Biomedical Imaging

OF was stated in the 1940s as the apparent movement of objects due to changes in the point of view. In the 1980s, numerical methods were introduced to link pixels of an image at time t with pixels of an image at time $t+1$ (Horn and Schunck 1981; Lucas and Kanade 1981), providing a basis for analyzing motile objects. Consider an image sequence, with space-time coordinates $\mathbf{x} = [x_1, \ldots, x_{d-1}, t]$, whose dimension d equals the number of spatial dimensions plus 1. The image series can then be represented by a function $I(\mathbf{x}) \in \mathbb{R}$ describing the brightness value at position $\mathbf{x}$. Assume further that we are imaging physical objects, whose motion between consecutive frames is described by a displacement field $U(\mathbf{x})$ mapping each point $\mathbf{x}$ at time t to a new position $\mathbf{x} + \mathbf{u}_x$ at time $t+1$, where $\mathbf{u}_x = [\Delta x_1, \ldots, \Delta x_{d-1}, 1]$. OF is then the problem of estimating $U(\mathbf{x})$ (Fig. 3b).

In the original formulation by Horn and Schunck (Horn and Schunck 1981), the problem is solved by assuming brightness constancy—i.e., a point in the image at time t retains the same brightness value as it moves in the image at time $t+1$:

$$I(\mathbf{x}) = I(\mathbf{x} + \mathbf{u}_x).$$

For sufficiently small displacements, the brightness constancy assumption can be approximated using a first-order Taylor expansion:

$$I(\mathbf{x} + \mathbf{u}_x) \approx I(\mathbf{x}) + \nabla I \cdot U(\mathbf{x})$$

where $\nabla I = \left[\frac{\partial I}{\partial x_1}, \ldots, \frac{\partial I}{\partial x_{d-1}}, \frac{\partial I}{\partial t}\right]^\top$ is the spatiotemporal gradient of I. Neglecting the higher-order terms of the Taylor expansion, this yields the *OF constraint*:

$$\nabla I \cdot U(\mathbf{x}) = 0. \tag{4}$$

Equation (4) provides a single constraint for each pixel, despite the $d-1$ unknowns in $U(\mathbf{x})$. For instance, in the case of 2D images over time ($d = 3$), the constraint reduces to:

$$\frac{\partial I}{\partial x}\Delta x + \frac{\partial I}{\partial y}\Delta y + \frac{\partial I}{\partial t} \cdot 1 = 0.$$

To obtain a unique solution, additional constraints and regularization terms must be introduced. These also allow to prefer desirable properties, such as smoothness. An optimal U^* is computed by minimizing an energy functional:

$$E = E_D + \lambda E_S, \tag{5}$$

where E_D represents the data term (Eq. 4), E_S is a regularization term, and λ controls the impact of regularization.

In Horn and Schunck (1981), E_S is defined as a quadratic penalty term on the gradient of the solution, which discourages large discontinuities in the flow:

$$E(u) = \int_{\Omega} \underbrace{(I(\mathbf{x}) - I(\mathbf{x} + U(\mathbf{x})))^2}_{\text{data term}} \, d\mathbf{x} + \lambda \int_{\Omega} \underbrace{\|\nabla U(\mathbf{x})\|^2}_{\text{smoothness term}} \, d\mathbf{x}. \tag{6}$$

More recent works (Zach et al. 2007; Sánchez et al. 2013) employ different regularization techniques to allow flow discontinuities and increase robustness to noise. However, Eq. 4 is only valid within a sufficiently small local neighborhood. Errors arise in scenarios involving large displacements, texture-less regions, or displacements parallel to image edges. Various methods address these challenges. For instance, additional features to brightness (e.g., SIFT) can be used to match points (Brox and Malik 2011), whereas coarse-to-fine approaches can improve accuracy for sequences with large displacements (Chang et al. 2013). These include pyramidal image decomposition with TV-L1 regularization (Sánchez et al. 2013) and temporal superpixel clustering (Chang et al. 2013).

Computing OF with time-lapse microscopy videos presents specific challenges. First, OF operates at pixel level, but biological interpretations often require object-level abstractions. Second, the assumption of constant pixel intensity between frames may not hold for fluorescence microscopy or transmitted light imaging due to photobleaching, changes in focal planes, or repositioning of the objective in high-throughput devices. Finally, the assumption of small displacements may not hold for large time intervals between frames and rapid moving cells. Coarse-to-fine methods combined with probabilistic frameworks (Amat et al. 2013) were proposed to address these issues, enabling the quantification of large cellular movements, but required segmentation masks to exclude background regions.

More recently, DL has revolutionized the robustness of OF estimation. By leveraging large synthetic and/or real-world data sets of high quality (high frame rates, no artifacts), networks can be first trained to reproduce classical OF formulations as "ground truth", then trained to reproduce these estimates after temporal downsampling and introduction of artifacts. Among these, FlowNet (Dosovitskiy et al. 2015) used convolutional neural networks for end-to-end OF estimation, and FlowNet2 and FlowNet3 (Ilg et al. 2018a,b) enhanced accuracy through multi-stream architectures to address both small and large displacements. DeepFlow (Weinzaepfel et al. 2013) combined deep feature extraction using convolutional neural networks with traditional OF formulations, and EpicFlow (Revaud et al. 2015) incorporated edge-preserving interpolation and geodesic distance from edges to improve flow estimation with large motions and occlusions. For further details, methods, and datasets on OF, see: Shah and Xuezhi (2021).

4.2 OF Applications to Intracellular, Cellular and Tissue Dynamics

OF can quantify cellular morphodynamics at various biological scales, including organelles, cytoskeleton, and tissues, especially when tracking is not possible. For example, trafficking of intracellular organelles and actin filament remodeling are difficult to track due to the limit of optical resolution and absence of a clear centroid in these complex structures—but OF can still provide quantitative, pixel-level information on movement orientation and magnitude. These complement morphological features used in digital pathology and were shown to offer additional insights, such as changes in motor proteins in healthy or pathological conditions (Huang et al. 2017; Drechsler et al. 2020).

At the cell level, morphodynamics result from the interaction of intracellular factors (i.e., ATP availability) and extracellular factors (i.e., extracellular matrix stiffness). At this level, OF can provide insights into the dynamic processes controlling cell migration, such as polarization towards a direction, revealing the effects of therapeutic agents, or modifying extracellular factors (Boric et al. 2013). Flow vectors can also be studied inside specific cells of interest, e.g., using binary masks or segmentations (when available) or an intracellular area defined through other means. Then, selected vectors can be described using the **histogram of optical flow** (HOF, Perš et al. 2010) and/or **histogram of oriented optical flow** (HOOF, Wang and Snoussi 2013) to distinguish different cellular behaviors, as discussed in Sect. 5.3 (Fig. 3c). In addition to cellular and subcellular applications, OF techniques have been applied to analyze tissue dynamics, including embryo development (Amat et al. 2013) and immune responses in lymphoid tissues (Pizzagalli et al. 2019b). In this case, OF provides insights at larger scale, such as flow fields (Schienstock et al. 2024) for identifying collective cell behavior, or tissue velocity heatmaps to identify areas with different motility patterns (Pizzagalli et al. 2019b) (Fig. 3d).

5 Action Recognition

Action recognition (AR) is a computer vision technique for identifying where and when specific actions in video or image sequences occur. An **action** is any meaningful interaction between a subject (human, animal, or cell) and its surroundings (Herath et al. 2017). AR has been applied in video surveillance, autonomous driving, and sports analysis (Vishwakarma and Agrawal 2013; Bukht et al. 2024; Karpathy et al. 2014). It has a huge potential for analyzing cell behavior in (patho)physiological processes like immune responses, cancer metastasis, and tissue regeneration, supporting the development of diagnostics and therapies (Pizzagalli et al. 2022b; Schienstock et al. 2022; Waite et al. 2011; Hanna et al. 2015; Singer and Clark 1999).

In microscopy videos, the types of actions that can be observed and quantified vary depending on cell type and imaging technique. At the single-cell level, key actions include cell cycle events (mitosis, cell death), biological functions (autophagy, phagocytosis, endocytosis, exocytosis), and adopting specific cell migration patterns (e.g., random patrolling in confinement, directed movement towards targets, or arrested states (Auffray et al. 2007; Arasa 2021; Friedl and Weigelin 2008)). Actions at the collective level involve interactions and communication between multiple cells—either through direct physical contact (essential for processes like diapedesis or precise delivery of stimuli to target cells via synapses), or indirectly through the secretion of soluble factors such as cytokines and chemokines (regulating collective behaviors like chemotaxis and swarm formation (Celli et al. 2007; Kienle et al. 2021)). We summarize here three main categories of AR techniques: track-based, segmentation-based, and video-based AR. For details, refer (Pizzagalli et al. 2022b; Vrigkas et al. 2015; Herath et al. 2017).

5.1 Track-based Action Recognition

Track-based methods (Fig. 4a) involve three main steps:

1. **Tracking**: trajectories of cells or particles are obtained as described in Sect. 2.1.
2. **Feature Extraction**: trajectories are described using features such as motility metrics (Sect. 2.2), derived from entire trajectories or tracklets (Sect. 3.1), or other features describing single or multiple objects.
3. **Classification**: trajectories can be classified using predefined thresholds on motility metrics (for example, on track straightness to distinguish directed migration from cell contacts (Pizzagalli et al. 2022b; Pécot et al. 2018)). Actions involving more complex patterns instead can be detected via supervised learning if sufficient annotations are available (Wagner et al. 2017).

Steps 2 and 3 can be combined with DL methods that automatically learn and classify features directly from trajectory data, represented as coordinates or image-based trajectories. For example, convolutional neural networks (CNNs) have been used to classify motility patterns by training on 2D image representations of trajectories (Nicolai et al. 2024).

5.2 Segmentation-based Action Recognition

Segmentation-based methods use the surfaces of objects and their temporal dynamics to enable AR (Fig. 4b). These methods generally follow four steps:

1. **Segmentation or Detection**: objects of interest are segmented from the background in the image, either manually or automatically (section "Cell Detection"). Automatic techniques range from classical approaches like thresholding to

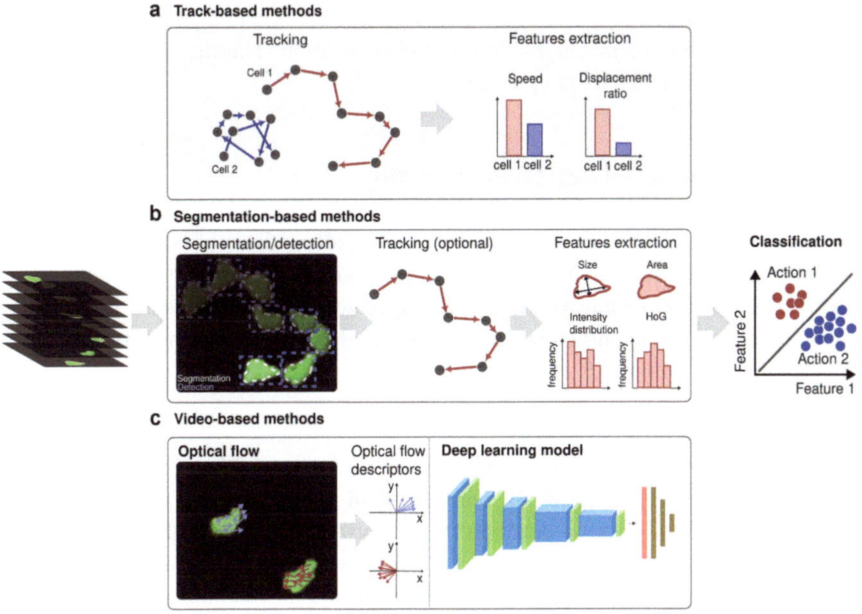

Fig. 4 Schematic representation of the three main classes of AR methods. Each method takes a microscope video as input and outputs the corresponding action class. (**a**) Track-based methods involve tracking individual cells and extracting relevant features. (**b**) Segmentation-based methods segment or detect cells, optionally track them, and extract features from the selected regions. (**c**) Video-based methods analyze the video directly, leveraging techniques such as OF (left panel) or DL approaches, exemplified here by the 3D CNN (right panel)

advanced DL methods, such as CNNs for robust object delineation (Liu et al. 2024; Lee et al. 2020; Cabini et al. 2024a, 2025a,b). In some cases, detection may involve generating bounding boxes around the objects instead of precise delineation, which is useful when exact boundaries are not needed (Pulfer et al. 2024; Anandakumaran et al. 2022; Delgado et al. 2024).

2. **(Optional) Tracking**: segmented objects are tracked across consecutive frames (see Sect. 2.1). This step is not necessary for all applications: for example, detecting contact between two cells in a microscopy video may not require tracking, but tracking is useful for identifying contact between a specific cell and its target.
3. **Feature Extraction**: once segmented (and optionally tracked), features are extracted from the objects. Single-object features describe morphological or statistical properties such as diameter, surface area, volume, intensity distribution, and OF vectors. Multiple-object features describe relationships between segmentations, such as the relative distance between surfaces or spatial arrangements, and can capture temporal changes if tracking is applied (Liu et al. 2024).
4. **Classification**: the extracted features are then classified using filters or supervised learning as outlined in Sect. 5.1.

Steps 3 and 4 can be combined using DL methods, where a single model simultaneously handles feature extraction and classification, learning relevant features directly from the data (Sillah et al. 2024).

5.3 Video-based Action Recognition

This category includes methods that process videos directly, without requiring preliminary steps such as tracking or segmentation (Fig. 4c). These methods are particularly useful in contexts where tracking or segmenting videos is impossible or prone to errors. While numerous video-based methods exist for other domains, their application for cellular AR is still limited.

5.3.1 Optical Flow Features

A major family of video-based methods relies on computing OF between consecutive frames (Fig. 4c, Sect. 4). One common OF-based approach to AR is to extract hand-crafted motion descriptors, such as HOF and HOOF (Uijlings et al. 2015) (Sect. 4), from OF vectors, which can then be used for classification using filtering-based approaches, supervised or unsupervised learning models. Another common approach is to track small image regions, called particles, over time based on OF vectors, generating trajectories (Wang et al. 2013). Features, such as chaotic invariants defined in (Wu et al. 2010; Ali and Shah 2007), are extracted from these trajectories and classified using ML models.

5.3.2 Video Classification

DL methods have become popular in AR due to their ability to automatically learn features from raw video data (Fig. 4c). Key DL architectures include **CNNs**, **multi-stream networks**, recurrent neural network (**RNN**)-based methods, and **transformer**-based methods (Herath et al. 2017; Zhu et al. 2020). CNNs use convolutional layers to capture both spatial and temporal information from raw video data, treating the input as volumetric data. These architectures typically consist of a stack of convolutional layers to learn spatiotemporal features, followed by pooling layers to reduce spatial and temporal dimensions. Convolutional layers can employ 3D convolutions, which operate simultaneously across spatial and temporal dimensions (Ji et al. 2013; Tran et al. 2014) (Fig. 4c), or a combination of 2D convolutions (for spatial dimensions) and 1D convolutions (for the temporal dimension), reducing the total number of parameters in the network (Sun et al. 2015).

Two-stream networks are a class of CNNs designed to perform object recognition and motion recognition separately. The spatial stream takes raw video frames as input to capture visual appearance information. The temporal stream takes a stack of OF images as input to capture motion information between video frames. The outputs from both networks are then fused to obtain the final prediction (Simonyan and Zisserman 2014; Tu et al. 2018).

RNN-based models exploit the capability of RNNs to handle temporal data, learning long-term relationships in the data. They consist of a 2D CNN that is simultaneously applied to each video frame separately, followed by an RNN that receives the extracted features as input and makes the final action prediction (Donahue et al. 2017; Baccouche et al. 2011). Some examples of RNNs include Long Short-Term Memory and Gated Recurrent Units.

Transformer-based methods have recently emerged as powerful tools for modeling temporal tasks, including AR. Similar to RNN-based methods, they operate on features extracted by a 2D CNN from individual video frames to produce the output prediction. However, transformers utilize a self-attention mechanism, allowing them to assess the importance of each part of the input data and capture long-range dependencies without the need for sequential computation, as in RNN-based methods (Ulhaq et al. 2022; Girdhar et al. 2019; Cabini et al. 2024b).

6 Conclusions and Future Perspectives

The development of methods for quantifying movement in biomedical videos is a current frontier of research both in computational and life sciences. In principle, the concepts and algorithms behind movement analysis methods can be applied to objects with different sizes and image modalities, quantifying movement across scales. In practice, however, many computer vision methods are currently more established for images and videos capturing natural scenes such as traffic and human behavior, and not all of them translate well to bioimaging. A more effective exchange of methods, datasets, and complementary needs between scientists in computer vision and life sciences will be crucial to overcome this gap in biomedical computer vision.

6.1 Overcoming the (Annotated) Data Gap

Modern computer vision owes its success in part to the availability of large, open, and well-curated datasets such as MNIST and ImageNet. However, a key challenge for the translation of bioimaging into the biomedical domain is that the subjects of bioimaging are substantially different from those in these classical datasets. For example, cells often lack stable features such as rigid edges or textures, ensuring that most general-purpose computer vision models trained on other types of data are not directly applicable to bioimaging applications unless trained on domain-specific data. This is especially relevant for DL methods, which typically require very high numbers of images for training. Even if models pre-trained on non-biomedical images can successfully be translated to the bioimaging domain using transfer learning, this still requires the availability of large, domain-specific and well-annotated datasets. And while these are increasingly available for medical imaging used in clinical practice (National Institute for Health Research 2025;

Stettler 2025), they are less widespread for the broad range in imaging techniques used in preclinical research.

Progress has been made in important initiatives such as the Cell Tracking Challenge (Maška et al. 2014, 2023; Ulman et al. 2017), as well as open data available in e.g., MotilityLab (Textor et al. 2015), the leukocyte tracking database (Pizzagalli et al. 2018), and Immunemap (McInnes et al. 2024), which effectively facilitate data exchange and support algorithm development. Yet annotated open datasets remain scarce given the large variety in imaging modalities, (fluorescent) labelling, and experimental settings—hampering the development of computer vision technology for more complex tasks. Especially in AR, annotations are critical for training action classifiers, and challenges arise from limited data availability and confounding motility patterns (i.e., mitosis vs. separation of two touching cells).

To overcome this issue, smart annotation strategies are likely required, e.g., by making annotations in an iterative manner: identifying cases yielding false positives / false negatives and adding them to the training dataset (Pulfer et al. 2024). A key direction of future research is how to use limited annotations more effectively, and/or how to develop active learning strategies to generate annotations where these are most needed. Likewise, an open question is how we can more effectively leverage simulations in computer vision for bioimaging, for example to generate large synthetic datasets for (pre-)training, or to generate annotated data for AR tasks where the ground truth is difficult and/or time-consuming to obtain (Brekke et al. 2019). In the past several decades, many computational models have been developed to simulate a wide range of biological dynamical processes—from cell sorting, to cell motility, to tissue development, collective migration, and many others (Graner and Glazier 1992; Beltman et al. 2007; Niculescu et al. 2015; Hirashima et al. 2017; Szabó and Merks 2013; Camley and Rappel 2017; Montagud et al. 2021). Their integration into computer vision development would be a promising direction of future research.

6.2 Overcoming Structural Barriers to Open Science Practices

A barrier to the availability of open data is the current lack of recognition and reward systems for publishing well-annotated datasets. Whereas taking and labelling pictures of cats and dogs is relatively cheap, generating specialized bioimaging data is typically (very) expensive. And even once raw data are available, providing annotations is often time-consuming as well. As long as the generation of data is not recognized as a meaningful contribution in and of itself, this provides an incentive to delay sharing of the data and/or metadata in order to capitalize on a dataset before making it available for the community to use. Solving this issue will require cultural changes as well as new perspectives on how we reward scientific contributions. Despite ongoing efforts (Devriendt et al. 2021), this area will remain in need of attention.

Aside from well-curated open datasets with rich metadata, the effective development and use of bio-image analysis methods also requires accessible (ideally open-source) analysis tools. Software should be accessible through key usability features including user interfaces, well-structured code, interoperability with different file formats and other tools, and a modular structure, and the analysis tools offered should be validated (Carpenter et al. 2021). Examples include larger platforms with community-supplied plugins such as Fiji (Schindelin et al. 2012) and Icy (Chaumont et al. 2012), but also web-based tools (Textor et al. 2015), or packages built on the existing ecosystems of popular programming languages such as R and python (Wortel et al. 2021). Despite ongoing efforts in the development of open-source software for bioimaging data, current research still often relies on custom scripts and/or proprietary software. Further supporting the transition to open-access will require the continued investment in Open Science practices such as open-access repositories, collaborative platforms, organization of scientific challenges, and release of open-source tools that can be customized or improved for different applications.

6.3 Learning Meaningful Representations of Dynamics

Finally, a promising area of ongoing research is to what extent ML can supplement hand-crafted motility metrics (Sect. 2.2) by facilitating the discovery of novel descriptors of dynamical processes, for example through self-supervised representation learning. An important caveat here is that machine-learned features will likely suffer from the same challenges and biases as hand-crafted ones (Sect. 3.3)—unless explicit care is taken to avoid issues such as confounding by track duration or bias introduced by the imaging window. A key challenge will therefore be to harness lessons learned from the application of hand-crafted features, ensuring that learned representations are meaningful.

6.4 Final Remarks

Scientific discoveries often go hand in hand with technological developments (Masters 2008). In the field of bioimaging, these developments have long been driven by the development of new microscopes and labelling techniques—but are increasingly driven by advances in the methodology used to process and analyze bioimaging data as well. Exciting developments in AI are working their way into the bioimaging domain, but some barriers remain to the optimal synergy between the fields of bioimaging and computer vision. Strengthening this synergy will be key to drive new discoveries in basic research and open up new avenues for clinical applications such as disease monitoring and precision therapeutics.

Acknowledgments DUP was supported by the Swiss National Science Foundation with grant 228512 and USI FIR grant, RFC was supported by USI FIR grant and swissuniversities with RE2VITAL CHORD B grant. IMNW was supported by AiNed Fellowship grant NGF.1607.22.020 from the Dutch Research Council (NWO).

Conflict of Interest Authors declare no conflict of interests.

References

Al-Kofahi O, Radke RJ, Goderie SK, Shen Q, Temple S, Roysam B (2006) Automated cell lineage construction: a rapid method to analyze clonal development established with murine neural progenitor cells. Cell Cycle 5(3):327–335

Ali S, Shah M (2007) A Lagrangian particle dynamics approach for crowd flow segmentation and stability analysis. In 2007 IEEE conference on computer vision and pattern recognition. IEEE, pp. 1–6

Amat F, Myers EW, Keller PJ (2013) Fast and robust optical flow for time-lapse microscopy using super-voxels. Bioinformatics 29(3):373–380

Anandakumaran PN, Ayers AG, Muranski P, Creusot RJ, Sia SK (2022) Rapid video-based deep learning of cognate versus non-cognate T cell-dendritic cell interactions. Sci Rep 12(1):559

Apostoloff N, Fitzgibbon A (2005) Learning spatiotemporal T-junctions for occlusion detection. In IEEE computer society conference on computer vision and pattern recognition, 2005 (CVPR 2005), vol. 2, IEEE, pp. 553–559

Apostoloff N, Fitzgibbon AW (2006) Automatic video segmentation using spatiotemporal T-junctions. Bmvc, pp. 1089–1098

Arasa J, Collado-Diaz V, Kritikos I, Medina-Sanchez JD, Friess MC, Sigmund EC, Schineis P, Hunter MC, Tacconi C, Paterson N et al (2021) Upregulation of VCAM-1 in lymphatic collectors supports dendritic cell entry and rapid migration to lymph nodes in inflammation. J Exp Med 218(7):e20201413

Arganda-Carreras I, Kaynig V, Rueden C, Eliceiri KW, Schindelin J, Cardona A, Sebastian Seung H (2017) Trainable weka segmentation: a machine learning tool for microscopy pixel classification. Bioinformatics 33(15):2424–2426

Ariotti S, Beltman JB, Borsje R, Hoekstra ME, Halford WP, Haanen JBAG, de Boer RJ, Schumacher TNM (2015) Subtle CXCR3-dependent chemotaxis of CTLs within infected tissue allows efficient target localization. J Immunol 195:5285–5295

Auffray C, Fogg D, Garfa M, Elain G, Join-Lambert O, Kayal S, Sarnacki S, Cumano A, Lauvau G, Geissmann F (2007) Monitoring of blood vessels and tissues by a population of monocytes with patrolling behavior. Science 317(5838):666–670

Baccouche M, Mamalet F, Wolf C, Garcia C, Baskurt A (2011) Sequential deep learning for human action recognition. In Human behavior understanding: second international workshop, HBU 2011, Amsterdam, The Netherlands, November 16, 2011. Proceedings 2. Springer, pp. 29–39

Banigan EJ, Harris TH, Christian DA, Hunter CA, Liu AJ (2015) Heterogeneous CD8+ T cell migration in the lymph node in the absence of inflammation revealed by quantitative migration analysis. PLOS Comput Biol 11:1–20

Beltman JB, Marée AF, Lynch JN, Miller MJ, de Boer RJ (2007) Lymph node topology dictates T cell migration behavior. J Exp Med 204:771–780

Beltman JB, Henrickson SE, von Andrian UH, de Boer RJ, Marée AF (2009a) Towards estimating the true duration of dendritic cell interactions with T cells. J Immunol Methods 347(1):54–69

Beltman JB, Marée AFM, de Boer RJ (2009b) Analysing immune cell migration. Nature Rev Immunol 9:789–798

Beltman JB, Allen CDC, Cyster JG, de Boer RJ (2011) B cells within germinal centers migrate preferentially from dark to light zone. Proc Natl Acad Sci 108(21):8755–8760

Berclaz J, Fleuret F, Turetken E, Fua P (2011) Multiple object tracking using k-shortest paths optimization. IEEE Trans Pattern Anal Mach Intell 33(9):1806–1819

Berg S, Kutra D, Kroeger T, Straehle CN, Kausler BX, Haubold C, Schiegg M, Ales J, Beier T, Rudy M et al (2019) Ilastik: interactive machine learning for (bio) image analysis. Nature Methods 16(12):1226–1232

Boric K, Orio P, Viéville T, Whitlock K (2013) Quantitative analysis of cell migration using optical flow. PloS One 8(7):e69574

Bousso P, Robey E (2003) Dynamics of CD8+ T cell priming by dendritic cells in intact lymph nodes. Nature Immunol 4:579–585

Brekke Å, Vatsendvik F, Lindseth F (2019) Multimodal 3D object detection from simulated pretraining. In Symposium of the Norwegian AI society. Springer, pp. 102–113

Brox T, Malik J (2011) Large displacement optical flow descriptor matching in variational motion estimation. IEEE Trans Pattern Anal Mach Intell 33(3):1–14

Bukht TFN, Rahman H, Shaheen M, Algarni A, Almujally NA, Jalal A (2024) A review of video-based human activity recognition: theory, methods and applications. Multimedia Tools Appl 1–47

Cabini RF, Tettamanti H, Zanella M (2024a) Understanding the impact of evaluation metrics in kinetic models for consensus-based segmentation. Preprint. arXiv:2412.03458

Cabini RF, Barzaghi L, Cicolari D, Arosio P, Carrazza S, Figini S, Filibian M, Gazzano A, Krause R, Mariani M et al (2024b) Fast deep learning reconstruction techniques for preclinical magnetic resonance fingerprinting. NMR Biomed 37(1):e5028

Cabini RF, Pichiecchio A, Lascialfari A, Figini S, Zanella M (2025a) A kinetic approach to consensus-based segmentation of biomedical images. Kinetic Relat Models 18(2):286–311

Cabini RF, Cozzi A, Leu S, Thelen B, Krause R, Del Grande F, Pizzagalli DU, Rizzo SMR (2025) Compositia: an open-source automated quantification tool for body composition scores from thoraco-abdominal CT scans. Eur Radiol Exp 9(1):12

Camley BA, Rappel W-J (2017) Physical models of collective cell motility: from cell to tissue. J Phys D Appl Phys 50(11):113002

Carpenter AE, Kamentsky L, Eliceiri KW (2021) A call for bioimaging software usability. Nature Methods 9(7):666–670

Celli S, Lemaître F, Bousso P (2007) Real-time manipulation of T cell-dendritic cell interactions in vivo reveals the importance of prolonged contacts for CD4+ T cell activation. Immunity 27(4):625–634

Chan TF, Vese LA (2001) Active contours without edges. IEEE Trans Image Process 10(2):266–277

Chang J, Wei D, Fisher JW (2013) A video representation using temporal superpixels. In Proceedings of the IEEE computer society conference on computer vision and pattern recognition, pp. 2051–2058

de Chaumont F, Dallongeville S, Chenouard N, Hervé N, Pop S, Provoost T, Meas-Yedid V, Pankajakshan P, Lecomte T, Le Montagner Y, Lagache T, Dufour A, Olivo-Marin J-C (2012) Icy: an open bioimage informatics platform for extended reproducible research. Nature Methods 9(7):690–696

Cheng Y, Li L, Xu Y, Li X, Yang Z, Wang W, Yang Y (2023) Segment and track anything. Preprint. arXiv:2305.06558

Chenouard N, Bloch I, Olivo-Marin J-C (2013) Multiple hypothesis tracking for cluttered biological image sequences. IEEE Trans Pattern Anal Mach Intell 35(11):2736–3750

Codling EA, Plank MJ, Benhamou S (2008) Random walk models in biology. J R Soc Interface 5(25):813–834

Crainiciuc G, Palomino-Segura M, Molina-Moreno M, Sicilia J, Aragones DG, Li JLY, Madurga R, Adrover JM, Aroca-Crevillén A, Martin-Salamanca S, del Valle AS, Castillo SD, Welch HCE, Soehnlein O, Graupera M, Sánchez-Cabo F, Zarbock A, Smithgall TE, Di Pilato M, Mempel TR, Tharaux P-L, González SF, Ayuso-Sacido A, Ng LG, Calvo GF, González-Díaz I, Díaz-de María F, Hidalgo A (2022) Behavioural immune landscapes of inflammation. Nature 601(7893):415–421

Davis DM (2009) Mechanisms and functions for the duration of intercellular contacts made by lymphocytes. Nature Rev Immunol 9:543–555 (2009)

Delgado-Rodriguez P, Sánchez RM, Rouméas-Noël E, Paris F, Munoz-Barrutia A (2024) Automatic classification of normal and abnormal cell division using deep learning. Sci Rep 14(1):14241

Devriendt T, Shabani M, Borry P (2021) Data sharing in biomedical sciences: a systematic review of incentives. Biopreservation Biobanking 19(3):219–227

Donahue J, Hendricks LA, Rohrbach M, Venugopalan S, Guadarrama S, Saenko K, Darrell T (2017) Long-term recurrent convolutional networks for visual recognition and description. IEEE Trans Pattern Anal Mach Intell 39(4):677–691

Dosovitskiy A, Fischer P, Ilg E, Hausser P, Hazirbas C, Golkov V, Van Der Smagt P, Cremers D, Brox T (2015) Flownet: learning optical flow with convolutional networks. In Proceedings of the IEEE international conference on computer vision, pp. 2758–2766

Drechsler M, Lang LF, Al-Khatib L, Dirks H, Burger M, Schönlieb C-B, Palacios IM (2020) Optical flow analysis reveals that kinesin-mediated advection impacts the orientation of microtubules in the drosophila oocyte. Molecular Biol Cell 31(12):1246–1258

Fürth R (1920) Die brownsche bewegung bei berücksichtigung einer persistenz der bewegungsrichtung. mit anwendungen auf die bewegung lebender infusorien. Z Phys 2(3):244–256

Friedl P, Weigelin B (2008) Interstitial leukocyte migration and immune function. Nature Immunol 9(9):960–969

Girdhar R, Carreira J, Doersch C, Zisserman A (2019) Video action transformer network. In Proceedings of the IEEE/CVF conference on computer vision and pattern recognition, pp. 244–253

Gorelik R, Gautreau A (2014) Quantitative and unbiased analysis of directional persistence in cell migration. Nature Protocols 9:1931–1943

Gröger M, Hirzinger G (2006) Optical flow to analyse stabilised images of the beating heart. In International conference on computer vision theory and applications, vol 3, pp. 237–244

Graner FMC, Glazier JA (1992) Simulation of biological cell sorting using a two-dimensional extended potts model. Phys Rev Lett 69:2013–2016

Guo D, Van de Ven AL, Zhou X (2013) Tracking and measurement of the motion of blood cells using optical flow methods. IEEE J Biomed Health Inf 18(3):991

Hanna RN, Cekic C, Sag D, Tacke R, Thomas GD, Nowyhed H, Herrley E, Rasquinha N, McArdle S, Wu R et al (2015) Patrolling monocytes control tumor metastasis to the lung. Science 350(6263):985–990

He S, Sillah M, Cole AR, Uboveja A, Aird KM, Chen Y-C, Gong Y-N (2024) D-mains: a deep-learning model for the label-free detection of mitosis, apoptosis, interphase, necrosis, and senescence in cancer cells. Cells 13(12):1004

Herath S, Harandi M, Porikli F (2017) Going deeper into action recognition: a survey. Image Vision Comput 60:4–21

Hirashima T, Rens EG, Merks RMH (2017) Cellular potts modeling of complex multicellular behaviors in tissue morphogenesis. Develop Growth Differ 59(5):329–339

Hirsch P, Epstein L, Guignard L (2022) Chapter 20 - mathematical and bioinformatic tools for cell tracking. In: Schnoor M, Yin L-M, Sun SX (eds) Cell movement in health and disease. Academic Press, Cambridge, pp. 341–361

Horn BK, Schunck BG (1981) Determining optical flow. Artif Intell 17(1–3):185–203

Hotelling H (1931) The generalization of student's ratio. Ann Math Stat 2(3):360–378

Huang Y, Hao L, Li H, Liu Z, Wang P (2017) Quantitative analysis of intracellular motility based on optical flow model. J Healthcare Eng 2017(1):1848314

Ibidi, Ibidi application guide: Chemotaxis assays. 2019. Accessed on Jan 10, 2025. https://ibidi.com/img/cms/resources/AG/FL_AG_035_Chemotaxis_150dpi.pdf

Ilg E, Saikia T, Keuper M, Brox T, Occlusions (2018a) motion and depth boundaries with a generic network for disparity, optical flow or scene flow estimation. In Proceedings of the European conference on computer vision (ECCV), pp. 614–630

Ilg E, Çiçek Ö, Galesso S, Klein A, Makansi O, Hutter F, Brox T (2018) Uncertainty estimates and multi-hypotheses networks for optical flow. In European conference on computer vision (ECCV)

Ji S, Xu W, Yang M, Yu K (2013) 3D convolutional neural networks for human action recognition. IEEE Trans Pattern Anal Mach Intell 35(1):221–231

Jiang Z, Rozgic V, Adali S (2017) Learning spatiotemporal features for infrared action recognition with 3D convolutional neural networks. In Proceedings of the IEEE conference on computer vision and pattern recognition, pp. 309–317

Kalman RE (1960) A new approach to linear filtering and prediction problems. J Basic Eng 82:35–45

Karpathy A, Toderici G, Shetty S, Leung T, Sukthankar R, Fei-Fei L (2014) Large-scale video classification with convolutional neural networks. In Proceedings of the IEEE conference on computer vision and pattern recognition, pp. 1725–1732

Kienle K, Glaser KM, Eickhoff S, Mihlan M, Knöpper K, Reátegui E, Epple MW, Gunzer M, Baumeister R, Tarrant TK et al (2021) Neutrophils self-limit swarming to contain bacterial growth in vivo. Science 372(6548):eabe7729

Krishna NM, Reddy RY, Reddy MSC, Madhav KP, Sudham G (2021) Object detection and tracking using YOLO. In 2021 third international conference on inventive research in computing applications (ICIRCA). IEEE, pp. 1–7

Krummel MF, Bartumeus F, Gérard A (2016) T cell migration, search strategies and mechanisms. Nature Rev Immunol 16:193–201

Lee M, Lee Y-H, Song J, Kim G, Jo Y, Min H, Kim CH, Park Y (2020) Deep-learning-based three-dimensional label-free tracking and analysis of immunological synapses of CAR-T cells. Elife 9:e49023

Letendre K, Donnadieu E, Moses ME, Cannon JL (2015) Bringing statistics up to speed with data in analysis of lymphocyte motility. PLOS ONE 10(5):1–18

Li F, Zhou X, Ma J, Wong STC (2010) Multiple nuclei tracking using integer programming for quantitative cancer cell cycle analysis. IEEE Trans Med Imaging 29(1):96–105

Liu B, Zhu Y, Yang Z, Yan HH, Leung SY, Shi J (2024) Deep learning–based 3D single-cell imaging analysis pipeline enables quantification of cell–cell interaction dynamics in the tumor microenvironment. Cancer Res 84(4):517–526

Lucas BD, Kanade T (1981) An iterative image registration technique with an application to stereo vision. In IJCAI'81: 7th international joint conference on artificial intelligence, vol. 2, pp. 674–679

Maška M, Ulman V, Svoboda D, Matula P, Matula P, Ederra C, Urbiola A, España T, Venkatesan S, Balak DM et al (2014) A benchmark for comparison of cell tracking algorithms. Bioinformatics 30(11):1609–1617

Maška M, Ulman V, Delgado-Rodriguez P, Gómez-de Mariscal E, Nečasová T, Guerrero Peña FA, Ren TI, Meyerowitz EM, Scherr T, Löffler K et al (2023) The cell tracking challenge: 10 years of objective benchmarking. Nature Methods 20(7):1010–1020

Magnusson KEG, Jaldén J (2012) A batch algorithm using iterative application of the Viterbi algorithm to track cells and construct cell lineages. In 2012 9th IEEE international symposium on biomedical imaging (ISBI). IEEE, pp. 382–385

Maiuri P, Rupprecht J-F, Wieser S, Ruprecht V, Bénichou O, Carpi N, Coppey M, De Beco S, Gov N, Heisenberg C-P, Lagev Crespo C, Lautenschlaeger F, Le Berre M, Lennon-Dumenil A-M, Raab M, Thiam H-R, Piel M, Sixt M, Voituriez R (2015) Actin flows mediate a universal coupling between cell speed and cell persistence. Cell 161(2):374–386

Markappa PSS, O'Leary C, Lynch C (2024) A review of YOLO models for soccer-based object detection. In 2024 sixth international conference on intelligent computing in data sciences (ICDS), pp. 1–7

Masters BR (2008) History of the optical microscope in cell biology and medicine. John Wiley & Sons, New York

McCutcheon M (1946) Chemotaxis in leukocytes. Physiol Rev 26(3):319–336. PMID: 20993553

McInnes L, Healy J, Melville J (2018) UMAP: Uniform manifold approximation and projection for dimension reduction. arXiv:1802.03426

Pizzagalli DU, Carrillo-Barbera P, Palladino E, Ceni K, Thelen B, Pulfer A, Moscatello E, Cabini RF, Textor J, Wortel I et al (2024) Systematic analysis of immune cell motility leveraging Immunemap, an open intravital microscopy atlas. bioRxiv

Mempel TR, Henrickson SE, von Andrian UH (2004) T-cell priming by dendritic cells in lymph nodes occurs in three distinct phases. Nature 427:154–159

Milan A, Rezatofighi SH, Dick A, Reid I, Schindler K (2017) Online multi-target tracking using recurrent neural networks. In Proceedings of the AAAI conference on Artificial Intelligence, vol. 31.

Miller MJ, Wei SH, Parker I, Cahalan MD (2002) Two-photon imaging of lymphocyte motility and antigen response in intact lymph node. Science 296(5574):1869–1873

Miller MJ, Wei SH, Cahalan MD, Parker I (2003) Autonomous T cell trafficking examined in vivo with intravital two-photon microscopy. Proc Natl Acad Sci 100(5):2604–2609

Moen E, Bannon D, Kudo T, Graf W, Covert M, Van Valen D (2019) Deep learning for cellular image analysis. Nature Methods 16(12):1233–1246

Mokhtari Z, Mech F, Zitzmann C, Hasenberg M, Gunzer M, Figge MT (2013) Automated characterization and parameter-free classification of cell tracks based on local migration behavior. PLOS ONE 8:1–20

Montagud A, de Leon MP, Valencia A (2021) Systems biology at the giga-scale: large multiscale models of complex, heterogeneous multicellular systems. Current Opinion Syst Biol 28:100385

Moore BR (1980) A modification of the Rayleigh test for vector data. Biometrika 67(1):175–180

National Institute for Health Research (2025) A collection of open source imaging data sets. Accessed on Mar 17, 2022

Nicolai M, Cabini RF, Pizzagalli DU (2024) Classification and regression of trajectories rendered as images via 2D convolutional neural networks. Preprint. arXiv:2409.18832

Niculescu I, Textor J, de Boer RJ (2015) Crawling and gliding: a computational model for shape-driven cell migration. PLOS Comput Biol 11(10):1–22

Oja H, Randles RH (2004) Multivariate nonparametric tests. Stat Sci 19(4):598–605

Pécot T, Zengzhen L, Boulanger J, Salamero J, Kervrann C (2018) A quantitative approach for analyzing the spatio-temporal distribution of 3D intracellular events in fluorescence microscopy. Elife 7:e32311

Perš J, Sulić V, Kristan M, Perše M, Polanec K, Kovačič S (2010) Histograms of optical flow for efficient representation of body motion. Pattern Recogn Lett 31(11):1369–1376

Pizzagalli DU, Farsakoglu Y, Palomino-Segura M, Palladino E, Sintes J, Marangoni F, Mempel TR, Koh WH, Murooka TT, Thelen F et al (2018) Leukocyte tracking database, a collection of immune cell tracks from intravital 2-photon microscopy videos. Sci Data 5(1):1–13

Pizzagalli DU, Gonzalez SF, Krause R (2019a) A trainable clustering algorithm based on shortest paths from density peaks. Sci Adv 5(10):eaax3770

Pizzagalli DU, Latino I, Pulfer A, Palomino-Segura M, Virgilio T, Farsakoglu Y, Krause R, Gonzalez SF (2019b) Characterization of the dynamic behavior of neutrophils following influenza vaccination. Front Immunol 10, 2621

Pizzagalli DU, Bordini J, Morone D, Pulfer A, Carrillo-Barberà P, Thelen B, Ceni K, Thelen M, Krause R, Gonzalez SF (2022a) Cancol, a computer-assisted annotation tool to facilitate colocalization and tracking of immune cells in intravital microscopy. J Immunol 208(6):1493–1499

Pizzagalli DU, Pulfer A, Thelen M, Krause R, Gonzalez SF (2022b) In vivo motility patterns displayed by immune cells under inflammatory conditions. Front Immunol 12:804159

Pulfer A, Pizzagalli DU, Gagliardi PA, Hinderling L, Lopez P, Zayats R, Carrillo-Barberà P, Antonello P, Palomino-Segura M, Grädel B, et al (2024) Transformer-based spatial–temporal detection of apoptotic cell death in live-cell imaging. Elife 12:RP90502

Rajič F, Ke L, Tai Y-W, Tang C-K, Danelljan M, Yu F (2023) Segment anything meets point tracking. Preprint. arXiv:2307.01197

Revaud J, Weinzaepfel P, Harchaoui Z, Schmid C (2015) Epicflow: edge-preserving interpolation of correspondences for optical flow. In Proceedings of the IEEE conference on computer vision and pattern recognition, pp. 1164–1172

Rodrigues V, Rodrigues PJ, Pereira AI, Lima R (2014) Automatic cell tracking method based on optical flow tuned by template matching. Virtual. ipb. pt

Rosen ME, Grant CP, Dallon JC (2021) Mean square displacement for a discrete centroid model of cell motion. PLOS ONE 16(12):1–19

Sánchez Pérez, J, Meinhardt-Llopis E, Facciolo G (2013) TV-L1 optical flow estimation. Image Processing On Line 3:137–150. https://doi.org/10.5201/ipol.2013.26

Schienstock D, Mueller SN (2022) Moving beyond velocity: opportunities and challenges to quantify immune cell behavior. Immunol Rev 306(1):123–136

Schienstock D, Hor JL, Devi S, Mueller SN (2024) Cecelia: a multifunctional image analysis toolbox for decoding spatial cellular interactions and behaviour. bioRxiv, pp. 2024–08

Schindelin J, Arganda-Carreras I, Frise E, Kaynig V, Longair M, Pietzsch T, Preibisch S, Rueden C, Saalfeld S, Schmid B, Tinevez J-Y, White DJ, Hartenstein V, Eliceiri K, Tomancak P, Cardona A (2012) Fiji: an open-source platform for biological-image analysis. Nature Methods 9(7):676–682

Schmidt U, Weigert M, Broaddus C, Myers G (2018) Cell detection with star-convex polygons. In Medical image computing and computer assisted intervention–MICCAI 2018: 21st international conference, Granada, Spain, September 16–20, 2018, Proceedings, Part II 11. Springer, pp. 265–273

Shah STH, Xuezhi X (2021) Traditional and modern strategies for optical flow: an investigation. SN Appl Sci 3(3):289

Simonyan K, Zisserman A (2014) Two-stream convolutional networks for action recognition in videos. Adv Neural Inf Process Syst 27

Singer AJ, Clark RA (1999) Cutaneous wound healing. New England J Med 341(10):738–746

Stettler MD (2025) Histopathology datasets for machine learning. Accessed on Mar 17, 2025. https://github.com/maduc7/Histopathology-Datasets

Stringer C, Wang T, Michaelos M, Pachitariu M (2021) Cellpose: a generalist algorithm for cellular segmentation. Nature Methods 18(1):100–106

Sun L, Jia K, Yeung D-Y, Shi BE (2015) Human action recognition using factorized spatio-temporal convolutional networks. In 2015 IEEE international conference on computer vision (ICCV), pp. 4597–4605

Szabó B, Szöllösi GJ, Gönci B, Jurányi Z, Selmeczi D, Vicsek T (2006) Phase transition in the collective migration of tissue cells: experiment and model. Phys Rev E 74:061908

Szabó A, Merks RMH (2013) Cellular Potts modeling of tumor growth, tumor invasion, and tumor evolution. Front Oncol 3

Textor J, Peixoto A, Henrickson SE, Sinn M, von Andrian UH, Westermann J (2011) Defining the quantitative limits of intravital two-photon lymphocyte tracking. Proc Natl Acad Sci 108(30):12401–12406

Textor J, Sinn M, de Boer RJ (2013) Analytical results on the Beauchemin model of lymphocyte migration. BMC Bioinf 14:S10

Textor J, Berry J, Miller MJ (2015) Motilitylab – a resource for cell migration analysis. Accessed on Jan 26, 2025

Thomas GL, Fortuna I, Perrone GC, Glazier JA, Belmonte JM, de Almeida RM (2020) Parameterizing cell movement when the instantaneous cell migration velocity is ill-defined. Phys A Stat Mech Appl 550:124493

Tinevez J-Y, Perry N, Schindelin J, Hoopes GM, Reynolds GD, Laplantine E, Bednarek SY, Shorte SL, Eliceiri KW (2017) Trackmate: an open and extensible platform for single-particle tracking. Methods 115:80–90

Tran D, Bourdev LD, Fergus R, Torresani L, Paluri M (2014) Learning spatiotemporal features with 3D convolutional networks. 2015 IEEE international conference on computer vision (ICCV), pp. 4489–4497

Tu Z, Xie W, Qin Q, Poppe R, Veltkamp RC, Li B, Yuan J (2018) Multi-stream CNN: learning representations based on human-related regions for action recognition. Pattern Recogn 79:32–43

Turetken E, Wang X, Becker CJ, Haubold C, Fua P (2016) Network flow integer programming to track elliptical cells in time-lapse sequences. IEEE Trans Med Imaging 36(4):942–951

Uijlings J, Duta IC, Sangineto E, Sebe N (2015) Video classification with densely extracted hog/hof/mbh features: an evaluation of the accuracy/computational efficiency trade-off. Int J Multimedia Inf Retrieval 4(1):33–44

Ulhaq A, Akhtar N, Pogrebna G, Mian A (2022) Vision transformers for action recognition: a survey. Preprint. arXiv:2209.05700

Ulman V, Maška M, Magnusson KE, Ronneberger O, Haubold C, Harder N, Matula P, Matula P, Svoboda D, Radojevic M et al (2017) An objective comparison of cell-tracking algorithms. Nature Methods 14(12):1141–1152

Vicsek T, Zafeiris A (2012) Collective motion. Phys Rep 517(3):71–140

Vishwakarma S, Agrawal A (2013) A survey on activity recognition and behavior understanding in video surveillance. Visual Comput 29:983–1009

Vrigkas M, Nikou C, Kakadiaris IA (2015) A review of human activity recognition methods. Front Robotics AI 2:28

Wagner T, Kroll A, Haramagatti CR, Lipinski H-G, Wiemann M (2017) Classification and segmentation of nanoparticle diffusion trajectories in cellular micro environments. PloS One 12(1):e0170165

Waite JC, Leiner I, Lauer P, Rae CS, Barbet G, Zheng H, Portnoy DA, Pamer EG, Dustin ML (2011) Dynamic imaging of the effector immune response to listeria infection in vivo. PLoS Pathogens 7(3):e1001326

Wang T, Snoussi H (2013) Histograms of optical flow orientation for abnormal events detection. In 2013 IEEE international workshop on performance evaluation of tracking and surveillance (PETS). IEEE, pp. 45–52

Wang H, Kläser A, Schmid C, Liu C-L (2013) Dense trajectories and motion boundary descriptors for action recognition. Int J Comput Vision 103:60–79

Weinzaepfel P, Revaud J, Harchaoui Z, Schmid C (2013) Deepflow: large displacement optical flow with deep matching. In Proceedings of the IEEE international conference on computer vision, pp. 1385–1392

Wortel IM, Liu AY, Dannenberg K, Berry JC, Miller MJ, Textor J (2021) CelltrackR: an R package for fast and flexible analysis of immune cell migration data. ImmunoInformatics 1–2:100003

Wortel IM, Postat J, Mihaylova M, Merino M, Bhagrath A, Harris M, Wouters L, Wiebke L, Parisi DR, Mandl JN, Textor J (2024) Cooperative motility emerges in crowds of T cells but not neutrophils. bioRxiv

Wu S, Moore BE, Shah M (2010) Chaotic invariants of Lagrangian particle trajectories for anomaly detection in crowded scenes. In 2010 IEEE computer society conference on computer vision and pattern recognition. IEEE, pp. 2054–2060

Zach C, Pock T, Bischof H (2007) A duality based approach for realtime tv-l 1 optical flow. In Pattern recognition: 29th DAGM symposium, Heidelberg, Germany, September 12–14, 2007. Proceedings 29. Springer, pp. 214–223

Zhu Y, Li X, Liu C, Zolfaghari M, Xiong Y, Wu C, Zhang Z, Tighe J, Manmatha R, Li M (2020) A comprehensive study of deep video action recognition. Preprint. arXiv:2012.06567

Zuraimi MAB, Zaman FHK (2021) Vehicle detection and tracking using YOLO and DeepSORT. In 2021 IEEE 11th IEEE symposium on computer applications & industrial electronics (ISCAIE). IEEE, pp. 23–29

Where Do We Stand in Preclinical Bioimaging: Synthesis and Perspectives

Ibrahim Fatih Cengiz, Joaquim Miguel Oliveira, and Rui L. Reis

Abstract

Recent advances in bioimaging have been transforming the way biological structures and events are visualized, quantified, and interpreted in preclinical research in bioengineering. Enabling analysis and visualization at structural, functional, and molecular scales, imaging technologies have been advancing with further integration with artificial intelligence, computational tools, and biological modeling. This chapter provides the author's opinion on the current state of the art of bioimaging in preclinical research focusing on emerging scientific directions, indicating the expansion of established modalities such as magnetic resonance imaging, micro-computed tomography, ultrasound, and optical techniques with novel concepts and analytical/computational approaches. This editorial chapter summarizes the current key trends that include tailoring imaging approaches to specific experimental needs to address the need for quantification/visualization with the use of hybrid/multimodal imaging and contributions of artificial intelligence computer vision.

Keywords

Multimodal imaging · Artificial intelligence · Translational research · Bioengineering · Image-based analysis

I. F. Cengiz (✉) · J. M. Oliveira · R. L. Reis
3B's Research Group, I3Bs – Research Institute on Biomaterials, Biodegradables and Biomimetics, University of Minho, Headquarters of the European Institute of Excellence on Tissue Engineering and Regenerative Medicine, Guimarães, Portugal

ICVS/3B's – PT Government Associate Laboratory, Braga/Guimarães, Portugal
e-mail: fatih.cengiz@i3bs.uminho.pt

I. F. Cengiz et al. (eds.), *Bioimaging Modalities in Bioengineering*,
Biomaterials, Bioengineering and Sustainability 9,
https://doi.org/10.1007/978-3-031-99410-4_30

1 Introduction

Bioimaging has been always important in the medical field to "see" any issue within the body, since seeing can be considered considered fundamental to diagnose, evaluate, and to understand (Hricak et al. 2025; Liao et al. 2023; Park et al. 2024; Schwenck et al. 2023). Moreover, visualization also opens the doors for quantification which is of critical importance. Similarly, in bioengineering, today we can see that bioimaging is one of the core tools which has been greatly progressed and allows us to visualize and quantify phenomena and processes which were not possible to be observed or even understood before (Lin and Wang 2022; Tao Liu et al. 2025). Bioimaging not only provides window into the structure, function, and dynamics of biological systems with increasing precision and scope, but also a connection point of several scientific domains to address outstanding scientific challenges in preclinical research. From the historical point of view, bioimaging is now deeply involved in preclinical research rather than just being a complementary/descriptive tool. Across preclinical bioengineering, bioimaging technologies enable the noninvasive study of biological materials, scaffolds, tissue de/regeneration, cellular behavior, and wide range complex disease models. In addition to the technological advances, the field is also responding to new demands in translational relevance, ethical responsibility, and data-driven discovery while also responding to and utilizing the advances in the wide field of artificial intelligence. Integration with computational techniques including deep learning and high-throughput analysis (Forsgren et al. 2022; McElliott et al. 2023; McGale et al. 2024; Oishi et al. 2024) further expands the potential of preclinical bioimaging, allowing for more comprehensive insights into biological and bioengineered preclinical research.

Over the years, we have been observing a big progress from single-modality imaging approaches toward hybrid/multimodal systems that deliver complementary information across spatial, temporal, and functional domains synergistically, even if there have been important advances within a single-modality (Bischof et al. 2024; Chen et al. 2024; Vermeulen et al. 2022). Herein, this chapter synthesizes the status quo as from these developments, examining cross-cutting themes and advances in imaging methods, analytical frameworks, and experimental strategies. The goal is to highlight the scientific progress that is shaping the next generation of bioimaging in life sciences and preclinical bioengineering.

2 Synthesis and Perspectives

The progress of bioimaging in bioengineering can be defined by both diversification and expansion. Imaging modalities such as micro-computed tomography, magnetic resonance imaging, ultrasound, optical coherence tomography, and light microscopy have matured, yet their scientific roles are being redefined and expanded through integration with molecular imaging, contrast-enhanced strategies, computational analytics, and artificial intelligence including computer vision. This convergence is not only technical just to see the results of an experiment but rather a dynamic tool involved in all stages of the research activities from hypothesis to outcomes. It is also exciting to see that scientific research trends are focused on application-specific design of imaging protocols; meaning that the trend is bioimaging has been moving from being performed as a routine endpoint towards actively tailored to address specific biological questions in preclinical research. This includes the use of multimodal platforms to assess scaffold integration, tissue regeneration, inflammation, vascularization, and tumor progression by in vitro, in vivo, and by ex vivo models. The merge of functional and structural imaging has provided opportunities for new levels of insight into the interaction between bio/tissue-engineered constructs and living systems from in vitro models to ex vivo models (Feng et al. 2024; Pi et al. 2024).

It is not only hardware advances, but also the integration of advanced software and computational methods have been reshaping how imaging data are interpreted and quantified. Artificial intelligence, machine learning, and computer vision are now the center to the extraction of quantitative results from high-resolution and large datasets (Archit et al. 2025). While these approaches may have limitations and challenges, they also have role in reducing subjectivity, enabling screening of large datasets, and supporting modeling of biological behaviors and events. At the same time, image-guided experimentation, namely using real-time feedback from imaging (Jiang et al. 2021) to adapt experimental conditions is also of critical importance in expanding the functional and applied role of bioimaging in preclinical research.

As discussed in the previous chapters of this book, the emerging modalities such as photon-counting CT, laser speckle contrast imaging, second harmonic generation, and mass spectrometry imaging contribute unique information about biochemical composition, microarchitecture, and dynamic physiological processes. Their integration into preclinical research enhances both the biological and clinical relevance of in vitro and ex vivo models, bridging the gap between preclinical research and clinics. It is also of critical importance for us to highlight that ethical and methodological considerations are also also influencing scientific practice and funding. The multi-dimensional interest to reduce/avoid animal use through improved imaging sensitivity, noninvasive monitoring, and high-content in vitro and ex vivo models, is a result of both technological advances and ethical consideration, especially in the EU and in some countries. Imaging is fundamental for the development of alternatives to conventional in vivo testing, enabling more humane and sustainable preclinical research with early-stage evaluation via in vitro models, and refinement of the experimental protocols. Another important dimension we would like to discuss is the scale. Bioimaging allows us to image from subcellular to organ-level resolution, supporting multiscale modeling and systems-level understanding (Park et al. 2023; Tolstik et al. 2024; Tuck et al. 2020). We can also see that the boundaries between imaging, modeling, and biological design are becoming weaker. Bioimaging will increasingly serve not just to observe biological outcomes, but to inform the development of more effective bioengineered systems. Real-time, multi-modal, and artificial intelligence-supported imaging and analysis will play an integral role in improving strategies in preclinical research, and advancing preclinical precision medicine.

3 Collaboration and Training on Bioimaging in Bioengineering

Scientific progress in bioimaging is also linked with collaboration between disciplines since bioimaging is no longer the domain of a single field, but a center where biology, engineering, physics, chemistry, and data science meet. Effective application of imaging technologies in preclinical bioengineering requires not only technical access, but also collaboration across domains. The bioimaging systems from acquisition hardware to reconstruction algorithms and post-processing steps demand specialized expertise and equipment. At the same time, the questions being addressed in bioengineering are naturally multidisciplinary, requiring imaging scientists to understand biological processes, and applied/experimental life science researchers to interpret and use complex image-derived data. To bridge this gap, to our opinion, collaboration and cross-training (Lai et al. 2025) seem to be essential: sharing core facilities, being funded by joint projects, shared experimental protocol, and integrated research teams that include but is not limited to bioimaging specialists, life scientists, biomedical engineers, and tissue engineers, artificial intelligence researchers.

Since it is not sufficient any more to teach bioimaging as a set of technical skills or analytical routines, the training approaches (Zylla et al. 2024) should be better include experimental design (considering statistical analysis, control groups, and reproducibility), as well as data treatment, storage, sharing, and interpretation in bioimaging-rich research. Training and staff-exchange programs that integrate hands-on imaging experience with exposure to regenerative medicine, tissue modeling, and translational goals are of critical importance. Especially, equipping early-career researchers with these cross-disciplinary competencies will not only enhance the impact of their work, but also strengthen the long-term integration of imaging into bioengineering research.

4 Take-Home Message

The literature supports that bioimaging has been a field undergoing a fast progress, as bioimaging technologies evolve from isolated modalities into integrated, focused on an application. The key scientific directions emerging from current preclinical imaging practices can be synthesized as:

(i) Bioimaging is increasingly application-centered, with imaging strategies tailored to address specific biological and engineering research questions and goals such as tissue regeneration, and disease modeling.

(ii) Multimodal and hybrid approaches are becoming foundational, allowing researchers to combine structural, functional, and molecular information.
(iii) Computational integration, including artificial intelligence and machine learning, is transforming how imaging data are analyzed, interpreted in preclinical studies.
(iv) Emerging imaging tools support scientific understanding to grow into previously inaccessible biological scales and biochemical processes, enabling insights into engineered tissues, models, and living systems.
(v) Noninvasive imaging and high-content in vitro models have great potential to contribute to a more reproducible, efficient, and ethical research with reduced in vivo experiments.
(vi) Future impact relies on interdisciplinary collaboration and training, ensuring that imaging technologies do not only exist technically but are actively used in scientific research.

Acknowledgments IFC thanks Portuguese Foundation for Science and Technology (FCT) and acknowledges the FCT distinction attributed under the Estímulo ao Emprego Científico program (2021.01969.CEECIND). IFC also thanks the TERM RES-Hub, Tissue Engineering and Regenerative Medicine Infrastructure project, funded by FCT. The authors acknowledge the financial support provided by the COMPETE2030 Program and the European Regional Development Fund (FEDER) (Project reference: COMPETE2030-FEDER-00682400). The authors acknowledge the support of CAN-TARGET – Decoding the Complexity of Cancer: Integrative Strategies to Model In vitro and Target the Multi-Cellular Tumor Micro-Environment, reference: NORTE2030-FEDER-02705300, funded by the CCDR-N – Northern Regional Development Coordination Committee, under the Norte 2030 Support System for the Creation of Scientific and Technological Knowledge – Integrated SR&TD Projects. This work has received funding from the European Union's Horizon Europe (2021–2027). The authors thank the financial support provided under UID/50026:3B's-Biomaterials, Biodegradables and Biomimetics Research Group, University of Minho (3B's Res. Group/UM), and the project "HEALTH-UNORTE: Setting-up biobanks and regenerative medicine strategies to boost research in cardiovascular, musculoskeletal, neurological, oncological, immunological and infectious diseases", reference: NORTE-01-0145-FEDER-000039, funded by the Norte Portugal Regional Coordination and Development Commission (CCDR-N), under the NORTE2020 Program.

Conflict of Interest The authors declare that they have no conflict of interest.

References

Archit A, Freckmann L, Nair S, Khalid N, Hilt P, Rajashekar V et al (2025) Segment anything for microscopy. Nat Methods:1–13

Bischof J, Fletcher G, Verkade P, Kuntner C, Fernandez-Rodriguez J, Chaabane L et al (2024) Multimodal bioimaging across disciplines and scales: challenges, opportunities and breaking down barriers. npj Imaging 2(1):5

Chen Z, Gezginer I, Zhou Q, Tang L, Deán-Ben XL, Razansky D (2024) Multimodal optoacoustic imaging: methods and contrast materials. Chem Soc Rev

Feng Q, Fatima K, Yang A, Li C, Chen S, Yang G et al (2024) Multi-modal imaging for dynamic visualization of osteogenesis and implant degradation in 3D bioprinted scaffolds. Bioact Mat 37:119–131
Forsgren E, Edlund C, Oliver M, Barnes K, Sjögren R, Jackson TR (2022) High-throughput widefield fluorescence imaging of 3D samples using deep learning for 2D projection image restoration. PLoS One 17(5):e0264241
Hricak H, Mayerhoefer ME, Herrmann K, Lewis JS, Pomper MG, Hess CP et al (2025) Advances and challenges in precision imaging. Lancet Oncol 26(1):e34–e45
Jiang Z, Li Z, Grimm M, Zhou M, Esposito M, Wein W et al (2021) Autonomous robotic screening of tubular structures based only on real-time ultrasound imaging feedback. IEEE Trans Ind Electron 69(7):7064–7075
Lai ACK, Yan WH, Wong CL, Yeung WY, Chan YK (2025) Enhancing interest and confidence of students in research through a cross-disciplinary undergraduate-initiated medical research program. Biomed Eng Educ:1–10
Liao S, Zhou M, Wang Y, Lu C, Yin B, Zhang Y et al (2023) Emerging biomedical imaging-based companion diagnostics for precision medicine. Iscience 26(8)
Lin L, Wang LV (2022) The emerging role of photoacoustic imaging in clinical oncology. Nat Rev Clin Oncol 19(6):365–384
McElliott MC, Al-Suraimi A, Telang AC, Ference-Salo JT, Chowdhury M, Soofi A et al (2023) High-throughput image analysis with deep learning captures heterogeneity and spatial relationships after kidney injury. Sci Rep 13(1):6361
McGale JP, Howell HJ, Beddok A, Tordjman M, Sun R, Chen D et al (2024) Integrating artificial intelligence and PET imaging for drug discovery: a paradigm shift in immunotherapy. Pharmaceuticals 17(2):210
Oishi H, Tabibzadeh N, Morizane R (2024) Advancing preclinical drug evaluation through automated 3D imaging for high-throughput screening with kidney organoids. Biofabrication 16(3):035003
Park B, Oh D, Kim J, Kim C (2023) Functional photoacoustic imaging: from nano-and micro-to macro-scale. Nano Converg 10(1):29
Park J, Choi S, Knieling F, Clingman B, Bohndiek S, Wang LV et al (2024) Clinical translation of photoacoustic imaging. Nat Rev Bioeng:1–20
Pi S, Ganjee R, Wang L, Arbuckle RK, Zhao C, Sahel JA et al (2024) Fully automated OCT-based tissue screening system. Opt Lett 49(16):4481–4484
Schwenck J, Sonanini D, Cotton JM, Rammensee H-G, la Fougère C, Zender L et al (2023) Advances in PET imaging of cancer. Nat Rev Cancer 23(7):474–490
Tao Liu JL, Li D, Tan S (2025) Bayesian deep-learning structured illumination microscopy enables reliable super-resolution imaging with uncertainty quantification. Nat Commun 16:5027. https://doi.org/10.1038/s41467-025-60093-w
Tolstik E, Lehnart SE, Soeller C, Lorenz K, Sacconi L (2024) Cardiac multiscale bioimaging: from nano-through micro-to mesoscales. Trends Biotechnol 42(2):212–227
Tuck M, Blanc L, Touti R, Patterson NH, Van Nuffel S, Villette S et al (2020) Multimodal imaging based on vibrational spectroscopies and mass spectrometry imaging applied to biological tissue: a multiscale and multiomics review. Anal Chem 93(1):445–477
Vermeulen I, Isin EM, Barton P, Cillero-Pastor B, Heeren RM (2022) Multimodal molecular imaging in drug discovery and development. Drug Discov Today 27(8):2086–2099
Zylla JL, Bomgni AB, Sani RK, Subramaniam M, Lushbough C, Winter R et al (2024) Convergence research and training in computational bioengineering: a case study on AI/ML-driven biofilm–material interaction discovery. Biomed Eng Educ 4(2):283–294

Correction to: Quantifying Biological Processes in Motion and Across Scales

Diego Ulisse Pizzagalli, Raffaella Fiamma Cabini, and Inge M. N. Wortel

Correction to:
Chapter 29 in: I. F. Cengiz et al. (eds.), *Bioimaging Modalities in Bioengineering*, Biomaterials, Bioengineering and Sustainability 9, https://doi.org/10.1007/978-3-031-99410-4_29

The chapter "Quantifying Biological Processes in Motion and Across Scales" written by Diego Ulisse Pizzagalli, Raffaella Fiamma Cabini, and Inge M. N. Wortel was inadvertently published without open access. Following the author's/authors' decision to opt for open access, the copyright of the chapter changed to © The Author(s) 2025 and the chapter is now distributed under the terms of the Creative Commons Attribution 4.0 International License http://creativecommons.org/licenses/by/4.0/, which permits use, duplication, adaptation, distribution, and reproduction in any medium or format, as long as appropriate credit is given to the original author(s) and the source, a link is provided to the Creative Commons licence, and any changes made are indicated. Open Access funded by Università della Svizzera italiana and Radboud University.

The updated version of this chapter can be found at
https://doi.org/10.1007/978-3-031-99410-4_29

I. F. Cengiz et al. (eds.), *Bioimaging Modalities in Bioengineering*,
Biomaterials, Bioengineering and Sustainability 9,
https://doi.org/10.1007/978-3-031-99410-4_31

Index

I. F. Cengiz et al. (eds.), *Bioimaging Modalities in Bioengineering*, Biomaterials, Bioengineering and Sustainability 9, https://doi.org/10.1007/978-3-031-99410-4

T

U

V

W

MIX
Papier aus verantwortungsvollen Quellen
Paper from responsible sources
FSC® C105338

If you have any concerns about our products,
you can contact us on
ProductSafety@springernature.com

In case Publisher is established outside the EU,
the EU authorized representative is:
Springer Nature Customer Service Center GmbH
Europaplatz 3, 69115 Heidelberg, Germany

Printed by Libri Plureos GmbH
in Hamburg, Germany